THE SCIENCE OF SOUND

THIRD EDITION

Thomas D. Rossing
Northern Illinois University

F. Richard Moore
University of California, San Diego

Paul A. Wheeler
Utah State University

Addison
Wesley

San Francisco   Boston   New York
Cape Town   Hong Kong   London   Madrid   Mexico City
Montreal   Munich   Paris   Singapore   Sydney   Tokyo   Toronto

Acquisitions Editor: Adam Black
Project Editor: Nancy Gee
Cover Designer: Blakeley Kim
Marketing Manager: Christy Lawrence
Managing Editor: Joan Marsh
Manufacturing Buyer: Vivian McDougal
Project Coordination: Integre Technical Publishing Co., Inc.

Library of Congress Cataloging-in-Publication Data

Rossing, Thomas D., 1929–
    The science of sound.—3rd ed. / Thomas D. Rossing, Paul Wheeler, and Richard Moore.
        p.   cm.
    Includes bibliographical references.
    ISBN 0-8053-8565-7
    1. Sound.   2. Music—Acoustics and physics.   I. Wheeler, Paul (Paul A.)   II. Moore, F. Richard.
III. Title.

QC225.15 .R67 2002
534—dc21   2001053994

ISBN 0-8053-8565-7

6 7 8 9 10 —MV— 08 07 06 05
www.aw.com/physics

Contents

12 ■ Woodwind Instruments   246

13 ■ Percussion Instruments   272

14 ■ Keyboard Instruments   310

21 ■ Digital Computers and Techniques   470

22 ■ Sound Recording   496

Preface to the Third Edition

During the twenty years since the first edition was published, this book has maintained its popularity. In fact, we are told that it has been the best selling textbook in introductory acoustics. But acoustics is a rapidly developing science, and it is necessary, once again, to revise it to include some of the newest developments and changes in this field. To do this, the senior author invited two eminent teachers to join him in the task. Professor F. Richard Moore is a musician and an engineer who is especially well known for his work in computer music, having written one of the first books on that subject. Professor Paul Wheeler is an electrical engineer, who has taught audio and acoustics to students of physics and music as well as engineering, using the Internet for course delivery in recent years.

It is relatively easy to add new material to a textbook; it is much more difficult to delete old material. In order to do justice to the subject, we have done both, and yet the book is about one hundred pages longer than the second edition, even though the number of chapters is one less than previous editions. The book is still organized into 8 parts or modules, so that the instructor can easily select the material that is most appropriate for a particular class. For example, Tom covers half the book in a one-semester course called Acoustics, Music and Hearing, and the balance of the book in a course entitled Audio and Electroacoustics. Richard covers Parts I, II, IV, V, and VI plus a chapter from Part VIII in a one-quarter course. Finally, Paul covers parts of all chapters in the book in a one-semester course by giving the students considerable choice in topics (which is possible by using the Internet). There are doubtless many other combinations of chapters and topics that can be put together to create courses that suit a wide variety of needs.

There is a strong emphasis on digital circuits and techniques and to computer music in this edition. This is in keeping with the times, of course. Although most students are quite familiar with digital computers these days, they may find Chapters 21, 23, 25 challenging in that they introduce a whole new language that pertains to digital sound and music. As in previous editions, the use of mathematics is limited to basic algebra, although some of the mathematical language used in Part VI on Electronic Music Technology may be new to some students. Brief discussions of logarithms and trigonometric functions are included in the Appendices.

A new feature in the third edition is the inclusion of Experiments for Home, Laboratory, and Classroom Demonstration. There is little doubt that the best way to learn any subject is by hands-on experiments, and so we have included suggestions for laboratory experiments that support each chapter. On the other hand, it is not always possible or convenient to schedule laboratory experiments, and so we briefly describe experiments for home and classroom demonstration. We urge instructors to do as many of these experiments as possible in class, and we urge students to do as many as possible at home. We generally award extra credit for demonstration experiments prepared at home by the students and demonstrated in class.

The authors thank their own students, as well as many users at other colleges and universities who have provided useful feedback during the past 20 years. We always welcome feedback from users!

T.D.R., F.R.M., and P.A.W.
October 2001

PART

I

Vibrations, Waves, and Sound

The first four chapters lay the groundwork for the rest of the book. After a brief introduction to sound in Chapter 1, we review some basic principles of motion. To understand sound we need to understand motion, including force and acceleration and how they are related by Newton's second law of motion. We need to understand work and energy, force, and pressure.

In Chapter 2 we consider vibrating systems, and in Chapter 3 we consider waves. We compare sound waves to light waves, ocean waves, and other types of waves. We note how all waves can undergo reflection, refraction, interference, and diffraction. Finally, we study resonance.

Readers who have had a good high school or college physics course may be familiar with most of the material covered in Chapters 1–4 and may wish to either skim over it quickly or go on the Chapter 5.

The metric system of units (in particular, the Système International, or SI) is used throughout this book. It is the preferred system for scientific work, and in the future (hopefully) it will come into general use in the United States as it has in the rest of the world. Other systems of units are listed in Appendix A.1.

1

What Is Sound?

The word *sound* is used to describe two different things:

1. An auditory sensation in the ear;
2. The disturbance in a medium that can cause this sensation.

By making this distinction, the age-old question, If a tree falls in a forest and no one is there to hear it, does it make a sound? can be answered.

The science of sound, which is called *acoustics*, has become a broad interdisciplinary field encompassing the academic disciplines of physics, engineering, psychology, speech, audiology, music, architecture, physiology, neuroscience, and others. Among the branches of acoustics are architectural acoustics, physical acoustics, musical acoustics, psychoacoustics, electroacoustics, noise control, shock and vibration, underwater acoustics, speech, physiological acoustics, etc.

This book is intended to be an introduction to acoustics, written in appropriate language, primarily for students without college physics and mathematics. A few basic mathematical ideas (such as logarithms) are introduced, and a brief review of algebra is included in the appendix for those who need it. A few basic concepts from physics, such as motion, energy, power, and waves, are introduced as needed.

In this chapter you should learn:

- About sound and sources of sound;
- About distance, speed, and velocity;
- How to represent motion graphically;
- About acceleration, force, and pressure;
- About energy and power.

1.1 ■ WHAT IS A SOUND WAVE?

The world is full of waves: sound waves, light waves, water waves, radio waves, X rays, and others. The room in which you are sitting is being crisscrossed by light waves, radio waves, and sound waves of many different frequencies. Practically all communication depends on waves of some type. Although sound waves are vastly different from radio waves or ocean waves, all waves possess certain common properties. One is that they carry information from one point to another. They also transport energy, as we will learn.

FIGURE 1.1
(a) Longitudinal motion of air molecules in a sound wave created by a loudspeaker; (b) transverse wave motion on a rope shaken up and down at one end.

Sound waves travel in a solid, liquid, or gas. Mostly we will focus our attention on longitudinal sound waves in air. Longitudinal means that the back-and-forth motion of air is in the direction of travel of the sound wave (as compared to waves on a rope, in which the back-and-forth motion is perpendicular to the direction of wave travel). As the wave travels through the air, the air pressure changes by a slight amount, and it is this slight change in pressure that allows our ears (or a microphone) to detect the sound. Longitudinal and transverse waves are compared in Fig. 1.1.

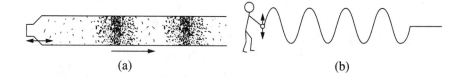

(a)   (b)

1.2 ■ SOURCES OF SOUND

Sound can be produced by a number of different processes, which include the following.

1. *Vibrating bodies*   When a drumhead or a piano soundboard vibrates, it displaces the air next to it and causes the local air pressure to increase and decrease slightly (Fig. 1.2(a)). These pressure fluctuations travel outward as a sound wave. Vibrating bodies are the most familiar sources of sound.

2. *Changing airflow*   When we speak or sing, our vocal folds (cords) alternately open and close so that the rate of air flow from our lungs increases and decreases, resulting in a sound wave. Similarly, a vibrating clarinet reed or the lips of a brass player cause a changing airflow into a clarinet or a trumpet. Air flowing through a screen or a grill produces a sort of hissing noise. An extreme case of changing airflow is a siren, in which holes on a rapidly rotating plate alternately pass and block air from a compressor, resulting in a very loud sound (Fig. 1.2(b)).

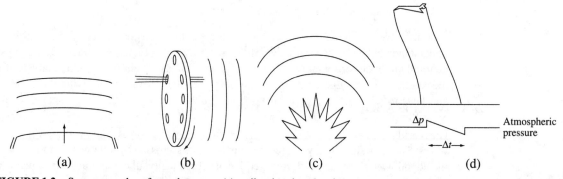

(a)   (b)   (c)   (d)

FIGURE 1.2   Some examples of sound sources: (a) a vibrating drumhead causes pressure changes in the air nearby; (b) holes in a rotating siren alternately allow and stop the flow of air; (c) an explosion rapidly heats the air nearby; (d) a supersonic airplane gives rise to shock waves.

3. *Time-dependent heat sources*   An electrical spark produces a crackle; an explosion produces a bang due to the expansion of the air caused by its rapid heating (Fig 1.2(c)). Thunder results from rapid heating of air by a bolt of lightning.

4. *Supersonic flow*   Shock waves result when a supersonic airplane or a speeding rifle bullet forces air to flow faster than the speed of sound (Fig 1.2(d)). We discuss sonic booms in Chapter 32.

How many sound sources can you think of? Do they fit one of the four categories above? Can you think of any sound sources that do not fit these categories? (There are a few.)

1.3 ■ WANTED AND UNWANTED SOUND

Our environment is filled with sound. Some of those sounds are made by humans, some by animals, some by machines, and some by natural causes, such as the weather. Some sounds, such as music and speech, are desirable (sometimes, at least); others, such as those of heavy vehicles on a street, are generally not. Unwanted sounds are often referred to as *noise*. Control of environmental noise is an important (but all too often overlooked) factor in maintaining environmental quality. Only recently has environmental noise been regulated by governmental agencies.

One difficult problem in controlling environmental noise is disagreement about what sounds are wanted and unwanted. Loud music may appeal to revelers at a party but not to their neighbors. The roar of a motorcycle engine may convey the feeling of power to the owner but insult the ears of bystanders.

Chapters 30–32 discuss environmental noise. In these chapters the reader will learn about the effect of noise on people and how to control noise. Although it is possible to isolate oneself from a noisy environment, it is difficult to do so (the term *soundproofing* could probably be described as a figment of our imagination). Control of noise is by far the best way to control environmental noise.

Before we go on to explore the science of sound and some of its applications, we should consider a few physical principles that we wish to apply in order to understand it. We want to understand such concepts as motion, force, kinetic and potential energy, pressure, and power and how these concepts apply to vibrating systems as well as to sound waves.

Not only do we wish to familiarize ourselves with these concepts, but sometimes we wish to solve problems that apply them, because doing physics is really the best way to understand physics. Physics involves more than solving problems, of course. Doing physics generally includes observation of natural phenomena, careful measurement, analysis of the measurements, formulating theories and laws to explain them, and applying these theories and laws to other situations or phenomena. Although it would be nice to make our own measurements on sound and formulate our own theories, this is not always practical, and so we make use of the measurements of many other scientists. We will, however, apply their theories to specific examples of interest, and that is the essence of most of the numerical exercises in this book.

In solving these numerical exercises or problems we will wish to apply some basic mathematics. Mathematics is the precise language of science, and when scientists talk to each other they often use this language quite freely. However, it is possible to translate most scientific concepts into a more familiar language, and that is what is intended in this book. We will, however, use the language of mathematics when it is especially appropriate.

1.4 ■ DISTANCE, SPEED, AND VELOCITY

For an object that travels at constant speed, the distance traveled is given by the simple formula

$$\text{distance} = \text{speed} \times \text{time}. \tag{1.1}$$

We realize that if we travel at a steady rate of 50 mi/h for 2 h, we will cover 100 mi, as the formula states. The same distance will be covered if our average speed is 50 mi/h for 2 h, even though our actual speed at different times may vary. In fact, the average speed can be defined as distance divided by time:

$$\text{speed} = \frac{\text{distance}}{\text{time}}. \tag{1.2}$$

Many different units are used to express distance: feet, inches, meters, yards, even furlongs. However, because most of the world uses metric units, it is prudent to emphasize the use of the metric system. The meter, then, will be our preferred unit of length, with occasional reference to feet and inches. Conversions between various units appear in the appendix.

The metric system uses prefixes to denote various powers of ten, as follows:

$$1 \text{ kilometer } = 1 \text{ km } = 10^3 \text{ m } = 1000 \text{ m}$$

$$1 \text{ centimeter } = 1 \text{ cm } = 10^{-2} \text{ m} = 1/100 \text{ m}$$

$$1 \text{ millimeter } = 1 \text{ mm } = 10^{-3} \text{ m} = 1/1000 \text{ m}$$

$$1 \text{ micrometer} = 1 \text{ } \mu\text{m } = 10^{-6} \text{ m} = 1/1,000,000 \text{ m}$$

These are the principal prefixes used for distance; others appear in the appendix. (The term *micron* is often used in place of *micrometer*.) Note that the first syllable is accented in all of these units, including "kil′ometer" and "mic′rometer."

Physicists often speak of *velocity* rather than speed. Velocity specifies the direction of motion as well as the speed. In describing motion in one direction, however, no distinction need be made, and in most of the vibrating systems we wish to discuss, the vibrations are in one direction. Speed is correctly defined as the magnitude of velocity (without regard to direction), and we will use the symbol v for speed, because it is customary to do so; furthermore, it reminds us that speed and velocity are closely related.

To find speed of an object, we must measure both the distance traveled and the time of travel. This is often done in the laboratory by photographing an object illuminated by a stroboscopic ("strobe") lamp that flashes at regular intervals. The photographs in Fig. 1.3 were taken with the lamp flashing ten times per second in a darkened room while the camera shutter remained open. Thus the position of the object was recorded at intervals of 0.1 s. It is easy to see that in Fig. 1.3(a) the speed stays the same, whereas in Fig. 1.3(b) it increases as the object moves from left to right.

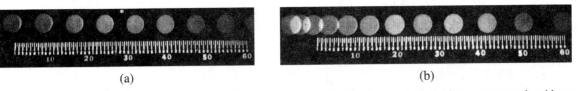

FIGURE 1.3 Stroboscopic pictures of motion: (a) constant speed; (b) changing speed. Both pictures were made with ten flashes per second.

We can easily determine the speed of the object in Fig 1.3(a). During each interval of 0.1 s, it appears to move 7.5 cm, so the speed is

$$v = \frac{7.5 \text{ cm}}{0.1 \text{ s}} = 75 \text{ cm/s} = 0.75 \text{ cm/s}.$$

Since the object moves with constant speed, the average speed and instantaneous speed are the same; this is not the case in Fig 1.3(b).

EXAMPLE 1.1   If the speed limit is posted at 30 mi/h, what is the corresponding limit in meters per second?

Solution

$$30\frac{\text{mi}}{\text{hr}} = \frac{(30 \text{ mi/h})(5280 \text{ ft/mi})(0.305 \text{ m/ft})}{3600 \text{ s/h}}$$

$$= 13.4 \text{ m/s}.$$

EXAMPLE 1.2   A motorist travels 100 mi. The first half of the distance takes one hour; the second half takes $1\frac{1}{2}$ h. What is the average speed for the journey?

Solution

$$v_{av} = \frac{d}{t} = \frac{100 \text{ mi}}{1 + 1.5 \text{ h}} = 40 \text{ mi/h}.$$

EXAMPLE 1.3   A motorist travels 300 mi. During the first half of the journey his average speed is 60 mi/h, and during the second half his average speed is 30 mi/h. What is his average speed for the entire journey?

Solution

$$t_1 = \frac{150 \text{ mi}}{60 \text{ mi/h}} = 2.5 \text{ h};$$

$$t_2 = \frac{150 \text{ mi}}{30 \text{ mi/h}} = 5.0 \text{ h}.$$

$$v_{av} = \frac{d}{t} = \frac{300 \text{ mi}}{2.5 + 5.0 \text{ h}} = 40 \text{ mi/h}.$$

(Note that the answer is *not* 45 mi/h.)

1.5 ■ GRAPHICAL REPRESENTATION OF MOTION

If a picture is worth a thousand words, a well-constructed graph must be worth at least five thousand, especially when it comes to describing motion. Suppose we wish to represent the changing position of the objects in Fig. 1.3 using two graphs. One coordinate is the position, represented by y, and the other coordinate is the time t. It does not really matter what point is selected as the zero or starting point (called the origin), but it is convenient to take it as the end of the meter stick; alternatively, it could have been the position of the object at the first flash. The graphs in Fig. 1.4 are the result.

Now we will use these graphs to help us determine average and instantaneous speed. To the graphs we add the useful constructions shown in Fig. 1.5. The distance FE is called Δy, a symbol read as "delta y," which means the "change in y." Similarly Δt (line DF) represents the "change in t." In Fig. 1.5(a), the average speed during the time interval from $t = 0.2$ s to $t = 0.4$ s is

$$v_{av} = \frac{\Delta y}{\Delta t} = \frac{15 \text{ cm}}{0.2 \text{ s}} = 0.75 \text{ m/s}.$$

In Fig. 1.5(b) it is

$$v_{av} = \frac{\Delta y}{\Delta t} = \frac{8.4 \text{ cm}}{0.2 \text{ s}} = 42 \text{ cm/s}.$$

FIGURE 1.4
Graphical representation of the motion shown in Fig. 1.3. The object in (a) moved at a constant speed; the object in (b) changed its speed.

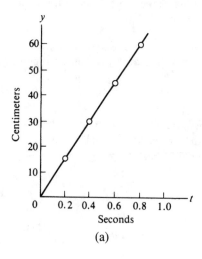

(a)

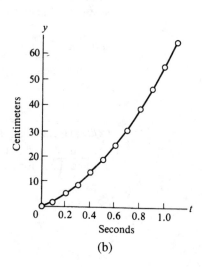

(b)

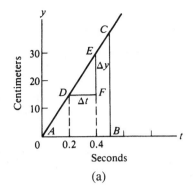
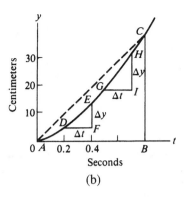

FIGURE 1.5
Curves of uniform and changing motion as in Fig. 1.4 with constructions added for determining speed.

(a)

(b)

In Fig. 1.5(a), the speed we calculate should be the same regardless of the time interval Δt selected. In fact, the larger time interval $\Delta t = 0.5$ s (represented by the line AB) should be used to improve accuracy:

$$v = \frac{\Delta y}{\Delta t} = \frac{37.5 \text{ cm}}{0.5 \text{ s}} = 75 \text{ cm/s}.$$

This is not true in the case of the changing motion in Fig. 1.5(b). Using the time intervals represented by DF, GI, and AB gives three different values of average speed, because the speed is changing:

$$v_1 = \frac{8.4}{0.2} = 42.0 \text{ cm/s} \qquad (DF),$$

$$v_2 = \frac{12.5}{0.2} = 62.5 \text{ cm/s} \qquad (GI),$$

$$v_3 = \frac{37.5}{0.8} = 46.9 \text{ cm/s} \qquad (AB).$$

If the instantaneous speed at a particular time is to be determined, the time interval Δt must be made very small. As Δt becomes smaller and smaller, the line (chord) DE more and more nearly approaches the slope of the curve at point P as shown in Fig. 1.6. Thus,

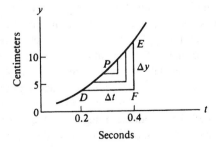

FIGURE 1.6
Shrinking Δy and Δt in order to obtain instantaneous speed.

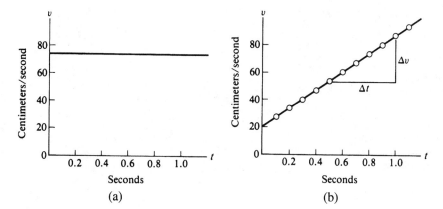

FIGURE 1.7
Speed as a function
of time for the
objects shown in
Figs. 1.3 and 1.4.

instantaneous speed can be interpreted as the *slope* (steepness) of the curve representing position y as a function of time t.

A graph showing speed as a function of time is also useful in describing motion. If the object photographed in Fig. 1.3 had a speedometer attached to it, we would have a photographic record of speed each time the strobe light flashed. These values of speed could then be plotted on a graph of v versus t. However, we can also determine' speed from the graphs already drawn, because speed at any time is represented by the slope of the curve showing y versus t at that time. Figure 1.7 shows speed as a function of time for each graph in Fig. 1.4.

1.6 ■ FORCE AND ACCELERATION

Force is a quantity with which we are all familiar; it can be described as a push or a pull. Practically all human activity involves forces: running, lifting, eating, writing, and even standing.

Applying a force to an object may distort the object, change its motion, or both. You may remember Newton's famous law of motion (his second law) tells us that the *acceleration* (change in motion) of an object is equal to the net force F divided by its mass m. Newton's second law of motion can be written as

$$F = ma \quad \text{or} \quad a = \frac{F}{m} \tag{1.3}$$

Clearly a greater force is required to obtain the same acceleration for an object of large mass as compared to an object of small mass.

A force applied to a movable object at rest causes it to accelerate (move with increasing speed) in the direction of the force. This is consistent with Newton's second law of motion as well as with our own experience. Applying a force in the direction of motion tends to increase its speed (we press the "accelerator" pedal to increase the speed of an automobile). Applying a force in a direction opposite to the direction of motion tends to decrease the speed (i.e., produces a negative acceleration, or a *deceleration*). Logically, we could call the brake pedal of an automobile the *decelerator* pedal.

Acceleration refers to a change in speed. When we push down the accelerator pedal in an automobile, we expect the speed to increase. In describing motion, acceleration is defined as the rate of change of speed (just as speed is the rate of change of position). *Average acceleration* is the ratio of change of speed Δv to time interval Δt:

$$a_{av} = \frac{\Delta v}{\Delta t}. \tag{1.4}$$

Instantaneous acceleration at a particular time can be determined by making Δt and Δv very small, just as we determined instantaneous speed v by making Δt and Δy very small.

In Fig. 1.7(a), the speed remains unchanged in time; in Fig. 1.7(b) it increases at a steady rate. The acceleration in Fig. 1.7(a) is therefore zero; in Fig. 1.7(b) it is

$$a = \frac{35 \text{ cm/s}}{0.5 \text{s}} = 70 \text{ cm/s}^2.$$

Note the units for acceleration; the unit *cm* appears to the first power, but the unit *s* appears squared.

An object in *free fall* in the gravitational field of the earth experiences a constant acceleration of 9.8 m/s^2. Thus if it begins with no initial speed (up or down), at the end of the first second it will have a speed of 9.8 m/s; at the end of 2 s its speed will be 19.6 m/s, and so on.

Note that acceleration does not always increase speed. If an object is thrown upward, the acceleration due to gravity acts to slow it down, or decelerates it. Figure 1.8 shows stroboscopic photographs of two objects in free fall. In Fig. 1.8(a), the object is dropped from rest; in Fig. 1.8(b), it is thrown upward, slows down, and then begins its descent.

Figure 1.9 shows a stroboscopic photograph of an object with an acceleration that changes its direction with time. The object, a mass attached to a spring, is executing a type of vibratory motion called *simple harmonic* motion, which will be discussed in Chap-

FIGURE 1.8
Stroboscopic photographs of two objects in free fall: (a) object dropped from rest; (b) object thrown upward. (Photographs by Christopher Chiaverina.)

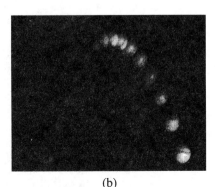

(a)   (b)

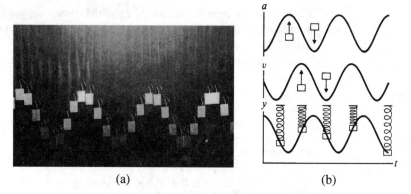

FIGURE 1.9
Vibratory motion in which y, v, and a all change with time.

(a)   (b)

ter 2. The camera has been turned during the exposure, so that multiple images of the object create a photographic record of position as a function of time. Note that position y, speed v, and acceleration a all change with time.

EXAMPLE 1.4   A bicyclist accelerates from 0 to 10 m/s in 20 s. He then applies the brakes and comes to a stop in 5 s. What is his average acceleration in each case?

Solution

$$a_{av} = \frac{\Delta v}{\Delta t} = \frac{10 - 0 \text{ m/s}}{20 \text{ s}} = 0.5 \text{ m/s}^2;$$

$$a_{av} = \frac{0 - 10 \text{ m/s}}{5 \text{ s}} = -2 \text{ m/s}^2.$$

EXAMPLE 1.5   A ball is thrown upward with a velocity of 15 m/s. How long does it take to reach its maximum height? How long does it take to fall back to the ground (neglect air resistance)?

Solution

$$a = \frac{\Delta v}{\Delta t}, \text{ so } \Delta t = \frac{\Delta v}{a} = \frac{0 - 15 \text{ m/s}}{-9.8 \text{ m/s}^2} = 1.53 \text{ s}.$$

1.7 ■ PRESSURE

Newton's second law of motion describes the way in which the net force acting on an object will set that object into motion. So far as that law is concerned, it does not matter whether the net force is due to a single force acting at a point or many forces distributed around the object. In weighing an object, we can consider the entire force of gravity acting at one point, which we appropriately refer to as the *center of gravity* or *center of mass*, even though gravity acts on every part of the object.

There are other times, however, when the distribution of forces is important. For example, you can walk on snow without sinking if you wear snowshoes; on the other hand, if you were to cross a wooden floor in spike heels, you would severely damage the floor. The difference in this case is the *area* over which the same total force is distributed.

It is useful to define a quantity called *pressure* as the force divided by the area over which it is distributed. To be more specific, it is the force acting perpendicular to a surface divided by the area of that surface:

$$p = F_\perp / A. \tag{1.5}$$

Because force is measured in newtons, pressure is measured in newtons/meter^2 (N/m^2) (or pounds/square inch in the British system). A 50-kg (110-lb) person standing on spike heels with an area of 20 mm^2 would exert a pressure of nearly 25 million N/m^2 (about 1.8 tons per square inch!) on the portion of the floor under the heel. A pascal (Pa) is often used to express pressure (1 Pa = 1 N/m^2). In most of the world, tire pressure is measured in kPa (kilopascals).

Fluids (liquids and gases) exert forces on the walls of their containers and anything immersed in them. One of the important properties of all fluids at rest, in fact, is that the pressure acts perpendicular to all surfaces (walls of the container as well as immersed objects). The pressure at any point in an open container of fluid (liquid or gas) is determined by the weight of the fluid above that point. For example, the weight of the atmosphere above us results in a pressure of about 10^5 N/m^2 (15 lb/in^2) at sea level; at an altitude of about 5.5 km (3.4 miles), the pressure is only one half as great. The *buoyancy* of an immersed object is due to the excess upward pressure on its bottom surface, as shown in Fig. 1.10.

FIGURE 1.10
Pressure in a container of fluid (a) acts on all surfaces; (b) is proportional to depth. Buoyant force (dashed arrow) on the immersed object is due to the excess upward force.

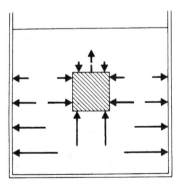

Slow variations in atmospheric pressure (as measured by a barometer) are indicative of changing weather. Sound waves consist of very small but rapid variations in pressure.

EXAMPLE 1.6   What is the total force on your chest wall due to the air outside?

Solution   Assume that your chest wall has an area of about 0.5 m^2,

$$F = pA = (10^5 \text{ N/m}^2)(0.5 \text{ m}^2) = 5 \times 10^4 \text{ N}$$

This is equal to about 11,000 lb; why doesn't your chest collapse?

1.8 ■ GRAPHICAL REPRESENTATION OF A SOUND WAVE

In Section 1.4, we learned how the motion of an object can be represented by graphs of distance, speed, or acceleration versus time. The graph tells us the position, speed, or acceleration of the object at each moment.

Similarly, it is useful to make a graph of sound pressure versus time. As the sound wave passes a certain point, the sound pressure rises and falls; a microphone measures this sound pressure. The graph of sound pressure versus time is called the *waveform* of the sound. Connecting the microphone to a *cathode-ray oscilloscope* or a computer allows us to display the waveform (graph of sound pressure versus time) on a screen.

Two sound waveforms are shown in Fig. 1.11. The one in Fig. 1.11(a) is a smoothly varying waveform; the one in Fig. 1.11(b) is a complex waveform (actually a musical sound made by a guitar). You may wish to speak or sing into a microphone connected to an oscilloscope so that you can observe the waveform of your golden voice.

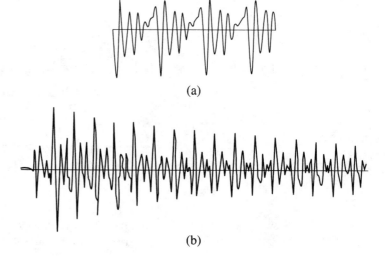

(a)

FIGURE 1.11
(a) Waveform (graph of sound pressure versus time) of a simple sound; (b) complex waveform of a musical sound made by a guitar.

(b)

1.9 ■ WORK AND ENERGY

The terms work and energy have various meanings in everyday life, but in the language of physics they have very definite meanings. *Work* is done when a force is applied to an object that moves. The work that is done is the product of the average force times the distance moved parallel to the force:

$$W = Fd. \tag{1.6}$$

Work is expressed in newton-meters, or *joules* (abbreviated J). If a force of one newton causes an object to move one meter, then one joule of work has been done.

The force of gravity on an object (its weight) is given by the formula mg where $g = 9.8$ m/s^2 is the acceleration of an object in free fall. Thus if an object falls a vertical distance h, the work done by gravity is:

$$W = mgh. \tag{1.7}$$

By the same token, raising an object of mass m to a height h requires an amount of work $W = mgh$.

Energy is perhaps the central idea underlying all branches of science. There are many forms of energy (e.g., mechanical, electrical, thermal, chemical, radiant, nuclear). A great deal of modern technology has as its goal the more efficient conversion of one of these forms of energy into another. For example, our future appears to depend on the development of the technology to convert solar (radiant) energy and the nuclear energy contained in sea water into electrical energy to replace our dwindling supply of oil and gas (chemical energy).

In our study of acoustics, we are concerned mainly with mechanical energy (and, to a lesser extent, electrical energy). Mechanical energy is closely related to work. Work is the transfer of energy. Systems with mechanical energy have the potential to do work. Vibrating systems have mechanical energy; mechanical energy is carried by the moving molecules in a sound wave. Energy, like work, is measured in joules. Sometimes a distinction is made between energy of motion, called *kinetic energy*, and stored energy, called *potential energy*, which is the capacity to do work (by virtue of position, for example).

Without going into a detailed discussion of energy, let us describe the energy of five completely different systems:

1. A baseball flying through the air has energy of motion, or *kinetic energy*. It is obvious that if it strikes another object (a bat, perhaps?) it can do work by virtue of its kinetic energy. The amount of kinetic energy is given by the formula

$$\mathrm{KE} = \tfrac{1}{2}mv^2, \tag{1.8}$$

where m is mass and v is its speed.

2. A block of wood lifted to a height h above the floor has *potential energy*, because if it were allowed to fall, it could do work. The amount of potential energy is given by the formula

$$\mathrm{PE} = mgh, \tag{1.9}$$

where h is its height above the floor.

3. A spring that has been stretched (or compressed) has potential energy, because if allowed to relax it can do work. If we use the constant K to denote its spring constant ("stiffness"), the potential energy when it is stretched an amount y from its relaxed length is given by the formula

$$PE = \tfrac{1}{2}Ky^2. \tag{1.10}$$

4. A bottle of gas of volume V whose pressure exceeds atmospheric pressure P_0 by a small amount p has potential energy

$$PE = \frac{1}{2}\frac{V}{P_0}p^2. \tag{1.11}$$

5. A guitar string displaced a small distance y at its midpoint has potential energy

$$PE = \frac{2T}{L}y^2, \tag{1.12}$$

where T is the tension in the string and L is its length. When the string is released, this energy is changed into kinetic energy, and thereafter the energy changes back and forth between kinetic and potential. The energy of vibrating systems will be discussed in Chapter 2.

Often the analysis of motion is facilitated by considering the way in which one form of energy is converted into another. For example, if we lift a heavy object we do work, and we give the object potential energy ($PE = mgh$). If we allow the object to free-fall, it acquires kinetic energy ($KE = \tfrac{1}{2}mv^2$). The speed it acquires in falling a distance h can easily be calculated by equating gain of KE to loss of PE without the need to calculate the time of fall:

$$\tfrac{1}{2}mv^2 = mgh$$
$$v^2 = 2gh$$
$$v = \sqrt{2gh}. \tag{1.13}$$

In describing vibrating systems in Chapter 2, we will make use of the fact that such systems continually convert potential energy to kinetic energy, and vice versa. Twice during each cycle of oscillation the energy is all kinetic, and twice it is all potential; at other times, the total energy is shared between potential and kinetic forms. As an oscillator slows down due to friction, the mechanical energy decreases because some of it is converted to another familiar form of energy: *heat*.

1.10 ■ POWER

Note that the definition of work (force × distance) says nothing about the time during which work is done. Raising a 2-kg mass to a height of 1 m requires 19.6 J of work whether the task is done in 1 s or 10. If the task were done in 4 s, for example, the average rate at which work is done would be 4.9 joules/second (J/s). The rate at which work is done is called *power*. Power is simply work divided by time:

$$P = \frac{W}{t}. \tag{1.14}$$

Power can be expressed in J/s, but this unit is used so frequently that it is given a special name, the *watt* (abbreviated W). Electrical equipment is rated according to the number of watts of electrical power it requires. A 100-W lamp converts electrical energy (to heat and light) at the rate of 100 J/s. The electric company, which sells electrical energy, installs meters that indicate how much electrical energy has been consumed. Instead of using joules (watt-seconds), however, the billing unit is the kilowatt-hour, the amount of work done by one kilowatt of power in one hour. One kilowatt-hour (kWh) equals $1000 \times 60 \times 60 = 3.6 \times 10^6$ J.

In the British system of units, work and energy are measured in *foot-pounds* and power in *horsepower*. One horsepower equals 550 ft = lb/s, which is roughly the rate at which a horse can do work. One horsepower equals 745.7 W.

EXAMPLE 1.7   A guitar string 65 cm long and having a tension of 55 N is displaced 8 mm at its midpoint. How much potential energy does it have?

Solution

$$\text{PE} = \frac{2(55\text{ N})}{(0.65\text{ m})}(8 \times 10^{-3}\text{ m})^2 = 0.11\text{ J}.$$

EXAMPLE 1.8   How much power is needed to raise a 2-kg mass to a height of 3 m in 15 s?

Solution

$$P = \frac{W}{t} = \frac{mgh}{t} = \frac{(2\text{ kg})(9.8\text{ m/s}^2)(3\text{ m})}{15\text{ s}}$$
$$= 3.92\text{ W}.$$

1.11 ■ UNITS

The preferred system of units for expressing physical quantities is the SI (Système International), or mks (meter-kilogram-second), system. Besides these three basic units, the system uses such units as newtons, joules, and watts, which are derived in a logical manner from the basic units. (For example, a newton equals kilograms × meters/seconds^2.) The SI, or mks, system is described in Appendix A.1.

Another metric system in use, the not as commonly as the mks system, is the cgs system, based on the centimeter, gram, and second. The cgs system is also described in Appendix A.1.

The fps (foot, pound, second) system of units is still very much in use in the United States, although its popularity is declining as we aim toward a conversion to metric units. Besides the basic units of foot, pound, and second, the fps system uses slugs, foot-pounds, horsepower, etc., as described in Appendix A.1.

1.12 ■ SUMMARY

Sound can refer to either an auditory sensation in the ear or the disturbance in a medium that causes this sensation. Sound is carried by waves in a solid, liquid, or gas. Unwanted sound is generally referred to as noise.

The motion of an object can be described by expressing its distance, speed, and acceleration as functions of time. A graphical representation of the motion consists of a plot of distance, speed, or acceleration as a function of time. Two other basic quantities, force and mass, are related to acceleration through Newton's second law of motion: $F = ma$. Pressure is the force per unit area, and is especially important in describing the behavior of objects immersed in a liquid or gas.

Another useful description of an object expresses its kinetic energy (energy of motion) and potential energy (stored energy) as functions of time. Power is the rate at which energy is expanded or the rate at which work is done. The preferred system of units is the mks (SI) system, which uses meters, kilograms, and seconds as its basic units, but also includes derived units such as newtons, joules, watts, and so forth.

REFERENCES AND SUGGESTED READINGS

Appel, K., J. Gastineau, C. Bakken, and D. Vernier (1998). *Physics with Computers*. Portland, Ore.: Vernier Software.

Hall, D. E. (1991). *Musical Acoustics*, 2nd ed. Pacific Grove, Calif.: Brooks/Cole.

Hewitt, P. G. (1993). *Conceptual Physics*, 7th ed. New York: Harper Collins.

Kirkpatrick, L. D., and G. F. Wheeler (1998). *Physics, A World View*, 3rd ed. Forth Worth: Saunders.

Rossing, T. D. (1982). *Acoustics Laboratory Experiments*. DeKalb: Northern Illinois University.

GLOSSARY

acceleration The rate of change of speed or velocity.

coordinates A set of numbers used to locate a point along a line or in space.

force An influence that can deform an object or cause it to change its motion.

gravity The force exerted by the Earth on all objects on or near it.

joule A unit of energy or work; one joule is equal to one newton-meter, also one watt-second.

kinetic energy Energy of motion; the capacity to do work by virtue of that motion; equal to one half mass times velocity (or speed) squared.

mass A measure of resistance to change in motion; equal to force divided by acceleration.

newton A unit of force.

potential energy Stored energy; the capacity to do work by virtue of position.

power The rate of doing work; equal to work or energy divided by time.

pressure Force divided by area.

speed The rate at which distance is covered; equal to distance divided by time.

stroboscope A light that flashes at a regular rate, making possible a photographic record of motion.

watt A unit of power; equal to one joule per second.

work The net force on an object times the distance through which the object moves.

Δ The Greek letter *delta*, denoting change in some quantity.

REVIEW QUESTIONS

1. What are two different meanings of the word *sound*?
2. What is the science of sound generally called?
3. What is the difference between a longitudinal and a transverse wave? Give an example of each.
4. What are four different processes that can produce sound? Give an example of each.
5. What is the difference between *speed* and *velocity*?
6. The slope of a graph of position versus time is equal to what quantity?
7. What three quantities are related by Newton's second law of motion?
8. Describe the motion of an object when no net force is applied.
9. Arrange the following in order from largest to smallest: 0.004 m, 0.4 mm, 4×10^{-5} km, 4×10^{-5} μm.

10. What is the difference between *pressure* and *force*?
11. Compare the pressure on the top and the bottom sides of a thin plate immersed in water.
12. What is the pressure of the atmosphere on our bodies?
13. What is a waveform of a sound?
14. What unit is used to express energy? work?
15. What is *kinetic energy*? *potential energy*?
16. Give a formula for the potential energy of a displaced guitar string and explain each symbol.
17. What is the difference between *power* and *energy*?
18. When you pay your electricity bill, are you paying for power used or for energy used?

QUESTIONS FOR THOUGHT AND DISCUSSION

1. At the same time a rifle is fired in an exactly horizontal position over level ground, a bullet is dropped from the same height. Both bullets strike the ground at the same time. Can you explain why?
2. What are some advantages of using the metric (SI) system of units rather than the English system?
3. In the sixteenth century, Galileo is said to have dropped objects of various weights from the Leaning Tower of Pisa. Since all objects in *free fall* accelerate at 9.8 m/s^2, one would expect them to reach the ground at the same time. Careful observation, however, indicates that an iron ball will strike the ground sooner than a baseball of the

same diameter. Can you explain why? Would the same be true on the moon? (The Apollo astronauts actually photographed a free-fall experiment on the moon using a hammer and a feather.)

4. Think of an object comparable in size to each of the following:
   (a) 10^7 m; (b) 10^3 m; (c) 1 m; (d) 10^{-3} m;
   (e) 10^{10} m.
5. Does shifting to a lower gear increase the power of an automobile? Explain.
6. Draw a diagram, similar to Fig. 1.10, showing how pressure acts on a floating object.

EXERCISES

1. Letting your classroom serve as the "origin" ($x = 0$, $y = 0$), express the approximate coordinates (x, y) of your place of residence. Let $x =$ the distance east and $y =$ the distance north, as on a map. Use any convenient unit of distance.

2. The speed of a bicycle increases from 5 mi/h to 10 mi/h in the same time that a car increases its speed from 50 mi/h to 55 mi/h. Compare their accelerations.

3. The density of water is 1.00 g/cm^3 and that of ice is 0.92 g/cm^3. What are the corresponding densities in SI units (kg/m^3)?

4. If the speed limit is posted as 55 mi/h, express this in km/h and in m/s (1 mi $= 1.61$ km).

5. A car accelerates from rest to 50 mi/h in 12 s. Calculate its average acceleration in m/s^2. Compare this to the acceleration of an object in free fall (1 mi/h $= 0.447$ m/s).

6. An object weighing 1 lb (English units) has a mass of 0.455 kg. Express its weight in newtons and thereby express a conversion factor for pounds to newtons.

7. Express your own mass in kilograms and your weight in newtons.

8. Calculate average speed in each of the following cases:

(a) An object moves a distance of 25 m in 3 s.

(b) A train travels 2 km, the first at an average speed of 50 km/h and the second at an average speed of 100 km/h. (*Note:* The average speed is not 75 km/h.)

(c) A runner runs 1 km in 3 min and a second kilometer in 4 min.

(d) An object dropped from a height of 75 m strikes the ground in 4 s.

9. Estimate the total force on the surface of your body due to the pressure of the atmosphere.

10. Calculate the kinetic energy of a 1500-kg automobile with a speed of 30 m/s. If it accelerates to this speed in 20 s, what average power has been developed?

11. An electric motor, rated at $\frac{1}{2}$ horsepower, requires 450 W of electrical power. Calculate its efficiency (power out divided by power in). What happens to the rest of the power?

12. Calculate the potential energy of:

(a) A 3-kg block of iron held 2 m above the ground;

(b) A spring with a spring constant $K = 10^3$ N/m stretched 10 cm from its equilibrium length;

(c) A 1-L bottle ($V = 1000$ cm^3) with a pressure 10^4 N/m^2 above atmospheric pressure ($P = 10^5$ N/m^2).

EXPERIMENTS FOR HOME, LABORATORY, AND CLASSROOM DEMONSTRATION

Home and Classroom Demonstration

1. *Longitudinal waves on a coiled spring (Slinky)*   For best results, suspend a Slinky from a long horizontal stick or rod by attaching several strings (about a meter in length). However, a giant Slinky will work satisfactorily on a smooth polished floor in spite of a small amount of friction. Jerk one end of the Slinky in the direction to increase its length and observe the pulse wave that propagates. Produce a small pulse and a large pulse in rapid succession. Does the distance between the two pulses change as they travel down the spring? What does this indicate about the relationship between amplitude (pulse size) and wave speed? Generate a series of waves by smoothly increasing and decreasing its length.

Repeat the experiment with transverse rather than longitudinal pulses and waves.

2. *Siren disk*   Blow air through a siren disk. If none is available, you can construct one by drilling regularly spaced holes in a wooden disk attached to a rotator. Note that the pitch of the tone depends upon the speed of rotation of the disk, whereas the loudness is determined by the rate of airflow.

3. *Moving object stroboscopically observed*   In a partially or totally dark room, observe a white ball in stroboscopic light. (If none is available, a hand stroboscope can be constructed by cutting slots around the circumference of a disc mounted on a dowel rod with a finger hole for rotating it). Roll the ball on a table or other horizontal surface and compare what you see to Fig. 1.3(a). Roll the ball down an incline and compare what you see to Fig. 1.3(b).

Observe a mass oscillating on the end of a spring, and see if you can make it appear to stand still by adjusting the rate of your stroboscope (either the flashing light or the hand stroboscope).

4. *Moving-object video capture*   Make a video recording of a moving object. Use a VCR with a single-frame player or a "frame grabber" to transfer single frames to a computer. Measure the distance the object has moved between successive frames.

5. *Falling object stroboscopically observed*   Observe a falling object in stroboscopic light (or with video capture) and compare what you see to Fig. 1.8(a). Toss a ball upward at an angle and compare what you see to Fig. 1.8(b).

6. *U-tube manometer*   Attach a length of rubber tubing to a U-shaped glass tube filled with colored water placed in front of a meter stick. The difference in heights of the water in the two sides of the U-tube represents the pressure in cm of water (a unit commonly used by organ builders). To convert cm of water to newtons/meter^2 or pascals (Pa), multiply by 100. Calibrate your lungs by blowing and sucking to obtain 100 cm of water (10^4 Pa) above and below atmospheric pressure. Which is easier to do?

7. *Deciding if pressure in a container depends upon the amount of water in the container*   Place the end of the tubing attached to a manometer at various depths in a cylinder of water and show that the pressure (in cm of water) is equal to the depth of the tube below the surface. Repeat with containers of varying size and shape to show that the pressure depends only of the depth below the surface, regardless of the shape of the container or how much water it holds.

8. *Force on a container wall*   Blowing a collapsed varnish can back to shape demonstrates the relationship of force to pressure. Measure the area of the large side of the can and multiply by the pressure difference inside and outside (indi-

cated on the manometer) to obtain the net force on the collapsed side of the can.

9. *Lifting a concrete block by blowing into a beach ball* A concrete block can be lifted by blowing air into a beach ball. A manometer indicates the pressure in the ball during and after inflation. (How much blowing pressure would be required to inflate an air mattress if someone is already lying on in?)

10. *Air pressure on a newspaper* Cover most of a thin board on a table with a sheet of newspaper. Strike the end of the board with the fist and note that the inertia of the air mass inhibits movement of the newspaper.

11. *Sound waveforms* Connect a microphone to a cathode-ray oscilloscope or a PC with a sound-input card and display sound waveforms (graphs of sound pressure versus time) on the screen. Speak and sing various vowel sounds. An "oo" sound sung in falsetto voice produces a very smooth waveform, for example. Try various musical instrument sounds.

12. *Human power* Run up a flight of stairs as fast as you can and time yourself with a stopwatch, a watch with a sweep second hand, or a digital watch. Measure the height of the stairway (number of steps times the height of each). Calculate the work done (using Eq. (1.7)) and the average power (using Eq. (1.14)). For your information, 1 hp equals 746 W (How does your power compare to that of a horse?)

Laboratory Experiments

Accelerated Motion (Experiment 1 in *Acoustics Laboratory Experiments*)

Graph Matching (Experiment 1 in *Physics with Computers*)

Picket Fence Free Fall (Experiment 5 in *Physics with Computers*)

Newton's Second Law (Experiment 9 in *Physics with Computers*)

CHAPTER

2

Vibrating Systems

Nature provides many examples of vibrating systems: trees swaying in the wind, atoms in a molecule of water, the motion of the tides, electric current in a flash of lightning, and so on. Although the motion is vastly different in each of these systems, they have several things in common: For one thing, the motion repeats in each regular time interval, which we call the *period* of the vibration; second, some type of *force* constantly acts to restore the system toward its point of equilibrium.

In this chapter you should learn:

- About simple vibrating systems;
- About vibrating systems with two or more masses;
- About vibrations in musical instruments;
- About vibration spectra.

2.1 ■ SIMPLE HARMONIC MOTION

Consider the very simple vibrating system in Fig. 2.1 consisting of a mass m attached to the end of a spring. We assume that the amount of stretch in the spring is proportional to the stretching force (which is true of most springs if they are not stretched too far), so that in order to stretch it a length l, a force Kl is required. The symbol K is the *spring constant*, or stiffness, of the spring.

Because the spring is vertical, the force of gravity on the mass stretches the spring by an amount that remains constant. In the equilibrium position shown in Fig. 2.1(b), the downward force of gravity on the mass (its weight) is just balanced by the upward force exerted by the spring; therefore, the system is in equilibrium. The description of the motion is simplified if we specify the displacement of y of the mass from the equilibrium position and the net force F that acts on the mass. The relationship between F and y is easily shown to be

$$F = -Ky. \tag{2.1}$$

The minus sign in this equations reminds us that when the mass is below its equilibrium position (y is negative), the net force will be upward (F is positive), as shown in Fig. 2.1(a). Thus the force F could be called a *restoring force*, which always acts in a direction to restore m to its equilibrium position. When the restoring force is proportional to the displacement, as it is in most vibrating systems we study, the motion is given a special name: *simple harmonic motion*. For a system in simple harmonic motion, the period

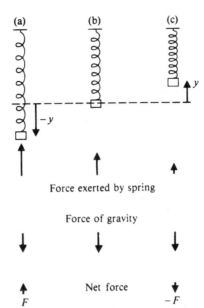

FIGURE 2.1
A simple vibrator consisting of mass and spring. In (b) the upward force exerted by the spring and the force of gravity balance each other, and the net force F on the mass is zero.

is independent of the amplitude (size) of the vibration. The *frequency f* of vibration is the number of oscillations per second, which is obviously the reciprocal of the period T of one vibration:

$$f = 1/T. \tag{2.2}$$

It is customary to use a unit called the *hertz* (abbreviated Hz) to denote cycles per second. In the case of the vibrating mass-spring system, the frequency of vibration is given by the formula

$$f = \frac{1}{2\pi}\sqrt{\frac{K}{m}}. \tag{2.3}$$

Note that to double the frequency of vibration, the mass m may be reduced to one-fourth its original size, or the spring constant K may be made four times larger.

EXAMPLE 2.1   Suppose a certain spring stretches 0.10 m when loaded with 2 kg. What is its spring constant? At what frequency will it vibrate when loaded with 2 kg? 0.5 kg?

Solution   At rest $Kl = mg$, so

$$K = \frac{mg}{l} = \frac{2(9.8)}{0.10} = 196 \text{ N/m}\quad \text{(newtons per meter)}.$$

Its frequency of vibration when $m = 2$ kg is

$$f = \frac{1}{2\pi}\sqrt{\frac{K}{m}} = \frac{1}{(2)(3.14)}\sqrt{\frac{196}{2}} = \frac{\sqrt{98}}{6.28} = 1.6 \text{ Hz.}$$

When loaded with 0.5 kg, the frequency is

$$f = \frac{1}{2\pi}\sqrt{\frac{K}{m}} = \frac{1}{(2)(3.14)}\sqrt{\frac{196}{0.5}} = \frac{\sqrt{392}}{6.28} = 3.2 \text{ Hz.}$$

A strobe photograph of a mass-spring system was shown in Fig. 1.9. Also shown was a graph of position y as a function of time. A graph of speed v versus time can be made by taking the slope of that graph at every time t. Graphs of y and v as functions of time are shown in Fig. 2.2. Mathematicians refer to curves shaped like these as *sinusoidal* or *sine* curves. The maximum value of y is called the *amplitude*.

2.2 ■ ENERGY AND DAMPING

The formula for the kinetic energy KE of a moving mass was given in Section 1.9:

$$\text{KE} = \tfrac{1}{2}mv^2, \tag{2.4}$$

where m is mass and v is speed. Similarly, the potential energy PE of a spring, stretched or compressed a distance y from its equilibrium length, was given as

$$\text{PE} = \tfrac{1}{2}Ky^2. \tag{2.5}$$

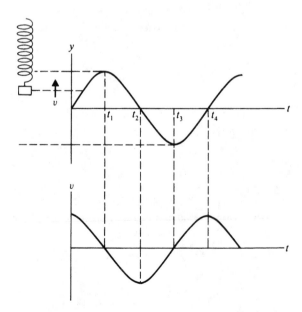

FIGURE 2.2
Graphs of simple harmonic motion:
(a) displacement versus time;
(b) speed versus time. Note that speed reaches its maximum when displacement is zero and vice versa.

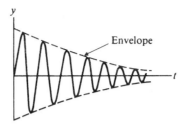

FIGURE 2.3
Displacement of a damped vibrator whose amplitude decreases with time.

From the graphs in Fig. 2.2, it is clear that v^2 reaches its maximum value when y^2 is zero, and vice versa. Thus, the total mechanical energy is constantly changing from kinetic to potential to kinetic. At times t_1, and t_3, potential energy is a maximum, and at t_2 and t_4, kinetic energy is a maximum.

Any real vibrating system tends to lose mechanical energy as a result of friction and other loss mechanisms. Unless the energy is renewed in some way, the amplitude of the vibrations will decrease with time, as shown in Fig. 2.3. In many vibrating systems, a certain fraction (usually small) of the energy is lost during each cycle of vibration; the result is a curve that decreases in amplitude in the manner shown in Fig. 2.3. The dashed curve, which indicates the change in amplitude with time, is called the *envelope*, or decay curve. A vibrating system whose amplitude decreases in this way is said to be *damped*, and the rate of decrease is the *damping constant*.

2.3 ■ SIMPLE VIBRATING SYSTEMS

Besides the mass-spring system already described, the following are examples of systems that vibrate in simple harmonic motion:

1. *Pendulum (small angle).*   A simple pendulum, consisting of a mass m attached to a string of length l (see Fig. 2.4), vibrates in simple harmonic motion, provided that $x \ll l$. Assume that the mass of the string is much less than m, the frequency of vibration is

$$f = \frac{1}{2\pi}\sqrt{\frac{g}{l}}, \tag{2.6}$$

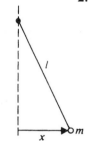

FIGURE 2.4
A simple pendulum.

where g is the acceleration due to gravity. Note that the frequency does not depend on the mass.

2. *A spring of air.*   A piston of mass m, free to move in a cylinder of area A and length l, vibrates in much the same manner as a mass attached to a spring (see Fig. 2.5). The spring constant of the air in the cylinder is determined by its compressibility and turns out to be $K = \gamma p A / l$, so the frequency is

$$f = \frac{1}{2\pi}\sqrt{\frac{\gamma p A}{ml}}, \tag{2.7}$$

where p is the gas pressure, A is the area, m is the mass of the piston, and γ is a constant that is 1.4 for air.

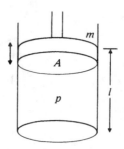

FIGURE 2.5
A piston free to
vibrate in a
cylinder.

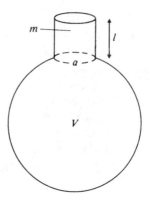

FIGURE 2.6
A Helmholtz
resonator.

3. *A Helmholtz resonator.*   Another common type of air vibrator, illustrated in Fig. 2.6, is often called a Helmholtz resonator, after H. von Helmholtz (1821–1894), who used it to analyze musical sounds. The mass of air in the neck now serves as the piston, and the air in the larger volume V as the spring. The frequency of vibration is

$$f = \frac{v}{2\pi}\sqrt{\frac{a}{Vl}}, \tag{2.8}$$

where a is the area of the neck, l is its length, V is the volume of the resonator, and v is the speed of sound ($v \simeq 344$ m/s).

The Helmholtz resonator can be thought of as having a mass m and a spring constant K that are

$$m = \rho a l \quad \text{and} \quad K = \frac{\rho a^2 v^2}{V}, \tag{2.9}$$

where ρ is the density of air.

Helmholtz resonators can have a variety of shapes and sizes. For example, blowing air across an empty pop bottle causes the air in its neck to vibrate at a fairly low frequency.

Note that the smaller the neck area a, the lower the frequency of vibration, which may seem a little surprising at first glance.

EXAMPLE 2.2   A small flask consists of a sphere 9.8 cm in diameter plus a neck 3 cm in diameter and 10 cm long. At what frequency will it resonate?

Solution

$$V = \frac{4}{3}\pi r^3 = \frac{4}{3}(3.14)(0.049 \text{ m})^3$$

$$= 4.93 \times 10^{-4} \text{ m}^3;$$

$$a = \pi r^2 = 3.14(0.015)^2 = 7.07 \times 10^{-4} \text{ m}^2.$$

$$f = \frac{v}{2\pi}\sqrt{\frac{a}{Vl}} = \frac{344 \text{ m/s}}{2(3.14)}\sqrt{\frac{7.07 \times 10^{-4} \text{ m}^2}{(4.93 \times 10^{-4} \text{ m}^3)(0.10 \text{ m})}}$$

$$= 207 \text{ Hz}.$$

2.4 ■ SYSTEMS WITH TWO OR THREE MASSES

The vibrating systems considered in the preceding section have one thing in common: A single coordinate is sufficient to describe their motion. In other words, they have one *degree of freedom*. In this section, we will consider vibrators with two or more degrees of freedom. Such systems have more than one natural *mode* of vibration, and the different modes will generally have different frequencies.

Consider the system consisting of two masses and three springs shown in Fig. 2.7. The system has two "normal," or independent, modes, as shown in Fig. 2.7(a) and 2.7(b). In one mode, the masses move in the same direction; in the other, they move in opposite directions. Assuming equal masses and springs with the same stiffness K, the frequencies of the two modes are

$$f_a = \frac{1}{2\pi}\sqrt{\frac{K}{m}}, \qquad f_b = \frac{1}{2\pi}\sqrt{\frac{3K}{m}}. \tag{2.10}$$

Note that mode (a) has the same frequency as the simple mass-spring system shown in Fig. 2.1, whereas mode (b) has a frequency that is 1.7 times that of mode (a).

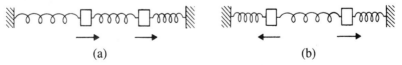

(a)   (b)

FIGURE 2.7   Modes of vibration of a two-mass vibrator. The mode shown in (a), in which the masses move in the same direction, will have the lower frequency.

(a)   (b)   (c)

FIGURE 2.8   Two-mass vibrators using (a) a linear air track; (b) an air table; (c) masses and springs hung from an overhead rod.

Modes (a) and (b) are virtually independent of each other. That is, the system can vibrate in mode (a) with minimal excitation of mode (b), and vice versa. If one sets the system into oscillation by giving the two masses a push or pull, the resulting motion will nearly always be a combination of modes (a) and (b). There are many recipes for combining these two modes in different proportions, and thus many ways in which the system can vibrate.

A great deal about the physics of vibration can be learned from watching the motion of a two-mass vibrator. Many physics laboratories have linear air tracks or air tables, on which objects move on a film of air with negligible friction. These are ideal for studying two-mass oscillators. Another convenient arrangement is to hang the masses on long cords from the ceiling or an overhead rod. The cords must be as long as possible to minimize the tendency of the masses to swing like pendulums. These three arrangements are shown in Fig. 2.8.

On the linear air track shown in Fig. 2.8(a), the masses are constrained to move in one direction only. In the systems shown in Fig. 2.8(b) and 2.8(c), however, the masses can move at right angles to the springs as well. Vibrations in this direction are called *transverse* vibrations, whereas vibrations in the direction of the springs are called *longitudinal* vibrations. Some systems vibrate only in transverse modes, some only in longitudinal. The air column of a musical wind instrument, for example, vibrates longitudinally, whereas the membrane of a drum vibrates transversely. A violin string normally vibrates transversely, although longitudinal vibrations (which sound like squeaks or squeals) are occasionally excited by the bowing of unskilled players.

In addition to their two modes of longitudinal vibration, the two-mass systems shown in Figs. 2.8(b) and 2.8(c) have two modes of transverse vibration, which are shown in Fig. 2.9. In the mode of lower frequency, the masses move in the same direction; in the

FIGURE 2.9
Modes of transverse vibration of a two-mass system. (a) In the mode of lower frequency, masses move in the same direction; (b) in the mode of higher frequency, masses move in opposite directions.

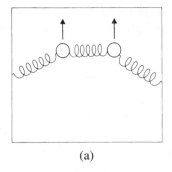
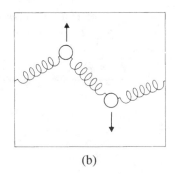

(a)   (b)

mode of higher frequency, they move in opposite directions. This behavior is similar to the longitudinal modes shown in Fig. 2.7.

Adding a third mass to the systems of Fig. 2.7 adds additional modes of vibration. In the case of the linear vibrator in Fig. 2.7(a), which vibrates only longitudinally, a third mode of longitudinal vibration appears. The three independent modes of vibration are those shown in Fig. 2.10. The systems in Figs. 2.7(b) and 2.7(c) can vibrate transversely as well; in addition to the three modes of longitudinal vibration in Fig. 2.10, they will have the three independent modes of transverse vibration shown in Fig. 2.11.

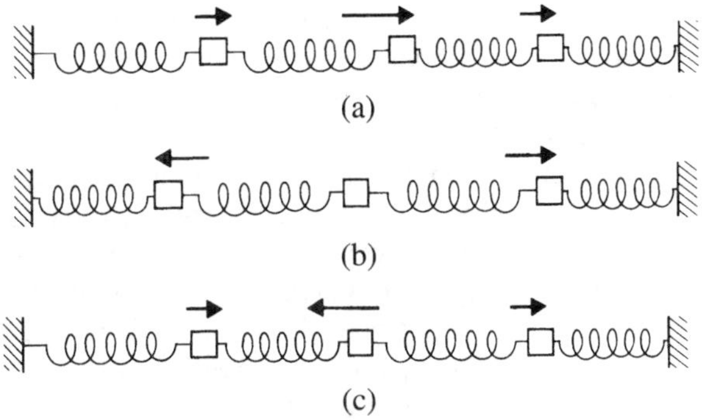

FIGURE 2.10
Independent modes
of longitudinal
vibrations of a
three-mass vibrator.

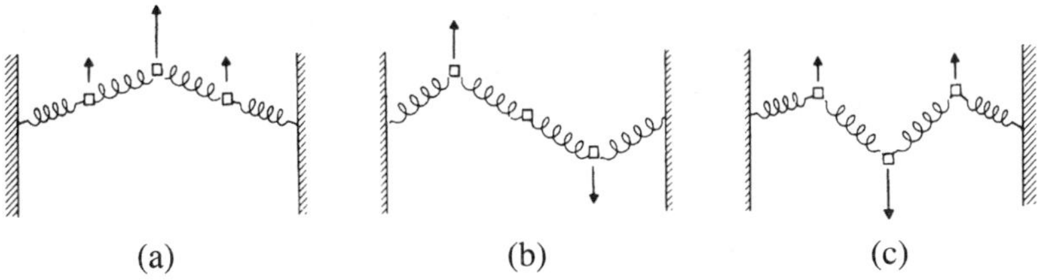

(a)

(b)

(c)

FIGURE 2.11
Independent modes
of transverse
vibration of a
three-mass
oscillator.

(a)   (b)   (c)

The independent modes shown in Figs. 2.7, 2.9, 2.10, and 2.11 are often called the *normal* modes of the vibrating systems. Getting the system to vibrate in a single normal mode requires special care. It is perhaps best done by driving the system at the frequency of the desired mode (this phenomenon, called *resonance*, will be discussed in Chapter 4). Carefully displacing the masses by the proper amounts and releasing them will also cause the system to vibrate in a single mode.

EXAMPLE 2.3   The vibrating system in Fig. 2.7 consists of two 0.5-kg masses and three springs having spring constants of 50 N/m. Find the frequencies of its vibrational modes.

Solution

$$f_a = \frac{1}{2\pi}\sqrt{\frac{K}{m}} = \frac{1}{2\pi}\sqrt{\frac{50 \text{ N/m}}{0.5 \text{ kg}}} = 1.59 \text{ Hz};$$

$$f_b = \frac{1}{2\pi}\sqrt{\frac{3(50)}{0.5}} = 2.76 \text{ Hz}.$$

2.5 ■ SYSTEMS WITH MANY MODES OF VIBRATION

In the case of the mass-spring vibrating systems, each new mass added one new mode of longitudinal vibration and one new mode of transverse vibration, if the system was able to vibrate it transversely. In general, a system of N masses of the type shown in Fig. 2.8(b) or 2.8(c) will have N longitudinal and N transverse modes of vibration. If the masses were free to move in all three coordinate directions, there would be $2N$ transverse modes of vibration and N longitudinal modes, where N is the number of masses. The number of frequencies associated with the transverse modes may be only N, however, because corresponding modes in two directions usually have the same frequency.

The transverse modes of vibration for 1, 2, 3, 4, 5, and 24 masses are sketched in Fig. 2.12. Note that in each case the number of transverse modes equals the number of masses. There are an equal number of longitudinal modes, but they are more difficult to represent in a diagram. In each case, the mode of highest frequency is the one in which adjacent masses move in opposite directions.

Note that as the number of masses increases, the system takes on a wavelike appearance. In fact, a vibrating guitar string can be thought of as a mass-spring system with very large N. The propagation of waves on a string will be considered in Chapter 3.

FIGURE 2.12
Modes of transverse vibration for mass-spring systems with different numbers of masses. A system with N masses has N modes.

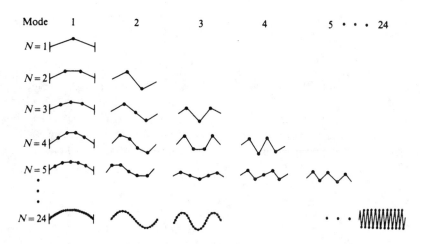

2.6 ■ VIBRATIONS IN MUSICAL INSTRUMENTS

All musical sound is generated by some type of vibrating system, whether it is a string on a violin, the air column of a trumpet, the head of a drum, or the voice coil of a loudspeaker. Often the vibrating system consists of two or more vibrators that work together, such as the reed and air column of a clarinet, the strings and sounding board of a piano, or the strings and body of a guitar. The acoustics of musical instruments is the subject of Part 3, but a brief description of several common musical vibrators will be made in closing this chapter on vibrating systems.

1. *Vibrating string.* The vibrating string can be thought of as the limit of the mass-spring system (see Fig. 2.11) when the number of masses becomes very large. The string itself has mass and elasticity or "springiness." There will be many modes of vibration, and their frequencies turn out to be very nearly whole-number multiples of the frequency of the lowest or *fundamental* mode. When the higher modes have frequencies that are whole number multiples of the fundamental frequency, we call them *harmonics*. Several modes of a vibrating string are illustrated in Fig. 2.13. The guitar (Fig. 2.14), for example, uses vibrating strings.

2. *Vibrating membrane.* Drumheads are membranes of leather or plastic stretched across some type of tensioning hoop or frame. A membrane can be thought of as a two-

FIGURE 2.13
(a) Modes of a vibrating string; (b) strobe picture of a string vibrating in its lowest two modes.

(a)   (b)

FIGURE 2.14
A guitar. Coupling between strings, wood plates, and enclosed air leads to many modes of vibration, which will be discussed in Chapter 10.

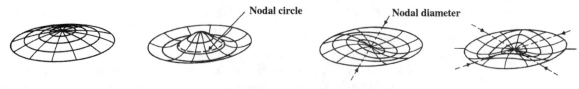

FIGURE 2.15   Modes of circular membrane. The first two modes have circular symmetry; the second two do not. (From Morse and Ingard, 1968.)

dimensional string in that its restoring force is due to tension applied from the edge. A membrane, like a string, can be tuned by changing the tension. Membranes, being two-dimensional, can vibrate in many modes that are not normally harmonic. Four modes of vibration of a circular membrane are illustrated in Fig. 2.15. The first two have circular symmetry; the second two have *nodal* lines (indicated by the arrows), which act as pivots for a rocking motion. Three familiar examples of drums that use vibrating membranes to produce sound are shown in Fig. 2.16.

3. *Vibrating bar.*   Many percussion instruments use vibrating bars as sound sources. The stiffness of a bar provides the restoring force when it bends, so no tension need be applied. Thus the ends may be free, as they are in most percussion instruments, or clamped (see Fig. 2.17). The frequencies of the vibrational modes of a uniform bar with free ends (as

FIGURE 2.16
(a) Bass drum and
snare drum;
(b) timpani.

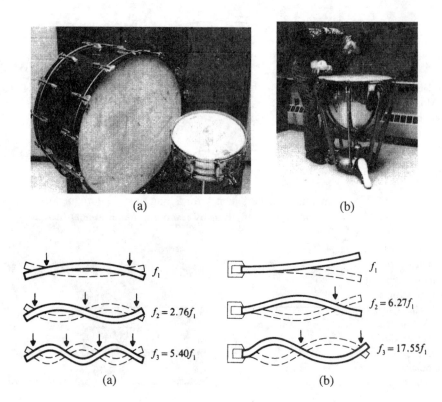

(a)   (b)

FIGURE 2.17
Nodes of vibrating
bars: (a) both ends
free; (b) one end
clamped. Arrows
locate the nodes.

FIGURE 2.18
A xylophone.
(Courtesy of
J. C. Deagan Co.)

the bars of a glockenspiel, for example) have the ratios 1 : 2.76 : 5.40 : 8.93, etc., which are nowhere near harmonic. The bars of marimbas, xylophones (see Fig. 2.18), and other instruments, however, have been shaped to have a quite different set of mode frequency ratios. Bars can also vibrate longitudinally, but longitudinal modes are not normally used in musical instruments.

4. *Vibrating plate.* Vibrating plates, like vibrating bars, depend on their own stiffness for the necessary restoring force. Plates have many modes of vibration, some exhibiting great complexity.

An interesting way to study the modes of vibration of plates is through the use of Chladni patterns, first described by E. F. F. Chladni in 1787. Particles of salt or sand are sprinkled on a vibrating plate, which is then excited to vibrate in one of its normal modes. The particles, agitated by the vibrations, tend to collect along nodal lines, where the vibrations are minimal. Chladni patterns of a circular plate are shown in Fig. 2.19.

Three musical instruments that use vibrating plates are shown in Fig. 2.20.

5. *Tuning fork.* A tuning fork consists of two bars joined together at one end. Thus the modes of vibration will resemble those of a bar clamped at one end, as illustrated in Fig. 2.17(b). Tuning forks are very convenient standards of frequency; once adjusted, they maintain their frequency for a long time. The frequency of a tuning fork may be raised by shortening its length or by removing material near the ends of the prongs. The frequency

FIGURE 2.19 Chladni patterns of a circular plate. The first four have two, three, four, and five nodal lines but no nodal circles: the second four have one or two nodal circles. (Rossing, 1977).

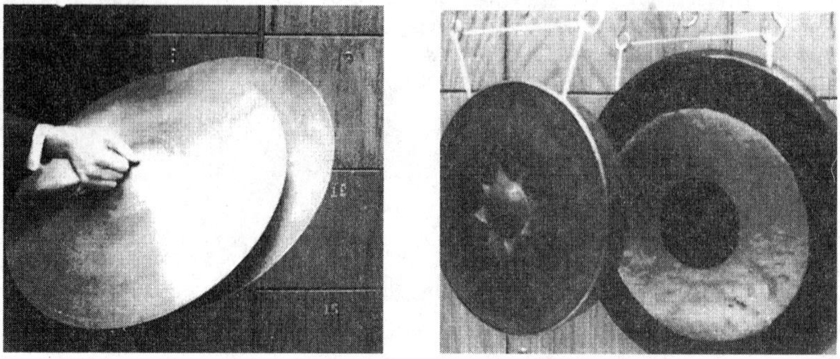

FIGURE 2.20
(a) Cymbals;
(b) gong (left) and
tam tam (right).

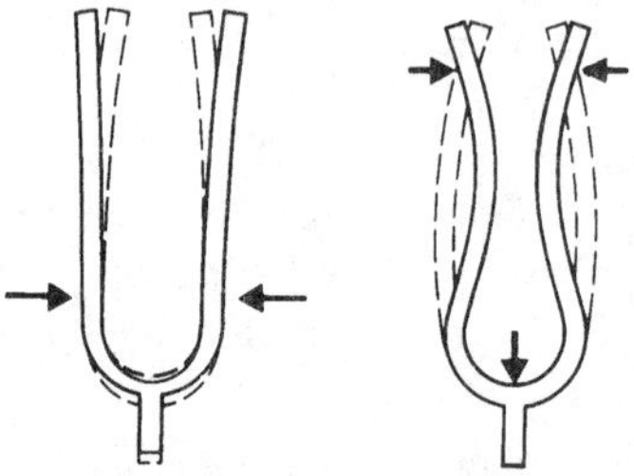

(a)   (b)

FIGURE 2.21
Vibrations of a
tuning fork:
(a) principal mode;
(b) "clang" mode,
which occurs at a
higher frequency
than the principal
mode.

(a)   (b)

can be lowered by removing material near the base of the prongs, which decreases the stiffness.

As shown in Fig. 2.21, two modes of vibration of the tuning fork are the principal mode and the "clang" mode, which occurs at a much higher frequency (nearly three octaves higher in a typical fork). In their normal motion, the bars pivot about two nodes marked by arrows, causing the handle to move up and down. Thus, if the handle is pressed against another object (e.g., a table top), it may cause that object to act as a sounding board. (In a noisy environment, the handle may be touched to one's forehead in order to conduct sound directly to the inner ear.)

6. *Air-filled pipes.* The vibrational behavior of a column of air, as found in an organ pipe or the bore of a trumpet, can be compared to that of the air spring we discussed in Section 2.3, but is better understood by considering the sound waves within it. Thus we leave the discussion of this type of musical vibrator to Chapter 4.

2.7 ■ COMPLEX VIBRATIONS: VIBRATION SPECTRA

In the preceding three sections, we have considered vibrating systems that can vibrate in several different modes. Each of these modes has a different frequency,* and hence it

---

*Occasionally two different modes of vibration will have the same natural frequency; they are then called *degenerate* modes. These are rare in musical instruments, however.

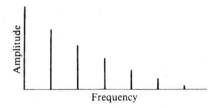

FIGURE 2.22 The vibration spectrum of a plucked string. The spectrum is a recipe that tells us the frequency and amplitude of each mode of vibration that is excited. In this case, the frequencies are harmonics (multiples) of the fundamental, but this will not be so in some vibrating systems.

can be excited individually by some type of driving force at that frequency, as shown in Figs. 2.9–2.19.

More commonly, however, when a vibrating system is excited, it vibrates in several modes at once. A description of its vibrational motion therefore requires a "recipe," which tells us the amplitude and frequency of each of the modes that have been excited. Such a recipe is called the *spectrum* of the vibration. A vibration spectrum of a plucked string is shown in Fig. 2.22. Spectra of this type will appear frequently throughout this book.

When we observe a vibrating system with several modes, we often wish to determine its vibration spectrum. Electronic instruments called *spectrum analyzers* enable us to do this in the laboratory. Spectrum analysis is also called Fourier analysis in honor of the mathematician Joseph Fourier (pronounced "four-yay"), who pioneered in the mathematics of spectrum analysis. The Fourier analysis of sound waves will be discussed in Chapter 7, and examples will be given in the chapters on musical instruments.

2.8 ■ SUMMARY

Vibrational motion repeats itself in a regular interval of time called the period. Vibrating systems have some type of force acting to restore the system toward its point of equilibrium. In the case of simple harmonic motion, this force is proportional to the displacement.

In a vibrating system, the total mechanical energy changes from kinetic to potential to kinetic during each cycle of vibration. The rate at which the total energy decreases depends on the damping forces. Some systems can vibrate in several independent modes. The actual vibratory motion may be a combination of these modes.

All musical sound is generated by some type of vibrating system. Common vibrators include strings, membranes, bars, plates, and air columns. The familiar tuning fork is combination of two bars vibrating in opposite directions, for example.

REFERENCES AND SUGGESTED READINGS

Appel, K., J. Gastineau, C. Bakken, and D. Vernier (1998). *Physics with Computers*. Portland, Ore.: Vernier Software.

Chladni, E. F. F. (1787). "Entdeckungen über die Theorie des Klanges," translated excerpts in R. B. Lindsay, *Acoustics:* *Historical and Philosophical Development*. Stroudsburg, Pa.: Dowden, Hutchinson and Ross, 1973.

Hewitt, P. G. (1993). *Conceptual Physics*. New York: Harper Collins.

Morse, P. M., and K. U. Ingard (1968). *Theoretical Acoustics*. New York: McGraw-Hill.

Rossing, T. D. (1982). *Acoustics Laboratory Experiments* (Northern Illinois University, DeKalb).

Rossing, T. D. (1976, 1977). "Acoustics of Percussion Instruments," *The Physics Teacher* **14**:546, and **15**:278.

Rossing, T. D., and N. H. Fletcher (1995). *Principles of Vibration and Sound*. New York: Springer-Verlag.

Rossing, T. D., D. A. Russell, and D. E. Brown (1992). "On the acoustics of tuning forks," *American J. Physics* **60**:620–626.

GLOSSARY

amplitude Maximum displacement from rest.

damping Loss of energy of a vibrator, usually through friction.

envelope Time variation of the amplitude (or energy) of a vibration.

frequency The number of vibrations per second; expressed in hertz (Hz).

fundamental mode The mode of lowest frequency.

harmonics Modes of vibration whose frequencies are whole-number multiples of the frequency of the fundamental mode.

Helmholtz resonator A vibrator consisting of a volume of enclosed air with an open neck or port.

longitudinal vibration Vibration in which the principal motion is in the direction of the longest dimension.

node, or nodal line A point or line where minimal motion takes place.

normal modes Independent ways in which a system can vibrate.

period The time duration of one vibration; the minimum time necessary for the motion to repeat.

simple harmonic motion Smooth, regular vibrational motion at a single frequency such as that of a mass supported by a string.

spectrum A "recipe" that gives the frequency and amplitude of each component of a complex vibration.

spring constant ("stiffness") The strength of a spring; restoring force divided by displacement.

transverse vibration Vibration in which the principal motion is at right angles to the longest dimension.

waveform Graph of some variable (e.g., position of an oscillating mass or sound pressure) versus time.

REVIEW QUESTIONS

1. What is meant by the period of a vibration? How is it related to the frequency?

2. In what units is the spring constant of a spring expressed?

3. What is meant by simple harmonic motion?

4. Doubling the distance a spring is stretched increases the restoring force by what factor?

5. Doubling the distance a spring is stretched increases its potential energy by what factor?

6. How does the frequency of a simple pendulum change when its mass is doubled?

7. How does the frequency of a Helmholtz resonator change when its volume is doubled? when the radius of its neck is doubled?

8. How many modes of longitudinal vibration does a two-mass system have? how many modes of transverse vibration?

9. How many modes of longitudinal vibration does a four-mass system have? how many modes of transverse vibration?

10. Describe the lowest mode of vibration in a vibrating string.

11. What is a node? Describe a node in a vibrating string. Describe a node in a vibrating membrane.

12. How many nodes are there in the lowest mode of a bar with free ends?

13. Describe the first two vibrational modes of a tuning fork.

14. What is a spectrum of vibration? In what sense is it a recipe?

QUESTIONS FOR THOUGHT AND DISCUSSION

1. Present an argument to show that the maximum kinetic energy of a mass-spring vibrator is equal to the maximum potential energy. Does the total mechanical energy remain constant throughout a cycle?

2. A damped vibrator is found to decrease its amplitude by one-half every 30 s. What is its amplitude at the end of 5 min? In theory will it every stop vibrating? Will it in practice? Explain. (*Hint:* $(\frac{1}{2})^{10} = \frac{1}{1024} \simeq 0.001$.)

3. With the help of Figs. 2.10 and 2.12, make a diagram of the four independent longitudinal modes of vibration for a four-mass vibrator.

4. To excite a tuning fork in its principal mode of vibration with a minimum of "clang" sound, where should you strike it? Of the four microphone positions $A, B, C,$

FIGURE 2.23

and D in Fig. 2.23, which will best pick up the sound of the fork? Why?

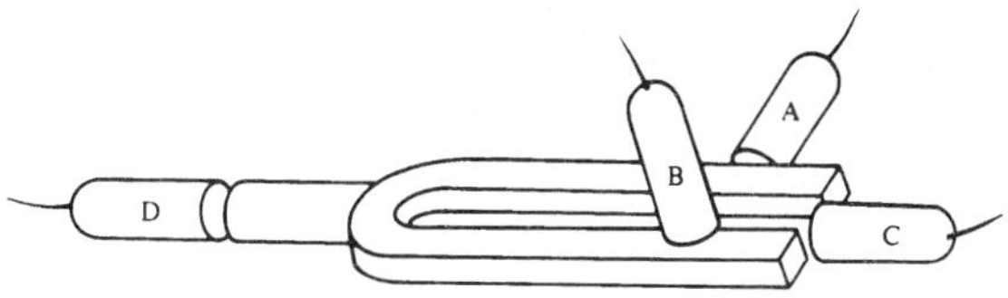

5. Why is the center arrow in Fig. 2.10(a) larger than the other two arrows?

6. Which Chladni patterns in Fig. 2.19 most nearly correspond to the second and fourth diagrams in Fig. 2.15? In what way are they different?

EXERCISES

1. Hanging a mass of 1 kg on a certain spring causes its length to increase 0.2 m.
   (a) What is the spring constant K of that spring?
   (b) At what frequency will this mass-spring system oscillate?

2. Copy the graphs of displacement and velocity shown in Fig. 2.2, and draw graphs of kinetic energy and potential to the same scale of time.

3. Most grandfather clocks have a pendulum that ticks (makes half a vibration) each second. What length of pendulum is required? (The value of g was given in Chapter 1 as 9.8 m/s^2.)

4. A bass-reflex loudspeaker enclosure (see Fig. 19.16) is essentially a Helmholtz resonator. Given the following parameters, what resonance frequency might be expected? $V = 0.5$ m^3, $a = 0.02$ m^2, $l = 0.05$ m, speed of sound $v = 343$ m/s at $T = 20°$C.

5. Calculate the maximum potential energy of the mass-spring system described in Problem 1 if its maximum displacement is 5 cm.

6. In the two-mass system shown in Fig. 2.7, each mass is 2 kg and each spring constant $K = 100$ N/m. Calculate the frequencies of modes (a) and (b).

7. Equation (2.3) for the frequency of a simple mass-spring vibrator assumes that the mass of the spring is much smaller than that of the load and thus can be neglected. This will not always be the case. The formula can be refined by letting m be the mass of the load plus one-third the mass of the spring. Suppose that the spring in the example in Section 2.1 has a mass of 100 g (K was found to be 196 N/m). Calculate the vibration frequencies with loads of 0.5 kg and 2 kg, and compare them to those given in the example.

EXPERIMENTS FOR HOME, LABORATORY, AND CLASSROOM DEMONSTRATION

Home and Classroom Demonstration

1. *Simple vibrating system: dependence of frequency on mass* Load a spring with several different masses and determine its frequency by counting oscillations during some appropriate time interval (such as a half-minute). What happens to the frequency when the mass is doubled? What happens when it is quadrupled?

2. *Simple vibrating system: dependence of frequency on spring constant* Determine the frequency of the simple vibrating system using several different spring constants. The spring constant can be determined by loading the spring with different masses and noting its static deflection, but this may not be necessary. Connecting two identical springs in series reduces the spring constant by half, whereas connecting them in parallel doubles the spring constant. What happens to the frequency when the spring constant is doubled? What happens when it is halved?

3. *Pendulum: dependence of frequency on mass*   Determine the frequency of a simple pendulum by counting oscillations during some appropriate time interval. What happens to the frequency when the mass is doubled? (Be careful not to change the length.) What happens when it is quadrupled?

4. *Pendulum: dependence of frequency on length*   Determine the frequency of a simple pendulum by counting oscillations. What happens to the frequency when the length is doubled? What happens when it is halved?

5. *Matching a pendulum to a mass-spring vibrator*   Begin with a spring having an unloaded length L_0. Load it with a mass M so that its length increases by L. Verify that a pendulum of length L has the same natural frequency as the mass-spring system. Can you show why this is so? (*Hint:* What is the spring constant?)

6. *Soda-bottle resonator*   Blow across the top of an empty soda bottle to determine its natural frequency (acting as a Helmholtz resonator). Fill the bottle half full of water and determine the frequency change.

7. *Modes of a vibrating string*   Pluck a stretched string and note its frequency (pitch). Touch the string lightly at its center for a moment to damp out the fundamental and note the frequency. Repeat by touching it at one-third its length.

8. *Chladni patterns*   Support a square plate at its center, sprinkle fine sand or salt on it, and bow the edge with a violin or cello bow. By touching the plate at different places and varying the bowing position, interesting vibration patterns can be obtained. Alternatively, a current-carrying coil can be positioned near a small permanent magnet attached to the plate in order to apply a sinusoidal force at a single frequency. Varying the frequency causes the plate to vibrate in its various modes. Repeat with a circular plate (the latter method was used to make the Chladni patterns in Fig. 2.19).

Laboratory Experiments

Simple harmonic motion (Experiment 2 in *Acoustics Laboratory Experiments*)

Vibrating strings (Experiment 4 in *Acoustics Laboratory Experiments*)

Vibrations of bars and plates (Experiment 5 in *Acoustics Laboratory Experiments*)

Stroboscopic measurements (Experiment 8 in *Acoustics Laboratory Experiments*)

Back and forth motion (Experiment 2 in *Physics with Computers*)

Pendulum periods (Experiment 14 in *Physics with Computers*)

Simple harmonic motion (Experiment 15 in *Physics with Computers*)

Energy in simple harmonic motion (Experiment 17 in *Physics with Computers*)

CHAPTER

3

Waves

Section 1.1 gave us a brief introduction to sound waves. We learned that sound waves in air are longitudinal waves; that is, the back-and-forth motion of the air is in the direction of travel of the sound wave. Many other types of waves, such as light waves and radio waves, are transverse waves. Although sound waves are very different from light waves or ocean waves, all waves possess certain common properties. In this chapter, some of these common properties will be discussed, along with some particular properties of sound waves.

In this chapter you should learn:

- About progressive waves and standing waves;
- About sound waves;
- About reflection of waves at a boundary;
- About refraction of waves;
- About interference and diffraction;
- About the Doppler effect.

3.1 ■ WHAT IS A WAVE?

One of the first properties noted about waves is that they can transport energy and information from one place to another through a medium, but the medium itself is not transported. A disturbance, or change in some physical quantity, is passed along from point to point as the wave propagates. In the case of light waves or radio waves, the disturbance is a changing electric and magnetic field; in the case of sound waves, it is a change in pressure and density. But in either case, the medium reverts to its undisturbed state after the wave has passed.

All waves have certain things in common. For example, they can be reflected, refracted, or diffracted, as we shall see later in this chapter. All waves have energy, and they transport energy from one point to another. Waves of different types propagate with widely varying speeds, however. Light waves and radio waves travel 3×10^8 m (186,000 mi) in 1 s, for example, whereas sound waves travel only 344 m/s. Water waves are still slower, traveling only a few feet in a second. Light waves and radio waves can travel millions of miles through empty space, whereas sound waves require some material medium (gas, liquid, or solid) for propagation.

3.2 ■ PROGRESSIVE WAVES

Suppose that one end of a rope is tied to a wall and the other end is held, as shown in Fig. 3.1. If the end being held is moved up and down f times per second, a wave with a frequency f will propagate down the rope, as shown. (When it reaches the tied end, a reflected wave will return, but we will ignore this for the moment.) The wave travels at a speed v that is determined by the mass of the rope and the tension applied to it.

If we were to observe the wave carefully (a photograph might help), we would note that the crests or troughs of the wave are spaced equally; we call this spacing the wavelength λ (Greek letter lambda).

It is not difficult to see that the wave velocity is the frequency times the wavelength:

$$v = f\lambda. \tag{3.1}$$

That is, if f waves pass a certain point each second and the crests are λ meters apart, they must be traveling at a speed of $f\lambda$ meters per second.

It is possible to propagate either transverse or longitudinal waves in solids, but in general only longitudinal waves propagate through gases and liquids. Figure 3.2 illustrates the propagation of longitudinal and transverse waves in a mass-spring system. This system is also a large-scale model (in one dimension) of a solid crystal and illustrates ways in which vibrations may propagate in a solid.

In a solid, longitudinal waves travel at a speed represented by the formula

$$v = \sqrt{\frac{E}{\rho}}, \tag{3.2}$$

where ρ is the density of the solid and E is called the elastic modulus (*Young's modulus*). Note that the speed of longitudinal waves in a solid bar is independent of its dimensions. This is not so for transverse waves, whose speed is dependent on the dimensions. For a wire or string, the transverse wave velocity (speed) is

$$v = \sqrt{\frac{T}{\mu}}, \tag{3.3}$$

FIGURE 3.1

A traveling wave generated by moving the end of a rope.

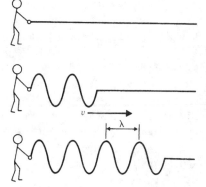

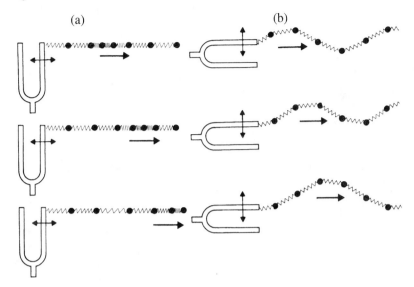

FIGURE 3.2
Wave motion in a
one-dimensional
array:
(a) longitudinal
waves;
(b) transverse
waves.

where T is the tension and μ is the mass per unit of length. In a stiff rod or bar, the velocity of transverse waves varies with frequency, and so a simple formula cannot be written. In general, longitudinal waves travel much faster than transverse waves do in solids. The speed of longitudinal (sound) waves in aluminum, for example, is 5000 m/s (about 3 mi/s).

EXAMPLE 3.1 The density of steel is 7700 kg/m^3 and Young's elastic modulus is 19.5×10^{10} N/m^2. What is the speed of longitudinal waves in a steel glockenspiel bar? How does this compare with the speed of longitudinal vibrations (sound waves) in air?

Solution

$$v = \sqrt{\frac{E}{\rho}} = \sqrt{\frac{19.5 \times 10^{10} \text{ N/m}^2}{7.7 \times 10^3 \text{ kg/m}^3}} = 5032 \text{ m/s}.$$

(This is $\frac{5032}{343} = 14.7$ times the speed of longitudinal (sound) waves in air.)

EXAMPLE 3.2 What tension would a steel wire 1 mm in diameter require in order that the transverse and longitudinal wave speeds are equal?

Solution

$$v = \sqrt{\frac{T}{\mu}}, \quad \text{so} \quad T = \mu v^2$$

$$\rho(\pi r^2)v^2 = 7700(3.14)(5 \times 10^{-4})^2(5032)^2$$
$$= 1.53 \times 10^5 \text{ N}$$

(far greater than the breaking force).

3.3 ■ IMPULSIVE WAVES; REFLECTION

Suppose that the rope in Fig. 3.1 is given a single impulse by quickly moving the end up and down. The impulse will travel at the wave speed v and will retain its shape fairly well as it moves down the rope, as illustrated in Fig. 3.3.

The question arises as to what happens when the incident pulse reaches the end of the rope. Careful observation shows that a pulse reflects back toward the sender. This reflected pulse is very much like the original pulse, except that it is upside down. If the end of the rope were left free to flop like the end of a whip, the reflected pulse would be right side up, as illustrated in Fig. 3.4(b). Photographs of impulsive waves on a long string with fixed and free ends are shown in Figs. 3.5 and 3.6. Note that the reflected pulse in Fig. 3.5 is upside down. This is called a *reversal of phase*. In Fig. 3.6 the *phase* of the reflected wave remains the same as that of the original wave.

It is instructive to tie the rope at the base of a mirror (see Fig. 3.7). Then the mirror image of the pulse (generated by your image) travels at the same speed and arrives at the end of the rope at the same time as the actual pulse, and appears to continue on as the reflected pulse. (To make the sense of the pulse correct, two mirrors can be used to form

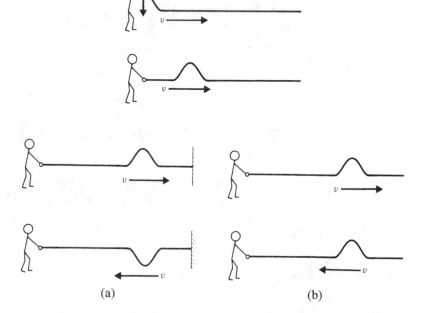

FIGURE 3.3   An impulsive wave generated by moving the end of a rope.

FIGURE 3.4 Reflection of an impulsive wave (a) at a fixed end; (b) at a free end.

(a)   (b)

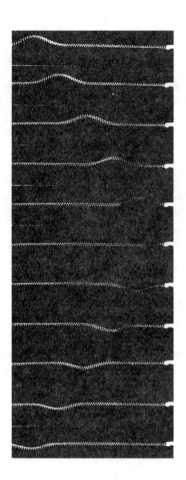

FIGURE 3.5 An impulsive wave in a long spring. The pulse travels left to right and reflects back to the left as in Fig. 3.4(a). (From *PSSC Physics*, 2nd ed., 1965, D. C. Heath & Co. with Education Development Center, Newton, Mass.)

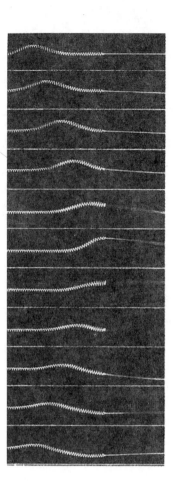

FIGURE 3.6 An impulsive wave in a spring showing reflection at a free end (actually a very light thread). Compare the reflected pulse to that of Fig. 3.5. (From *PSSC Physics*, 2nd ed., 1965, D. C. Heath & Co. with Education Development Center, Newton, Mass.)

FIGURE 3.7 The mirror image of an impulsive wave approaching a point of reflection P. In a plane mirror, the two pulses have the same sense, but in a corner mirror the two pulses have opposite sense (just as the incident and reflected pulses on the rope with a fixed end).

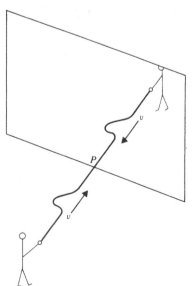

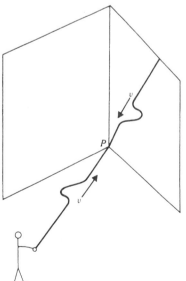

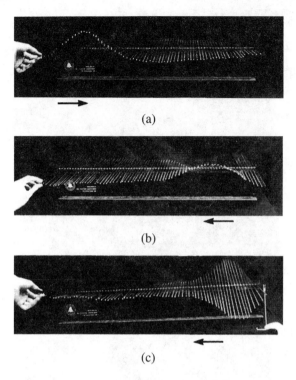

(a)

(b)

(c)

FIGURE 3.8
Wave propagation
on a "wave
machine":
(a) incident pulse;
(b) reflection at a
free end;
(c) reflection at a
fixed end.
(Photographs by
Craig Anderson.)

a corner reflector, but this is really not necessary to achieve the sensation of the reflected pulse coming from a virtual source and meeting the original pulse at the point of reflection.)

One can think of the reflected wave on the rope as coming from an imaginary source, whether or not a mirror is there to show a reflected image. If the rope is tied to a solid object at its far end, the deflection must be zero at all times, even when the pulse arrives; this requires that the pulse be met by a pulse of opposite sense, as shown in Fig. 3.4(a). If the end is free, it snaps like a whip, momentarily doubling its displacement when the pulse arrives. This is equivalent to the arrival of a pulse with the same sense, which then continues as the reflected pulse shown in Fig. 3.4(b).

Several interesting properties of waves can be studied with a wave machine developed at the Bell Laboratories, which consists of a long array of rods attached to a wire. Waves travel slowly on this machine; hence they can be observed rather easily. Reflection of a pulse at free and fixed ends is illustrated in the photographs of the wave machine in Fig. 3.8.

3.4 ■ SUPERPOSITION AND INTERFERENCE

An interesting feature of waves is that two of them, traveling in opposite direction, can pass right through each other and emerge with their original identities. The *principle of linear superposition* describes this behavior. For wave pulses on a rope or spring, for example, the displacement at any point is the sum of the displacements due to each pulse by itself. The wave pulses shown in Fig. 3.9 illustrate the principle of superposition. If the pulses have the

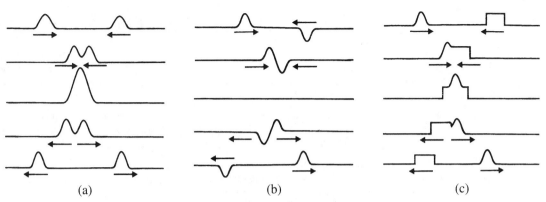

FIGURE 3.9 The superposition of wave pulses that travel in opposite directions: (a) pulses in the same direction; (b) pulses in opposite directions; (c) pulses with different shapes.

same sense, they add; if they have the opposite sense, they subtract when they meet. These are examples of *interference* of pulses. The addition of two similar pulses (3.9a) is called *constructive interference*; the subtraction of opposing pulses (3.9b) is called *destructive interference*.

Suppose that both ends of a rope (or the wave machine shown in Fig. 3.8) are shaken up and down at the same frequency, so that continuous waves travel in both directions. Continuous waves interfere in much the same manner as the impulsive waves we have just considered. If two waves arrive at a point when they have opposite sense, they will interfere destructively; if they arrive with the same sense, they will interfere constructively. Under these conditions, the waves do not appear to move in either direction, and we have what is called a *standing wave*.

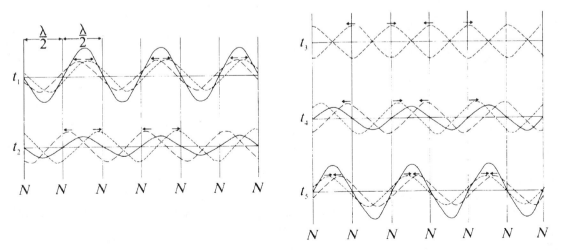

FIGURE 3.10 Interference of two identical waves in a one-dimensional medium. At times t_1 and t_5 there is constructive interference, and at t_3 there is destructive interference. Note that at points marked N, the displacement is always zero.

In the case of two identical waves (same frequency and amplitude) traveling in opposite directions on a rope or a spring, there will be alternating regions of constructive and destructive interference, as shown in Fig. 3.10. The points of destructive interference that always have zero displacement are called nodes; they are denoted by N in Fig. 3.10. Between the nodes are points of constructive interference, where displacement is a maximum; these are called antinodes. At the antinodes, the displacement oscillates at the same frequency as in the individual waves; the amplitude is the sum of the individual wave amplitudes.

Note that the antinodes in Fig. 3.10, formed by the interference of two identical waves, are one-half wavelength apart. Because these points of maximum displacement do not move through the medium, the configuration is called a standing wave. Standing waves result whenever waves are reflected back to meet the oncoming waves. The case illustrated in Fig. 3.10, in which the forward and backward waves have the same amplitude, is a special case that leads to total interference. If the two incident waves do not have the same amplitude, the nodes will still be points of minimum but not zero displacement.

3.5 ■ SOUND WAVES

Sound waves are longitudinal waves that travel in a solid, liquid, or gas. To aid in your understanding of sound waves, consider a large pipe or tube with a loudspeaker at one end. Although sound waves in this tube are similar in many respects to waves on a rope, they are more difficult to visualize, because we cannot see the displacement of the air molecules as the sound wave propagates down the tube.

Suppose we consider first a single pulse, as we did in the case of the rope. An electrical impulse to the loudspeaker causes the speaker cone to move forward suddenly, compressing the air directly in front of it *very* slightly (even a very loud sound results in a pressure increase of less than 1/10,000 atmospheric pressure). This pulse of air pressure travels down the tube at a speed of about 340 m/s (more than 700 mi/h). It may be absorbed at the far end of the tube, or it may reflect back toward the loudspeaker (as a positive pulse of pressure or a negative one), depending on what is at the far end of the tube.

Reflection of a sound pulse for three different end conditions is illustrated in Fig. 3.11. If the end is open, the excess pressure drops to zero, and the pulse reflects back as a negative

FIGURE 3.11
Reflection of a sound pulse in a pipe: (a) incident pulse; (b) reflection at an open end; (c) reflection at a closed end; (d) no reflection from absorbing end.

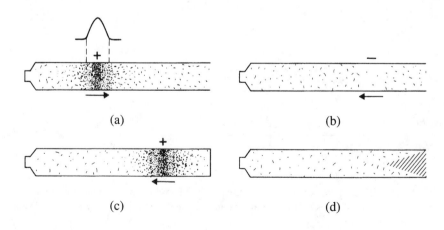

pulse of pressure, as shown in Fig. 3.11(b); this is analogous to the fixed-end condition illustrated in Figs. 3.5 and 3.8(b).* If the end is closed, however, the pressure builds up to twice its value, and the pulse reflects back as a positive pulse of pressure; this condition, shown in Fig. 3.11(c), is analogous to the free-end reflection of Figs. 3.6 and 3.8(c). If the end is terminated with a sound absorber, there is virtually no reflected pulse. Such a termination is called *anechoic*, which means "no echo."

The speed of sound waves in an ideal gas is given by the formula

$$v = \sqrt{\frac{\gamma R T}{M}},\tag{3.4}$$

where T is absolute temperature, M is the molecular weight of the gas, and γ and R are constants for the gas. For air, $M = 2.88 \times 10^{-2}$, $R = 8.31$, and $\gamma = 1.4$, so $v = 20.1\sqrt{T}$. The absolute temperature T is found by adding 273 to the temperature on the Celsius scale. At $t = 21°C$, for example, $T = 294$ K, so $v = 344$ m/s. At Celsius zero, $v = 331$ m/s. Over the range of temperature we normally encounter, the speed of sound increases by about 0.6 m/s for each Celsius degree, and an approximate formula for the speed of sound is sufficiently accurate:

$$v = 331.3 + 0.6t \text{ m/s},\tag{3.5}$$

where t is the temperature in degrees on the Celsius scale. The speed of sound in an ideal gas is independent of atmospheric pressure. In air the change in speed with pressure change is generally too small to measure.

Sound waves travel much faster in liquids and solids than they do in gases. The speed of sound in several materials is given in Table 3.1.

TABLE 3.1   Speed of sound in various materials

Substance	Temperature (°C)	Speed (m/s)	Speed (ft/s)
Air	0	331.3	1,087
Air	20	343	1,127
Helium	0	970	3,180
Carbon dioxide	0	258	846
Water	0	1,410	4,626
Methyl alcohol	0	1,130	3,710
Aluminum	—	5,150	16,900
Steel	—	5,100	16,700
Brass	—	3,480	11,420
Lead	—	1,210	3,970
Glass	—	3,700–5,000	12–16,000

*In an actual tube with an open end, a little of the sound will be radiated; most of it, however, will be reflected as shown.

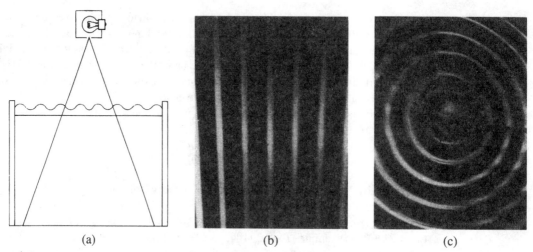

(a)   (b)   (c)

FIGURE 3.12   (a) A ripple tank for projecting an image of water waves; (b) straight waves on a ripple tank; (c) circular waves on a ripple tank. (Photographs by Christopher Chiaverina.)

3.6 ■ WAVE PROPAGATION IN TWO AND THREE DIMENSIONS

Thus far we have considered only waves that travel in a single direction (along a rope or in a pipe, for example). One-dimensional waves of this type are a rather special case of wave motion. More often, waves travel outward in two or three dimensions from a source.

Water waves are a familiar example of two-dimensional waves. Many wave phenomena, in fact, can be studied conveniently by means of a ripple tank in the laboratory. A ripple tank uses a glass-bottom tray filled with water; light projected through the tray forms an image of the waves on a large sheet of paper, as shown in Fig. 3.12.

Three-dimensional waves are difficult to make visible. An ingenious technique has been used to photograph three-dimensional wave patterns (though not the actual waves) at the Bell Laboratories and elsewhere. A tiny microphone and a neon lamp together scan the sound field in a dark room while a camera lens remains open in a time exposure. The brightness of the neon lamp is controlled by an amplifier, so that bright streaks appear at wave crests, as shown in Fig. 3.13.

FIGURE 3.13
The pattern of sound waves from a loudspeaker produced by scanning with a microphone and neon lamp. (From Kock 1971.)

FIGURE 3.14
Sound wave
patterns from (a) a
single small
loudspeaker (*point
source*); (b) a
column of
loudspeakers
(*cylindrical
source*).

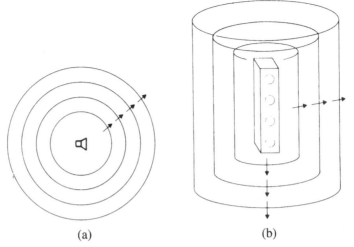

(a)   (b)

Different types of sources radiate different kinds of patterns. A point source or a source
that is spherically symmetric radiates spherical waves. A line source or a source with cylindrical symmetry radiates cylindrical waves. A large flat source radiates plane waves. Real
sound sources are never true point sources, line sources, or flat sources, however; what we
may have in real life are sources that approximate one of these geometries.

A source that is very small compared to a wavelength of sound approximates a point
source and emits nearly spherical waves. A small enclosed loudspeaker will radiate nearly
spherical waves at low frequency, as shown in Fig. 3.14(a). A column of small loudspeakers
may resemble a line source at low frequency and emit cylindrical waves, as shown in
Fig. 3.14(b). Symmetrical radiation patterns of this type can be observed outside, away
from reflecting objects, or in an anechoic (echo-free) room.

3.7 ■ THE DOPPLER EFFECT

Ordinarily, the frequency of the sound waves that reach the observer is the same as the
frequency of vibration of the source. There is a notable exception, however, if either the
source or the observer is in motion. If they are moving toward each other, the observed
frequency is greater than f_s; if they are moving apart, the observed frequency is lower than
f_s. This apparent frequency shift is called the *Doppler effect*.

The Doppler effect is explained quite simply with the aid of Fig. 3.15. Suppose that a
source S emits 100 waves per second. An observer at rest, O, will count 100 waves per
second passing him or her. However, an observer O' moving toward the source will count
more waves because he or she "meets" them as he or she moves, just as the driver of an
automobile meets oncoming traffic. The apparent frequency (the rate at which the observer
meets waves) will be

$$f' = f_s \frac{v + v_o}{v}, \tag{3.6}$$

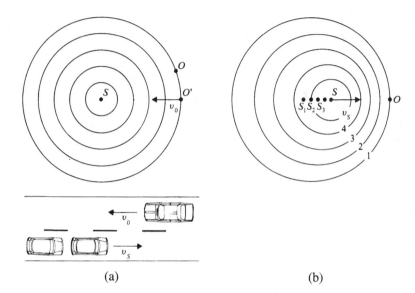

FIGURE 3.15
The Doppler effect:
(a) observer
moving toward the
sound source;
(b) source moving
toward the
observer.

(a)   (b)

where f_s is the frequency of the source, v_o is the speed of the observer, and v is the speed of sound. Note that after the observer passes the sound source, v_o must be subtracted from v. Thus the frequency drops abruptly as the observer passes the source.

There is also a Doppler effect if the source is in motion. You have probably observed a drop in pitch or frequency of the noise as a truck or car passes by while you are standing at the side of the road. The case of the moving source is shown in Fig. 3.15(b). The source emitted the wave numbered 1 when it was at position S_1, number 2 when at S_2, etc. The wave fronts resemble spheres with centers constantly shifting to the right as the source moves. Thus the observer O receives waves at a greater rate than he or she would from a stationary source. If the speed of the source is v_s, the apparent frequency will be

$$f = f_s \frac{v}{v - v_s}. \tag{3.7}$$

Note that if the source moves directly toward the observer, the frequency will drop *abruptly*, not gradually, as the source passes by.

The Doppler effect is also used by astronomers to determine the motion of a distant star. By carefully analyzing the spectra of starlight, astronomers can detect shifts in frequency (Doppler shifts) due to the motion of the stars toward or away from us. Because the universe appears to be expanding, most distant starts exhibit a "red shift"; that is, their frequency is shifted toward the low-frequency (red) end of the visible spectrum. It is remarkable how much we know about our universe just by analyzing the light from distant stars, some of which was radiated millions of years ago and has been traveling through space ever since!

EXAMPLE 3.3   An automobile horn emits a tone with a frequency of 440 Hz. What is the apparent frequency when the automobile approaches an observer at 55 mi/h (25 m/s) and what is the apparent frequency when it recedes at this same speed?

Solution

$$f' = f_s \frac{v}{v - v_s} = 440 \frac{343}{343 - 25} = 475 \text{ Hz};$$

$$f' = 440 \frac{343}{343 + 25} = 410 \text{ Hz}.$$

Note that the pitch drops by 14% (more than two semitones on the musical scale).

EXAMPLE 3.4 A police radar speed gun transmits microwaves having a frequency of 9600 MHz. What is the upward shift in frequency for waves reflected from an automobile traveling at 55 mi/h (25 m/s)?

Solution View the automobile as a mirror moving at 25 m/s; the "image" of the source appears to move at twice this speed, or 50 m/s.

$$\Delta f = f' - f_s = f_s \frac{v}{v - v_s} - f_s$$

$$= 9.6 \times 10^9 \frac{3 \times 10^8}{3 \times 10^8 - 50} - 9.6 \times 10^9$$

$$\cong 9.6 \times 10^9 \left(1 + \frac{50}{3 \times 10^8} - 1 \right)$$

$$= 1600 \text{ Hz}.$$

(Although this frequency shift is only one part in six million, it can readily be measured—as many speeding motorists know—by mixing together the transmitted and reflected microwaves.)

3.8 ■ REFLECTION

The reflection of wave pulses of one dimension on a rope or in a tube was discussed in Sections 3.3 and 3.5. Waves of two or three dimensions undergo similar reflections when they reach a barrier. The reflection of light waves by a mirror is a phenomenon familiar to all of us, as is the echo that results from clapping one's hands some distance away from a large wall, which reflects the sound waves back to the source. Figure 3.16(a) shows the reflection of water waves from a straight barrier. Note that the spherical reflected waves appear to come from a point behind the barrier. This point, which is called the *image* is denoted by S' in Fig. 3.16(b). It is the same distance from the reflector as the source S is.

Reflection of waves from a curved barrier can lead to the *focusing* of energy at a point, as shown in Fig. 3.17. A curious case of sound focusing, which occurs in "whispering galleries," is shown in Fig. 3.17(b). Sound originating from a source S is reflected by a curved barrier, beamed to a second curved barrier, and focused at O.

FIGURE 3.16
Reflection of waves
from a barrier:
(a) waves on a
ripple tank from a
point source;
(b) reflected waves
appear to originate
from image S'.

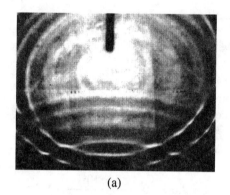

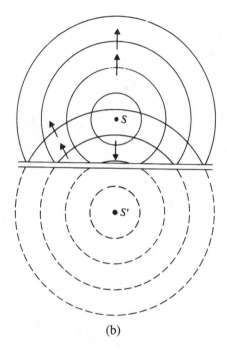

(a)   (b)

Well-known examples of whispering galleries are found in the Museum of Science and
Industry in Chicago and in the National Capitol in Washington. The curved ceilings of
certain auditoriums, such as the Mormon Tabernacle in Salt Lake City, make it possible to
transmit whispers between selected spots. However, the focusing of sound by curved walls
is frequently detrimental to the acoustics of auditoria, as we will discuss in Chapter 23.

FIGURE 3.17
Reflection of waves
by a curved barrier:
(a) incoming waves
are focused at F by
a curved reflector;
(b) whispering
gallery in which
two curved
reflectors beam
sound from source
S to observer O
with great
efficiency.

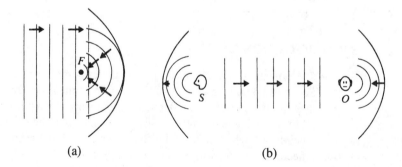

(a)   (b)

3.9 ■ REFRACTION

When the speed of waves changes, a phenomenon called *refraction* occurs, which can
result in a change in the direction of propagation or a bending of the waves. The change
of speed may occur abruptly as the wave passes from one medium to another, or it may

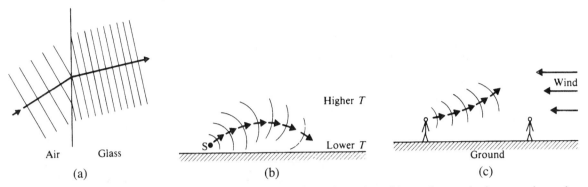

FIGURE 3.18 Refraction of waves: (a) light waves passing from air to glass; (b) sound waves in the atmosphere when temperature varies with height; (c) sound traveling against the wind.

change gradually if the medium changes gradually. These two situations are illustrated in Fig. 3.18.

The situation illustrated in Fig. 3.18(b), which sometimes occurs during the cool evening hours, causes sounds to be heard over great distances. Because the speed of sound increases with temperature (Section 3.5), the sound travels faster some distance above the ground where the temperature is greater. This results in a bending of sound downward as shown. Sound that would ordinarily be lost to the upper atmosphere is refracted back toward the ground.

Figure 3.18(c) shows why it is difficult to be heard when yelling against the wind. (It is *not* because the wind blows the sound waves back; even a strong wind has a speed much less than that of sound). Refraction results because the wind speed is less near the ground than it is some distance above it. Because the speed of sound with respect to the air (in this case, moving air) remains the same, the ground speed of the sound changes with altitude. The resulting refraction causes some of the sound to miss its target.

3.10 ■ DIFFRACTION

When waves encounter an obstacle, they tend to bend around the obstacle. This is an example of a phenomenon known as *diffraction*. Diffraction is also apparent when waves pass through a narrow opening and spread out beyond it. Examples of the diffraction of water waves, light waves, and sound waves are shown in Figs. 3.19, 3.20, and 3.21.

An important point to remember is that it is the size of the opening in relation to the wavelength that determines the amount of diffraction. A loudspeaker 0.2 m (8 in.) in diameter, for example, will distribute sound waves of 100 Hz ($\lambda = 3.4$ m) in all directions, but waves of 2000 Hz ($\lambda = 0.2$ m) will be much louder directly in front of the speaker than at the sides, because diffraction will be minimal.

3.11 ■ INTERFERENCE

In Section 3.4, we pointed out that interference between incident and reflected waves leads to standing waves. Standing waves exist in a room due to interference between waves

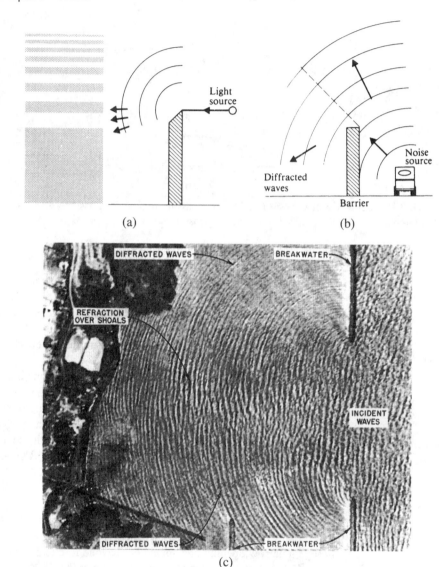

FIGURE 3.19
Diffraction of
waves by a barrier:
(a) shadow of a
straight edge
magnified to show
diffraction of light;
(b) diffraction of
sound waves allows
noise to "leak"
around a wall;
(c) ocean waves in
a harbor.
(Photograph (c)
courtesy of
University of
California at
Berkeley.)

reflected from the ceiling, walls, and other surfaces; these can be observed by moving one's head around while a pure tone is played through a loudspeaker.

Waves from two identical sources provide another example of interference. Constructive and destructive interference lead to minima and maxima in certain directions, as shown in Fig. 3.22. The interference patterns are determined by the spacing of the two sources compared to a wavelength.

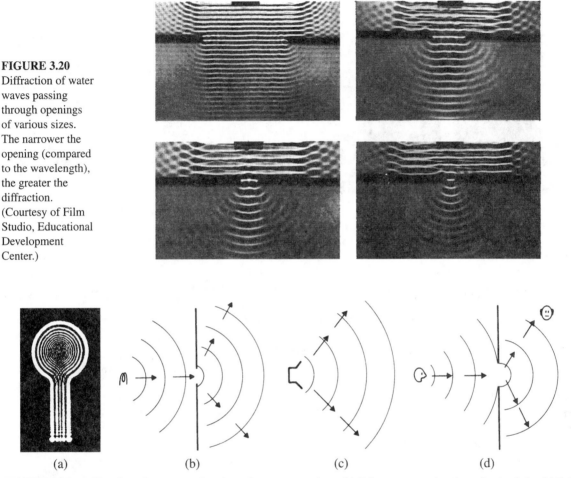

FIGURE 3.20
Diffraction of water
waves passing
through openings
of various sizes.
The narrower the
opening (compared
to the wavelength),
the greater the
diffraction.
(Courtesy of Film
Studio, Educational
Development
Center.)

(a)   (b)   (c)   (d)

FIGURE 3.21   Diffraction of waves passing through narrow openings: (a) light waves passing through a keyhole; (b) light waves through a very narrow slit; (c) sound waves from a loudspeaker; (d) diffraction allows sound to be heard behind a doorway.

3.12 ■ SUMMARY

We are surrounded by waves of many types (light waves, radio waves, sound waves, water waves, etc.). These quite different types of waves have many properties in common. All carry energy; all can be reflected, refracted, and diffracted; interference leads to regions of minimum and maximum amplitude. However, the speeds at which these waves travel varies widely. Waves can be classed as transverse or longitudinal depending on the direction of the vibrations. Sound waves are longitudinal vibrations of molecules that result in pressure fluctuations. The speed of sound waves in air increases with temperature.

Some wave phenomena can be understood best by considering wave propagation in one dimension and extending the ideas to two- and three-dimensional waves.

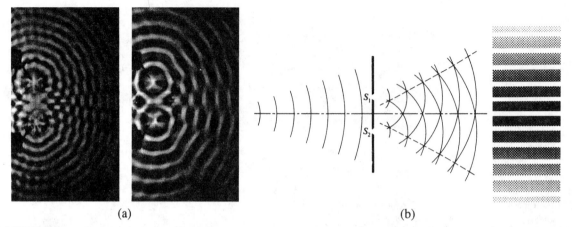

(a)   (b)

FIGURE 3.22   Interference of waves from two identical sources: (a) water waves in a ripple tank; (b) light waves from two slits illuminated by the same light source. (From *PSSC Physics*, 2nd ed., 1965, D. C. Heath & Co. with Education Development Center, Newton, Mass.)

REFERENCES AND SUGGESTED READINGS

Appel, K., J. Gastineau, C. Bakken, and D. Vernier (1998). *Physics with Computers*. Portland, Ore.: Vernier Software.

Baez, A. V. (1967). *The New College Physics*. San Francisco: Freeman.

Haber-Schaim, U., J. H. Dodge, R. Gardner, and E. A. Shore (1991). *PSSC Physics*, 7th ed. New York: Kendall/Hunt.

Hewitt, P. G. (1993). *Conceptual Physics*, 7th ed. New York: Harper Collins.

Kirkpatrick, L. D., and G. F. Wheeler (1995). *Physics: A World View*, 2nd ed. Philadelphia: Saunders.

Kock, W. E. (1971). *Seeing Sound*. New York: Wiley.

Physical Science Study Committee (1965). *Physics*, 2nd ed. Boston: D. C. Heath.

Project Physics—Text (1975). New York: Holt, Rinehart and Winston.

Rossing, T. D. (1982). *Acoustics Laboratory Experiments* DeKalb: Northern Illinois University.

GLOSSARY

absolute temperature   The temperature (in kelvins) on a scale that has its zero at the lowest attainable temperature ($-273°C$); absolute temperature is found by adding 273 to the Celsius temperature.

amplitude   The maximum displacement from equilibrium in a wave or a vibrating system.

anechoic   Echo free; an anechoic room is one whose walls, ceiling, and floor are covered with sound-absorbing material, usually in the shape of wedges.

diffraction   The spreading out of waves when they encounter a barrier or pass through a narrow opening.

Doppler effect   The shift in apparent frequency when the source or observer is in motion.

impulsive wave   A brief disturbance or pressure change that travels as a wave.

interference   The interaction of two or more identical waves, which may support (constructive interference) or cancel (destructive interference) each other.

longitudinal wave   A wave in which the vibrations are in the direction of propagation of the wave; *example:* sound waves in air.

reflection   An abrupt change in the direction of wave propagation at a change of medium (by waves that remain in the first medium).

refraction A bending of waves when the speed of propagation changes, either abruptly (at a change of medium) or gradually (e.g., sound waves in a wind of varying speed).

standing wave A wavelike pattern that results from the interference of two or more waves; a standing wave has regions of minimum and maximum amplitude called nodes and antinodes.

superposition The motion at one point in a medium is the sum of the individual motions that would occur if each wave were present by itself without the others.

transverse wave A wave in which the vibrations are at right angles to the direction of propagation of the wave; *example:* waves on a rope.

waveform The graph of some variable (e.g., wave displacement, sound pressure) versus time.

wavelength The distance between corresponding points on two successive waves.

Young's modulus An elastic modulus of a solid; the ratio of force per unit area to the stretch it produces.

REVIEW QUESTIONS

1. How many times faster do light waves travel as compared to sound waves?

2. Compare the speed of longitudinal (sound) waves in an aluminum rod 10 mm in diameter with a similar rod 5 mm in diameter.

3. Doubling the frequency of a sound wave multiplies the wavelength by what factor?

4. Compare the phases of an incident and reflected pulse on a rope with a fixed end and then with a free end.

5. What must be true of the sizes and shapes of two pulses that meet on a rope in order for them to cancel each other (interfere destructively)?

6. Compare the speed of sound in air, water, and steel.

7. How does the speed of sound in air change with temperature?

8. How does the speed of sound in air change with atmospheric pressure?

9. Describe the sound of a car horn as the moving car passes an observer standing at roadside.

10. What is the cause of the "red shift" observed in the spectra of distant stars?

11. Why are curved walls sometimes detrimental to concert hall acoustics?

12. Why is it difficult to be heard when you shout into a strong wind?

13. Why is it possible to hear around a corner but not to see around a corner?

14. Why do you see a flash of lightning seconds before you hear the thunder associated with it?

15. Why do low frequency sound waves from a subwoofer spread out in all directions, but high-frequency sound waves from a tweeter travel pretty much straight ahead?

QUESTIONS FOR THOUGHT AND DISCUSSION

1. Although ocean waves are often described as transverse waves, the motion of a small bit of water is actually in a circle. Why could strictly transverse waves not exist on a water surface?

2. Mine operators carefully select the right atmospheric conditions for blasting operations in order to minimize community disturbance. What atmospheric conditions would be optimum?

3. **(a)** Will a larger pulse (with more energy) overtake a smaller pulse as they travel down a rope?

   (b) Will a baseball thrown with more energy overtake a baseball with less energy?

   (c) Does a loud sound travel faster than a softer sound?

4. A camera lens is made of glass in which light travels slower than it does in air. Could you construct a lens for sound by filling a balloon with carbon dioxide?

5. Would you expect interference effects from two loudspeakers in a room?

6. Compare the speed of sound at altitude of a jet airplane with the speed at ground level.

7. At some high altitude the density of air will equal the density of helium gas at sea level. What would you expect the speed of sound to be at this altitude?

EXERCISES

1. Electromagnetic waves travel through space at a speed of 3×10^8 m/s. Find the frequency of the following. (1 nm $= 10^{-9}$ m)

   (a) radio waves with $\lambda = 100$ m

   (b) waves of red light ($\lambda = 750$ nm)

   (c) waves of violet light ($\lambda = 500$ nm)

   (d) microwaves with $\lambda = 3$ cm (used in police radar)

2. Two trumpet players tune their instruments to exactly 440 Hz. Find the difference in the apparent frequencies due to the Doppler effect if one plays his or her instrument while marching away from an observer and the other plays while marching toward the observer. Is this enough to make them sound out of tune? (Assume 1 m/s as a reasonable marching speed.)

3. How much will the velocity of sound in a trumpet change as it warms up (from room temperature to body temperature, for example)? If the wavelength remains essentially the same (the expansion in length will be very small), by what percentage will the frequency change?

4. At what frequency does the wavelength of sound equal the diameter of the following? (1 in. $= 0.0254$ m)

   (a) a 15-in. woofer

   (b) a 3-in. tweeter

5. A nylon guitar string has a mass per unit length of 8.3×10^{-4} kg/m and the tension is 56 N. Find the speed of transverse waves on the string.

6. The audible range of frequencies extends from approximately 50 to 15,000 Hz. Determine the range of wavelengths of audible sound.

7. The distance from the bridge to the nut on a certain guitar is 63 cm. If the string is plucked at the center, how long will it take the pulse to propagate to either end and return to the center? (Use the speed calculated in Problem 5.)

8. Find the speed of sound in miles per hour at $0°C$. This is called Mach 1. A supersonic airplane flying at Mach 1.5 is flying at 1.5 times this speed. Find its speed in miles per hour.

9. A thunderclap is heard 3 s after a lightning flash is seen. Assuming that they occurred simultaneously, how far away did they originate?

10. The density of aluminum is 2700 kg/m^3 and Young's elastic modulus is 7.1×10^{10} N/m^2 (Pa). Compare the speed of longitudinal waves in aluminum to those in steel (See Example 3.1).

11. Compare the speed of sound calculated from Eqs. (3.4) and (3.5) when $t = 30°C$.

EXPERIMENTS FOR HOME, LABORATORY, AND CLASSROOM DEMONSTRATION

Classroom Demonstrations

Waves in one dimension

1. *Waves on a rope*   A rope is stretched across the front of the room and one end is fastened to a door handle or other fixed point. The other end is held in one hand, and the other hand strikes it quickly to create a pulse. The speed of this pulse is shown to increase as the tension increases. The phase of the reflected pulse on the rope is seen to be reversed (see Fig. 3.4(a)).

2. *Wave machine*   A pulse is sent down a wave machine of the type developed at Bell Laboratories (see Fig. 3.8). Reflection at a fixed end again reverses the impulse, whereas reflection at a free end maintains the same orientation. When the wave machine is terminated with a dashpot, no reflection occurs.

3. *Standing waves on a rope*   The rope (in Experiment 1) is moved up and down rhythmically to produce a standing wave pattern. Vary the frequency to produce one, two, and three loops. Have a student grab the rope at a nodal point to show that waves still propagate "through" his or her hand. An elastic rope attached to an electromagnetic wave driver (Pasco SE9409 and WA9753, for example) is particularly convenient for this demonstration. Projecting a transparency similar to Fig. 3.10 (preferably with the three waves in different colors) at the same time is a great help to the students in understanding standing waves.

4. *Standing waves on a wave machine*   Standing waves are generated on the wave machine by moving the hand up and down rhythmically, by attaching a motorized driver, or by using an electromagnetic driver. Compare the standing waves that result from a fixed end and a free end.

5. *Wave reflection at an interface*   Wave machines generally include two sections having different wave speeds. Attach two

sections together, and show the difference when a pulse originates in the slower medium and the faster medium.

6. *Waves in a ripple tank* Wave images can be projected onto a screen by means of an overhead projector or a large mirror. In a small class, projection onto the ceiling or a white paper on the floor may be satisfactory. The class should be shown how single pulses, plane waves, and circular waves propagate, reflect (at straight and curved reflectors), and refract. Diffraction at a slit and two-source interference should also be demonstrated. More complex phenomena may be demonstrated by means of film or videotape.

7. *Standing waves in a room* A three-dimensional pattern of standing waves can be demonstrated by driving a loudspeaker with a sine wave of about 1000 Hz and having the students move their heads. Ask them to estimate the distance between adjacent maxima (these will be only approximately a half-wavelength apart in a three-dimensional standing wave). Repeat the experiment at other frequencies. They may be surprised to discover a maximum (of sound pressure) in each corner of the room.

Little is to be gained by using two loudspeakers. Two-source interference should probably not be demonstrated in a classroom.*

8. *Diffraction by a slit* Diffraction can best be demonstrated with light waves. A vertical line source (straight-filament lamp or fluorescent tube) is viewed through a slit formed by two fingers; as the spacing is narrowed a diffraction pattern appears. Slits of varying size can be ruled on smoked glass or an old photographic negative and passed around. Best of all are photographic negative slits made by photographing black lines of various widths.

Students should be reminded that the visible spectrum extends less than one octave, whereas the audible spectrum extends more than seven octaves; also, acoustics wavelengths are 30 thousand to 20 million times larger than optical wavelengths.

9. *Diffraction grating* Although multislit diffraction gratings for sound are rare, the diffraction of light by an inexpensive diffraction grating is so impressive that it is worth distributing such gratings in class so that students can view a vertical filament white lamp (showcase bulb), a bare fluorescent tube, and various gas discharge lamps (neon is particularly striking). Inexpensive diffraction glasses of various types are also very striking. (The Arbor Scientific catalog, for example, lists three different types.)

10. *Doppler effect with a moving source* A "Doppler ball," consisting of a small sound source and battery in a Styrofoam ball, can be thrown about the room to illustrate the Doppler effect. Twirling a small loudspeaker on the end of its electrical cable creates an impressive change in sound, but it is a little confusing because the Doppler effect is often overpowered by variation in sound-pressure level, generation of beats, etc.

11. *Recorded Doppler effect* Make a tape recording of trucks passing on an expressway or a train whistle on a passing train (better yet, ask students to make them). Try to estimate the speed of the passing vehicle from the change in pitch as it passes. Emphasize that the change is abrupt as the vehicle passes rather than gradual as the distance to the observer changes. The record set "The Science of Sound" (Folkways) has a good recording of racing cars, as does the video "Physics at the Indy 500" (AAPT).

Laboratory Experiments

Sinusoidal motion: The oscilloscope (Experiment 3 in *Acoustics Laboratory Experiments*)

Wave propagation: The ripple tank (Experiment 6 in *Acoustics Laboratory Experiments*)

Speed of sound (Experiment 9 in *Acoustics Laboratory Experiments*)

Speed of sound (Experiment 24 in *Physics with Computers*)

*See T. D. Rossing, "Acoustic demonstrations in lecture halls: A note of caution," *American J. Physics* **44**, 1220 (1976).

CHAPTER

4

Resonance

Consider a simple mechanical system: a child in a swing (Fig. 4.1). The swing has a natural frequency that is determined by its length (as the pendulum described in Section 2.3). If the swing is given a small push at the right time in each cycle, its amplitude gradually increases. This is an example of *resonance*. The swing receives only a small amount of energy during each push, but provided this amount is larger than the energy lost during each cycle (due to friction and air drag), the *amplitude* of swing increases.

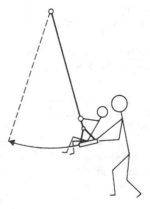

FIGURE 4.1
An example of resonance: a child in a swing.

In this chapter you should learn:

- About resonance in vibrating systems;
- How amplitude and phase change near a resonance;
- About standing waves as resonances in a pipe or a vibrating string;
- About partials, harmonics, and overtones;
- About acoustic impedance.

4.1 ■ RESONANCE OF A MASS-SPRING VIBRATOR

Consider a mass-spring system similar to the one discussed in Section 2.1. Suppose that the spring is attached to a crank, as shown in Fig. 4.2. Let the crankshaft revolve at a frequency f and let the natural frequency of the mass-spring system be f_0. If f is varied slowly, the amplitude A of the mass is observed to change, reaching its maximum $A_{\max}$ when $f = f_0$.

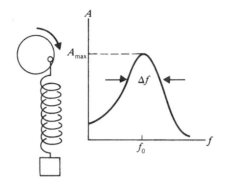

FIGURE 4.2
Resonance of
mass-spring
vibrator driven at a
frequency f.

The mass is forced to vibrate at the frequency f of the crank, but when f matches f_0, the natural frequency of the system, resonance occurs. At resonance the maximum transfer of energy occurs, and the amplitude builds up to a value A_{max} determined by the friction in the system. The graph of amplitude A as a function of frequency f, shown in Fig. 4.2, is nearly symmetrical around its peak, with a width Δf often called the *linewidth*. The linewidth is usually measured at an amplitude of 71% of $A_{max}(A_{max}/\sqrt{2})$.

Just as A_{max} depends on the rate of energy loss (due to friction or *damping*), so Δf also depends on energy loss. For a heavily damped system, Δf is large, and A_{max} is small. For a system with little loss, a "sharp" resonance with small Δf and large A_{max} occurs. Engineers define a quantity $Q = f_0/\Delta f$ to characterize the sharpness of a resonance. (The use of the letter Q comes from the term "quality factor" used to describe electrical circuits. A high-Q circuit is one with a sharp resonance; a low-Q circuit has a broad resonance curve.)

The linewidth Δf and the Q associated with a resonance are intimately related to the damping constant and the decay curve of a vibrator described in Section 2.2. A vibrator that loses its energy slowly will have a sharp resonance, and a vibrator that loses its energy rapidly will have a broad resonance. If the vibrator is set into motion and left to vibrate freely, its decay time is directly proportional to the Q of its resonance.

4.2 ■ PHASE OF DRIVEN VIBRATIONS

If we carefully observe the direction of motion of the crank and the mass, we note an interesting phenomenon. At low frequencies, far below resonance, the two move in the same direction. At frequencies far above resonance, however, they move in opposite directions.

We describe this phenomenon by using the term *phase*, which may be thought of as a specification of the starting point of a vibration. At low frequencies, the entire system follows the motion of the crank, and the spring hardly stretches at all. As the frequency of the crank increases, however, it is more difficult to move the mass, and thus it begins to lag behind the driving force supplied by the crank. At resonance, the mass is one-fourth cycle behind the crank, although its amplitude builds up to its maximum value. As the crank frequency is increased still further, the phase difference becomes greater until finally the mass is one-half cycle behind the crank; that is, the mass and the crank move in opposite directions, as shown at the right in Fig. 4.3. The higher the Q, the more abrupt is this

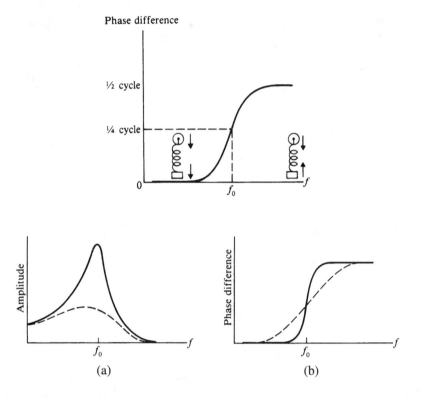

FIGURE 4.3
Phase difference
between crank and
mass in a driven
mass-spring
system. At
resonance they
differ in phase by
one-fourth of a
cycle.

FIGURE 4.4   (a)
Response and (b)
phase difference for
vibrators with more
(dashed curve) and
less (solid curve)
damping.

transition from *in phase* to *opposite phase*. A vibrator with a lot of damping, on the other hand, exhibits a gradual change of phase, as shown in Fig. 4.4.

4.3 ■ STANDING WAVES ON A STRING

In Section 3.4, we learned how interference between two waves traveling in opposite directions leads to standing waves. We also learned how reflection occurs when waves or pulses reach the boundary of the medium in which they propagate. We now combine these two ideas, and show how the modes of vibration or resonances of acoustical systems can be interpreted as waves propagating back and forth between the boundaries.

A simple and familiar example of such a system is a string of length L with both ends fixed as shown in Fig. 4.5. In its fundamental mode (that is, the standing wave with the lowest frequency and the longest wavelength), the string vibrates as shown in Fig. 4.5(a). The wavelength λ can be seen to be twice the string length, so the frequency is $f_1 = v/2L$. In the second mode, shown in Fig. 4.5(b), the wavelength λ equals the string length, so $f_2 = v/L = 2f_1$. Continuing to higher modes, we find that they have frequencies $3f_1$, $4f_1$, etc. The frequency of the nth mode will be

$$f_n = n\frac{v}{2L} = nf_1.$$

(4.1)

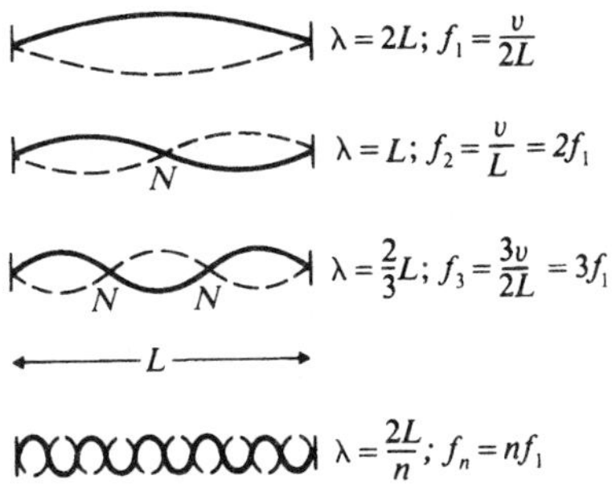

FIGURE 4.5
Modes of vibration or resonances of a vibrating string as standing waves. The nodes are denoted by N. Note that the frequencies are harmonics of the fundamental frequency f_1.

(a) $\lambda = 2L; \; f_1 = \dfrac{v}{2L}$

(b) $\lambda = L; \; f_2 = \dfrac{v}{L} = 2f_1$

(c) $\lambda = \dfrac{2}{3}L; \; f_3 = \dfrac{3v}{2L} = 3f_1$

$\longleftarrow L \longrightarrow$

(d) $\lambda = \dfrac{2L}{n}; \; f_n = nf_1$

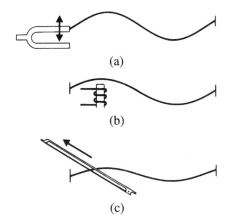

(a)

(b)

(c)

FIGURE 4.6
Three ways to drive a string at one of its resonances: (a) a tuning fork; (b) an electromagnet; (c) a violin bow.

Substituting the expression for wave speed given in Section 3.2 gives the following expression for the modes of a vibrating string:

$$f_n = \frac{n}{2L}\sqrt{\frac{T}{\mu}}, \tag{4.2}$$

where T is tension and μ is mass per unit length.

If a string is driven at the frequency of one of its natural modes, resonance can occur. There are many ways in which to apply the driving force, three of which are illustrated in Fig. 4.6. The magnetic drive shown in Fig. 4.6(b) works only for a string of steel or other magnetic material. The moving violin bow applies a rather complicated driving force (to be described in Chapter 10), which has components at several different frequencies. The string may also be driven by *electromagnetic force*, even when the string is made of a nonmagnetic metal, by placing a permanent magnet near the string and passing an alternating current of the desired frequency through the string.

EXAMPLE 4.1   A steel guitar string with a diameter of 0.3 mm and 65 cm long has a tension of 100 N. Find the frequencies of its first three modes of vibration. The density of steel is 7700 kg/m³.

Solution

$$\mu = \pi r^2 \rho = \pi (1.5 \times 10^{-4})^2 (7700)$$
$$= 5.44 \times 10^{-4} \text{ kg/m};$$

$$f_1 = \frac{1}{2L}\sqrt{\frac{T}{\mu}} = \frac{1}{2(0.65)}\sqrt{\frac{100}{5.44 \times 10^{-4}}} = 330 \text{ Hz};$$

$$f_2 = \frac{2}{2L}\sqrt{\frac{T}{\mu}} = \frac{2}{2(0.65)}\sqrt{\frac{100}{5.44 \times 10^{-4}}} = 660 \text{ Hz} \ (= 2f_1);$$

$$f_3 = 3f_1 = 990 \text{ Hz}.$$

4.4 ■ PARTIALS, HARMONICS, AND OVERTONES

It is appropriate at this point to clarify nomenclature in order to avoid confusion. We will use the term *harmonics* to refer to modes of vibration of a system that are whole-number multiples of the fundamental mode and also to the sounds that they generate. (It is customary to stretch the definition a bit so that it includes modes that are *nearly* whole-number multiples of the fundamental: 2.005 times the fundamental rather than 2, for example.) Thus we say that the modes of an ideal vibrating string are harmonics of the fundamental, but the modes of a real string are usually so close to being whole-number multiples that we also speak of them as harmonics. Note that the term *first harmonic* refers to the fundamental.

Many vibrators do not have modes that are whole-number multiples of the fundamental frequency, however, and the term *overtones* is used to denote their higher modes of vibration. Harmonics are therefore described as overtones whose frequencies are whole-number multiples of the fundamental frequency. Minor confusion arises, however, from the fact that the term harmonics includes the fundamental, but the term overtones does not. Thus the second harmonic is the first overtone, the third harmonic is the second overtone, and so forth.

There is another term in common use that refers to modes of vibration of a system or the components of a sound: *partials*. Partials include all the modes or components, the fundamental plus all the overtones, whether they are harmonics or not. The term *upper partials* excludes the fundamental and thus is a synonym of overtones, but use of the former will be avoided in this book.

The actual motion of a vibrating string is a combination of the various modes of vibration. The way in which these modes or partials combine is given by the *spectrum* of the vibration. A vibration spectrum is like a recipe that specifies the relative amplitudes of the partials. Similarly, the spectrum of a sound specifies the amplitudes of its partials, as we will discuss in Chapter 7.

4.5 ■ OPEN AND CLOSED PIPES

The reflection of sound pulses at open and closed ends of pipes was described in Section 3.5. At an open end, a pulse of positive pressure reflects back as a negative pulse; at a closed end, it reflects as a positive pulse. These two end conditions can be used to arrive at the resonances for open and closed pipes.

The motion of the vibrating air in a pipe is a little harder to visualize than the transverse vibrations of a string, because the motion of the air is longitudinal. The displacement of the air is greatest at an *open* end, but the pressure variation is maximum at a *closed end*. A pressure-sensitive microphone inserted into the tube will pick up the most sound at the points where the pressure variations above and below atmospheric pressure are maximum. Thus, in Figs. 4.7 and 4.8, both the air motion and the pressure variations are shown.

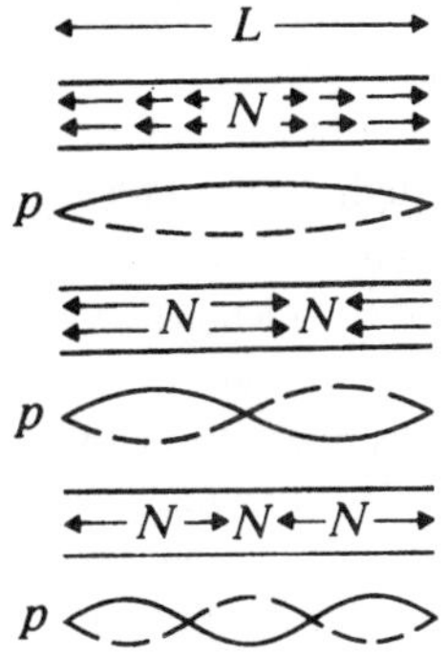

$$\lambda = 2L; \ f_1 = \frac{v}{2L}$$

$$\lambda = L; \ f_2 = \frac{v}{L} = 2f_1 \qquad (4.3)$$

$$\lambda = \frac{2}{3}L; \ f_3 = \frac{3v}{2L} = 3f_1$$

$$f_n = nf_1$$
$$(n = 1, 2, 3 \ldots)$$

FIGURE 4.7   Modes of vibration or resonances of an open pipe. At the open ends the pressure is equal to atmospheric pressure. The resulting modes include both odd-numbered and even-numbered harmonics. Minimum displacement occurs at the nodes denoted by N.

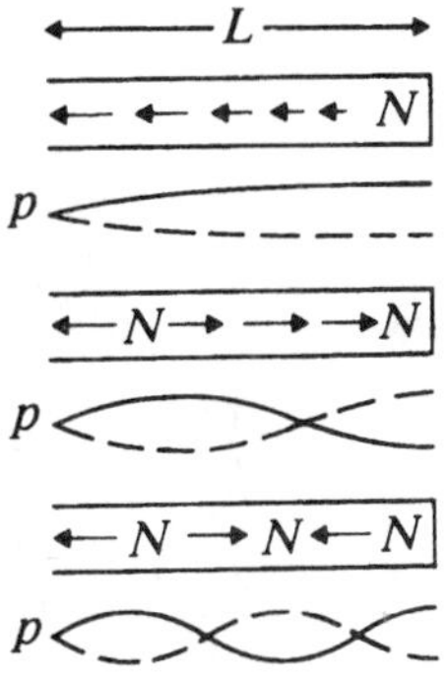

$$\lambda = 4L; \ f_1 = \frac{v}{4L}$$

$$\lambda = \frac{4}{3}L; \ f_3 = \frac{3v}{4L} = 3f_1 \qquad (4.4)$$

$$\lambda = \frac{4}{5}L; \ f_5 = \frac{5v}{4L} = 5f_1$$

$$f_n = nf_1$$
$$(n = 1, 3, 5 \ldots)$$

FIGURE 4.8   Modes of vibration or resonances of a closed pipe. At the closed end, the air motion is minimum but the pressure is maximum. The resulting modes include odd-numbered harmonics only. Minimum displacement occurs at the nodes denoted by N.

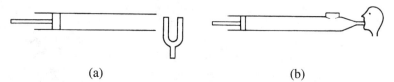

(a)   (b)

FIGURE 4.9   Resonance of a closed tube demonstrated by (a) holding a tuning fork near one end; (b) blowing on an organ pipe.

In an actual pipe, the pressure variations do not drop to zero right at the open end of the pipe, but rather a small distance beyond. Thus, the pipe appears to have an acoustic length that is slightly greater than its physical length. For a cylindrical pipe of radius r, the additional length, called the *end correction*, is $0.61r$. Twice this amount should be added to the length of a pipe with two open ends to obtain its acoustic length.

Resonance of a tube can be demonstrated by placing a tuning fork near one end of it, as shown in Fig. 4.9(a). A piston at the closed end makes it possible to change the resonance frequency. Blowing through a closed ("stopped") organ pipe may excite several of its resonances. Gentle blowing excites the lowest mode, but blowing much harder causes the pipe to vibrate in its first overtone, which for a closed pipe is the third harmonic ($f_3 = 3f_1$), a musical twelfth above the lowest mode.

In Chapter 15, which deals with speech production, we will be interested in resonances of the human vocal tract that allow us to enunciate various vowel sounds. There, we will consider not only reflections from open and closed ends but from constrictions and changes in pipe diameter as well. At every such discontinuity, a portion of the sound wave is reflected, thus leading to standing waves and resonances. If reflections occur at several places along the pipe, the resonances (in the vocal tract, they are called *formants*) can become rather complex.

EXAMPLE 4.2   Find the first three modes of vibration of a pipe 0.75 m long with open ends (neglect end corrections).

Solution

$$f_1 = \frac{v}{2L} = \frac{343}{2(0.75)} = 229 \text{ Hz};$$

$$f_2 = \frac{2v}{2L} = \frac{2(343)}{2(0.75)} = 457 \text{ Hz};$$

$$f_3 = \frac{3v}{2L} = \frac{3(343)}{2(0.75)} = 658 \text{ Hz}.$$

EXAMPLE 4.3   Find the first three modes of vibration of a pipe 0.75 m long with one open end and one closed end (neglect end correction).

Solution

$$f_1 = \frac{v}{4L} = \frac{343}{4(0.75)} = 114 \text{ Hz};$$

$$f_3 = \frac{3v}{4L} = \frac{3(343)}{4(0.75)} = 343 \text{ Hz};$$

$$f_5 = \frac{5v}{4L} = \frac{5(343)}{4(0.75)} = 572 \text{ Hz}.$$

4.6 ■ ACOUSTIC IMPEDANCE

A quantity that acoustical engineers find very useful is *acoustic impedance* Z_A. It is defined as the ratio of sound pressure p to volume velocity U and is measured in acoustic ohms:

$$Z_A = p/U. \tag{4.5}$$

The volume velocity U is the amount of air that flows through a specified area per second due to passage of a sound wave. In the case of sound propagating in a tube, the specified area would be the cross-sectional area of the tube.

In the case of plane waves propagating in a tube, the acoustic impedance can be found by using the formula $Z_A = \rho v/S$, where ρ is the density of the air (1.15 kg/m^3 at room temperature), v is the speed of sound, and S is the cross-sectional area of the tube. Hence $Z_A \simeq 400/S$, with S measured in square meters. We will find the numerical value of Z_A much less important than the fact that it varies inversely with area S. Thus when there is a constriction, a change in diameter, or a sidebranch in a tube, the impedance change at that point leads to reflection of sound waves.

Acoustic impedance is analogous to electrical impedance, which is the ratio of voltage to electrical current (see Chapter 18). In this case, the voltage is the forcing function that causes current to flow. In the case of a sound wave, sound pressure is the forcing function that causes air flow with volume velocity U. In the chapters on wind instruments, use will be made of acoustic input impedance, which is the ratio of pressure to volume velocity at the input (mouthpiece) of a wind instrument.

4.7 ■ HELMHOLTZ RESONATOR

The Helmholtz resonator was described as a vibrating system in Section 2.3. As a resonator, it has many applications in acoustics. Only a few of them will be described.

1. Before the invention of microphones, amplifiers, and spectrum analyzers, Helmholtz resonators were used to study vibrating objects and analyze complex sounds. In Fig. 4.10(a) the modes of vibration of a bell are being probed by a Helmholtz resonator having a second opening through which resonant sound can be heard by the investigator.

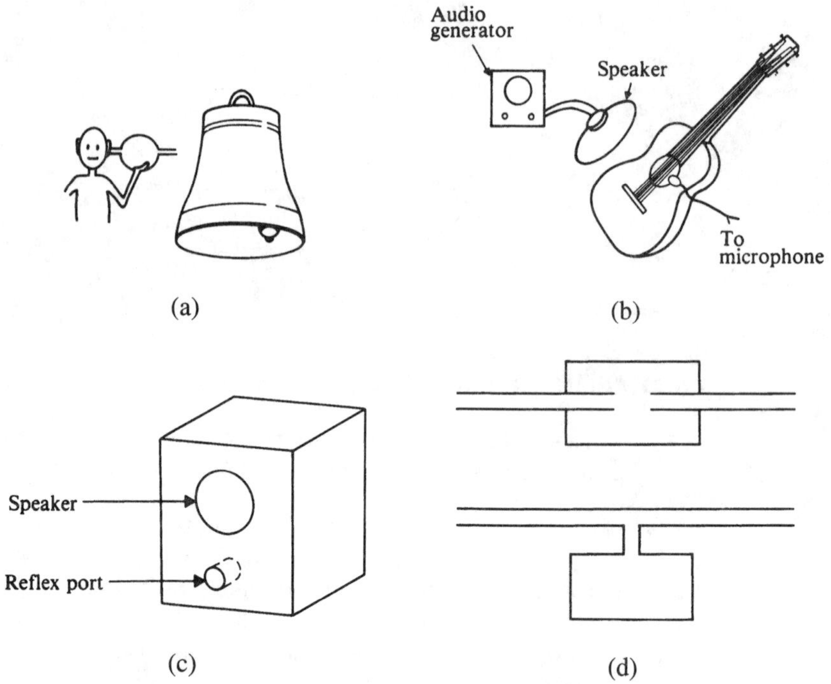

FIGURE 4.10
Some examples of
Helmholtz
resonators:
(a) analysis of
vibrations of a bell;
(b) air resonance of
a guitar body;
(c) bass reflex
loudspeaker
cabinet; (d) two
types of
side-branch
mufflers.

(a)

(b)

(c)

(d)

2. The main air resonance of a violin or guitar is essentially a Helmholtz resonance. The frequency can be determined by blowing across the f-holes of a violin or the sound hole of a guitar and listening for the pitch of the resonance. A more precise method is to insert a microphone inside the instrument and generate sound outside by means of a loudspeaker and audio generator.

3. Bass reflex loudspeaker cabinets are designed so that radiation from the back of the speaker cone excites the Helmholtz resonance of the cabinet and appears at the reflex port in phase with the front of the speaker.

4. Some automobile mufflers make use of side branches that absorb sound at their resonance frequency, as shown in Fig. 4.10(d).

4.8 ■ SINGING RODS AND WINEGLASSES

In addition to the bending modes of a bar shown in Fig. 2.17, there are several other ways in which a bar or rod can vibrate. These include longitudinal vibrations (similar to sound waves) and torsional (or twisting) modes.

Sound waves traveling in air are longitudinal waves; the air molecules vibrate back and forth in the same direction the wave is propagating. Similar longitudinal (compressional) waves can propagate in solids and liquids, resulting in longitudinal standing waves or normal modes of vibration. In a bar or rod, longitudinal waves travel at a speed $v = \sqrt{\frac{E}{\rho}}$, where E is Young's elastic modulus for the material and ρ is its density. Note that the wave speed does not depend on frequency or the diameter of the rod.

FIGURE 4.11
Stroking an aluminum rod with the fingers to excite longitudinal resonances.

In a bar or rod with free ends, the fundamental mode will have a node at its center, and the maximum vibration occurs at the ends just like the open pipe in Fig. 4.7. Its frequency is the wave speed divided by the wavelength (twice the rod length): $f_1 = \frac{n}{2L}\sqrt{\frac{E}{\rho}}$. The next mode has two nodes at $\frac{1}{4}L$ and $\frac{3}{4}L$ like the second mode in Fig. 4.7, the third mode has three nodes, and so on. Note that only the odd-numbered modes have nodes at the center. The modal frequencies are

$$f_n = \frac{n}{2L}\sqrt{\frac{E}{\rho}} \tag{4.6}$$

In most bars or rods, the longitudinal modes of vibration occur at much higher frequencies than the transverse modes.

Stroking an aluminum rod with the fingers to excite longitudinal resonances, as shown in Fig. 4.11, can create rather loud sounds, and stroke rods are actually used as percussion instruments. Further discussion of longitudinal and transverse vibrations in rods is given in Chapter 13.

Can a singer break a wineglass? We have never seen it happen, but perhaps it is possible. If the singer sings a note at precisely the frequency of the lowest mode of vibration of the wineglass (which can be determined by tapping it lightly on the rim or by running one's finger around the rim), it will certainly excite a resonance in the glass. The problem is that the sound power of the singing voice is small to start with, and only a small part of the total power is contained in the lowest partial or fundamental. Breaking a wineglass with sound (as in the TV commercials) is usually done with an amplifier and a loudspeaker to produce ample sound at the resonance frequency.

FIGURE 4.12   Collapse of the Tacoma Narrows Bridge (1940) is a dramatic example of self-excited oscillation. The wind excited the bridge in one of its vibrational modes.

4.9 ■ SELF-EXCITATION

A self-excited oscillator is one in which a steady motion, flow of air, or electric current is modulated and a part of the modulated flow is then fed back in such a way as to excite the oscillator. One example, a bowed violin string, will be discussed in Chapter 10. A dramatic example is the Tacoma Narrows Bridge, which collapsed in 1940. A torsional (twisting) mode that the designers had apparently overlooked (and for which they failed to provide adequate damping) was excited by the wind, and it just kept building up in amplitude until the bridge broke into pieces, as shown in Fig. 4.12.

4.10 ■ SYMPATHETIC VIBRATIONS: SOUNDBOARDS

The amount of sound radiated by a vibrating system depends on the amount of air it displaces as it moves (the volume velocity defined in Section 4.6). A vibrating string or the narrow prongs of a tuning fork displace very little air as they vibrate; thus they radiate a small amount of sound. The moving cone of a loudspeaker and the vibrating membrane of a drum, on the other hand, radiate sound rather efficiently.

It is possible to increase the sound radiation from a tuning fork by pressing its handle against a wood plate or tabletop, so that the tuning fork forces the large wood area to vibrate. The vibrations of the wood, called *sympathetic vibrations*, may or may not occur at a frequency near a resonance of the wood plate, but nevertheless they amplify the sound because of the large surface set into vibration.

Violins, guitars, cellos, lutes, and other string instruments depend almost completely on sympathetic vibrations of the wood sounding box for radiation of their sound. Most of the sound radiation in these instruments comes from sympathetic vibrations of the top plate, which is driven by the vibrating strings through the bridge. The top plate has many resonances of its own distributed throughout the playing range, and these resonances, to a large part, determine the quality of the instrument. Sympathetic vibration of the wood also sets the air inside the instrument into vibration, so that sound is radiated from the f-holes (violin) or sound hole (guitar). String instruments will be discussed in Chapter 10.

Pianos and harpsichords have large soundboards with many resonances of their own closely spaced throughout the playing range. Vibrations are transmitted from the string to the soundboard through the bridge as in the violin or guitar. We will discuss pianos and harpsichords in Chapter 14.

4.11 ■ SUMMARY

Resonance occurs when a force applied to a vibrating system oscillates with a frequency at or near the natural frequency of the system. Linewidth, Q, and maximum response are ways to describe the sharpness of the resonance, which in turn depends on the amount of damping in the system. The phase difference between the vibrating object and the driving force changes near the frequency resonance.

Normal modes of vibration of strings, pipes, and similar systems may be described as standing waves. Standing waves, which consist of waves propagating back and forth between the boundaries, lead to resonances in vibrating systems. Acoustic impedance, which is the ratio of sound pressure to volume velocity, is a quantity useful in the analysis of acoustical systems. Since sound radiation depends on air displacement, radiation from a vibrating string is greatly enhanced by the sympathetic vibration of a soundboard.

REFERENCES AND SUGGESTED READINGS

Carpenter, D. R., Jr., and R. B. Minnix (1993). *The Dick and Rae Physics Demo Notebook*. Lexington, Va.: Dick and Rae, Inc.

French, A. P. (1966). *Vibrations and Waves*. New York: Norton. (See Chapter 4.)

Hewitt, P. G. (1993). *Conceptual Physics*, 7th ed. New York: Harper Collins.

Kirkpatrick, L. D., and G. F. Wheeler (1995). *Physics: A World View*, 2nd ed. Philadelphia: Saunders.

Meiners, H. F. (1985). *Physics Demonstration Experiments*. Malabar, Fla.: R. E. Krieger.

Lord Rayleigh (J. W. Strutt) (1894). *The Theory of Sound*, Vol. 2, 2nd ed. London: Macmillan. Reprinted by Dover, New York, 1945. Section 253b.

Rossing, T. D. (1982). *Acoustics Laboratory Experiments*. DeKalb: Northern Illinois University.

Rossing, T. D., and N. H. Fletcher (1995). *Principles of Vibration and Sound*. New York: Springer-Verlag.

GLOSSARY

acoustic impedance A measure of the difficulty of generating flow (in a tube, for example); it is the ratio of the sound pressure to the volume velocity due to a sound wave.

amplitude The height of a wave; the maximum displacement of a vibrating system from equilibrium.

damping Energy loss in a system that slows it down or leads to a decrease in amplitude.

electromagnetic force The force that results from the interaction of an alternating electric current with a magnetic field.

fundamental The mode of vibration (or component of sound) with the lowest frequency.

harmonic A mode of vibration (or a component of a sound) whose frequency is a whole-number multiple of the fundamental frequency.

Helmholtz resonator A vibrator consisting of a volume of enclosed air with an open neck or port.

linewidth The width Δf of a resonance curve, usually measured at 71% of its maximum height; a measure of the sharpness of a resonance (a sharp resonance is characterized by a small linewidth).

overtone A component of a sound with a frequency greater than the fundamental frequency.

partial A component of a sound; includes the fundamental plus the overtones.

phase difference A measure of the relative positions of two vibrating objects at a given time; also the relative positions, in a vibration cycle, of a vibrating object and a driving force.

Q A parameter that denotes the sharpness of a resonance; $Q = f_0/\Delta f$, where f_0 is the resonance frequency and Δf is the linewidth.

resonance When a vibrator is driven by a force that oscillates at a frequency at or near the natural frequency of the vibrator, a relatively large amplitude results.

soundboard A sheet of wood or other material that radiates a substantial amount of sound when it is driven in sympathetic vibration by a vibrating string or in some other manner.

spectrum A recipe for vibratory motion (or sound) that specifies the relative amplitudes of the partials.

sympathetic vibration One vibrator causing another to vibrate at the same frequency (which may or may not be a resonance frequency). An example is a piano string causing the bridge and soundboard to vibrate at the string's frequency.

REVIEW QUESTIONS

1. Write a definition of *resonance* and give several examples.

2. In Fig. 4.4, do the solid curves or the dashed curves represent a higher Q? Explain.

3. What does n equal in Fig. 4.5(d)? How many wavelengths equal L?

4. Distinguish between partials, harmonics, and overtones.

5. If you blow over the ends of two pipes, one with the other end closed and one with it open, which pipe will give the tone of lower pitch? approximately how much lower?

6. A pipe with one open and one closed end has its lowest resonance at 200 Hz. What are the frequencies of its next two resonances?

7. What is acoustic impedance?

8. In order to lower the Helmholtz resonances of a guitar, would you make the sound hole larger or smaller?

9. To excite a singing rod in its fundamental mode, where should you hold it? Where should you stroke it?

10. Can a singer break a wineglass by singing loudly?

11. What is the main function of a piano soundboard?

QUESTIONS FOR THOUGHT AND DISCUSSION

1. If a child in a swing is pushed with the same impulsive force in each cycle, will the amplitude increase by the same amount in each cycle?

2. List as many examples of Helmholtz resonators as you can other than those given in Section 4.7. Are the resonances sharp or broad?

3. Attach a mass to a spring, as in Fig. 2.1 or 4.2, and determine the approximate resonance frequency by moving the top of the spring up and down by hand. Then move it at frequencies below and above resonance, and carefully describe the force exerted on your hand in each case.

4. Does the end correction given in Section 4.5 lower all harmonics of a pipe proportionally, or does it result in the overtones going out of tune? An exact expression for the end correction shows that it varies slightly with wavelength. Does that change your answer?

EXERCISES

1. A particular vibrator has a resonance frequency of 440 Hz and a Q of 30. What is the linewidth of its resonance curve?

2. Sketch a waveform that represents the displacement of the mass in Fig. 4.2 as a function of time. Then carefully sketch a second wave one-fourth cycle in advance of the first to represent the driving force at resonance. Label each curve correctly.

3. Determine the frequencies of the fundamental and first overtone (second partial) for the following. Neglect end corrections.
   (a) A 16-ft open organ pipe
   (b) a 16-ft stopped organ pipe (one open end, one closed end)

4. Extend Figs. 4.7 and 4.8 to include two more modes each.

5. Find the difference in the fundamental frequency, calculated with and without the end correction, of an open organ pipe 2 m long and 10 cm in diameter.

6. A nylon guitar string 65 cm long has a mass of 8.3×10^{-4} kg/m and the tension is 56 N. Find the frequencies of the first four partials.

7. A steel bar 1 m long is held at the center and tapped on one end. Because its ends are free to move, its modes of longitudinal vibration will be similar to those of the air in a pipe open at both ends. Using the speed of sound given in Table 3.1, calculate the frequencies of the first three longitudinal modes.

8. Determine the frequencies of the pipes in Problem 3 if helium is substituted for air. (The speed of sound in helium is given in Table 3.1.)

EXPERIMENTS FOR HOME, LABORATORY, AND CLASSROOM DEMONSTRATION

Home and Classroom Demonstration

1. *Resonance of hand-held oscillator* It is easy to demonstrate resonance by moving the top of a spring, to which a mass is attached, up and down at the correct frequency. When the hand is moved up and down slowly, the mass is seen to move in phase with the hand; when the hand is moved quite rapidly, it can be seen that the mass and the hand move in opposite phase, as shown in Fig. 4.3.

2. *Resonance of a driven oscillator* Quantitative data require a sinusoidal drive with variable frequency. Although it is easy to demonstrate that the amplitude of a mass-spring oscillator has maximum value at the resonance frequency (Fig. 4.2), it is more difficult to show the important phase change at resonance (Fig. 4.3). Attaching markers as shown at the right is fairly effective. In the Pasco ME9210A harmonic motion analyzer (now discontinued), a flashing LED showed the phase relationship between the driver and the oscillating mass.

3. *Resonance of a wire string* The resonances of a thin wire string should be demonstrated, preferably both by electromagnetic excitation and by bowing with a violin bow (see Fig. 4.6).

4. *Resonance of a closed tube* A tuning fork is held above one end of a glass tube whose other end is immersed in a large reservoir of water. The tube is raised or lowered in the water until resonance occurs.

Another way to change the length of a closed-end tube is to connect it to a reservoir by means of a hose. The water height in the tube is adjusted by raising or lowering the reservoir.

5. *Resonance of a tube by a loudspeaker* A loudspeaker L driven by an audio generator sets up standing waves in a large cardboard or Plexiglas tube. A small microphone M can be moved up and down to locate the pressure maxima for each resonance.

6. *Tubing fork resonator* The Ames tube, sold by Riverbank Laboratories, is a tuning fork and open-end resonance tube combined. Their Ames tube kit includes materials for several interesting demonstrations. Choirchimes, made by Malmark, Inc. (a handbell manufacturer), similarly combine a tuning fork with a closed-end resonator. Choirchimes, which are popular with handbell choirs in churches and schools, include a clapper with which to set the tuning fork into vibration.

7. *Smoke alarm vibrator* A vibrator of the type used in smoke alarms is attached to one end of a tube, which is adjustable in length. When powered by a battery, the vibrator generates a tone with several harmonics, and the tube can be adjusted to resonate with individual harmonics.

8. *Kundt's tube* In the classical Kundt's tube, a brass rod is stroked with a rosin-soaked cloth so that it vibrates (longitudinally) at its resonance frequency. A plunger is adjusted until standing waves appear in cork dust in the glass tube. Note that two resonances are being demonstrated: one in the brass rod and one in the air column.

9. *Electrified Kundt's tube* In the electrified Kundt's tube, a horn driver or small loudspeaker drives the tube. In this case, it is easy to obtain striations in the cork dust, which fascinated Lord Rayleigh (*Theory of Sound*, Section 253b) and others.

10. *Impedance of an air column* If the loudspeaker in Experiment 9 is driven at a nearly constant current (by adding a resistor of 200 Ω or more in series), the voltage across the speaker (and hence the electrical impedance) can be shown to rise at each resonance. (If the voltage and current are displayed on an oscilloscope, a shift in phase can also be shown.)

11. *Resonances of a tin whistle* Some tin whistles (or plastic whistles) will sound several notes with the end closed as well as open. With a closed end the fundamental will be an octave lower and only the odd-numbered harmonics will sound ($f, 3f, 5f$, etc.). With a closed end, all the harmonics of the higher note will sound ($2f, 4f, 6f$, etc.). Thus by alternately opening and closing the end, a complete harmonic series is obtained and bugle calls can be played.

12. *Open and closed organ pipes* Many cylindrical organ pipes will sound several notes with the end closed as well as open. With a closed end, the fundamental will be an octave lower and only the odd-numbered harmonics will sound. With both ends open, both odd and even harmonics will sound.

13. *Flutes and clarinets* Flutes and recorders behave like open pipes and thus sound both odd and even harmonics of the fundamental (Fig. 4.7). Clarinets behave like closed-end pipes and thus play only the odd-numbered harmonics (Fig. 4.8).

14. *Rubens flame tube* Flame tubes with regularly spaced holes (Meiners 1985, 19–3.5) are widely used to demonstrate standing waves, but the explanation of why variations in average gas pressure occur requires a little thought (see T. D. Rossing, "Average pressure in standing waves," Phys. Teach. *15*, 260 (1977)).

15. *Pipe excited with a Meeker burner* A pipe about 1 to 2 m in length will resonate when lowered to optimum position over a Meeker (Fisher) grid-top burner. Large-diameter tubes are easiest to excite. Adjust the burner until blue tips are formed about 1 mm above the grid.

16. *Rijke tube* The Rijke tube has one or more layers of wire gauze at about one-fourth of its length to produce turbulence. If you heat the gauze cherry red, the tube sings when removed from the flame (as long as it is near vertical). To amuse the class, sound can be "poured out" into a bucket and then "poured back" into the tube.

17. *Singing corrugated hose* Plastic corrugated hose emits tones when held near one end and twirled in a circle. Sump pump hose (1 m long × 2.5–3.5 cm in diameter) from building supply stores works well, as do "hummers," sold in toy stores. Some spiral-wound vacuum cleaner hose will not work, nor will hose with closely spaced corrugations.

By measuring the frequencies of several harmonics, the speed of sound in the hose can be determined. It is up to 10% slower than the speed of sound in air, due to the periodic loading effect of the corrugations.

18. *Soda-straw reed instrument* Flatten one end of a soda straw and cut an arrow-shaped end with scissors. Place well into your mouth and blow hard. You should get a tone as the end vibrates in the manner of an oboe or bassoon reed. While blowing, shorten the straw in small steps with the scissors to hear the frequency rise. Alternatively, you can cut some tone holes in the straw that can be covered or uncovered with your fingers. Or, you can add a second straw of slightly greater or lesser diameter, so that the tube can be lengthened by sliding the outer (or inner straw) to create a sort of "kazoo."

19. *Longitudinal resonances in a singing rod* An aluminum rod about 1.3 cm (0.5 in.) in diameter and 2 m long can be held at the center and stroked longitudinally to give a rather loud sound. When held at the center, the rod vibrates at a frequency such that it is one-half a wavelength. Holding the rod at one-fourth of its length causes it to vibrate in its second harmonic. The rod can be sprayed with a tacky material (such as Firm Grip, sold in sporting goods stores) or the fingers can be coated with a little rosin to enhance the stick-slip action. Some demonstrators prefer to stroke the rod with a paper towel soaked in rubbing alcohol (Carpenter and Minnix 1993).

20. *Longitudinal and transverse vibrations of a rod* An aluminum rod about 0.6 cm (0.25 in.) in diameter and 2 m long can be used to demonstrate a great difference in speed between longitudinal and transverse waves. Hold the rod at the center and stroke it (as in Experiment 19) or tap it on the floor to excite the lowest longitudinal resonance. Then hold it at about one-fourth of its length and strike the center with the other hand to excite the lowest transverse resonance. The vibrations are slow enough to be readily visible. Formulas for transverse and longitudinal wave speeds are given in Table 13.1.

21. *Breaking a beaker (or wineglass)* Tap a beaker (or wineglass) to determine the frequency of its fundamental

mode. Supply a signal of that frequency to an audio amplifier and drive a loudspeaker positioned close to the beaker (a compression horn driver works well, because the sound output is concentrated in a small area). The beaker must have a high Q, and the sound source must have ample power in order for the beaker to break.

Laboratory Experiments

Resonance (Experiment 7, *Acoustics Laboratory Experiments*)

Vibrating strings (Experiment 4, *Acoustics Laboratory Experiments*)

Vibrations of rods: longitudinal and transverse (Experiment 45, *Acoustics Laboratory Experiments*)

PART
II

Perception and Measurement
of Sound

Psychoacoustics is the science that deals with the perception of sound. This interdisciplinary field overlaps the academic disciplines of physics, biology, psychology, music, audiology, and engineering, and utilizes principles from each of them. Our understanding of sound perception has increased substantially in recent years.

Loudness, pitch, timbre, and duration are four attributes used to describe sound, especially musical sound. These attributes depend in a rather complex way on measurable quantities such as sound pressure, frequency, spectrum of partials, duration, and envelope. The relationship of the subjective attributes of sound to physical quantities is the central problem of psychoacoustics and it has received a great deal of attention in recent years. Another interesting subject is the method of using rather subtle clues to localize the direction of a sound source and also draw surprisingly accurate conclusions about the nature of the acoustic environment. (In a dark, unfamiliar room, for example, we could probably point to a speaker and also conclude whether the room is large or small.)

The following chapters introduce some of the important topics of psychoacoustics. Chapter 5 briefly describes the human auditory system and the hearing process. Chapters 6 and 7 discuss three important attributes of sound: loudness, pitch, and timbre. Chapter 8 discusses several phenomena having to do with combination tones and how they relate to musical sound. Finally, Chapter 9 discusses musical scales and temperaments.

CHAPTER

5

Hearing

The human auditory system is complex in structure and remarkable in function. Not only does it respond to a wide range of stimuli, but it precisely identifies the pitch and the timbre (quality) of a sound and even the direction of the source. Much of the hearing function is performed by the organ we call the ear, but recent research has emphasized how much hearing depends on the data processing that occurs in the central nervous system as well.

In this chapter you should learn:

- About the human auditory system (ear) and how it functions;

- How large and small numbers are expressed as powers of ten and on a logarithmic scale;

- About critical bands in hearing;

- About binaural hearing and localization;

- How *subjective* attributes of sound and music relate to *physical* parameters.

5.1 ■ RANGE OF HEARING

The range of sound *intensity* (pressure) and the range of *frequency* to which the ear responds, as shown in Fig. 5.1, is remarkable indeed. The intensity ratio between the sounds that bring pain to our ears and the weakest sounds we can hear is more than 10^{12} (1,000,000,000,000). The frequency ratio between the highest and lowest frequencies we can hear is nearly 10^3 (1000) times, or more than nine octaves (each octave represents a doubling of frequency).

Human vision is remarkable, too, but the frequency range does not begin to compare to that of human hearing. The frequency range of vision is about one octave (4×10^{14} to 7×10^{14} Hz, corresponding to wavelengths of 400 to 750 nanometers). Within this one octave range of frequency we can identify more than 7 million different colors (Rossing and Chiaverina 1999). Given that the frequency range of the ear is nine times greater than that of the eye, you can begin to imagine how many sound "colors" might be possible.

In Chapter 1, we learned that *power* is the rate at which work is done; it is equal to work or energy divided by time. We also learned that *pressure* is force per unit area (or force divided by area). In dealing with sound waves (or light waves), it is useful to talk about *intensity*, which is the power per unit area carried by the wave. It is expressed in

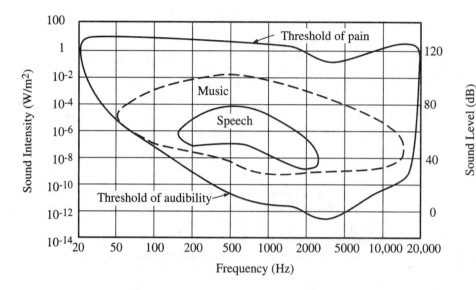

FIGURE 5.1   Range of frequencies and intensities to which the auditory system (ear) responds.

watts/square meter (W/m^2). The intensity of a sound wave multiplied by the area of our eardrum, for example, expresses the amount of power that the sound wave transmits to each ear. It is a small fraction of a watt even for the loudest of sounds.

Because it is rather difficult to measure sound intensity, we generally measure sound pressure instead, as we will discuss in Chapter 6. A microphone gives us an electrical signal proportional to sound pressure. The intensity is proportional to sound pressure squared. We will discuss this in more detail in Chapter 6.

The ear is extremely sensitive to small pressure changes. The pressure change in a very loud sound is still only 10^{-5} normal atmospheric pressure. At some sound frequencies, the vibrations of the eardrum may be as small as 10^{-8} mm, about one-tenth the diameter of the hydrogen atom. It is estimated that the vibrations of the very fine membrane in the inner ear that transmit this stimulus to the auditory nerve are nearly 100 times smaller yet in amplitude (Békésy 1960).

The frequency range of hearing varies greatly among individuals; a person who can hear over the entire audible range of 20–20,000 Hz is unusual. The ear is relatively insensitive to sounds of low frequency; for example, its sensitivity at 100 Hz is roughly 1000 times less than its sensitivity at 1000 Hz. Sensitivity to sounds of high frequency is greatest in early childhood and decreases gradually throughout life, so that an adult may have difficulty hearing sounds beyond 10,000 or 12,000 Hz. (This deterioration of perception of high frequencies, termed *presbycusis*, is compared in Chapter 31 to noise-induced hearing loss.)

Another remarkable quality of the auditory system is its selectivity. From the blended sounds of a symphony orchestra, a listener can pick out the sound of a solo instrument. In a noisy room crowded with people, it is possible to pick out a single speaker. Even during sleep the conditioned ear of a mother can respond to the cry of an infant. We can train ourselves to sleep through the noise of city traffic but to awaken at the sound of an alarm clock or unusual noise.

5.2 ■ STRUCTURE OF THE EAR

For convenience of description it is usual to divide the ear into three sections: the outer ear, the middle ear, and the inner ear (see Fig. 5.2). The *outer ear* consists of the external *pinna* and the *auditory canal* (meatus), which is terminated by the *eardrum* (tympanum). The pinna helps, to some extent, in collecting sound and contributes to our ability to determine the direction of origin of sounds of high frequency. The auditory canal acts as a pipe resonator that boosts hearing sensitivity in the range of 2000 to 5000 Hz.

The *middle ear* begins with the eardrum, to which are attached three small bones (shaped like a hammer, an anvil, and a stirrup) called *ossicles*. The eardrum, which is composed of circular and radial fibers, is kept taut by the tensor tympani muscle. The eardrum changes the pressure variations of incoming sound waves into mechanical vibrations to be transmitted via the ossicles to the inner ear.

The ossicles perform a very important function in the hearing process. Together they act as a lever, which changes the very small pressure exerted by a sound wave on the eardrum into a much greater pressure (up to 30 times) on the oval window of the inner ear. This function, which an engineer might call a mechanical transformer, is illustrated in Fig. 5.3. The lever action of the ossicles provides a factor of about 1.5 in force multiplication, whereas the remaining factor of about 20 in pressure comes from the difference in the areas of the eardrum and round window (the same force distributed over a smaller area results in a greater pressure, as explained in Section 1.6).

Another function of the small bones is to protect the inner ear from very loud noises and sudden pressure changes. Loud noise triggers two sets of muscles; one tightens the eardrum and the other pulls the stirrup away from the oval window of the inner ear. This response to loud sounds, called the *acoustic reflex*, will be discussed in Chapter 6.

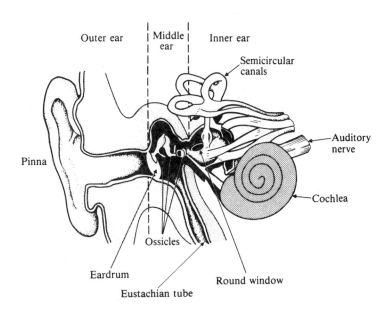

FIGURE 5.2
A schematic diagram of the ear, showing outer, middle, and inner regions. This drawing is not to scale; for purposes of illustration, the middle ear and inner ear have been enlarged.

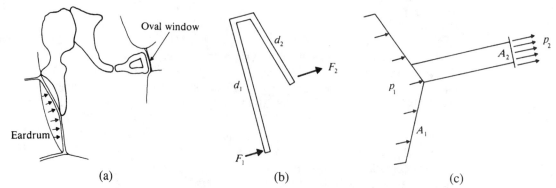

FIGURE 5.3 Pressure amplification by the ossicles. (a) Three bones link the eardrum to the inner ear. (b) Lever action: A smaller force acts through a larger distance, resulting in a larger force acting through a smaller distance. (c) Pressure multiplication by piston action: A small pressure on a large area produces the same force as a large pressure on a small area.

Because the eardrum makes an airtight seal between the middle and outer parts of the ear, it is necessary to provide some means of pressure equalization. The *Eustachian tube*, which connects the middle ear to the oral cavity, is such a safety device. If the Eustachian tube is slow to open, a "popping" may be heard in the ears when the outside air pressure changes, for example, during a rapid change in altitude. It is remarkable that all these middle ear functions take place in a space approximately the size of an ordinary sugar cube!

The marvelously complex *inner ear* contains the *semicircular canals* and the *cochlea*. The semicircular canals contribute little or nothing to hearing; they are the body's horizontal-vertical detectors necessary for balance. The spiral cochlea, a masterpiece of miniaturization, contains all the mechanism for transforming pressure variations into properly coded neural impulses.

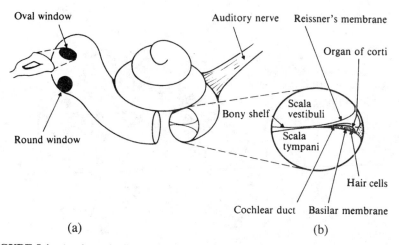

FIGURE 5.4 A schematic diagram of (a) the cochlea; (b) a section cut from the cochlea.

The cross section of the cochlea in Fig. 5.4 shows three distinct chambers that run the entire length: the *scala vestibuli*, the *scala tympani*, and the *cochlear duct*.

> The cochlea is filled with liquid and surrounded by rigid bony walls. Actually there are two different liquids, called perilymph (in the canals or scala) and endolymph (in the cochlear duct); the total capacity of the cochlea is only a fraction of a drop. Perilymph is similar to spinal fluid, whereas endolymph is similar to the fluid within cells. The two liquids are kept separate by two membranes: *Reissner's membrane* and the *basilar membrane*. Reissner's membrane is exceedingly thin, approximately two cells thick.

Resting on the basilar membrane is the delicate and complex *organ of Corti*, a gelatinous mass about $1\frac{1}{2}$ in. long. This "seat of hearing" contains several rows of tiny *hair cells* to which are attached nerve fibers. A single row of inner hair cells contains about 4000 cells, whereas about 12,000 outer hair cells occur in several rows. Each hair cell has many hairs, or *stereocilia*, that are bent when the basilar membrane responds to a sound. The bending of the stereocilia stimulates the hair cells, which in turn excite neurons in the auditory nerve.

> Modern auditory research has shown that the inner and outer hair cells function quite differently. The inner hair cells are mainly responsible for transmitting signals to the auditory nerve fibers. The more numerous outer hair cells apparently act as biological amplifiers. When their stereocilia are bent in response to a sound wave, the cell changes in length. This pushes against the tectoral membrane, selectively amplifying the vibration of the basilar membrane. It is estimated that the outer hair cells add about 40 dB of amplification, so that hearing sensitivity decreases by a considerable amount when these delicate cells are destroyed by overexposure to noise.

In order to understand how the basilar membrane vibrates, we can see the cochlea uncoiled and simplified in Fig. 5.5. The cochlea then appears as a tapered cylinder divided into two sections by the basilar membrane. (Because the cochlear duct is quite thin, we can ignore it—as a first approximation—and consider the two sections separated by a single membrane.) At the larger end of the cylinder are the oval and round windows, each closed by a thin membrane, and near the far end of the basilar membrane is a small hole called the *helicotrema* connecting the two sections. The basilar membrane terminates just short of the smaller end of the cylinder, so that fluid can transmit pressure waves around the end of the membrane.

When the stapes (stirrup) vibrates against the oval window, hydraulic pressure waves are transmitted rapidly down the scala vestibuli, inducing ripples in the basilar membrane. High tones create their greatest amplitude in the region near the oval window where the basilar membrane is narrow and stiff. On the other hand, low tones create ripples of greatest amplitude where the membrane is slack at the far end (see Fig. 5.6). Thus the initial frequency analysis takes place in the cochlea, although we will see in Chapter 7 that much

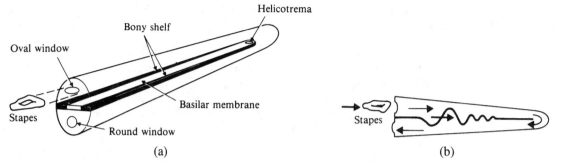

(a)   (b)

FIGURE 5.5   (a) A schematic diagram of uncoiled cochlea showing the basilar membrane and oval and round windows. (b) When the stapes (stirrup) presses against the oval window, a pressure pulse propagates through the cochlear fluid toward the round window, causing ripples to occur in the basilar membrane.

FIGURE 5.6
Basilar membrane displacement amplitude as a function of distance for several different frequencies. (After von Békésy 1960.)

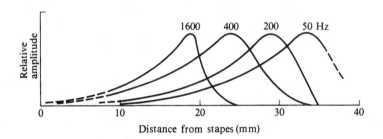

of the sense of pitch is determined in the central nervous system, where the data from the auditory nerve are processed.

The conversion of the mechanical vibrations of the basilar membrane into electrical impulses in the auditory nerve is accomplished in the organ of Corti. When the basilar membrane vibrates, the "hairs" of the hair cells are bent, thus generating nerve impulses that travel to the brain. The impulse rate on the auditory nerve depends on both the intensity and the frequency of the sound.

The overall hearing mechanism is illustrated in Fig. 5.7. Sound waves propagate through the ear canal, excite the eardrum, and cause mechanical vibrations in the middle ear. The

FIGURE 5.7
A schematic representation of the ear, illustrating the overall hearing mechanism. Sound waves in the outer ear cause mechanical vibrations in the middle ear, and eventually nerve impulses that travel to the brain to be interpreted as sound.

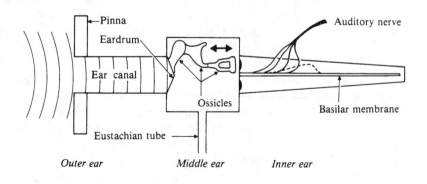

stapes vibrating against the oval window causes pressure variations in the cochlea, which in turn excite mechanical vibrations in the basilar membrane. These vibrations of the basilar membrane cause the hair cells to transmit electrical impulses to the brain via the auditory nerve.

Some sounds are heard through vibrations of the skull that reach the inner ear. Hearing by bone conduction plays an important role in speaking. The sounds of humming or clicking one's teeth are heard almost entirely by bone conduction. (If you stop your ears with your fingers, thus interfering with the air path, the humming may actually sound louder.) During speaking or singing, two different sounds are heard, one by bone conduction and one by air conduction. The recorded sound of your own voice sounds very unnatural to you because only the airborne sound is received by the microphone, whereas you are used to hearing both components in your own voice.

Many researchers have contributed to our understanding of the hearing process, but two scientists deserve special mention: Hermann von Helmholtz and Georg von Békésy.

Hermann Ludwig Ferdinand von Helmholtz (1821–1894) was a physician and a man of many sciences. He did pioneering work in the field of physiology, mathematics, thermodynamics, optics, and acoustics. He invented the ophthalmoscope used to study the interior of the eye and formulated an important theory of color perception. In 1862 he published his monumental book *On the Sensations of Tone as a Physiological Basis for the Theory of Music*, which has been reprinted many times and is useful even today to researchers in psychoacoustics. Helmholtz envisioned the fibers of the basilar membrane as selective resonators tuned, like the strings of a piano, to different frequencies. Thus a complex sound would be analyzed into its various components by selectively exciting fibers tuned to the frequency of one of the components. It turned out that Helmholtz was nearly, but not quite, correct in this assumption, as we shall learn in Chapter 7.

Georg von Békésy, a communications engineer in Budapest, Hungary, became interested in the mechanism of hearing while studying ways to improve telephones. In order to carry out his studies, Békésy carefully removed cochleas from the ears of animal and human cadavers. For his careful and extensive research, he was awarded a Nobel prize in 1961.

In order to illustrate vibrations of the basilar membrane, Békésy built several mechanical models of the cochlea, one of which is shown in Fig. 5.8. A brass tube with a slit at the top is covered by a plastic of varying thickness with a raised ridge. The tube is closed with a piston at one end and a fixed plate at the other, and filled with water. The elasticity of the plastic varies along its length in much the same manner as the basilar membrane. Thus when the piston is driven at various frequencies, the point of maximum excitation moves up and down the tube, which can be felt by placing one's forearm in gentle contact with the ridge of the plastic (Békésy 1960, 1970).

FIGURE 5.8
(a) Cochlear model constructed by Békésy. (b) The observer notes that the point of maximum sensation moves up and down the forearm as the frequency of the sound changes.

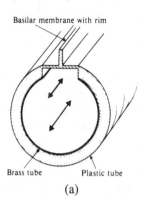

Basilar membrane with rim

Brass tube   Plastic tube

(a)   (b)

The cochlear model, as well as other instruments used by Békésy, can be seen and operated at a small museum at the University of Hawaii.

Much of Békésy's success was due to the careful techniques he developed for removing the cochleas of fresh cadavers. Working under a microscope with micro-tools of his own design, he was able to lay open a part of the basilar membrane. The cochlear fluid was drained and replaced by a salt solution with a suspension of pow-dered aluminum and coal. By observing light scattered from the suspended powder, he discovered an undulation in the basilar membrane when the cochlea was excited by sound.

Békésy studied the ears of many different mammals. An amusing story is told about his excitement in learning that an elephant had died in the Budapest zoo. He traced the carcass to a local glue factory where he was able to recover the elephant's cochleas. To Békésy's delight, traveling waves were observable also in the basilar membrane of the elephant (Stevens and Warshofsky 1965).

5.3 ■ SIGNAL PROCESSING IN THE AUDITORY SYSTEM

Signal processing in the auditory system can be divided into two parts: that done in the peripheral auditory system (ears themselves), and that done in the auditory nervous system (brain). The ears process an acoustic pressure signal by first transforming it into a mechanical vibration pattern on the basilar membrane, as shown in Figs. 5.5 and 5.6, and then representing this pattern by a series of pulses to be transmitted by the auditory nerve. Perceptual information is extracted at various stages of the auditory nervous system.

It is possible, by inserting a tiny electrode into the auditory nerve, to pick up the elec-trical signals traveling in a single fiber of the auditory nerve from the cochlea to the brain (Tasaki 1954). The signal consists of a series of voltage spikes, each spike corresponding to the stimulation of a hair cell attached to the basilar membrane. The spikes are found to be closely correlated to the mechanical vibration pattern on the basilar membrane up to frequencies of about 4000 or 5000 Hz.

Each auditory nerve fiber responds over a certain range of frequency and sound pressure. Each nerve fiber has a characteristic frequency (CF) at which it has maximum sensitivity. Fibers with a high CF show a rapid rolloff in sensitivity above their CF but a long "tail" below it. A 90-dB stimulus at 500 Hz, for example, causes spikes to appear on all six fibers. By sophisticated techniques such as probing with laser light (Khanna and Leonard 1982) and using the Mössbauer effect (Johnstone and Boyle 1967), it has been found that basilar membrane displacements in live animals show a much sharper frequency response than those of Fig. 5.6 in the cochlea of a dead animal. Rhode and Robles (1974) found that within several hours after death, the basilar membrane response decreases 10–15 dB, the frequency of maximum response shifts downward, and the response curve broadens. In fact, the mechanical frequency response of the basilar membrane in live cochleas is quite comparable to the tuning curves observed in nerve fibers. There is some evidence for sharpening of neural tuning curves further along the neurological pathway, however.

If we were to observe the spikes on a nerve fiber when the stimulus is a tone of a single frequency, we would note that the time between spikes almost always corresponds to one or two or more periods of the tone. Although the nerve fiber does not fire at the peak of every vibration cycle in the basilar membrane, it rarely fires at any other time. The situation is a little more complicated when the stimulus is a complex tone, but still we find that the pattern of spikes on the auditory nerve carries accurate information about the frequency spectrum of the stimulus tone.

Consider a stimulus consisting of the pure tones C_4 (523 Hz) and C_5 (1046 Hz), spaced one octave apart. Their neural tuning curves (or frequency response curves) shown in Fig. 5.9(a) show very little overlap, so very few hair cells respond to both frequencies. Processing of the one component in the brain is only slightly affected by the presence of the other one.

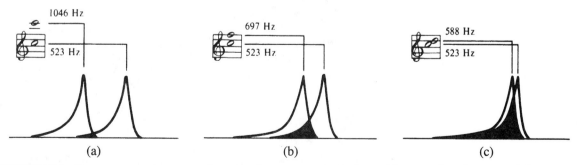

FIGURE 5.9 Frequency response curves for pairs of pure tones. As the interval between them decreases, their response curves show increasing overlap.

As the interval between the two components decreases, the situation changes. Their amplitude envelopes show more and more overlap, as in Fig. 5.9(b) and (c), so an increasing number of hair cells are stimulated by both components. This leads to many interesting auditory phenomena, some of which will be discussed in Chapter 6, 7, and 8.

5.4 ■ CRITICAL BANDS

When two pure tones are so close in frequency that there is considerable overlap in their amplitude envelopes on the basilar membrane, they are said to lie within the same *critical band*. Critical bands are of great importance in understanding many auditory phenomena, such as loudness, pitch, and timbre. They have been defined and measured in a variety of ways (Fletcher 1940; Plomp 1976; Zwicker, Flottorp, and Stevens 1957).

Each critical band may be regarded as a data collection unit on the basilar membrane. About 24 critical bands span the audible frequency range, and the regions on the basilar membrane to which each of these corresponds is about 1.3 mm long and embraces about 1300 neurons (Scharf 1970). The *critical bandwidth* varies with center frequency, as shown in Fig. 5.10, having nearly a constant value at low frequency and being roughly proportional to frequency at high frequency. Bandwidths are found to vary substantially, depending upon the type of experiment.

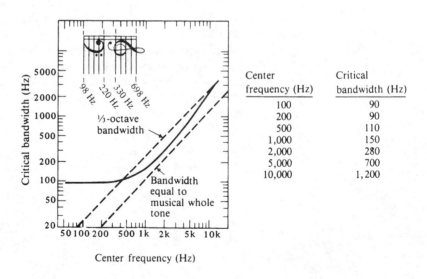

FIGURE 5.10
Critical bandwidth as a function of the critical band center frequency. Bandwidths are typical of those reported in various experiments.

Center frequency (Hz)	Critical bandwidth (Hz)
100	90
200	90
500	110
1,000	150
2,000	280
5,000	700
10,000	1,200

Critical Bands as Musical Intervals

Over much of the audible range, critical bands are slightly less than $\frac{1}{3}$ octave in width, as indicated in Fig. 5.10. An octave is the musical interval between two tones whose frequencies are in the ratio 2 : 1. The ratio of frequencies of two tones that are

$\frac{1}{3}$ octave apart is $\sqrt[3]{2} = 1.26$. In musical language, $\frac{1}{3}$ octave equals four semitones or a major third. Sound analyzers that measure sound pressure in each of about 30 $\frac{1}{3}$-octave bands are quite common (30 such bands are required to span the audible range as compared to only 24 critical bands because critical bands are substantially greater than $\frac{1}{3}$ octave at low frequency; see Fig. 5.10).

5.5 ■ BINAURAL HEARING AND LOCALIZATION

"Nature," said the ancient Greek philosopher Zeno, "has given us one tongue, but two ears, that we might hear twice as much as we speak." Excellent advice.

The most important benefit we derive from binaural hearing is the sense of localization of the sound source. Although some degree of localization is possible in monaural listening, binaural listening greatly enhances our ability to sense the direction of the sound source.

Lord Rayleigh, who contributed so much to our understanding of acoustics, was one of the first to explain binaural localization of sound. In 1876 Rayleigh performed experiments (which, unknown to him, had been performed nearly a century earlier by Giovanni Venturi, an Italian scientist remembered for his work on fluid dynamics) to determine his ability to localize sounds of different frequencies. He found that sounds of low frequency were more difficult to locate than those of high frequency. According to Rayleigh's explanation, a sound coming from one side of the head produces a more intense sound in one ear than in the opposite ear, because the head casts a "sound shadow" for sounds of high frequency, as shown in Fig. 5.11(a). At low frequency, however, the shadow effect is small because sound waves of long wavelength diffract around the head. At 1000 Hz, the sound level is about 8 decibels greater at the ear nearest the source, but at 10,000 Hz the difference could be as great as 30 decibels.

FIGURE 5.11
Localization of a sound source. (a) At frequencies above 4000 Hz, localization is due to intensity difference at two ears. (b) At frequencies below 1000 Hz, localization is due to an interaural time difference between sound traveling paths L_1 and L_2.

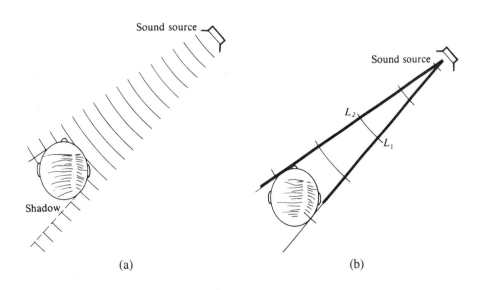

(a)   (b)

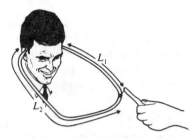

FIGURE 5.12   An experiment illustrating the sensitivity of the ear to interaural time difference. Tapping the tube so that $L_1 = L_2$ causes the sound to appear centered. When $L_2 > L_1$ the sound appears to come from the left.

Sounds of low frequency can be localized, although with slightly less accuracy than those of higher frequency. In 1907, Rayleigh offered a second theory of localization to explain low-frequency effects. A sound coming from the side strikes one ear before the other, and thus the sounds in the two ears will be slightly out of phase, as shown in Fig. 5.11(b). He confirmed this theory by experiments with two tuning forks tuned to slightly different frequencies, so that their relative phases constantly changed. The sound of the beating tone moved from right to left and back again.

Several experiments have confirmed the fact that for frequencies up to about 1000 Hz, localization occurs mainly through detection of the phase difference at the two ears (for steady sounds) or the difference in arrive time (for clicks), as illustrated in Fig. 5.12. Above 4000 Hz, localization by intensity difference takes over. Between 1000 and 4000 Hz, the accuracy of localization declines, with a high error rate around 3000 Hz demonstrating that the two mechanisms do not overlap appreciably.

At high frequencies (about 5000 Hz and upward), the pinna aids in localization of a sound, particularly in distinguishing between sound coming from the front or the back, because it receives sound with slightly greater efficiency from the front. Some animals have the ability to aim their pinnas toward sounds of interest, but human beings must turn the entire head to change pinna orientation.

An important corollary to sound localization is the *precedence effect* (sometimes referred to as the *Haas effect*), which applies to efforts to localize a sound source in a room. If similar sounds arrive within about 35 ms (0.035 s), the apparent direction of the sound source is the direction from which the first arriving sound comes. The ear automatically assumes this to be the direct sound and successive sounds to have been reflected one or more times. This effect will be discussed further in Chapter 23.

5.6 ■ MEASURING SENSATIONS: PSYCHOPHYSICS

Information about the world around us comes from our senses: vision, taste, smell, touch, and hearing. Each of our sensory organs responds to a particular type of stimulus over a limited range of energies. Our eyes, for example, respond to electromagnetic waves over an extremely narrow range of frequency compared to the wide range of electromagnetic radiations all around us.

Perception involves not only the reception of information by the appropriate sensory organ, but the coding, transmission, and processing of this information by the central nervous system. Our understanding of how this is accomplished has advanced remarkably in recent years, but still remains only fragmentary. (This may be due partly to the fact that research in this area involves several disciplines: physics, psychology, physiology, speech and hearing, engineering, mathematics, etc.) An excellent source of information about perception is a collection of articles from *Scientific American* (1972). It appears that many perceptual abilities are intrinsic; others are acquired or developed through experience and training.

The study of the relationships between stimuli and the subjective sensations they produce is the basis of *psychophysics*, so named by a pioneer in the field, G. T. Fechner. Inspired by earlier work on the subject by Ernst Weber, Fechner spent many years trying to determine quantitative relationships between stimulus and perceived sensation, and in 1860 published many of his findings in a monumental book entitled *Elements of Psychophysics*. He summed up much of this work in a simple mathematical law relating sensation to stimulus, which is often referred to today as *Fechner's law*. It expresses the relationship between stimulus and sensation rather simply: As stimuli are increased by *multiplication*, sensations increase by *addition*. For example, as the intensity of a sound is doubled, its loudness increases by one step on a scale. Mathematicians call such a relationship logarithmic; Fechner's law states that sensation grows as the logarithm of the stimulus.

Fechner argued that the same relationship applies to any stimulus and its corresponding sensation: to light and vision, etc. Recent investigations have pointed out its inadequacies; nevertheless, Fechner's law served as a basis for psychophysical theory for nearly a century thereafter. Fechner answered his early critics by saying "The Tower of Babel was never finished because the workers could not reach an understanding on how they should build it; my psychophysical edifice will stand because the workers will never agree on how to tear it down" (Stevens and Warshofsky 1965, 82). We will return to the subject in the following chapters.

5.7 ■ LOGARITHMS IN SOUND AND MUSIC

Although this book employs a minimum of mathematics, there are some times when the use of a little mathematics actually makes things easier to understand. One mathematical tool that should be familiar to everyone who wishes to understand the science of sound (and that should include serious musicians, students of speech and hearing, and anyone interested in sound recording, reproduction, and amplification) is the *logarithm*.

The logarithm to the base 10 of a number x is the power to which 10 must be raised in order to equal x. For example, $100 = 10^2$, so the logarithm of 100 (to base 10) is 2 ($\log 100 = 2$). There are other numbers besides 10 that can serve as a base, of course, but in this book $\log x$ will always mean the logarithm of x to the base 10. Fortunately pressing the log key on most calculators gives the logarithm to base 10.

Why are logarithms so useful in the study of sound? Three applications come to mind:

1. Decibel scales used to express such things as sound level and amplifier gain are based on logarithms.

2. Frequency response of the ear or audio devices are generally expressed on a compressed or logarithmic scale (Fig. 5.1 or 5.11, for example).

3. The keyboard on a piano or other musical instruments is logarithmic.

4. The musical scale is logarithmic (that is, each step is a certain ratio of frequencies).

At this time, therefore, it is appropriate to review or introduce (depending on the background of the reader) some properties of logarithms and logarithmic scales.

Logarithms and Powers of Ten

It is inconvenient to write out numbers such as 1,530,000,000 and 0.000087. These same numbers can be written better as 1.53×10^9 and 8.7×10^{-5}, respectively, because $10^9 = 1,000,000,000$ and $10^{-5} = 0.00001$. Other powers of ten are

$$10^3 = 1000,$$
$$10^2 = 100,$$
$$10^1 = 10,$$
$$10^0 = 1,$$
$$10^{-1} = 0.1,$$
$$10^{-2} = 0.01, \text{ etc.}$$

On some electronic calculators, the scientific notation used to display very large and very small numbers expresses 10^3 as E3 and 10^{-3} as E−3, so 1,530,000,000 would be expressed as 1.53 E9 and 0.000087 as 8.7 E−5 (E is an abbreviation for *exponent*, which means the power to which 10 is raised). Other calculators omit the letter E but leave a space between the first part of the number and its exponent (e.g., 1531 becomes 1.531 03 in scientific notation).

To multiply two numbers in scientific or exponential notation, we *add* the exponents; to divide, we *subtract* exponents. Thus

$$(10^3)(10^4) = 10^7;$$
$$(5 \times 10^2)(3 \times 10^5) = 15 \times 10^7 = 1.5 \times 10^8;$$
$$(3 \times 10^{-3})(2 \times 10^5) = 6 \times 10^2;$$
$$\frac{10^4}{10^2} = 10^2;$$
$$\frac{6 \times 10^5}{3 \times 10^3} = 2 \times 10^2.$$

In general, then, $(10^A)(10^B) = 10^{A+B}$, and $10^A/10^B = 10^{A-B}$.

Closely related to exponents and scientific notation are logarithms. As we stated, logarithms are defined as follows: The logarithm to the base 10 of a number x is equal to the power to which 10 must be raised in order to equal x. That is, if $x = 10^y$, then $y = \log x$.

For example, $100 = 10^2$, so $2 = \log 100$ (here $x = 100$, $y = 2$); or $1000 = 10^3$, so $3 = \log 1000$ ($x = 1000$, $y = 3$).

Although logarithms to other bases are used in mathematics, we nearly always use base 10 in acoustics. Most calculators use *log x* to denote the logarithm of x to base 10, and *ln x* to denote the logarithm to the base 2.7183. On some calculators, the inverse logarithm is computed by pressing INV and then log; on others a 10^x key is used.

The following identities are useful for performing calculations with logarithms:

$$\log AB = \log A + \log B;$$

$$\log A/B = \log A - \log B;$$

$$\log A^n = n \log A.$$

The logarithms of some numbers are as follows:

x	$\log x$	x	$\log x$
1	0	6	0.778
2	0.301	7	0.845
3	0.477	8	0.903
4	0.602	9	0.954
5	0.699	10	1.000

Using this table and the identities listed above, we can compute the logarithms of many numbers. For example:

$$\log 400 = \log 4 + \log 100 = 0.602 + 2 = 2.602 \text{ (first identity)};$$

$$\log 2.5 = \log 5 - \log 2 = 0.699 - 0.301 = 0.398 \text{ (second identity)};$$

$$\log 25 = 2 \log 5 = (2)(0.699) = 1.398 \text{ (third identity)}.$$

If one remembers that $\log 2 = 0.3$, many logarithms can be estimated closely. For example:

$$\log 4 = \log 2 \times 2 = 0.3 + 0.3 = 0.6;$$

$$\log 5 = \log \frac{10}{2} = 1 - 0.3 = 0.7;$$

$$\log 8 = \log 2 \times 2 \times 2 = 0.3 + 0.3 + 0.3 = 0.9.$$

FIGURE 5.13
Linear and logarithmic scales. On the linear scale, moving one unit to the right adds an increment of one; on the logarithmic scale, moving one unit to the right multiplies by a factor of ten.

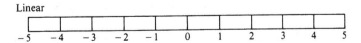

Linear

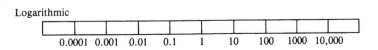

Logarithmic

A *logarithmic scale* is one on which equal distances represent the same factor anywhere along the scale (in contrast to a *linear scale*, on which equal distances represent equal increments). Logarithmic and linear scales are shown in Fig. 5.13.

Sound frequencies are usually represented on a logarithmic scale for reasons that will become clear later on. In Fig. 5.14, the distance from 20 to 200 Hz is the same as from 200 to 2000 Hz or from 2000 to 20,000 Hz.

FIGURE 5.14
Graph paper with a logarithmic scale of frequencies. Such graph paper is called *semilog*, because only one axis is logarithmic. On *log-log* graph paper, both axes are logarithmic.

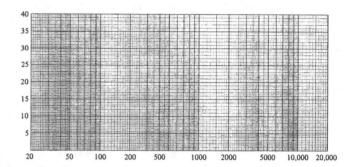

Logarithmic Scales in Music

Musical scales are discussed in Chapter 9. At this time, however, we will mention the *scale of equal temperament*, which is a logarithmic scale of frequency. The most common scale of equal temperament divides an octave (a frequency ratio of 2 : 1) into 12 equal steps. To do this, the octave is divided into frequency ratios of $2^{1/12} = 1.05946$. Going up an octave means traversing 12 such steps; the initial frequency is multiplied by $2^{1/12}$ 12 times, which is equivalent to multiplying it by 2. Most other scales that are used in music use steps of slightly different sizes as one goes up the scale.

Piano tuners, composers of electronic music, researchers of musical instruments, etc., often find it convenient to divide the octave into 1200 equal steps called *cents*. Each cent is 1/100 of a semitone, just as each cent of money is 1/100 of a dollar. Raising the pitch by 1 cent, then, means multiplying the frequency by $2^{1/1200} = 1.000578$. The formula for converting from a frequency ratio to cents (or visa versa), given in Section 9.5, uses logarithms, naturally enough.

5.8 ■ SUBJECTIVE ATTRIBUTES OF SOUND

Four attributes are frequently used to describe sound, especially musical sound. They are loudness, pitch, timbre, and duration. Each of these subjective qualities depends on one or more physical parameters that can be measured. Loudness, for example, depends mainly on sound pressure but also on the spectrum of the partials, the physical duration, etc. Pitch depends mainly on frequency but also shows less dependence on sound pressure, envelope, etc. Timbre is a sort of catchall, including all those attributes that serve to distinguish sounds with the same pitch and loudness. Table 5.1 relates subjective qualities to physical parameters, and is presented here as an introduction to the next three chapters, which discuss loudness, pitch, and timbre in more detail.

TABLE 5.1   Dependence of subjective qualitites of sound on physical parameters

Physical Parameter	Subjective Quality			
	Loudness	Pitch	Timbre	Duration
Pressure	+ + +	+	+	+
Frequency	+	+ + +	+ +	+
Spectrum	+	+	+ + +	+
Duration	+	+	+	+ + +
Envelope	+	+	+ +	+

+ = weakly dependent; + + = moderately dependent; + + + = strongly dependent.
Note: Spectrum refers to the frequencies and amplitudes of all the partials (components) in the sound. The physical duration of a sound and its perceived (subjective) duration, though closely related, are not the same. Envelope includes the attack, the release, and variations in amplitude. These parameters will be discussed in Chapters 6, 7, and 8.

5.9 ■ SUMMARY

The human auditory system responds to pressure stimuli over a range of a million times. The frequency range of hearing extends from 20 to 20,000 Hz for some individuals, substantially less for others. Like other sensations, hearing tends to follow *logarithmic* relationships. The *outer ear* boosts hearing sensitivity in the middle frequency range and aids in determining the direction of a sound. The middle ear contains three small bones, called *ossicles*, which transmit sound pressure from the eardrum to the inner ear. The main part of the inner ear is the *cochlea*, which transforms pressure variations into neural impulses. Much of our ability to determine the direction of a sound source depends on *binaural* hearing, with a different mechanism of localization being dominant at high and low frequencies.

REFERENCES AND SUGGESTED READINGS

Békésy, G. von (1960). *Experiments in Hearing.* New York: McGraw-Hill.

Békésy, G. von (1970). "Enlarged Mechanical Model of the Cochlea with Nerve Supply," in *Foundations of Modern Auditory Theory*, Vol. I. Ed., J. V. Tobias. New York: Academic Press.

Denes, P. B., and E. N. Pinson (1973). *The Speech Chain.* Garden City, N.Y.: Anchor.

Flannagan, J. L. (1972). *Speech Analysis, Synthesis and Perception.* New York: Springer-Verlag.

Fletcher, H. (1940). "Auditory Patterns," *Revs. Modern Physics* **12**: 47.

Helmholtz, H. von (1877). *On the Sensation of Tone as a Psychological Basis for the Study of Music*, 4th ed. Trans., A. J. Ellis. New York: Dover, 1954.

Johnstone, B. M., and A. J. F. Boyle (1967). "Basilar Membrane Vibration Examined with Mössbauer Technique," *Science* **158**: 389.

Khanna, S. M., and D. G. B. Leonard (1982). "Basilar Membrane Tuning in the Cat Cochlea," *Science* **215**: 305.

Plomp, R. (1976). *Aspects of Tone Sensation.* London: Academic Press.

Readings from *Scientific American* (1972). *Perception: Mechanisms and Models.* San Francisco: W. H. Freeman. This collection of readings contains reprints of 38 articles published during 1950–1971 with introductions to each section by R. Held and W. Richards. "The Ear" by Békésy (1957) is included.

Rhode, W. S., and L. Robles (1974). "Evidence from Mössbauer Experiments for Nonlinear Vibration in the Cochlea," *J. Acoust. Soc. Am.* **55**: 588.

Rossing, T. D., and C. J. Chiaverina (1999). *Light Science: Physics and the Visual Arts.* New York: Springer-Verlag.

Scharf, B. (1970). "Critical Bands" in *Foundations of Modern Auditory Theory,* ed. J. Tobias. New York: Academic Press.

Scharf, B., and S. Buus (1986). "Audition I. Stimulus, Physiology, Thresholds" in *Handbook of Perception and Human Performance.* Ed. K. R. Boff, L. Kaufman, and J. P. Thomas. New York: J. Wiley.

Stevens, S. S., and F. Warshofsky (1965). *Sound and Hearing.* New York: Time Life.

Tasaki, I. (1954). "Nerve Impulses in Individual Auditory Nerve Fibres of Guinea Pigs," *J. Neurophysiology* **17**: 97.

Teas, D. C. (1970). "Cochlear Processes," in *Foundations of Modern Auditory Theory.* Ed., J. V. Tobias. New York: Academic Press.

Zwicker, E., G. Flottorp, and S. S. Stevens (1957). "Critical Bandwidth in Loudness Summation," *J. Acoust. Soc. Am.* **29**: 548.

GLOSSARY

auditory canal A tube in the outer ear that transmits sound from the external pinna to the eardrum.

basilar membrane A membrane in the cochlea that separates the cochlear duct from the scala tympani and to which the organ of Corti is attached.

cochlea The spiral organ of the inner ear containing the sound-sensing mechanism.

critical band Frequency band within which two or more tones excite many of the same hair cells on the basilar membrane and thus are difficult to distinguish as separate tones.

eardrum (tympanum) The fibrous membrane that terminates the auditory canal and is caused to vibrate by incoming sound waves.

envelope Time history of the amplitude.

Eustachian tube A tube connecting the middle ear to the oral cavity that allows the average pressure in the middle ear to equal atmospheric pressure.

exponent The number expressing the power to which 10 or some other number is raised.

Fechner's (Weber's) law An empirical law expressing the way in which sensation varies with stimulus.

hair cells The tiny sensors of sound in the cochlea.

intensity Power per unit area. The intensity of a sound wave is proportional to the square of the sound pressure.

linear scale A scale in which moving a given distance right or left adds or subtracts a given increment.

localization The ability to determine the location or direction of a sound source.

logarithm (of a number) The power to which 10 (or some other base) must be raised to give the desired number.

logarithmic scale A scale on which moving a given distance right or left multiplies or divides by a given factor.

organ of Corti The part of the cochlea containing the hair-cells; the "seat of hearing."

ossicles Three small bones of the middle ear that transmit vibrations from the eardrum to the cochlea.

pinna The external part of the ear.

precedence effect If similar sounds arrive within about 35 ms, the apparent direction is the direction from which the first arriving sound comes.

psychoacoustics The study of the relationships between sound and the sensations it produces. The psychophysics of sound.

psychophysics The study of the relationship between stimuli and the sensations they produce.

Reissner's membrane A membrane in the cochlea that separates the cochlear duct from the scala vestibuli.

scala vestibuli A canal in the ear that transmits pressure variations from the oval window to the cochlear duct.

stereocilia The tiny fibers attached to hair cells that bend and cause electrical signals to be transmitted on the auditory nerve fibers.

REVIEW QUESTIONS

1. What is the intensity ratio between the threshold of pain and the threshold of audibility?

2. What is the frequency ratio between the highest and lowest frequencies we can hear?

3. What is the frequency ratio between the highest and lowest light frequencies we can see?

4. What membrane terminates the outer ear?

5. What are the bones in the middle ear called?

6. What tube connects the middle ear to the oral cavity?

7. What is the main function of the semicircular canal in our inner ear?

8. What is the spiral organ in the inner ear called?

9. What happens to the cilia when the basilar membrane responds to a sound?

10. What part of the basilar membrane responds the most to low-frequency vibrations?

11. In addition to transmission through the outer ear, how else can sound reach the inner ear?

12. At low frequency, the critical bandwidth remains nearly constant. (T or F)

13. How are sounds of low frequency localized?

14. How are sounds of high frequency localized?

15. What is the precedence effect?

16. According to Fechner's law, how do sensations increase as stimuli are increased by multiplication?

17. Pitch depends mainly on what physical parameter?

18. What other physical parameters does pitch also depend on?

19. What is the envelope of a sound?

20. What is a logarithm?

QUESTIONS FOR THOUGHT AND DISCUSSION

1. If everyone's hearing sensitivity were reduced by 10 dB, in what ways would our lives probably be different?

2. What advantage is there in having our various senses respond on a (nearly) logarithmic rather than a linear scale?

3. Before the development of radar, a device used to determine the direction of aircraft consisted of two sound-receiving horns, each of which transmitted sound to one ear. Comment on the effectiveness of such a device.

4. Listen to a tape recording of your own voice and compare its sound to what you hear when you speak and sing. Try to describe the difference in terms of relative balance between high and low frequency components, etc.

5. Compare following distances on the horizontal scale in Fig 5.14.
   (a) 20 to 100
   (b) 100 to 500
   (c) 2000 to 10,000

EXERCISES

1. Assume that the outer ear canal is a cylindrical pipe 3 cm long, closed at one end by the eardrum. Calculate the resonance frequency of this pipe (see Fig. 4.8). Our hearing should be especially sensitive for frequencies near this resonance.

2. At what frequency does the wavelength of sound equal the distance between your ears? What is the significance of this with respect to your ability to localize sound?

3. The effective area of the eardrum is estimated to be approximately 0.55 cm^2. During normal conversation, the sound pressure variations of about 10^{-2} N/m^2 reach the eardrum. What force is exerted on the eardrum (force = pressure × area)?

4. Pressure is force per unit area. Calculate the pressure when a force of 500 N (approximate weight of a 110-lb person) is supported by:
   (a) Spike heels having an area of 10^{-5} m^2 each;

   (b) Standard heels having an area of 10^{-2} m^2 each.
   Comment on the likelihood of denting the floor in each case.

5. Measure the distance between your ears. Divide this distance by the speed of sound (Table 3.1) to find the maximum difference in arrival time $\Delta t = (L_2 - L_1)/v$ that occurs when a sound comes directly from the side.

6. Calculate the difference in arrival time at the two ears for a sound that comes from a 45° direction (from the northwest, for example, when the listener faces north).

7. Perform the following arithmetic operations.
   (a) $(1.6 \times 10^{-8})(5.0 \times 10^3)$
   (b) $\frac{4.5 \times 10^{-2}}{1.5 \times 10^{-3}}$
   (c) $1.3 \times 10^3 + 4.3 \times 10^2$
   (d) $4.2 \times 10^2 - 5.4 \times 10^{-2}$

8. Find the following logarithms using the logarithms of the numbers 1–10 and the three identities given.

(a) log 50

(b) log 0.5

(c) $\log 2 \times 10^{10}$

(d) log 16

9. Given log x, find the number x in each case.

(a) $\log x = 0.3$

(b) $\log x = 3.0$

(c) $\log x = 1.3$

(d) $\log x = -0.3$

EXPERIMENTS FOR HOME, LABORATORY, AND CLASSROOM DEMONSTRATION

Home and Classroom Demonstration

1. *Frequency range of audible sound*   Use a wide-range loudspeaker, an audio generator, and an amplifier to determine the frequency range of audible sound. (Be careful to avoid harmonic distortion at low frequency, because the harmonics may be heard when the fundamental is inaudible.)

2. *Loudness and sound level*   Listen to how much a sound changes in loudness when the sound pressure is doubled and tripled (6-dB and 9.5-dB increase on a sound-level meter). Is it different for a pure (single frequency) tone as compared to broadband noise or music?

3. *Sensitivity to interaural time difference*   Place each end of a rubber hose (about 2 to 3 m long) in your ears (see Fig. 5.9). Close your eyes and have someone tap the hose near its center and at varying distances to the left and right of center as you point in the apparent direction of the sound source. How far from the center must the tap occur in order to identify the sound source as *left* or *right*? What, then, is the minimum interaural time difference you can detect?

4. *Time resolution of the ear*   Repeat Joseph Henry's experiment (T. D. Rossing, "Joseph Henry and Acoustics," *Physics Teacher* **16**, 600 (1978)). Clap your hands as you move back from a large wall, and note the minimum distance at which a distinct echo can be heard. Determine the time resolution by dividing the distance the sound wave traveled (twice your distance from the wall) by the speed of sound.

5. *Critical bands by masking*   Listen to Demonstration 2 on the *Auditory Demonstrations* CD.

6. *Critical bands by loudness comparison*   Listen to Demonstration 3 on the *Auditory Demonstrations* CD.

Laboratory Experiments

Critical bands by masking or critical bands by loudness comparison (*Auditory Demonstrations* CD).

Critical bands by loudness comparison (*Auditory Demonstrations* CD).

Binaural localization (*Auditory Demonstrations* CD).

CHAPTER

6

Sound Pressure, Power, and Loudness

In this chapter, we will discuss the quality of loudness and the physical parameters that determine it. The principal such parameter, we learned in Table 5.1, is sound pressure. Related to the sound pressure are the sound *power* emitted by the source and the sound *intensity* (the power carried across a unit area by the sound wave). The output signal of a microphone is generally proportional to the sound pressure, so sound pressure can be measured with a microphone and a voltmeter.

In this chapter, you should learn:

- About sound pressure level, sound intensity level, and sound power level;
- About the decibel scale for comparing sound levels;
- How to combine sound levels from several sources;
- What determines the perceived loudness of a sound;
- How one sound can mask another.

6.1 ■ DECIBELS

Decibel scales are widely used to compare two quantities. We may express the power gain of an amplifier in decibels (abbreviated dB), or we may express the relative power of two sound sources. We could even compare our bank balance at the beginning with the balance at the end of the month. ("My bank account decreased 27 decibels last month.") The decibel difference between two power levels, ΔL, is defined in terms of their power ratio W_2/W_1:

$$\Delta L = L_2 - L_1 = 10 \log W_2/W_1. \tag{6.1}$$

Although decibel scales always compare two quantities, one of these can be a fixed reference, in which case we can express another quantity in terms of this reference. For example, we often express the *sound power level* of a source by using $W_0 = 10^{-12}$ W as a reference. Then the sound power level (in decibels) will be

$$L_W = 10 \log W/W_0 \tag{6.2}$$

EXAMPLE 6.1 What is the sound power level of a loudspeaker that radiates 0.1 W?

Solution $L_W = 10 \log W/W_0 = 10 \log(0.1/10^{-12}) = 10(11) = 110$ dB.

Although L_W is the preferred abbreviation for sound power level, one often sees it abbreviated as *PWL*.

EXAMPLE 6.2   What is the decibel gain of an amplifier if an input of 0.01 W gives an output of 10 W?

Solution $L_2 - L_1 = 10 \log W_2/W_1 = 10 \log 10/0.01 = 10(3) = 30$ dB.

One number fact worth remembering is that the logarithm of 2 is 0.3 (actually 0.3010, but 0.3 will do). Why is this worthwhile? Because $10 \log 2 = 3$, doubling the power results in an increase of 3 dB in the power level. In rating the frequency response of audio amplifiers and other devices, one often specifies the frequencies of the 3-dB points, the upper and lower frequencies at which the power drops to one-half its maximum level.

Actually, remembering that $\log 2 = 0.3$ can be used to estimate other decibel levels. We know that $\log 10 = 1$, so a power gain of 10 represents a power level gain of 10 dB. For a power gain of 5, note that $5 = 10/2$, so if we gain 10 dB (multiplying by 10) and lose 3 dB (dividing by 2), multiplying power by 5 results in a gain of 7 dB. Multiplying by 4 should give 6 dB of gain, because $4 = 2 \times 2$. If multiplication by 2 is equivalent to 3 dB and multiplication by 4 is equivalent to 6 dB, we can probably guess that multiplying by 3 would give about 5 dB (actually 4.8, but 5 is often close enough).

Also, $100 = 10 \times 10$, so a power gain of 100 should represent 20 dB (two 10-dB increases).

Here, then, is a summary of what we have just figured out.

Power ratio:	2	3	4	5	10	100
Decibel gain:	3	5	6	7	10	20

EXAMPLE 6.3   What is the decibel gain when the power gain is 400?

Solution $400 = 2 \times 2 \times 10 \times 10$, so the decibel gain is $3 + 3 + 10 + 10 = 26$ dB.

6.2 ■ SOUND INTENSITY LEVEL

We have just seen how the strength of a sound source can be expressed in decibels by comparing its power to a reference power (nearly always $W_0 = 10^{-12}$ W). Similarly, the sound intensity level at a point some distance from the source can be expressed in decibels by comparing it to a reference intensity, for which we generally use $I_0 = 10^{-12}$ W/m². Thus the *sound intensity level* at some location is defined as

$$L_I = 10 \log I/I_0. \tag{6.3}$$

EXAMPLE 6.4   What is the sound intensity level at a point where the sound intensity is 10^{-4} W/m^2?

Solution $L_I = 10 \log I/I_0 = 10 \log 10^{-4}/10^{-12} = 10(8) = 80$ dB.

Even though they are both expressed in decibels, do not confuse sound power level, which describes the sound source, with sound intensity level, which describes the sound at some point. The relationship between the sound intensity level at a given distance from a sound source and the sound power level of the source depends upon the nature of the sound field. In the following boxes, we consider two cases.

Free Field

When a point source (or any source that radiates equally in all directions) radiates into free space, the intensity of the sound varies as $1/r^2$ (and the sound pressure varies as $1/r$), where r is the distance from the source S. This may be understood as a given amount of sound power being distributed over the surface of an expanding sphere with area $4\pi r^2$ (see Fig. 6.1). Thus the intensity is given by

$$I = W/4\pi r^2, \tag{6.4}$$

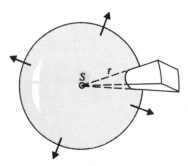

FIGURE 6.1
Spherical sound waves in a free field. The power from source S is distributed over a spherical surface $4\pi r^2$ in area.

where W is the power of the source. An environment in which there are no reflections is called a *free field*. In a free field, the sound intensity level decreases by 6 dB each time the distance from the source is doubled. The sound intensity level (or sound pressure level) at a distance of 1 m from a source in free field is 11 dB less than the sound power level of the source. This is easily shown as follows:

$$I = \frac{W}{4\pi r^2} = \frac{W}{4\pi (1)};$$

$$L_I = 10\log\frac{I}{10^{-12}} = 10\log\frac{W}{10^{-12}} - 10\log 4\pi = L_W - 11 \simeq L_p.$$

Similarly, it can be shown that at a distance of two meters, L_I is 17 dB less than L_W.

Hemispherical Field

More common than a free field is a sound source resting on a hard, sound-reflecting surface and radiating hemispherical waves into free space (see Fig. 6.2). Under these conditions, the sound intensity level L_I and the sound pressure level L_p at a distance of one meter at 8 dB less than the sound power level, once again diminishing by 6 dB each time the distance is doubled. In actual practice, few sound sources radiate sound equally in all directions, and there are often reflecting surfaces nearby that destroy the symmetry of the spherical or hemispherical waves.

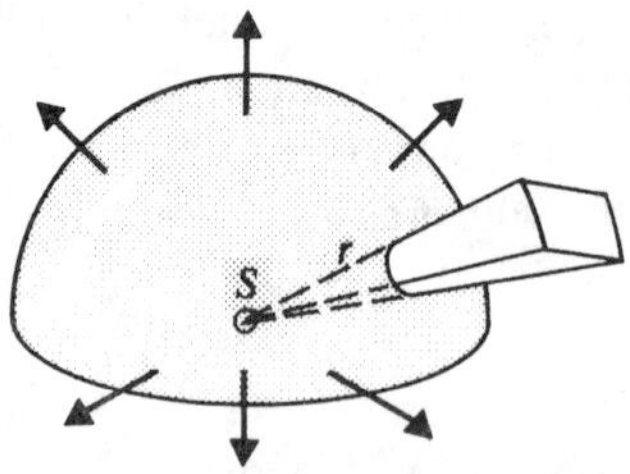

FIGURE 6.2
Hemispherical sound waves from a source S on a hard reflecting surface. The power is distributed over a surface $2\pi r^2$ in area.

EXAMPLE 6.5   If a trombone bell has an area of 0.1 m^2 and the power radiated from the bell during a very loud note is 1.5 W, what is the average intensity and sound intensity level at the bell?

Solution

$$I = \frac{W}{A} = \frac{1.5}{0.1} = 15 \text{ W/m}^2;$$

$$L_I = 10\log\frac{I}{I_0} = 10\log\frac{15}{10^{-12}}$$

$$= 10\log 15 \times 10^{12} = 132 \text{ dB}.$$

EXAMPLE 6.6   The sound pressure level 1 m from a noisy motor resting on a concrete floor is measured to be 95 dB. Find the sound power and the sound power level of the source.

Solution 1

$$L_I = 10 \log \frac{I}{I_0} \simeq L_p = 95 \text{ dB};$$

$$I = I_0 \text{ INV} \log \frac{95}{10} = 3.16 \times 10^{-3} \text{ W/m}^2;$$

$$W = 2\pi r^2 I = 2\pi (1)^2 (3.16 \times 10^{-3}) = 1.98 \times 10^{-2} \text{ W}.$$

Solution 2   For a hemispherical field,

$$L_W = L_p(1 \text{ m}) + 8 = 95 + 8 = 103 \text{ dB};$$

$$W = W_0 \text{ INV} \log \frac{103}{10} = 1.98 \times 10^{-2} \text{ W}.$$

Demonstration 4 on the *Auditory Demonstrations* CD (Houtsma, Rossing, and Wage-naars 1987) provides test tones to "calibrate" your hearing. Broadband noise is reduced in steps of 6 dB, 3 dB, and 1 dB. You probably noticed that the 1-dB steps are about the smallest steps in which you can notice a difference. The demonstration also records free-field speech at distances of 0.25, 0.5, 1, and 2 m from a microphone. Doubling the distance from the source also reduces the sound level in 6-dB steps, as we have just discussed.

6.3 ■ SOUND PRESSURE LEVEL

In a sound wave there are extremely small periodic variations in air pressure to which our ears respond. The minimum pressure fluctuation to which the ear can respond is less than 1 billionth (10^{-9}) of atmospheric pressure. This threshold of audibility, which varies from person to person, corresponds to a sound pressure amplitude of about 2×10^{-5} N/m^2 at a frequency of 1000 Hz. The threshold of pain corresponds to a pressure amplitude approximately 1 million (10^6) times greater, but still less than 1/1000 of atmospheric pressure.

The intensity of a sound wave is proportional to the pressure squared. In other words, doubling the sound pressure quadruples the intensity. The actual formula relating sound intensity I and sound pressure p is

$$I = p^2/\rho c \tag{6.5}$$

where ρ is the density of air and c is the speed of sound. The density ρ and the speed of sound c both depend on the temperature (see Section 3.5). At normal temperatures the product ρc is around 410 to 420, but for ease of calculation, we often set it equal to 400.

It is useful to substitute for I from Eq. 6.5 setting $\rho c = 400$) in Eq. 6.3: $L_I = 10 \log p^2/400 I_0 = 10 \log p^2/4 \times 10^{-10} = 20 \log p/2 \times 10^{-5}$. The latter expression is defined as the *sound pressure level* L_p (sometimes abbreviated *SPL*, although L_p is pre-ferred).

$$L_p = 20 \log p/p_0, \tag{6.6}$$

TABLE 6.1   Typical sound levels one might encounter

Jet takeoff (60 m)	120 dB	
Construction site	110 dB	*Intolerable*
Shout (1.5 m)	100 dB	
Heavy truck (15 m)	90 dB	*Very noisy*
Urban street	80 dB	
Automobile interior	70 dB	*Noisy*
Normal conversation (1 m)	60 dB	
Office, classroom	50 dB	*Moderate*
Living room	40 dB	
Bedroom at night	30 dB	*Quiet*
Broadcast studio	20 dB	
Rustling leaves	10 dB	*Barely audible*
	0 dB	

where the reference level $p_0 = 2 \times 10^{-5}$ N/m^2 = 20μPa (a pascal (Pa) is an alternative name for N/m^2). Note that Eq. 6.6 is the definition of sound pressure level, which is equal to sound intensity level only when $\rho c = 400$ (which would happen at 30° C and 748 mm Hg, for example). However, at ordinary temperatures, the two are so close to each other that they are often considered to be equal and called merely *sound level*. For precise measurement, a distinction should be made, however.

Sound pressure levels are measured by a sound-level meter, consisting of a microphone, an amplifier, and a meter that reads in decibels. Sound pressure levels of a number of sounds are given in Table 6.1. If you have access to a sound-level meter, it is recommended that you carry it with you to many locations to obtain a feeling for different sound pressure levels.

EXAMPLE 6.7   What sound pressure level corresponds to a sound pressure of 10^{-3} N/m^2?

Solution $L_p = 20 \log \dfrac{10^{-3}}{2 \times 10^{-5}} = 34.0$ dB.

EXAMPLE 6.8   How much force does a sound wave at the pain threshold ($L_p \simeq 120$ dB) exert on an eardrum having a diameter of 7 mm?

Solution

$$L_p = 120 = 20 \log \frac{p}{p_0};$$

$$p = p_0 \text{ INV} \log \frac{120}{20} = 20 \text{ N/m}^2;$$

$$F = pA = 20\pi (3.5 \times 10^{-3})^2 = 1.54 \times 10^{-3} \text{ N}.$$

6.4 ■ MULTIPLE SOURCES

Very frequently we are concerned with more than one source of sound. The way in which sound levels add may seem a little surprising at first. For example, two uncorrelated sources, each of which would produce a sound level of 80 dB at a certain point, will together give 83 dB at that point. Figure 6.3 gives the increase in sound level due to additional equal sources. It is not difficult to see why this is the case, because doubling the sound power raises the sound power level by 3 dB and thus raises the sound pressure level 3 dB at our point of interest. Under some conditions, however, there may be interference between waves from the two sources, and this doubling relationship will not hold true.

When two waves of the same frequency reach the same point, they may interfere constructively or destructively. If their amplitudes are both equal to A, the resultant amplitude may thus be anything from zero up to 2 A. The resultant intensity, which is proportional to the amplitude squared, may thus vary from 0 to 4 A^2. If the waves have different frequencies, however, these well-defined points of constructive and destructive interference do not occur. In the case of sound waves from two noise sources (as in the case of light from two light bulbs), the waves include a broad distribution of frequencies (wavelength), and we do not expect interference to occur. In this case, we can add the energy carried by each wave across a surface or, in other words, the intensities.

In the case of independent (uncorrelated) sound sources, what we really want to add are the mean-square pressures (average values of p^2) at a point. Because intensity is proportional to p^2, however, we can add the intensities. For example, two sources that by themselves cause $L_I = 40$ dB at a certain location will cause $L_I = 43$ dB at the same location when sounded together. (This result is also obtained from the graph in Fig. 6.3.)

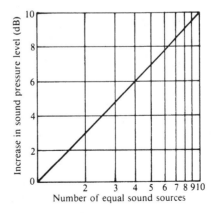

FIGURE 6.3
Addition of equal
(uncorrelated)
sound sources.

EXAMPLE 6.9   With one violin playing, the sound pressure level at a certain place is measured to be 50 dB. If three violins play equally loudly, what will the sound pressure level most likely be at the same location?

Solution

$$L_p = 10 \log \frac{p_1^2 + p_2^2 + p_3^2}{p_0^2} = 10 \log \frac{I_1 + I_2 + I_3}{I_0}$$

$$= 10 \log \frac{I_1}{I_0} + 10 \log 3$$

$$= 50 + 4.8 = 54.8 \text{ dB}$$

(This result could also be determined from Fig. 6.3.)

EXAMPLE 6.10   If two sound sources independently cause sound levels of 50 and 53 dB at a certain point, what is L_I at that point when both sources contribute at the same time?

Solution

$$50 = 10 \log \frac{I_1}{I_0}$$

so $I_1 = I_0 \text{ INV} \log \dfrac{50}{10} = (10^{-12})(10^5) = 10^{-7} \text{ W/m}^2$;

likewise $I_2 = 2 \times 10^{-7} \text{ W/m}^2$;

$$L_I = 10 \log \frac{I_1 + I_2}{I_0} = 10 \log \frac{10^{-7} + 2 \times 10^{-7}}{10^{-12}}$$

$$= 10 \log 3 \times 10^5 = 54.8 \text{ dB}.$$

(Note that the answer is *not* $50 + 53 = 103$ dB.)

6.5 ■ LOUDNESS LEVEL

Although sounds with a greater L_I or L_p usually sound louder, this is not always the case. The sensitivity of the ear varies with the frequency and the quality of the sound. Many years ago Fletcher and Munson (1933) determined curves of equal *loudness level* (L_L) for pure tones (that is, tones of a single frequency). The curves shown in Fig. 6.4, recommended by the International Standards Organization, are quite similar to those of Fletcher and Munson. These curves demonstrate the relative insensitivity of the ear to sounds of low frequency at moderate to low intensity levels. Hearing sensitivity reaches a maximum between 3500 and 4000 Hz, which is near the first resonance frequency of the outer ear canal, and again peaks around 13 kHz, the frequency of the second resonance.

The contours of equal loudness level are labeled in units called *phons*, the level in phons being numerically equal to the sound pressure level in decibels at $f = 1000$ Hz. The phon is a rather arbitrary unit, however, and is not widely used in measuring sound. It is

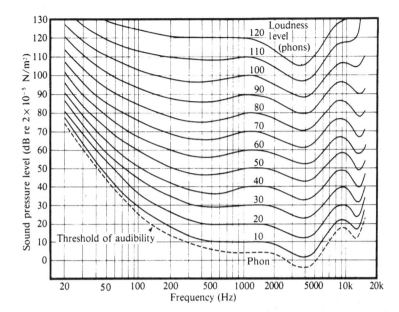

FIGURE 6.4
Equal-loudness
curves for pure
tones (frontal
incidence). The
loudness levels are
expressed in phons.

important, however, to note the relative insensitivity of the ear to sounds of low frequency, which is one reason why weighting networks are used in sound-measuring equipment.

Sound level meters have one or more weighting networks, which provide the desired frequency responses. Generally three weighting networks are used; they are designated A, B, and C. The C-weighting network has an almost flat frequency response, whereas the A-weighting network introduces a low-frequency rolloff in gain that bears rather close resemblance to the frequency response of the ear at low sound pressure level. A sound level meter is shown in Fig. 6.5, along with the frequency responses of A-, B-, and C-weighting networks.

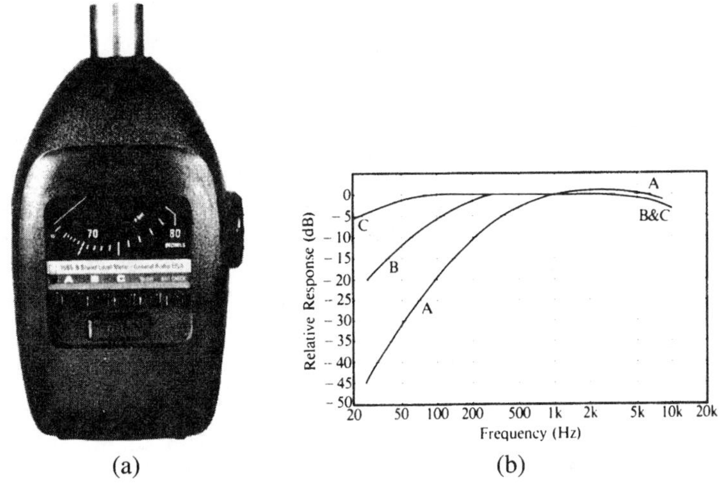

FIGURE 6.5
Sound-level meter
with the frequency
response of its A-,
B-, and
C-weighting
networks.
(Photography
courtesy of
GenRad, Inc.)

Measurements of sound level are usually made using the A-weighting network; such measurements are properly designated as $L_p(A)$·or $SPL(A)$ in dB, although the unit dBA or dB(A) is often used to denote A-weighted sound level. Inside a building, the C-weighted sound level may be substantially higher than the A-weighted sound level, because of low-frequency machinery noise, to which the ear is quite insensitive. Many sound-level meters have both fast and slow response, the slow response measuring an "average" level.

Although it is difficult to describe a sound environment by a single parameter, for many purposes the A-weighted sound level will suffice. At low to medium sound levels, it is reasonably close to the true loudness level so that dBA (easily measured with a sound-level meter) may be substituted for phons without too much error.

There are many examples of interesting sound environments to measure. In the classroom, one can ask the entire class to shout loudly, then half the class to do so, one-fourth of the class, etc. The sound level should drop about 3 dB in each step. One can also measure traffic noise, noise near a construction site, sound level at a concert, noise in an automobile, and so on. In each case the A-weighted sound level should be measured, although it may be interesting to measure the C-weighted level (which places more emphasis on sounds of low frequency) as well.

6.6 ■ LOUDNESS OF PURE TONES: SONES

In Chapter 5, we mentioned Fechner's law, relating sensation to stimulus. The logarithmic relationship in that law was found to provide only a rough approximation to listeners' estimates of their own sensations of loudness. In an effort to obtain a quantity proportional to the loudness sensation, a loudness scale was developed in which the unit of loudness is called the *sone*. The sone is defined as the loudness of a 1000-Hz tone at a sound level of 40 decibels (a loudness level of 40 phons).

For loudness levels of 40 phons or greater, the relationship between loudness S in sones and loudness level L_L in phons recommended by the International Standards Organization (ISO) is

$$S = 2^{(L_L - 40)/10}. \tag{6.7}$$

A graph of Eq. 6.7 is shown in Fig. 6.6. An equivalent expression for loudness S that avoids the use of L_L is

$$S = Cp^{0.6}, \tag{6.8}$$

where p is the sound pressure and C depends on the frequency.

Equations 6.7 and 6.8 are based on the work of S. S. Stevens, which indicated a doubling of loudness for a 10-dB increase in sound pressure level. Some investigators, however, have found a doubling of loudness for a 6-dB increase in sound pressure level (Warren 1970). This suggests the use of a formula in which loudness is proportional to sound pressure (Howes 1974):

$$S = K(p - p_0), \tag{6.9}$$

where p is sound pressure and p_0 is the pressure at some threshold level.

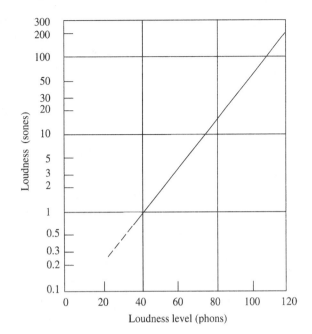

FIGURE 6.6
The relationship between the loudness (in sones) and the loudness level (in phons) from Eq. 6.7.

EXAMPLE 6.11   Find the loudness level and the loudness of a 500-Hz tone with $L_p = 70$ dB.

Solution   From Fig. 6.4, the loudness level is $L_L = 74$ phons.
   The loudness is: $S = 2^{(74-40)/10} = 10.6$ sones.

6.7 ■ LOUDNESS OF COMPLEX TONES: CRITICAL BANDS

As pointed out in Table 5.1, loudness depends mainly on sound pressure, but it also varies with frequency, spectrum, and duration. We have already seen how loudness depends on frequency; now we will consider its dependence on the spectrum of the sound.

   If we were to listen to two pure tones having the same sound pressure level but with increasing frequency separation, we would note that when the frequency separation exceeds the *critical bandwidth*, the total loudness begins to increase. Broadband sounds, such as those of jet aircraft, seem louder than pure tones or narrowband noise having the same sound pressure level. Figure 6.7 illustrates the dependence of loudness on bandwidth with fixed sound pressure level and center frequency. Note that loudness is not affected until the bandwidth exceeds the critical bandwidth, which is about 160 Hz for the 1-kHz center frequency shown.

   One way to estimate the critical bandwidth is to increase the bandwidth of a noise signal while decreasing the amplitude in order to keep the power constant. When the bandwidth is greater than a critical band, the subjective loudness increases above that of a reference noise

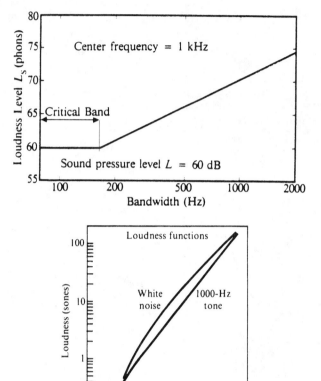

FIGURE 6.7
The effect of bandwidth on loudness.

FIGURE 6.8
Loudness of white noise compared to that of a 1000-Hz tone at the same sound pressure level. Fifteen subjects judged the noise presented binaurally through headphones (Scharf and Fishken 1970).

signal because the stimulus now extends over more than one critical band (Demonstration 3 in Houtsma, Rossing, and Wagenaars 1988).

The perceived loudness of broadband (*white*) noise is compared to that of a 1000-Hz tones have the same SPL in Fig. 6.8. At a sound pressure level of 55 dB, the white noise is judged to be about twice as loud as the 1000-Hz tone, but at higher and lower levels the difference is substantially less (Scharf and Houtsma 1986).

The dependence of loudness on stimulus variables, such as sound pressure, frequency, spectrum, duration, etc., appears to be about the same whether the sound is presented to one ear (monaurally) or to both ears (binaurally). However, a sound presented to both ears is judged nearly twice as loud as the same sound presented to one ear only (Scharf and Houtsma 1986).

6.8 ■ LOUDNESS OF COMBINED SOUNDS

The loudness of combined sounds is a subject of considerable interest. How many violins must play together, for example, in order to double the loudness? Or, how does the loudness of traffic noise depend on the number of vehicles? We stated in Section 6.4 that the intensities (or mean-square pressures) from two or more uncorrelated sound sources add together to give a total intensity. The loudness is not necessarily additive, however. Accepted methods for combining loudness are given in the following box.

When two or more tones are mixed, the way in which their individual loudnesses combine depends on how close they are to each other in frequency. We can have three different situations:

1. If the frequencies of the tones are the same or fall within the critical bandwidth, the loudness is calculated from the total intensity $I = I_1 + I_2 + I_3 + \cdots$. If the intensities I_1, I_2, I_3, etc., are equal, the increase in sound level is as shown in Fig. 6.3. The loudness may then be determined from the combined sound level.

2. If the bandwidth exceeds the critical bandwidth, the resulting loudness is greater than that obtained from simple summation of intensities. As the bandwidth increases, the loudness approaches (but remains less than) a value that is the sum of the individual loudnesses:

$$S = S_1 + S_2 + S_3 + \cdots. \tag{6.10}$$

3. If the frequency difference is very large, the summation becomes complicated. Listeners tend to focus primarily on one component (e.g., the loudest or the one of highest pitch) and assign a total loudness nearly equal to the loudness of that component (Roederer 1975).

To determine the loudness of sones of a complex sound with many components, it is advisable to measure the sound level in each of the ten standard octave bands (or in thirty $\frac{1}{3}$-octave bands). Octave bands are frequency bands one octave wide (that is, the maximum frequency is twice the minimum frequency). Octave-band analyzers, available in many acoustic laboratories, usually have a filter that allows convenient measurement of the sound level in standard octave bands with center frequencies at 31, 63, 125, 250, 500, 1000, 2000, 4000, 8000, and 16,000 Hz. Once these levels have been measured, a suitable chart (see ISO Recommendations No. 532) can be used to find the loudness in sones.

Is this seemingly complicated procedure necessary? For precise determination of loudness, yes. For estimating loudness, no. A pretty fair estimate of loudness can be made by using an ordinary sound level meter to measure the A-weighted sound level. To estimate the number of sones, let 30 dBA correspond to 1.5 sones and double the number of sones for each 10-dBA increase, as shown in Table 6.2. This procedure works quite well at low to moderate levels, because the A-weighting is a reasonable approximation to the frequency response of the ear.

Because the previous paragraphs have dealt with numbers, formulas, and graphs, it is appropriate to make a few comments on how they apply to music, environmental noise,

TABLE 6.2　Chart for estimating loudness in sones of complex sounds from A-weighted sound levels

$L_p(A)$	30	40	50	55	60	65	70	75	80	85	90	dB
S	1.5	3	6	8	12	16	24	32	48	64	96	sones

and audiometric measurements. It should be emphasized that loudness is subjective, and its assessment varies from individual to individual. On the average, a sound of four sones sounds twice as loud as a sound of two sones, but some listeners may regard it as three times louder or one and a half times louder.

Interesting examples illustrating the importance of loudness phenomena in music appear throughout the literature. Roederer (1975) discusses the selection of combinations of organ stops. Benade (1976) describes how a saxophone was made to sound louder at the same sound pressure level by a change in timbre.

6.9 ■ MUSICAL DYNAMICS AND LOUDNESS

Variations in loudness add excitement to music. The range of sound level in musical performance, known as the *dynamic range*, may vary from a few decibels to 40 dB or more, depending on the music (loud peaks and pauses may cause the instantaneous level to exceed this range). The approximate range of sound level and frequency heard by the music listener is shown in Fig. 6.9.

Composers use dynamic symbols to indicate the appropriate loudness to the performer. The six standard levels are shown in Table 6.3.

Measurements of sound intensity of a number of instrumentalists have shown that seldom do musical performers actually play at as many as six distinguishable dynamic levels, however. In one study, the dynamic ranges of 11 professional bassoonists were found to vary from 6 to 17 decibels with an average of 10 dB (Lehman 1962). A 10-dB increase

FIGURE 6.9
Approximate range of frequency and sound level of music compared to the total range of hearing.

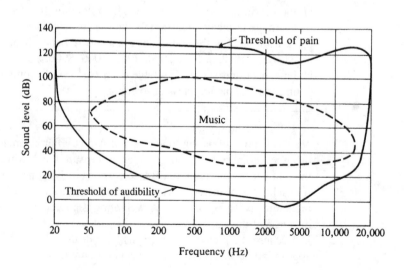

TABLE 6.3   Standard levels of musical dynamics

Name	Symbol	Meaning
Fortissimo	*ff*	Very loud
Forte	*f*	Loud
Mezzo forte	*mf*	Moderately loud
Mezzo piano	*mp*	Moderately soft
Piano	*p*	Soft
Pianissimo	*pp*	Very soft

TABLE 6.4   Dynamic ranges of musical instruments

Instrument	Average dynamic range (dB) (Clark and Luce 1965)	Maximum dynamic range (dB) (Patterson 1974)
Violin	14	40
Viola	16	
Cello	14	
String bass	14	30
Recorder		10
Flute	7	30
Oboe	7	
English horn	5	
Clarinet	8	45
Bassoon	10	40
Trumpet	9	
Trombone	17	38
French horn	18	
Tuba	13	

in sound level, you will recall from Section 6.4, is usually said to double the loudness (expressed in sones). Most listeners would have considerable difficulty identifying six different levels within a dynamic range of 10 dB. Dynamic ranges of several instruments are given in Table 6.4.

The dynamic ranges in Table 6.4 are for single notes played loudly and softly. Several instruments have much more sound power near the top of their playing range than near the bottom. (Fortissimo on a French horn, for example, is found to be nearly 30 dB greater at C_5 than at C_2, although the difference between *ff* and *pp* on any note of the scale may be 20 dB or less.)

Measurement of the dynamic ranges of various musical instruments and players is an instructive and relatively easy experiment for the reader to perform. The dynamic ranges of most players we have measured fall close to those reported by Clark and Luce (1965).

6.10   ■ MASKING

When the ear is exposed to two or more different tones, it is a common experience that one may mask the others. *Masking* is probably best explained as an upward shift in the hearing threshold of the weaker tone by the louder tone and depends on the frequencies of the two tones. Pure tones, complex sounds, narrow and broad bands of noise all show differences in their ability to mask other sounds. Masking of one sound can even be caused by another sound that occurs a split second after the masked sound.

Some interesting conclusions can be made from the many masking experiments that have been performed:

1. Pure tones close together in frequency mask each other more than tones widely separated in frequency.
2. A pure tone masks tones of higher frequency more effectively than tones of lower frequency.
3. The greater the intensity of the masking tone, the broader the range of frequencies it can mask.
4. If the two tones are widely separated in frequency, little or no masking occurs.
5. Masking by a narrow band of noise shows many of the same features as masking by a pure tone; again, tones of higher frequency are masked more effectively than tones of lower frequency than the masking noise.
6. Masking of tones by broadband ("white") noise shows an approximately linear relationship between masking and noise level (that is, increasing the noise level 10 dB raises the hearing threshold by the same amount). Broadband noise masks tones of all frequencies.
7. *Forward masking* refers to the masking of a tone by a sound that ends a short time (up to about 20 or 30 ms) before the tone begins. Forward masking suggests that recently stimulated cells are not as sensitive as fully rested cells.
8. *Backward masking* refers to the masking of a tone by a sound that begins a few milliseconds later. A tone can be masked by noise that begins up to ten milliseconds later, although the amount of masking decreases as the time interval increases (Elliott 1962). Backward masking apparently occurs at higher centers of processing where the later-occurring stimulus of greater intensity overtakes and interferes with the weaker stimulus.

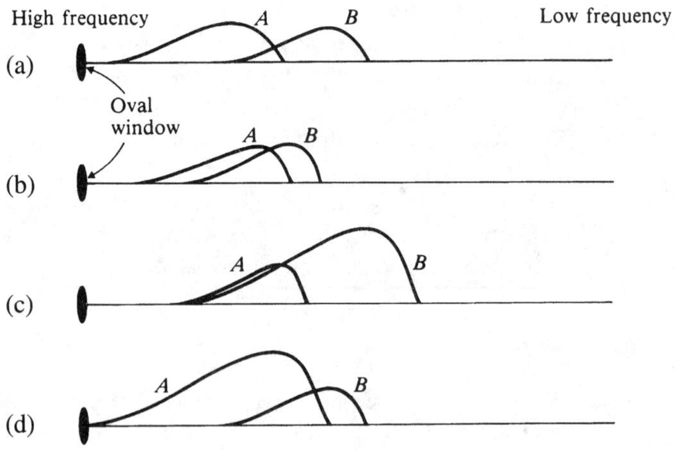

FIGURE 6.10
Simplified response of the basilar membrane for two pure tones *A* and *B*. (a) The excitations barely overlap; little masking occurs. (b) There is an appreciable overlap; tone *B* masks tone *A* and somewhat more than the reverse. (c) The more intense tone *B* almost completely masks the higher-frequency tone *A*. (d) The more intense tone *A* does not completely mask the lower-frequency tone *B*.

9. Masking of a tone in one ear can be caused by noise in the other ear, under certain conditions; this is called *central masking*.

Some of the conclusions about masking just stated can be understood by considering the way in which pure tones excite the basilar membrane (see Fig. 5.6). High-frequency tones excite the basilar membrane near the oval window, whereas low-frequency tones create their greatest amplitude at the far end. The excitation due to a pure tone is asymmetrical, however, having a tail that extends toward the high-frequency end as shown in Fig. 6.10. Thus it is easier to mask a tone of higher frequency than one of lower frequency. As the intensity of the masking tone increases, a greater part of its tail has amplitude sufficient to mask tones of higher frequency.

6.11 ■ LOUDNESS REDUCTION BY MASKING

Sounds are seldom heard in isolation. The presence of other sounds not only raises the threshold for hearing a given sound but generally reduces its loudness as well (this is sometimes called *partial masking*).

Figure 6.11 shows how white noise reduces the apparent loudness of a 1000-Hz tone. Compared to the tone in quiet (see Fig. 6.8), the loudness functions in white noise are steeper. Rising form an elevated threshold, the partially masked tone eventually comes to its full unmasked loudness when the noise level is less than 80 dB. In more intense noise, the loudness does not reach its full unmasked value, but the function approaches the same slope as the function without masking noise (Scharf and Houtsma 1986).

6.12 ■ LOUDNESS AND DURATION: IMPULSIVE SOUNDS AND ADAPTATION

How does the loudness of an impulsive sound compare to the loudness of a steady sound at the same sound level? Numerous experiments have pretty well established that the ear averages sound energy over about 0.2 s (200 ms), so loudness grows with duration up to this

FIGURE 6.11
Loudness functions for a 1000-Hz tone partially masked by white noise at various sound pressure levels. Subjects adjusted the level of the tone in quiet so that it sounded as loud as the tone with noise. (After Scharf 1978.)

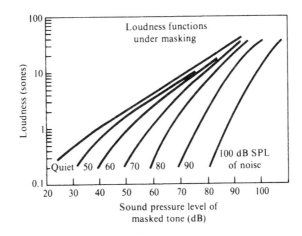

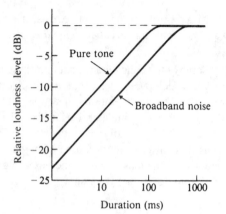

FIGURE 6.12
Variation of
loudness level with
duration. (After
Zwislocki 1969.)

value. Stated another way, loudness level increases by 10 dB when the duration is increased by a factor of 10. The loudness level of broadband noise seems to depend somewhat more strongly on stimulus duration than the loudness level of pure tones, however. Figure 6.12 shows the approximate way in which loudness level changes with duration.

The ear, being a very sensitive receiver of sounds, needs some protection to avoid injury by very loud sounds. Up to 20 dB of effective protection is provided by muscles attached to the eardrum and the ossicles of the middle ear. When the ear is exposed to sounds in excess of 85 dB or so, these muscles tighten the ossicular chain and pull the stapes (stirrup-shaped bone) away from the oval window of the cochlea. This action is termed the *acoustic reflex*.

Unfortunately the reflex does not begin until 30 or 40 ms after the sound overload occurs, and full protection does not occur for another 150 ms or so. In the case of a loud impulsive sound (such as an explosion or gunshot), this is too late to prevent injury to the ear. In fact a tone of 100 dB or so preceding the loud impulse has been proposed as a way of triggering the acoustic reflex to protect the ear (Ward 1962). It is interesting to speculate what type of protective mechanism, analogous to eyelids, might have developed in the auditory system had the loud sounds of the modern world existed for millions of years (earlids, perhaps?).

Like most other sensations, loudness might be expected to decrease with prolonged stimulation. Such a decrease is called *adaptation*. Under most listening conditions, however, loudness adaptation appears to be very small. A steady 1000-Hz tone at 50 dB causes little adaptation, although the loudness of a tone that alternates between 40 and 60 dB appears to decrease in loudness over the first two or three minutes, as do tones within about 30 dB of threshold (Scharf and Houtsma 1986).

Exposure to a loud sound affects our ability to hear another sound at a later time. This is called *fatigue* and may result in both a temporary loudness shift (TLS) and a temporary threshold shift (TTS). TLS and TTS appear to be greatest at a frequency a half octave higher than that of the fatiguing sound. Noise-induced TTS is discussed in Chapter 31.

6.13 ■ SUMMARY

Each of the quantities sound pressure, sound power, and sound intensity has an appropriate decibel level that expresses the ratio of these quantities to appropriate reference levels.

Sound pressure can be measured directly by a sound level meter, which may offer one to three different frequency weightings. Loudness level (in phons) expresses the sound pressure level of an equally loud 1000-Hz tone, whereas the loudness (in sones) expresses a subjective rating of loudness. Expressing the loudness of complex tones is fairly subtle, involving critical bandwidth, masking one tone by another, etc. The loudness of impulsive sounds increases with their duration up to about 0.2 s. The dynamic range of music covers about 40 dB, although individual instruments have dynamic ranges considerably less than this. Composers use six standard levels to indicate loudness. The ear is partially protected from loud sounds by the acoustic reflex.

REFERENCES AND SUGGESTED READINGS

Benade, A. H. (1976). *Fundamentals of Musical Acoustics.* New York: Oxford.

Clark, M., and D. Luce (1965). "Intensities of Orchestral Instrument Scales Played at Prescribed Dynamic Markings," *J. Audio Eng. Soc.* **13**: 151.

Egan, J. P., and H. W. Hake (1950). "On the Masking Pattern of a Simple Auditory Stimulus," *J. Acoust. Soc. Am.* **22**: 622.

Elliott, L. L. (1962). "Backward and Forward Masking of Probe Tones of Different Frequencies," *J. Acoust. Soc. Am.* **34**: 1116.

Fletcher, H., and W. A. Munson (1933). "Loudness, Definition, Measurement and Calculation," *J. Acoust. Soc. Am.* **6**: 59.

Hawkins, J. E., Jr., and S. S. Stevens (1950). "Masking of Pure Tones and Speech by Noise," *J. Acoust. Soc. Am.* **22**: 6.

Houtsma, A. J. M., T. D. Rossing, and W. M. Wagenaars (1988). *Auditory Demonstrations* (Philips Compact Disc #1126-061 and text). Woodbury, N.Y.: Acoustical Society of America.

Howes, W. L. (1974). "Loudness Function Derived from Data on Electrical Discharge Rates in Auditory Nerve Fibers," *Acustica* **30**: 247.

International Standards Organization, Standards R131, 26, R357 and R454 (International Organization for Standardization, 1 Rue de Narembé, Geneva, Switzerland).

Lehman, P. R. (1962). "The Harmonic Structure of the Tone of the Bassoon," Ph.D. thesis, University of Michigan, Ann Arbor.

Littler, T. S. (1965). *The Physics of the Ear.* New York: Macmillan.

Patterson, B. (1974). "Musical Dynamics," *Sci. Am.* **231**(5): 78.

Plomp, R. (1976). *Aspects of Tone Sensation.* London: Academic.

Roederer, J. G. (1975). *Introduction to the Physics and Psychophysics of Music,* 2nd ed. New York: Springer-Verlag.

Scharf, B. (1978). "Loudness," in *Handbook of Perception,* Vol. 4. Ed., E. C. Carterette and M. P. Friedman. New York: Academic Press.

Scharf, B., and D. Fishken (1970). "Binaural Summation of Loudness," *J. Experimental Psych.* **86**: 374–79.

Scharf, B. and A. J. M. Houtsma (1986). "Audition II. Loudness, Pitch, Localization, Aural Distortion, Pathology," in *Handbook of Perception and Human Performance.* Ed., K. R. Boff, L. Kaufman, and J. P. Thomas. New York: J. Wiley.

Stevens, S. S., and G. Stevens (1975). *Psychophysics Introduction to Its Perceptual, Neural and Social Prospects.* New York: Wiley.

Ward, W. D. (1962). "Studies on the Aural Reflex III. Reflex Latency as Inferred from Reduction of Temporary Threshold Shift from Impulses," *J. Acoust. Soc. Am.* **34**: 1132.

Warren, R. M. (1970). "Elimination of Biases in Loudness Judgements for Tones," *J. Acoust. Soc. Am.* **48**: 1397.

Zwislocki, J. J. (1969). "Temporal Summation of Loudness: An Analysis," *J. Acoust. Soc. Am.* **46**: 431.

GLOSSARY

acoustic reflex Muscular action that reduces the sensitivity of the ear when a loud sound occurs.

auditory fatigue Change in loudness of a sound that follows a loud sound.

critical bandwidth The frequency bandwidth beyond which subjective loudness increases with bandwidth (see also definition in chapter 5).

decibel A dimensionless unit used to compare the ratio of two quantities (such as sound pressure, power, or intensity), or to express the ratio of one such quantity to an appropriate reference.

intensity Power per unit area; rate of energy flow.

intensity level $L_I = 10 \log I/I_0$, where I is intensity and $I_0 = 10^{-12}$ W/m^2 (abbreviated *SIL* or L_I).

loudness Subjective assessment of the "strength" of a sound, which depends on its pressure, frequency, and timbre; loudness may be expressed in sones.

loudness level Sound pressure of a 1000-Hz tone that sounds equally loud when compared to the tone in question; loudness level is expressed in phons.

masking The obscuring of one sound by another.

phon A dimensionless unit used to measure loudness level; for a tone of 1000 Hz, the loudness level in phons equals the sound pressure in decibels.

sone A unit used to express subjective loudness; doubling the number of sones should describe a sound twice as loud.

sound power level $L_W = 10 \log W/W_0$, where W is sound power and $W_0 = 10^{-12}$ W (abbreviated *PWL* or L_W).

sound pressure level $L_p = 20 \log p/p_0$, where p is sound pressure and $p_0 = 2 \times 10^{-5}$ N/m^2 (or 20 micropascals) (abbreviated *SPL* or L_p).

white noise Noise whose amplitude is constant throughout the audible frequency range.

REVIEW QUESTIONS

1. In what units is sound intensity measured?
2. What reference level is used to measure sound intensity level?
3. What is meant by a free field?
4. How large is the "just noticeable difference" in sound level?
5. How much does the sound level decrease in a free field when the distance from the source is doubled?
6. In air, how does ρ change as the temperature increases?
7. In air, how does c change as the temperature increases?
8. In air, how does ρc change with temperature?
9. What is the approximate sound level in normal conversation?
10. If each of two sound sources alone produces a sound level of 55 dB at a certain point, what will the level most likely be at that point if both sources are active?
11. In what units is loudness level expressed?
12. In what units is loudness expressed?
13. What generally happens to loudness as the bandwidth of a noise source is increased while the sound level stays constant?
14. By approximately how many decibels must the A-weighted sound level increase in order to double the loudness of a complex tone?
15. What is the average dynamic range of a single note played on a musical instrument?
16. What is backward masking?
17. Is it easier for a tone of lower frequency to mask a tone of a higher frequency, or vice versa?
18. Does a given tone generally sound louder or less loud against a background noise as compared to the same tone in a quiet setting?
19. How does loudness depend upon duration at constant sound level?

QUESTIONS FOR THOUGHT AND DISCUSSION

1. Which will sound louder, a pure tone of $L_p = 40$ dB, $f = 2000$ Hz, or a pure tone of $L_p = 65$ dB, $f = 50$ Hz?
2. If two identical loudspeakers are driven at the same power level by an amplifier, how will the sound levels

due to each combine? Does it make a difference whether the program source is stereophonic or monophonic?

3. How is it possible for one sound to mask a sound that has already occurred (backward masking)? Speculate what might happen in the human nervous system when such a phenomenon occurs.

4. How low long must a burst of broadband (white) noise be in order to be half as loud as a continuous noise of the same type?

5. Why do community noise laws generally specify maximum $L_p(A)$ rather than $L_p(C)$?

EXERCISES

1. What sound pressure level is required to produce minimum audible field at 50, 100, 500, 1000, 5000, and 10,000 Hz?

2. What sound pressure level of 100-Hz tone is necessary to match the loudness of a 3000-Hz tone with $L_p = 30$ dB? What is the loudness level (in phons) of each of these tones?

3. With one violin playing, the sound level at a certain place is measured as 50 dB. If four violins play equally loudly, what will the sound level most likely be at this same place?

4. If two sounds differ in level by 46 dB, what is the ratio of their sound pressures? their intensities?

5. A loudspeaker is supplied with 5 W of electrical power, and it has an efficiency of 10% in converting this to

sound power. What is its sound power level? If we assume that the sound radiates equally in all directions, what is the sound pressure level at a distance of 1 m? at a distance of 4 m?

6. A 60-Hz tone has a sound pressure level of 60 dB measured with C-weighting on a sound level meter. What level would be measured with A-weighting?

7. Find the sound pressure and the intensity of a sound with $L_p = 50$ dB.

8. What is the decibel gain when the power gain is 30? when it is 50?

9. According to Fig. 6.6, what is the loudness level that produces a loudness of 10 sones? 100 sones?

EXPERIMENTS FOR HOME, LABORATORY, AND CLASSROOM DEMONSTRATION

Home and Classroom Demonstration

1. *The decibel scale* Demonstration 4 on the *Auditory Demonstrations* CD (Houtsma, Rossing, and Wagenaars 1988). Broadband is reduced in steps of 6 dB, 3 dB, and 1 dB. This is followed by free-field speech, recorded at distances of 0.25, 0.5, 1, and 2 m from a microphone.

2. *Frequency response of the ear* Demonstration 6 on the *Auditory Demonstrations* CD. Tones having frequencies of 125, 250, 500, 1000, 2000, 4000, and 8000 Hz are decreased in 10 steps of −5 dB each in order to determine thresholds of audibility at each frequency.

3. *Loudness scaling* Demonstration 7 on the *Auditory Demonstrations* CD. Listeners are asked to rate the loudness of 20 test tones in comparison to a reference tone. These ratings are plotted against the sound level of each test tone to establish an average loudness scale. If done as a class demonstration, better statistics are obtained by combining all the responses on a single graph.

4. *Critical bands by loudness comparison* Demonstration 3 on the *Auditory Demonstrations* CD. The bandwidth of a

noise burst is increased while its amplitude is decreased to keep the power constant. When the bandwidth is greater than a critical band, the subjective loudness increases above that of a reference noise burst, because the stimulus now extends over more than one critical band.

5. *Critical bands by masking* Demonstration 2 on the *Auditory Demonstrations* CD. A 2000-Hz tone is masked by spectrally flat (white) noise of different bandwidths. You expect to hear more steps in the 2000-Hz tone staircase when the noise bandwidth is reduced below the critical bandwidth.

6. *Temporal integration* Demonstration 8 on the *Auditory Demonstrations* CD. Bursts of broadband noise having durations of 1000, 300, 100, 30, 10, 3, and 1 ms are presented at eight decreasing levels. A graph of a number of steps heard as a function of duration should give an indication of integration time (see Fig. 6.12).

7. *Asymmetry of masking* Demonstration 9 on the *Auditory Demonstrations* CD. This demonstration compares the mask-

ing of a 2000-Hz tone by a 1200-Hz tone with the masking of a 1200-Hz tone by a 2000-Hz tone.

8. *Backward and forward masking*   Demonstration 10 on the *Auditory Demonstrations* CD. This demonstration of masking by nonsimultaneous tones compares forward masking (masking tone before the test tone) with backward masking (masking tone after the test tone). Forward masking is more robust, of course, but the amazing thing is that backward masking occurs at all!

9. *Familiarity with sound levels*   Use an inexpensive sound-level meter to determine the sound level of as many different sounds as possible (music, broadband noise, traffic noise, conversation, etc.).

10. *Asymmetry of masking with two oscillators*   Two audio generators can be used to show that a tone of lower frequency masks a tone of higher frequency more effectively than the other way around.

Laboratory Experiments

Sound level (Experiment 10 in *Acoustics Laboratory Experiments*)

Loudness level and audiometry (Experiment 12 in *Acoustics Laboratory Experiments*)

CHAPTER

7

Pitch and Timbre

Pitch has been defined as that characteristic of a sound that makes it sound high or low or that determines its position on a scale. For a pure tone, the pitch is determined mainly by the frequency, although the pitch of a pure tone may also change with sound level. The pitch of complex sounds also depends on the spectrum (timbre) of the sound and its duration. In fact, the pitch of complex sounds has been one of the most interesting objects of study in psychoacoustics for several years.

In this chapter you should learn:

- About pitch scales and pitch discrimination;
- How pitch depends on frequency, sound level, duration, timbre, and competing sounds;
- About theories of pitch;
- About pitch of complex tones and virtual pitch;
- About absolute pitch;
- About timbre;
- About spectral analysis of complex tones;
- About analytical and synthetic listening.

7.1 ■ PITCH SCALES

The American National Standards Institute (1960) defines *pitch* as "that attribute of auditory sensation in terms of which sounds may be ordered on a scale extending from low to high." This definition probably leads most of us to think of a musical scale. Are there other pitch scales besides musical scales? Is there a subjective scale of pitch similar to the sone scale of loudness discussed in the previous chapter?

Pitch is a subjective sensation. Two persons hearing the same sound may assign it different may assign it different positions on a pitch scale. In fact, some listeners may assign a different pitch to a sound depending upon whether it is presented to the right or left ear (this is called *binaural diplacusis*).

The basic unit in most musical scales is the *octave*. Notes judged an octave apart have frequencies nearly (but not always exactly, as we will see) in the ratio 2:1. As early as the sixth century B.C., according to legend, Pythagoras of Athens noted that if one segment of a string is half as long as the other, the pitches produced by plucking the two segments have a special similarity. Errors of one octave are frequently made in judging the pitch of a musical note. (If you don't believe this, ask a musician to whistle a note and then to name the octave in which the note lies.)

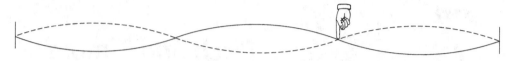

Pythagoras discovers the octave (ca 600 B.C.).

In music, the octave is subdivided in different ways, as we shall see in Chapter 9. Western music normally divides the octave into 12 intervals called *semitones;* these are given note names (A through G with sharps and flats) and designated on musical staves.

Psychophysical Pitch Scales

Various attempts have been made to establish a psychophysical pitch scale. If an average listener were allowed to listen to a tone of 4000 Hz followed by a tone of low frequency and then asked to tune an oscillator to a pitch halfway between, a likely choice would be something around 1000 Hz. On a scale of pitch, then, 1000 Hz is judged as halfway between 0 and 4000 Hz. The unit used for subjective pitch is the *mel;* the scale is arranged so that doubling the number of mels doubles the pitch. From 0 to 2400 mels spans the frequency range 0 to 16 kHz; the correspondence between mels and hertz is shown in Fig. 7.1.

Another psychophysical scale is based on critical bands of hearing. A critical bandwidth is designated one *bark*. Interestingly enough, it turns out that one bark is very nearly equal to 100 mels, so the two scales are actually quite similar.

A numerical scale of pitch (in mels) is not nearly so useful as a numerical scale of loudness (in sones), however. Pitch is usually related to a musical scale where the octave, rather than the critical bandwidth, is the "natural" pitch interval.

FIGURE 7.1
Pitch scale versus frequency scale. On the pitch scale, 100 mels is close to the width of the critical band, which is 160 Hz at a center frequency of 1000 Hz (dashed lines). (From Wightman and Green 1974. Reprinted by permission of the Am. Inst. of Physics.)

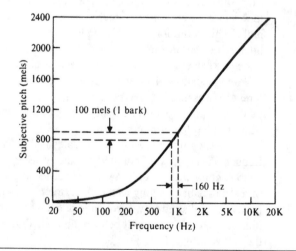

FIGURE 7.2
Just-noticeable
difference (jnd) in
frequency
determined by
modulating the
frequency of a tone
at 4 Hz. Note that
the jnd at each
frequency is nearly
a constant
percentage of the
critical bandwidth.
(From Zwicker,
Flottorp, and
Stevens, 1957.)

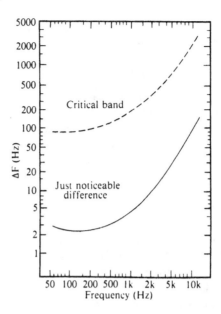

7.2 ■ PITCH DISCRIMINATION

The ability to distinguish between two nearly equal stimuli is often characterized, in psychophysical studies, by a *difference limen* or *just-noticeable difference* (jnd). Two stimuli will be judged the same if they differ by less than the jnd.

The jnd for pitch has been found to depend on the frequency, the sound level, the duration of the tone, and the suddenness of the frequency change. It also depends on the musical training of the listener and to some extent on the method of measurement. Figure 7.2 shows the average (of four subjects) for pure tones at a sound level of 80 dB. From 1000 to 4000 Hz, the jnd is approximately 0.5 percent of the pure tone frequency, which is about one-twelfth of a semitone. Sometimes the term *frequency resolution* is used to denote the jnd divided by the frequency ($\Delta f/f$).

By comparing the upper and lower curves in Fig. 7.2, we can see that critical bandwidth is roughly equal to 30 difference limens or jnd's at all center frequencies. This remarkable result suggests that the same mechanism in the ear is responsible for critical bands and for pitch discrimination. It is quite likely related to regions of excitation along the basilar membrane (see Section 5.4).

It is interesting to compare pitch discrimination to color discrimination. Whereas the visible spectrum extends over one octave (violet light has roughly twice the frequency of red) and includes 128 just noticeable differences (distinguishable hues or colors), the auditory spectrum covers about 10 octaves with 5000 jnds.

7.3 ■ PITCH OF PURE TONES

We have already noted that pitch depends mainly on frequency; pitch scaling with frequency was discussed in Section 7.1. We now consider the pitch dependence of pure tones on other physical quantities such as sound pressure, duration, envelope, and the presence of other sounds.

Pitch and Sound Level

Early experiments on pitch versus sound level reported substantially larger pitch dependence on sound level than more recent studies do. Early work by Stevens (1935) indicated shifts in pitch as large as two semitones (apparent frequency changes of 12%) as the sound level of pure tones increased from 40 to 90 dB. Tones of low frequency were found to fall in pitch with increasing intensity; tones of high frequency rise in pitch with increasing intensity, and tones of middle frequency (1–2 kHz) show little change. (This has sometimes been referred to as *Stevens' rule*.) Stevens found the maximum downward shift with sound level at 150 Hz and the largest upward shift with sound level around 8000 Hz.

It now appears that the effect is small, even for pure tones, and varies from observer to observer; in one experiment, for example, five musically trained subjects hear pitch lowerings that varied from 0 to 75 cents (75 cents = $\frac{3}{4}$ semitone) when a 250-Hz tone increased from 40 to 90 dB (Ward 1970). Whereas pitch changes for individuals tend to follow Stevens' rule, then, averaging over a group of observers makes the changes less significant. Figure 7.3 shows the pitch shifts of pure tones with frequencies from 200 Hz to 6000 Hz averaged over 15 subjects.

The small pitch changes shown in Fig. 7.3, as well as the larger changes described by early investigators, are for pure tones. Less is known about the effect for complex tones. Studies with musical instruments have generally shown very small pitch change with intensity (around 17 cents for an increase from 65 to 95 dB, for example). Whether the pitch of a complex tone rises or falls with increasing intensity appears to depend on which partials (above or below 1000 Hz) are predominant (Terhardt 1979).

In contrast with the results in Fig. 7.3, however, increasing the amplitude of short tone bursts causes a downward shift in pitch over a wide range of frequency. Similar results are found in experiments using 12-ms bursts (Doughty and Garner 1948) and 40-ms bursts (Rossing and Houtsma 1986).

A phenomenon of pitch change that has been observed during reverberant decay may be due in part to pitch change with sound level, although other effects appear to contribute as well. This phenomenon is quite apparent when one is listening to pipe organ music in

FIGURE 7.3
Pitch shift of pure tones as a function of sound pressure level. Shifts are shown in both percent and cents (100 cents = 1 semitone). The curves are based on data from 15 subjects. (After Terhardt 1979.)

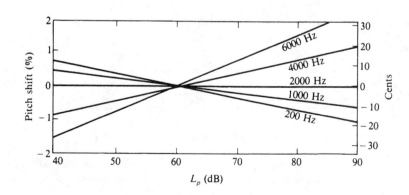

churches with substantial reverberation; the pitch often appears to rise as the sound level diminishes after a loud chord ends (Parkin 1974).

It is fortunate for performing musicians and listeners alike that the change in pitch with sound level for complex tones is much less than was reported from early experiments with pure tones. Musical performance would be very difficult if substantial changes of pitch occurred during changes in dynamic level.

Pitch and Duration

How long must a tone be heard in order to have an identifiable pitch? Although early experiments by Savart (1840) indicated that a sense of pitch develops after only two cycles, most subsequent investigations indicated that a longer duration is required (see Fig. 7.4). Very brief tones are described as *clicks*, but as the tones lengthen, the clicks take on a weak sensation of pitch, increasing in strength upon further lengthening.

The transition from click to tone depends on sound level; if the tone does not begin abruptly, but rather with a soft onset, tone-recognition times as short as 3 ms are possible, which is shorter than the attack time of most musical instruments (Winckel 1967).

It has been suggested that the dependence of pitch on duration follows a sort of "acoustical uncertainty principle" $\Delta f \Delta t = K$, where Δf is the uncertainty in frequency and Δt is the duration of a tone burst. Under optimum conditions, K can be less than 0.1 (Majerník and Kalužný 1979). When the tone duration falls below 25 ms, the pitch may appear to change, although slightly different results are reported by various investigators (Rossing and Houtsma 1986).

The ear has an especially high sensitivity for detecting frequency changes of pure tones. The jnd for frequency change in pure tones Δt is less than for noise, provided that the amplitude of the pure tone remains constant. Even with a band of noise as narrow as 10 Hz at a center frequency of 1500 Hz (which sounds like a pure tone of varying amplitude), Δt will be six times greater than for a pure tone of 1500 Hz (Zwicker 1962).

Pitch and Envelope

The perceived pitch of a short exponentially decaying sinusoidal tone is found to be consistently higher than a simply gated sine tone with the same frequency and energy (Hartmann 1978). Rossing and Houtsma (1986) found the same effect for tones with rising exponen-

FIGURE 7.4 The duration required for a given tone to produce a definite pitch. The solid line is from the data of Bürck, Kotowski, and Lichte (1935); the dashed line is the duration of two cycles (Savart 1840).

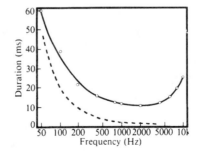

tial envelopes, and found that the pitch shift depends on the sound pressure level as well as the rate of rise or fall of the tone envelope.

The reason for the envelope dependence of pitch is not clear, but it appears to be related to the pitch shift with intensity discussed earlier in this section. It certainly is an effect that musicians should take into account when dealing with the pitch of percussion instruments.

Effect of Interfering Sounds

Sounds are seldom heard in isolation. Another factor that influences the pitch of pure tone is the presence of other interfering sounds. Experiments both with a second interfering tone and with interfering noise can be summarized as follows:

1. If the interfering tone has a frequency below that of the test tone, an *upward* shift always occurs.
2. If the interfering tone frequency is above that of the test tone, a *downward* shift is observed at low frequencies.
3. Interfering noise always causes an upward pitch shift if it has a lower frequency than the test tone (but if it has a higher frequency, the shift can occur in either direction).
4. The pitch shift increases with the amount by which the interfering tone or noise amplitude exceeds that of the test tone (Terhardt and Fastl 1971).

7.4 ■ PITCH OF COMPLEX TONES: VIRTUAL PITCH

When the ear is presented with a tone composed of exact harmonics, it is easy to predict what pitch will be heard. It is simply the lowest common factor in these frequencies, which is the fundamental. The ear identifies the pitch of the fundamental, even if the fundamental is very weak or missing altogether. For example, if the ear hears a tone having partials with frequencies of 600, 800, 1000, and 1200 Hz, the pitch will nearly always be identified as that of a 200-Hz tone, the *missing fundamental*. This is an example of what is called *virtual pitch*, since the pitch does not correspond to any partial in the complex tone. The ability of the ear to determine a virtual pitch makes it possible for the undersized loudspeaker of a portable radio to produce bass tones and also forms the basis for certain mixture stops on a pipe organ.

If a strong fundamental is not essential for perceiving the pitch of a musical tone, the question arises as to which harmonics are most important. Experiments have shown that for a complex tone with a fundamental frequency up to about 200 Hz, the pitch is mainly determined by the fourth and fifth harmonics. As the fundamental frequency increases, the number of the dominant harmonics decreases, reaching the fundamental itself for $f_0 = 2500$ Hz and above (Plomp 1967). Consider, for example, a tone A_3 with a frequency $f_0 = 200$ Hz; if the fourth and fifth harmonics were raised in frequency, the pitch of the tone would most likely appear to rise even though the fundamental remained at 220 Hz.

When the partials of the complex tone are not harmonic, however, the determination of virtual pitch is more subtle. According to current theories of pitch, the ear picks out a series of nearly harmonic partials somewhere near the center of the audible range, and determines the pitch to be the largest near-common factor in the series (Goldstein 1973).

Several demonstrations of virtual pitch are presented by Houtsma, Rossing, and Wagenaars (1987).

Musical examples of the ability of the auditory system to formulate a virtual pitch from near harmonics in a complex tone are the sounds of bells and chimes. In each case the pitch of the *strike note* is determined mainly by three partials that have frequencies almost in the ratio 2 : 3 : 4, as we shall see in Chapter 13. In the case of the bell, there is usually another partial with a frequency near that of the strike note, which reinforces it. In the case of the chime, however, there is none: The pitch is purely subjective.

The following two sections discuss theories of pitch perception and their historical development, including some important experiments that led to our present understanding of pitch of complex tones (which are so important in music and speech). You may read them carefully, skim them, or skip directly to Section 7.7.

7.5 ■ SEEBECK'S SIREN AND OHM'S LAW: A HISTORICAL NOTE

About the middle of the eighteenth century, A. Seebeck performed a series of experiments on pitch perception that produced some significant, if surprising, results. As a source of sound, Seebeck used a siren consisting of a rotating disc with periodically spaced holes that created puffs of compressed air at regular intervals, as shown in Fig. 7.5(a). Seebeck noted that this siren produced sound with a very strong pitch corresponding to the time between puffs of air. Doubling the number of holes, as shown in Fig. 7.5(b), raised the pitch exactly an octave, as expected.

However, using a disk with unequal spacing of holes, as shown in Fig. 7.5(c), produced an unexpected result: the pitch now heard matched that heard with the siren in (a). This may be understood by studying the corresponding waveforms (amplitude versus time) and spectra (amplitude versus frequency) shown in Fig. 7.5. In (a) the spectrum has components at the fundamental frequency $1/T$ (where T is the period) and its harmonics ($2/T$, $3/T$, etc.). In (b) the fundamental frequency is twice as great ($2/T$), and the harmonics occur at

FIGURE 7.5
Three different
sirens used by
Seebeck along with
the waveforms and
spectra of sound
they generate.
(After Wightman
and Green 1974.)

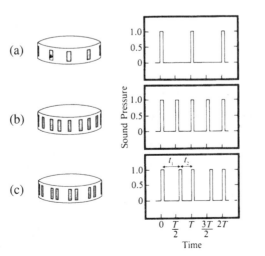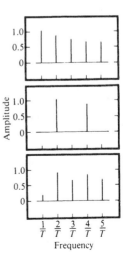

$4/T$, $6/T$, etc.; thus, the pitch is an octave higher. In (c) where the spacing between puffs is alternately t_1 and t_2, the period of repetition is $T = t_1 + t_2$; thus, harmonics occur at the same frequencies as in (a), although the fundamental is very much weaker. The pitch therefore matches that of case (a), although the quality or timbre of the sound is quite different.

It is quite easy and instructive to repeat Seebeck's experiment using an electronic pulse generator to generate the wave forms shown in Fig. 7.5. What one hears in the case of waveform (c) is two tones an octave apart, the lower tone becoming softer as $t_2 \to t_1$, disappearing rather abruptly when $t_2 = t_1$, whereas the upper tone remains relatively constant in loudness.

About the time Seebeck was performing his experiment, G. S. Ohm adapted Fourier's theorem on spectrum analysis (see Section 7.10) to acoustics and formulated what is often known as *Ohm's acoustical law* (or *Ohm's second law*, his first law having dealt with electric circuits). Ohm believed that a pitch corresponding to a certain frequency could be heard only if the acoustic wave contained power at that frequency. Thus, he criticized Seebeck's interpretation of his siren experiment that periodicity, rather than fundamental frequency, determines pitch. In the case of the waveform shown in Fig. 7.5(c), however, the sensation of pitch is far too strong to be explained on the basis of the weak component or partial at the fundamental frequency, and thus Ohm's law is contradicted. Ohm finally suggested that the phenomenon was due to an acoustical illusion (Wightman and Green 1974).

In his monumental work *On the Sensations of Tone as a Physiological Basis for the Theory of Music*, H. von Helmholtz (1877) supported Ohm's position, adding the important idea of distortion products generated in the ear. For pure tones, these distortion products would be primarily harmonics of the pure tone (harmonic distortion). For the waveforms shown in Fig. 7.5(a) and (c), however, distortion would produce sum and difference tones, resulting in the generation of a strong fundamental, since difference tones between all the adjacent partials would be at this frequency.

Experiments with filtered sound by H. Fletcher (1934) and others appeared to support Helmholtz. When the lower harmonics of a complex tone are filtered out, the pitch remains the same. This phenomenon can be demonstrated by recording the sound of a musical instrument and playing it back through a high-pass filter to remove the fundamental (and even the lower harmonics). The missing fundamental is supplied by the ear of the listener.

7.6 ■ THEORIES OF PITCH: PLACE PITCH VERSUS PERIODICITY PITCH

Two major theories of pitch perception have gradually developed on the basis of numerous experiments in many different laboratories. They are usually referred to as the place (or frequency) theory and the periodicity (or time theory). Before discussing these theories, let us briefly review the relationship between frequency and period.

A periodic waveform is one that repeats itself after a certain interval of time, called the period T. The reciprocal of the period is the fundamental frequency f_1. If the waveform is complex, it can be resolved into a spectrum of partials with frequencies $2f_1$, $3f_1$, etc., called the harmonics (see Section 2.7). A periodic waveform need not have energy at its fundamental frequency f_1, as will become apparent later in this chapter. In a pulse waveform, the fundamental *frequency* is not necessarily the same as the pulse *rate*. The waveform in Fig. 7.5(c), for example, has $2/T$ pulses per second, although its fundamental frequency $f_1 = 1/T$ is only half as great. In determining pitch, the ear apparently performs *both a time analysis and a frequency analysis* of the sound wave and reaches its final decision after a considerable amount of computation!

The idea that vibrations of different frequencies excite resonant areas on the basilar membrane is often referred to as the *place theory* of hearing. According to this theory, the cochlea converts a vibration in time into a vibration pattern in space (along the basilar membrane), and this in turn excites a spatial pattern of neural activity. The place theory explains many aspects of auditory perception but fails to explain others.

Helmholtz regarded the basilar membrane as a frequency analyzer, with transfer fibers "tuned" to resonate at frequencies determined by their length, mass, and tension. A complex wave of sound pressure would excite regions of the basilar membrane corresponding to the frequencies of its components or partials, the higher frequencies acting on regions near the oval window, and the lower frequencies acting closer to the far end where the membrane is thick and loose. (Helmholtz was nearly correct; later investigations showed that the individual fibers are not free to resonate, but the membrane as a whole can create the effect of resonances.)

In his experiments with cochleas removed from human cadavers, Békésy provided support for the place theory of pitch perception. By ingenious and careful experiments, he directly observed wavelike motions of the basilar membrane caused by sound stimulation. Just as Helmholtz had suggested, the place of maximum vibration moved up and down the basilar membrane as the frequency of the sound wave changed (see Figs. 5.6 and 5.8).

More recent experiments have pointed to limitations in the place theory of pitch perception, however. One difficulty is in explaining fine frequency discrimination. In order to respond to rapid changes in frequency, a resonator must have considerable damping. But damping decreases selectivity, that is, the ability to discriminate between small differences in frequency. Another difficulty arises in attempting to explain why we hear a complex tone as one entity with a single pitch.

According to the *periodicity theory* of pitch, the ear performs a *time* analysis of the sound wave. Presumably the time distribution of the electrical impulses carried by the auditory nerve has encoded into it information about the time distribution of the sound wave. This information is decoded by a process called *autocorrelation* (to be discussed in Section 8.13) in the central nervous system.

In the late 1930s, J. F. Schouten and his colleagues in the Netherlands performed experiments that supported the periodicity theory of pitch Schouten studied stimuli, such as those shown in Fig. 7.6, in which the pitch corresponds to the repetition rate of the pulses, 200 Hz. In the waveform shown in Fig 7.6(b), the fundamental component has been canceled out by addition of an out-of-phase signal of 200 Hz; the pitch remains unchanged at 200 Hz, the frequency of the missing fundamental. Schouten then added a pure tone of 206 Hz. If a distortion product of 200 Hz were actually present in the ear, as suggested by the hypothesis of Helmholtz, beats should be heard at a frequency of six per second. No beats were heard.

Schouten continued his experiments with a type now called pitch-shift experiments. Using amplitude modulation, he produced complex waveforms in which the frequencies of individual components could be shifted by the same amount, thus leaving the spacing between components undisturbed. For instance, a carrier frequency of 1200 Hz modulated by a 200-Hz signal produces components at 1000 Hz and 1400 Hz (called *sidebands*) along with the 1200-Hz component. Such a waveform, shown in Fig. 7.7, has a clear pitch of 200 Hz. If the carrier frequency is changed to 1240 Hz, however, the components are shifted to 1040, 1240, and 1440 Hz. The pitch is now found to shift to about 207 Hz, even though the difference frequency remains at 200 Hz. This experiment can be repeated in the laboratory using a generator with provision for amplitude modulation or an electronic music synthesizer. (See Demonstration 21, Houtsma, Rossing, and Wagenaars 1987).

FIGURE 7.6

Cancellation of the fundamental frequency of a complex signal. Part (a) shows a periodic pulse train and its spectrum. By appropriate adjustment of phase and amplitude, the fundamental may be canceled as shown in (b). In both cases, however, the pitch of the signal corresponds to the fundamental. (After Schouten 1940.)

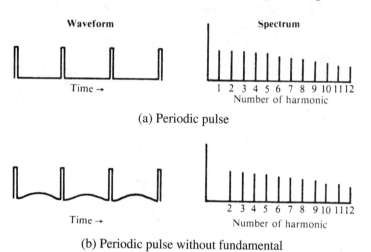

(a) Periodic pulse

(b) Periodic pulse without fundamental

The virtual pitch can be estimated by dividing the component frequencies by successive integers 5, 6, and 7 to obtain a "nearly common factor." In this case $1040/5 = 208$, $1240/6 = 206.7$, and $1440/7 = 205.7$. Averaging these three factors together gives 207 Hz, which the auditory system accepts as the frequency of the missing fundamen-

FIGURE 7.7

Waveforms for pitch-shift experiments of the Schouten type: (a) carrier of 1200 Hz modulated at 200 Hz; (b) carrier at 1240 Hz modulated at 200 Hz.

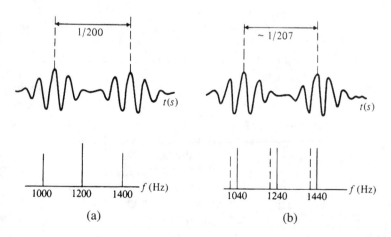

tal. Using 4, 5, and 6 or 6, 7, and 8 leads to less consistent trial factors, so the auditory system prefers the 207-Hz factor.

Schouten explained the pitch-shift phenomena as due to synchronous firing in the auditory nerve due to an unresolved "residue" of high-frequency components. These components, too close in frequency to be resolved on the basilar membrane, retain the periodicity of the original tone envelope. Schouten's residue theory of pitch provided a reasonable alternative to the distortion hypothesis of Helmholtz, but subsequent experiments (e.g., Plomp (1967), Ritsma (1967)) showed that the pitch of complex tones is determined by the low-frequency (resolved) components rather than by the high-frequency (unresolved) residue. An excellent historical review of the subject is given by Plomp (1967).

The importance of some sort of *central* pitch processor in the nervous system was illustrated by experiments in which a single harmonic of a missing fundamental was presented to one ear and a different harmonic to the other ear (Houtsma and Goldstein 1972). The resulting virtual pitch heard this way (*dichotic* presentation) appeared to be as strong as when both harmonics were presented to the same ear (*monotic* presentation). In both monotic and dichotic presentations, the virtual pitch tends to deteriorate with increasing harmonic number.

One might correctly conclude from the foregoing discussion that both the place and periodicity theories of pitch have validity. Clues from both frequency and time analyses of the sound are used to determine pitch, although one or the other may predominate under certain conditions. For low-frequency tones, the time (periodicity) analysis appears to be more important, whereas at high frequencies, the frequency analysis in the basilar membrane (*place clues*) plays a more important role. The relative importance of each type of clue and the frequency range over which the clues predominate are still under study.

Modern Theories of Pitch

Modern theories of pitch, given such names as *optimum processor theory* (Goldstein 1973), *virtual pitch theory* (Terhardt 1974), and *pattern transformation theory* (Wightman 1973), describe how the ear-brain processor determines the pitch of complex tones. Each of them has attractive features. A detailed discussion of them is beyond the scope of this book.

Quite a few experiments have been conducted to evaluate the predictions of these theories (references are given in Scharf and Houtsma (1986) and in Houtsma and Rossing (1987)). Some of these experiments compare the observed pitch shifts in the complex tone with those observed in the partials due to masking noise, amplitude envelope change, intensity change, etc. Others compare complex tones made up of high and low partials, partials of unequal amplitude, etc. The general conclusion appears to be that none of the current pitch theories is completely successful in explaining all the experiments.

Repetition Pitch: A Demonstration of Pitch

In 1693, astronomer Christiaan Huygens, standing at the foot of a staircase at the castle at Chantilly de la Cour in France, noticed that sound from a nearby fountain produced a certain pitch. He correctly concluded that the pitch was caused by periodic reflections of the sound against the steps of the staircase. *Repetition pitch*

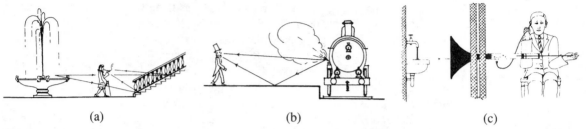

(a)   (b)   (c)

FIGURE 7.8   Examples of repetition pitch (from Bilsen and Ritsma 1969/1970): (a) Huygens (1693) observed periodic reflections of the noise of a fountain against the steps of a staircase; (b) Minnaert (1941) observed the interference of the hissing sound from a locomotive with its reflection from a platform; (c) Hermann (1912) observed interference between the noise of running water and its reflection in a tube of adjustable length.

due to interference between noise and its delayed repetition is discussed by Bilsen and Ritsma (1969/1970), who describe several historical examples, including those shown in Fig. 7.8.

Repetition pitch can be demonstrated in a number of ways, including the two shown in Fig. 7.9. In both cases, broadband noise is combined with identical noise

FIGURE 7.9

(a) Two ways to demonstrate repetition pitch by combining noise with similar noise delayed by time $T = L/v$; (b) Pitch change with time delay using the second arrangement in (a). (After Rossing and Hartmann 1975.)

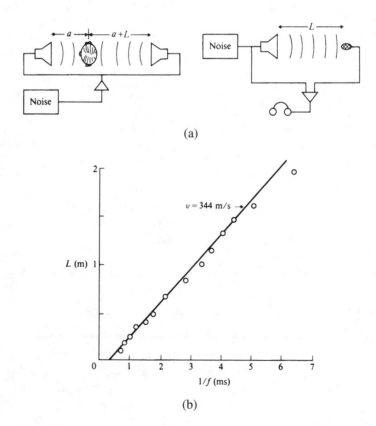

that has traveled a distance L farther and thus is delayed by a time $T = L/v$, where v is the speed of sound. The perceived pitch corresponds to a frequency $f = 1/T = v/L$ and is easy to identify for time delays of 1 to 7 ms. Some blind persons make use of the phenomenon to locate obstructions by observing the interference between the direct and reflected sound.

7.7 ■ ABSOLUTE PITCH

A subject that has held considerable fascination, but also causes no small amount of controversy, is *absolute pitch*. The term refers to the ability to recognize and define (e.g., by naming or singing) the pitch of a tone without the use of a reference tone. This ability is often compared to absolute recognition of color (e.g., green) without any comparison to a standard spectrum. Whereas absolute color recognition is possessed by about 98% of the population (only 2% being partially or totally colorblind), absolute pitch recognition is rare (less than 0.01% of the population appears to have it).

Absolute pitch contrasts with relative pitch, which most persons have to some degree. Nearly all persons can tell whether one tone is higher than another; persons with some musical experience or training can recognize intervals between tones with varying degrees of precision. Someone with a well-trained ear can tell when the frequency of a second tone deviates a little as one percent from the expected interval, although these judgments are not as accurate as they are consistent. For example, the frequency that a person judges, with great consistency, to be an octave above a 1000-Hz tone may actually be 2060 Hz. Relative pitch, in fact, is a remarkable sensory ability that has no counterpart in our other senses. We cannot judge a color that has twice the frequency of a reference color; the only comparable judgment in the visual domain might be selection of a complementary color, and few people develop the ability to do that with great accuracy.

Psychologists have studied absolute pitch for at least 75 years, and during that time there has been considerable discussion and some controversy concerning its origin. In particular, there is less than unanimous agreement as to whether absolute pitch is inherited, acquired, or possibly both. At least four different theories about absolute pitch have developed (Ward 1963):

1. *Heredity theory.* The faculty for developing absolute pitch is inherited, just as the ability for color identification is (unless one inherits colorblindness). The child, so gifted, learns pitch names in early life just as color names are learned.

2. *Learning theory.* The opposite point of view, that absolute pitch can be acquired by almost anyone by diligent and constant practice, is not too widely held.

3. *Unlearning theory.* The ability to develop absolute pitch is nearly universal, but is simply trained out of most people at an early age (by emphasis on relative pitch, for example).

4. *Imprinting theory.* Imprinting is a term used to describe rapid irreversible learning that takes place at a specific developmental stage (used to explain, for example, why duck-

lings will follow for the rest of their life the first moving object they see after hatching). Proponents of this theory feel that nearly all children could be taught absolute pitch at the appropriate age of development.

Bachem (1955) distinguishes between *chroma* and *tone height* as two separate components of pitch. All A's up and down the scale have the same chroma, or quality, but differ in tone height (possessors of absolute pitch frequently make octave errors in identifying tones). Above about 5000 Hz, chroma tends to become fixed, whereas tone height continues to increase, so that absolute pitch identification is not possible.

At least one person with absolute pitch has reported a change in his internal pitch standard with time (Vernon 1977). At age 52 he noted a tendency to identify keys one semitone higher than they should be. He was troubled because he heard the overture to Wagner's "Die Meistersinger" in the "effeminate" key of C$^\sharp$ rather than the "strong and masculine" key of C. By age 71, however, it had moved still further into the sturdier key of D! The shift of internal pitch standard may have been due to a change in elasticity of the basilar membrane with age; in other words a tone of a given frequency was invoking maximum activity at a different place on the basilar membrane than in earlier years.

Speakers of *tone languages*, in which a speech sound can take on several different meanings depending on its tone (see Section 15.8), appear to have a knack for absolute pitch. Vietnamese and Mandarin speakers repeat words on different days with pitches within a semitone, demonstrating a remarkably precise and stable absolute pitch template in producing words (Deutsch, Henthorn, and Dolson 1999).

However it develops, absolute pitch is a remarkable ability. Absolute pitch (inherited or acquired) may continue to be a controversial subject for some time to come, because of the obvious difficulty of experimenting with human subjects in isolation. If one really wants a child to acquire absolute pitch (it has disadvantages as well as advantages!), one should probably begin as early as possible to play find-the-note games on the piano.

7.8 ■ PITCH STANDARDS

The advantages of a universal pitch standard are so obvious that it is quite remarkable that for so many years there was none. Pipe organs were built with A's tuned all the way from 374 to 567 Hz (Helmholtz 1877). In 1619, Praetorius suggested a pitch of 424 Hz; Handel's tuning fork reportedly vibrated at 422.5 Hz. This pitch standard prevailed, more or less, for two centuries, and it is the pitch standard for which Hayden, Mozart, Bach, and Beethoven composed.

Early in the nineteenth century pitch began to rise, probably due to an increased use of brass instruments, which were found to sound more brilliant at the higher pitch. In 1859 a commission appointed by the French government (which included Berlioz, Meyerbeer, and Rossini) selected 435 Hz as a standard. Early in the twentieth century a *scientific pitch*, with all the C's being powers of 2 (128, 256, 512, and so on), appeared; this leads to about 431 Hz for A. Unfortunately, tuning forks made to this standard are still being distributed by scientific and medical supply houses.

In 1939 an International Conference in London unanimously adopted 440 Hz as the standard frequency for A$_4$, and this is almost universally used by musicians. A few orchestras have once again begun a "pitch-raising" game by tuning to 442 or even 444 Hz for greater brightness. This is unfortunate, however, because instruments designed to play well at one pitch may not retain their tone or intonation at another (this is especially true of woodwinds). Singers of today sing the arias of Mozart and Beethoven about a semitone above the pitch for which they were written; most violins of the old masters have already had to be strengthened by adding stouter bass bars and necks to accommodate the increased string tension of today's pitch standard.

Tuning forks have served as convenient pitch standards since the time of Handel. More recently, quartz crystals have provided us with a more precise standard for measuring frequency as well as time. Electronic frequency counters and stroboscopic tuners have made it possible for every physics laboratory as well as every band or orchestra to have precise and dependable frequency standards. The United States Bureau of Standards broadcasts an exceedingly precise 440-Hz tone on its short wave radio station WWV for checking local standards.

The frequency of most musical instruments changes with temperature, and those using wood and gut also change with humidity. The velocity of sound increases about 0.6 m/s for each degree Celsius, so the pitch of a wind instrument rises about 3 cents ($\frac{3}{100}$ of a semitone) per degree of temperature rise (the slight lowering of pitch due to expansion in length is negligible). String instruments generally fall in pitch due to relaxing tension as temperature rises.

7.9 ■ TIMBRE OR TONE QUALITY

The word *timbre*, borrowed from French, is used to denote the tone quality or tone color of a sound. The American National Standards Institute (1960) defines it: "Timbre is that attribute of auditory sensation in terms of which a listener can judge two sounds similarly presented and having the same loudness and pitch as dissimilar." An explanatory note is added: "Timbre depends primarily on the spectrum of the stimulus, but it also depends upon the waveform, the sound pressure, the frequency location of the spectrum, and the temporal characteristics of the stimulus." This definition suggests that judgment of timbre must take place under conditions of equal loudness and pitch (and probably equal duration as well), and so Pratt and Doak (1976) have suggested an alternative definition: "Timbre is that attribute of auditory sensation whereby a listener can judge that two sounds are dissimilar using any criteria other than pitch, loudness or duration."

Timbre may be described as a multidimensional attribute of sound (Plomp 1970); it is impossible to construct a single subjective scale of timbre of the type used for loudness (sones) and pitch (mels), for example. Two recent attempts to construct subjective scales, by asking listeners to rate various verbal attributes of steady sounds, are illustrated in Fig. 7.10. Each investigator found the dull–sharp (brilliant) scale the most significant.

In discussing timbre, and especially in reading about the many experiments on timbre described in the literature, it is important to distinguish between the timbre of *steady* complex tones and those that include *transients* or other variations with time. Plomp (1970) suggested the possibility of using *tone color* to refer to the perceptual differences between steady complex tones; this suggestion has not been widely accepted, however.

FIGURE 7.10
Subjective rating
scales for timbre:
(a) Pratt and Doak
(1976); (b) von
Bismarck (1974).

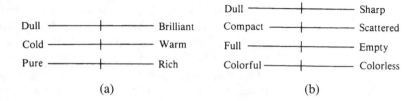

(a)

(b)

A thorough investigation of the timbre of steady tones was carried out by Helmholtz (1877). Helmholtz demonstrated that the sounds of most musical instruments (including the vocal folds or cords) consist of series of harmonics that determine the timbre. Further-more, he carefully described a way in which the ear could comprehend timbre. On the basis of his experiments, he formulated the following general rules:

1. Simple tones, such as those of tuning forks and widely stopped organ pipes, have very soft, pleasant sound, free from roughness but dull at low frequencies.
2. Musical tones with a moderately loud series of harmonics up to the sixth (such as those produced by the piano, the French horn, and the human voice) sound richer and more musical than simple tones, yet remain sweet and soft if the higher harmonics are absent.
3. Tones consisting of only odd harmonics (narrow stopped organ pipes, clarinet) sound hollow and, if many harmonics are present, nasal. When the fundamental predominates, the quality of tone is rich; when the fundamental is not sufficiently strong, the quality of tone is poor.
4. Complex tones with strong harmonics above the sixth or seventh are very distinct, but the quality of tone is rough and cutting.

Helmholtz continued with careful experiments to determine the dependence of timbre on the relative phases of the harmonics. Using electrically driven tuning forks and tuned resonators (of the type we now call Helmholtz resonators), he concluded that timbre does not depend on phase differences between the harmonics. Unfortunately, Helmholtz could detect only very slow changes in phase in his experiments (a limitation that he appar-ently recognized), and thus some interesting dynamic phase effects were overlooked. So thorough were the studies of Helmholtz, that until 1950 very little new information of significance appeared in the literature.

Before continuing the discussion of timbre, we will investigate the Fourier analysis of a tone.

7.10 ■ FOURIER ANALYSIS OF COMPLEX TONES

The determination of the harmonic components of a periodic waveform is called *Fourier analysis*, after the mathematician Joseph Fourier (1768–1830), who formulated an impor-tant mathematical theorem: *Any periodic vibration, however complicated, can be built up from a series of simple vibrations, whose frequencies are harmonics of a fundamental fre-quency, by choosing the proper amplitudes and phases of these harmonics.* Constructing a

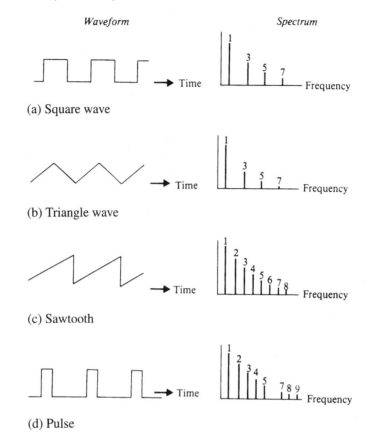

Waveform   *Spectrum*

(a) Square wave

(b) Triangle wave

(c) Sawtooth

(d) Pulse

FIGURE 7.11
Spectra of complex
waveforms. The
square wave and
triangle wave are
missing all the
even-numbered
harmonics.

complex tone from its harmonics (the opposite of Fourier analysis) is called *Fourier synthesis*. The terms *spectrum analysis*, *harmonic analysis*, and *sound analysis* are sometimes used to describe Fourier analysis applied to sound. A specification of the strengths of the various harmonics (usually in the form of a graph) is called a *spectrum*.

Spectra of four different complex waveforms are shown in Fig. 7.11. Although they sound rather harsh and unmusical (with the exception of the flutelike triangle wave), these waveforms are frequently used to create sound in electronic music synthesizers (see Chapter 27). The square wave, for example, is composed of only odd-numbered harmonics with amplitudes in the ratio $1/n$. Thus, if the fundamental has frequency f and amplitude A, the other harmonics in the spectrum will have frequencies of $3f, 5f, 7f, \ldots$, and amplitude $A/3, A/5, A/7, \ldots$. The triangle wave has odd harmonics with amplitudes in the ratio $1/n^2$ (that is, $A, A/9, A/25, \ldots$). The sawtooth wave, on the other hand, has both odd-numbered and even-numbered harmonics with amplitudes in the ratio $1/n$ ($A, A/2, A/3, \ldots$).

Figure 7.12 illustrates how Fourier analysis works. The first six harmonics of a sawtooth wave are shown individually and collectively. Note that when combined *in the proper*

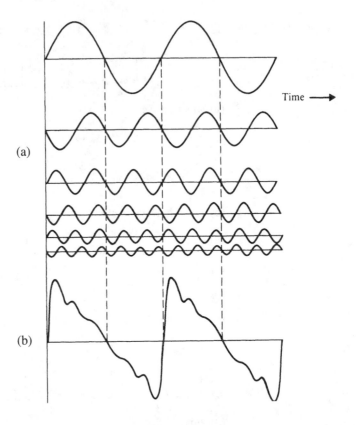

Time ⟶

(a)

(b)

FIGURE 7.12
Fourier synthesis of
a sawtooth wave:
(a) first six
harmonics; (b) sum
of the first six
harmonics.

phase, the first six harmonics approximate a sawtooth wave, although "wiggles" occur that diminish with the addition of higher harmonics.

Textbooks present many "typical" spectra of musical instruments. It should be emphasized, however, that sound spectra from a given instrument vary widely according to the way in which the instrument is played (soft, loud, high, low, or midrange) and how the sound is recorded (near field, far field, reverberant field, direction of microphone from the instrument, etc.).

One way to determine the spectrum of harmonics is by direct computation from the recorded waveform. One of the earliest instruments developed for recording waveforms was the *phonodeik* designed by D. C. Miller (1916), which used a vibrating mirror to direct a beam of light onto a moving film. Most of the sound spectra in early publications were calculated from phonodeik recordings.

Modern spectrum analyzers are of two types: digital and analogue. Digital-spectrum analyzers begin by sampling one period of the wave at regular intervals and feeding these samples into a digital computer. The computer then calculates the amplitude and phase of each harmonic.

Analogue spectrum analyzers use filters or other electronic circuits to isolate the harmonics one after another. If this is done very rapidly (in a few milliseconds), the analyzer

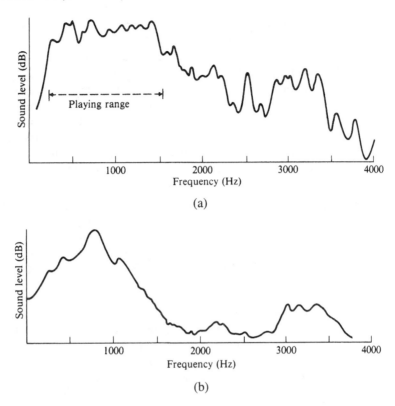

FIGURE 7.13
Time-averages of spectra; (a) clarinet; (b) tenor singing "ah."

is called a *real-time* spectrum analyzer, which is very useful for studying changing sounds or spectra during attack and decay of sounds.

Some interesting information about timbre can be obtained by averaging many spectra (from the same instrument, for example). Figure 7.13 shows averages of 512 spectra of a clarinet and a male voice. In each case, the pitch is varied by playing (singing) up and down the scale during the recording. The significance of the various maxima will be become clear after reading about woodwind instruments (Chapter 12) and voice formants (Chapter 15).

FIGURE 7.14
Long-time average spectra of a violin with and without a mute. (From Jansson and Sundberg 1975.)

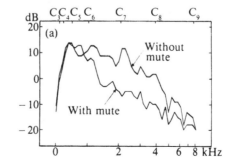

Long-time-average spectra have been used extensively at the Royal Institute of Technology in Stockholm to study musical instruments and the singing voice. A long-time-average spectrum contains information on the written music, the performance, the musical instrument, and the room in which it is played. The effect of varying any one of these factors can be studied by holding the others constant. Figure 7.14 shows the long-time-average spectra of a violin played with and without a mute, for example.

It should be mentioned that whereas the effects of phase on timbre are small for steady tones, the ear is in fact quite sensitive to *changes* in phase, especially if they take place at a regular rate. This is illustrated by the phenomenon described as *second-order beats*, to be discussed in Chapter 8.

7.11 ■ TIMBRE AND DYNAMIC EFFECTS: ENVELOPE AND DURATION

In Sections 7.9 and 7.10, the discussion focused on the timbre of steady complex tones. Transients and other dynamic effects, however, play an important role in determining the timbre of musical and speech sounds, as you can prove to yourself by two simple experiments.

Record the sounds of a number of different musical instruments. In the first experiment, play the tape backward (so that the attack transient occurs at the end). You will hear some curious effects. For example, a piano played backward sounds like a reed organ or a harmonium. (This is illustrated in Demonstration 29, Houtsma, Rossing, and Wagenaars 1987). For a second experiment, cut and splice the tape so that the attack transient is removed. Without attack transients, a remarkable similarity is noted between dissimilar pairs of instruments, such as a French horn and a saxophone and even a trumpet and an oboe.

Berger (1963) performed an experiment in which the sounds of various instruments were presented with the first and last half seconds removed; using 30 band students as a jury of listeners, he obtained the *confusion matrix* shown in Table 7.1. Note that with the transients removed, the sound of an alto saxophone was correctly identified by only four

TABLE 7.1 Listener judgments of recorded wind-instrument tones presented with first and last half seconds removed (Berger 1963)

						Response					
Stimulus	Flute	Oboe	Clarinet	Tenor saxophone	Alto saxophone	Trumpet	Cornet	French horn	Baritone	Trombone	No answer
Flute	1	2		1	6	5	4			4	7
Oboe		28									2
Clarinet	1	1	20	4	3						1
Tenor saxophone			25	2	1						2
Alto saxophone				3	4		1	11	5	5	1
Trumpet	8				6	2	3	4	1	3	3
Cornet		1				12	15				2
French horn	1			2	3			5	6	6	7
Baritone			1	1	2	3	2	4	7	3	7
Trombone	2	1		5	3			1	5	9	4

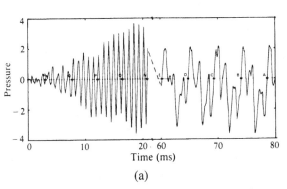

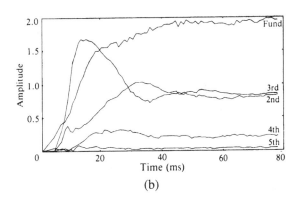

(a)   (b)

FIGURE 7.15   (a) The waveform of an attack transient. (b) Amplitudes of the first five harmonics of the attack transient of a 110-Hz diapason organ pipe. (From Keeler 1972.)

jurists, whereas eleven jurists thought it was a French horn. Also surprising is the confusion of tenor saxophone with clarinet, because the "woody" tone of a clarinet emphasizes odd-numbered harmonics.

During attack, the various partials of a musical sound may develop at different rates. Figure 7.15 shows the attack waveform of an organ pipe tones, along with the onset of the first five harmonics. During the attack transient, the waveform is not exactly periodic; note the difference from cycle to cycle. Note that the second harmonic of the organ pipe develops slightly faster than the others; in other wind instruments, the fundamental is often found to lead.

Strong and Clark (1967) performed some interesting experiments in which they interchanged spectra and time envelopes of wind instrument tones. They synthesized many tones, each time using the envelope characteristic of one instrument with the spectrum of another, and asked listeners to identify the instrument. They found that in the cases of the oboe, clarinet, bassoon, tuba, and trumpet, the spectrum is much more important than the envelope; in the case of the flute, the envelope is more important than the spectrum; in the cases of the trombone and French horn, spectrum and envelope appear to be of comparable importance. The general principle seems to be that the spectrum takes on greatest importance when it has a maximum in a unique location within its playing range.

7.12 ■ VIBRATO

Vibrato is widely used to enhance musical performance, both instrumental and vocal. In order to avoid misunderstanding, it is important to carefully define vibrato.

The definition recommended by the American National Standards Institute (1960) is "The vibrato is a family of tonal effects in music that depend on periodic variations of one or more characteristics in the sound wave." The important note is added: "When the particular characteristics are known, the term 'vibrato' should be modified accordingly: e.g., frequency vibrato, amplitude vibrato, phase vibrato and so forth." In keeping with this recommendation, we use the term *frequency vibrato* to refer to frequency modulation (FM) and *amplitude vibrato* to refer to amplitude modulation (AM). In practice it is virtually impossible to have frequency vibrato without amplitude vibrato because of the effect of room resonances and resonances in the source instrument. Amplitude vibrato without frequency vibrato is possible (in the case of a vibraphone with resonators that open and close periodically, for example), but would be the exception rather than the rule.

Unfortunately, some texts use the term *vibrato* to refer to frequency vibrato but the term *tremolo* to refer to amplitude vibrato. This is unfortunate, not only because frequency vibrato and amplitude vibrato nearly always coexist in musical performance, but also because the term *tremolo* is generally used in music to refer to something else: rapid back-and-forth strokes of a violin bow or rapid alternation between two notes.

Vibrato was studied extensively some 40 to 50 years ago by C. E. Seashore and colleagues at the University of Iowa, and many of their findings are confirmed by more recent experiments (Ward 1970). Vibrato appears to vary with individual performers, an "average" rate for both singers and instrumentalists being around 7 Hz. Singers seem to use a slightly greater depth of frequency vibrato than instrumentalists do, however.

You can perform interesting experiments on vibrato using an audio generator with provision for frequency modulation (many generators can be frequency modulated by a second oscillator), or with an electronic music synthesizer. Try varying both the *rate* and the *depth* of frequency modulation. You will probably find that with modulation rate in the rage of 1 to 5 Hz, you can recognize the periodicity of pitch change (most clearly around 4 Hz). Beginning at about 6 Hz, however, the tone takes on a single average pitch with intensity fluctuations at the frequency of the vibrato. At a still higher rate (around 12 Hz), the sound becomes a rather unpleasant confusion of more than one tone. It is not difficult to see why performers choose a vibrato rate around 7 Hz.

The parameters of a natural vibrato fluctuate slightly during the duration of a tone. Tones from electronic instruments, which have a fixed rate and depth of vibrato, sound artificially rigid. Analyses of the vibrato used by opera singers Maria Callas and Dietrich Fischer-Dieskau show that both singers use deep vibratos (Winckel 1975). The rates of vibrato and trill used by Callas were the same, and in fact her transition from vibrato to trill was made with no change of phase. When trained singers sing duets, they reportedly adjust their vibratos to have identical rate and phase (but not necessarily depth); the adjustment is most likely subconscious (Winckel 1975).

Vibrato is said to cover up small errors in frequency. Fletcher, Blackham, and Geertsen (1965) found that the vibrato of many violinists apparently centers 15 to 20 cents above the target pitch. Vibrato makes identification of vowel sounds more difficult and tends to conceal formant frequencies of singers that may deviate substantially from the corresponding formant frequencies of normal speech (Sundberg 1975).

7.13 ■ BLEND OF COMPLEX TONES

Our auditory system has the ability to listen to complex sounds in different modes. When we listen *analytically*, we hear the different partials separately; when we listen *synthetically* or holistically, we focus on the whole sound and pay little attention to the partial sounds. Listeners differ in the degree to which they listen analytically or synthetically. If a two-tone complex of 800 and 1000 Hz is followed by one of 750 and 1000 Hz, for example, an analytic listener will hear one partial go down in pitch; a synthetic listener will hear a virtual pitch rising a major third from 200 to 250 Hz (Demonstration 25, Houtsma, Rossing, and Wagenaars 1987).

A tone with several harmonic partials, whose frequencies and relative amplitudes remain steady, is generally heard as a single tone, even if the total intensity changes. However, when one of the harmonics is turned off and on, it stands out clearly (Demonstration 1, Houtsma, Rossing, and Wagenaars 1987). The same is true if one of the harmonics is given a vibrato (i.e., its frequency, its amplitude, or its phase is modulated at a slow rate).

One of the most remarkable feats of our auditory system is its ability to single out complex tones from a complex background, such as the sounds of different instruments in a symphony orchestra or conversation at a cocktail party, for example. In the former case, the ear interprets certain partial tones as belonging to one particular instrument, other partials as belonging to another instrument. In other words, it looks for familiar or likely sets of partial tones and fuses these together into a single complex tone at the same time it hears a blend of many instrument sounds. As we have seen, the mechanism for this analysis is partially understood, but much research remains to be done.

Erickson (1975) addresses this subject from the standpoint of a composer. In an enlightening chapter, "Some Territory Between Timbre and Pitch," he discusses three ways in which a complex sound can be heard: (1) as a chord; (2) as a pitch (with timbre); (3) as a sound (an unpitched sound without definite pitch or pitches such as the sound of a bass drum). These three concepts can be represented as the apexes of a triangle (see Fig. 7.16) with the grey areas between them represented by the sides of the triangle.

Transformation from a chord to the fused condition described as a sound, for example, is illustrated by the music of Edgard Varese. A pitch (with timbre)-to-chord transformation occurs in the unusual chanting of Tibetan lamas recorded and described by Smith, Stevens, and Tomlinson (1967). The chanting is done in such a way that certain harmonics of the voice become audible as separate pitches, giving the effect of one person singing a continuous chord.

It is well known that the partials in a piano tone are stretched further apart than partials in a true harmonic series (see Chapter 14). Stretching the partials even further apart causes the sounds to become bell-like or chime-like. More surprising, perhaps, is the observation that compression of the partials also produces bell-like timbres (Slaymaker 1970). Individual partials, in both cases, can be singled out more easily than the harmonic partials of the usual musical tone; the transformation can be described as going from pitch (with timbre) to an inharmonic chord as the partials are stretched or compressed beyond certain limits.

Inharmonicity in the partials of a complex tone appears to be detected in a different way for low and high harmonics. For low harmonics, the inharmonic partial appears to "stand

FIGURE 7.16
Three ways in which a complex sound can be heard. (From Erickson 1975.)

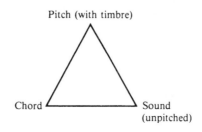

out" when it is mistuned by an amount that varies from 1 to 3% in different subjects. For high harmonics, on the other hand, the mistuning is detected as a kind of beat, or roughness, presumably reflecting a sensitivity to changing phase of the mistuned harmonic relative to the other harmonics (Moore, Peters, and Glasberg 1985).

7.14 ■ SUMMARY

Pitch has been defined as the characteristic of a sound that gives it the sensation of high or low. It is determined mainly by the frequency of a tone, but sound level, spectrum, and duration also influence pitch. Early models for pitch perception regarded the basilar membrane as a frequency analyzer of high resolution (place theory), but more recent studies have shown that much of the determination of pitch is contributed by a temporal analysis in the central nervous system (periodicity pitch). The ear is able to assign a pitch to complex sounds composed of inharmonic partials and even to some presentations of wideband noise. Some persons have the ability to identify pitch independent of a reference pitch (absolute pitch).

Timbre or tone quality depends on the frequency of a tone, its time envelope, its duration, and the sound level at which it is heard. Any complex waveform that is periodic can be constructed from simple tones with the right frequency and phase; determination of the spectrum of simple tones is called *spectrum analysis* or *Fourier analysis*. Under most conditions, the timbre of a complex sound is insensitive to the phase of its components. Periodic variation of the frequency and amplitude, called *vibrato*, lends warmth and blend to musical tones. A vibrato rate of about 7 Hz is common in musical performance.

REFERENCES AND SUGGESTED READINGS

American National Standards Institute (1960). "USA Standard Acoustical Terminology." S1.1–1960.

Appel, K., J. Gastineau, C. Bakken, and D. Vernier (1998). *Physics with Computers*. Portland, Ore.: Vernier Software.

Bachem, A. (1955). "Absolute Pitch," *J. Acoust. Soc. Am.* **27**: 1180.

Bassett, I. G. (1982). "Vibration and Sound of the Bass Drum," *Percussive Notes* **19**(3): 50.

Berger, K. W. (1963). "Some Factors in the Recognition of Timbre," *J. Acoust. Soc. Am.* **36**: 1888.

Bilsen, F. A., and R. J. Ritsma (1969/1970). "Repetition Pitch and Its Implication for Hearing Theory," *Acustica* **22**: 63.

Bismarck, G. von (1974). "Timbre of Steady Sounds: A Factorial Investigation of Its Verbal Attributes," *Acustica* **30**: 146.

Bürck, W., P. Kotowski, and H. Lichte (1935). "Die Horbarkeit von Laufzeitdifferenzen," *Elektrotechn. Nachr.-Techn.* **12**: 355.

Deutsch, D., T. Henthorn, and M. Dolson (1999). "Absolute Pitch Demonstrated in Speakers of Tone Languages," *J. Acoust. Soc. Am.* **106**: 2267 (abstract).

Doughty, J. M., and W. M. Garner (1948). "Pitch Characteristics of Short Tones II: Pitch as a Function of Duration," *J. Exp. Psychol.* **38**: 478.

Erickson, R. (1975). *Sound Structure in Music*. Berkeley: University of California. (See especially Chapter 2.)

Evans, E. F., and J. P. Wilson, eds. (1977). *Psychophysics and Psychology of Hearing*. London: Academic Press.

Fletcher, H. (1934). "Loudness, Pitch and Timbre of Musical Tones and Their Relation to the Intensity, the Frequency, and the Overtone Structure," *J. Acoust. Soc. Am.* **6**: 153. See also *Phys. Rev.* **23**: 427 (1924).

Fletcher, H., E. D. Blackham, and O. N. Geertsen (1965). "Quality of Violin, Viola, Cello and Bass-Viol Tones," *J. Acoust. Soc. Am.* **37**: 851.

Goldstein, J. L. (1973). "An Optimum Processor Theory for the Central Information of Complex Tones," *J. Acoust. Soc. Am.* **54**: 1496.

Grey, J. M. (1977). "Multidimensional Perceptual Scaling of Musical Timbres," *J. Acoust. Soc. Am.* **61**: 1270.

Hartmann, W. M. (1978). "The Effect of Amplitude Envelope on the Pitch of Sinewave Tones," *J. Acoust. Soc. Am.* **63**: 1105.

Helmholtz, H. L. F. von (1877). *On the Sensations of Tone as a Physiological Basis for the Theory of Music*, 4th ed. Trans. A. J. Ellis. New York: Dover, 1954.

Houtsma, A. J. M., and J. L. Goldstein (1972). "The Central Origin of the Pitch of Complex Tones: Evidence from Musical Interval Recognition," *J. Acoust. Soc. Am.* **51**: 520.

Houtsma, A. J. M., and T. D. Rossing (1987). "Effects of Signal Envelope on the Pitch of Short Complex Tones," *J. Acoust. Soc. Am.* **81**: 439.

Houtsma, A. J. M., T. D. Rossing, and W. M. Wagenaar (1987). *Auditory Demonstrations* (Philips Compact Disc #1126-061 and text). Melville, N.Y.: Acoustical Society of America.

Jansson, E. V., and J. Sundberg (1975). "Long-Time-Average Spectra Applied to Analysis of Music. Part I: Method and General Applications," *Acustica* **34**: 15.

Keeler, J. S. (1972). "Piecewise-Periodic Analysis of Almost-Periodic Sounds and Musical Transients," *IEEE Trans. Audio and Electroacoustics* **AU-20**: 338.

Majerník, V., and J. Kalužný (1979). "On the Auditory Uncertainty Relations," *Acustica* **43**: 132.

Miller, D. C. (1916). *Science of Musical Sounds*. New York: Macmillan.

Moore, D. C. J., R. W. Peters, and B. R. Glasberg (1985). "Threshold for the Detection of Inharmonicity in Complex Tones," *J. Acoust. Soc. Am.* **77**: 1861.

Parkin, P. H. (1974). "Pitch Change During Reverberant Decay," *J. Sound and Vibration* **32**: 530.

Plomp, R. (1967). "Pitch of Complex Tones," *J. Acoust. Soc. Am.* **41**: 1526.

Plomp, R. (1970). "Timbre as a Multidimensional Attribute of Complex Tones," in *Frequency Analysis and Periodicity Detection in Hearing*. Eds., R. Plomp and G. Smoorenburg. Leiden: Sijthoff.

Pratt, R. L., and P. E. Doak (1976). "A Subjective Rating Scale for Timbre," *J. Sound and Vibration* **45**: 317.

Ritsma, R. (1967). "Frequencies Dominant in the Perception of the Pitch of Complex Sounds," *J. Acoust. Soc. Am.* **42**: 191.

Rossing, T. D., and W. M. Hartmann (1975). "Perception of Pitch from Time-Delayed White Noise," AAPT meeting, Anaheim, California (January 1975).

Rossing, T. D., and A. J. M. Houtsma (1986). "Effects of Signal Envelope on the Pitch of Short Sinusoidal Tones," *J. Acoust. Soc. Am.* **79**: 1926.

Savart, F. (1840). "Uber die Ursachen der Tonhohe," *Ann. Phys. Chem.* **51**: 555.

Scharf, B., and A. J. M. Houtsma (1986). "Audition II: Loudness, Pitch, Localization, Aural Distortion, Pathology," in *Handbook of Perception and Human Performance*, Vol. 1. Eds., K. R. Boff, L. Kaufman, and J. P. Thomas. New York: J. Wiley.

Schouten, J. F. (1940). "The Perception of Pitch," *Philips Tech. Rev.* **5**: 286. (This article summarizes much of the contents of several articles in *Proc. Ned. Acad. Wet.* **41** and **43**.)

Slaymaker, F. H. (1970). "Chords from Tones Having Stretched Partials," *J. Acoust. Soc. Am.* **47**: 1569–1571.

Smith, H., K. N. Stevens, and R. S. Tomlinson (1967). "On an Unusual Mode of Chanting by Tibetan Lamas," *J. Acoust. Soc. Am.* **41**: 1262.

Stevens, S. S. (1935). "The Relation of Pitch to Intensity," *J. Acoust. Soc. Am.* **6**: 150.

Strong, W., and M. Clark, Jr. (1967). "Perturbations of Synthetic Orchestral Wind-Instrument Tones," *J. Acoust. Soc. Am.* **41**: 277.

Sundberg, J. (1975). "Vibrato and Vowel Identification," in Report STL-QPSR 2-3/1975, Speech Technology Laboratory, Royal Institute of Technology, Stockholm.

Terhardt, E. (1974). "Pitch, Consonance, and Harmony," *J. Acoust. Soc. Am.* **55**: 1061.

Terhardt, E. (1979). "Calculating Virtual Pitch," *Hearing Research* **1**: 155.

Terhardt, E., and H. Fastl (1971). "Zum Einfluss von Störtönen und Störgeräuschen auf die Tonhöhe von Sinustönen," *Acustica* **25**: 53.

Vernon, P. E. (1977). "Absolute Pitch: A Case Study," *Br. J. Psychol.* **68**: 485.

Ward, W. D. (1963). "Absolute Pitch," *Sound* **2**(3): 14; **2**(4): 33.

Ward, W. D. (1970). "Musical Perception," in *Foundations of Modern Auditory Theory*, Ed. J. V. Tobias. New York: Academic.

Wightman, F. L. (1973). "The Pattern-Transformation Model of Pitch," *J. Acoust. Soc. Am.* **54**: 407.

Wightman, F. L., and D. M. Green (1974). "The Perception of Pitch," *Am. Scientist* **62**: 208.

Winckel, F. (1967). *Music, Sound and Sensation.* Trans., T. Binkley. New York: Dover.

Winckel, F. (1975). "Measurement of the Acoustic Effectiveness and Quality of Trained Singers' Voices," 90th meeting of Acoustical Society of America, San Francisco.

Zwicker, E. (1962). "Direct Comparisons between the Sensations Produced by Frequency Modulation and Amplitude Modulation," *J. Acoust. Soc. Am.* **34**: 1425.

Zwicker, E., G. Flottorp, and S. S. Stevens (1957). "Critical Bandwidth in Loudness Summation," *J. Acoust. Soc. Am.* **29**: 548.

GLOSSARY

absolute pitch The ability to identify the pitch of any tone without the aid of a reference.

analytic listening Listening to a complex tone in a way that individual components or partial tones are heard as separate entities.

bark An interval of frequency equal to a critical bandwidth.

critical band The frequency bandwidth at which subjective response (to loudness, pitch, etc.) changes rather abruptly (see Chapter 6).

distortion An undesired change in waveform. Two common examples are harmonic distortion and intermodulation distortion. *Harmonic distortion* means that harmonics are generated by altering the waveform in some way ("clipping" the peaks, for example). *Intermodulation distortion* refers to the generation of sum and difference tones.

envelope The amplitude of a tone as a function of time.

Fourier analysis, or **spectrum analysis** The determination of the component tones that make up a complex tone or waveform.

Fourier synthesis The creation of a complex tone or waveform by combining its spectral components.

fundamental The lowest common factor in a series of harmonic partials. The fundamental frequency of a periodic waveform is the reciprocal of its period.

harmonic A partial whose frequency is a multiple of some fundamental frequency.

inharmonic partial A partial that is not a harmonic of the fundamental.

just noticeable difference (jnd) or **difference limen** The minimum change in stimulus that can be detected.

mel The unit of subjective pitch; doubling the number of mels doubles the subjective pitch for most listeners. The critical band is about 100 mels wide.

octave The basic unit in most musical scales. Notes judged an octave apart have frequencies nearly in the ratio 2:1.

partial tone (or **partial**) One of the components in a complex tone (it may or may not be a harmonic of the fundamental).

period The smallest increment of time over which a waveform repeats itself.

periodic quantity One that repeats itself at regular time intervals.

periodicity pitch Pitch determination on the basis of the period of the waveform of a tone.

phase The fractional part of a period through which a waveform has passed, measured from a reference.

pitch An attribute of auditory sensation by which sounds may be ordered from low to high.

place theory of pitch A view of the basilar membrane as a frequency analyzer of high resolution; pitch is determined by sensing the place on the basilar membrane that has maximum excitation.

repetition pitch Pitch sensation created by the interference of a sound with a time-delayed repetition.

residue theory of pitch A view that components of a tone that cannot be resolved by the basilar membrane (the residue) are analyzed in time by the central nervous system.

semitone One step on a chromatic scale. Normally $\frac{1}{12}$ of an octave.

spectral dominance A view that certain partials dominate in the determination of the pitch of a complex tone.

spectrum The "recipe" for a complex tone that gives the amplitude and frequency of the various partials.

strike note Note heard when a bell or chime is struck.

subjective pitch Pitch determined to have a frequency that does not correspond to that of any partial.

synthetic (holistic) listening Listening to a complex tone in a way that focuses on the whole sound rather than the individual partials.

timbre An attribute of auditory sensation by which two sounds with the same loudness and pitch can be judged dissimilar.

transient A sound that does not reoccur, at least on a regular basis.

tristimulus diagram A way of representing timbre graphically in terms of the relative loudness of three different parts of the spectrum.

vibrato Tonal effect in music resulting from periodic variation of amplitude, frequency, and/or phase.

virtual pitch Subjective pitch created by two or more partials in a complex tone (two examples are the "missing fundamental" of a filtered tone and the strike note of a bell).

REVIEW QUESTIONS

1. What is the basic unit in most musical scales and what frequency ratio does it represent?

2. How does the jnd for pitch generally compare to the critical bandwidth at the same frequency?

3. On what physical parameter(s) does pitch depend?

4. How does the pitch of a 200-Hz tone depend on sound level?

5. How does the pitch of a tone change if noise of a lower frequency is added?

6. What pitch will generally be heard when tones of 800, 1000, and 1200 Hz are sounded together?

7. What is meant by *absolute pitch*?

8. What is the frequency of A_4, according to the International pitch standard?

9. Does the timbre of a complex tone depend on the relative phases of its harmonics?

10. What frequencies would appear in the spectrum of a 100-Hz square wave?

11. Does playing a tone backward change its spectrum?

12. Does playing a tone backward change its timbre?

13. Vibrato is generally defined as a periodic change in
    (a) frequency
    (b) amplitude
    (c) phase
    (d) timbre
    (e) all of these

14. A preferred vibrato rate is
    (a) 3 Hz
    (b) 7 Hz
    (c) 15 Hz
    (d) 100 Hz
    (e) depends upon the frequency of the tone

15. A single harmonic in a complex tone can be made to stand out by
    (a) turning it on and off
    (b) modulating its frequency
    (c) modulating its amplitude
    (d) mistuning it
    (e) any of these

QUESTIONS FOR THOUGHT AND DISCUSSION

1. A "tonic" chord in the key of A consists of tones with frequencies of 440, 550, and 660 Hz. When such a chord is played on the piano or by three instruments, why is this not heard as a single tone with a pitch of 110 Hz (the "missing fundamental")?

2. Have you ever experienced the pitch change during reverberation described by Parkin? Would this effect be apparent on a recording with reverberation?

3. Discuss the advantages and disadvantages to a performing musician of possessing absolute pitch.

4. Try to account for the most prevalent "confusions" in Berger's experiment (Table 7.1) in the identification of instrument tones without the transients.

5. Why is it virtually impossible to have frequency vibrato in a musical instrument without amplitude vibrato?

EXERCISES

1. At what point would you divide a 65-cm guitar string (as Pythagoras did) so that the two segments sound pitches one octave apart?

2. From Fig. 7.2, find the jnd at frequencies 200, 1000, and 5000 Hz.

3. By referring to Fig. 7.2, show that the critical band comprises roughly 30 jnds. (Compare them at 200, 1000, 5000, and 10,000 Hz, for example.)

4. According to Fig. 7.3, how many cents does the pitch of a 200-Hz tone fall when the sound pressure level is changed from 50 to 90 dB?

5. In Fig. 7.5(c) let $t_1 = 7$ ms and $t_2 = 3$ ms. Determine $1/T$, $2/T$, and $3/T$. What is the frequency of the pitch that would be heard? What is the pulse rate?

6. From Fig. 7.15(b), determine the approximate rise times of the first and second harmonics of a diapason organ pipe.

7. From Fig. 7.1, determine the number of mels in an octave from

   (a) C_3 (131 Hz) to C_4 (262 Hz);

   (b) C_4 to C_5 (523 Hz);

   (c) C_5 to C_6 (1046 Hz).

8. If a pure tone with a frequency of 800 Hz is modulated at 150 Hz, what sidebands are produced? According to the theory discussed in Section 7.6, what virtual pitch will probably be heard? (Try dividing by various sets of integers such as 4, 5, 6 and 5, 6, 7, etc.)

9. If the steps in Fig. 7.8(a) are 30 cm deep, what pitch would most likely be heard?

10. Compare the tension in a violin string tuned to a standard A (440 Hz) with the tension in the same string tuned to match Handel's tuning fork (422 Hz). (See Section 3.2.)

11. What are the frequencies of the first four partials in a 300-Hz square wave?

12. What is the frequency of the maximum sound level in the spectrum of Fig. 7.13(b)?

EXPERIMENTS FOR HOME, LABORATORY, AND CLASSROOM DEMONSTRATION

Home and Classroom Demonstration

1. *Dependence of pitch on intensity*   Demonstration 12 on the *Auditory Demonstrations* CD (Houtsma, Rossing, and Wagenaars 1987). Tone bursts having frequencies of 200, 500, 1000, 3000, and 4000 Hz are presented at two levels 30 dB apart. Does the pitch of the second tone sound lower, higher, or the same as the first pitch?

2. *Dependence of pitch on intensity*   An audio generator is connected to an amplifier and loudspeaker so that pure tones (sine waves) of different frequencies (200 to 4000 Hz) can be heard different sound levels. Does the pitch rise, fall, or stay the same when the intensity is increased?

3. *Pitch salience and tone duration*   Demonstration 13 on the *Auditory Demonstrations* CD. Tones of 300, 1000, and 3000 Hz are presented in bursts of 1, 2, 4, 8, 16, 32, 64, and 128 periods. How many periods are necessary to establish a sense of pitch?

4. *Influence of masking noise on pitch*   Demonstration 14 on the *Auditory Demonstrations* CD. A 1000-Hz tone, 500 ms in duration and partially masked by noise low-pass filtered at 900 Hz, alternates with an identical tone presented without masking noise. The tone partially masked by noise of lower frequency generally appears slightly higher in pitch. Do you agree?

5. *Octave matching*   Demonstration 15 on the *Auditory Demonstrations* CD. A 500-Hz tone alternates with another tone that varies from 985 to 1035 Hz in steps of 5 Hz. Which one sounds like a correct octave? Most listeners select a tone somewhere around 1010 Hz, which illustrates our preference for "stretched" octaves.

6. *Stretched and compressed scales*   Demonstration 16 on the *Auditory Demonstrations* CD. Another demonstration illustrating preference for stretched intonation. A melody is played in a high register with an accompaniment in a low register.

7. *Difference limen, or jnd*   Demonstration 17 on the *Auditory Demonstrations* CD. Ten groups of four tone pairs are presented. In each tone pair the second tone pair may be higher or lower than the first (write down which you hear). The frequency difference decreases with each group.

8. *Difference limen, or jnd*   The difference limen for frequency is conveniently demonstrated by a two-tone switching generator (such as the Automated Industrial Electronics 2TSG-1) or a computer by switching back and forth between tones of frequency f and $f + \Delta f$.

9. *Seebeck's siren*   The waveforms shown in Fig. 7.1 are generated electronically (with a pulse generator (see T. D. Rossing, "Seebeck's Siren," *Phys. Teach.* **17**: 352 (1959)) or with a computer), displayed on an oscilloscope, and fed to an audio amplifier and loudspeaker. The abrupt disappearance of the $1/T$ tone when $t_2 - t_1 = 0$ is rather dramatic.

10. *Virtual pitch*   Demonstration 20 on the *Auditory Demonstrations* CD. A complex tone consisting of 10 harmonics of 200 Hz is presented, followed by the same tone without the fundamental, with the two lowest harmonics, etc. Does the pitch of the complex tone change?

11. *Shift of virtual pitch*   Demonstration 21 on the *Auditory Demonstrations* CD. The 800-, 1000-, and 1200-Hz partials in

a complex tone are shifted upward in steps of 20 Hz. The virtual pitch is heard to rise. Shifting to 850, 1050, and 1250 Hz, for example, generally produces a shift in virtual pitch from 200 to 210 Hz, as can be determined by matching to a spectral pitch.

12. *Shift of virtual pitch* Schouten's pitch-shift experiment (Fig. 7.3) can be done with an amplitude-modulated audio signal. Some signal generators may require a band-reject filter to eliminate leakage of the modulation tone into the output.

13. *Masking spectral and virtual pitch* Demonstration 22 on the *Auditory Demonstrations* CD. The Westminster chime melody is played with pairs of tones. The first tone of each pair is a pure tone, the second a complex tone with the same pitch. Low-pass noise masks only the pure-tone notes, whereas high-pass noise masks only the virtual pitch of the complex tone.

14. *Virtual pitch with random harmonics* Demonstration 23 on the *Auditory Demonstrations* CD. The Westminster chime melody is presented with various harmonics of a missing fundamental.

15. *Strike note of a chime* Demonstration 24 on the *Auditory Demonstrations* CD. An orchestral chime is struck eight times, each time preceded by cue tones equal to the first eight partials of the chime. In most orchestral chimes the virtual pitch of the strike note lies between the second and third partial.

16. *Analytic versus synthetic pitch* Demonstration 25 on the *Auditory Demonstrations* CD. A two-tone complex of 800 and 1000 Hz is followed by one of 750 and 1000 Hz. Do you hear the pitch go up or down? If you listen analytically, you will hear one partial go down in pitch; if you listen synthetically you will hear the virtual pitch go up a major third (from 200 to 250 Hz).

17. *Scales with repetition pitch* Demonstration 26 on the *Auditory Demonstrations* CD. Repetition pitch can be demonstrated by playing scales or melodies with pairs of pulses having appropriate time delays between members of a pair.

18. *Repetition pitch* Repetition pitch can be demonstrated using a tape recording made with a movable microphone (see Fig. 7.9). As a home experiment, moving the head between two loudspeakers (first drawing in Fig. 7.9) works well.

19. *Circularity in pitch judgment* Demonstration 27 on the *Auditory Demonstrations* CD. The *Shepherd scale*, which demonstrates circularity in pitch judgment, is an auditory analog to the ever-ascending staircase visual illusion.

20. *Spectrum analysis* Use a tunable bandpass filter to present each of the harmonics of a square wave, both on an oscilloscope and aurally through headphones or through an audio amplifier and loudspeaker. Do the same for sustained tones from musical instruments recorded on a loop of tape in order to play back continuously.

21. *Fourier analysis* An FFT analyzer or a PC with an FFT card can be used to display the spectra of various waveforms and musical instrument sounds.

22. *Fourier synthesis* Fourier synthesis using separate oscillators plus a mixer (the Pasco 9300 combines them in one unit) or a computer is entertaining as well as instructive. Observe the synthesized waveforms both visually (on an oscilloscope) and audibly (on headphones or loudspeaker).

23. *Effect of spectrum on timbre* Demonstration 28 on the *Auditory Demonstrations* CD. A carillon bell and a guitar tone are synthesized in eight steps by adding successive partials. How many partials are necessary to make the sound source recognizable?

24. *Effect of tone envelope on timbre* Demonstration 29 on the *Auditory Demonstrations* CD. Piano tones, heard backward, do not sound like piano tones, even through the spectrum remains unchanged. This demonstrates the significant influence of temporal envelope (including attack and decay) on timbre.

25. *Change in timbre with transposition* Demonstration 30 on the *Auditory Demonstrations* CD. A three-octave scale, synthesized by transposing the highest note of a bassoon, sounds different from a scale played on a bassoon.

26. *Canceled harmonics* Demonstration 1 on the *Auditory Demonstrations* CD. When the amplitudes of all 20 harmonics in a tone remain steady, we tend to hear the tone holistically (as a single, complex tone). When a harmonic is canceled and restored, it calls attention to itself and we tend to listen analytically. This is demonstrated for harmonics 1 through 10.

Laboratory Experiments

Perception of pitch (Experiment 13 in *Acoustics Laboratory Experiments*)

Sound spectra (Experiment 11 in *Acoustics Laboratory Experiments*)

Tones, vowels, and telephones (Experiment 22 in *Physics with Computers* by Appel et al.)

The demonstration experiments on the *Auditory Demonstrations* CD can be used as laboratory experiments.

CHAPTER

8

Combination Tones and Harmony

In this chapter we will focus our discussion on the various effects that can occur when two or more tones reach the ear simultaneously. We begin with the simplest case, two pure tones with the same frequency, then proceed to two pure tones of different frequency, and finally to complex tones and chords.

In this chapter you should learn:

- About combining two harmonic motions;
- About beats;
- About difference tones;
- About consonance and dissonance;
- About signal processing in the central nervous system.

8.1 ■ LINEAR SUPERPOSITION

A *linear* system is one in which doubling the driving force doubles the response. If two driving forces are applied to a linear system simultaneously, the response will be the sum of the responses to the driving forces individually. That is, the response of the system to one driving force is not affected by the presence of the second one.

More than likely, you have heard the terms linear or linearity applied to components of a high-fidelity sound system. A loudspeaker with a linear response, for example, can reproduce the sound of a clarinet and, at the same time quite independently, can reproduce a violin sound when presented with (electrical) driving signals characteristic of each instrument. If the loudspeaker were not completely linear in its response, one signal would influence (i.e., modulate) the other, and intermodulation distortion would be the result. This will be discussed in Section 19.4.

To determine the response of a linear system to two simple harmonic (pure tone) driving forces, we simply add together the two individual responses at each point in time. If two simple harmonic motions at the same frequency are superimposed, the resultant will also be a simple harmonic motion at that frequency. The amplitude of the resultant will depend not only on the amplitude of the components but also on the fractional part of the period through which each component has passed. This fraction of the period is known as the *phase*. Figure 8.1 illustrates two cases as examples.

In the first case, the two motions are identical in phase and the displacements add. In the second case, they are opposite in phase and the displacements subtract (if $A = B$, total cancellation is the result). The concept of phase is a useful one in the comparison of two

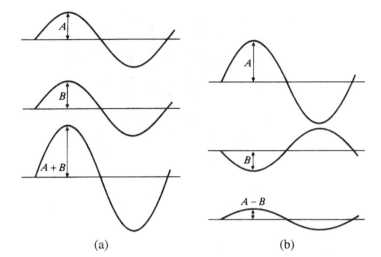

FIGURE 8.1
Linear
superposition of
two simple
harmonic motions
at the same
frequency: (a) same
phase; (b) opposite
phase.

(a)   (b)

simple harmonic motions as well as in their superposition. It merits discussion in some detail.

8.2 ■ PHASE ANGLE

Simple harmonic motion can be described as the projection of uniform circular motion onto an axis, as shown in Fig. 8.2. As the circle rotates, the projection of point P on the y-axis moves in simple harmonic motion. The angle ϕ indicates how far the circle has turned, so as ϕ increases, the corresponding point P moves to the right on the graph is displacement versus time. During one complete revolution, ϕ increases by 360°, and the point moves to the right a distance T on the time axis (T is the period of the motion in seconds).

The representation of simple harmonic motion as the projection of circular motion can be demonstrated in several ways, one of which is shown in Fig. 8.3. The shadow of a wheel with a crank is projected alongside a mass vibrating in simple harmonic motion at the end of a spring (see Section 2.1). If the speed of the wheel is adjusted so that the time required for one revolution is the same as the period of the mass-spring vibrator, its shadow will

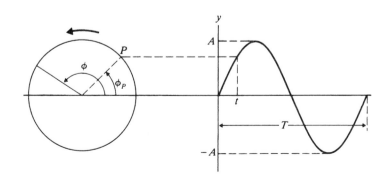

FIGURE 8.2
Simple harmonic
motion represented
as the projection on
a vertical axis of
the point P moving
around a circle at a
uniform rate.

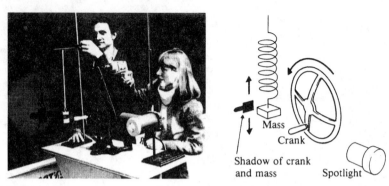

FIGURE 8.3   Demonstration of simple harmonic motion as the projection of circular motion. By rotating the wheel at the proper rate, the shadow of a crank can be made to move up and down in synchronism with a mass-spring vibrator. (Photograph by Christopher Chiaverina.)

move up and down in synchronism with the mass. Both the mass and the shadow of the crank move up and down in simple harmonic motion.

Simple harmonic motion is often referred to as *sinusoidal* motion, because the projection of circular motion on the y-axis is given analytically by a trigonometric function called the sine; that is,

$$y = A \sin \phi = A \sin 360 \frac{t}{T},$$

where the amplitude of the vibration A also equals the radius of the circle projected. Also, 360 is the number of degrees in a complete circle, t/T is the fraction of a complete circle subtended by angle ϕ in time t, T is a complete period, and sin is the abbreviation for sine. The language of trigonometry is not generally used in this book, however (see Appendix A.10).

We now add a second point in simple harmonic motion, such as point Q in Fig. 8.4. The projection of point Q moves with the same period T and amplitude A as point P, but it

FIGURE 8.4
Two points P and Q move with the same period T and amplitude A and maintain a constant phase difference $\phi_P - \phi_Q$.

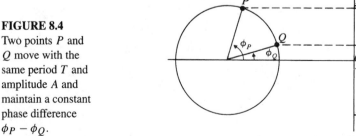

obviously has a different motion. At any given time t, points P and Q will be at different points on the circle. Thus the projected points reach their maximum and minimum values at different times. At any given time t, the positions of P and Q on the two curves result from different angular positions ϕ_P and ϕ_Q of the rotating points. The difference $\phi_P - \phi_Q$, which remains constant, is called the *phase difference* between the two simple harmonic motions.

The complete description of a given simple harmonic motion requires three parameters: the *period* (or frequency), the *amplitude*, and the initial *phase*. In the case of sound, phase has significance only when it is used to compare two or more waves or vibrations.

8.3 ■ COMBINATION OF TWO SIMPLE HARMONIC MOTIONS

In order to radiate a pure tone, a loudspeaker cone moves in and out with a motion that is essentially simple harmonic motion. How does a loudspeaker cone move when two pure tones are radiated? If its response is linear, its motion at any time will be the linear combination or superposition of two simple harmonic motions. When it radiates a complex tone, its motion will be a combination of all the spectral components of the complex tone (see Section 7.10).

We consider first the combination of two simple harmonic motions having the same period (frequency) but different amplitudes and initial phases. Two cases were illustrated already in Fig. 8.1: same phase ($\phi_B - \phi_A = 0$) and opposite phase ($\phi_B - \phi_A = 180°$). Another important case is the one in which the phase difference $\phi_B - \phi_A = 90°$. In this case the displacement curve of one simple harmonic motion reaches its maximum when the other curve is at zero, as shown in Fig. 8.5. We would expect the resultant curve obtained by linear superposition to reach its maximum value somewhere between the maxima of curves A and B, which indeed it does. The amplitude of the resultant curve can be shown to be $\sqrt{A^2 + B^2}$ in this case.

The amplitude of the resultants for the values of phase difference between simple harmonic motions A and B can be calculated. The results are summarized in Table 8.1.

FIGURE 8.5
Linear superposition of simple harmonic motions with the same period (frequency) but with a phase difference $\phi_B - \phi_A = 90°$. (Compare with Fig. 8.1.)

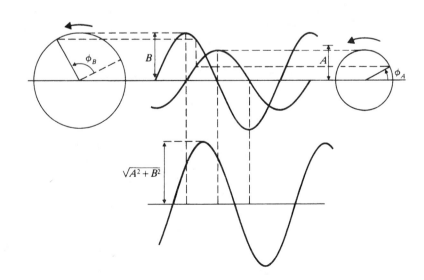

TABLE 8.1   Amplitude of resultant obtained by superposition of two simple harmonic motions with the same frequency

Phase difference $\phi_B - \phi_A$	0	45°	90°	135°	180°	270°
Amplitude of resultant	$A + B$	$\sqrt{A^2 + B^2 + 1.4AB}$	$\sqrt{A^2 + B^2}$	$\sqrt{A^2 + B^2 - 1.4AB}$	$A - B$	$\sqrt{A^2 - B^2}$

8.4 ■ PURE TONES WITH SLIGHTLY DIFFERENT FREQUENCIES: BEATS

A *pure tone* is a sound wave with a single frequency. The superposition of two pure tones proceeds in the same way as the superposition of two simple harmonic motions, which we discussed in Section 8.3. If the tones have the same frequency, the resultant amplitude will be somewhere between $A + B$ and $A - B$ (or $B - A$, if B is larger than A), depending on their phase difference (see Table 8.1). A pure tone is often referred to as a *sine wave* for the same reason that simple harmonic motion is called sinusoidal (see the box on p. 152).

If two pure tones have slightly different frequencies, f_1 and $f_1 + \Delta f$, the phase difference $\phi_B - \phi_A$ changes continually with time, and so the amplitude of the resultant tone changes also. The amplitude of the resultant varies between $A + B$ and $A - B$ at a frequency Δf. These slow periodic variations in amplitude at frequency Δf are called *beats*. If the amplitudes A and B are equal, the resultant amplitude varies between $2A$ and 0, but this condition is not necessary for beats to occur.

In the case of two pure tones of slightly different frequency, linear superposition at our ears leads to a sensation of audible beats at the *difference frequency* Δf. These beats are heard as a pulsation in the loudness of the tone having the *average frequency* $f = \frac{1}{2}(f_1 + f_2)$. An example of beats is shown in Fig. 8.6.

So long as the frequency difference Δf is less than about 10 Hz, the beats are easily perceived. When Δf exceeds 15 Hz, the beat sensation disappears, and a characteristic roughness appears. As Δf increases still further, a point is reached at which the "fused" tone at the average frequency gives way to two tones, still with roughness. The respective resonance regions on the basilar membrane are now separated sufficiently to give two distinct pitch signals, but the excitations corresponding to the two pitches still overlap to give an effect of roughness (see Fig. 5.9). When the separation Δf exceeds the width of the critical band (see Section 5.4), the roughness disappears, and the two tones begin to blend. This process is illustrated by the graph (not to scale) in Fig. 8.7.

Probably the easiest way to illustrate beats is by connecting two audio generators to an amplifier and a loudspeaker. A pair of resistors (about 1000 Ω) can serve as a *mixer*. One

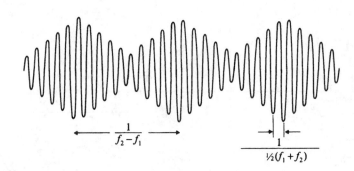

FIGURE 8.6
Waveform with beats due to pure tones with frequencies f_1 and $f_2 = f_1 + \Delta f$.

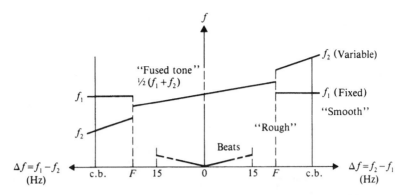

FIGURE 8.7   Schematic representation of frequencies heard when pure tones of frequencies f_1 and f_2 are superimposed. Note that the disappearance of beats occurs around $\Delta f = 15$ Hz regardless of the values of f_1 and f_2; the critical bandwidth (c.b.) and the fusion frequency (F) increase with f_2 and f_1. (After Roederer 1975.)

should keep the frequency of one generator constant as f_1 and vary the other slowly above and below f_1. Note how the frequency of beats Δf changes as f_2 changes, and note also the frequency f_2 at which the beats disappear and two separate tones can be distinguished. It is instructive to view the resultant waveform on an oscilloscope at the same time that it is heard by using an arrangement like that shown in Fig. 8.8.

The limit of pitch discrimination, or *fusion frequency*, the point at which the single fused tone changes to two tones, varies with center frequency in a manner somewhat like the critical band. It varies from about a semitone (at 500 Hz) to more than a whole tone* (below 200 Hz and above 4000 Hz) (Plomp 1965), but is always less than the critical bandwidth. At the same time it is 7 to 30 times larger than the just noticeable difference (jnd) for frequency (see Fig. 7.2). In other words, we can detect a very small change in frequency of a pure tone, but two tones sounded together may have to differ by a semitone or more in order to be heard as separate tones.

Beats can also be heard when tones of slightly different frequency are presented separately to our two ears; these are called *binaural beats*. They are difficult to detect and are best heard with f_1 around 500 Hz and Δf in the range of 5 to 20 Hz (Perrott and Nelson

FIGURE 8.8
Demonstration of beats with electrical signals from two audio generators.

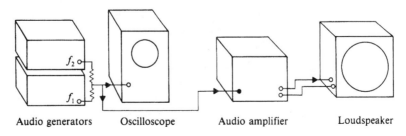

Audio generators   Oscilloscope   Audio amplifier   Loudspeaker

*For a semitone interval $\Delta f/f \simeq 0.06$, and for a whole-tone interval $\Delta f/f \simeq 0.12$. These and other musical intervals will be discussed in Chapter 9.

1969). They have been described as a *muffled* sound. When Δf is less than 5 Hz, the beats change to a *rotating tone*, a single tone that appears to move around inside the head. Binaural beats can be heard over a wide range of intensity of the stimulus tones; they apparently originate in the central auditory processor, not the individual ears. The rotating tone sensation is probably a result of the continuous change in phase, which makes it appear that the sound source is changing its direction (see Section 5.5).

Beats may be used to tune two audio generators or musical instruments to precisely the same pitch. Although we have discussed beats between pure tones, beats with a slightly different timbre occur between complex tones, and we will discuss beats between tones with various ratios of frequency in Section 8.12. (It is these *second-order beats* that are counted by piano tuners to achieve just the right interval between two notes.)

8.5 ■ THE MUSICAL STAFF: MUSICIANS' GRAPH PAPER

In Section 5.7 we discussed linear and logarithmic scales and the way in which they may be used to represent quantities on graphs. Graph paper that has one logarithmic axis is called semilogarithmic graph paper, and we use it frequently for representing quantities that are functions of frequency (see Fig. 5.14). When frequency is represented on a logarithmic scale, octaves (frequency ratio of 2:1) and other familiar musical intervals become equidistant anywhere on the scale (the piano keyboard approximates a logarithmic scale; an octave requires a reach of the same distance anywhere on the keyboard).

Music is written on staves which approximate logarithmic frequency scales. Normally a musical staff consists of five lines, and a clef sign is placed at the beginning of each staff to show the exact location of a particular note (G, F, or C). Piano music is normally written on two staves (connected by a brace) which carry treble and bass clef signs. The five lines of a musical staff are separated either by three or four semitones, as shown in Fig. 8.9.

Although the lines on a normal musical staff are evenly spaced, we have drawn the staves in Fig. 8.9 with two different spacings, so that they exactly fit a logarithmic frequency scale and also indicate clearly whether the lines are separated by an interval of a minor third (three semitones) or a major third (four semitones). Note that the two staves in Fig. 8.9 (used in most piano and vocal scores) span a frequency range of just under three octaves from 98 to 698 Hz. This is a small part of the total $7\frac{1}{2}$-octave range of a piano, so ledger lines are added above and below the staves. Also, the notation 8va is used to indicate that notes on the staff are to be played one octave higher than normal (as in Fig. 8.11). Several years ago, we proposed the use of three new clefs to extend written music over a full 10 octaves, the range of audible sound (see Appendix A.7). The musical world has not rushed to adopt our "invention," however.

The horizontal time axis in music is linear. But instead of marking off the axis in seconds, the musical staff uses bar lines (the distance between two bar lines is sometimes called a *measure*). The rate at which the music is to be played is denoted in several ways: by the time signature (e.g., 2/4, 3/4, 6/8, etc.), by the meter (e.g., ♩ = 64, which means 64 quarter notes per minute), and by designation for tempo (e.g., adagio, allegro, vivace, etc.). In 4/4 time at a meter of ♩ = 64, there will be four beats per bar, or measure, each quarter note receiving one beat with a duration of 1/64 min, or 60/64 s. Thus each bar of four beats should equal 240/64 = 3.75 s. The two passages shown in Fig. 8.10 would be

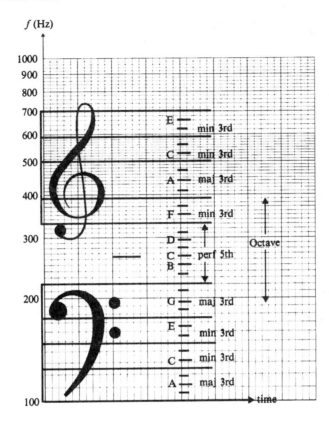

FIGURE 8.9 The musical staff as semilogarithmic graph paper. Some lines are spaced three semitones (minor third) apart, and some are spaced four semitones (major third) apart. The scale of time on the linear horizontal axis indicates the tempo.

FIGURE 8.10 Two passages that would appear as straight lines on a graph of log frequency versus time.

represented by straight lines on a graph of frequency (logarithmic scale) versus time (linear scale). This is because semitone intervals are written as quarter notes (♩) and whole-tone intervals as half notes (♪).

8.6 ■ COMBINATION TONES

When two tones are sounded together, a lower tone is frequently heard. This undertone is called a *difference tone*, or *Tartini tone*, after the Italian violinist Tartini, who reportedly discovered it around 1714. If the two tones have frequencies f_1 and f_2, this difference tone, which is an example of a *combination tone*, occurs at a frequency $f_2 - f_1$ (or $f_1 - f_2$).

Difference tones can be demonstrated by using two audio generators, two flutes, or even two soprano voices. The notes shown in Fig. 8.11 should produce a difference tone melody;

another such passage is given by Stickney and Englert (1975). We often demonstrate difference tones with a plastic whistle (originally obtained from a box of breakfast cereal) that emits loud tones with frequencies of 1727, 1896, and 2081 Hz. Difference tones at 169, 185, and 374 Hz are clearly audible, and in fact fast beats occur between the 169- and 185-Hz difference tones (see Rossing 1974). Police in London use whistles of this type.

FIGURE 8.11
Playing these notes on a piccolo or descant recorder should produce a difference-tone melody about three octaves lower, as shown on the lower clef.

Other combination tones that can be heard have frequencies given by $2f_1 - f_2$ and $3f_1 - 2f_2$ (Plomp 1965). The $2f_1 - f_2$ tone, called the cubic difference tone, may be detected at stimulus levels as low as 15 dB but is heard only in a limited range of frequency below f_1 (Smoorenburg 1972). Other members of the class given by $f_1 - k(f_2 - f_1)$ can also be detected but with some difficulty ($4f_1 - 3f_2$, for example).

Suppose that we continue the experiment with two tones described in Section 8.4 by increasing f_2 until it reaches twice the value of f_1. The difference tone $f_2 - f_1$ would be heard over most of the range, and the cubic difference tone $2f_1 - f_2$ would be heard over much of the range as well. These two tones are shown in Fig. 8.12, as well as the primary tones f_2 and f_1 and the less prominent difference tones $3f_1 - 2f_2$ and $4f_1 - 3f_2$. The quartic and higher-order difference tones are best heard in the region just above $f_2/f_1 = 1.1$. The two most prominent difference tones are also shown on musical staves in Fig. 8.13, along with the primary tones.

Combination tones have many applications in music, several of which are discussed by Hindemith (1937) in his classic book on musical composition. For example, very low tones

FIGURE 8.12
Most prominent combination tones for a two-tone presentation consisting of one tone of fixed frequency f_1 and one of variable frequency f_2. Stronger and weaker tones are indicated by solid and dashed lines, respectively. (After Plomp 1965.)

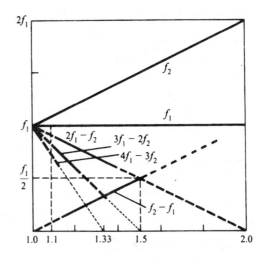

FIGURE 8.13 Combination tones on a musical staff. As in Fig. 8.12, f_1 is fixed and f_2 is variable. Note that the difference tone ($f_2 - f_1$) and the cubic difference tone ($2f_1 - f_2$) cross at $f_2 = 1.5 f_1$, just as they do in Fig. 8.12. This ratio of frequencies corresponds to an interval called a *perfect fifth*.

can be produced on a small organ by the combination tone from two smaller pipes instead of requiring a large pipe tuned to the desired low pitch. (The reader should be warned, however, that not all of Hindemith's ideas are consistent with more recent experimental results in perception and psychoacoustics).

8.7 ■ MODULATION OF ONE TONE BY ANOTHER

Closely akin to combination tones is the phenomenon of *amplitude modulation* (AM), which will be further discussed in Chapter 20 on components for high-fidelity sound reproduction. When vibrations occur at two different frequencies in a nonlinear system, various *sidebands* are generated, having frequencies equal to the various difference tones and summation tones as well. If one of the frequencies is considerably less than the other, we describe the process as modulation of the higher frequency by the lower one. That is, the amplitude of the high-frequency component changes at the frequency of the low-frequency component, as shown in Fig. 8.14.

Note that the principal sidebands occur at $f_2 + f_1$ and $f_2 - f_1$ with much smaller sidebands at $f_2 \pm 2f_1$, etc. In the case of pure tones, they are not usually heard as separate tones but as components of a complex tone. One example of audible sidebands occurs in the pitch-shift experiments described in Section 7.6. Intermodulation distortion products generated when several frequencies interact in some sound system component having a slight nonlinearity (such as a loudspeaker) are another example.

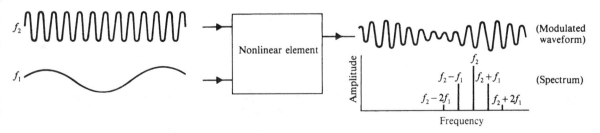

FIGURE 8.14 Amplitude modulation of one vibration by another in a nonlinear element. The spectrum of the modulated wave shows sidebands at $f_2 - f_1$ and $f_2 + f_1$.

8.8 ■ OTHER NONLINEAR EFFECTS: AURAL HARMONICS AND SUMMATION TONES

A single tone of frequency f, if it is sufficiently loud, may produce additional pitch sensations at $2f, 3f, 4f$, etc. These are called *aural harmonics*, and they are due to nonlinear behavior in the auditory system (analogous to nonlinearity in an amplifier or loudspeaker).

Aural harmonics were observed many years ago by Fletcher (1929), who suggested a simple power series for the response of the ear:

$$x = a_0 + a_1 p + a_2 p^2 + a_3 p^3 + \cdots, \tag{8.1}$$

where p is sound pressure and the a_s are constants that can be determined by experiment. Such a response predicts that for each 1-dB increase in signal level the second harmonic will increase by 2 dB and the third harmonic by 3 dB. This is consistent with the results of experiments by Clack (1977) and others. At a signal level of 70 dB the second and third aural harmonics have sound levels in the neighborhood of 25 and 15 dB but with wide variation between individuals.

If the response of the ear is nonlinear, summation as well as difference tones ought to be generated, but no one has presented convincing evidence that even simple sum tones $(f_1 + f_2)$ can be heard, not to mention other summation tones $(2f_1 + f_2, f_1 + 2f_2$, etc.). According to the theory of Helmholtz (1877), the sum and difference tones, which have frequencies of $f_1 + f_2$ and $f_1 - f_2$, respectively, ought to have amplitudes in the ratio

$$\left(\frac{f_1 - f_2}{f_1 + f_2} \right)^2 .$$

Because $f_1 - f_2$ is usually much smaller than $f_2 + f_1$, this ratio is quite small. Lying above f_1 and f_2 in frequency, the weaker summation tones may be masked by the primary tones.

8.9 ■ ORIGIN OF DIFFERENCE TONES

Difference tones have been studied extensively, not only because of their importance in the perception of musical sound, but also because they provide us with a window into the functioning of the auditory system. Intensities of difference tones can be measured by adjusting the amplitude and phase of a "cancellation tone" until it just cancels the difference tone of interest (Zwicker 1955).

The difference tone with frequency $f_2 - f_1$ (properly called the quadratic difference tone) has the behavior we would expect from a quadratic distortion product (i.e., one that results from the quadratic term in Eq. 8.1). If either one of the primary tones increases 3 dB in level, the difference tone also increases 3 dB; if both primary tones increase 3 dB, the difference tone increases 6 dB. If the cubic difference tone were a result of the cubic term in Eq. 8.1 (i.e., a cubic distortion product), it would be expected to increase 9 dB when each primary tone increased 3 dB. Instead, the cubic difference tone is observed to increase more nearly 3 dB, or about half as much as the quadratic difference tone.

Intensities of the cubic (and higher-order) difference tones decrease rapidly with increasing frequency ratio f_2/f_1 (Goldstein 1967). They are best heard in the region just above $f_2/f_1 = 1.1$ (see Fig. 8.12). The quadratic difference tone $f_2 - f_1$, on the other hand, can be heard over a wider range of frequency, and its intensity varies much less with the frequency ratio f_2/f_1. These results, among others, seem to indicate that the quadratic and cubic difference tones are *not* produced in the same way in the ear. The cubic and higher order difference tones appear to be produced in the frequency selective inner ear, whereas the quadratic difference tone results from a nonlinearity without frequency selectivity, probably in the middle ear.

It would be well to point out that the beats that occur when $f_2 - f_1$ is small (less than about 15 Hz) are distinctly different from difference tones that occur at greater frequency separation. Beats, which are heard as periodic variations in intensity, do *not* require nonlinearity in the ear; audible difference tones *do*. Hall (1981) demonstrates this by presenting tones f_1 and f_2 with and without nonlinear distortion (and also shows the spectra of the audible signals). Without the distortion, masking noise that spans the region of f_1 and f_2 can mask the difference tones, but adding distortion generates a signal at the difference tone frequency that can be heard in the presence of noise.

8.10 ■ CONSONANCE AND DISSONANCE: MUSICAL INTERVALS

Pythagoras of ancient Greece is considered to have discovered that the tones produced by a string vibrating in two parts with simple length ratios such as 2 : 1, 3 : 2, or 4 : 3 sound harmonious. These ratios define the so-called perfect intervals of music, which are considered to have the greatest consonance. Galileo (1638) observed: "Agreeable consonances are pairs of tones which strike the ear with a certain regularity; this regularity consists in the fact that the pulses delivered by the two tones, in the same time, shall be commensurable in number, so as not to keep the eardrum in perpetual torment, bending in two different directions in order to yield to the ever-discordant impulses." The most consonant intervals of music are generally considered to be the following (in descending order of consonance):

2:1	octave	(C/C),
3:2	perfect fifth	(G/C),
4:3	perfect fourth	(F/C),
5:3	major sixth	(A/C),
5:4	major third	(E/C),
8:5	minor sixth	(A$^\flat$/C),
6:5	minor third	(E$^\flat$/C).

Why are some intervals more consonant than others? Helmholtz (1877) explained *consonance* by referring to Ohm's acoustical law (see Section 7.5), which stated that the ear performs a spectral (Fourier) analysis of sound, separating a complex sound into its various partials. Helmholtz concluded that *dissonance* occurs when partials of the two tones produce 30–40 beats per second. The more the partials of one tone coincide with the partials of the other, the less chance that beats in this range will produce roughness (dissonance). This explains why simple frequency ratios define the most consonant intervals.

More recent studies have shown that Helmholtz was on the right track. A more accurate picture of consonance and dissonance should be based on critical bands in hearing (which, of course, were unknown to Helmholtz). We will briefly consider the consonance of two pure tones, two complex tones, and chords of three or four notes.

Two Pure Tones

Research by Plomp and Levelt in The Netherlands and by Kameoka and Kuriyagowa in Japan led to the same conclusion: The consonance or dissonance of two pure tones sounded together depends upon their frequency difference rather than on their frequency ratio. If the frequency difference between two pure tones is greater than a critical band, they sound consonant; if it is less than a critical band they sound dissonant. Plomp and Levelt (1965) found that maximum dissonance occurs when the frequency difference is approximately one-fourth of a critical bandwidth, as shown in Fig. 8.15. Kameoka and Kuriyagowa (1969), however, suggested a slightly more complicated dependence on frequency and also found a dependence on sound pressure level. Their empirical expression giving the frequency difference for greatest dissonance Δf_d is

$$\Delta f_d = 2.27 \left(1 + \frac{L_p - 57}{40} \right) f^{0.447},\tag{8.2}$$

where L_p is sound pressure level and f is the frequency of the primary tone. Note that around 500 Hz, the maximum dissonance predicted in both Eq. 8.2 and Fig. 8.15 corresponds reasonably well with the 30- to 40-Hz criterion of Helmholtz.

It is instructive to examine a few intervals between pure tones that occur in different octaves. If C_4 (262 Hz) and G_4 (392 Hz) are sounded together, the difference frequency is 130 Hz, which is 40% greater than the critical bandwidth (approximately 90 Hz in this octave). Thus, they sound consonant. However, an octave lower on the scale, a perfect fifth is not quite so consonant, because the frequency difference between C_3 (131 Hz) and G_3 (196 Hz) is 65 Hz, which is less than the critical bandwidth. Another octave or so lower, the frequency difference approaches one-fourth of the critical bandwidth, the criterion for maximum roughness or dissonance. So the degree of dissonance of the interval between two pure tones is strongly dependent on their location on the musical scale. The higher on the scale, the closer two pure tones can be to each other in pitch and still sound consonant when sounded together.

FIGURE 8.15
Consonance of two
pure tones as a
function of
frequency
separation relative
to the critical
bandwidth (Plomp
and Levelt 1965).

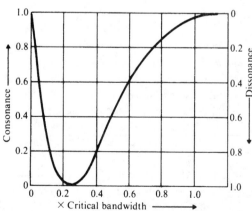

TABLE 8.2 Interactions between harmonics of two tones separated by different intervals

Perfect fifth C_4:G_4 ($f_2/f_1 = 3/2$)		Major sixth C_4:A_4 ($f_2/f_1 = 5/3$)		Minor third C_4:E_4 ($f_2/f_1 = 6/5$)		Major whole tone C_4:D_4 ($f_2/f_1 = 9/8$)	
mf_1	nf_2	mf_1	nf_2	mf_1	nf_2	mf_1	nf_2
262	392	262	436	262	314	262	294
523		523		523 R	628	523 R	589
785 =	785	785 R	872	785		785 R	883
1047 R*1177		1047		1047 R	942	1047 R	1177
1308		1308 =	1308	1308 R	1256	1308	
1570 =	1570	1570		1570 =	1570	1570 R	1472
1831 R		1831 R	1744	1831 R	1884	1831 R	1766
2093 R		2093 R	2180	2093 R	2198	2093 R	2060

*R denotes roughness due to frequency difference within the critical bandwidth.

Two Complex Tones

In the case of musical tones, which nearly always have several harmonics, the situation is quite different, however. Roughness can occur between the harmonics of the tones as well as between the fundamentals. Herein lies the reason why certain intervals are inherently more consonant than others. Fewer harmonics of the more consonant intervals have frequency differences within the roughness range.

In Table 8.2, harmonics of four musical intervals are tabulated. We have used frequency ratios from the just scale rather than the scale of equal temperament (see Section 9.1). In the case of the perfect fifth, two of the lower harmonics coincide and two produce frequency differences within the critical bandwidth but well away from the range of maximum roughness. In the case of the minor third, however, there are many interactions that produce roughness, and in the case of the major second, nearly all the harmonics thus interact.

Figure 8.16 shows the dissonance to be expected between two tones, each having six harmonics, making the assumption that dissonances between various harmonics are addi-

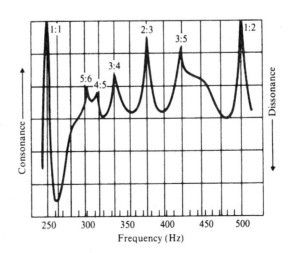

FIGURE 8.16
Consonance expected from tone A with $f = 250$ Hz and tone B with varying frequency, both A and B having six harmonics. (From Plomp and Levelt 1965.)

tive. Maxima in consonance occur when tone *B* forms consonant intervals with tone *A*. The major sixth (5:3) appears more consonant than the perfect fourth (5:4) under these assumptions.

It certainly should not implied that consonant intervals are good and dissonant intervals bad. Music written with consonant intervals alone would be exceedingly dull; musicians make a clear distinction between pleasantness and consonance. Sixths (5:3, 8:5) and thirds (5:4, 6:5) are generally found to be pleasant intervals, as are the fourth (5:4) and minor seventh (9:5), even though some of these intervals are not particularly consonant.

Chords

When three or more musical (complex) tones are sounded together in a chord, there are many opportunities for roughness-producing interactions between the various partials that lie within critical bandwidths of one another. Musicians generally consider major and minor chords to be more consonant than diminished and augmented chords, which sound dissonant and need to be resolved (see Fig. 8.17). In a psychoacoustic study of three- and four-note chords, Roberts (1986) found major chords to be the most consonant, followed (in order) by minor, diminished, and augmented chords. Chords in root position were found to be more consonant than in first or second inversion, and chords in equal temperament were found to be more consonant than chords in just or Pythagorean tuning (see Chapter 9). Chords were judged to be more consonant when heard in a traditional musical context.

FIGURE 8.17   Examples of: (a) major triad; (b) minor triad; (c) diminished triad; (d) augmented triad; (e–h) same chords in first inversion.

An interesting indication of the preferences of musicians resulted from a statistical analysis by Plomp and Levelt (1965) of chords in two musical compositions. They analyzed a movement of J. S. Bach's *Trio Sonata for Organ, No. 3* and a movement of A. Dvořák's *String Quartet, Op. 51*, and determined the interval width that is not exceeded 25%, 50%, and 75% of the time. Figure 8.18 shows the results for each piece, taking into account only the fundamental ($n = 1$) and also the first nine harmonics ($n = 9$). In other words, Fig. 8.18(a) indicates the distribution in the size of intervals in which each of seven notes appears along with another note of higher frequency. Figure 8.18(b) includes many more cases in which one of 14 notes appears as a fundamental or as a harmonic of the fundamental.

Note that in the Bach composition, at least half of the intervals between fundamentals exceed the critical band and thus would be judged as consonant. Inclusion of the harmonics places many intervals in the musically interesting region of dissonance, but seldom is the maximum degree of dissonance approached. Dvořák's composition, however, shows a much larger percentage of dissonant intervals.

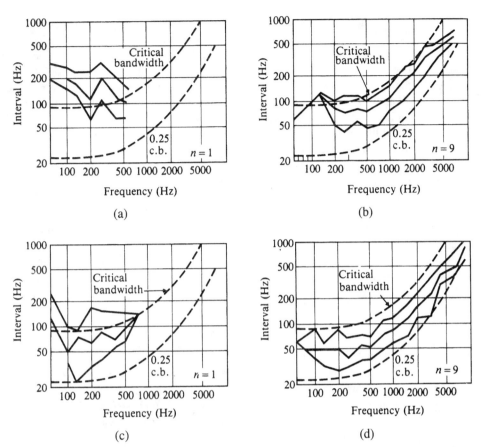

FIGURE 8.18   (a) and (b) Statistical analysis of the chords in Bach's *Trio Sonata for Organ, No. 3* (C-minor). (c) and (d) Similar analysis of chords in Dvořák's *String Quartet, Op. 51* (E♭-major). Shown in each graph are the interval widths (in Hz) not exceeded 25%, 50%, and 75% of the time. The dotted lines indicate the critical bandwidth and one-fourth of the critical bandwidth. In (a) and (c) only the fundamental is analyzed; in (b) and (d) the first nine harmonics are included. (From Plomp and Levelt 1965.)

Papers by Houtsma and Goldstein (1972) and by Terhardt (1974) discuss the role of the central processor in the perception of musical intervals. Terhardt feels that tonal meaning plays at least as important a role as roughness in determining consonance and dissonance.

8.11 ■ EFFECT OF PHASE ON TIMBRE

A good chef knows that the quality of the end product depends not only on using the right amount of the various ingredients but also combining them in a prescribed way. (Likewise, the chemistry student is cautioned to pour acid into water, not water into acid!) In the same way, building up complex tones using the same spectrum with different phases between the harmonics can lead to totally different waveforms. An interesting question to consider is: Do these different waveforms, which have the same harmonic spectrum, sound different?

Helmholtz (1877) stated that "the quality of the musical portion of a compound tone depends solely on the number and relative strength of its partial simple tones, and in no respect on their differences of phase." Although Helmholtz felt the influence of phase between various harmonics to be negligibly small in general, he did recognize exceptions in the case of "unmusical" sounds. Experiments by R. König using siren discs, performed about the same time, indicated that timbre has some dependence on phase. Using a generator that controlled both the phase and the amplitude of sixteen harmonics, Licklider (1959) noted that changing the phase of a high-frequency component has more effect on timbre than changing the phase of a low-frequency component.

Plomp (1970) summarizes the most important results of his experiments conducted with H. J. M. Steeneken on phase and timbre:

1. The maximum effect of phase on timbre is the difference between a complex tone in which the harmonics are in phase and one in which alternate harmonics differ in phase by 90°.
2. The effect of lowering each successive harmonic by 2 dB is greater than the maximum phase effect described above.
3. The effect of phase on timbre appears to be independent of the sound level and the spectrum.

Effect of Phase on Waveform

Four of the waveforms used by Plomp and Steeneken in their experiments on phase and timbre are shown in Fig. 8.19. All four of them consist of ten harmonics with amplitudes proportional to $1/n$ but with different phases. In the first two, the harmonics are in phase; in waveforms 3 and 4, alternate harmonics differ in phase by 90°. Mathematical expressions for the waveforms are:

$$\sin \omega t + \tfrac{1}{2} \sin 2\omega t + \tfrac{1}{3} \sin 3\omega t + \cdots + \tfrac{1}{10} \sin 10\omega t; \tag{1}$$

$$\cos \omega t + \tfrac{1}{2} \cos 2\omega t + \tfrac{1}{3} \cos 3\omega t + \cdots + \tfrac{1}{10} \cos 10\omega t; \tag{2}$$

$$\sin \omega t + \tfrac{1}{2} \cos 2\omega t + \tfrac{1}{3} \sin 3\omega t + \cdots + \tfrac{1}{10} \cos 10\omega t; \tag{3}$$

$$\cos \omega t + \tfrac{1}{2} \sin 2\omega t + \tfrac{1}{3} \cos 3\omega t + \cdots + \tfrac{1}{10} \sin 10\omega t. \tag{4}$$

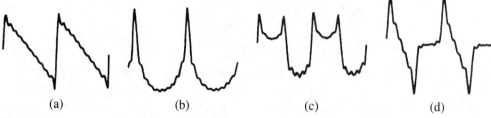

(a)   (b)   (c)   (d)

FIGURE 8.19   Four waveforms consisting of ten harmonics in different phase patterns. Waveforms 3 and 4 sound different from waveforms 1 and 2. (Plomp and Steeneken 1969.)

> Plomp and Steeneken (1969) found that their subjects could easily distinguish waveforms 3 and 4 from 1 and 2, but it was much more difficult to distinguish 1 from 2 and 3 from 4. Additional experiments established that the timbre difference between the two groups is the greatest possible using this spectrum of harmonics; hence the first conclusion above.

The nature of the amplitude envelope can have an effect on timbre. When all the harmonics are in phase, for example, the resulting "spikes" tend to make the tone sound rougher than a complex tone with the same harmonics in random phase (Patterson 1973). Phase shift that increases linearly with frequency (or with harmonic number) delays the tone but leaves the amplitude envelope unchanged; most observers find the timbre unchanged (an observation of importance to designers of high fidelity sound reproducing equipment, as we will see in Chapter 22).

Adding distortion can produce noticeable changes in the timbre of a complex tone when phase angles are changed (Demonstration 33, Houtsma, Rossing, and Wagenaars 1987), because the distortion products now interfere constructively or destructively with the harmonics, depending upon their relative phases.

The mechanism for phase perception is not well understood at present. Goldstein (1967) constructed a theory that explains certain experiments on phase perception on the basis of the ear's limited frequency resolution. When two or more spectral components lie within a critical band, the ear is not able to resolve them, so it seeks clues from the time envelope.

It should be mentioned that whereas the effects of phase on timbre are small for steady tones, the ear is in fact quite sensitive to *changes* in phase, especially if they take place at a regular rate. This is illustrated by the phenomenon known as *second-order beats*, to be described in Section 8.12. Thus, in the dynamic case, phase can have an appreciable effect on timbre.

8.12 ■ BEATS OF MISTUNED CONSONANCES

A sensation of beats occurs when the frequencies of two tones f_2 and f_1 are nearly, but not quite, in a simple ratio. If $f_2 = 2f_1 + \delta$, beats are heard at a frequency δ. In general, when $f_2 = \frac{n}{m} f_1 + \delta$, $m\delta$ beats occur per second. These are sometimes called *second-order beats*, and have been discussed by Helmholtz (1877), Plomp (1967), and many others. They are described by various observers as variations in timbre, changes in loudness of one or both constituent tones, etc., and they are quite prominent even at low sound levels. They are clearly visible on an oscilloscope as periodic changes in pattern of the type shown in Fig. 8.20. These changes in pattern occur in synchronism with the beat sensations heard. Second-order beats between mistuned consonances can be heard up to about 1500 Hz or so, depending on the sound level.

In the case of a mistuned octave, we observe envelope changes as large as 13% that correspond to variations in sound level of about 1.3 dB, too small to account for the prominent beats that are observed, however. Apparently the beats of mistuned consonances are related to periodic variations of the waveform. The ear, which is a poor detector of static phase, appears to be quite sensitive to cyclical variations in phase. In the light of modern

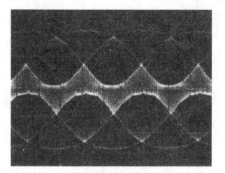

FIGURE 8.20
Oscilloscope
presentation of
beats of a mistuned
fifth (pure tones);
$f_1 = 332$ Hz,
$f_2 = 220$ Hz.

auditory theory, it would appear that the beats are due to the periodicity of nerve impulses. Nerve impulses are evoked when the displacement of the basilar membrane passes a certain critical value, and thus slow variations in the waveform, corresponding to beats, result in slow variations in the time pattern of the impulses.

Beats of mistuned consonances have long been used by piano tuners, for example, to tune fifths and fourths or even octaves on the piano. Violinists also make use of them in tuning their instruments.

When the two interacting tones include harmonics of the fundamental, second-order beats generally are stronger than when they are pure tones. Most musical tones have harmonics, and in this case ordinary first-order beats can occur between the various harmonics. For example, a mistuned fifth consisting of tones of 220 and 332 Hz, the third harmonic of f_1 and the second harmonic of f_2 have frequencies of 660 and 664 Hz, and they will produce four beats per second. However, the beats between complex tones do not always appear louder than between pure tones of the same frequency (Rossing and Dols 1976); this phenomenon needs further study.

8.13 ■ THE CENTRAL NERVOUS SYSTEM: AUTOCORRELATION AND CROSS-CORRELATION

In Chapters 5–8, we discussed many aspects of the modern theory of hearing. We have not described how auditory data is processed in the central nervous system, however. In Sections 8.13 and 8.14, only a brief summary will be given of this rather complex subject, which has been the focus of much recent research in psychoacoustics.

The basic building block of the nervous system is the *neuron* or nerve cell, which both transmits and processes neural impulses. The neuron has receptors called *dendrites*, which receive information from other neurons, and *axons*, which transmit that information to selected neurons. The "wiring scheme" among our ten billion neurons, which determines which neurons receive pulses from which others, is the key to human intelligence. Although much of the wiring scheme is fixed, in the cerebral cortex the interconnection of neurons is the result of repeated stimulation patterns, which leads to *learning*.

It is possible to attach very small electrodes to nerve fibers and thus observe the rate of neural impulses. Some neurons generate pulses spontaneously at a certain rate that can be either increased or decreased in accordance with the pulses received from other neurons. When several nerve fibers receiving stimuli from one region of the basilar membrane are tied together, the sum of their impulses constitutes a *volley*, which is synchronous with the auditory stimulus.

A given auditory nerve fiber carries two types of information (Roederer 1975):

1. The fact that it is firing at all indicates that the basilar membrane has been excited at a particular place; this gives information on primary pitch at all frequencies (*place theory*).
2. The time of distribution of impulses carries information on repetition rate or periodicity (*periodicity pitch*) and possibly on the vibration pattern itself; this works only in the lower-frequency range.

Two very important processing functions that the nervous system appears able to perform are autocorrelation and cross-correlation. *Autocorrelation* is the comparison of a pulse train with previous pulse trains in order to pick out repetitive features. Autocorrelation could account for the pitch of delayed noise in Section 7.6, for example. *Cross-correlation*, on the other hand, describes a comparison between signals on two different nerve fibers (from the cochleas of our two ears, for instance). Cross-correlation could account for the localization of sound at low frequencies by measuring the time delay between signals from our two ears, for example.

8.14 ■ CEREBRAL DOMINANCE

Information from our left ear passes preferentially to the right side of the brain (right cerebral cortex) and information from our right ear to the left side. (This is consistent with the bilateral symmetry exhibited in nearly all our sensory and motor processing.) There are, of course, pathways from right ear to right cerebral cortex and strong ties between the right and left cortexes, so that both sides of the brain receive processed information from both ears.

Clinical and experimental evidence has shown that the dominant hemisphere of the brain (the left side in 97% of the population) is specialized for speech processing and the minor hemisphere for nonlinguistic functions such as music. This is apparently related to the way in which most people process speech and music. Speech processing requires analytic and serial processing of incoming information for which the dominant hemisphere is best suited. Most aspects of musical perception, on the other hand, require holistic or synthetic processing, which is better done in the minor hemisphere.

Recognition of melodies apparently requires some of each. It has been found that musically experienced listeners recognize melodies better in their right ear than their left, while the reverse is true for nonmusicians (Bever and Chiarello 1974). This suggests that musicians learn to process melodies as they do speech in the dominant hemisphere. Patients with severe traumatic speech impediments are sometimes able to sing a song that had been learned before the trauma had arisen, even though they could not speak the same words.

8.15 ■ SUMMARY

Linear superposition of two pure tones with nearly the same frequency produces *beats*, or variations in loudness at the frequency difference. As the frequency difference increases, a sensation of roughness or *dissonance* develops, finally giving way to *consonance* when the interval between the tones exceeds the critical bandwidth.

Combination tones and aural harmonics are indicative of the nonlinear behavior of the auditory system. The most audible combination tones are the difference tone ($f_2 - f_1$) and the cubic difference tone ($2f_1 - f_2$). Differences tones are important in music.

Musical intervals appear to be consonant when their frequency ratios are simple numbers or when the differences between the frequencies of most of the various harmonics exceed the critical bandwidth. Maximum dissonance occurs when frequency differences are as small as one-fourth the critical bandwidth. Skilled use of dissonant intervals adds interest to musical composition and performance. Beats between slightly mistuned consonances are evidence of the ear's ability to detect changes in waveform.

Although the ear is a relatively poor detector of phase, the relative phase of various harmonics can affect the timbre of musical sound. Cyclical variations in pattern or phase, which occur in the case of mistuned consonances, can be heard as beats. The central nervous system makes extensive use of both autocorrelation and cross-correlation in processing sound.

REFERENCES AND SUGGESTED READINGS

Bever, T. G., and R. J. Chiarello (1974). "Cerebral Dominance in Musicians and Non-Musicians," *Science* **185**: 537.

Clack, T. D. (1977). "Growth of the Second and Third Aural Harmonics of 500 Hz," *J. Acoust. Soc. Am.* **62**: 1060.

Fletcher, H. (1929). *Speech and Hearing*. New York: Van Nostrand.

Galileo Galilei (1638). *Dialogues Concerning Two New Sciences*, Trans., H. Crew and A. de Salvio. New York: Dover, 1952.

Goldstein, J. L. (1967). "Auditory Nonlinearity," *J. Acoust. Soc. Am.* **41**: 676.

Hall, D. E. (1981). "The Difference Between Difference Tones and Rapid Beats," *Am. J. Phys.* **49**: 632.

Helmholtz, H. L. F. von (1877). *On the Sensations of Tone as a Physiological Basis for the Theory of Music*, 4th ed. Trans., A. J. Ellis. New York: Dover, 1954.

Hindemith, P. (1937). *The Craft of Musical Composition*. Trans., A. Mendel. New York: Associated Music Publishers, 1945.

Houtsma, A. J. M., and J. L. Goldstein (1972). "Perception of Musical Intervals: Evidence for the Central Origin of the Pitch of Complex Tones," *J. Acoust. Soc. Am.* **51**: 520.

Houtsma, A. J. M, T. D. Rossing, and W. M. Wagenaars (1987). *Auditory Demonstrations* (Philips Compact Disc #1126–061 and text). Melville, N. Y.: Acoustical Society of America.

Kameoka, A., and M. Kuriyagawa (1969). "Consonance Theory I: Consonance of Dyads," *J. Acoust. Soc. Am.* **45**: 1451.

Licklider, J. C. R. (1959). "Three Auditory Theories," in *Psychology: A Study in Science*, Vol. 1. Ed., S. Koch. New York: McGraw-Hill.

Oster, G. (1973). "Auditory Beats in the Brain," *Sci. Am.* **229**(4): 94.

Patterson, R. D. (1973). "The Effects of Relative Phase and the Number of Components on Residue Pitch," *J. Acoust. Soc. Am.* **53**: 1565.

Perrott, D. R., and M. A. Nelson (1969). "Limits for the Detection of Binaural Beats," *J. Acoust. Soc. Am.* **46**: 1477.

Plomp, R. (1965). "Detectability Threshold for Combination Tones," *J. Acoust. Soc. Am.* **37**: 1110.

Plomp, R. (1967). "Beats of Mistuned Consonances," *J. Acoust. Soc. Am.* **42**: 462.

Plomp, R. (1970). "Timbre as a Multidimensional Attribute of Complex Tones," in *Frequency Analysis and Periodicity Detection in Hearing*. Eds., R. Plomp and G. Smoorenburg. Leiden: Sijthoff.

Plomp, R. (1976). *Aspects of Tone Sensation*. London: Academic Press.

Plomp, R., and W. J. M. Levelt (1965). "Tonal Consonance and Critical Bandwidth," *J. Acoust. Soc. Am.* **38**: 548.

Plomp, R., and H. J. M. Steeneken (1969). "Effect of Phase on the Timbre of Complex Tones," *J. Acoust. Soc. Am.* **46**: 409.

Roberts, L. A. (1986). "Consonance Judgments of Musical Chords by Musicians and Untrained Listeners," *Acustica* **62**: 163.

Roederer, J. G. (1975). *Introduction to the Physics and Psychophysics of Music*, 2nd ed. New York: Springer-Verlag.

Rossing, T. D. (1974). "Subjective Tones from a Toy Whistle," *Am. J. Physics* **42**: 616.

Rossing, T. D. (1979). "Physics and Psychophysics of High-fidelity Sound," *Phys. Teach.* **17**: 563.

Rossing, T. D., and K. Dols (1976). "Second-Order Beats Between Complex Tones," *Catgut Acoust. Soc. Newsletter* **25**: 17.

Smoorenburg, G. F. (1972). "Audibility Region of Combination Tones," *J. Acoust. Soc. Am.* **52**: 603; "Combination Tones and Their Origin," *J. Acoust. Soc. Am.* **52**: 615.

Stickney, S. E., and T. J. Englert (1975). "The Ghost Flute," *Phys. Teach.* **13**: 518.

Terhardt, E. (1974). "Pitch, Consonance and Harmony," *J. Acoust. Soc. Am.* **55**: 1061.

Wightman, F. L., and D. M. Green (1974). "The Perception of Pitch," *Am. Scientist* **62**: 208.

Zwicker, E. (1955). "Der ungewöhnliche Amplitudengang der nichtlinearen Verzerrungen des Ohres," *Acustica* **5**: 67.

GLOSSARY

aural harmonic A harmonic that is generated in the auditory system.

autocorrelation The comparison of a signal with a previous signal in order to pick out repetitive features.

axon That part of a neuron or nerve cell that transmits neural pulses to other neurons.

beats Periodic variations in amplitude that result from the superposition or addition of two tones with nearly the same frequency.

combination tones A secondary tone heard when two primary tones are received. Combination tones are usually difference tones.

consonance Tones presented together with a minimum of roughness.

critical band The range of frequencies over which tones simply add in loudness; the critical bandwidth appears to determine consonance or dissonance.

cross-correlation The comparison of two signals to pick out common features.

dendrite That part of a neuron that receives neural pulses from other neurons.

difference tone When two tones having frequencies f_1 and f_2 are sounded together a difference tone with frequency $f_2 - f_1$ is often heard. (Properly this should be called the quadratic difference tone to distinguish it from the cubic and other difference tones.)

dissonance Roughness that results when tones with appropriate frequency difference are presented simultaneously.

linear superposition Addition of two waves applied simultaneously to a linear system.

major triad A chord of three notes having intervals of a major third and a minor third, respectively (as C : E : G).

minor triad A chord of three notes having intervals of a minor third and a major third, respectively (as C : E$^\flat$: G).

musical staff (pl: **staves**) A five-line graph on which musical notes are written. A clef sign shows the exact location of some particular note.

modulate To change some parameter (usually amplitude or frequency) of one signal in proportion to a second signal.

neuron, or nerve cell Building block of the nervous system that both transmits and processes neural pulses.

phase The fractional part of a period through which a waveform has passed. Phase is often expressed as an angle that is an appropriate fraction of 360°.

phase difference The difference in phase angle between two simple harmonic motions or waves. (If the phase difference is zero, they are in phase; if it is 180°, they are in opposite phase.)

second-order beats Beats between two tones whose frequencies are nearly but not quite in a simple ratio; also called **beats between mistuned consonances**.

sidebands Sum and difference tones generated during modulation.

sine wave A waveform that is characteristic of a pure tone (that is, a tone without harmonics or overtones) and also simple harmonic motion.

REVIEW QUESTIONS

1. What is a linear system?

2. Describe the motion of the shadow of a crank on a rotating wheel.

3. When two simple harmonic motions with frequency f are added, what is the nature of the resulting motion?

4. What is the maximum number of beats between two sounds that can be heard?

5. In what way is a musical staff semilogarithmic graph paper?

6. What are the frequencies of the most prominent difference tones when pure tones f_1 and f_2 are sounded together?

7. What are the two most consonant musical intervals?

8. What are second-order beats?

9. Changing the relative phase of high- and low-frequency components of sound (by moving the tweeter, for example) may change the timbre heard by the listener. What connection does this have to distortion in the sound system?

10. What is autocorrelation?

11. In what part of the brain are signals from the left ear processed?

12. Learning takes place when neurons become interconnected by repeated stimulation patterns. (T or F)

QUESTIONS FOR THOUGHT AND DISCUSSION

1. Is there any technical difference between 4/4 time with $\quarternote = 96$ and 2/2 time (also designated as *cut time* $\mathbf{\phi}$) with $\halfnote = 48$? Is there any implied difference in mood, interpretation, etc.?

2. The lowest fifth on a piano is from A_0 (27.5 Hz) to E_1 (41.2 Hz). By applying the criteria for roughness of Plomp and Levelt, show why it sounds less pleasant than the fifth from A_4 (440 Hz) to E_5 (660 Hz), for example.

3. Pianos are tuned to a *tempered scale*, for which $A_4 = 440$ Hz and $E_5 = 659$ Hz, whereas a *perfect fifth* would require $A_4 = 440$ Hz and $E_5 = 660$ Hz. Describe how a piano tuner could first set A to exactly 440 Hz by using a 440-Hz tuning fork and then set E_5 to 659 Hz by listening for second-order beats. How many beats should the tuner hear? How can he or she be sure that he or she has not set E to 661 Hz rather than 659 Hz?

4. The strings of a violin are tuned at intervals of fifths. Is it more likely that they will be tuned to perfect fifths or to the fifths found on a piano? (You may wish to discuss this with a violinist.)

EXERCISES

1. Make a linear superposition of a square-wave motion and a simple harmonic (sine-wave) motion with half the amplitude and twice the frequency of the square wave. This can be done by making a graph of each waveform to scale and then adding the ordinates point by point. (Alternatively, of course, it could be done by computation.)

2. Pure tones with frequencies of 440 and 448 Hz are sounded together. Describe what is heard (pitch of the fused tone, frequency of beats). Do the same for tones with frequencies of 440 and 432 Hz.

3. Verify each of the six amplitudes in Table 8.1 when $A = 3$ and $B = 2$ by drawing to scale vectors A and B (with correct angles between them) to represent the two harmonic motions, and a vector $A + B$ to represent the resultant. Compare the lengths of the resultant to those calculated from Table 8.1.

4. Calculate the first three difference frequencies that result from $f_1 = 900$ Hz and $f_2 = 1000$ Hz.

5. Make a list of possible subjective tones (combination tones, aural harmonics) that might result when tones of 240 and 300 Hz are sounded together. Indicate the most prominent tones by an asterisk (*). Enclose in parentheses those that occur only for loud tones.

6. Square waves of 200 and 301 Hz are sounded together. How many beats are heard? Write out the harmonics of each tone and indicate harmonics that might beat with harmonics of the other tone. Do the same for sawtooth waves at these frequencies. Can second-order beats be explained as beats between harmonics of the two tones?

7. Using frequencies of the first three notes in Fig. 8.11 ($B_2^\flat$, $B_5^\flat$, and C_6) as given in Table 9.2, verify that the lowest note nearly corresponds to the difference tone produced by the upper two. What is the melody shown in Fig. 8.11?

8. When an AM radio station amplitude-modulates a carrier wave of 800 kHz with a single pure tone of 1000 Hz, what are the principal sideband frequencies?

9. Redraw Fig. 8.9 with a linear frequency scale.

EXPERIMENTS FOR HOME, LABORATORY, AND CLASSROOM DEMONSTRATION

Home and Classroom Demonstration

1. *Simple harmonic motion* Project the shadow of a crank on a rotating wheel on the wall and on a mass-spring oscillator as shown in Fig. 8.3.

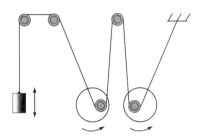

2. *Combination of two simple harmonic motions* A mechanical device for showing the combination of two simple harmonic motions is shown above. The wheels can be rotated at the same rate to illustrate the resultant amplitudes in Table 8.1 or they can be turned at slightly different rates to show beats.

3. *Beats* Connect the outputs of two audio generators in series (if the ground terminals are internally connected to the power-line ground, you will have to use a mixer) to the input of an audio amplifier and loudspeaker and to an oscilloscope as well. Vary the frequency of one of them and note the beats that you hear and see. Illustrate the three critical frequencies in Fig. 8.7.

4. *Difference tones* Leaving one of the audio generators in Demonstration 3 at a constant frequency, vary the other over a range wide enough to track both the difference tone $f_2 - f_1$ and the cubic difference tone $2f_1 - f_2$. One should rise in frequency while the other falls; they cross when $f_2 = 1.5f_1$. Do the demonstration both with the sum of the two tones versus time displayed on an oscilloscope and with one voltage versus the other to obtain a Lissajous pattern.

5. *Difference tones from a whistle* A two-tone whistle, such as a London police whistle, produces an audible difference tone. A three-tone whistle (from a toy store or a box of breakfast cereal) produces two difference tones, which can be heard to beat with one another (see Section 8.6). A four-tone model train whistle produces several difference tones, but they are difficult to hear.

6. *Difference tones between two flutes* Two piccolos (or flutes), two recorders, or two sopranos can produce a difference tone melody (see Fig. 8.11).

7. *Primary and secondary beats* Demonstration 32 on the *Auditory Demonstrations* CD. Tones of 1000 and 1004 Hz are presented together to give (primary) beats at a 4-Hz rate. Tones of 2004 Hz, 1502 Hz, and 1334.67 Hz are combined with a 1000-Hz tone to give second-order (secondary) beats, also at a 4-Hz rate.

8. *Aural combination tones* Demonstration 34 on the *Auditory Demonstrations* CD. Tones of 1000 and 1200 Hz are presented. When an 804-Hz probe tone is added, it beats with the 800-Hz combination (difference) tone. When the frequency of the higher tone is increased from 1200 to 1600 Hz, the moving difference tone and cubic difference tone are heard.

9. *Effect of distortion on combination tones* Demonstration 33 on the *Auditory Demonstrations* CD. A 440-Hz tone distorted by a symmetrical compressor shows a strong third harmonic, whereas a 440-Hz tone distorted asymmetrically by a half-wave rectifier shows a strong second harmonic. A 440-Hz tone is added to its second harmonic with a phase varying from $-90°$ to $90°$. The phase change is practically inaudible until the tone is distorted.

10. *Binaural beats* Demonstration 36 on the *Auditory Demonstrations* CD. (This demonstration must be heard through headphones.) When sinusoidal (pure) tones of slightly different frequency are presented to the two ears, binaural beats are sometimes heard.

Laboratory Experiments

Musical intervals, beats, and combination tones (Experiment 16 in *Acoustics Laboratory Experiments*)

9 Musical Scales and Temperament

This chapter describes musical scales, tunings, and temperaments, including a little about their mathematical basis and their history. It should be of interest not only to those seriously interested in music theory, history, and performance, but to serious music listeners as well. Computers and music synthesizers now allow us to explore any kind of scale and tuning imaginable, and considerable interest in scales and tuning has been aroused.

A number of musical terms are used, but we have attempted to explain them as we go and to keep the mathematics simple, so that it can be readily understood by readers regardless of their musical or mathematical experience.

In this chapter you should learn:

- About musical scales;
- About tunings and temperaments;
- About musical and mathematical bases for these scales and temperaments;
- Advantages and disadvantages of various scales.

The word *scale* is derived from a Latin word (*scala*) meaning "ladder" or "staircase." A musical scale is a succession of notes arranged in ascending or descending order. Most musical composition is based on scales, the most common ones being those with five notes (*pentatonic*), twelve notes (*chromatic*), or seven notes (major and minor *diatonic*, Dorian and Lydian modes, etc.). Western music divides the octave into 12 steps called *semitones*. All the semitones in an octave constitute a *chromatic scale* or 12-tone scale. However, most music makes use of a scale of seven selected notes, designated as either a *major* scale or a *minor* scale and carrying the note name of the lowest note. For example, the C-major scale is played on the piano by beginning with any C and playing white keys until another C is reached.

Other musical cultures use different scales. The pentatonic or five-tone scale, for example, is basic to Chinese music and also appears in Celtic and Native American music. (The familiar old Scottish melody "Auld Lang Syne," for example, uses a pentatonic scale.) A pentatonic scale can be played on the piano by beginning with C♯ and playing only black keys, or by playing the notes C, D, F, G, A, C. The scales of Indian music are often said to be quite complex because of the abundance of microtonal intervals. However, Benade (1976) points out that Indian music is based on a seven-tone scale quite similar to our own major scale, and that the microtonal decorations are a matter of style, not unlike that of the American jazz player.

9.1 ■ SCALE, TUNING, TEMPERAMENT, AND INTONATION

Before proceeding to our discussion of scales, we should distinguish between the following terms, which are sometimes confused: scale, tuning, temperament, and intonation. We have already defined a *scale* as a succession of notes arranged in ascending or descending order. In Chapter 7 we learned that although pitch can be expressed (in mels) on a psychophysical scale, we generally use a musical scale. There are many ways to construct musical scales. The construction of scales has fascinated mathematicians as well as musicians since the time of the Greeks. We will discuss four important scales: the Pythagorean scale, the just scale, and the scales of meantone and equal temperament.

A *tuning* may be defined as an adjustment of pitch in any instrument so that it corresponds to an accepted norm. That norm will generally be one of the accepted musical scales, but contemporary composers experiment with other tunings as well. *Temperament* is a system of tuning in which the intervals deviate from acoustically pure (Pythagorean) intervals, whereas *intonation* refers to the degree of accuracy with which pitches are produced.

On the standard keyboard in Fig. 9.1, successive keys produce tones that are a semitone apart. The white keys are labeled with letters A–G, and the black keys are denoted by a letter followed by ♯ (sharp) or ♭ (flat). Each note has a frequency that is approximately 6% greater than the one below it; in other words, the frequencies follow a logarithmic scale, not a linear one (see Fig. 5.13). Notes are represented on a musical staff, so two equal frequency ratios represent two equal distances on the logarithmic scale of frequency. Intervals on a musical scale represent frequency *ratios*, so when you go up or down by a musical interval, you multiply or divide frequency by the corresponding number.

Touching a guitar or violin string at its center causes the frequency to double and the pitch to rise one octave (see Fig. 4.5). Nearly every listener would agree that the new pitch bears a close similarity to the old. This close similarity is one reason that the octave has such a special place in music. Another reason is that two notes an octave apart sound very harmonious or consonant when sounded together (see Section 8.10). The principle of *octave equivalence*, well known to musicians, is based on the fact that the octave comprises only the even harmonics, so a given tone with many harmonics already contains the octave as a subset of its partials. It is not surprising that the octave appears in nearly all musical cultures. In a sense, we could say that the different musical scales are different ways of dividing up the octave.

FIGURE 9.1
Standard keyboard. White keys are labeled with letters from A–G, and black keys are denoted by a letter followed by ♯ (sharp) or ♭ (flat).

9.2 ■ THE PYTHAGOREAN SCALE

Tradition has it that Pythagoras was born about 580 B.C. on the island of Samos, although he traveled and studied at all the Greek centers of learning around the Mediterranean. He established a school at Crotona where he taught arithmetic, geometry, astronomy, and music. It took in women students as well as men, which was remarkably progressive for that time (Johnston 1989).

Although no written records exist, Pythagoras is given credit for learning through experimentation that dividing a string into halves, thirds, or fourths yielded segments that vibrated and emitted harmonious tones. We now know that if one string is divided into halves and an identical string is divided into thirds, the two strings will vibrate with frequencies in the ratio 3 : 2. We call this interval a *perfect fifth*, and it is second only to the octave in consonance (see Section 8.10). Similarly, the frequency ratio of a string divided into fourths to a string divided into thirds is 4 : 3 and the interval between them is a *perfect fourth*, another very consonant interval.

The Pythagorean scale is one that creates the greatest number of perfect fourths and fifths. In constructing the Pythagorean scale based on fourths and fifths, two facts become apparent:

1. An octave is a fourth plus a fifth ($3/2 \times 4/3 = 2$); recall that in adding intervals, we multiply the corresponding frequency ratios. Therefore, going up a fourth leads to the same letter as going down a fifth, and vice versa.
2. All notes on the scale (sharps and flats included) can be reached by going up or down 12 successive fifths or 12 successive fourths.

The second of these facts can be represented on the *circle of fifths* shown in Fig. 9.2. Beginning at C, one can follow the outer circle by going up a fifth (or down a fourth) at each step. Twelve such steps bring one back to C... almost. If 3/2 is multiplied by itself 12 times, one obtains 129.7, whereas $2^7 = 128$. This means that going up 12 perfect fifths takes one up 7 octaves plus one-fourth of a semitone extra. By the same token, going down

FIGURE 9.2

The circle of fifths. The outer circle visits all 12 notes on the chromatic scale by going up by fifths (or down by fourths). The inner circle goes down by fifths (or up by fourths).

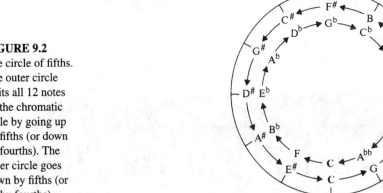

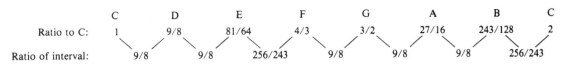

FIGURE 9.3 Frequency ratios of notes in the Pythagorean scale. Numbers in the bottom row give intervals between two adjacent notes.

12 fifths (or up 12 fourths), one follows the note names along the inner circle, encounters flats instead of sharps, and arrives back at C one-fourth of a semitone flat.

To determine the notes of the C-major scale, we can see clearly from Fig. 9.2 that we should go down a fifth (or up a fourth) to determine F as 4/3 and then go up five successive fifths to determine the other five notes. This gives G = 3/2, D = 3/2 × 3/2 = 9/4, A = $(3/2)^3$ = 27/8, E = 81/16, and B = 243/32. Putting them into the proper octave gives the following frequency ratios:

$$\begin{array}{cccccccc} C & D & E & F & G & A & B & C \\ 1 & 9/8 & 81/64 & 4/3 & 3/2 & 27/16 & 243/128 & 2 \end{array}$$

(Note that if we had omitted the last two fifths, E and B would not appear, and we would have the pentatonic scale C–D–F–G–A–C with appropriate intervals.)

Next the intervals between successive notes are calculated in Fig. 9.3. Note that there are two different intervals: whole tones and semitones.

> Continuing around the circle of fifths to obtain the frequencies of the flats and sharps leads to another interval. The ratio of F$^\sharp$ to F turns out to be 2187/2048 = 1.068, which is called a *chromatic semitone*. It is larger than the diatonic semitone (256/243 = 1.053) previously determined. So the chromatic Pythagorean scale has one size of whole tone but two different semitones. The ratio of the chromatic to the diatonic semitone is 1.0136, which defines an interval called the *Pythagorean comma*. This is the same interval between 12 fifths and 7 octaves noted on the circle of fifths. It is also the interval between pairs of enharmonic notes (e.g., A$^\flat$ and G$^\sharp$) facing each other on the inner and outer circles.

The great advantage of the Pythagorean scale is the emphasis on perfect fifths and fourths. A great disadvantage, however, is the poor tuning of thirds. The major thirds exceed the ratio 5 : 4 by a ratio of 1.0125, and the minor thirds are smaller than the ratio 6 : 5 by the same ratio. This ratio is called the *syntonic comma*. Pythagorean thirds sound quite out of tune. Nevertheless, studies have shown that many concert violinists tend to favor Pythagorean intonation in their performance, which points out the great importance of fifths and fourths in music.

EXAMPLE 9.1 Compare the frequency ratios of the major thirds C : E, F : A, and G : B in the Pythagorean scale.

Solution

$$C : E \quad \frac{81}{64} \div 1 = \frac{81}{64}$$

$$F : A \quad \frac{27}{16} \div \frac{4}{3} = \frac{81}{64}$$

$$G : B \quad \frac{243}{128} \div \frac{3}{2} = \frac{81}{64}$$

They are the same.

9.3 ■ MEANTONE TEMPERAMENT

Because Pythagorean thirds sound out of tune, numerous alterations to the Pythagorean scale have been developed. Nearly all of them flat the third (E in the C-major scale) so that the major third (C to E) and minor third (E to G) are close to the corresponding just intervals, to be discussed in Section 9.4. Similar adjustments are made in other notes. Such compromises form the bases of *meantone temperament* and its variants.

Meantone temperament (also called *quarter-comma meantone temperament*) raises or lowers various notes by 1/4, 1/2, 3/4, or 5/4 of the syntonic comma δ, as shown in Fig. 9.4.* (The syntonic comma, you will recall from Section 9.2, is the amount by which the major third and minor third differ from the ratios 5/4 and 6/5.) Note that the fourths and fifths are no longer perfect intervals, although they are not too far from it. The real problems with meantone temperament, however, occur when playing instruments in keys with many sharps or flats. If the instrument is set to meantone temperament in C, it becomes increasingly out of tune as sharps and flats are added to the key signature.

Pythagorean:	C	D	E	F	G	A	B	C
Meantone:	C	D − 1/2 δ	E − δ	F + 1/4 δ	G − 1/4 δ	A − 3/4 δ	B − 5/4 δ	C

FIGURE 9.4 Quarter-comma meantone temperament. The diatonic scale is compared to the Pythagorean. δ is the syntonic comma and represents a frequency ratio of 1.0125.

9.4 ■ THE SCALE OF JUST INTONATION

The *scale of just intonation* (or just diatonic scale) is based on the *major triad*, a group of three notes that sound particularly harmonious (for example, C : E : G). The notes of the major triad are spaced in two intervals: a major third (C : E) and a minor third (E : G). When these intervals are made as consonant as possible, the notes in the major triad are found to have frequencies in the ratios 4 : 5 : 6.

The psychophysical basis of consonance and dissonance was discussed in Section 8.9, where the most consonant musical intervals were given as:

*Raising a note by $\frac{1}{4}$ δ means the frequency is multiplied by $(1.0125)^{1/4}$.

$$
\begin{array}{ll}
2:1 & \text{octave,} \\
3:2 & \text{perfect fifth,} \\
4:3 & \text{perfect fourth,} \\
5:3 & \text{major sixth,} \\
5:4 & \text{major third,} \\
8:5 & \text{minor sixth,} \\
6:5 & \text{minor third.}
\end{array}
$$

The numbers above are the frequency ratios of the just intervals. These intervals are consonant because the dissonant combinations between harmonics of the two tones are minimal, especially in the case of the so-called perfect intervals (octave, fifth, and fourth), which have the simplest ratios.

The three major triads in a major scale are the tonic, subdominant, and dominant chords (also called the I, IV, and V chords, since they are built on the first, fourth, and fifth notes of the major scale). The frequencies of all seven notes in the just diatonic scale can be determined by letting these three triads consist of notes with the frequency ratios 4 : 5 : 6.

Frequency Ratios in the Just Scale

This may be illustrated in the key of C as follows. First we let the notes of the tonic chord (C, E, G) have the ratios 4 : 5 : 6 and set C = 1; we have now determined C = 1, E = 5/4, G = 6/4 = 3/2. Next we let the notes of the dominant chord (G, B, D) be in the ratio 4 : 5 : 6; this determines B = 5/4 × 3/2 = 15/8 and D = 3/2 × 3/2 = 9/4. Dropping D down into the same octave as C makes D = 9/8. Now we require that the subdominant chord (F, A, C) likewise be in the ratio 4 : 5 : 6; in this case we work down from C an octave up (C = 2), and obtain F = 2 ÷ 3/2 = 4/3 and A = 2 ÷ 6/5 = 5/3. Putting this all together, we obtain the frequency ratios for the just scale in C-major:

$$
\begin{array}{ccccccc}
C & D & E & F & G & A & B & C \\
1 & 9/8 & 5/4 & 4/3 & 3/2 & 5/3 & 15/8 & 2
\end{array}
$$

Note that E is major third above C, F is a perfect fourth above C, and G is a perfect fifth above C. Next consider the intervals between successive notes: D to E is 5/4 ÷ 9/8 = 10/9; E to F is 4/3 ÷ 5/4 = 16/15; G to F is 3/2 ÷ 4/3 = 9/8, and so on. In fact, if we write all the ratios as in Fig. 9.5, we observe that there are only three different intervals and they have ratios 9/8, 10/9, and 16/15. The interval

---

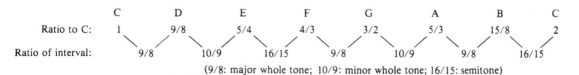

(9/8: major whole tone; 10/9: minor whole tone; 16/15: semitone)

FIGURE 9.5   Frequency ratios of notes in the just diatonic scale. Numbers in the bottom row give intervals between two adjacent notes.

corresponding to 9/8 is called a *major whole tone*, the 10/9 interval, a *minor whole tone*, and the 16/15 interval, a *semitone*. In the just scale, there are three major whole tones, two minor whole tones, and two semitones.

Besides the three major triads, the just scale has two triads (E, G, B and A, C, E) with frequencies in the ratio 10 : 12 : 15 that are called *minor triads*. Minor triads, like major triads, include one major and one minor third, but with the order reversed. That is, the lower interval (12/10 = 6/5) is a minor third and the upper interval (15/12 = 5/4) is a major third. So far so good; there are three major triads and two minor triads, all in "just" intonation. But further examination reveals some shortcomings of the just scale.

Problems with the Just Scale

If the fifths in the just scale are examined, five are seen to be perfect, but the other one, D : A is imperfect (that is, the frequencies are not in the ratio 3 : 2). The same is true of the fourths in the scale; five are perfect but A : D is not.

Further difficulties appear when sharps and flats are added. Requiring that E to G♯ be a major third sets G♯ at 5/4 × 5/4 = 25/16, but requiring A♭ to C to be a major third sets A♭ at 2 ÷ 5/4 = 8/5. Thus A♭ should be a little higher in pitch than G♯ to maintain the justness of the above intervals. (A♭ and G♯ are called *enharmonic* notes; on the piano they are the same note.)

The just scale has never been of much practical use. Because of the two different whole-tone intervals (major and minor), retuning of an instrument would be required at each change of key. The interval C to D, for example, would have the ratio 9/8 in the key of C, but in the key of F, this interval would need to be tuned to the ratio of 10/9. Pipe organs have been built with complicated keyboards that would allow playing in just intonation in several keys, but few are in existence today. (For example, a keyboard devised by Salinas in the early eighteenth century had 24 notes to the octave, which made it possible to play all major triads in the circle of fifths from G♭ to G♯ and all the minor triads from E♭ to E♯ (Campbell and Greated 1987).) An orchestra composed of instruments with just intonation would approach musical chaos.

EXAMPLE 9.2   Find the frequency ratio for C : F♯ in the scale of just intonation by requiring that the triads D–F♯–A and B–D♯–F♯ have frequency ratios 4 : 5 : 6.

Solution   (a) In order to be a major third above D, F♯ must have the ratio 9/8 × 5/4 = 45/32 with C. (b) In order to be a perfect fifth above B, F♯ must have the ratio 15/16 × 3/2 = 45/32 with C.

9.5 ■ EQUAL TEMPERAMENT

The ultimate compromise is equal temperament, which makes all semitones the same. The *scale of equal temperament* (often referred to simply as the *tempered scale*) consists of five equal whole tones and two semitones; the whole tones are twice the size of the semitones. Twelve equal semitones make up an octave.

> To determine the frequency ratio of the interval that is exactly 1/12 of an octave is a mathematical rather than a musical exercise. The square root of 2 (written $\sqrt{2}$, or $2^{1/2}$) is a number that can be multiplied by itself to give the product 2. Likewise, the twelfth root of 2 (written $\sqrt[12]{2}$, or $2^{1/12}$) is a number that multiplied by itself 12 times will give the product 2. Many pocket calculators have a y^x key; if you have access to one, you can verify that $2^{1/12} = 1.05946$. This is the semitone of equal temperament. A whole tone is $(1.05946)^2$, or 1.12246. A fifth is 1.498 and a fourth is 1.335, both very close to the perfect (beatless) intervals 1.500 and 1.333. A major third is 1.260 and a minor third is 1.189, not very close to the just intervals 1.250 and 1.200, but not as far sharp and flat as the thirds on the Pythagorean scale.

Rather than deal with ratios, it is customary to compare tones by using cents. One *cent* is 1/100 of a semitone in equal temperament. Thus an octave is 1200 cents, a tempered fifth is 700 cents, and so forth. An equal *difference* in cents implies an equal *ratio* of frequencies. One cent has the ratio $2^{1/1200} = 1.000578$. When we use cents, the comparison between various scales becomes more meaningful. Such a comparison is made in Fig. 9.6.

Note the rather large differences in the third (E) and sixth (A) notes of the scale. A rather surprising feature is the great difference in the sharps and flats in the three scales. Whereas $C^\sharp$ lies above $D^\flat$ in the Pythagorean scale, for example, it lies well below $D^\flat$ in the just scale. A performer attempting to play in either scale might very well tend toward some meantone tuning of the sharps and flats.

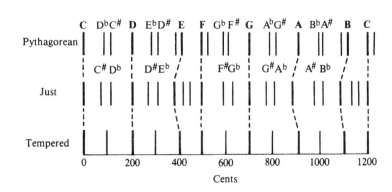

FIGURE 9.6
A comparison of Pythagorean, just, and equally tempered scales on a scale of cents (see Table 9.2).

Conversion from Cents to Frequency Ratio

To convert from an interval in cents $I(\cent)$ to a frequency ratio R:

$$R = 10^{(I \log 2)/1200}.$$ (9.1)

To convert from frequency ratio R to an interval in cents, $I(\cent)$:

$$I(\cent) = \frac{1200}{\log 2} \log R.$$ (9.2)

To convert 45¢ to a frequency ratio, for example, you would press the following keys on a typical calculator: 2, log, ÷, 1200, ×, 45, =, INV log (or 10^x); you should obtain 1.026.

Three tables of intervals and tunings follow. Table 9.1 compares the main intervals, and Table 9.2 has the frequencies of nine octaves of notes in equal temperament with A_4 at 440 Hz. Table 9.3 gives the tunings for all the notes of one octave.

TABLE 9.1   Musical intervals in various tunings

Interval	Equal tempered Ratio	Equal tempered Cents	Pythagorean Ratio	Pythagorean Cents	Meantone Ratio	Meantone Cents	Just Ratio	Just Cents
Octave	2.000	1200	2.000	1200	2.000	1200	$2/1 = 2.000$	1200
Fifth	1.498	700	1.500	702	1.495	697	$3/2 = 1.500$	702
Fourth	1.335	500	1.333	498	1.337	503	$4/3 = 1.333$	498
Major third	1.260	400	1.266	408	1.250	386	$5/4 = 1.250$	386
Minor third	1.189	300	1.185	294	1.196	310	$6/5 = 1.200$	316
Major sixth	1.682	900	1.687	906	1.672	890	$5/3 = 1.667$	884
Minor sixth	1.587	800	1.580	792	1.600	814	$8/5 = 1.600$	814

TABLE 9.2   Frequencies of notes in tempered scale

C_0	16.352	C_3	130.81	C_6	1046.5
	17.324		138.59		1108.7
D_0	18.354	D_3	146.83	D_6	1174.7
	19.445		155.56		1244.5
E_0	20.602	E_3	164.81	E_6	1318.5
F_0	21.827	F_3	174.61	F_6	1396.9
	23.125		185.00		1480.0
G_0	24.500	G_3	196.00	G_6	1568.0
	25.957		207.65		1661.2
A_0	27.500	A_3	220.00	A_6	1760.0
	29.135		233.08		1864.7
B_0	30.868	B_3	246.94	B_6	1975.5
C_1	32.703	C_4	261.63	C_7	2093.0
	34.648		277.18		2217.5
D_1	36.708	D_4	293.66	D_7	2349.3
	38.891		311.13		2489.0
E_1	41.203	E_4	329.63	E_7	2637.0
F_1	43.654	F_4	349.23	F_7	2793.8
	46.249		369.99		2960.0
G_1	48.999	G_4	392.00	G_7	3136.0
	51.913		415.30		3322.4
A_1	55.000	A_4	440.00	A_7	3520.0
	58.270		466.16		3729.3
B_1	61.735	B_4	493.88	B_7	3951.1
C_2	65.406	C_5	523.25	C_8	4186.0
	69.296		554.37		4434.9
D_2	73.416	D_5	587.33	D_8	4698.6
	77.782		622.25		4978.0
E_2	82.407	E_5	659.26	E_8	5274.0
F_2	87.307	F_5	698.46	F_8	5587.7
	92.499		739.99		5919.9
G_2	97.999	G_5	783.99	G_8	6271.9
	103.83		830.61		6644.9
A_2	110.00	A_5	880.00	A_8	7040.0
	116.54		932.33		7458.6
B_2	123.47	B_5	987.77	B_8	7902.1

TABLE 9.3   Notes of scales based on C

Note	Tempered Cents	Just Ratio	Just Cents	Pythagorean Ratio	Pythagorean Cents	Meantone Ratio	Meantone Cents
C	1200	2.000	1200	2.000	1200	2.000	1200
B$\sharp$	1200	1.953	1159	2.027	1223	1.953	1159
C$\flat$	1100	1.920	1129	1.873	1086	1.914	1124
B	1100	1.875	1088	1.898	1110	1.869	1083
B$\flat$	1000	1.800	1018	1.778	996	1.789	1007
A$\sharp$	1000	1.758	977	1.802	1020	1.747	966
A	900	1.667	884	1.688	906	1.672	890
A$\flat$	800	1.600	814	1.580	792	1.600	814
G$\sharp$	800	1.563	773	1.602	816	1.562	773
G	700	1.500	702	1.500	702	1.495	697
G$\flat$	600	1.440	631	1.405	588	1.431	621
F$\sharp$	600	1.406	590	1.424	612	1.398	579
F	500	1.333	498	1.333	498	1.337	503
E$\sharp$	500	1.302	457	1.352	522	1.306	462
F$\flat$	400	1.280	427	1.249	384	1.280	427
E	400	1.250	386	1.266	408	1.250	386
E$\flat$	300	1.200	316	1.185	294	1.196	310
D$\sharp$	300	1.172	275	1.201	318	1.168	269
D	200	1.125	204	1.125	204	1.118	193
D$\flat$	100	1.067	112	1.053	90	1.070	117
C$\sharp$	100	1.042	71	1.068	114	1.045	76
C	0	1.000	0	1.000	0	1.000	0

Tuning to Equal Temperament

The tuning of most keyboard instruments is based on equal temperament. The exact tuning of keyboard instruments, however, depends on how nearly harmonic the overtones are. Organs, which have harmonic overtones, can be tuned almost exactly to equal temperament; piano tuners take the inharmonicity of the strings into account, however. Pianos will sound better if the intervals are stretched, especially in the upper octaves; small upright pianos require a greater amount of stretch than concert grands do. This subject will be discussed in Chapter 14.

We now present a brief outline of a procedure for *laying the temperament*. The first step in tuning an instrument consists in tuning a series of intervals slightly sharp or flat so that each interval beats at a certain rate. This can be done by working around the circle of fifths shown in Fig. 9.2 and setting each one 2¢ (cents) flat (a frequency ratio of 1.4983 rather than 1.5000). Twelve such flatted fifths equal exactly 7 octaves, whereas 12 perfect fifths, you will recall, come to 7 octaves plus 23.5¢ (about one-fourth of a semitone).

Suppose that C_4 (middle C) is first set to exactly 261.63 Hz by means of a tuning fork (one can start with A_4 if a 440-Hz fork is more readily available). Then G_4 is set at 392.00 Hz by counting beats. Recall from Section 8.11 that in the case of the mistuned fifth, where $f_2 = (3/2)f_1 + \delta$, 2δ beats are heard each second. A little arithmetic shows that if $f_2/f_1 = 1.4983$, this leads to $\delta = 0.0017 f_1$, or $0.0034 f_1$ beats per second. With $f_1 = 261.63$ Hz, the rate of beats is 0.89 Hz (one beat each 1.12 s), so one might begin with a near-perfect fifth and lower it until 8.9 beats are heard each 10 s.

Next, D_5 would be tuned to give $(0.0034) \times (392) = 1.33$ beats per second (this beat rate can also be calculated as the difference between the second harmonic of D_5 and the third harmonic of G_4). After setting D_5, it is best to tune D_4 exactly an octave below and then proceed up a fifth to A_4. This procedure eventually tunes the entire octave C_4 to C_5 to equal temperament. The number of beats for each of the 12 intervals would be:

Interval	Beat frequency
C_4 to G_4	0.89
G_4 to D_5	1.33
D_4 to A_4	1.00
A_4 to E_5	1.50
E_4 to B_4	1.12
B_3 to $F_4^{\sharp}$	0.84
$F_4^{\sharp}$ to $C_5^{\sharp}$	1.26
$C_4^{\sharp}$ to $G_4^{\sharp}$	0.94
$G_4^{\sharp}$ to $D_5^{\sharp}$	1.41
$D_4^{\sharp}$ to $A_4^{\sharp}$	1.06
$A_4^{\sharp}$ to F_5	1.58
F_4 to C_5	1.19

If one is tuning an organ, the other octaves will probably be tuned to match the C_4 to C_5 octave. In the case of a piano, however, the octaves will no doubt be stretched, so the tuning procedure becomes more complicated.

9.6 ■ COMPARISON OF SCALES

Tunings of all notes (including enharmonics) in the scale of equal temperament, the just scale, and the Pythagorean scale are compared in Table 9.3. Cents above C are given for each note.

In comparing different scales, it is helpful to return to the circle of fifths. In the case of equal temperament, each tone is of the same size and exactly equal to two semitones, so the circle of fifths will close, as shown in Fig. 9.7(a). In meantone temperament (quarter-comma), there is only one size of whole tone and one size of semitone, so the fifth is 3.5 cents short of the fifth used in equal temperament. Therefore, the circle of fifths fails to close by $12 \times 3.5 = 42$ cents, as shown in Fig. 9.7(b). In Pythagorean tuning, the fifths are again equal but 2 cents greater than in equal temperament, so the circle overlaps itself by $12 \times 2 = 24$ cents, as shown in Fig. 9.7(c).

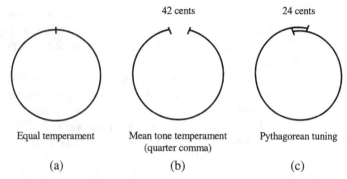

FIGURE 9.7   The circle of fifths showing (a) closure for equal temperament; (b) a gap for mean-tone temperament (quarter-comma); and (c) an overlap for Pythagorean tuning. (After Campbell and Greated 1987.)

9.7 ■ OTHER MEANTONE TEMPERAMENTS

The quarter-comma meantone temperament is the only true meantone temperament because it is derived from the mean of the major and minor tones having frequency ratios of 5/4 and 4/3. It is possible to have other meantone temperaments, however. An example is the 1/3-comma meantone temperament devised by Salinas (Barbour 1951):

Note:	C	D	E	F	G	A	B	C
Cents:	0	190	379	505	695	884	1074	1200

This is described as a *regular* temperament, because the whole tones and fifths both have fixed values. Equal temperament is regular; it could be called 1/11-comma meantone.

In equal temperament, the Pythagorean comma (which is approximately 24 cents) is equally distributed among the 12 fifths of the circle of fifths, so each fifth is contracted from 702 to 700 cents. It is possible to distribute the Pythagorean comma in different ways among the 12 fifths, and a number of such temperaments developed in the late seventeenth and eighteenth centuries. These are described as *circular* temperaments because the circle of fifths closes exactly (as it does for equal temperament). Barnes (1979) points out that Bach's 24 preludes and fugues for *The Well-Tempered Clavier* were probably written for a circular temperament similar to the one known as Werckmeister III. In this temperament, the C–G, G–D, D–A, and B–F♯ fifths are narrowed by 6 cents each. In this temperament, the keys with few sharps or flats sound better than those with many, though even the worst ones remain usable (Campbell and Greated 1987).

Characteristics of Tempered Scales

Equal temperament is generally considered to be the most convenient for pianos and other modern keyboard instruments because this is the only temperament that allows the complete freedom of musical modulation (gradual key change) or transposition (key substitution) demanded by modern music. However, there is a feeling among many musicians that music of earlier periods, particularly the Renaissance and Baroque, should be played on authentic instruments of the period with authentic tunings.

In equal temperament all keys should sound alike because the interval relationships between notes is unchanged in moving from one key to another. This is not the case with other temperaments, and many musicians through the ages have associated emotional characteristics with different keys. Beethoven described the key of $D^\flat$ major as "majestic" and C major as "triumphant." In the nineteenth century, equal temperament was strongly criticized by some musicians, but over the ages our ears have become more accustomed to equal temperament, just as they have to the more dissonant harmonies of contemporary music.

N-note Scales of Equal Temperament

Musicians and scientists have explored many interesting scales of equal temperament. We will mention only a few of those that divide the octave into a integral number N of equal steps. Some of the more interesting are those for which $N = $ 15, 19, 31, 22, and 53. Among the recorded compositions by composer Easley Blackwood (whose name is often associated with the bridge convention invented by his father, Easley Blackwood) are pieces for synthesizer and guitar in 15-note and 19-note tunings. Paul Rapoport has written six "Songs of Fruits and Vegetables" using 31-note temperament.

In 1691, astronomer Christiaan Huygens (pronounced "how-kens") described a 31-note system in his *Lettre touchant le cycle harmonique*. Two and a half centuries later, physicist Adrian Fokker read this and asked the Dutch organ builder Pels and van Leeuwen to build a 31-note-per-octave organ for the Teylers Museum in Haarlem. It is now usually called the Fokker organ and is heard regularly in concert. It has a main console with two 31-tone matrix keyboards and a pedal keyboard and an additional console with 12-tone keyboards on which portions of the 31-tone can be played. Pictures of the instrument and its unique keyboard can be seen at www.xs4all.nl/~huygensf/instrum.html.

9.8 ■ INTONATION

Research on the intervals played or sung by skilled musicians shows substantial variability in the intonation from performance to performance. These variations are frequently larger than the differences between the various tunings. Studies of large numbers of performances have shown that deviations from equal temperament are usually in the direction of Pythagorean intervals (Ward 1970).

Several investigators have found that performers tend to stretch their intervals (including octaves), even when unaccompanied by a piano. Many choral conductors prefer the third in a chord slightly raised, especially in sustained chords or cadences, to avoid any suggestion of "flatting" the chord, a particular nemesis of choirs. How much of the apparent preference for sharpened intervals is due to constant exposure to pianos with stretched tuning is difficult to determine.

9.9 ■ SUMMARY

Most musical composition makes use of scales. Four different ways to construct scales have been discussed; these lead to Pythagorean tuning, just tuning, meantone temperament, and equal temperament. Each has its own advantages and disadvantages:

Pythagorean tuning is based on perfect (beat-free) fifths and fourths which are preferred for tuning string instruments. An instrument tuned in this way can play well in many keys, but thirds and sixths tend to sound bad, so composers avoid them.

Just tuning (intonation) has beat-free thirds and sixths as well as fifths and fourths, which is what musicians strive for if they can adjust the notes on their instruments. A fixed-pitch keyboard instrument plays well in a few keys but sounds out of tune in many other keys.

Meantone temperament adjusts the Pythagorean scale so that thirds and sixths sound much better while fifths and fourths deviate slightly from perfect (beat-free) ratios. When optimized in C, meantone temperament works well in keys without too many sharps and flats but deteriorates as more sharps and flats are added.

Equal temperament divides the octave into 12 equal semitone intervals. Fourths and fifths are quite good, but thirds and sixths are wider than the beat-free intervals on the just scale. Modulation from key to key is easy.

REFERENCES AND SUGGESTED READINGS

Backus, J. (1969). *The Acoustical Foundations of Music*. New York: Norton. (See Chapter 8.)

Barbour, J. M. (1951). *Tuning and Temperament*. East Lansing: Michigan State College Press.

Barnes, J. (1979). "Bach's keyboard temperament," *Early Music* **7**: 236.

Benade, A. H. (1976). *Fundamentals of Musical Acoustics*. New York: Oxford University Press.

Burns, E. M., and W. D. Ward (1982). "Intervals, Scales, and Tuning," in *The Psychology of Music*. Ed. D. Deutsch. New York: Academic Press.

Campbell, M., and C. Greated (1987). *The Musician's Guide to Acoustics*. New York: Schirmer Books. (See Chapter 4.)

Houtsma, A. J. M., T. D. Rossing, and W. M. Wagenaars (1987). *Auditory Demonstrations* (Philips Compact Disc # 1126-061 and text). Melville, N.Y.: Acoustical Society of America.

Johnston, I. (1989). *Measured Tones*. Bristol: Adam Hilger.

Martin, D. W. (1962). "Musical Scales Since Pythagoras," *Sound* **1**: 3, 22.

Ward, W. D. (1970). "Musical Perception," in *Foundations of Modern Auditory Theory*. Ed. J. Tobias. New York: Academic.

White, B. W. (1943). "Mean-Tone Temperament," *J. Acoust. Soc. Am.* **15**: 12.

Wood, A. (1975). *The Physics of Music*, 7th ed. Rev., J. M. Bowsher. London: Chapman and Hall. (See Chapter 11.)

GLOSSARY

cent 1/100 of a semitone.

chromatic scale An ascending or descending sequence of twelve tones, each separated by a semitone.

diatonic scale A scale of seven whole tones and semitones appropriate to a particular key.

enharmonic notes Two different notes that sound the same on keyboard instruments (e.g., G♯ and A♭).

equal temperament A system of tuning in which all semitones are the same; namely, a frequency ratio of $2^{1/12} = 1.059$.

inharmonic overtones Overtones whose frequencies are not multiples of the fundamental (i.e., harmonics).

intonation Refers to the degree of accuracy with which pitches are produced.

just intonation A system of tuning that attempts to make thirds, fourths, and fifths as consonant as possible; it is based on major triads with frequency ratios 4 : 5 : 6.

major diatonic scale A scale of seven notes with the following sequence of intervals: two whole tones, one semitone, three whole tones, one semitone.

meantone temperament A system of tuning which raises or lowers various notes from their Pythagorean values by quarters of the syntonic comma.

microtone Any interval smaller than a semitone.

minor scale A scale with one to three notes lowered a semitone from the corresponding major scale. In the key of C-minor, the three minor scales are:

> natural: C D E$^\flat$ F G A$^\flat$ B$^\flat$ C;
> harmonic: C D E$^\flat$ F G A$^\flat$ B C;
> melodic (ascending): C D E$^\flat$ F G A B C;
> melodic (descending): C D E$^\flat$ F G A$^\flat$ B$^\flat$ C.

pentatonic scale A scale of five notes used in several musical cultures, such as the Chinese, Native American, and Celtic cultures.

Pythagorean comma The small difference between two kinds of semitones (chromatic and diatonic) in the Pythagorean tuning; a frequency ratio 1.0136 corresponding to 23.5(¢).

Pythagorean tuning A system of pitches based on perfect fifths and fourths.

scale Succession of notes arranged in ascending or descending order.

semitone A half step; in equal temperament, a semitone corresponds to 100(¢) or to a frequency ratio of 1.059.

stroboscopic tuner A tuning device that makes use of a rotating pattern illuminated by flashing lights.

syntonic comma The small difference between a major or minor third in the Pythagorean and just tunings.

temperament System of tuning in which intervals deviate from acoustically "pure" (Pythagorean) intervals.

triad A chord of three notes; in the just tuning, a major triad has frequency ratios 4 : 5 : 6, and a minor triad has ratios 10 : 12 : 15.

tuning An adjustment of pitch in any instrument so that it corresponds to an accepted norm.

REVIEW QUESTIONS

1. Define *musical scale*.
2. How many notes are in a pentatonic scale?
3. How many notes are in a diatonic scale?
4. How many notes are in a chromatic scale?
5. Name the three most common musical scales.
6. What frequency ratio gives an octave interval?
7. What frequency ratio gives an interval of a perfect fifth?
8. What frequency ratio gives an interval of a perfect fourth?
9. What is the *circle of fifths*?
10. What is the great disadvantage of the Pythagorean scale?
11. What is the syntonic comma?
12. What two intervals are found in a major triad?
13. What two intervals are found in a minor triad?

14. What is the main problem with the just scale?
15. What frequency ratio gives an interval that is exactly 1/12 of an octave?
16. Define *cent*.
17. How many cents are there in an octave?
18. How many cents are there in a perfect (Pythagorean) fifth? (See Table 9.3.)
19. How many cents are there in a just major third? (See Table 9.3.)
20. In what temperament is the circle of fifths a closed circle?
21. In what temperament is there a gap in the circle of fifths?
22. In what tuning is there overlap in the circle of fifths?

QUESTIONS FOR THOUGHT AND DISCUSSION

1. Why is the fingerboard of a guitar fretted, whereas the fingerboard of a violin is not?
2. To play "Auld Lang Syne" using only black keys, in what key should it be played?
3. Handel was one of several composers who felt that the key in which a piece of music is written does much to set its mood. (For example, F major sounds pastoral; F minor and F$^\sharp$ minor are tragic keys; C major expresses vigor or military discipline.) Think of acoustical reasons why this might be true. Do any of these reasons remain valid in this day of tuning to equal temperament?

EXERCISES

1. Verify by direct multiplication that a major third in equal temperament has the ratio of 1.26 and a minor third the ratio 1.19.

2. From your knowledge of equal temperament, show that if you invest money at an interest rate of 5.9% compounded annually, your investment doubles in 12 years!

3. An octave-band sound analyzer measures the sound level in 10 octave bands with center frequencies 31.5, 63, 125, 250, 500, 1000, 2000, 4000, 8000, and 16,000 Hz. What are the closest notes on the musical scale?

4. The sounds used in a touch-tone telephone have the following frequencies: 697, 770, 850, 941, 1209, 1337, and 1477 Hz. What are the closest notes on the musical scale?

5. Verify by multiplication that a fifth plus a fourth equals an octave in any tuning, as does a major sixth plus a minor third.

6. According to the data in Fig. 7.2, the jnd in frequency up to about 700 Hz (roughly the lower half of the piano keyboard) is about 3 Hz. Convert $\Delta f / f = 3/400$ to cents. Then refer to Table 9.2 (or Fig. 9.6) and answer the following.

    (a) Can one normally hear a difference between A_4 on the just, Pythagorean, and tempered scales based on C?

    (b) How about E_4, F_4, and A_4?

7. Using the frequency ratios given in Fig. 9.5, verify that the intervals C : G, E : B, F : C, G : D, and A : E are perfect fifths in the just diatonic scale. Determine the frequency ratios for the imperfect fifth D : A.

8. Find the frequency ratio that corresponds to $25(\cancel{c})$. What are the frequencies of the note $A_4 + 25(\cancel{c})$? $A_4 - 25(\cancel{c})$?

9. Some tuning forks are designed to a scale in which the C's have frequencies that are powers of 2 (128, 256, 512 Hz, etc.). How many cents flat are they compared to the international standard frequencies given in Table 9.2?

EXPERIMENTS FOR HOME, LABORATORY, AND CLASSROOM DEMONSTRATIONS

Home and Classroom Demonstration

1. *Best octave*   Alternate between two tone generators, and tune them so that they sound an octave apart. Determine the frequency ratio that sounds like the "best" octave. Try this in different frequency ranges.

2. *Octave matching*   Listen to Demonstration 15 on the *Auditory Demonstrations* CD.

3. *Stretched and compressed scales*   Listen to Demonstration 16 on the *Auditory Demonstrations* CD.

4. *Linear and logarithmic tone scales*   Listen to Demonstration 18 on the *Auditory Demonstrations* CD.

5. *Tones and tuning with stretched partials*   Listen to Demonstration 31 on the *Auditory Demonstrations* CD.

6. *Tone sequences in two ears: an auditory illusion*   Listen to Demonstration 39 on the *Auditory Demonstrations* CD.

7. *Comparison of different tunings*   Several recordings have been made of keyboard instruments with different tunings (see, for example, *AAPT Announcer*, Dec. 1980, Paper LL6). One of the best demonstrations is on Philips record 6831-013, side 1 (see the article by Franssen and van der Peet in *Philips Tech. Rev.* **31** (11/12) 1970).

8. *N-note scales of equal temperament*   With a computer, it is possible to divide the octave into N equal steps. Try different values of N, including 15, 19, 31, 22, and 53. Do any of these have nearly beat-free (just) intervals close to a major third (1.25), perfect fourth (1.333), perfect fifth (1.5), or major sixth (1.667)?

Laboratory Experiments

Mathematics of music (Experiment 23 in *Physics with Computers*)

Comparison of different tunings (see 7 above)

N-note scales of equal temperament (see 9 above)

PART
III

Musical Instruments

Musical instruments are found in nearly all cultures of the world, past and present. Considerable variation is found in the sophistication of design and the level of craftsmanship in construction, as would be expected. Musical instruments have been the main source of musical sound and the most practical means of musical expression through the ages.

There are many ways to classify musical instruments. One of the most common methods places them in three familiar families: string, wind, and percussion. For convenience we describe the wind family in two different chapters: brasses in one and woodwinds in the others. We have also gathered the discussions of various keyboard instruments from different families into one chapter.

There will probably be a tendency for the reader to read first the chapter that discusses a favorite family of instruments and to spend less time with the others. This is quite understandable, although the reader should be reminded that an understanding of one's own instrument is greatly enhanced by comparing and contrasting it with the acoustical behavior of other families of instruments.

CHAPTER

10

String Instruments

String instruments have held a special place in music for many years. Bowed string instruments are the backbone of the symphony orchestra, and plucked string instruments fulfill a similar roll in folk music as well as rock music. The violin has been the most popular instrument for solo performances as well as chamber music for several generations.

In this chapter you should learn:

- About the vibrations of plucked and bowed strings;
- About modes of vibration in violin bodies;
- About families of bowed string instruments, old and new;
- About the guitar as a system of coupled oscillators;
- About how guitars radiate sound;
- About electric guitars.

Bowed string instruments date back to medieval times, and perhaps even earlier. The *viola de gamba*, or *viol*, was developed in the late Middle Ages and reached its zenith in the seventeenth century. Of the once large viol family, the four sizes in general use today have six strings and fretted fingerboards. The six strings are tuned in fourths except for the middle two, which are separated by a major third (e.g., the soprano viol tunes to D_3, G_3, C_4, E_4, A_4, D_5).

The modern violin was developed in Italy in the sixteenth century, largely by Gasparo da Salo and the Amati family. In the eighteenth century, Antonio Stradivari, a pupil of Nicolo Amati, and Guiseppi Guarneri created instruments with great brilliance that have set the standard for violin-makers since that time.

Violins have probably had more attention from historians, musicians, and scientists than any other musical instrument. Outstanding contributions to our understanding of violin acoustics have been made by such distinguished scientists as Felix Savart (France, 1791–1841), Hermann von Helmholtz (Germany, 1821–1894), Lord Rayleigh (England, 1842–1919), C. V. Raman (India, 1888–1970), Frederick Saunders (United States, 1875–1963), and Lothar Cremer (Germany, 1905–1990). More recently, the work of Professor Saunders has been continued by members of the Catgut Acoustical Society, an organization that publishes an excellent journal describing current research on the acoustics of string instruments.

Interesting questions still remain unanswered: Do the best instruments of today compare to the great instruments of the Italian masters? Did Stradivari, Guarneri, and the Amatis possess secrets still undiscovered today? Most experts think not, but the question can arouse some lively discussion.

FIGURE 10.1
Bowed string
instruments:
(a) violin and viola;
(b) string bass and
cello.

(a)   (b)

It is impossible to discuss in detail all the members of the string family, even the principal bowed string instruments shown in Fig. 10.1. We will first describe how plucked and bowed strings vibrate, applying principles learned in Chapters 3 and 4, and then focus on some of the acoustical features of the violin and the guitar. Brief mention will be made of other string instruments.

Bowed string instruments are discussed in Chapter 10 of Fletcher and Rossing (1998) and guitars, in Chapter 9. *The Physics of the Violin* (Cremer 1984) is the most authoritative work on violin physics, whereas Hutchins and Benade (1997) have reprinted many recent papers on violin research.

10.1 ■ CONSTRUCTION OF THE VIOLIN

A violin has four strings, of steel or gut, tuned to G_3, D_4, A_4, and E_5 (196, 294, 440, and 660 Hz). Strings of small diameter displace very little air as they vibrate; therefore, they radiate very little sound. The vibrating strings, however, set the body of the violin into vibration, which results in radiated sound of considerable strength. The body of the violin is designed to vibrate over a very wide range of frequency, and in this playing range it will have a great many resonances, a few of which will be discussed in Section 10.4.

An exploded view of a violin is shown in Fig. 10.2. The top plate, or *belly*, of the body is usually made of spruce, the ribs and back plate, of curly maple. Two openings, called *f-holes*, are cut in the top plate, and a *bass bar* is glued to the top plate directly under one foot of the *bridge*. Near the other foot of the bridge is a short wooden stick called the *sound post*, which extends from the top to the back plate. The strings are attached to the tail piece, pass over the bridge, along the fingerboard, and over the *nut*, and finally are attached to wooden pegs inserted in the peg box.

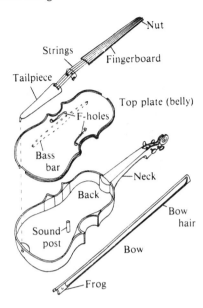

FIGURE 10.2
Parts of a violin.

In order to vibrate properly, the top plate must be made very thin (its thickness is typically in the range of 2 to 4 mm). When tuned to normal pitch, the combined tension of the strings is about 220 N (50 lb). The downward force exerted on the top plate by the bridge is nearly 100 N. Without the sound post and the bass bar, the fragile top plate would not be able to support this load for long. These two important structural members also serve important acoustic functions, as will be discussed in Section 10.4.

The bow consists of many strands of horsehair held under tension by a wood stick (pernambuco, a very dense, stiff wood, is the preferred material). Violin bows typically have a length of about 73 cm and a mass of about 60 grams (Reder 1970). The tension of the hair is adjusted by moving the frog with a screw. Although the violin players attach a great deal of importance to the bow, its properties have not been studied nearly as carefully as those of the violin body.

10.2 ■ VIBRATIONS OF A PLUCKED STRING

When the string of a musical instrument is excited by bowing, plucking, or striking, the resulting vibration can be considered to be a combination of the normal modes of vibration or resonances. For example, if the string is plucked at its center, the resulting vibration will consist of the fundamental plus the odd-numbered harmonics.

Figure 10.3 illustrates how the modes associated with the odd-numbered harmonics, when each is present in the right proportion, can add up at one instant in time to give the initial shape of the string. Modes 3, 7, 11, etc., must be opposite in phase from modes 1, 5, and 9 in order to give a maximum in the center.

Because all the modes shown in Fig. 10.3 have different frequencies of vibration, they will quickly get out of phase, and the shape of the string changes rapidly after plucking. What happens is that two identical pulses propagate in opposite directions away from the

FIGURE 10.3
Odd-numbered
modes of vibration
add up in
appropriate
amplitude and
phase to the shape
of a string plucked
at its center.

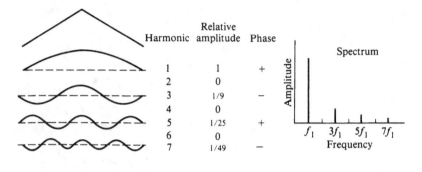

Harmonic	Relative amplitude	Phase
1	1	+
2	0	
3	1/9	−
4	0	
5	1/25	+
6	0	
7	1/49	−

FIGURE 10.4
The motion of a
string plucked at its
midpoint through
one half cycle.
Motion can be
thought of as due to
two pulses traveling
in opposite
directions.

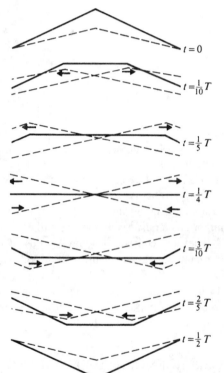

center, as shown in Fig. 10.4. The shape of the string at each moment in time can still
be obtained by adding up the modes in the proportions shown in Fig. 10.4, but it is more
difficult to do so, because each of the modes will be at a different point in its cycle. The
resolution of the string motion into two pulses, shown by the dotted lines in Fig. 10.4,
results in a simpler analysis.

    If the string is plucked at a point other than its center, the recipe of the constituent modes
is different, of course. For example, if the string is plucked one-fifth the distance from one
end, the recipe of mode amplitudes is as shown in Fig. 10.5. Note that the fifth harmonic is
missing. Plucking it one-fourth the distance from the end suppresses the fourth harmonic,

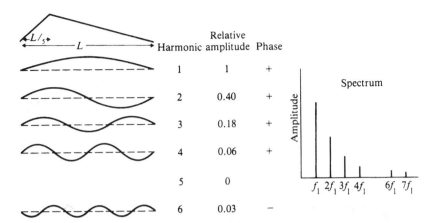

FIGURE 10.5   The addition of modes to obtain the shape of a string plucked at one-fifth its length. It should be noted that the spectra in Fig. 10.3 and the figure above show the relative amplitudes of the different modes of vibration. The spectra of the radiated sound will have the same frequencies but their relative amplitudes will be quite different due to the acoustical properties of the instrument.

and so on. (In Fig. 10.3 it can be noted that plucking it at one-half the distance eliminated the second harmonic, as well as other even-numbered ones.)

10.3 ■ VIBRATIONS OF A BOWED STRING

As the bow is drawn across the string of a violin, the string appears to vibrate back and forth smoothly between two curved boundaries, like a free string vibrating in its fundamental mode (see Fig. 4.4 or the second curve in Fig. 10.3). But this appearance of simplicity is deceiving. If we took a high-speed photograph of the bowed violin string, we would find that the string is nearly straight with a sharp bend at one point; at certain times it resembles the initial shapes of the plucked strings shown in Figs. 10.3 and 10.5. Over a hundred years ago, Hermann von Helmholtz (who contributed so much to our understanding of physics, anatomy, physiology, and the arts) discovered what really happens. The sharp bend racing along the bowed string follows the curved path that we see; because of its speed, our eye sees only the curved envelope.

During the greater part of each vibration, the string is carried along by the bow. Then it suddenly detaches itself and moves rapidly back until it is caught by the moving bow again. The motion of the string at the point of bowing is shown in Fig. 10.6.

Up to the point of release a and again from c to i, the string moves at the constant speed of the bow (recall that speed is represented by the slope of the displacement curve). From a to c, the string makes a rapid return until caught up by a different point on the bow. The displacement and velocity (speed) of a very flexible string during bowing are shown in the oscillograms in Fig. 10.7. The oscillogram on the left, which shows the displacement versus time, is almost identical to Fig. 10.6, whereas the curve of velocity versus time (on the right) shows rather narrow spikes, which represent the large velocity of the string when it slides rapidly along the bow (a-b-c in Fig. 10.6).

FIGURE 10.6
Displacement of
bow and string at
the point of contact
with the bow. Note
that the midpoint of
the vibration is
displaced slightly
from the rest
position of the
string.

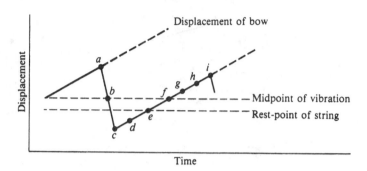

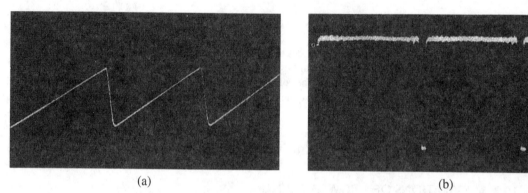

FIGURE 10.7   (a) Displacement and (b) velocity of a bowed string. The velocity (b) at every point in time equals the slope of the displacement curve. (From Schelleng 1974.)

The connection between the sawtoothlike motion shown in Figs. 10.6 and 10.7 and the motion of the bends racing back and forth on the string can be understood by referring to Fig. 10.8, which shows the entire string at successive times in the vibration cycle. At the moment of release, shown in Fig. 10.8(a), the bend has just passed the bow. In (b), the bend has reached the bridge, from which it will reflect back down the string in (c), (d), and (e), until it reaches the nut in (f) and is again reflected. At point c in Fig. 10.6 and also in frame (c), in Fig. 10.8, the string is captured by the bow and once again moves upward at the speed of the bow. In the set of diagrams in the right-hand column of Fig. 10.8, the bow has been deleted, and the velocity of the string at several points is indicated by the arrows.

The *stick* and *slip* of the string against the moving bow are determined partly by the friction between the horsehair of the bow and the string. It is well known that the force of friction between two objects is less when they are sliding past each other than when they move together without slippage. Once the string begins to slip, it moves rather freely until it is once again captured by the bow. It is important to note, however, that the beginning and the end of the slipping are triggered by the arrival of the bend (slipping begins when the bend arrives from the nut and ends when it arrives again from the bridge). Because the time required for one round trip depends only on the string length and wave velocity (which, in turn, depends on tension and mass), the vibration frequency of the string remains the same

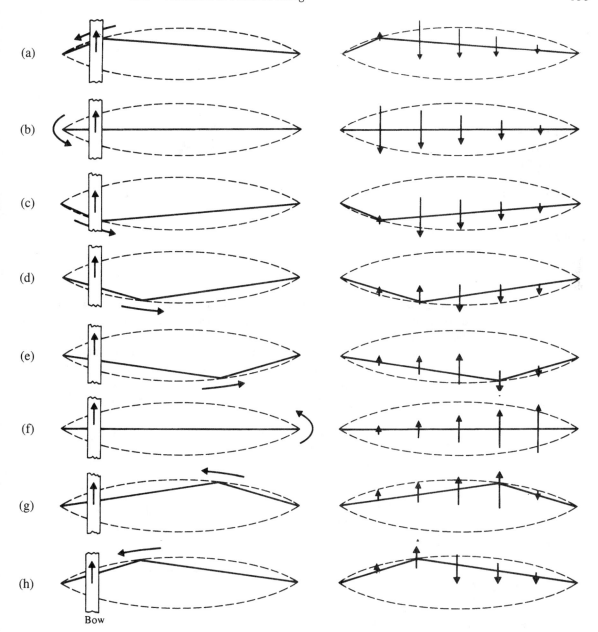

Bow

FIGURE 10.8 The motion of a bowed string at successive times during the vibration cycle. The configurations shown in (a)–(h) correspond to the points *a–h* in Fig. 10.6. The bend races around a curved boundary, which appears to be the profile of the string viewed during bowing.

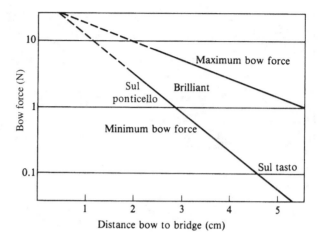

FIGURE 10.9
Range of bowing
force for different
bow-to-bridge
distances for a cello
bowed at 20 cm/s.
(After Schelleng
1974.)

under widely varying bowing conditions. If friction alone determined the beginning and the end of slipping, the vibrations would be irregular rather than regular.

The limits on the bowing conditions are the limits on the conditions at which the bend can trigger the beginning and the end of slippage between bow and string. For each position of the bow, there is a maximum and minimum bowing force, as shown in Fig. 10.9. The closer to the bridge the instrument is bowed, the less leeway the violinist has between minimum and maximum bowing force. Bowing close to the bridge (*sul ponticello*) gives a loud, bright tone, but requires considerable bowing force and the steady hand of an experienced player. Bowing further from the bridge (*sul tasto*) produces a gentle tone with less brilliance.

Further study of Fig. 10.8 will help us understand how amplitude of the vibration is determined by the speed and position of the bow. Because the speed of the bend around its curved path is essentially independent of the speed and position of the bow, the amplitude of vibration can increase either by increasing the bow speed or by bowing closer to the bridge.

Figures 10.7 and 10.8 illustrate the motion of the string at the point of contact with the bow. The displacement and velocity at two other points on the string are shown as functions

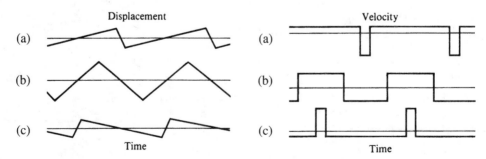

FIGURE 10.10   Displacement and velocity of a bowed string as a function of time at three positions: (a) near the bridge; (b) at the center; (c) near the nut. For bowing in the opposite direction, exchange (a) and (c).

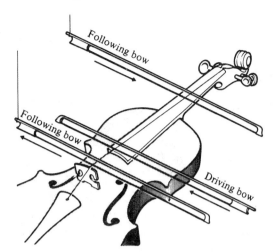

FIGURE 10.11
The following-bow experiment. A second bow hanging by a string indicates the direction of string motion during the longer part of each vibration cycle.

of time in Fig. 10.10. At the center of the string, the velocity (speed) of the string (slope of the displacement curve) is the same in both directions.

When the bow moves in the opposite direction, the curves in Figs. 10.10(a) and (c) are exchanged. Also, the bend moves clockwise, rather than counterclockwise as shown in Fig. 10.8.

> The difference in motion at the three points can be demonstrated by a "following-bow" experiment (Schelleng 1974). (See Fig. 10.11.) With the violin supported horizontally, a second bow is suspended by a string at its heavy end, the other end resting lightly on the bowed string. As the string is bowed loudly, the hanging bow moves in the direction of string motion during the longer part of each cycle. Placed near the bridge, it moves in the same direction as the driving bow; near the nut, it moves in the opposite direction. At the center, there is little motion in either direction.
>
> The behavior of the following bow can be understood by examining the velocity of the string near the bow and the nut as shown in Fig. 10.10. The suspended bow follows the slow motion of the string in one direction but is unable to follow the rapid return of the string in the other direction; thus it undergoes a net motion during each cycle of vibration.

The vibrating string exerts a sideways force on the bridge, which in turn transmits this force to the top plate. In the ideal case of a completely flexible string vibrating between two fixed end supports, this force has a sawtooth waveform with a spectrum of harmonics varying in amplitude as $1/n$, as shown in Fig. 10.12. In actual practice, the waveform of the force is modified by the string stiffness, mechanical properties of the bridge (see Hacklinger 1978), and other factors.

The motion of the top plate, which is the source of most of the violin sound, is the result of a complex interaction between the driving force from the bridge and the various resonances of the violin body. These will now be considered.

FIGURE 10.12
(a) Idealized
waveform of the
force exerted on the
bridge by a bowed
string; (b) spectrum
of the force
showing harmonics
decreasing as $1/n$.

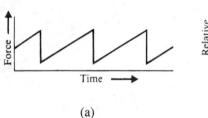

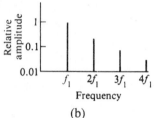

(a)

(b)

10.4 ■ VIBRATIONS OF THE VIOLIN BODY

An important factor in the sound quality and playability of a violin is the vibrational behavior of its body. We often describe this in terms of normal modes of vibration, which
consist of coupled motions of the top plate, the back plate, and the enclosed air. Smaller
contributions are made by the ribs, neck, and fingerboard.

To determine the normal modes of vibration, it is customary to apply an oscillating
force to the bridge and observe the motion of the various parts of the violin. This may
be done electrically, by means of accelerometers or optically, by means of holographic
interferometry.

Six low-frequency modes of vibration of a violin are shown in Fig. 10.13. The C_1 and
C_2 body (*corpus*) modes radiate little sound, although they contribute to the "feel" of the
violin. The other four modes are indicated on the frequency response curve in Fig. 10.14,
where the vertical coordinate is the velocity divided by the force (called *mechanical admittance*, or *mobility*).

The A_0 (*f-hole* or *air*) mode and the T_1 (*top*) mode both involve considerable motion of
air in and out of the f-holes. Both modes radiate sound efficiently, and they dominate the
low-frequency sound spectra of most violins.

Above 1 kHz, the vibrational modes or resonances are bunched together, as can be seen
in Fig. 10.14, and it is difficult to identify the individual modes. Of considerable interest is
the concentration of resonances around 2 to 3 kHz, which resembles the "singer's formant"
found in the spectra of most opera singers (see Chapter 17). Dünnwald (1983) found this
concentration to be particularly characteristic of violins made by the old Italian masters
(Stradivari, Guarneri, etc.).

10.5 ■ TUNING THE TOP AND BACK PLATES

The carving of the top and back plates is one of the most critical steps in constructing a
violin. Since both plates have an arch, they are carved from blocks of wood thicker at the
center, as shown at O–O in Fig. 10.15.

As the plates near their desired shapes, the violin-maker tests them by listening to *tap
tones*. The hear these, the plate is held in a particular way and tapped at certain spots. The
trained ear of the violin maker can extract useful information by noting the pitch and the
decay time of the tap tones.

In recent years, the analysis of the vibrations of the plates has been refined considerably
by the application of electronics and optics. One rewarding technique has been the obser-

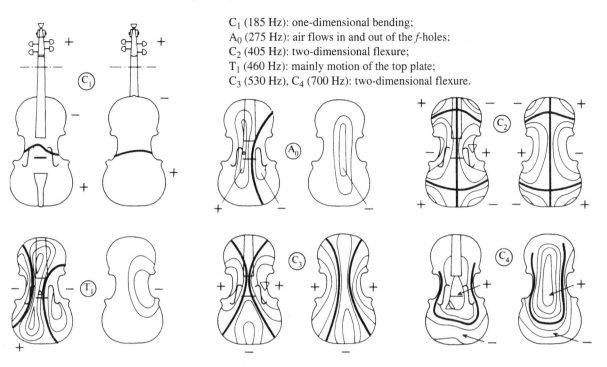

C_1 (185 Hz): one-dimensional bending;
A_0 (275 Hz): air flows in and out of the f-holes;
C_2 (405 Hz): two-dimensional flexure;
T_1 (460 Hz): mainly motion of the top plate;
C_3 (530 Hz), C_4 (700 Hz): two-dimensional flexure.

FIGURE 10.13 Modal shapes for six modes in a violin. Top plate and back plate are shown for each mode. The heavy lines are nodal lines; the direction of motion is indicated by + or −. The drive point is indicated by a small triangle. (After Alonso Moral and Jansson 1982a.)

FIGURE 10.14
Input admittance (driving point mobility) of a Guarneri violin driven on the bass bar side (Alonso Moral and Jansson 1982b).

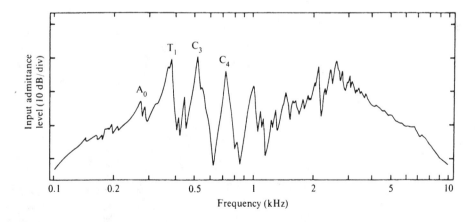

vation of individual modes of vibration by *Chladni patterns* of the type shown in Fig. 2.19. The plate is usually mounted horizontally and excited by a loudspeaker. The two most important modes of vibration are those shown in Fig. 10.16. (The plates shown are for a viola, but violin and cello plates show similar patterns.) If the frequency of the lower mode (a) in the top plate matches the back plate, then tuning the upper modes (b) to the same

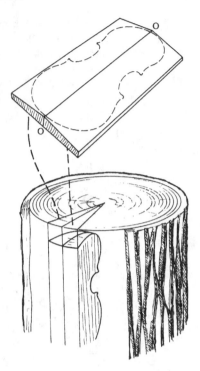

FIGURE 10.15
Wood for a top or back plate. The top plate is spruce; the back is generally maple.

frequency or a semitone different will usually produce plates of high quality (Hutchins 1977). Of course, the shape of the nodal lines is equally as important as the frequency of the modes.

Other methods of vibration analysis give rather detailed information on the modes but are less convenient to set up in the violin shop. One such method involves setting a plate into vibration and scanning it with a beam of light. The light reflected from the plate falls on a photocell, where it creates an electrical signal that indicates the character of vibration

FIGURE 10.16
Chladni patterns showing the modes of vibration in the viola top and back plates. (From Hutchins 1977. Reprinted by permission.)

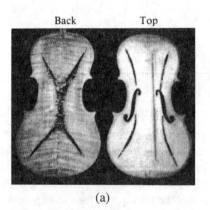

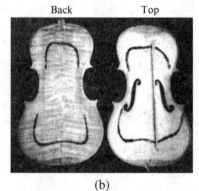

(a)   (b)

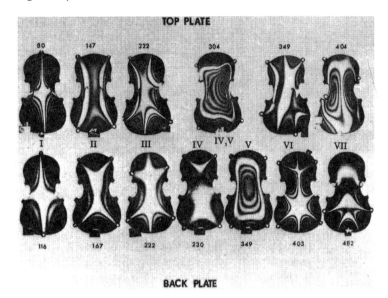

FIGURE 10.17
Modes of vibration of a violin top and back plates visualized by hologram interferometry. (Hutchins, Stetson, and Taylor 1971.)

at that point. An elegant method, called *hologram interferometry*, employs laser light. Figure 10.17 is a photograph of the vibrational modes of a violin top plate and back plate using this technique (Hutchins, Stetson, and Taylor 1971). The dark lines indicate areas that vibrate with the same amplitude; along the center of each large white area is a nodal line. Mode II in Fig. 10.17 is the mode shown in Fig. 10.16(a), whereas mode V corresponds to the mode in Fig. 10.16(b).

The vibrational characteristics of wood plates depend not only on the size, shape, thickness, and arching of wood but also on the density, stiffness, and internal damping of the wood. These properties vary from sample to sample and even from day to day, as the temperature and humidity change. Varnish fortifies the wood to some extent against the effects of changing humidity, but the varnish also affects the wood vibrations. Varnish adds some mass and stiffness to the wood plates (resonances in a violin plate top may shift by as much as a quarter tone), but its greatest effect is to increase the damping. The best advice on varnish may be "not too soft, not too hard, not too much" (Schelleng 1968).

One particular property of wood that deserves mention is its *anisotropy*; its stiffness along the grain is much greater than that across the grain. Spruce has a particularly high ratio (more than 10:1), which is one of the reasons it is used in a number of musical instruments (for top plates in violins and guitars, for soundboards in pianos and harpsichords, etc.). This anisotropy obviously has a large influence on its modes of vibration.

Attempts have been made to develop a synthetic material that would substitute for spruce. One of the most successful materials has been a sandwich with an inner core of cardboard overlaid with graphite fibers set in epoxy. The fibers are carefully aligned in one direction to achieve the desired anisotropy (like wood grain). A violin and a guitar with top plates of this material have been constructed and demonstrated at meetings of the Acoustical Society of America (Haines, Chang, and Hutchins 1975).

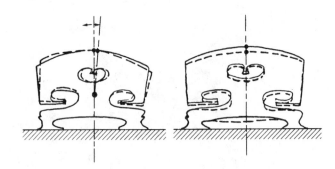

FIGURE 10.18
First two vibrational modes (resonances) of a violin bridge. (After Reinecke 1973.)

10.6 ■ THE BRIDGE

The primary role of the bridge is to transform the motion of the vibrating strings into periodic driving forces applied by its two feet to the top plate of the instrument. However, generations of violin makers have discovered that shaping the bridge is a convenient way to alter the frequency response of a violin as well.

Violin bridges typically have strong resonances around 3000 and 6000 Hz. The lower resonance is due to a rocking-bending motion, as shown in Fig. 10.18, whereas the upper one is due to a symmetrical up-and-down motion. Clearly, these resonance frequencies can easily be changed by cutting away wood at the appropriate places.

In order to "darken" the sound of a violin, the player may attach a mute to the bridge. The additional mass of the mute shifts the bridge resonances to lower frequencies, changing the sound spectrum of the instrument. Typical violin mutes have masses of about 1.5 g.

10.7 ■ OTHER BOWED STRING INSTRUMENTS

The other string instruments of the orchestra are the viola, cello (violoncello), and string bass (also known as contrabass, double bass, or bass viol). The general shape and construc-

FIGURE 10.19
Tunings and resonances of violin, viola, cello, and string bass. The main air resonance is denoted by A, the plate resonances, by T and C. The strings are given by o.

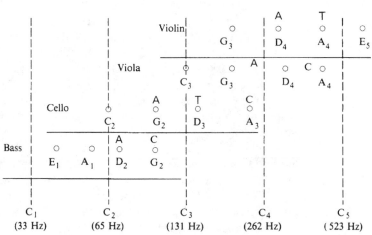

FIGURE 10.20
The eight instruments in the new fiddle family. (Photograph by John Castronovo. Reprinted by permission.)

tion of the viola and cello are similar to those of the violin. The strings, like those of the violin, are tuned in intervals of fifths: C_3, G_3, D_4, A_4 in the viola and C_2, G_2, D_3, A_3 in the cello. Although the viola is tuned a fifth below the violin, its main air resonance is typically less than a third below that of the violin, and the main plate resonance is a little over a third lower. Thus the lowest notes on a viola lack carrying power. Tunings and resonances of the violin, viola, and cello are shown in Fig. 10.19.

The lower resonances of a cello are related to its strings in about the same way as those of the viola. The tall bridge of the cello, however, results in a large driving force and a strong response near the main wood resonance. Also, the second air resonance is nearly an octave above the main air resonance (Bynum and Rossing 1997) and thus reinforces this resonance by strengthening the second harmonic.

The string bass is tuned in fourths rather than fifths, which reduces the distances that the hand must travel along the long fingerboard. It has relatively narrow sloping shoulders, and its flat back saves a substantial amount of wood, as compared to what would be needed for an arched back like that of the violin or cello.

10.8 ■ MUSIC AND PHYSICS: A NEW FAMILY OF FIDDLES

In 1958, Carleen Hutchins and other members of the Catgut Acoustical Society set about to develop a new family of fiddles, with resonances scaled to those of a fine violin. Their research led to a family of eight instruments shown in Figs. 10.20 and 10.21. The family has two instruments pitched above the violin: the treble and the soprano. The alto, which is tuned like the viola, is held either under the chin (if one has long arms) or played vertically on a peg in the fashion of a cello. The tenor tunes between the viola and cello, and the baritone has cello tuning. Finally there are two basses, with their strings tuned in fourths.

These new instruments have appeared in several concerts and have aroused considerable interest, both musically and scientifically. The alto violin, especially, is considered to have increased power and tone quality over the viola. "That is the sound I have always wanted

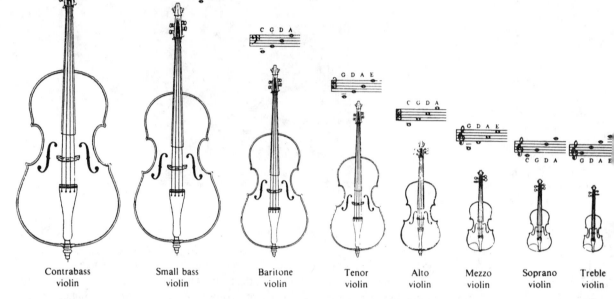

FIGURE 10.21   Comparative sizes and tuning of the new violin family. (Reprinted from *The Instrumentalist*, June 1967, p. 41. © The Instrumentalist Company 1989. Used by permission of The Instrumentalist Co.)

from the violas in my orchestra," declared conductor Leopold Stokowski upon hearing the alto violin (Hutchins 1967).

10.9 ■ CONSTRUCTION OF THE GUITAR

The guitar has its origins in antiquity, probably more than 3000 years ago in Egypt. Its ancestors include the lute and the vihuela. Although the guitar appeared in several countries of Europe (Antonio Stradivari constructed about a dozen guitars), it was in Spain that it developed great popularity. By the end of the thirteenth century, guitar making was a developing art in Andalusia in southern Spain. Perhaps the most celebrated of the Spanish luthiers was Antonio de Torres Jurado, whose innovations established a Spanish school of guitar making in the late nineteenth century.

It is in the twentieth century, however, that the guitar has become one of the most popular of all musical instruments. In the United States alone, there are an estimated 15 to 20 million guitars. Guitar music has developed along at least five different lines: classical, flamenco, folk, jazz, and rock.

Throughout the years, the guitar has undergone many changes in design, including the size and shape of the body and the number of strings. Torres developed the larger body, fan-braced sound board, and 65-cm string length that are popular today. The modern guitar, shown in Fig. 10.22, has six strings tuned to E_2, A_2, D_3, G_3, B_3, and E_4 (82, 110, 147, 196, 247, and 330 Hz). The strings, which lie in a single plane, are fastened directly to

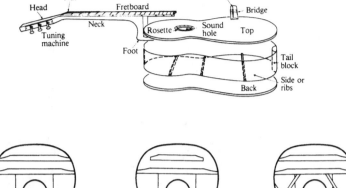

FIGURE 10.22
An exploded view of a guitar, showing details of construction.

 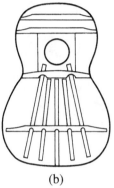  

| (a) | (b) | (c) | (d) |

FIGURE 10.23 Various designs for bracing guitar sound board: (a) traditional fan bracing; (b) Bouchet (France); (c) Ramirez (Spain); (d) crossed bracing.

the bridge. The long fingerboard is fitted with frets, which greatly simplify the playing of chords.

The top is usually cut from spruce, planed to a thickness of about 2.5 mm (3/32 in.). The back is usually of a hardwood, such as rosewood, mahogany, or maple, also about 2.5 mm thick. Both top and back are braced, the bracing of the top being one of the critical parameters in guitar design. Braces strengthen the fragile plate and also transmit vibrations of the bridge to various parts of the sound board. Figure 10.23 shows several different designs of bracing used in guitars.

Acoustic guitars usually fall into one of four families of design: classical, flamenco, flattop (or folk), and archtop. Classical and flamenco guitars have nylon strings; flattop and archtop guitars have steel strings. Since steel strings are under greater tension, flattop guitars usually have a steel rod imbedded inside the neck, and their sound boards are provided with crossed bracing. The fingerboards of flattop and archtop guitars are narrower than those of classical and flamenco guitars. Steel strings tend to give a louder sound than nylon strings. Classical, flattop, and archtop guitars are shown in Fig. 10.24.

Some flattop guitars carry 12 strings, which are positioned and played in pairs. In the two pairs of highest pitch (B, E), the strings are tuned in unison, but in the other four pairs, one string is tuned an octave higher than normal for greater brilliance (occasionally one of the low E strings is tuned two octaves higher than the other).

FIGURE 10.24
Styles of acoustic
guitars:
(a) classical;
(b) flattop;
(c) archtop.
(Courtesy of
Gibson Guitar
Corp., Nashville,
Tenn. Reprinted by
permission.)

(a)   (b)   (c)

10.10 ■ THE GUITAR AS A VIBRATING SYSTEM

The guitar can be considered to be a system of coupled oscillators. The plucked strings, which were discussed in Section 10.2, radiate only a small amount of sound directly, but they excite the bridge and top plate, which in turn transmit vibrational energy to the air cavity and back plate. Sound is radiated efficiently by the vibrating plates and through the sound hole.

Figure 10.25 is a simple schematic of a guitar. At low frequency the top plate transmits energy to the back plate and the sound hole via the air cavity; the bridge essentially acts as part of the top plate. At high frequency, however, most of the sound is radiated by the top plate, and the mechanical properties of the bridge becomes significant. We will expand this simple model in the following sections.

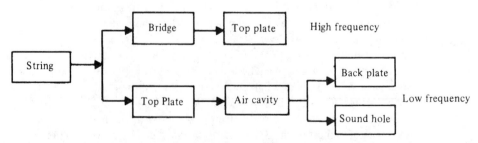

FIGURE 10.25   Simple schematic of a guitar. At low frequency sound is radiated by the top and back plates and the sound hole. At high frequency most of the sound is radiated by the top plate.

The waveform of the actual bridge force is strongly influenced by the stiffness and damping of the string and the manner in which it is plucked. If the string is plucked with the finger or a soft plectrum, the spectrum of the force will have less prominent high harmonics.

10.11 ■ VIBRATIONS OF THE TOP PLATE, BACK PLATE, AND AIR CAVITY

Before considering the way in which the guitar body vibrates, it is well to consider the vibrational modes of the individual parts. In describing these modes, it is important to specify the exact conditions under which they are measured. A top plate with a free edge, for example, has completely different vibrational modes than a top plate attached to guitar ribs.

The first four modes of a guitar top plate are illustrated in Fig. 10.26, using the technique of hologram interferometry. (These interferograms were made with the strings and back plate removed and with the sides clamped in place.) The second mode (Fig. 10.26(b)) has a nodal line running in the direction of the grain, and the next mode (Fig. 10.26(c)) has anode just above the bridge. For the top plate shown in Fig. 10.26, the modal frequencies were 185, 287, 460, and 508 Hz.

In our laboratory, we usually observe the top plate modes with the back plate in place but heavily damped in a sandbox and with the sound hole closed. This changes the mode frequencies slightly, but the mode shapes remain nearly the same as in the backless guitar of Fig. 10.26.

Vibrational modes of the top plate, the back plate, and the air cavity of the folk guitar are shown in Fig. 10.27. Note that in this folk guitar with crossed bracing (as in Fig. 10.23(d)), the cross-grain bending mode (designated as the $(1, 0)$ mode in Fig. 10.27) is observed at a higher frequency than the long-grain $(0, 1)$ bending mode. This behavior, opposite to that observed in Fig. 10.26, occurs because the crossed bracing in the folk guitar adds considerable stiffness across the grain.

The lowest mode of the air cavity is the familiar Helmholtz resonance (see Fig. 4.10(b)), whose frequency is determined by the cavity volume and sound hole diameter. Higher modes resemble the standing waves in a rectangular box.

FIGURE 10.26
Holographic interferograms of the first four modes of vibration of the top plate of a guitar without back and strings. Along each line, the amplitude of the vibrational motion is a constant. (From Jansson 1971.)

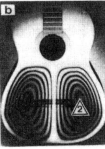
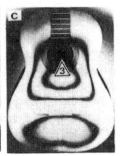
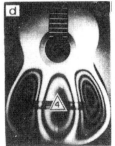

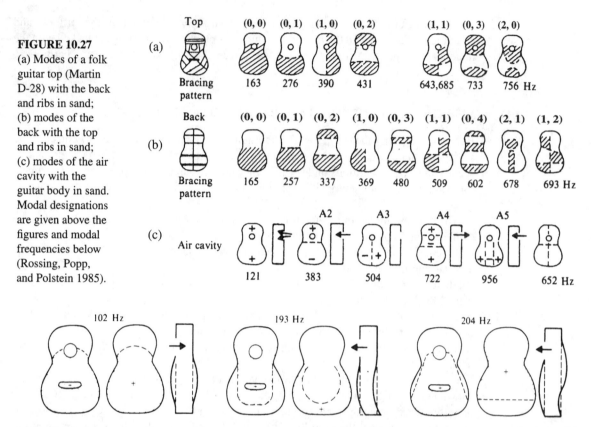

FIGURE 10.27
(a) Modes of a folk guitar top (Martin D-28) with the back and ribs in sand; (b) modes of the back with the top and ribs in sand; (c) modes of the air cavity with the guitar body in sand. Modal designations are given above the figures and modal frequencies below (Rossing, Popp, and Polstein 1985).

FIGURE 10.28 Vibrational motion of a freely supported Martin D-28 folk guitar at three strong resonances in the low frequency range.

10.12 ■ RESONANCES OF THE GUITAR BODY

Most of the prominent low-frequency resonances of a guitar can be attributed to coupled motion of the top plate, the back plate, and the enclosed air. Coupling between the lowest modes of each of these in the guitar shown in Fig. 10.27 leads to resonances at 102, 193, and 204 Hz, as shown in Fig. 10.28.

At the lowest of these three resonances, the top and back plates move in opposite directions, causing the guitar to breathe in and out of the sound hole. At the second resonance, the plates move in the same direction. At the third resonance, they again move in opposite directions, but the air in the sound hole moves in a direction opposite to its motion in the first mode. This is somewhat like the highest mode of the three-mass vibrator in Fig. 2.10(c).

Note that the resonance frequencies in Fig. 10.28 are given for a guitar freely supported on rubber bands. Fixing the ribs lowers the second resonance from 193 to 169 Hz, but the first and third resonances remain essentially unchanged because they involve but little motion of the ribs. This illustrates the dependence of the vibrational modes on the method

FIGURE 10.29
Vibrational
configurations of a
Martin D-28 guitar
at two resonances
resulting from
seesaw motion of
the (1, 0) type.

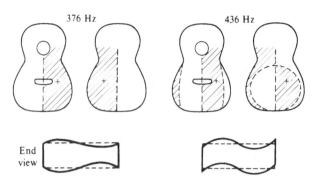

of support, and suggests that the timbre of the instrument depends upon the way it is held
by the player.

The (1, 0) modes in the top plate, back plate, and air cavity combine to give at least one
strong resonance between 250 and 300 Hz in a classical guitar but closer to 400 Hz in a
cross-braced folk guitar. Motion of the plates at two such resonances in a Martin D-28 folk
guitar are shown in Fig. 10.29.

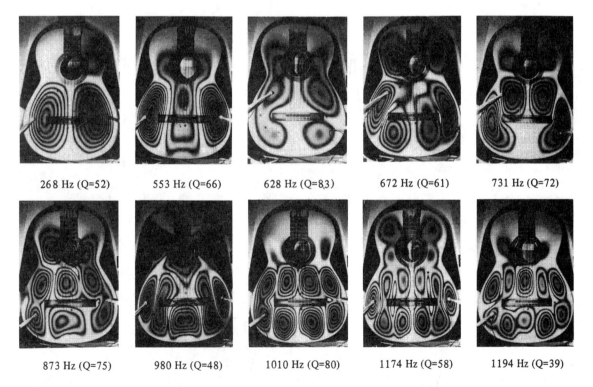

FIGURE 10.30   Holographic interferograms of a classical guitar top plate at several resonances. Resonance frequencies and
Q-values (a measure of the sharpness of the resonance) are given. (Richardson and Roberts 1985.)

Above 400 Hz, the coupling between top and back plate modes appears to be weaker, so the observed resonances are due mainly to resonances in one or the other of the two plates. A fairly prominent (2, 0) top plate resonance is generally observed in classical guitars around 550 Hz, but this mode is generally less prominent in folk guitars. Vibrational configurations of a classical guitar top plate at several resonances are illustrated by the holographic interferograms in Fig. 10.30. Q-values are a measure of the sharpness of each resonance.

10.13 ■ SOUND RADIATION

At low frequencies, considerable sound is radiated by the top plate, the back plate, and the sound hole. The sound spectrum depends upon the direction from the guitar in which it is observed. Figure 10.31 shows how the radiated sound varies with frequency in one particular direction (straight ahead of the sound hole) in an anechoic (echo-free) room. Figure 10.32 shows how the sound level varies with direction at the frequencies of four

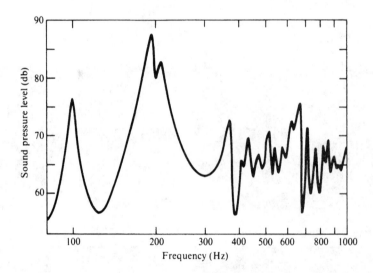

FIGURE 10.31
Sound pressure level one meter in front of a folk guitar (Martin D-28) driven by a force of constant amplitude applied to the treble side of the bridge.

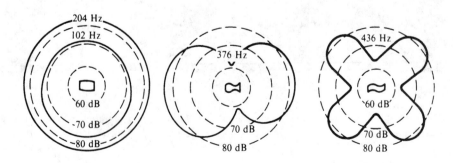

FIGURE 10.32
Sound radiation patterns at four resonance frequencies in a Martin D-28 folk guitar (compare with Figs. 10.28 and 10.29, which show the corresponding vibrational configurations).

of the resonances shown in Figs. 10.28 and 10.29. At 102 and 204 Hz, the guitar radiates well in all directions, but at 376 and 436 Hz it does not. (At 376 Hz, the radiation pattern takes on a dipole character, not unlike the field of a bar magnet with two poles; at 436 Hz a quadrupole character is apparent.)

In an ordinary room, the directionality of the sound radiation pattern is well obscured by reflections from the walls, ceiling, and other surfaces. Nevertheless, a different sound spectrum occurs at every location in the room. It is clear why many guitarists take great care to hold their instruments in such a way that the back is free to vibrate.

For the second spectra and radiation patterns in Figs. 10.31 and 10.32, a sinusoidal force was applied to the treble side of the bridge perpendicular to the top plate. When a guitar string is plucked, the force on the bridge has spectral components at many frequencies (see Figs. 10.3 and 10.5), and furthermore the force is not exactly perpendicular to the top plate. Thus the sound spectra become more complicated.

One way to study the playing characteristics of a guitar is to record the sound pressure for each note of the scale plucked with as consistent an initial displacement as possible. Such a curve for a classical guitar is shown in Fig. 10.33. Playing curves of this type will be quite different from frequency response curves, such as Fig. 10.31, because plucking the string excites a large number of harmonics. Driving the guitar sinusoidally at 87 Hz, for example, gives a weak response, but the note F_2 comes through fairly well because its harmonics are near enough to guitar resonances. There is some difference in the sound levels due to playing the same note on different strings. The heavier strings tend to drive the guitar a little more strongly, as might be expected.

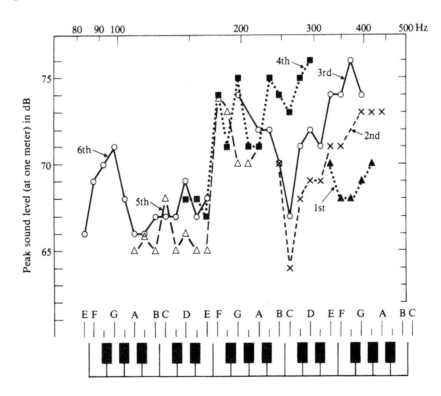

FIGURE 10.33
Playing curve for a classical guitar. Each note is plucked with as consistent a displacement as possible, and the sound level is recorded one meter away in a fairly reverberant room.

A folk guitar with steel strings will produce sound levels that are typically 5 to 10 dB greater than those of a nylon-strung classical guitar. Steel strings normally have approximately twice the tension and mass of the corresponding nylon strings.

10.14 ■ THE GUITAR: A DEVELOPING INSTRUMENT

The violin is a mature musical instrument, the product of more than 300 years of development by skilled craftspeople. It is unlikely that any major innovations will be seen. The guitar, by contrast, is a rather young instrument and many guitar builders (luthiers) are constantly experimenting. We will briefly mention some relatively recent developments.

Asymmetric Bodies

Although most classical guitars tend to be symmetrical around their center plane, a number of luthiers (e.g., Hauser in Germany and Ramirez in Spain) have had considerable success by introducing varying degrees of asymmetry into their designs. Most asymmetrical designs use shorter but thicker struts on the treble side, thus making the plate stiffer. Three such top-plate designs are shown in Fig. 10.34. Asymmetric radial bracing in the top plate, such as that shown in Fig. 10.34(c), appears to offer some advantages over the more traditional fan bracing (Rossing and Eban 1999).

Australian luthier Greg Smallman, who builds guitars for John Williams, has enjoyed considerable success by using lightweight top plates supported by a lattice of braces whose thicknesses are tapered away from the bridge in all directions, as shown in Fig. 10.35. Smallman generally uses carbon fiber epoxy struts (typically 3 mm wide and 8 mm high at their tallest point) in order to achieve high stiffness-to-mass ratio and, hence, high-resonance frequencies, or "lightness" (Caldersmith and Williams 1986).

A Family of Scaled Guitars

Members of guitar ensembles (trios, quartets) generally play instruments of similar design, but Australian physicist/luthier Graham Caldersmith has created a new family of guitars especially designed for ensemble performance. (Actually, he has created two such families:

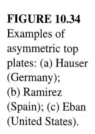

FIGURE 10.34
Examples of asymmetric top plates: (a) Hauser (Germany); (b) Ramirez (Spain); (c) Eban (United States).

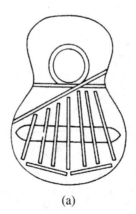

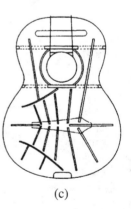

(a)   (b)   (c)

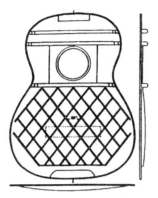

FIGURE 10.35
Lattice bracing of a
guitar top plate
used by Australian
luthier Greg
Smallman.

one of classical guitars and one of steel-string folk guitars.) His classical guitar family, including a treble guitar, a baritone guitar, and a bass guitar in addition to the conventional guitar—which becomes the tenor of the family—has been played and recorded extensively by Australian quartet Guitar Trek (Caldersmith 1989).

Use of Synthetic Materials

Traditionally guitars have top plates of spruce or redwood, with backs and ribs of rosewood or some comparable hardwood. Partly because traditional woods are sometimes in short supply, luthiers have experimented with a variety of other woods, such as cedar, pine, mahogany, ash, elder, and maple. Bowls of fiberglass, used to replace the wood back and sides of guitars, were developed and patented by the Kaman company in 1966; their Ovation guitars have become very popular, partly because of their great durability.

One of the first successful attempts to build a guitar mostly of synthetic materials was described by Haines, Chang, and Hutchins (1975). The body of this instrument, built to the dimensions of a Martin folk guitar, used composite sandwich plates with graphite-epoxy facings around a cardboard core. In listening tests, the guitar of synthetic material was judged equal to the wood standard for playing scales but inferior for playing chords.

In France, Charles Besnainou and his colleagues have constructed lutes, violins, violas, cellos, double basses, harpsichords, and guitars using synthetic materials (Besnainou 1995). A graphite-epoxy guitar specifically designed to be played outside in inclement weather is the "Rainsong" guitar developed by John Decker and his associates (Decker 1995).

10.15 ■ THE ELECTRIC GUITAR

Although it is possible to attach a contact microphone to the body of an acoustic guitar, an electric guitar nearly always uses magnetic pick-ups in which the vibrating strings induce electric signals directly.

Electric guitars may have a solid wood body or a hollow body, the solid design being the more common. Vibrations of the body have much less influence on tone in the electric guitar than in its acoustic cousin. The solid guitar, although heavier, is less susceptible

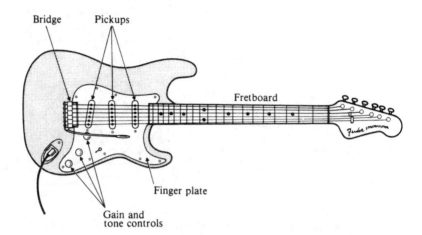

FIGURE 10.36
An electric guitar.

to acoustic feedback (from the loudspeaker to the guitar) and it also allows the strings to continue vibrating for a slightly longer time. Figure 10.36 shows the main features of an electric guitar.

The vibrations of strings are influences by their end supports. As a result of nonrigid end supports, energy can flow from the strings to the body of an instrument, causing the string vibrations to decay faster than in the case of rigid supports. In an electric guitar, this mechanism can lead to "dead spots" at certain locations on the fretboard (Fleischer and Zwicker 1998).

The electromagnetic pickup consists of a coil with a magnetic core. The vibrating steel string causes changes in the magnetic flux through the core, thus inducing an electrical signal in the coil. The principle of the guitar pickup is similar to the magnetic phonograph pickup (see Fig. 22.2). Most electric guitars have at least two pickups for each string, some have three. These pickups, located at different points along the string, sample different strengths of the various harmonics, as shown in Fig. 10.37(b). The front pickup (nearest

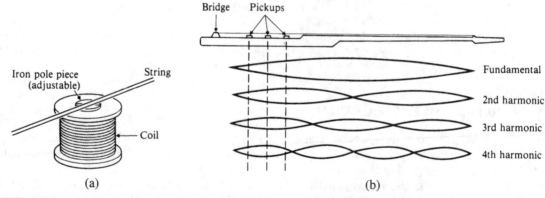

FIGURE 10.37   (a) Electromagnetic pickup for a guitar; (b) Arrangement of multiple pickups to sample various modes of vibration of the string.

FIGURE 10.38
An electric bass.
(Photo courtesy of
Calvin Rose.)

the fretboard) generates the strongest signal at the fundamental frequency, whereas the rear pickup (nearest the bridge) is most sensitive to the higher harmonics (the resulting tones are sometimes characterized as "mellow" and "gutsy," respectively). Switches or individual gain controls allow the guitarist to mix together the signals from the pickups as desired.

Most pickups have a threaded pole piece that can be adjusted in height by screwing it in or not. Adjusting the pole piece closer to the string will usually increase the volume; if it is too close to the string, however, distortion will result due to the force exerted on the string by the magnet. The distortion becomes especially noticeable when fingering beyond the twelfth fret, which brings the string down close to the pickup.

Electrical circuits for guitar pickups may incorporate a wide variety of different features. Many guitars include tone controls that adjust the high-frequency response. Others have a switch to reverse the phase of the signal from one pickup with respect to the others. *Humbucking pickups* have two coils wound in such a way that stray magnetic fields (from power cords, lights, etc.) will induce opposing electrical signals in the two coils; thus the hum they produce will be minimized.

A special type of electric guitar is the *bass guitar* or *electric bass* widely used in rock and jazz bands. Besides being tuned lower (E_1, A_1, D_2, G_2), it differs from the ordinary electric guitar in that it has only four strings and a longer fretboard (90 cm instead of 65 cm). An electric bass is shown in Fig. 10.38.

10.16 ■ STRINGS, FRETS, AND COMPENSATION

Strings for the modern classical and flamenco guitars are made of nylon, replacing the gut strings used in the past. The three highest strings are usually monofilament nylon, while the three lowest strings have nylon cores wrapped with a metal winding. Some flamenco players substitute second and third strings with plastic windings around nylon cores for a slightly more brilliant sound.

Flattop or folk guitars use steel wire for the highest two strings and sometimes the third, whereas the remaining strings have steel cores wrapped with steel, nickel-steel, or bronze. Usually the wrapping is composed of a fine wire of circular cross section (*round-wound* strings), but sometimes a flat ribbon of stainless steel is used for the wrapping (*flatwound* strings). Other variations are the *flatground* string (wound with round wire that is then

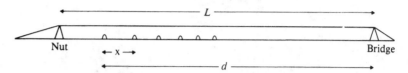

FIGURE 10.39   Fret placement. According to the rule of eighteen, each fret is placed 1/18 of the remaining distance d to the bridge saddle, or $x = d/18$. (Greater accuracy is obtained by using 17.817 rather than 18.)

ground flat) and compound strings with a winding of silk between the steel core and metal outer windings.

Spacing the frets on the fretboard presents some interesting design problems. Semitone intervals on the scale of equal temperament correspond to frequency ratios of 1.05946 (see Section 9.4). This is very near the ratio $18:17$, which has led to the well-known *rule of eighteen*. This rule states that each fret should be placed 1/18 of the remaining distance to the bridge, as shown in Fig. 10.39. Obviously the fret spacing x decreases as one moves down the fingerboard.

Since the ratio 18/17 equals 1.05882 rather than 1.05946 (an error of about 0.06%), each semitone interval will be slightly flat if the rule of eighteen is used to locate the frets. By the time the twelfth fret is reached, the octave will be 12 cents (12/100 semitone) flat, which is noticeable to the ear. Thus for best tuning the exact figure 17.817 should be used in place of 18; in other words, each fret should be placed 0.05613 of the remaining distance to the bridge.

Another problem in guitar design is the fact that pressing down a string against a fret increases the tension slightly. This effect is much greater in steel strings than nylon, since a much greater force is required to produce the same elongation. Fretted notes will tend to be sharp compared to open ones. The greater the clearance between strings and frets, the greater this sharpening effect will be.

To compensate for this change in tension in fingering fretted notes, the actual distance from the nut to the saddle is made slightly greater than the scale length used to determine the fret spacings. This small extra length is called the string *compensation*, and it usually ranges from between 1–5 mm on acoustic guitars to several centimeters on an electric bass. Bass strings require more compensation than treble strings, and steel strings require considerably more than nylon strings. A guitar with high action (larger clearance between strings and frets) requires more compensation than one with a lower action. Some electric guitars have bridges that allow easy adjustment of compensation for each individual string.

10.17 ■ SUMMARY

The vibrating strings of a violin radiate very little sound directly, but they transmit their vibrations to the violin body. Plucking a string at the center tends to excite modes that are odd-numbered harmonics of the fundamental. Bowing a string causes a bend to propagate along the string, and the displacement of each point on the string describes a sawtooth curve as a function of time. As the bend passes the point of contact, it initiates "sticking" and "slipping" of the string and bow.

The violin body has two strong resonances, called the main air and main wood resonances. Tuning the top and back plates during construction is made much easier by using Chladni patterns to observe the modes of vibration. The resonances of violas and cellos lie somewhat higher in relationship to their open strings than those of a violin. Recently a new family of eight fiddles has been designed to have resonances scaled to those of the violin.

The guitar also has two resonances called air and main wood, which largely determine the low-frequency characteristics. The frequencies of these two resonances are determined by the frequencies of fundamental vibration mode of the top plate and the Helmholtz air resonance plus the amount of coupling between them. The vibration modes of the top plate depend on the plate stiffness and bracing. Electric guitars have pickups, which sense the vibrations of the strings directly. Guitars may have either steel or nylon strings, plain or wrapped. Strings are made slightly longer than the calculated scale length in order to compensate for the change in tension when a string is pressed against a fret.

REFERENCES AND SUGGESTED READINGS

Alonso Moral, J., and E. Jansson (1982a). "Eigenmodes, Input Admittance, and the Function of the Violin," *Acustica* **50**: 329.

Alonso Moral, J., and E. Jansson (1982b). "Input Admittance, Eigenmodes, and Quality of Violins," Report STL-QPSR 2-3/1982, Speech Transmission Laboratory, Royal Inst. of Tech. (KTH), Stockholm, p. 60.

Besnainou, C. (1995). "From Wood Mechanical Measurements to Composite Materials for Musical Instruments: New Technology for Instrument Makers." *MRS Bulletin* **20**(3): 34.

Bynum, E., and T. D. Rossing (1997). "Holographic Studies of Cello Vibrations," *Proc. Inst. Acoustics* **19**(5): 155.

Caldersmith, G. (1977). "Low Range Guitar Function and Design," *Catgut Acoust. Soc. Newsletter* **23**: 19. A more complete but more mathematical paper appears in *J. Acoust. Soc. Am.* **63**: 1566 (1978).

Caldersmith, G., and J. Williams (1986), "Meet Greg Small," *American Lutherie* (8): 30.

Cremer, L. (1984). *The Physics of the Violin*, trans. J. S. Allen. Cambridge: MIT Press.

Decker, J. A. (1995). "Graphite-epoxy Acoustic Guitar Technology," *MRS Bulletin* **20**(3): 37.

Dünnwald, H. (1983). "Auswertung von Geigenfrequenzgangen," *Proc. 11th ICA, Paris*, **4**: 373.

Fleischer, H., and T. Zwicker (1998). "Mechanical Vibrations of Electric Guitars," *Acustica* **84**: 758.

Fletcher, N. H., and T. D. Rossing (1998). *The Physics of Musical Instruments*, 2nd ed. New York: Springer Verlag. (See Chapters 9 and 10.)

Hacklinger, M. (1978). "Violin Timbre and Bridge Frequency Response," *Acustica* **39**: 323.

Haines, D. W., N. Chang, and C. M. Hutchins (1975). "Violin with Graphite-Epoxy Top Plate" (abstract), *J. Acoust. Soc. Am.* **57**: S21.

Hutchins, C. M. (1967). "Founding a Family of Fiddles," *Physics Today* **20**(2): 23.

Hutchins, C. M. (1973). "Instrumentation and Methods for Violin Testing," *J. Audio Eng. Soc.* **21**: 563.

Hutchins, C. M. (1977). "Another Piece of the Free Plate Tap Tone Puzzle," *Catgut Acoust. Soc. Newsletter* **28**: 22.

Hutchins, C. M., and V. Benade, Eds. (1997). *Research Papers in Violin Acoustics 1975–1993*, vols. 1, 2. Woodbury, N.Y.: Acoustical Society of America.

Hutchins, C. M., K. A. Stetson, and P. A. Taylor (1971). "Clarification of 'Free Plate Tap Tones' by Hologram Interferometry," *Catgut Acoust. Soc. Newsletter* **16**: 15. Also reprinted in *Musical Acoustics, Part 2*. Ed., C. M. Hutchins. Stroudsburg, Penn.: Dowd, Hutchinson and Ross, 1976.

Jansson, E. V. (1971). "A Study of Acoustical and Hologram Interferometric Measurements of the Top Plate Vibrations of a Guitar," *Acustica* **25**: 95.

Reder, O. (1970). "The Search for the Perfect Bow," *Catgut Acoust. Soc. Newsletter* **13**: 21.

Reinecke, W. (1973). "Ubertragungseigenschaften des Streichinstrumentenstegs," *Catgut Acoust. Soc. Newsletter* **19**: 26.

Richardson, B. E., and G. W. Roberts (1985). "The Adjustment of Mode Frequencies in Guitars: A Study by Means of Holographic Interferometry and Finite Element Analysis," *Proc. SMAC 83* (Royal Swedish Academy of Music, Stockholm).

Rossing, T. D., and G. Eban (1999). "Normal Modes of a Radially Braced Guitar Determined by Electronic TV Holography," *J. Acoust. Soc. Am.* **106**: 2991.

Rossing, T. D., J. Popp, and D. Polstein (1985). "Acoustical Response of Guitars," *Proc. SMAC 83* (Royal Swedish Academy of Music, Stockholm).

Schelleng, J. C. (1968). "Acoustical Effects of Violin Varnish," *J. Acoust. Soc. Am.* **44**: 1175.

Schelleng, J. C. (1974). "The Physics of the Bowed String," *Sci. Am.* **230**(1): 87.

Sloane, I. (1966). *Classic Guitar Construction*. New York: Dutton.

GLOSSARY

anisotropy The difference in some property when measured in different directions (such as the stiffness of wood along and across the grain).

bass bar The wood strip that stiffens the top plate of a violin or other string instrument and distributes the vibrations of the bridge up and down the plate.

belly The top plate of a violin.

bridge The wood piece that transmits string vibrations to the sound board or top plate.

Chladni pattern A means for studying vibrational modes of a plate by making nodal lines visible with powder.

compensation (string) An extra length of string added because tension changes when a string is pressed against a fret.

f-holes The openings in the top plate of a string instrument shaped like the letter f.

Helmholtz resonator A vibrator consisting of a volume of enclosed air with an open neck or port (see Sections 2.3 and 4.7).

humbucking pickup A magnetic pickup with two coils designed to minimize hum caused by stray magnetic fields.

mobility (mechanical admittance) The ratio of velocity to force (called input admittance or driving point mobility if velocity and force are measured at the same point).

nut The strip of hard material that supports the strings at the head end.

purfling The thin wood strip near the edge of the top or back plate of a string instrument.

saddle The strip of hard material (ivory or bone) that supports the string at the bridge of a guitar.

sinusoidal force A smoothly varying force with a single frequency; the waveform is described as a sine wave.

sound hole (rose hole) The round hole in the top plate of a guitar that plays an important role in determining the lower resonances of the body.

sound post The short round stick (of spruce) connecting the top and back plates of a violin or other string instrument.

sul ponticello Bowing near the bridge.

sul tasto Bowing near the fingerboard.

viol An early bowed string instrument usually having six strings and a fretted fingerboard.

viola da gamba A viol played in an upright position.

REVIEW QUESTIONS

1. What musical interval separates the open strings of a violin?

2. In what century did Antonio Stradivari and Guiseppi Guarneri make fine violins?

3. Describe the spectrum of a string plucked at its center.

4. Describe the spectrum of a string plucked at one-fifth of its length.

5. During how much of the bowing cycle does the string move with the bow?

6. How does a player increase the amplitude of vibration of a bowed string?

7. How are Chladni patterns created in a free violin top?

8. What is meant by *anisotropy* in a sheet of wood?

9. What is the approximate frequency of the *f-hole* (A_0) *resonance* of a violin?

10. What are the approximate frequencies of the first two resonances of a violin bridge?

11. How much below violin strings are viola strings tuned?

12. How much below violin strings are cello strings tuned?

13. What musical intervals separate the open strings of a guitar?

14. What is the approximate frequency of the lowest resonance of a guitar?

15. How does string tension in a folk guitar with steel strings compare to the tension in those of a classical guitar?

16. Why do most electric guitars have more than one set of pickups?

17. Each fret on a guitar is placed at what fraction of the remaining distance to the bridge?

18. What is meant by string compensation in a guitar?

QUESTIONS FOR THOUGHT AND DISCUSSION

1. Does increasing the force on a violin bow increase the loudness of the tone? Explain why.

2. Why does a folk guitar with steel strings play more loudly than the same guitar with nylon strings?

3. How might the resonances of a hollow-body electric guitar affect the tonal output? (Remember that the pickups sense string motion only.)

4. Electric guitars (especially those with a hollow body) are susceptible to acoustic feedback (see Section 24.6), even though they have no microphone. Explain why this occurs and how it can be prevented.

5. Why do classical guitarists hold the back of the guitar away from their body when they play?

EXERCISES

1. Continue the sketches in Fig. 10.4 to show the shape of the plucked string during the next half cycle from $t = \frac{1}{2}T$ to $t = T$.

2. In Figs. 10.3 and 10.5, note that plucking a string one-fifth the distance from one end suppresses the fifth harmonic, and plucking it at the midpoint (one-half the distance) suppresses the second harmonic. Also note that the phase of the harmonics (indicated by + or −) changes in going through a zero. Using this information, draw a similar diagram to show the addition of modes to obtain the shape of a string plucked at one-third its length.

3. Assuming a frequency of 440 Hz and a bow speed of 0.2 m/s in Fig. 10.7, what is the displacement of the string at its midpoint? What is the speed of the string when it leaves the bow and "snaps" back? (*Hint*: First determine the time during which the string moves at the speed of the bow.)

4. Note the similarity between the Chladni patterns in Fig. 10.16 and the holograms in Fig. 10.17 for modes II and V. Sketch a Chladni pattern that one might expect

for mode I in Fig. 10.17.

5. If the main resonances (A and T in Fig. 10.19) of the alto violin in the new family of fiddles are scaled to those of the conventional violin, near what notes will they lie? (The alto is tuned a fifth below the violin.) Compare these resonances to those of the viola.

6. Determine the musical intervals between the strings of the guitar (see Section 10.9) and those of the electric bass (see Section 10.15).

7. Calculate the two lowest resonance frequencies of a pipe 46 cm long closed at both ends (they are the same as those a pipe open at both ends; see Section 4.5). Do the same for a pipe 33 cm long. Now compare these frequencies to those given for the resonances A_2 and A_4 (longer pipe) and A_3 and A_5 (shorter pipe) in Fig. 10.27(c). Discuss the significance of the similarity.

8. Carefully measure the distance of each fret of a guitar from the saddle of the bridge, and determine how closely the rule of eighteen has been followed. If done carefully, you can determine how much compensation is included for each string.

EXPERIMENTS FOR HOME, LABORATORY, AND CLASSROOM DEMONSTRATION

Home and Classroom Demonstration

1. *Motion of a plucked string* Straddle a steel string (on a guitar or a monocord) with a magnet and display the voltage induced by the moving string on an oscilloscope. Move the magnet to different locations on the string.

2. *String force*   Place a guitar force transducer on the saddle of a guitar bridge and note the force waveform when the string is plucked. An inexpensive force transducer can also be placed under a monocord string.

3. *Motion of a bowed string*   Orient a magnet with its magnetic field passing vertically through a steel string so that it senses the horizontal motion. Bow the string at various points and note the waveform on an oscilloscope. Move the magnet to different locations on the string.

4. *Following bow*   Set up Schelleng's following-bow demonstration (see Fig. 10.11).

5. *Plate vibrations*   Chladni patterns on square and rectangular plates of wood and metal can be displayed by supporting the plate on pieces of foam over a loudspeaker. Alternatively, the plate can be mounted on an electromagnetic vibrator. Of particular interest are the modes with nodal lines forming patterns such as +, ×, and ○.

6. *Violin and guitar plates*   Chladni patterns of free violin and guitar plates can be made in the same manner as in Demonstration 5. (Apparatus suppliers sell a flat metal plate cut in the shape of a violin plate, but a carved plate is better.)

7. *Tap tone frequencies*   If an FFT spectrum analyzer is available, the tap-tone frequencies of violin and guitar plates can be determined. A guitar top plate without bracing typically has its lowest resonance around 50 Hz; with bracing this rises to around 80 or 90 Hz. A complete guitar has its lowest resonance around 100 Hz.

8. *Fret locations*   Measure the distance L_0 from the bridge saddle to the nut and the distance L_1 from the saddle to the first fret. The ratio L_1/L_0 should be close to 1.06. The ratio L_{12}/L_0 for the twelfth fret should be close to 2 (an octave).

9. *Electric guitar output*   Feed the output waveform of an electric guitar to an oscilloscope as well as into the guitar amplifier. Show the difference between plucking in the horizontal plane and in the vertical plane. Show the different signals from each of the pickups and relate each one to the velocity waveform of the string at that point.

10. *Longitudinal string vibrations*   Rather faint "scratch" tones can be excited on wrapped strings of a guitar by scratching them longitudinally with a fingernail. Stopping the strings at the frets raises the pitch of these scratch tones (in fact you can play a tune), but changing the tension on the string does not.

11. *Bowing at different positions*   Show the effect of bowing a violin at different positions along the string.

12. *Violin mute*   Show the effect of loading a violin bridge with a mute or other added mass.

13. *Fret buzz*   Show that a strong upward pluck of a guitar near the sound hole causes the string to "buzz" against the frets, whereas a downward pluck does not.

14. *Partials of a guitar string*   Second part of Demonstration 28 in the *Auditory Demonstrations CD*.

15. *Harmonics of touching the string*   When a plucked guitar string is lightly touched at its center to damp out the fundamental, you hear the second harmonic an octave higher. Similarly, touching it at one-third of its length produces tones with frequencies 1.5 times and 3 times the fundamental, which sound a fifth and a twelfth (octave plus a fifth) above the fundamental.

Laboratory Experiments

Acoustics of a guitar (Experiment 17 in *Acoustics Laboratory Experiments*).

Acoustics of bowed string instruments (Experiment 18 in *Acoustics Laboratory Experiments*).

Vibrations of plates (Experiment 7 in *Acoustics Laboratory Experiments*).

Brass Instruments

In Chapters 11 and 12, we will discuss a wide variety of wind instruments, or aerophones, which differ widely in their construction and in their acoustical properties. It is customary to classify them into two families, the brasses and the woodwinds. When a brass instrument is played, the player's lips act as a valve, introducing puffs of air at just the right time to maintain oscillations of the air column. When a woodwind is played, an oscillating air stream or an oscillating reed excites the air column. The brass instruments radiate sound from a flared end of the tube, called the *bell*; woodwinds usually radiate sound from several holes in the sides of the air column. Other differences will become clear in these next two chapters.

In this chapter you should learn:

- About the acoustic impedance of various brass instruments;
- How a player's lips function as a pressure-controlled valve;
- About functions of the mouthpiece, bell, and valves;
- About sound spectra of brass instruments.

In ancient times, people blew animal horns, which served as sound sources for religious ceremonies and as warnings of danger. Attempts to use these horns to play music, however, apparently began during the Middle Ages. Later, wood and metal tubing were substituted for the horns of animals, and mouthpieces were added to simplify playing. Among the brass instruments that have survived from the Renaissance and Baroque eras are the valveless posthorn; the cornett, played with side holes; and the sackbut, forerunner of the trombone.

11.1 ■ INSTRUMENTS OF THE BRASS FAMILY

The principal members of the brass family are the trumpet, French horn, trombone, and tuba (see Fig. 11.1(a)). Their playing ranges are indicated in Fig. 11.1(b). Other important brass instruments are the cornet, fluegelhorn, bugle, and baritone horn.

Brass instruments have four sections: a *mouthpiece*, a tapered *mouthpipe*, a *cylindrical* section, and a *bell*, as shown in Fig. 11.2. The trumpet, French horn, and trombone have cylindrical sections of considerable length, as indicated in Table 11.1, whereas the fluegelhorn, baritone, and tuba, often referred to as instruments of conical bore, are tapered throughout much of their length.

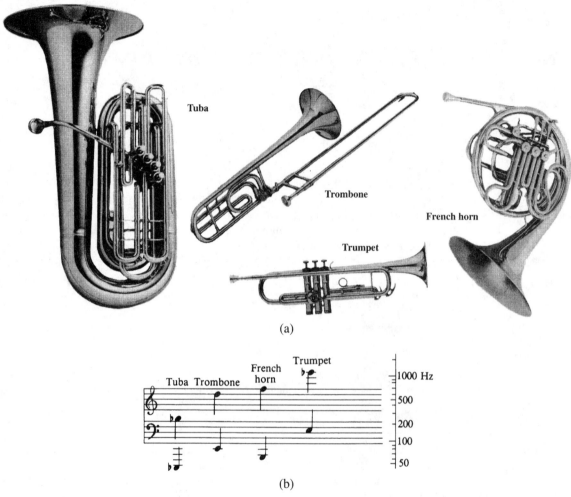

(a)

(b)

FIGURE 11.1   (a) Tuba, trombone, French horn, and trumpet; (b) their playing ranges. (Courtesy of C. G. Conn, Ltd.)

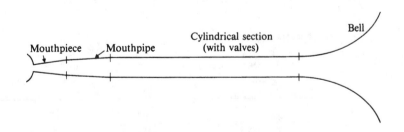

FIGURE 11.2
A cross section of a
brass instrument.

TABLE 11.1   Comparison of brass instruments

	Trumpet	French horn	Trombone	Tuba	Baritone
Fundamental	B^b_2	F_1	B^b_1	B^b_0	B^b_1
Lowest open note	B^b_3	F_2*	B^b_2	B^b_1	B^b_2
Length	140 cm	375 cm	275 cm	536 cm	264 cm
Bore (diameter) of main tube	1.1 cm	1.1 cm	1.2 cm	1.8 cm	1.3 cm
Cylindrical portion	53 cm	193 cm	170 cm		
Bell diameter	11 cm	32 cm	18 cm	35–60 cm	25 cm

*Many French horns are "double horns" composed of two horns tuned to F_2 and B^b_2.

11.2 ■ OSCILLATIONS IN A PIPE

The air column of a B^b trumpet has a length of about 140 cm. As we discussed in Section 4.5, the resonances of a closed cylindrical pipe of this length should be

$$f_n = n\frac{v}{4L} = n\frac{343}{4(1.4)} = 61.3n \qquad (n = 1, 3, 5, 7, \ldots)$$
$$= 61, 184, 306, 429, \ldots \text{ Hz.}$$

If we blow softly on the end of the pipe in the manner of a brass player (this may be easier if a smooth ring is attached to the end), the frequencies obtainable are very near those calculated here. (The fundamental, which is difficult to blow with the lips, can be sounded by attaching a clarinet mouthpiece.)

One way to study the resonances of a pipe (or a wind instrument) is to make a graph of its *acoustic impedance* as a function of frequency. (Acoustic impedance, defined in Sec-

FIGURE 11.3
Apparatus for graphing acoustic impedance of a wind instrument, as used by Benade (1973), Backus (1976), and other investigators. (Copyright ©1976 by Virginia Benade. Reprinted by permission.)

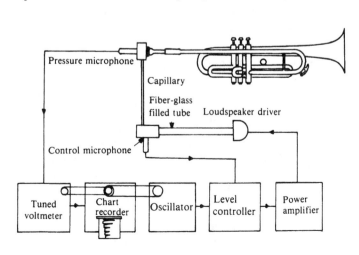

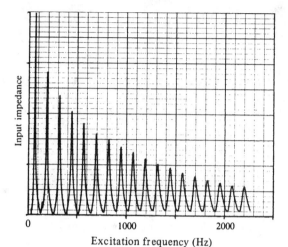

FIGURE 11.4
Impedance curve
for a cylindrical
pipe 140 cm long
(about the length of
a trumpet). (From
Benade 1976)

tion 4.6, is the sound pressure divided by the volume velocity.) An apparatus for doing so is shown in Fig. 11.3. A loudspeaker driver forces sound through a capillary tube designed to produce a volume velocity of constant amplitude. A pressure microphone next to the mouthpiece detects pressure variations; because the velocity amplitude is kept essentially constant, the pressure peaks and valleys will be a replica of the impedance peaks and valleys.

Figure 11.4 shows the impedance peaks of a trumpet-length cylindrical tube 140 cm long.* Note that the frequencies are the odd-numbered harmonics of the fundamental, as shown in Fig. 4.8. At the frequency of one of these pressure (impedance) peaks, the tube cooperates with the lips, because the excess pressure during a portion of the cycle helps to open the lips to admit air at the right time. Energy is thus supplied to the vibrating air column, which makes up for various losses (including the radiation of sound) and sustains the oscillations. The blowing pressure in the mouth must be maintained high enough so that when the lips open momentarily and emit a burst of air pressure, this burst will be strong enough after one round-trip to the bell and back to again open the lips in the same way.

11.3 ■ PRESSURE-CONTROLLED VALVES

Consider what happens when a brass player's lips open for a moment to admit a puff of air into the mouthpiece. A standing wave builds up in the horn, but it immediately begins to die out, like a plucked guitar string, as the energy is dissipated at the walls of the tube or radiated as sound from the bell. In order to sustain the oscillation, the player must continue to supply puffs of air at appropriate times, not unlike the regular pushes applied to a child in a swing.

A steady flow of air into the instrument will not sustain the oscillation any more than a steady force applied to a swing, because energy would be added during half the cycle and

*Graphs showing impedance versus frequency, called *impedance curves*, should not be confused with spectra, which show sound output or vibration amplitude versus frequency (as in Fig. 7.11 or Fig. 10.3, for example).

removed during the other half. To maintain oscillation, air must be added at the appropriate part of each cycle: when the mouthpiece pressure is high. It would be difficult, if not impossible, for the player to synchronize his or her lip opening by muscular action alone; fortunately, this is not necessary. Pressure pulses reflected back from the horn tend to force the player's lips open at the right time during the cycle of oscillation. This *regenerative* or *positive feedback* is a characteristic of all oscillators, mechanical, acoustical, or electrical.

The cooperation between lips and air column to sustain oscillations is similar to the electrical feedback that sustains oscillations in an audio generator (see Chapter 18). It is equivalent to having a pressure-controlled valve that admits air whenever the pressure is high, as shown in Fig. 11.5(a). An analysis of the cooperation between a pressure-controlled valve and a vibrating air column is given by Helmholtz (1877), but a much more lucid discussion is given by Benade (1973, 1976).

Benade's "water trumpet," shown in Fig. 11.5(b), illustrates the principle of pressure-controlled feedback with sloshing water substituted for the longitudinally vibrating column of air. When the water level rises at the mouthpiece end, a valve opens to admit more water, and the water piles up into a larger wavecrest. A steady water pressure in the supply pipe has been used to add energy to the water waves in the trough and thus compensate for the losses incurred due to friction, leakage at the bell end, and so on.

We will not go into the mathematical details of the pressure feedback that controls the player's lips. It will suffice to point out that stable oscillation can be maintained when the horn oscillation frequency is *above* the natural resonance frequency of the lips. The player's lips, which were described by Helmholtz (1877) as an "outward-striking reed," have a natural resonance frequency determined by their mass and tension (which can be adjusted by the player).

A very important principle in describing the cooperation of a pressure-controlled valve with a vibrating air column is the following.

FIGURE 11.5
(a) A pressure-controlled valve (player's lips) admits air at the time when air pressure is a maximum, thus sustaining the oscillation. (b) A water trumpet, which illustrates the same principle with sloshing water in a trough. (From Benade 1976.)

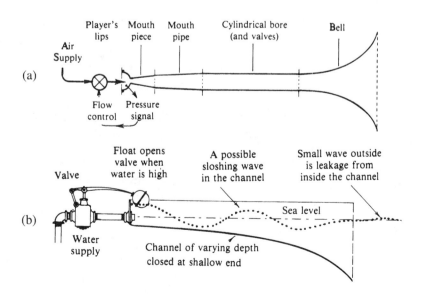

> If the flow rate is not proportional to the pressure (that is, if the valve is nonlinear), oscillation is favored when the air column has one or more resonances that correspond to the partials (overtones) of the tone being produced.

Except at the softest level of playing, the vibrating air column of a brass instrument will have many harmonics. If the pressure peaks of each of these partials reach the lips at different times, cooperation between the lips and the air column will be difficult. However, if these pressure peaks arrive at the same time, they add together, and the oscillation is stabilized. This is equivalent to the statement in the box above: that oscillations are favored when these partials correspond to the resonances of the air column itself. The oscillation of an air column with nonlinear excitation has been carefully studied by several investigators and is summarized in Chapter 20 of Benade (1976), by Fletcher (1979), and by Fletcher and Rossing (1998).

11.4 ■ THE BELL AND MOUTHPIECE

All brass instruments have flared bells. The bell has several important effects on the acoustics of the brass instruments:

1. It changes both the frequencies and the heights of the impedance peaks.
2. It changes the radiation pattern of the horn, making it more directional at high frequencies.
3. It changes the spectrum of the radiated sound.
4. It allows more efficient radiation of sound by matching the high pressure inside the horn to the lower pressure outside.

Figure 11.6 shows the impedance curve of a length of trumpet tubing with a trumpet bell attached. Comparison with Fig. 11.4 indicates that the peaks are shifted toward lower frequencies, but in a slightly different way than would occur from a mere lengthening of the tube. The higher peaks have been lowered proportionately more than the lower peaks. This is due to a basic property of flared horns: The effective length increases with frequency. The manner in which the effective length changes with frequency depends on the flare of the bell (Pyle, 1975). The *turning point** of the wave can be observed by noting the sound level at a small microphone as it is slowly inserted into the bell; at sufficiently loud playing levels, one can feel the turning point by inserting a finger into the bell.

Another experiment is to disconnect the bell by removing the tuning slide of a trumpet or trombone. The instrument now plays out of tune, and would do so even if enough tubing were added to match the effective length of the instrument including the bell.

Also apparent in Fig. 11.6 is the virtual disappearance of peaks above 1500 Hz when a bell is added to the pipe (compare with Fig. 11.4). At the same time, the higher frequencies become more prominent in the radiated sound. These two related effects both result from the more efficient radiation of higher frequencies by the bell. At the frequencies of the

---

*The point of furthest penetration of the standing sound wave into the bell.

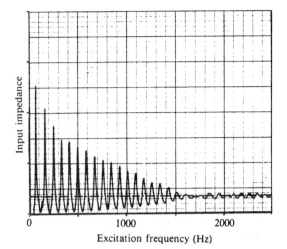

FIGURE 11.6
Impedance curve for a 140-cm length of trumpet tubing with a trumpet bell attached. Compare this with Fig. 11.4. (From Benade 1976.)

lower peaks, most of the sound is reflected back from the bell, and the amplitude builds up; at higher frequencies, however, a substantial part is radiated into the room.

The mouthpiece also shifts the frequencies of the impedance peaks, although it has a more dramatic effect on the peak heights. Peaks (as well as valleys) in the vicinity of the lowest resonance of the mouthpiece are greatly enhanced. The frequency of the lowest resonance of the mouthpiece, which usually occurs in the range 750 to 850 Hz, is often called its *popping frequency*, because it can be determined by slapping the rim with the palm of the hand and noting the frequency of the resulting pop.

The impedance curve of a $B^\flat$ trumpet is shown in Fig. 11.7. Comparisons with Figs. 11.4 and 11.6 point out the enhancement of impedance (pressure) peaks in the vicinity of the mouthpiece resonance. These large pressure peaks greatly enhance the ability of the air column to control the rather massive lips of the player. Skillful design of the bell and

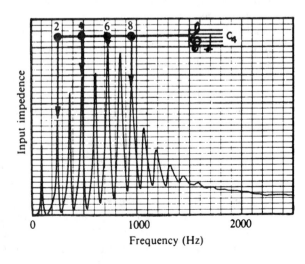

FIGURE 11.7
Impedance curve of a $B^\flat$ trumpet, showing peaks that form a "regime of oscillation" for C_4. (From Benade 1976.)

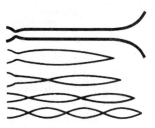

FIGURE 11.8 Approximate pressure distribution for the first four modes in a trumpet. Note that the turning point moves outward in the bell as the frequency increases. Mode frequencies are nearly in the ratios 0.8 : 2 : 3 : 4.

the mouthpiece have now brought the peak frequencies, except for the lowest one, very nearly into a harmonic relationship, so that the air column resonances will correspond to the partials of the desired tone.

Also indicated in Fig. 11.7 are the peaks (2, 4, 6, and 8) that correspond to the partials of B^b_3 ($f = 233$ Hz; this is the written C_4 for a B^b instrument). When the note is blown very softly, the player's lips vibrate nearly sinusoidally, and only peak 2 is of importance. As the level of loudness increases, more harmonics are produced, which excite peaks 4, 6, and 8. The tone takes on increasing stability as the lips become subject to control by pressure impulses corresponding to all four peaks.

Note that the lowest peak in Fig. 11.7 lies below the fundamental frequency (about 117 Hz, corresponding to B^b_2). It is possible to play the "missing" fundamental note, sometimes called the pedal tone, by relying on feedback from other peaks having harmonically related frequencies. It is even possible to play at the frequency of the lowest peak, but that is not a useful note at all on a trumpet. Figure 11.8 shows the approximate pressure distribution for each of the first four resonances in Fig. 11.7. Note the way in which the *turning point* in the bell changes with frequency.

It is clear from Fig. 11.7 why the notes above F_5 (written G_5) or so become increasingly difficult to play. The peaks that correspond to the harmonics are small, which indicates that the corresponding resonances are too weak to be of much help in controlling the lips and stabilizing the tone. F_5 is quite easy to play softly, because peak 6 is very tall, indicating a strong resonance; little increase in stability is noted during crescendo, however, because peak 12, which lies near the second harmonic, is small. Above B^b_6 (written C_6), even the fundamental is weakly supported by the air column, and the lips must provide their own stability, which makes great demands on the player's ability to control his or her lips. This control makes use, to a small extent, of a velocity-dependent force called the *Bernoulli* force.

The Bernoulli Force

When a fluid flowing through a pipe enters a region in which the area of the pipe decreases, the speed of the fluid increases. To thus accelerate the fluid requires a net force, so the pressure in the larger pipe behind must be greater than in the smaller pipe ahead, as shown in Fig. 11.9(a). The reduced pressure in the center section of the tube causes the liquid in the U-tube to stand at a higher level. This effect (i.e.,

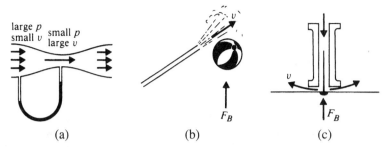

large p
small v small p
              large v

v

F_B v F_B

(a)   (b)   (c)

FIGURE 11.9   The Bernoulli effect. (a) The reduced pressure in the center section of the tube causes the liquid in the U-tube to stand at a higher level. (b) The reduced pressure in the air jet from a vacuum cleaner hose gives rise to a net upward force F_B, which can support a beach ball. (c) Blowing downward through the spool causes air to flow outward, supporting a card and pin by means of the Bernoulli force.

reduction in pressure when flow velocity increase) was described by Daniel Bernoulli in 1738 and now carries his name.

Simple experiments that illustrate the Bernoulli effect are shown in Fig. 11.9(b) and (c). In each case, the reduced pressure in the moving air stream gives rise to a net upward force F_B, which is often referred to as the *Bernoulli* force.

11.5 ◼ VALVES AND SLIDES: FILLING IN BETWEEN THE MODES

The instrument we have been discussing thus far is a valveless trumpet like the familiar military bugle. It is quite possible to play eight or more notes, corresponding to the peaks in Fig. 11.7, and a few more may be added by skillful playing. To play the remaining notes of the scale, however, the acoustical length of the instrument must be varied. This is done by moving a slide (trombone) or by inserting lengths of additional tubing by means of valves (trumpet, French horn, baritone, tuba, etc.). A French horn normally uses rotary valves, but most other brass instruments in the United States use piston valves of the type shown in Fig. 11.10. In the raised position, a hole in the piston, called the *windway*, extends

FIGURE 11.10
Action of a piston
valve: (a) piston up;
(b) piston down.

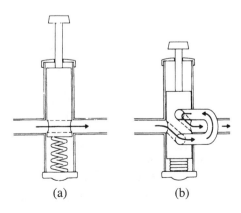

(a)   (b)

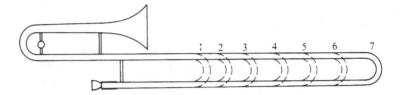

FIGURE 11.11
Playing positions of
a trombone slide.

the air column straight through the valve. When the valve is depressed, however, two other windways connect the air column to an additional length of tubing.

The playing positions of a trombone slide are shown in Fig. 11.11. Seven positions of the slide are needed to play all desired notes of the scale; note that the seven positions are not spaced equally. Going down a semitone decreases the frequency by about 6%, so the instrument should be lengthened by about this same amount. But as the instrument gets longer, a 6% increase in length also becomes progressively longer.

Another complication arises in determining the lengths of tubing to be added. The additional length of tubing added by moving a trombone slide or by depressing a trumpet piston valve must be cylindrical. Thus the average flare of the entire horn is reduced. The result is that the addition of a piece of tubing makes a larger percentage change in the lower modes than in the higher ones.

Another fundamental problem arises in designing the additional lengths of tubing for each valve. If the lengths of tubing are designed so that the first and second valves can lower the pitch by a whole tone and a semitone, respectively, then the combination will be inadequate to lower the pitch by three semitones. This can be illustrated by the following simple arithmetic.

To lower the pitch a whole tone, the first valve must add a length of tubing 12.2% or the total length L, or $0.122L$. To lower the pitch a semitone requires the addition of 5.9% of the total length, or $0.059L$. Thus the two valves together increase the length by 18.1%, or $0.181L$. But three semitones requires an increase of 18.9% (see Table 9.1).

The same conclusion can be reached by examining Fig. 11.10. The length of tubing needed to go one semitone down from the second to the third position is greater than that needed to go from the first to the second position.

Various compromises have been proposed to deal with this tuning problem (Young 1967). The valve slides for the first and second valves may be made slightly longer than the optimum length, thus accepting a slight flattening of the first two semitones in order to reduce the sharpening of the third one. Similarly, the third valve slide may be designed to give a shift slightly more than three semitones. Since the discrepancy is usually greatest when the third valve is used in combination with the others, most trumpets have a provision for moving the third valve slide with the left hand while the instrument is being played. Beyond that, the brass player must "lip" the various notes into tune, that is, make small corrections in frequency by changing the tension of the lips.

The valve problem becomes increasingly difficult in the larger brass instruments. A fourth valve is frequently added to the tuba; it lowers the pitch by a fourth, and thus substitutes for the troublesome combination of valves 1 and 3. Even a fifth or sixth valve is sometimes added, their function varying in the different tubas.

11.6 ■ THE FRENCH HORN

The French horn in common use is a double horn composed of two horns tuned to F_2 and $B^\flat{}_2$, which share the same mouthpiece and bell. The three main valves each have two sets of windways and valve slides, and a fourth valve is used to switch from one horn to the other. The $B^\flat$ horn is preferred for playing high notes, because it has stronger resonances at the higher frequencies, and they are spaced further apart.

The length of the F-horn is about 375 cm (more than 12 ft), 30% longer than the trombone. It has many resonances, and much of the time the horn is played at a pitch corresponding to one of the higher resonances. This means that many notes on a French horn lack the stability of notes normally played on a trumpet or trombone, where there are several prominent resonances corresponding to the overtones of the note played.

Placing the hand in the bell of the horn makes it easier to play the higher notes. By inhibiting the radiation and increasing the reflection of the higher frequencies, it allows standing waves to build up and leads to more usable resonances at the higher frequencies. Figure 11.12 shows the impedance curves of a French horn with and without a hand in the bell. The additional resonances in Fig. 11.12(a) stabilize the higher notes by providing feedback at the overtone frequencies.

Placing the hand in the bell also lowers the pitch. If the hand is inserted as far as possible, a *stopped* tone is produced, which appears to be about a semitone higher than normal. What actually happens is that all the resonances are lowered to the point where the next higher resonance is only a semitone above the desired note. This can be confirmed by holding a

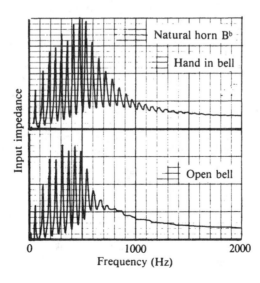

FIGURE 11.12
Impedance curves of a $B^\flat$ horn with and without a hand in the bell. (From Benade 1976.)

note as the hand is slowly inserted to the stopped position. If no attempt is made to hold the original pitch with the lips, the pitch will fall slowly. If the player tries to hold the pitch, however, the note jumps up to the next higher resonance at some point as the hand is inserted (Backus 1977).

11.7 ■ THE TROMBONE

The trombone, like the trumpet, has a bore that is predominantly cylindrical. Its tubing has twice the length of a trumpet, so its main resonances occur at frequencies that are approximately harmonics of a missing fundamental near $B^\flat_1$ (58 Hz). The lowest note that is played in "first" position (i.e., with the slide fully retracted, see Fig. 11.11) is normally $B^\flat_2$, although the $B^\flat_1$ pedal note in a trombone is a more useful note than in a trumpet, and occasionally is called for in musical scores.

The playing range of the standard tenor trombone extends up to D_5 (the tenth harmonic of the fundamental $B^\flat_1$). The bass trombone has a slightly larger bore and a mouthpiece of larger volume than the tenor, making it easier to play the lower notes. Most bass trombones, and some tenor trombones as well, include a rotary valve and additional tubing to lower the fundamental to F_1.

Valve trombones use valves to change the effective length rather than a slide, but they are much less common.

11.8 ■ TUBA, BARITONE, AND FLUEGELHORN

The tuba, baritone horn, and fluegelhorn are examples of instruments that are essentially conical throughout their entire length. The baritone horn has roughly the same playing range as a trombone, but with a conical bore, it has a different timbre.

The small tuba in $B^\flat_2$ has much the same playing range as the bass trombone. Larger tubas tune to $E^\flat_2$ and $B^\flat_1$. One version of the large tuba, popular in marching bands, is called a *sousaphone* in honor of bandsman John Philip Sousa. It coils around the player's body, which makes it easier to carry.

The fluegelhorn is the soprano of the conical brass instrument family. It has a playing range similar to that of the trumpet, and its mellow tone has made it a popular solo instrument in jazz ensembles.

11.9 ■ THE SPECTRA OF BRASS INSTRUMENTS

"Typical" sound spectra of the various instruments frequently appear in books and articles about musical instruments. These can be a little misleading, however, because the spectrum of an instrument changes substantially with changes in pitch and loudness. This statement is especially true of the brass instruments.

We have already discussed (in Section 11.3), how pianissimo playing on a trumpet uses mainly the resonance that matches the fundamental frequency of the played note, but increasing the sound level brings additional resonances into prominence. Thus the spectrum of the sound inside the instrument takes on more and more harmonics as the playing level increases. However, the internal spectrum is not what we hear.

In any wind instrument, the spectrum of the radiated sound depends both on the spectrum of the standing waves within the instrument and the portion of the sound energy that leaks to the outside. In a brass instrument, this portion is determined by the radiation efficiency of the bell, and it changes markedly with frequency.

> For a cylindrical pipe with an open end, the radiation efficiency increases with f^2 up to a certain frequency f_c and then remains more or less constant. The value f_c is called the *cutoff frequency* and is approximately $f_c = c/\pi a$, where c is the speed of sound and a is the radius of the pipe. The radiation efficiency of a small pipe thus quadruples each time the frequency goes up one octave, giving a treble boost of 6 dB per octave.
>
> For a pipe equivalent to the trumpet bore the cutoff frequency will be nearly 20 kHz. Adding a bell lowers the cutoff frequency several octaves, however.

Figure 11.13 illustrates how the internal spectrum of a trumpet may be combined with the radiation curve (sometimes called the spectrum transformation function) to obtain the spectrum of the radiated sound. Note that the internal and external spectra move up and down the frequency axis as the pitch changes; the radiation curve does not, because it is determined by the shape of the instrument. Thus the various harmonics receive different amounts of treble boost for different notes on the scale. The curves in Fig. 11.13(a) and (c) are for $B^\flat_4$ (466 Hz); they would be quite different for notes in other parts of the playing range. The curve in Fig. 11.13(b) would not change, however.

Recording spectra of the brass instruments is further complicated by the fact that sounds of high frequency are radiated straight ahead in a relatively narrow beam, whereas sounds of low frequency are radiated in all directions, due to the phenomenon of diffraction (see Section 3.10). In addition, the acoustics of the room have a strong influence on the sound spectrum if the microphone is any appreciable distance away from the bell. In spite of these uncertainties, the increasing prominence of the higher harmonics in the louder sounds is clear.

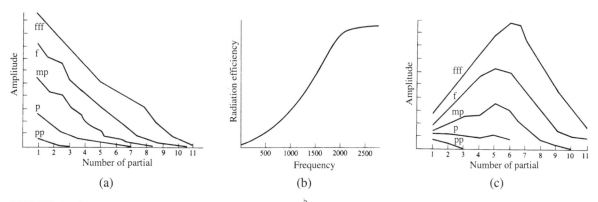

(a)   (b)   (c)

FIGURE 11.13   (a) Spectrum of sound inside a trumpet for $B^\flat_3$ (233 Hz). (b) Radiation curve for a trumpet. (c) Spectrum of radiated sound obtained by combining (a) and (b). (From Benade 1976.)

11.10 ■ SHOCK WAVES

Only a small part of the sound energy that reaches the bell of a brass instrument is radiated externally. The rest of it is reflected back into the instrument, where the sound builds up to ear-shattering levels. Sound levels as high as 175 dB have been measured inside a trumpet (Long 1947). At such levels, acoustic shock waves can form. The formation of shock waves is rather complicated, but a simple description takes into account that pressure maxima travel at a slightly higher speed than pressure minima (just as the crests of an ocean wave travel faster than the troughs, leading to breaking waves).

Hirschberg et al. (1996) concluded that shock waves generated in the long cylindrical part of the trombone bore are responsible for the change in tone color in extreme fortissimo playing. They show that this phenomenon is associated with progressive sharpening of an already large pressure gradient in the sound wave and that a long cylindrical tube is required in order for a shock wave to form. For this reason, shock waves are more likely to be found in instruments such as a trombone and a trumpet than in more conical brass instruments, such as the cornet or saxhorn.

11.11 ■ TRANSIENTS

Starting transients are an important identifying feature in all musical instruments, as we learned in Section 7.11. In brass instruments, the lip generator can operate autonomously with the assistance of pipe resonances (it is largely this feature that makes it possible to play reliably the notes in the upper registers). However, we would expect a different transient behavior for partials above and below the cutoff frequency of the instrument. Indeed, Luce and Clark (1967) found that the partials below the cutoff frequency of the bell build up together and reach their steady-state values simultaneously. Partials above cutoff, on the other hand, build up more slowly and the time to reach steady state increases at higher frequency.

When a player starts to blow a note, his or her lips are more or less on their own until the first reflected wave comes back to initiate the pressure-feedback process (Section 11.3). Normally the wave will travel all the way to the bell, so the time for the round trip is about equal to one period of the first mode, irrespective of the note actually being played. For a high-register note, the lips will have to go through several oscillation periods before they receive help from the horn. To attack the note correctly right from the start takes well-trained lips, as brass players well know. This is especially true of a French horn, which is almost always played in a high register.

Of course smaller reflections occur at discontinuities, especially at a partially closed valve. This is one possible advantage cited for European-style brass instruments with rotary valves located quite close to the mouthpiece.

11.12 ■ MUTES

Mutes of varying designs are used to alter the timbre of brass instruments. Three types of trumpet mutes are shown in Fig. 11.14. The straight mute is a truncated cone closed at the large end with three narrow cork pads spaced around the cone so that an air space of

FIGURE 11.14
Three types of
trumpet mute:
(a) straight mute;
(b) cup mute;
(c) Harmon, or
wah-wah, mute.

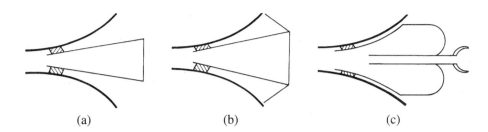

(a)   (b)   (c)

about 3 mm in thickness is maintained between the bell and the mute. The cup mute adds a second cone at the large end of the original cone. The Harmon, or "wah-wah," mute is a metal enclosure of a special shape with a small adjustable bell at one end.

Backus (1976) has studied the acoustical effects of trumpet mutes. His results can be summarized as follows:

1. The mute changes the radiation characteristics of the bell, reducing the radiation at low frequencies much more than the radiation at high frequencies.
2. An extra peak is added to the impedance curve around 100 Hz, associated with the resonance of the mute itself. Because this peak is below the playing range of the instrument, it has little or no effect on its playing characteristics.
3. The straight mute acts as a high-pass filter, letting through frequencies above 1800 Hz; the cup mute passes a band of frequencies in the range 800 to 1200 Hz; the Harmon mute passes a band of frequencies from 1500 to 2000 Hz.
4. Several tiny peaks in the impedance curve above 1300 Hz are enhanced; by reducing the radiation from the bell, which is usually large at these frequencies, build-up of standing waves is encouraged.

The spectra of trumpet tones with and without mutes in Fig. 11.15 were recorded by Richard Ross in our laboratory. They show substantial, but not complete, agreement with Backus's results.

11.13 ■ WALL MATERIAL

The question of how the playing qualities of a brass instrument depend on the material used in its construction has been debated for many years. Brass instruments are nearly always made from thin-walled brass tubing, although bells have been crafted from sterling silver and other materials. Large tubas (sousaphones) of synthetic material are used in marching bands. Experimental brass instruments made of other metals and even wood are reported to be indistinguishable from instruments of brass. It is probably safe to say that the quality of the craftsmanship in a brass instrument is more important than the type of material used.

The walls of the brass tubing vibrate during playing, and whereas the walls radiate in a negligible amount of sound, wall vibrations can contribute to damping. The friction of air against the walls is also a factor. However, if the walls of the tubing were made thick enough to minimize the effect of wall vibrations, and the inside of the tubing were smooth

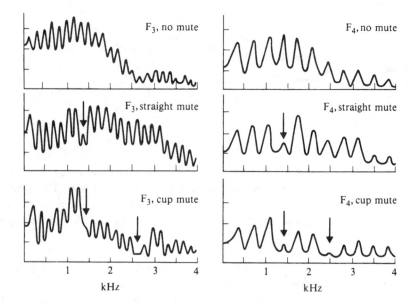

FIGURE 11.15
Spectra of trumpet tones with and without mutes. Adding a mute reduces the strengths of the low harmonics slightly, but gives a boost to harmonics in the range 1500 to 3000 Hz. Note the filtering by the resonances (indicated by arrows).

enough to minimize friction, it is doubtful whether the nature of the material would be of importance.

Careful studies by Smith (1978), by Watkinson and Bowsher (1982), and by other investigators have shown how the bells of brass instruments vibrate at high playing amplitudes. Smith (1978) found that the vibration amplitude increases rapidly when the wall thickness is reduced below 0.4 mm. These vibrations have little effect on the radiated sound along the axis of the bell, but the sound radiated laterally (which is heard mainly by the player) is augmented. Thus the player probably would notice a difference even though the audience would not. Vibrations of a 0.3-mm thick trombone bell at 240 and 630 Hz are illustrated by the holographic interferograms in Fig. 11.16.

11.14 ■ PERFORMANCE TECHNIQUE

To control loudness, the player must control blowing pressure. The actual blowing pressures used for brass instruments vary widely. A soft, low-pitched note on a French horn can be produced with an air pressure as low as 300 Pa (3-cm water gauge), whereas a high note played loudly on the same instrument may require more than 6000 Pa. Trumpet players use even higher blowing pressures, up to 10 or 15 kPa, near the limit set by normal systolic blood pressure in the arteries of the neck. (This can often be noted by the reddening face of a trumpet player playing fortissimo.) Attempts to play extremely loudly may lead to dizziness or collapse!

The maximum sound power output from a brass instrument is about 1 W. The dynamic range is typically about 30 dB, but the higher harmonics increase rapidly in level at higher playing levels. In fact, the change in spectrum is probably a more important part of a crescendo than the change in sound level. From measurements of blowing pressure and

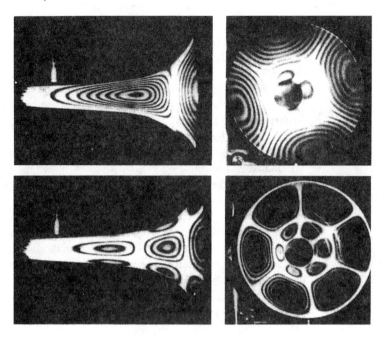

FIGURE 11.16
Holographic
interferograms
showing the
vibrations of a
0.3-mm-thick
trombone bell
driven acoustically
at 240 (top) and
630 Hz (bottom).
(Courtesy of
Richard
Smith/Crown
copyright, NPL.
Reprinted by
permission.

estimates of flow rates, the input power can be estimated to vary from about 30 mW to 10 W (Fletcher and Rossing 1998). The maximum efficiency, which is achieved for very loud playing, can be as high as 10%, but an average efficiency may be more in the range of 0.1%. Most of the power loss occurs in the vibrating lips, a lesser amount being lost to the instrument walls.

One of the most critical techniques is the ability to set the lips into vibration at the correct frequency before feedback is received from the instrument, as discussed in Section 11.10. Attacking high notes requires considerable muscle tension, muscle memory, and a great deal of skill.

The extent to which the sound of a wind instrument can be modified by changing the configuration of the player's vocal tract is a topic of considerable interest. The effect is strongest when the instrument air-column cross section is approximately the same as the human airway, so it is more important in woodwinds than in brasses. Some players associate particular vowels with certain notes and shape their vocal tracts (including the mouth cavity) accordingly. Mukai (1989) has observed that beginning players hold their vocal folds (cords) wide open, but in more experienced players they are nearly closed. The effect of the vocal tract resonances are probably more apparent to the player as change in stability and response than to the audience as a change in the sound.

11.15 ■ SUMMARY

Brass instruments have a mouthpiece, various lengths of straight and tapered tubing, and a flared bell. The player's lips act as a valve, controlled to some extent by pressure (impedance) peaks due to the various resonances of the air column. Cooperation between the lips and the resonances of the air column sustains oscillations at a stable pitch, and the stability is greatest when the air column has strong resonances at several partials of the tone being produced.

The bell changes the frequencies of the pressure peaks of the air column and also determines the radiation characteristics of the instrument. The mouthpiece also shifts the frequencies of the pressure peaks, and it increases the heights of peaks in the vicinity of its own resonance frequency. Valves or slides are used to change the lengths of brass instruments to fill in notes of the chromatic scale that lie between modes in the open instrument. The radiated spectrum of a brass instrument is determined by the internal spectrum of standing waves and the radiation characteristic of its bell. The resonances of a French horn can be adjusted by placing a hand in the bell. Mutes act as high-pass or band-pass filters and thus alter the timbre and also the loudness of brass instruments.

Partials below the cutoff frequency build up to their steady state together, but partials above cutoff build up more slowly the higher their frequency. A player's lips are more or less on their own until the reflected wave returns to initiate pressure feedback. Thus a player develops the ability to set the lips into vibration at the frequency before feedback is received from the instrument. The extend to which the sound can be modified by changing the vocal tract is perhaps less in brasses than in woodwinds.

REFERENCES AND SUGGESTED READINGS

Backus, J. (1976). "Impedance Curves for the Brass Instruments," *J. Acoust. Soc. Am.* **60**: 470.

Backus, J. (1977). *The Acoustical Foundations of Music.* 2nd ed. New York: Norton. (See Chapter 12.)

Benade, A. H. (1973). "The Physics of Brasses," *Sci. Am.* **229**(1): 24.

Benade, A. H. (1976). *Fundamentals of Musical Acoustics.* New York: Oxford. (See Chapter 20.)

Campbell, M., and C. Greated (1987). *The Musician's Guide to Acoustics.* New York: Schirmer Books. Chapter 9.

Fletcher, N. H. (1979). "Air Flow and Sound Generation in Musical Wind Instruments," *Ann. Rev. Fluid Mech.* **11**: 123. (See also *Acustica* **43**: 63.)

Fletcher, N. H., and T. D. Rossing (1998). *The Physics of Musical Instruments.* 2nd ed. New York: Springer Verlag. Chapters 13 and 14.

Helmholtz, H. von (1877). *On the Sensations of Tone.* Trans., A. Ellis. New York: Dover, 1954. (See Appendix VII.)

Hirschberg, A., J. Gilbert, R. Msallam, and A. P. J. Wijnands (1996). "Shock Waves in Trombones," *J. Acoust. Soc. Am.* **99**: 1754.

Long, T. H. (1947). "The Performance of Cup-Mouthpiece Instruments," *J. Acoust. Soc. Am.* **19**: 892.

Luce, D., and M. Clark (1967). "Physical Correlates of Brass Instrument Tones," *J. Acoust. Soc. Am.* **42**: 1232.

Martin, D. W. (1942). "Lip Vibrations in a Cornet Mouthpiece," *J. Acoust. Soc. Am.* **13**: 305.

Mukai, S. (1989). "Laryngeal Movement During Wind Instrument Play," *J. Otolaryngol. Jpn.* **92**: 260.

Pratt, R. L., S. J. Elliott, and J. M. Bowsher (1977). "The Measurement of the Acoustic Impedance of Brass Instruments," *Acustica* **38**: 242.

Pyle, R. W., Jr. (1975). "Effective Length of Horns," *J. Acoust. Soc. Am.* **57**: 1309.

Smith, R. A. (1978). "Recent Developments in Brass Design," *International Trumpet Guild Journal* **3**: 27.

Smithers, D., K. Wogram, and J. Bowsher (1986). "Playing the Baroque Trumpet," *Sci. Am.* **254**(4): 108.

Watkinson, P. S., and J. M. Bowsher (1982). "Vibration Characteristics of Brass Instrument Bells," *J. Sound and Vibration* **85**: 1.

Young, R. W. (1967). "Optimum Lengths of Valve Tubes for Brass Wind Instrument," *J. Acoust. Soc. Am.* **42**: 224.

GLOSSARY

acoustic impedance The ratio of sound pressure to volume velocity (see Section 4.6). A graph of acoustic impedance of a musical instrument as a function of frequency shows peaks that correspond to the resonances of the air column.

bell The flared section that terminates all brass instruments and determines their radiation characteristics.

Bernoulli effect The pressure in a fluid is decreased when the flow velocity is increased.

cutoff frequency The frequency above which an instrument radiates so efficiently that standing waves inside the instrument are weak. In brass instruments the cutoff frequency is determined mainly by the shape of the bell.

feedback The addition of a part of the output of a system to the input; positive feedback is used to sustain oscillations in wind instruments, audio generators, etc. (See Chapter 18.)

filters (high-pass and band-pass) Acoustic elements that allow certain frequencies to be transmitted while attenuating others. A high-pass filter allows all components above a cut-off frequency to be transmitted; a band-pass filter allows frequencies within a certain band to pass. Electrical filters are discussed in Chapter 18.

mouthpiece The part of a brass instrument that couples the vibrating lips to the air column.

mouthpipe Tapered tubing that connects the mouthpiece to the main section of a brass instrument.

mute An acoustic device that alters the timbre and loudness of a musical instrument.

popping frequency The lowest resonance of a brass instrument mouthpiece.

radiation curve, or characteristic A graph showing what portion of the internal sound is radiated by the bell or other part of the instrument.

turning point The point at which reflection of a wave occurs at the open end of the bell or tubing.

volume velocity The rate of air flow in a tube, expressed in units of volume per unit of time (such as m^3/s).

REVIEW QUESTIONS

1. What are the four main sections of a brass instrument?
2. Which has the largest playing range: trumpet, French horn, trombone, or tuba?
3. What is meant by acoustic impedance?
4. Is it easier or more difficult to blow air into an instrument at the frequency of an impedance peak?
5. What are four effects of the bell on the acoustics of a brass instrument?
6. What is the *turning point* of a sound wave in a musical instrument?
7. At what frequency does the lowest resonance of a trumpet mouthpiece occur?
8. What is the Bernoulli effect?
9. Which has the greater length of tubing: a French horn or a trombone?
10. What is the purpose of placing your hand in the bell of a French horn?
11. What are the maximum sound levels observed inside a trumpet?
12. What four acoustical effects does a mute have on a trumpet?
13. Why do wind-instrument players shape their vocal tract in a particular way when playing certain notes?
14. What is the *popping frequency* of a mouthpiece?

QUESTIONS FOR THOUGHT AND DISCUSSION

1. A trombone plays one octave lower than a trumpet. Why is its bore length not exactly twice that of a trumpet?
2. A French horn has a greater bore length than a trombone. Why are its played notes generally higher than those of a trombone?
3. A slide trombone can be played in perfect intonation, whereas a valve-operated instrument requires adjustments in the pitch by "lipping." Explain why slide trumpets, having this advantage of a slide trombone, are not used in an orchestra.
4. Make a tracing of the trumpet impedance curve shown in Fig. 11.7. Indicate which peaks form regimes of oscillation for F_4 ($f = 349$ Hz) and F_5 ($f = 698$ Hz).

Comment on the stability of these two notes played *pp* and *ff*.

5. The trumpet spectrum in Fig. 11.13(c) is typical of what would be heard immediately in front of the trumpet bell. What change would occur in the spectrum heard at some distance (say, 50 ft) away:

   (a) in the direction in which the bell is pointed;

   (b) at right angles to this direction (assume no room reflections).

6. Dents in the tubing of a brass instrument will cause the reflection of sound waves. How will these show up the impedance curves? How will they affect the playing characteristics of the instrument?

EXERCISES

1. Assume that the length of a trombone is 275 cm in first ("open") position. How far should the slide have to be moved to lower the pitch one semitone? (Remember the length is increased by twice this amount.) If possible, compare this with the slide motion of an actual trombone.

2. Calculate the frequencies of the first three resonances of closed tubes having lengths of 140 cm and 375 cm. Compare these to the resonances shown in the impedance curves of the trumpet and French horn.

3. Determine the frequencies of the trumpet resonances as accurately as you can from Fig. 11.7. How closely do they correspond to the bugle notes: $B^b{}_3$, F_4, $B^b{}_4$, D_5, F_5, $A^b{}_5$, and $B^b{}_5$ (written C_4, G_4, C_5, E_5, G_5, $B^b{}_5$, and C_6)?

4. Pressing the first valve of a trumpet or tuba increases the acoustic length by 12.2% and lowers the pitch a whole tone. How long must the first valve slide be in each of these instruments? (In other words, how many cm of tubing should be added to produce the pitch change?) If possible, compare your answers to the measured lengths in actual instruments.

5. Find the frequency ratios of the first five peaks in Fig. 11.4 to the corresponding peaks in Fig. 11.6.

EXPERIMENTS FOR HOME, LABORATORY, AND CLASSROOM DEMONSTRATION

Home and Classroom Demonstration

1. *Resonances of a pipe* Attach a rubber membrane to a 140-cm length of tubing about 1 cm in diameter, and excite the membrane with a loudspeaker. Show that the resonances of the pipe are those of a closed tube. Then have a brass player excite these same resonances by blowing the pipe.

2. *Resonances of a pipe and mouthpiece* Attach a trumpet mouthpiece to a 140-cm length of tubing with approximately the diameter of a trumpet (1.1 cm) and note the resonances. Compare these to the resonances of a trumpet having the same length. (The lowest resonance of a trumpet is quite difficult to sound with a trumpet mouthpiece, but attaching a clarinet mouthpiece by means of a small adapter makes it relatively easy.)

3. *Trombone slide* Measure the length of a trombone with the slide in several positions (see Fig. 11.10). Show that the acoustical length increases by 6% between successive positions.

4. *Hand stopping* Have a skilled French horn player demonstrate the effect of hand-stopping the bell. Slowly insert the hand as a note is sustained. If no attempt is made with the lips to hold the original pitch, the pitch will fall as the hand is inserted. If the player tries to hold the pitch, however, the note jumps to the next higher resonance at some point as the hand is inserted (see Section 11.5).

5. *Tuning effect of the bell* Have a trombone or trumpet player play all the notes which can be sounded with one slide position or fingering. Then disconnect the bell by removing the tuning slide, and show that the resonances are no longer harmonic in frequency.

6. *Bernoulli effect* Demonstrate the Bernoulli effect by means of the simple experiments shown in Fig. 11.9.

7. *Internal sound field* By means of a miniature microphone or a probe tube coupled to a microphone, record the sound inside a trumpet and compare it to a tape recording of the radiated sound (see Fig. 11.13).

8. *Turning point* Have a trumpeter or trombonist play a sustained note, and observe the turning point by inserting a finger into the bell. Alternatively, insert a small microphone into

the bell and observe the sound pressure on an oscilloscope. It may be possible for the player to observe the disruption of the turning point as a broomstick or rod is inserted into the bell.

9. *Mutes*   Ask a trumpet or trombone player to demonstrate several mutes.

10. *Internal and external sound spectra*   Place a miniature microphone or a probe tube coupled to a microphone inside a trumpet or trombone and a second microphone $\frac{1}{2}$ m or so in front of the bell. By means of a two-channel spectrum analyzer, compare the two internal and external sound spectra (see Fig. 11.13).

11. *Sound spectra at various dynamic levels*   Record sound spectra of various brass instruments at different dynamic levels ranging from *pp* to *ff*. Compare them to Fig. 11.13(c).

Laboratory Experiments

Demonstrations 1, 2, 3, 5, 7, 8, 10, and 11 can be combined to make an interesting laboratory experiment.

CHAPTER
12

Woodwind Instruments

The woodwinds are a large family of instruments with great diversity in size, shape, and design. The principal woodwinds in the symphony orchestra are the flute, clarinet, oboe, and bassoon; auxiliary instruments include the piccolo, English horn, bass clarinet, and contrabassoon. The saxophone is the principal woodwind in jazz bands and also plays an important role in marching and concert bands and, occasionally, in symphony orchestras. Old instruments, such as recorders and crumhorns, have recently experienced a revival in interest.

There are basically two types of woodwinds: those in which the air column is excited by a vibrating reed and those in which it is excited by an oscillating air stream. The behavior of these two types of woodwinds is very different, as we shall see. Like most musical instruments, woodwinds have evolved over the centuries.

In this chapter you should learn:

- How a pipe-reed system vibrates;
- About the function and design of tone holes;
- About resonances of cylinders and cones;
- About single- and double-reed instruments with cylindrical and conical bores;
- How a flow-controlled air stream collaborates with the air column in a flute.

Woodwinds were originally constructed from wood, with few exceptions, and hence their family name. Modern woodwinds may be either wood, metal, or plastic, although wood is still the preferred material for woodwinds other than flutes, piccolos, and saxophones. Like the brass instruments, woodwinds make use of feedback from an oscillating air column to control the flow of air input and maintain the oscillations. The flow-control valve may be either a vibrating reed or an oscillating stream of air. The feedback mechanisms in these two cases are quite different, as will be discussed in this chapter.

In woodwinds, the resonances of the air column are tuned by opening and closing tone holes with the fingers and with mechanical keys. Sound is radiated from the open tone holes, so that the radiation pattern becomes more complex than that of the brass instruments, which radiate virtually all their sound from the bell.

Woodwinds are often classified into three groups: the single reeds, the double reeds, and the air reeds, or flute-type instruments. Some modern woodwind instruments are shown in Fig. 12.1.

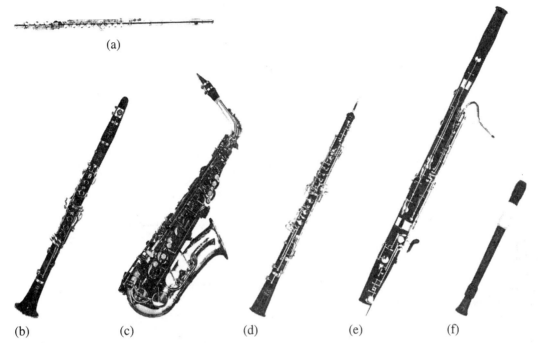

FIGURE 12.1   Woodwind instruments: (a) flute; (b) clarinet; (c) saxophone; (d) oboe; (e) bassoon; (f) recorder. (Courtesy of Selmer, Elkhart, Indiana.)

12.1 ■ HOW A PIPE-REED SYSTEM VIBRATES

Imagine a clarinet mouthpiece attached to a cylindrical pipe of a length such that the total acoustical length of the pipe plus mouthpiece is L. Now as blowing pressure is applied, the reed valve allows a puff of air to flow into the instrument and at the same time starts the reed swinging shut toward the mouthpiece, as shown in Fig. 12.2(a).

The puff of air (or pulse of positive pressure) travels down the pipe until it comes to the open end, where the excess pressure (above atmospheric pressure) rather abruptly drops to zero (see Fig 3.11). This causes a negative pressure pulse (pressure below atmospheric) to propagate back up the pipe toward the mouthpiece, as shown in Fig. 12.2(b) and (c). When it arrives, the reed is just completing its swing toward the mouthpiece, and the negative pressure pulse pulls the reed valve a little farther shut (Fig. 12.2(d)). Because the reed valve is now closed or nearly closed, very little air enters, so a negative pressure pulse starts back down the tube toward the open end (Fig. 12.2(e)).

Now we reverse the chain of events described in the preceding paragraph. The negative pressure pulse arrives at the open end, the pressure suddenly rises to zero (actually, to normal atmospheric pressure), and a positive pressure pulse begins its journey back toward the mouthpiece (Fig. 12.2(f) and (g)). When it arrives, the reed is swinging open (h) and the pressure pulse pushes it farther open, so that a new puff of air can be introduced from the player's mouth (i).

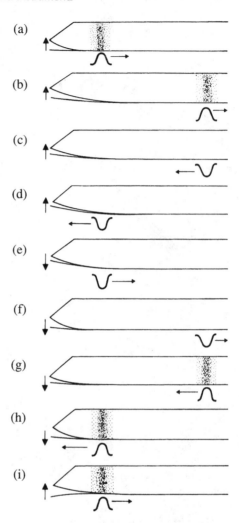

FIGURE 12.2
Vibration of a
pipe-reed system.
A pulse of excess
pressure propagates
down the pipe (a, b)
and is reflected as a
pulse of
underpressure
(c, d), which
returns and helps to
draw the reed valve
shut. In (e)–(h) the
process repeats
with this negative
pulse, which
reflects as a positive
pressure pulse.

Note that the reed has received one pull and one push as the pulse makes two trips up and down the pipe (once as a positive pulse, once as a negative pulse). This is analogous to giving brief pushes to a child in a swing to make up for energy lost during a cycle of oscillation. Often called *regenerative feedback* or *positive feedback*, it is analogous to the positive feedback in an electrical oscillator. Actually, the pulses are spread out in time, and the air pressure at any point in the pipe varies continuously from positive to negative, measured with respect to atmospheric pressure.

The air in the pipe has appreciable mass, and is able to pretty much force the vibrating reed to lock in on the natural frequency of the air column. Thus the reed in a woodwind has little to say about the selection of a frequency at which the system will vibrate, at least when compared to brass instruments, where the vibrating lips of the player have considerable mass, allowing the player a substantial amount of lip control of frequency. One has only to observe the sound of a reed and mouthpiece blown by themselves to realize that the reed is forced by the air column to vibrate at an unfamiliar frequency.

The reed could also have "locked in" with the air column by vibrating three times as fast as in the mode just described. In this case, the reed would close, open, and close again during one round trip of the pressure pulse, so that once again it would be ready to receive a pull when the negative pulse returned. If the embouchure pressure is great, there is a tendency to jump into this mode, which has a frequency three times that of the fundamental mode (provided the pipe is cylindrical). Normally, reed instruments shift into this mode of higher frequency only when a register key is opened, however (see Section 12.5).

The frequencies of the modes of oscillation can be calculated easily. In the fundamental mode, the pulse or wave travels down and back the length of the pipe during each half-cycle of the reed vibration, so the period of the lowest mode is

$$T_1 = \frac{4L}{v} \quad \text{and} \quad f_1 = \frac{1}{T_1} = \frac{v}{4L},$$

where T and f represent the period and frequency of vibration, L is the acoustical length of the tube plus mouthpiece, and v is the speed (or velocity) of sound. The second mode has a frequency $f_3 = 3f_1$, and the modes of higher frequency are the odd-numbered harmonics of the fundamental frequency (see Section 4.5):

$$f_n = nf_1 = n\frac{v}{4L}, \quad n = 1, 3, 5, \ldots. \tag{12.1}$$

A Note on Reeds

Musical instrument reeds are often described as either free or striking. Woodwind reed instruments and most organ reed pipes use *striking* reeds, so named because they "strike" against some surface as they vibrate. *Free* reeds, which do not strike a surface, are used in the harmonium (reed organ), the harmonica (mouth organ), and some organ pipes. The lips of the brass player are sometimes referred to as a *lip reed*, and the oscillating air jet of a flute as an *air reed*. (The brass player's lips would be classified as a striking reed.)

The reeds of the clarinet, oboe, bassoon, and organ pipe *strike inward*; that is, they close by moving with the flow of air. The lips of the brass player, on the other hand, *strike outward*; that is, they move against the flow of air in order to close. In the clarinet, the maximum flow of air outward must coincide with the maximum displacement of the reed inward (Helmholtz 1877). In this respect, woodwind reed instruments differ from brass instruments, where the maximum flow of air outward coincides with the maximum displacement of the lip reed outward.

A detailed discussion of the flow-control characteristics and elastic properties of reeds is beyond the scope of this book. Figure 12.3 shows the way in which the flow velocity varies with the pressure across the reed. Because the clarinet reed strikes inward, the desired flow control occurs on the descending slope of the curve. The entire curve may be lowered by increasing embouchure pressure, because the reed opening is thus decreased.

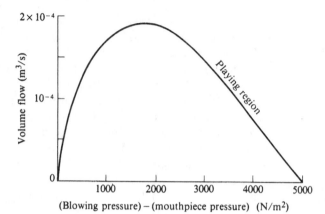

FIGURE 12.3
Flow velocity as a
function of air
pressure for a
clarinet. (After
Worman 1971.)

The compliance of the reed in this graph is about 8×10^{-8} m^3/N, which is typical of clarinet reeds. (Compliance is the inverse of stiffness.) The maximum Bernoulli force (see Section 11.3) is estimated to be 1.2×10^{-2} N, which is less that two percent of the force necessary to close the reed, and in fact moves the reed only 7×10^{-6} m (Worman 1971). Thus the Bernoulli force can be ignored in the clarinet reed, although it plays a fairly important role in brass instruments and apparently also in double reeds (see the discussion of the Bernoulli effect in Section 11.4).

12.2 ■ TONE HOLES

A pipe fitted with a clarinet mouthpiece and having a length and a bore comparable to those of a clarinet produces a sound resembling that of a clarinet. However, as the pipe is shortened, it sounds less clarinetlike. The difference lies in the tone holes of the clarinet. Both the open and closed tone holes of the clarinet affect its acoustical behavior, as we shall see.

One function of the tone holes is to change the effective or acoustical length of the clarinet. In the case of a single tone hole, the larger the hole, the more the effective length is shortened, as shown in Fig. 12.4. When the size of the tone hole matches the bore, the

FIGURE 12.4
Effective length of
a pipe with open
tone holes of
different diameters.

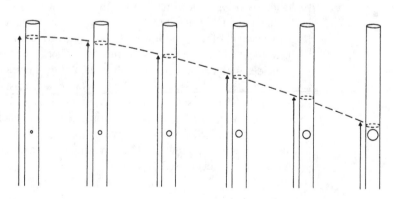

pipe effectively ends at the open tone hole. Opening and closing tone holes permits the clarinetist to change the pitch of his or her instrument.

When a pipe has more than one open hole, its acoustical behavior exhibits several interesting features. If the open holes are regularly spaced, they constitute a *tone-hole lattice*, not unlike a lattice of atoms in a crystal or a line of beads spaced equally on a string. The open tone-hole lattice acts as a filter that transmits waves of high frequency but reflects those of low frequency. The critical frequency above which sound waves can propagate through a lattice of tone holes is called the *cutoff frequency* of the lattice, which has been found to be an important factor in determining the timbre of a woodwind instrument.

FIGURE 12.5
An open tone-hole lattice indicating the parameters to be used in calculating the cutoff frequency: a = the radius of the bore, b = the radius of the tone hole, $2s$ = the tone hole spacing, and t = the tone-hole height.

The cutoff frequency of a lattice of tone holes depends on their size, shape, and spacing. The formula for calculating the cutoff frequency (Benade 1976) is

$$f_c = 0.11 \frac{b}{a} \frac{c}{\sqrt{s(t + 1.5b)}},$$

where c is the speed of sound (344 m/s), and a, b, s, and t (expressed in meters) are physical parameters, shown in Fig. 12.5.

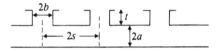

The effective length of a pipe with open tone holes is found to vary with frequency in much the same manner as a tube terminated with a trumpet bell (see Section 11.3). As the frequency increases, the turning point moves further down the pipe; the pipe acts longer at

FIGURE 12.6
Impedance curves for a piece of cylindrical clarinet tubing with and without tone holes. (From Benade, 1976.)

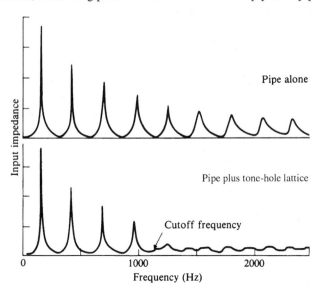

high frequency than at low frequency. Thus the upper resonances are lowered slightly with respect to the lowest one, as can be seen in the impedance curves shown in Fig. 12.6.

More important, however, is the fact that the resonances above the cutoff frequency become very weak. Above the cutoff frequency, the open tone holes radiate so well that very little sound is reflected back toward the mouthpiece. The spectrum of the sound recorded by a small microphone inside the mouthpiece will have peaks that resemble those found on the impedance curve. The spectrum measured outside the instrument does not show the same weakening of upper partials, however, because of the greater efficiency of radiation from the open tone holes above the cutoff frequency.

Closed tone holes also affect the acoustics of a woodwind air column. The increased volume of air at each closed tone hole reduces the velocity of the sound wave down the tube, and lowers the resonance frequencies slightly. Another way of expressing this is to say that a pipe with regularly spaced bumps appears longer acoustically than a smooth pipe of the same length.

12.3 ■ BORE TYPES

Woodwinds are designed with cylindrical or conical bores, because these shapes have resonance frequencies that are harmonically related to the fundamental. The bores of the flute and clarinet are essentially cylindrical; most other woodwinds (oboe, English horn, bassoon, saxophones, etc.) are essentially conical. The cone angles of the oboe and bassoon are small (1.4° and 0.8°), whereas those of the saxophone are quite large (3° to 4°) (Nederveen 1998).

The resonances of a cone have essentially the *same frequencies as an open pipe of the same length*. This statement is true even if the cone is truncated, which may seem paradoxical at first glance. As a sound wave travels toward the small end of a cone, its pressure must increase. Pressure distributions for the first three modes of a cone are shown in Fig. 12.7, along with the corresponding modes of a pipe.

The four lowest-mode frequencies of a cylindrical tube (closed at one end), a cone, and a flared horn are compared in Table 12.1. The mode frequencies for a cylindrical tube have the ratios 1 : 3 : 5 : 7, whereas those of a cone are nearly in the ratios 1 : 2 : 3 : 4. Wavefronts in the cone are spherical sections, but they travel at the same speed as the plane waves in the cylindrical tube. When the cone takes on a flare, however, the wave speed depends on frequency, and thus the mode frequencies are no longer harmonic.

The input impedance for a conical pipe is different from that of a cylindrical pipe (as shown in Fig. 12.6). The impedance peaks occur at frequencies having the ratios 1 : 3 : 5 : 7 : . . . , as in Table 12.1, but the impedance minima do not lie midway between the peaks, as in a cylindrical pipe. Impedance curves for cylindrical and conical pipes are compared by Strong and Plitnik (1983) and by Ayers, Eliason, and Mahgerefteh (1985).

12.4 ■ THE CLARINET

A cross-sectional view of a clarinet is shown in Fig. 12.8. The cylindrical bore has a diameter of about 15 mm. A single cane reed is clamped by the ligature against a specially

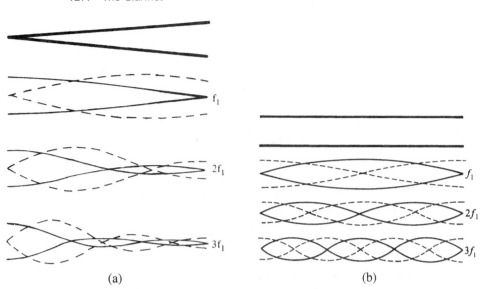

FIGURE 12.7   Pressure (solid line) and velocity (dashed line) distributions for the first three modes of a cone (a) and an open pipe (b). (For modes of a closed pipe see Fig. 4.8.)

TABLE 12.1   Mode frequencies in cylindrical, conical, and flared tubes (Strong and Plitnik 1983)

Mode	Cylindrical	Conical	Flared
1	136.4 Hz	255.6 Hz	281 Hz
2	409.2	512.0	520
3	682.0	771.0	767
4	954.9	1032.0	1021

FIGURE 12.8
A cross-sectional
view of a clarinet.

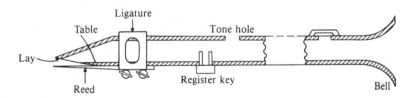

designed surface of the mouthpiece called the *table*. The vibrating portion of the reed is shaved down to a wedge shape with a thickness of about 0.1 mm.

The opening between the tip of the mouthpiece and the reed is typically about 1 mm. When the instrument is played, the lower lip pushes the reed in to about half this distance, and it vibrates around this position. For soft tones, the tip of the reed does not touch the mouthpiece; therefore, the flow of air is not interrupted. As the blowing pressure is increased, the amplitude of the reed increases, until for loud tones the tip of the reed touches the mouthpiece during about half of the cycle (Backus 1961). Thus the flow of air, which

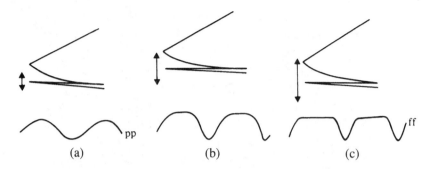

FIGURE 12.9
Vibration of a clarinet reed at different dynamic levels: (a) pp; (b) mf; (c) ff.

is more or less sinusoidal (single frequency) for small amplitude, takes on more harmonics as the blowing pressure increases, as shown in Fig. 12.9.

A clarinet reed has a natural frequency of its own at around 2000–3000 Hz; oscillations at this frequency are prevented largely by the damping action of the lower lip pressed against the reed. If the clarinet is blown with the teeth against the reed rather than the lip, unpleasant squeaks and squeals appear.

The clarinet is an important instrument in the orchestra, but it plays an even more important role in the band, where B^b clarinets form a principal choir much as violins do in the orchestra. The B^b clarinet has a wide playing range of $3\frac{1}{2}$ octaves divided into three registers, as shown in Fig. 12.10. The low or, *chalumeau*, register extends from D_3 to E_4, and a second register, the *clarion*, plays a twelfth above this, because the first overtone of a cylindrical closed tube is the third harmonic of the fundamental. Notes above B^b_5 can be played in the *altissimo* register, which uses the third mode (fifth harmonic) of the air column. Often the *throat tones* (G_4 to $A^\sharp_4$), which open keys in the throat of the clarinet, are included in the chalumeau register.

The bass clarinet, which plays an octave lower than the B^b clarinet, is also used in orchestras and bands. Orchestral clarinetists alternate between clarinets tuned in A and B^b, the choice usually depending on the key signature of the music. Other clarinets are the E^b soprano and E^b alto. The C clarinet, used in the Classical period, is rarely seen today.

The open tone-hole cutoff frequency in most good clarinets is around 1500 Hz. A clarinet with a higher cutoff frequency tends to have a "bright" tone; an instrument with a lower cutoff frequency has a "dark" tone. In an experiment in which two matching clarinets were reworked to raise the cutoff frequency slightly in one and to lower it in the other, it was found that players of classical music preferred the one with the lower cutoff frequency, whereas jazz clarinetists chose the one with the higher cutoff and bright tone (Benade 1976).

Impedance curves are shown for several clarinet fingerings in Fig. 12.11. In fingering E_3, all tone holes are closed, and the resonances are not greatly different from those of the

FIGURE 12.10
Three registers of the clarinet. Shown are the notes fingered; notes sounded on a B^b clarinet are one whole tone lower.

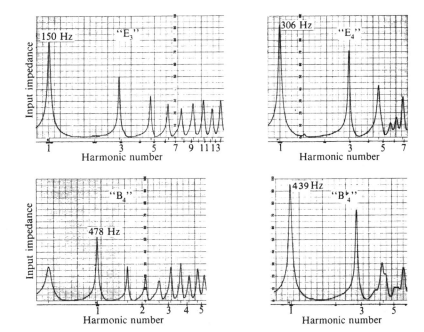

FIGURE 12.11
Impedance curves for several clarinet fingerings: E$_3$ and B$_4$ have the same fingering except for the open register hole. Actual notes on a B$^\flat$ clarinet are a whole note lower than the fingered note. (From Backus 1974.)

pipe with a clarinet mouthpiece shown in Fig. 12.6. The differences are mainly due to the bell and the effect of the closed tone holes. In E$_4$ and B$^\flat$$_4$ cases, a cutoff due to the open tone-hole lattice is apparent. B$_4$ has the same fingering as E$_3$, except for an open register hole, which "spoils" the lowest resonance.

12.5 ■ REGISTERS AND REGISTER HOLES

All the woodwinds play in at least two different registers. In a clarinet, corresponding notes in the lower two registers differ by a musical twelfth (see Fig. 12.10), whereas in many other woodwinds there is an octave difference. In the lower register, the note sounded corresponds to the first resonance. On the impedance curve of a clarinet, this peak is the tallest one, although in the case of the conical instruments, it is usually not. Nevertheless to shift to a high register, it is necessary to reduce the strength of this lowest resonance or to shift its frequency in a way that discourages cooperation with the other modes.

The easiest way to reduce the strength of a resonance is to leak a little air at a point of maximum pressure for that mode, which can be done with a register hole. This function of a register hole can be demonstrated by fitting a clarinet mouthpiece to a length of pipe; at about one-third the distance from the reed end of the pipe (considering the mouthpiece as part of the pipe), a tiny hole is drilled. Uncovering this hole will cause the oscillation to shift up a musical twelfth from the first to the second register (i.e., to the third harmonic). A similar hole at about one-fifth the distance down the pipe will cause it to shift up an additional sixth to the fifth harmonic, because this mode of oscillation has a pressure node at that point, whereas the two lower modes do not, as shown in Fig. 12.12.

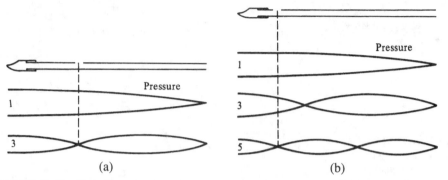

FIGURE 12.12   The effect of register holes on the modes of a pipe attached to a clarinet mouthpiece. (a) The hole one-third of the way down the pipe damps the first mode, and encourages the third harmonic. (b) The hole one-fifth of the way down the pipe damps the first and third harmonics, and encourages oscillation at the fifth harmonic.

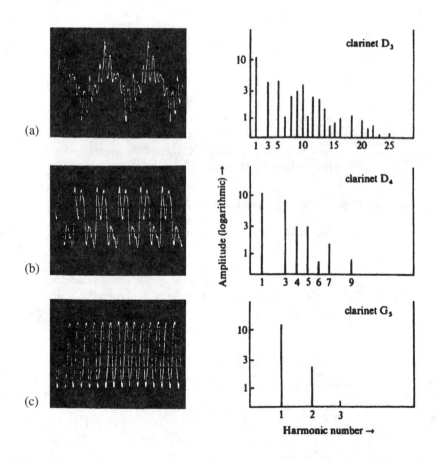

FIGURE 12.13
Radiated sound
pressure waveform
and corresponding
spectrum for three
notes played on a
clarinet (Fletcher
1976).

A practical difficulty arises when an entire register is considered. In order to be able to open a register hole near a pressure node for every note in the register, a large number of register holes would be needed. An oboe effectively has three register holes; a saxophone has two; but a clarinet has a single register hole for transitions between the chalumeau and clarion registers. It becomes somewhat of a challenge to design a single register hole that functions well throughout the entire register. An open tone hole near the top end of the clarinet (ordinarily closed by the first finger) serves as a register hole for the altissimo register.

12.6 ■ RADIATED SOUND

The sound of a clarinet, like all reed woodwinds, is rich in harmonics. In the low (chalumeau) register, the second harmonic is almost completely absent, as we would expect from the absence of this harmonic in the first impedance curve in Fig. 12.11. The fourth harmonic is also missing, although even-numbered harmonics begin to appear starting with the sixth. This is clearly seen in the radiated sound-pressure waveform and spectrum for this same note in Fig. 12.13(a) (D_3 is the sound of the fingered E_3 in Fig. 12.11). The missing second and fourth harmonics give the clarinet its distinctive "woody" tone in the low register.

Note the change in waveform and spectrum from (a) to (b) to (c) in Fig. 12.13. Not only do the even-numbered harmonics grow stronger, but the higher harmonics become weaker. Clarinet sound takes on a completely different timbre in its three registers. Furthermore, the timbre changes with dynamic level. As the vibration amplitude of the reed increases, so do the amplitudes of all the harmonics, the nth harmonic increasing by n dB for every 1 dB increase in the level of the fundamental.

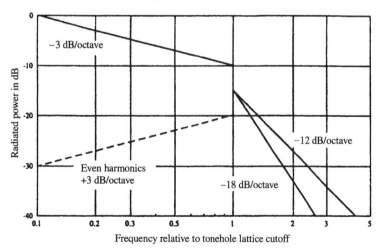

FIGURE 12.14   Spectral envelope for a typical reed-woodwind instrument played loudly. Below the tone-hole cutoff frequency, the radiated power in the harmonics falls off at about -3 dB/octave, whereas above the cutoff it falls at -12 dB/octave if the reed beats and at -18 dB/octave if it just fails to beat. In an instrument with a cylindrical bore, the power in the even harmonics rises at about 3 dB/octave below the cutoff frequency while behaving like the odd harmonics above cutoff (Fletcher and Rossing 1998).

In mezzoforte playing, the spectral envelope of the clarinet pretty much follows the generalized reed-woodwind curve shown in Fig. 12.14. Below the tone-hole lattice cutoff frequency (typically around 1500 Hz), the radiated power in the harmonics falls off at about -3 dB/octave (amplitudes proportional to $1/\sqrt{f}$), whereas above the cutoff frequency it falls off at about -18 dB/octave (amplitudes proportional to $1/f^3$). For louder playing, so that the reed "beats" against the lay of the mouthpiece, the radiated power falls off at about -12 dB/octave (Fletcher and Rossing 1998). Playing spectra are shown by Benade and Kouzoupis (1988).

12.7 ■ THE DOUBLE REEDS

The family of orchestral double reeds include the *oboe, English horn, bassoon,* and *contrabassoon.* Basically, they are conical tubes with the tip of the cone cut off and a reed attached. All these instruments use a double reed consisting of two halves of cane beating against each other. Small mouthpieces with a single reed have been used, but have not become very popular because they change the quality of the instrument. Cross sections of the oboe and bassoon are shown in Fig. 12.15.

The oboe bore is a nearly straight cone about 60 cm in length. Since the resonances of a cone are an octave apart, the registers of the oboe are an octave apart, extending from D_4 to C_5 and D_5 to C_6. Additional keys and cross fingerings extend the playing range from B^b_3 to G_6. Oboes are usually constructed from three pieces of wood: a top section, a bottom section, and a bell, all three sections having tone holes.

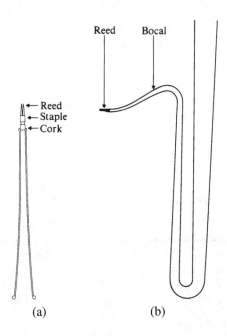

Reed   Bocal

← Reed
← Staple
← Cork

(a)   (b)

FIGURE 12.15
Cross section of
(a) an oboe; (b) a
bassoon.

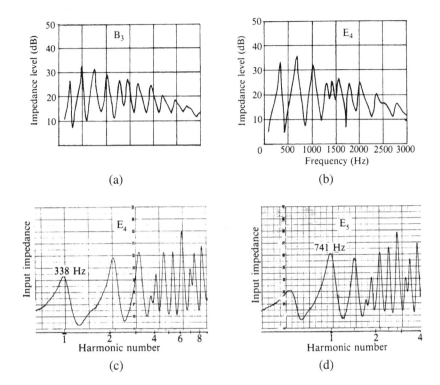

FIGURE 12.16
Impedance curves
for oboes. (Parts (a)
and (b) from Plitnik
and Strong 1979;
parts (c) and (d)
from Backus 1974.)

Impedance curves for four different oboe notes by two different investigators are shown in Fig. 12.16. Curve (b) for E_4 shows the effect of an open tone-hole cutoff at about 1200 Hz; in curve (c) a cutoff is not apparent. Comparison of (c) and (d) shows how opening the register hole weakens the first resonance.

Quality oboes have cutoff frequencies that are nearly constant throughout their playing range, and for different instruments vary from about 1100 to 1500 Hz (Benade 1976). A higher cutoff frequency results in a bright tone and a lower cutoff frequency, in a dark tone, just as in the case of clarinets. The spectra of oboe sounds show substantial amounts of the higher harmonics (see Fig. 12.17) along with broad resonances, or *formants*, in

FIGURE 12.17
Spectra of a 250-Hz
oboe tone played at
dynamic markings
of p, mf, and ff.
These may be
compared to
Fig. 12.14(a),
which is the
impedance curve
for B_3. Note the
formants around
1000 and 3000.
(From Strong and
Plitnik 1983.)

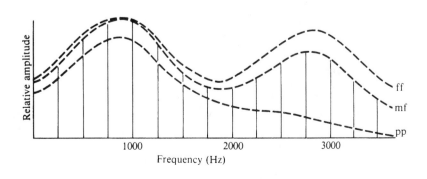

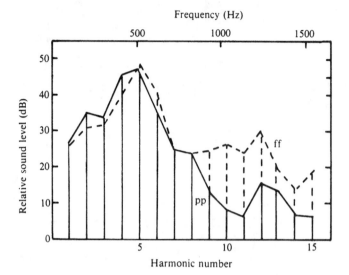

FIGURE 12.18
Bassoon spectra for
fortissimo and
pianissimo playing.
(After Lehman
1962.)

the neighborhood of 1000 and 3000 Hz. Fransson (1967) attributes these formants to the mechanical properties of the reed.

The *English horn*, or cor Anglais, is an alto version of the oboe with a pear-shaped bell that gives a distinctive tone quality to certain notes near its resonance. Composers have taken note of this distinctive tone quality in writing solos for this instrument. The English horn is tuned in F, a fifth below the oboe.

The *bassoon* has a nearly conical bore with a total length of about 254 cm. The tube makes several bends, as shown in Fig. 12.15(b) in order to be of a manageable size. The playing range of the bassoon extends from $B^\flat_1$ to about C_5 (58–523 Hz), whereas the *contrabassoon* plays an octave lower. Open tone-hole cutoff frequencies for quality bassoons range from about 350 to 500 Hz (Benade 1976). Like the oboe, the bassoon has many sharp resonances, and the tone is rich in harmonic overtones.

An extensive study of the bassoon played by eleven professionals revealed a rather strong formant extending from about 440 to 494 Hz (Lehman 1962). Presumably this is attributable to the reed, as in the oboe. Spectra of the bassoon for pianissimo and fortissimo playing are shown in Fig. 12.18. (These spectra, recorded in an anechoic (echo-free) room with a single microphone, are therefore influenced by variation in the radiation pattern of the various harmonics; nevertheless, the comparison between spectra at the two dynamic levels is probably significant.)

12.8 ■ THE SAXOPHONE

The saxophone is a conical-bore single-reed instrument invented in 1846 by Adolphe Sax. Although used occasionally in symphony orchestras, its main use has been in all types of bands and in ensembles that play jazz and popular music. The four main members of the saxophone family are the soprano in $B^\flat$, the alto in $E^\flat$, the tenor in $B^\flat$, and the baritone in $E^\flat$. Saxophones are characterized by a large bore diameter, a low input impedance, and

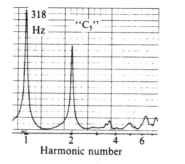 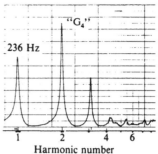

FIGURE 12.19
Impedance curves
for an E♭ alto
saxophone. (From
Backus 1974.)

a louder sound than the other woodwinds. The impedance curves for two fingerings on an alto saxophone (see Fig. 12.19) show few modes of high frequency.

Since the saxophone has a conical bore, its spectrum has even-numbered harmonics as well as odd-numbered ones. Its cone angle is greater than that of the oboe, however, so saxophone tone has fewer prominent harmonics. Its registers are an octave apart, and so it uses oboelike fingerings.

In order for the modes of the instrument to be as nearly harmonic as possible, the saxophone mouthpiece should mimic the acoustic behavior of the missing apex of the cone. At low frequencies, this match is achieved if the internal volume of the mouthpiece is equal to that of the missing conical apex, so the mouthpiece has a slightly bulbous internal shape. The high-frequency match is achieved by arranging the constriction where it joins the main part of the instrument so that the Helmholtz resonance frequency (Section 2.3.3) of the mouthpiece is the same as the missing conical apex. The mouthpiece cavity gives the spectral envelope an extra rise of 6 dB/octave below its resonance and a fall of −6 dB/octave above. The Helmholtz resonance frequency of the mouthpiece is typically comparable to the tone-hole lattice cutoff frequency (about 850 Hz for an alto saxophone), so the spectral envelope rises 3 dB/octave below cutoff and falls −18 dB/octave below cutoff. Playing spectra are shown by Benade and Lutgren (1988).

12.9 ■ EFFICIENCY AND PERFORMANCE TECHNIQUE

Typical blowing pressures for the clarinet, alto saxophone, oboe, and bassoon are shown in Fig. 12.20 for *forte* and *piano* playing. In all four instruments there is an increase in blowing pressure with dynamic level, and in the oboe and bassoon, there is also an increase in pressure with frequency.

Given that a player can sustain a note for about 30 s, the volume flow is around 1000 mL/s (10^{-4} m^3/s), which, when combined with the blowing pressures in Fig. 12.20, gives a power input of roughly 0.3 to 0.5 W. Yet the sound power from most woodwind instruments rarely exceeds 1 mW, so the efficiency is clearly less than 1%. Where does all the input power go? Most of it is dissipated in the flow resistance of the reed valve, somewhat less being lost in the walls of the instrument itself. Wall losses, although less than reed losses, exceed the radiation by a factor of 10 (Fletcher and Rossing 1998).

Blowing pressure gives some measure of control over loudness, but this is coupled to a change in timbre. Lip pressure on the reed also affects the playing level and timbre.

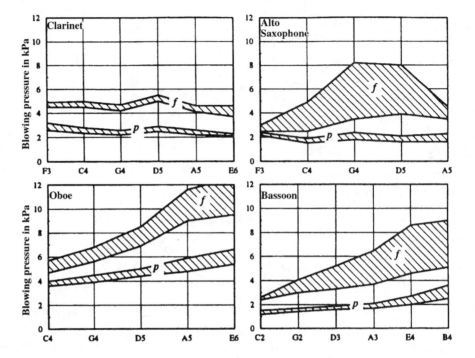

FIGURE 12.20
Typical blowing pressures in the clarinet, alto saxophone, oboe, and bassoon. (Fuks and Sundberg 1996.)

Squeezing the reed reduces the acoustic flow and it also increases the damping of the reed vibration.

12.10 ■ OSCILLATING AIR STREAMS AND WHISTLES

In nearly all musical instruments, sound production depends in some way on the vibration of some mechanical element. In a string or percussion instrument, the mechanical properties of the vibrating element determine its vibration frequency. The vibrating reed of the clarinet or oboe and the brass player's lips, on the other hand, are strongly influenced by the resonances of the vibrating air column of the instrument, as we have seen.

In flutes, recorders, flue organ pipes, and toy whistles, the vibrating element is a jet of air. The oscillating air stream in these aerodynamic whistles is sometimes called an *air reed*, because its behavior bears some resemblance to the cane reed of the clarinet or the *lip reed* of the trumpet. There is an important difference, however: The input flow or air reed is controlled not by the feedback of pressure pulses from the air column but by the direction of air flow due to standing waves in the air column. We say that the input is *flow*-controlled rather than *pressure*-controlled.

A familiar example of a flow-controlled valve is the excitation of oscillations in a bottle by blowing across its mouth. If the blowing is done at the proper angle and pressure, air in the neck of the bottle will be set in motion and sound is radiated (see Section 4.7). Flow control by blowing across a bottle is illustrated in Fig. 12.21. When the air in the neck of the bottle is flowing inward, a part of the air jet is directed inward also. When the air in the neck is flowing outward, the jet of blown air is directed away from the bottle. Thus the

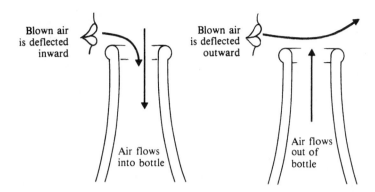

Blown air
is deflected
inward

Air flows
into bottle

Blown air
is deflected
outward

Air flows
out of
bottle

FIGURE 12.21
Flow control of air
being blown across
a bottle.

energy supplied from the lungs via the steady jet of air is converted to energy of oscillation
through the mechanism of flow control.

In most aerodynamic whistles, the frequency of oscillation varies almost linearly with
flow speed (Chanaud 1970). Examples are the *aeolian tones* generated by a wind blowing
past telephone wires or through trees, the *edge tones* generated when a jet of air encounters
a wedge-shaped obstacle, and the *hole tone* of a whistling teakettle.

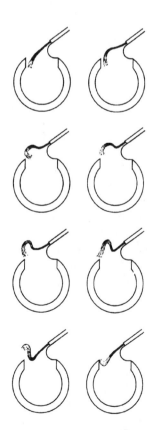

FIGURE 12.22
Oscillation of air
stream in flute
embouchure hole.
Compare the
oscillation of the jet
stream in and out of
the hole with the
flow of air in and
out of the bottle in
Fig. 12.17. (After
Coltman 1968.)

The manner in which a stream of air interacts with the embouchure hole in a flute was studied several years ago by Coltman (1968). Using an artificial lip, he injected smoke into the center of the jet stream and photographed it by using a strobe light. The sketches in Fig. 12.22 are based on his photographs.

12.11 ■ THE FLUTE

A flow-controlled air stream, such as that which occurs in a blown bottle or a flute, collaborates with an air column to oscillate at frequencies at which the air column has dips in its impedance curve (that is, where the pressure is at a minimum and, therefore, the flow-controlling jet deflection is at a maximum). This is opposite to the woodwind reed or brass player's lips, which are controlled by pressure peaks.

The flute is basically a cylindrical pipe, about 1.9 cm in diameter, open at both ends. (See Fig. 12.23.) Its vibration modes thus form a series of frequencies that includes all the harmonics of the fundamental. The resonance frequencies of the pipe are influenced by the presence of the stream, however. In particular, the frequency of the lowest resonance is raised when the blowing pressure increases, whereas the second resonance remains more or less unchanged. The range of the flute is normally B_3 to D_7.

The narrowing of the distance between these peaks with increasing pressure must be counteracted by shaping the air column, if the flute is to play in tune at all dynamic levels. In most modern flutes, this is accomplished by tapering the first section of the air column, called the *head joint*. Baroque flutes, on the other hand, have a cylindrical head and a slightly tapered bore.

The flute plays in three registers, and the player must select the desired register, without the aid of register keys, by adjusting the technique of blowing. The three parameters at the player's control are the blowing pressure, the length of the air jet, and the area of lip opening. The technique used by most flute players includes adjustments in all three of these (Fletcher 1974).

The most efficient excitation of the fundamental takes place when the time for the air jet to travel across the embouchure hole is about half the period of an oscillation. If the jet travel time is shortened much below this, the fundamental will not sound. Thus to sound the second register, the player moves the lips forward to decrease the jet length and/or increases the blowing pressure. At the same time the blowing pressure is increased, the size of the lip opening is decreased to maintain the loudness and tuning.

Studies of a number of flute players by Fletcher (1974) show that to shift to a register an octave higher, a flute player will typically double the blowing pressure, reduce the jet length by 20%, and reduce the lip opening by about 30%. (Doubling the blowing pressure

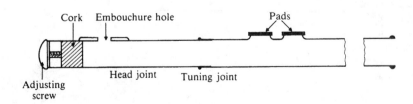

FIGURE 12.23
Construction of the flute.

increases the jet speed by 40%, because jet speed is approximately proportional to the square root of pressure.)

The angle of blowing depends on the shape of the player's lips but typically ranges from 25° to 40° below the horizontal. Most players use a shallower angle for low notes and for loud playing.

The narrow end of the head joint is closed by a cork plug whose position can be adjusted by a screw. Changing the length of the small cavity between the cork and the embouchure hole can have a substantial effect on the tuning and "playability" of certain notes.

The *piccolo* is about one-half the length of the flute, and it sounds one octave higher. An alto flute, which plays a fourth lower than the standard flute, is occasionally seen.

Pressure standing waves in a flute are shown in Fig. 12.24. The pressure, which is maximum at the cork, dips down at the open embouchure hole. With the Boehm tapered head joint, the cork is set about 17 mm (or one tube diameter) from the embouchure hole. If the cork is pulled out, the speaking length is increased and all notes are flattened. However, the flattening effect on the upper modes is greater than on the lower ones, because for the higher modes the effective position of the pressure node at the blowing end is beyond the cork.

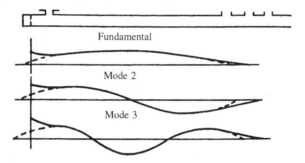

FIGURE 12.24 Pressure standing waves in a flute (schematic). (Campbell and Greated 1987.)

Moving the cork also changes the relative pitch of different notes. Pulling out the cork, for example, will flatten the higher notes more than the lower ones, because their tube length is shorter, so they are more sensitive to small changes in length.

12.12 ■ OTHER FLUTES

The *piccolo* is about one-half the length of the standard flute, and it sounds one octave higher. Piccolos are made with either cylindrical or conical bores, and the bodies may be of metal, wood, or plastic. Orchestral players generally prefer conical-bored wooden piccolos, whereas band musicians use cylindrical metal instruments because of their brighter tone. The tone-hole lattice cutoff frequency of a piccolo, because of its smaller size, is nearly twice as high as that of the flute, and thus the piccolo sounds lighter and brighter than a flute playing the same note. Also, its smaller size limits the radiation efficiency for low partials in the sound (Coltman 1991).

The alto flute is a transposing instrument sounding a fourth below the standard flute. Its mechanism is identical with that of the standard flute, except that the position of the finger keys has been shifted for convenience. Boehm likened the tone of an alto flute to that of a female contralto voice. Bass flutes, sounding an octave below the standard flute, are also made but are not commonly played.

12.13 ■ THE RECORDER

The *recorder*, also known as the Blockflöte, flauto dolce, and English flute, is an early member of the flute family, now experiencing a revival. The Baroque recorder has a reverse conical bore, tapering inward toward the foot, and a whistle-type mouthpiece with a fixed windway, as shown in Fig. 12.25. The three most commonly played recorders are the descant (soprano) in C, the treble (alto) in F, and the tenor in C. A bass in F, a sopranino in F, and a great bass in C complete the family. Each instrument has a two-octave normal playing range.

The fixed windway makes the recorder easy to sound but somewhat difficult to play in tune at different dynamic levels. Without the flexibility to adjust the embouchure in the manner of the flautists, the recorder player changes to the upper register by half-closing the thumb hole and increasing the blowing pressure slightly. Wind pressure in an also recorder varies from about 100 N/m^2 for the lowest note to about 500 N/m^2 for the highest (Herman 1959). Figure 12.26 compares the blowing pressures of the alto recorder with those of the flute. In both instruments, the blowing pressure is approximately proportional to frequency, but deviations from this rule are necessary to keep the notes in tune (Martin 1994).

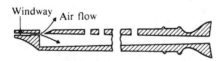

Windway, Air flow

FIGURE 12.25
(a) Cross section of a Baroque recorder.
(b) An alto recorder.

(a)   (b)

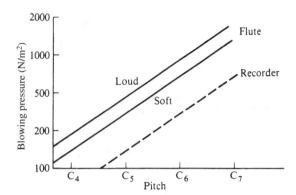

FIGURE 12.26
Variation in
blowing pressure
with frequency for
the flute and
recorder (1000
N/m^2 is 1/100 of
atmospheric
pressure).

The recorder has only eight tone holes. Therefore, in order to play a full chromatic scale, liberal use is made of cross-fingerings. Cross-fingerings leave one or more tone holes open and close other tone holes below these. In general, cross-fingered notes are not as stable as other notes, and the recorder player often changes the fingering slightly from instrument to instrument. Since the pitch produced by a given fingering may rise as much as 100 cents from low to high blowing pressure (Bak 1969), a skilled recorder player often changes fingerings with dynamic level.

12.14 ■ ORGAN PIPES

Flue organ pipes are similar to flute-type woodwind instruments. The tone-producing mechanism is similar to that of a recorder. A narrow stream of air impinges on the upper lip and oscillates back and forth, sometimes directed into the pipe, sometimes into the room. A variety of pipe resonators are used, including closed and open cylinders and cones. The pipe organ will be discussed in Chapter 14.

12.15 ■ SUMMARY

Woodwind instruments use feedback from an oscillating air column to control the flow of air into the instrument. The feedback control may be applied to a vibrating reed (clarinet, oboe) or to an air jet (flute). In woodwinds with reeds, the feedback is pressure-controlled, and the instruments play at the frequencies of one or more impedance (pressure) peaks. In flute-type woodwinds, the input is flow-controlled, and the instruments play at impedance (pressure) minima. The open tone holes radiate sound, and their size and spacing determine the cutoff frequency of the instrument.

The air column of a clarinet is essentially a closed cylinder, that of a flute an open cylinder, whereas the oboe, bassoon, and saxophone are conical (closed). The registers of a clarinet are thus spaced a twelfth apart, whereas those of the other instruments are an octave apart. In all instruments except the flute, one or more register keys help the player change registers. The flute player changes registers by adjusting blowing pressure and the position and shape of the lips.

The recorder is a flute-type instrument with a reverse conical bore and a fixed windway. The blowing pressure is substantially less than that of the flute; blowing pressure increases with frequency in both instruments. Cross-fingerings are used to play a chromatic scale on the recorder.

REFERENCES AND SUGGESTED READINGS

Ayers, R. D., L. J. Eliason, and D. Mahgerefteh (1985). "The Conical Bore in Musical Acoustics," *J. Acoust. Soc. Am.* **53**: 528.

Backus, J. (1961). "Vibrations of the Reed and the Air Column in the Clarinet," *J. Acoust. Soc. Am.* **33**: 806.

Backus, J. (1974). "Input Impedance Curves for the Reed Woodwind Instruments," *J. Acoust. Soc. Am.* **56**: 1266.

Bak, N. (1969/70). "Pitch, Temperature and Blowing Pressure in Recorder Playing. Study of Treble Recorders," *Acustica* **22**: 295.

Benade, A. H. (1976). *Fundamentals of Musical Acoustics.* New York: Oxford. (See Chapters 21–22.)

Benade, A. H., and S. N. Kouzoupis (1988). "The Clarinet Spectrum: Theory and Experiment," *J. Acoust. Soc. Am.* **83**: 292.

Benade, A. H., and S. J. Lutgren (1988). "The Saxophone Spectrum," *J. Acoust. Soc. Am.* **83**: 1900.

Campbell, M., and C. Greated (1987). *The Musician's Guide to Acoustics.* London: Oxford.

Chanaud, R. C. (1970). "Aerodynamic Whistles," *Sci. Am.* **222**(1): 40.

Coltman, J. W. (1968). "Acoustics of the Flute," *Physics Today* **21**(11): 25.

Coltman, J. W. (1991). "Some Observations on the Piccolo," *The Flutist Quarterly*, Winter 1991: 17.

Fletcher, N. H. (1974). "Some Acoustical Principles of Flute Technique," *The Instrumentalist* **28**(7): 57. A more mathematical version appears in *J. Acoust. Soc. Am.* **57**: 233 (1975).

Fletcher, N. H. (1976). *Physics and Music.* Melbourne: Heinemann.

Fletcher, N. H. (1979). "Air Flow and Sound Generation in Musical Wind Instruments," *Ann. Rev. Fluid Mech.* **11**: 123.

Fletcher, N. H., and T. D. Rossing (1998). *The Physics of Musical Instruments*, 2nd ed. New York: Springer Verlag. Chapters 15 and 16.

Fransson, F. (1967). "The Source Spectrum of Double-Reed Woodwind Instruments," Report STL-QPSR 1, 25, Royal Institute of Technology, Stockholm.

Fuks, L., and J. Sundberg (1996). "Blowing Pressures in Reed Woodwind Instruments," *Quart. Prog. Status Report, Speech Music and Hearing, No. 3/1996.* Stockholm: Royal Inst. Tech., p. 41.

Helmholtz, H. L. F. von (1877). *On the Sensations of Tone as a Physiological Basis for the Theory of Music*, 4th ed. Trans., A. J. Ellis. New York: Dover, 1954.

Herman, R. (1959). "Observations on the Acoustical Characteristics of the English Flute," *Am. J. Physics* **27**: 22.

Lehman, P. R. (1962). "The Harmonic Structure of the Tone of the Bassoon." Ph.D. dissertation, University of Michigan, Ann Arbor.

Martin, J. (1994). *The Acoustics of the Recorder.* Celle, Germany: Moeck-Verlag.

Nederveen, C. J. (1998). *Acoustical Aspects of Woodwind Instruments*, DeKalb, Ill.: Northern Illinois University Press.

Plitnik, G. R., and W. J. Strong (1979). "Numerical Method for Calculating Input Impedances of the Oboe," *J. Acoust. Soc. Am.* **65**: 816.

Strong, W. J., and G. R. Plitnik (1983). *Music, Speech, and High Fidelity*, 2nd ed. Provo, Utah: Soundprint. Chapter 6.

Worman, W. E. (1971). "Self-Sustained Nonlinear Oscillations of Medium Amplitude in Clarinet-like Systems." Ph.D. dissertation, Case Western Reserve University, Cleveland.

GLOSSARY

aerodynamic Having to do with the flow of air and its interaction with other bodies.

Bessel horn A family of horns of different shapes, including cylindrical and conical.

cutoff frequency The frequency above which the sound loss due to radiation through a lattice of tone holes is large, so that the resonances of an air column are weak.

embouchure The lip position used in playing a wind instrument. The embouchure hole in a flute is the hole through which the lips blow air.

feedback Use of an output signal to control or influence the input. Positive feedback, if great enough, can cause a system to oscillate.

formant A range of frequency to which a system responds preferentially or which is emphasized in its output.

impedance The ratio of the pressure to the velocity in a sound wave.

lay The slightly curved portion of a clarinet or saxophone mouthpiece that faces the reed.

register A group of related notes on a musical instrument; one register, for example, may include all notes whose pitch corresponds to the lowest resonance of an air column.

register hole A hole that can be opened in order to cause an instrument to play in a higher register.

sinusoidal Pertaining to a sine wave; thus a pure tone or single frequency of vibration.

tone hole A hole that can be opened to raise the pitch of an instrument.

turning point The point in a musical instrument at which most of the sound wave is reflected back toward the mouthpiece.

REVIEW QUESTIONS

1. If a trumpet player can play a melody using only the mouthpiece, why can't a clarinet player?

2. Which of the following employ inward-striking reeds?
   (a) Clarinet
   (b) Oboe
   (c) Bassoon
   (d) Organ reed pipe
   (e) All of these

3. List three instruments that have conical bores.

4. What instrument has a reverse conical bore?

5. Are the resonance frequencies of a cone higher, lower, or the same as an open cylindrical pipe of the same length?

6. Do modern flutes have a tapered or a cylindrical head joint?

7. What is the approximate natural frequency of a clarinet reed?

8. Missing second and fourth harmonics give the clarinet its hollow, "woody" sound in which register?

9. Which of the following instruments requires the highest blowing pressure?
   (a) Clarinet
   (b) Saxophone
   (c) Oboe
   (d) Bassoon
   (e) Recorder

10. Flutes play at frequencies where the acoustic impedance is (a) maximum; (b) minimum; (c) either a maximum or minimum depending on other factors.

11. Do modern flutes have a tapered or a cylindrical bore?

12. What is the approximate efficiency of woodwind instruments in converting wind power to sound power?

13. What are two techniques a clarinet player uses to increase the loudness of a note?

QUESTIONS FOR THOUGHT AND DISCUSSION

1. Why must a clarinet have more keys than a flute?

2. Discuss the acoustical implications of playing the clarinet with a stiff reed rather than a soft one.

3. If a clarinet were played in a pure helium atmosphere, what would its lowest note be? (See Table 3.1.) Would you expect this same pitch if a clarinet player completely filled his or her lungs with helium before blowing?

4. Can you "overblow" a bottle to obtain a higher note in a different register? Try it.

5. Why do saxophone players not use a mute (as trumpet players do)?

6. Clarinets and flutes both have nearly cylindrical bores of about the same length. Why is their timbre so different?

EXERCISES

1. A B$^\flat$ clarinet is about 67 cm long.

   (a) What is the lowest resonance frequency of a closed pipe of that length?

   (b) The lowest note on a B$^\flat$ clarinet is D$_3$. Compare its frequency to your answer in (a).

   (c) Can you explain the difference between these two frequencies?

2. It is possible to fit a flute-type head joint to a clarinet, so that it plays in the manner of a flute. What would you expect the lowest note to be?

3. (a) Making use of the fact that the speed of an air jet is approximately proportional to the square root of the blowing pressure, show that doubling the blowing pressure increases the jet speed by 40%.

   (b) How much does the jet speed increase when the blowing pressure is tripled?

4. Show that the frequencies of the notes in the different clarinet registers that are fingered nearly the same are in the ratio of 1 : 3 : 5 (see Fig. 12.10; the frequencies of the notes can be found in Table 9.2).

5. The length of an alto recorder (from the windway to the bell) is about 42 cm.

   (a) What is the lowest resonance frequency of an open pipe of that length?

   (b) The lowest note on the recorder is F$_4$. Compare its frequency to your answer in (a).

   (c) Explain the difference between these two frequencies.

6. The length of a flute (from the embouchure hole to the open end) is about 60 cm. What do you expect the frequency of the lowest note to be?

EXPERIMENTS FOR HOME, LABORATORY, AND CLASSROOM DEMONSTRATION

Home and Classroom Demonstration

1. *Register holes*   Attach a clarinet mouthpiece to a length of tubing (electrical conduit is ideal), as shown in Fig. 12.12. Drill small holes one-third and one-fifth of the way down the pipe and cover them with tape. The conduit has a clarinetlike sound. Uncovering the lower hole will make the pipe sound its third harmonic and uncovering the upper hole will make it sound its fifth harmonic.

2. *Register holes in a clarinet*   Follow Demonstration 1 by showing that any set of three notes in Fig. 12.10 are the first, third, and fifth harmonics of the closed-pipe fundamental.

3. *Overblowing a flute or recorder*   Show that increasing the air flow or opening a register hole (recorder) encourages the open pipe to sound its second harmonic.

4. *Comparing open and closed end pipes*   Show that placing a finger over the end of a plastic tube with a whistle causes the fundamental to go down one octave and also makes it impossible to sound the second harmonic. By alternately opening and closing the end of the pipe, bugle calls can be played (by using both the open and closed pipe harmonic series).

5. *Comparing open and closed end organ pipes*   Show that capping an open organ pipe causes the fundamental to go down one octave and also makes it impossible to sound the second harmonic.

6. *Soda-straw double reed*   Flatten one end of a soda straw and cut an arrow-shaped end with scissors. Place the straw well into your mouth and blow hard to obtain a double-reed sound. Use the scissors to cut off more and more of the straw while blowing, resulting in a rising pitch.

7. *Instruments with conical bore*   An oboe or saxophone can be used to demonstrate that conical pipes sound even-numbered harmonics even though the pipe has a closed end.

8. *Clarinet bell*   Demonstrate that removing the bell of a clarinet has little effect except on the lowest note of each register (fingered E$_3$ and B$_5$) and a note or two above them. Likewise, removing the lower half of the bore has little effect on notes such as fingered E$_4$ and B$_5$ because the open tone-hole lattice immediately beyond them is left intact.

9. *Clarinet with flute head joint*   A clarinet fitted with a flute head joint will sound even-numbered as well as odd-numbered harmonics.

10. *PVC pipe instruments*   A variety of inexpensive woodwind instruments can be constructed from PVC pipe. Pan pipes are the easiest. An alpen horn is described in *Physics Teacher* **28**: 459 (1990).

11. *Change in timbre in clarinet bore*   By means of a miniature microphone or a probe tube coupled to a microphone record the sound at various points along a clarinet bore, in particular at each of the first three open-tone holes. Note the difference in timbre (spectrum) due to the frequency-dependent transfer characteristic of the open tone-hole lattice.

Laboratory Experiments

Acoustics of woodwind instruments (Experiment 18 in *Acoustics Laboratory Experiments*)

Percussion Instruments

Praise Him with the sounding cymbals.
Praise Him with the clanging cymbals...

Psalm 150

Percussion instruments may be our oldest musical instruments (with the exception of the human voice), but recently they have experienced a new surge in interest and popularity. Many novel percussion instruments have been developed recently and more are in the experimental stage. What is often termed "contemporary sound" makes extensive use of percussion instruments. Yet, relatively little material has been published on the acoustics of percussion instruments.

In this chapter you should learn:

- About vibrations of bars, rods, or tubes in marimbas, xylophones, vibes, chimes, and triangles;
- About vibrations of membranes in drums;
- About vibrations of plates in cymbals, gongs, and steelpans;
- About the acoustics of bells.

There are many instruments in the percussion family and a number of ways in which to classify them. Sometimes they are classified into four groups: idiophones (xylophone, marimba, chimes, cymbals, gongs, etc.); membranophones (drums); aerophones (whistles, sirens); and chordophones (piano, harpsichord). Another system divides them into two groups: those that have definite pitch and those that do not. In this chapter we will discuss various membranophones and idiophones; chordophones will be discussed in Chapter 14.

Percussion instruments generally use one or more of the following basic types of vibrators: strings, bars, membranes, plates, air columns, or air chambers. The first four are mechanical, and the latter two are pneumatic. Two of them (the string and the air column) tend to produce harmonic overtones; the others, in general, do not. Bars, membranes, and plates are three classes of vibrators whose modes of vibration are not related harmonically. Thus the overtones they sound will not be harmonics of the fundamental tone. The inharmonic overtones of these complex vibrators give percussion instruments their distinctive timbres.

13.1 ■ VIBRATIONS OF BARS

Vibrations of bars were discussed briefly in Section 2.6. A bar (or rod or tube) can vibrate either longitudinally (by expanding and contracting in length) or transversely (by bending at right angles to its length). In percussion instruments, the transverse modes of vibration are nearly always used (one exception being the aluminum stroke rods, which are excited longitudinally by stroking with a rosined cloth or gloves). Longitudinal vibrations are much higher in frequency than transverse vibrations, and the various longitudinal modes are related harmonically. The frequency of longitudinal vibration depends on the length of the rod and on the elasticity of the material from which it is fabricated but is independent of the thickness, surprisingly enough. In fact, the frequency of longitudinal vibration of a rod is expressed by the simple formula of $f_n = n v_L / 2L$, where v_L is the speed of sound in the rod, L is its length, and $n = 1, 2, 3, \ldots$, denotes the number of the harmonic (beginning with $n = 1$ for the fundamental frequency). The velocity $v_L = \sqrt{E/\rho}$, where E is Young's elastic modulus and ρ is the density.

When a bar vibrates longitudinally, its motion is almost identical to the movement of the air in a pipe that is open at both ends (a flute, for example). Maximum movement occurs at the ends of the bar or air column (called antinodes), and one or more points in between have a minimum of movement (called nodes). For the fundamental mode, there is one node, and that occurs at the center. For the first overtone (harmonic number 2), there are two nodes with another antinode at the center. One can selectively excite any desired mode of vibration in a bar by clamping it at the location of one of the nodes for that particular mode. If the fundamental mode is desired, the bar should be supported at the center only.

Transverse vibrations in a bar are a little more complicated. First of all, there are three possible end conditions for the bar: clamped, simply supported, and free. However, nearly all percussion instruments use bars with free ends, so only the free bar will be considered. (One exception is the electronic carillon, which sometimes uses bars or rods that are free at one end and clamped at the other.)

The frequency of a bar in either longitudinal or transverse vibration depends on its length and the density and elasticity of the material, but in transverse vibration, the frequency depends on the thickness of the bar as well. The frequency of transverse vibrations in a bar with free ends is given by the formula $f_n = (\pi v_L K / 8L^2) m^2$, where v_L is the speed of sound, L is the length of the bar, m is a sequence number to be defined in a moment, and K is the *radius of gyration*, which is related to the size and shape of the bar. (For a flat bar, K is the thickness divided by 3.46; values for other shapes are given in Table 13.1.) Note the following differences from the longitudinally vibrating bar.

1. The frequency depends on L^2 rather than L.
2. The mode frequencies are not harmonic, but increase as m^2.
3. The frequency depends on the shape of the bar through the factor K.

The frequencies of the modes are in proportion to the squares of the odd integers—almost. The number m begins with 3.0112 and then continues with the simple values $5, 7, 9, \ldots, (2n + 1)$. Transverse vibrations of a bar are illustrated in Fig. 2.17.

TABLE 13.1   Formulas for vibration frequencies
of strings and bars

String

　　Transverse vibration: $f_n = n\frac{v}{2L}$

　　　　v = speed of waves in the string = $\sqrt{T/\mu}$
　　　　L = length of the string
　　　　$n = 1, 2, 3, \ldots$
　　　　T = tension
　　　　μ = mass per unit length

　　Longitudinal vibration: $f_n = n\frac{v_L}{2L}$

　　　　v_L = speed of sound in the string = $\sqrt{E/\rho}$
　　　　　　(does not change with tension)
　　　　E = Young's modulus of elasticity
　　　　ρ = density

Bar with free ends

　　Transverse vibration: $f_n = \frac{\pi v_L K}{8L^2}m^2$

　　　　v_L = speed of sound = $\sqrt{E/\rho}$
　　　　E = Young's modulus of elasticity
　　　　ρ = density
　　　　L = length
　　　　$m = 3.0112, 5, 7, \ldots, (2n + 1)$
　　　　K = radius of gyration
　　　　　　$K = t/\sqrt{12} = t/3.46$ for a rectangular bar
　　　　　　t = thickness
　　　　　　$K = \frac{1}{2}\sqrt{a^2 + b^2}$ for a tube
　　　　　　a = inner radius
　　　　　　b = outer radius

　　Longitudinal vibration: $f_n = n\frac{v_L}{2L}$

　　　　v_L = speed of sound = $\sqrt{E/\rho}$
　　　　L = length
　　　　$n = 1, 2, 3, \ldots$

　　　The frequencies of the transverse modes of vibration of a uniform bar have the ratios
1.0 : 2.76 : 5.40 : 8.90, and so on, which are anything but harmonic and, in fact, match
no intervals on the musical scale. They give a distinctive timbre to instruments such as
chimes, orchestra bells, and triangles, which are nearly uniform bars. (The bars of marimbas, xylophones, and related instruments are not uniform; they have been cut to have quite
a different set of mode frequency ratios and thus a different timbre.)

　　　Longitudinal and transverse vibrations may be strikingly demonstrated with aluminum
rods about 1 cm in diameter (see Section 4.8). Longitudinal vibrations can be excited by
stroking the rod with a rosined cloth or tapping lightly on the end with a hammer; transverse vibrations, which in a long rod occur at a much lower frequency, can be excited by

hitting the rod with one's hand near the center. Since the frequency of transverse vibrations varies with $1/L^2$ but the frequency of longitudinal vibrations only varies with $1/L$ (see Table 13.1), it is possible to cut a rod to a length such that the two fundamental frequencies coincide. Note that the nodes for longitudinal and transverse vibration occur at different places, so one's grip must be changed. Aluminum stroke rods are now being used as percussion instruments in the performance of contemporary music.

13.2 ■ RECTANGULAR BARS: THE GLOCKENSPIEL

The glockenspiel, or orchestra bells, use rectangular steel bars 2.5 to 3.2 cm (1 to $1\frac{1}{4}$ in.) wide and 0.61 to 1 cm ($\frac{1}{4}$ to $\frac{3}{8}$ in.) thick. Its range is customarily from G_5 ($f = 784$ Hz) to C_8 ($f = 4186$ Hz), although it is scored two octaves lower than it sounds. The glockenspiel is usually played with brass or hard plastic mallets. The bell lyra, a portable version that uses aluminum bars and usually covers the range A_5 ($f = 880$ Hz) to A_7 ($f = 3520$ Hz), is sometimes used in marching bands. Glockenspiel and bell lyra are shown in Figure 13.1.

When struck with a hard mallet, a glockenspiel bar produces a crisp, metallic sound, which quickly gives way to a clear ring at the designated pitch. Because the overtones have very high frequencies and die out quickly, they are of relatively less importance in determining the timbre of the glockenspiel than are the overtones of the marimba or xylophone, for example. For this reason, no effort is made to bring the inharmonic overtones of a glockenspiel into a harmonic relationship through overtone tuning.

The frequencies for transverse vibrations in a bar with free ends, given in Section 13.1, were shown to be inharmonic; thus the glockenspiel has no harmonic overtones. However, for even the lowest bar the first overtone has a frequency of 2160 Hz, which is getting into the range in which the pitch discrimination of human listeners is diminished. (This will not be the case for other bar percussion instruments, however, as we will see.)

The vibrational modes of a glockenspiel bar are shown in Fig. 13.2. These modes can be excited individually by passing an electric current from an audio amplifier through the bar while it is in the field of a magnet. Besides the transverse modes, which are labeled 1, 2, 3, 4, and 5, there are torsional, or "twisting," modes labeled a, b, c, d; a longitudinal

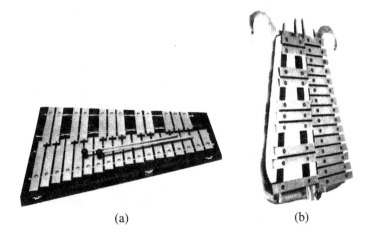

FIGURE 13.1
(a) Orchestra bells
or glockenspiel;
(b) bell lyra.

(a)   (b)

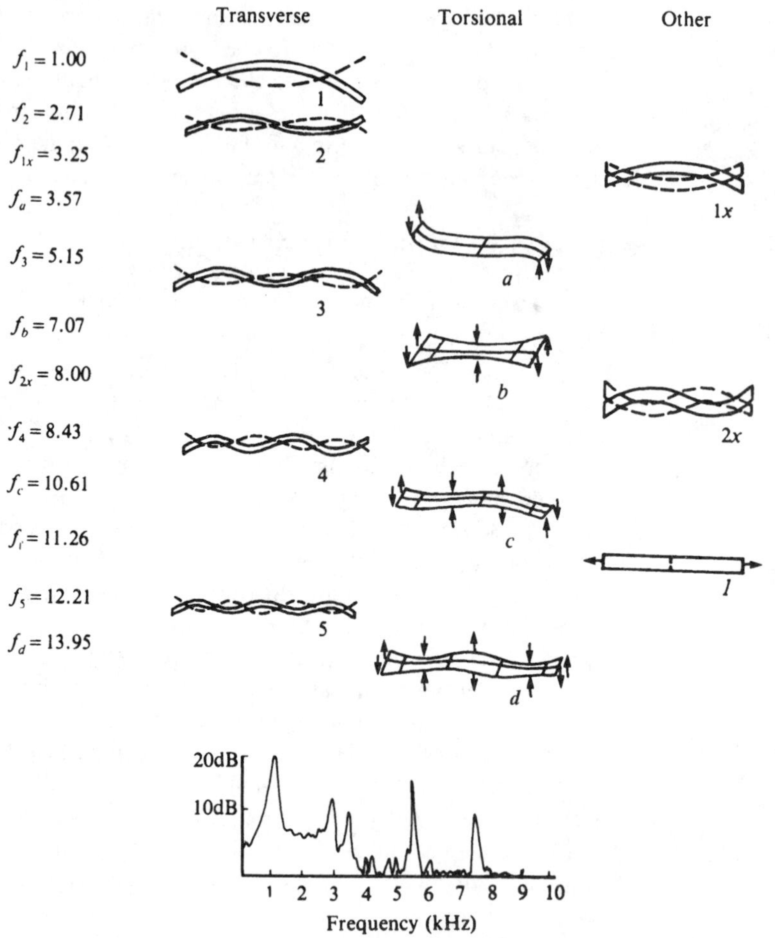

Transverse
Torsional
Other
$f_1 = 1.00$
$f_2 = 2.71$
$f_{1x} = 3.25$
$f_a = 3.57$
$f_3 = 5.15$
$f_b = 7.07$
$f_{2x} = 8.00$
$f_4 = 8.43$
$f_c = 10.61$
$f_l = 11.26$
$f_5 = 12.21$
$f_d = 13.95$
1
2
3
4
5
a
b
c
d
1x
2x
l
20dB
10dB
1 2 3 4 5 6 7 8 9 10
Frequency (kHz)

FIGURE 13.2 Modes of C_6 glockenspiel bar, which lie within the audible range of frequency. The transverse modes are labeled 1, 2, 3, 4, and 5; *a*, *b*, *c*, and *d* are torsional modes; *l* is a longitudinal mode; and 1*x* and 2*x* are transverse modes excited edgewise. Mode frequencies are given as ratios to the fundamental. The sound spectrum above shows the relative amplitudes of the partials associated with these modes. (From Rossing 1976.)

mode *l*; and transverse modes in the plane of the bar, 1*x* and 2*x*. Recording a spectrum of the strike sound requires a real-time spectrum analyzer.

The transverse modes of vibration of the glockenspiel bar are spaced a little closer together than predicted by the simple theory of thin bars. The more sophisticated theory for thick bars (called Timoshenko bars in engineering textbooks) takes into account other factors such as the moment of inertia of the bar and the shear stresses in the bar.

13.3 ■ THE MARIMBA, XYLOPHONE, AND VIBES

The three most common bar percussion instruments are the marimba, the xylophone, and the vibraphone or vibraharp (commonly called *vibes*). All three instruments consist of tuned bars with tubular resonators. They are played with mallets of varying hardness.

The *marimba* typically includes 3 to 5 octaves of tuned bars of rosewood or fiberglass synthetic (typical tradenames: Kelon, Klyperon), graduated in width from about 4.5 to

6.4 cm ($1\frac{3}{4}$ to $2\frac{1}{2}$ in.). Beneath each bar is a tubular resonator tuned to the fundamental frequency of that bar (see Fig. 13.3(a)). When the marimba is played with soft mallets, it produces a rich mellow tone. The playing range of a large concert marimba is A_2 to C_7 ($f = 110$ to 2093 Hz), although bass marimbas extend to C_2 ($f = 65$ Hz).

A deep arch is cut in the underside of marimba bars, particularly in the low register. This arch serves two useful purposes: It reduces the length of bar required to reach the low pitches, and it allows tuning of the overtones (the first overtone is nominally tuned two octaves above the fundamental). Figure 13.3(a) shows a scale drawing of a marimba bar and also indicates the positions of the nodes for each of the first seven modes of vibration. (This may be compared to Fig. 13.2, which shows the nodes of a glockenspiel bar without a cut arch.) The ratios of the frequencies of these modes are also indicated. Note that the second partial (first overtone) of this bar has a frequency 3.9 times that of the fundamental, which is close to a two-octave interval (a ratio of 4.0).

Modal tuning of the bars is discussed in Chapter 19 in Fletcher and Rossing (1998). In general, removing material from a location where the bending moment for a given mode is large will lower the frequency of that mode more than others. It is more difficult to raise the frequency of a given mode, but removing material near the end of the bar will raise the frequencies of all the modes.

Marimba resonators are cylindrical pipes tuned to the fundamental mode of the corresponding bars. A pipe with one closed end and one open end resonates when its acoustical length is one-fourth of a wavelength of the sound. The purpose of the tubular resonators is to emphasize the fundamental and also to increase the loudness, which is done at the expense of shortening the decay time of the sound. We have measured the decay time (60 dB) of a rosewood bar in the low register (E_3) to be 1.5 s with the resonator and 3.2 s without it. Decay times in the upper register are generally shorter; we measured 0.4 s and 0.5 s for an E_6 bar with and without the resonator, respectively. The corresponding decay times for synthetic bars are somewhat longer.

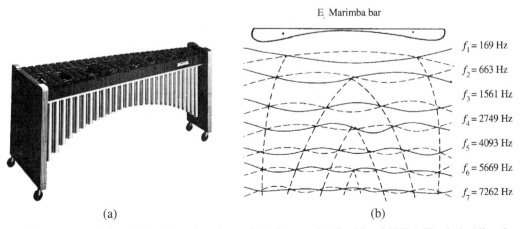

E Marimba bar

$f_1 = 169$ Hz
$f_2 = 663$ Hz
$f_3 = 1561$ Hz
$f_4 = 2749$ Hz
$f_5 = 4093$ Hz
$f_6 = 5669$ Hz
$f_7 = 7262$ Hz

(a)   (b)

FIGURE 13.3   (a) A marimba. (b) Scale drawing of a marimba bar tuned to E_3 ($f = 265$ Hz). The dashed lines locate the nodes of the first seven modes. (Photograph courtesy of J.C. Deagan Co.)

It is not difficult to test the overtone tuning of a marimba by noting the position of the nodes as shown in Fig. 13.3(b). To suppress the fundamental mode and emphasize the first overtone, one should touch the bar firmly at the center and strike the bar either at one end or at a point about one-third of the way from the center to either end.

As an experiment, the author designed a marimba with variable timbre which has a second set of resonators tuned to the first overtone of the bars. Each resonator is equipped with a vane that can partially or completely close the mouth of the tube; thus the timbre can be varied by adjusting the amount of closure. (This instrument was first played at a meeting of the Acoustical Society of America.)

The *xylophone* is a close cousin to the marimba (or perhaps "uncle" is a better term, since the xylophone apparently has a longer history). Xylophones typically cover a range of 3 to $3\frac{1}{2}$ octaves extending from F_4 or C_5 to C_8 ($f = 349$–4186 Hz) and may have bars of synthetic material or rosewood. Modern xylophones are nearly always equipped with tubular resonators to increase the loudness of the tone (see Fig. 2.18).

Xylophone bars are also cut with an arch on the underside, but the arch is not as deep as that of the marimba, since the first overtone is tuned to a musical twelfth above the fundamental (that is, three times the frequency of the fundamental). Since a pipe closed at one end can also resonate at three times its fundamental resonant frequency, the twelfth will also be reinforced by the resonator. This overtone boost, plus the hard mallets used to play it, give the xylophone a much crisper, brighter sound than the marimba. We have found that careful overtone tuning is usually ignored in the upper register, as in the case of the marimba.

The vibrational frequencies of a xylophone bar tuned to $F_4^\#$ ($f = 370$ Hz) are shown in Fig. 13.4. The frequencies of the bending and torsional modes have been lowered from those of a bar with free ends (as shown in Table 13.1) by cutting an arch on the underside; this lowering is greatest in the lowest modes in each type of vibration. Nevertheless, the longitudinal and torsional mode frequencies are more or less proportional to the mode

FIGURE 13.4
Modal frequencies of an $F_4^\#$ xylophone bar. Solid lines have slopes of 1 or 2. The mode number is $2n + 1$ for bending modes (Rossing and Russell 1990).

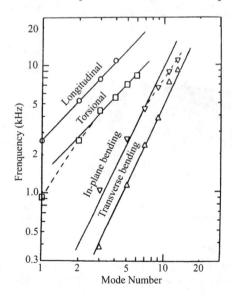

FIGURE 13.5
Sound level
recording for A_4
bar ($f = 440$ Hz)
with a vibe rate of
about 4 Hz. The
sound level
fluctuates about
6 dB, and the decay
time (60 dB) is
seven seconds.

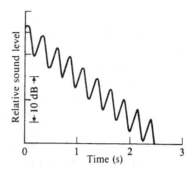

number, and the bending modes are more or less proportional to the square of mode number as in a uniform bar (Table 13.1).

The xylomarimba is a large xylophone with a $4\frac{1}{2}$- to 5-octave range (C_3 or F_3 to C_8), and is occasionally used in solo work or in modern scores. Bass xylophones and keyboard xylophones have also been constructed (Brindle 1970).

A very popular bar percussion instrument is the *vibraphone* or *vibraharp*, as they are designed by different manufacturers. *Vibes*, as they are popularly called, usually consist of aluminum bars tuned over a three-octave range from F_3 to F_6 ($f = 175 - 1397$ Hz). The bars are deeply arched so that the first overtone has four times the frequency of the fundamental, as in the marimba. The aluminum bars tend to have a much longer decay time than the wood or synthetic bars of the marimba or xylophone, and so vibes are equipped with pedal-operated dampers.

The most distinctive feature of vibes, however, is the vibrato introduced by motor-driven discs at the top of the resonators, which alternately open and close the tubes. The vibrato produced by these rotating discs or pulsators consists of rather substantial fluctuation in amplitude (*intensity vibrato*) and a barely detectable change in frequency (*pitch vibrato*). The speed of rotation of the discs may be adjusted to produce a slow vibe or a fast vibe. Often vibes are played without vibrato by switching off the motor. They are usually played with soft mallets or beaters, which produce a mellow tone, although some passages call for harder beaters.

Because vibraphone bars have a much longer decay time than do marimba and xylophone bars, the effect of the tubular resonators on decay time is more dramatic. At 200 Hz (A_3), for example, we measure a decay time (60 dB) of 40 s without the resonator and 9 s with the tube full open. For A_5, we measure 24 s with the resonator closed and 8 s with it open. In the recording of sound level shown in Fig. 13.5, the intensity modulation, as well as the slow decay of the sound, can be clearly seen.

13.4 ■ MALLETS

The modern percussion player can select from a wide variety of mallets, which differ in mass, shape, and hardness. Through intelligent selection of a mallet, a player can greatly influence the timbre of the instrument being played. Striking a marimba or xylophone with a hard mallet, for example, produces a sound rich in overtones that emphasizes the woody character of the instrument. A soft mallet, on the other hand, which excites mainly the harmonically tuned lower partials, gives a dark sound to the instrument.

A mallet whose mass nearly equals the dynamic mass of the struck bar (typically about 30% of the total mass for a marimba bar in its fundamental mode) transfers the maximum amount of energy to the bar. A lighter mallet rebounds after a short contact time. A heavier mallet remains in contact for a longer time, which results in considerable damping of the higher partials. This is a desirable effect in drums, where a large amplitude and short decay time are desired, but may be only a special effect in bar percussion instruments.

According to Hertz's law, the impact force is proportional to the $\frac{3}{2}$ power of the mallet deformation. From measurements of impact force and impact time, it appears that Hertz's law describes the impact over a fairly wide range of mallet hardness and impact time (Chaigne and Doutaut 1997).

13.5 ■ CHIMES

Chimes, or tubular bells, are usually fabricated from lengths of brass tubing $1\frac{1}{4}$ to $1\frac{1}{2}$ in. in diameter. The upper end of each tube is partially or completely closed by a brass plug with a protruding rim. This rim forms a convenient and durable strike point.

One of the interesting characteristics of chimes is that there is no mode of vibration with a frequency at, or even near, the pitch of the strike tone one hears. The frequencies excited when a chime is struck are very nearly those of a free bar described earlier. Modes 4, 5, and 6 appear to determine the strike tone. This can be understood by noting that these modes for a free bar have frequencies nearly in the ratio $9^2 : 11^2 : 13^2$, or $81 : 121 : 169$, which are close enough to the ratio $2 : 3 : 4$ for the ear to consider them nearly harmonic and to use them as a basis for establishing a pitch (see Section 7.4). The largest near-common factor in the numbers 81, 121, and 169 is 41.

Figure 13.6 is a graph of the frequencies of the G-chime as a function of m, along with those predicted by the thin-bar theory of Lord Rayleigh (1894) and also the more detailed thick-bar theory (Flugge 1962). Also shown are the frequencies of vibration with the end plug removed. Note that the end plug lowers the frequencies of the first few modes but has little effect on the higher modes. The strike tone, which lies one octave below the fourth mode, is also indicated.

The ratios of the modal frequencies of a chime tube with and without a load at one end are given in Table 13.2. Also given are the ratios considered desirable for a tuned carillon bell. Note the similarity between the partials of a chime and those of a carillon bell. Adding a load to one end of a chime lowers the frequencies of the lower modes more than the higher ones (see also Fig. 13.6) and thus "stretches" the modes into a more favorable ratio. The end plug also adds to the durability of the chime and helps to damp out the very high modes.

The bell-like quality of chimes is well known, of course, and has been used in many compositions for band and orchestra (Tchaikovsky's *1812 Overture*, for example). This bell-like timbre can be maximized by selecting the optimum size of end plug for each chime. For the G-chime in Table 13.2, the 193-g plug, with which the chime was originally fitted, is quite near the optimum. Most chime-makers use the same size plug throughout the entire set of chimes, however, so the timbre changes up and down the scale, typically being

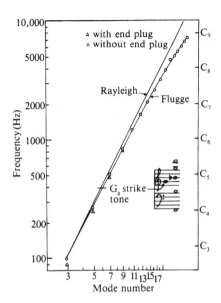

FIGURE 13.6 Frequencies of a G-chime as a function of mode number m, along with those predicted by thin-bar theory (Rayleigh) and thick-bar theory (Flugge). The first five modes are shown on musical staffs, along with the subjective strike note. (From "Acoustics of Percussion Instruments, Part I" *Phys. Teach.* 14:546.)

TABLE 13.2 Ratios of mode frequencies for loaded and unloaded chime tube (compared with the strike tone)

n	Thin rod	Tube	Loaded with 193 g	435 g	666 g	Tuned bell
1	0.22	0.24	0.23	0.22		0.5
2	0.61	0.64	0.63	0.62	0.61	1
3	1.21	1.23	1.22	1.22	1.22	1.2 or 1.25 (1.5)
4	2	2	2	2	2	2 (2.5)
5	2.99	2.91	2.93	2.95	2.94	3
6	4.17	3.96	4.01	4.04	4.03	4
7	5.56	5.12	5.21	5.21	5.18	5.33
8	7.14	6.37	6.50	6.43	6.37	6.67
Strike tone		416 Hz	393 Hz (G_4)	383 Hz	381 Hz	

(All modes are compared with the strike tone.)

Source: Rossing (1976).

optimum only near the center. A set of chimes "scaled" to have the same timbre throughout would have some advantages.

A well-tuned chime not only has its overtones tuned to resemble those of a carillon bell, but also is free of beats between modes with nearly, but not quite, the same frequencies of vibration. These beating modes occur when the chime tube is not perfectly round or its wall thickness is not perfectly uniform. As a result, the transverse vibrations will have slightly different frequencies in two different transverse directions, resulting in beats when both modes are excited. These undesired beats can be eliminated by squeezing the chime in a vise or thinning the wall slightly on one side to bring the modes into tune.

13.6 ■ TRIANGLES

Because of their many modes of vibration, triangles are characterized as having an indefinite pitch. They are normally steel rods bent into a triangle (usually, but not always, equilateral) with one open corner. Triangles are suspended by a cord from one of the closed corners, and are struck with a steel rod or hard beater.

Triangles are typically available in 15-, 20-, and 25-cm (6-, 8-, and 10-in.) sizes, although other sizes are also used. Sometimes one end of the rod is bent into a hook, or the ends may be turned down to smaller diameters than the rest of the triangle to alter the modes of vibration. The sound of the triangle depends on the strike point as well as the harness of the beater. Single strokes are usually played on the base of the triangle, whereas the rapid strokes of a tremolo are made on the inside of the triangle near the upper angle.

The modes of the triangle are many, and they are not harmonically related. Generally, they can be characterized by vibrations in the plane of the triangle and perpendicular to the plane. Normal strokes will tend to emphasize vibrations in the plane of the triangle. Figure 13.7 shows sound spectra for a 10-in. triangle at two different strike points with strokes parallel and perpendicular to the plane of the triangle.

Because triangles are made of steel, it is easy to excite single modes of vibration magnetically. A choke coil with a ferrite core or a ferrite antenna from a radio can be driven by an audio amplifier to produce an oscillating magnetic field. At each resonance frequency, a small microphone can be used to locate the nodes and antinodes and thus identify the mode. Some of the modes of vibration of a 10-in. triangle are shown in Fig. 13.8, along

FIGURE 13.7
Sound spectra for a 25-cm steel triangle (a) struck in the plane; (b) struck perpendicular to the plane. Two frequency and amplitude ranges are shown in each case. (Reprinted by permission of the Percussive Arts Society.)

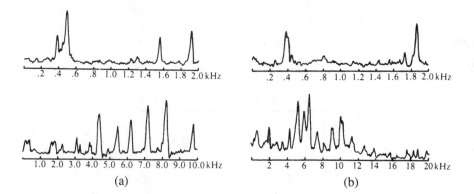

(a)   (b)

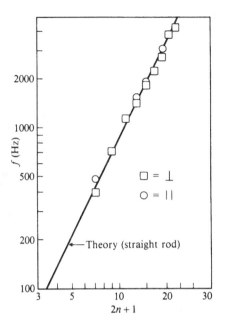

FIGURE 13.8
Mode frequencies
for a 25-cm steel
triangle driven in
the plane and
perpendicular to the
plane. The line
gives the predicted
frequencies for a
steel rod of the
same diameter and
length. (Reprinted
by permission of
the Percussive Arts
Society.)

with those calculated for a steel rod of the same length and diameter. The triangle modes show a surprisingly close correspondence to those of a straight rod.

13.7 ■ VIBRATIONS OF MEMBRANES

FIGURE 13.9
Modes of an ideal
membrane,
showing radial and
circular nodes and
the customary
mode designation
(the first number
gives the number of
radial modes, and
the second number
the circular nodes,
including the one at
the edge). The
number below each
mode diagram
gives the frequency
of that mode
compared to the
fundamental (0, 1)
mode.

The vibrations of an ideal membrane were discussed briefly in Section 2.6. We pointed out that a membrane may be thought of as a two-dimensional string, in that the restoring force necessary for it to vibrate is supplied by tension applied from the edge. A membrane, like a string, can be tuned by changing its tension. One major difference between vibrations in the membrane and in the string, however, is that the mode frequencies in the ideal string are harmonics of the fundamental, but in the membrane they are not. Another difference is that in the membrane, nodal lines replace the nodes that occur along the string. These nodal lines are circles and diameters, as shown in Fig. 13.9.

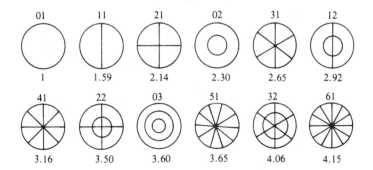

The formula for the frequencies of the various vibrational modes of an ideal membrane is

$$f_{mn} = \frac{1}{2a}\sqrt{\frac{T}{\sigma}}\beta_{mn},$$

a = the radius of the membrane (m), T = the surface tension applied to the membrane (N/m), σ = the area density (kg/m^2), and β_{mn} = values for which a certain mathematical function (the Bessel function) becomes zero.

The formula itself is not particularly important to the musician, except that it shows how the pitch of a vibrating membrane varies with radius, tension, and thickness. The values m and n for each mode of vibration indicate the number of nodal diameters and the number of nodal circles, respectively.

The first 12 modes of vibration of a circular membrane are shown in Fig. 13.9. Above each sketch are given the values of m and n, and below it is the frequency of vibration for that mode divided by the frequency of the lowest (01) mode. For example, the (31) mode has 3 nodal diameters and 1 nodal circle (around the edge) and vibrates at a frequency 2.65 times that of the lowest mode. The 4 modes shown in Fig. 2.14 are the (01), (02), (11), and (21) modes.

13.8 ■ TIMPANI

Drums consist of membranes of animal skin or synthetic material stretched over some type of air enclosure. Some drums (e.g., timpani, tabla, boobams) sound a definite pitch; others convey almost no sense of pitch at all. Drums are important in nearly all musical cultures and constitute one of the most universally used types of musical instruments throughout history.

The timpani or kettledrums are the most important drums in the orchestra, one member of the percussion section usually devoting attention exclusively to them. During the last century, various mechanisms were developed for changing the tension to tune the drumheads rapidly. Most modern timpani have a pedal-operated tensioning mechanism in addition to six or eight tensioning screws around the rim of the kettle. The pedal typically allows the player to vary the tension over a range of 2 : 1, which corresponds to a tuning range of about a musical fourth. A modern pedal-equipped kettledrum is shown in Fig. 13.10.

At one time all timpani heads were calfskin, but this material has gradually given way to Mylar (polyethylene terephthalate). Calfskin heads require a great deal of hand labor to prepare and great skill to tune properly. Some orchestral timpanists prefer them for concert work under conditions of controlled humidity, but use Mylar when touring. Mylar is insensitive to humidity and easier to tune, due to its homogeneity. A thickness of 0.19 mm (0.0075 in.) is considered standard for Mylar timpani heads. Timpani kettles are roughly hemispherical; copper is the preferred material, although fiberglass and other materials are frequently used.

FIGURE 13.10
A kettledrum with
suspended bowl
and pedal-operated
mechanism for
tuning.

Although the modes of vibration of an ideal membrane are not harmonic, a carefully tuned kettledrum will sound a strong principal note plus two or more harmonic overtones. Rayleigh (1894) recognized the principal note as coming from the (11) mode and identified overtones about a perfect fifth ($f : f_1 = 3 : 2$), a major seventh (15 : 8), and an octave (2 : 1) above the principal tone. Taylor (1964) identified a tenth (octave plus a third of $f : f_1 = 5 : 2$) by humming near the drumhead, a technique some timpanists use to fine-tune their instruments.

How are the inharmonic modes of the ideal membrane coaxed into a harmonic relationship? Three effects contribute: (1) the membrane vibrates in a "sea of air," and the mass of this air "sloshing" back and forth lowers the frequency of the principal modes of vibration; (2) the air enclosed by the kettle has resonances of its own that will interact with the modes of the membrane that have similar shapes; (3) the stiffness of the membrane, like the stiffness of piano strings, raises the frequencies of the higher overtones. Our studies show that the first effect (air loading) is mainly responsible for establishing the harmonic relationship of kettledrum modes; the other two effects only "fine tune" the frequencies but may have considerable effect on the rate of decay of the sound (Rossing 1982a).

Stiffness in a two-dimensional membrane has a little different meaning than in a one-dimensional string. A plastic membrane, like a sheet of paper, offers little resistance to bending along a line. However, it strongly resists the type of distortion needed to wrap it around a ball without wrinkling, for example. This resistance is characterized as *stiffness to shear*. Stiffness to shear affects the frequencies and the shapes of vibrational modes in the same way that stiffness to bending affects the modes of a vibrating string (see Section 14.2).

We should point out that the frequencies of the fundamental (01) and other symmetrical modes (02, 03, etc.) will be raised by the "stiffness" of the enclosed air in the kettle. (This is not unlike the increase in the resonance frequency of a loudspeaker when it is mounted in an airtight enclosure; see Chapter 19.) The other modes are not thus affected, however, because their net displacement of air is zero.

TABLE 13.3 Vibration frequencies of a kettledrum, a drumhead without the kettle, and an ideal membrane

Mode	Kettledrum		Drumhead alone		Ideal membrane
	f	f/f_{11}	f	f/f_{11}	f/f_{11}
01	127 Hz	0.85	82 Hz	0.53	0.63
11	150	1.00	155	1.00	1.00
21	227	1.51	229	1.48	1.34
02	252	1.68	241	1.55	1.44
31	298	1.99	297	1.92	1.66
12	314	2.09	323	2.08	1.83
41	366	2.44	366	2.36	1.98
22	401	2.67	402	2.59	2.20
03	418	2.79	407	2.63	2.26
51	434	2.89	431	2.78	2.29
32	448	2.99	479	3.09	2.55
61	462	3.08	484	3.12	2.61
13	478	3.19	497	3.21	2.66
42			515	3.32	2.89

Table 13.3 shows the vibration frequencies of a 65-cm (26-in.) kettledrum and an identical drum without the kettle. In both cases, the ratios of the frequencies to the principal (11) mode are given. Note that the (21) and (31) modes will radiate overtones a fifth and an octave above the fundamental, as noted by Rayleigh.

The sound spectra obtained by striking the drum in its normal place (about one-fourth of the way from edge to center) and at the center are shown in Fig. 13.11. Note that the fundamental mode (01) appears much stronger when the drum is struck at the center, as do the other symmetrical modes (02, 03). These modes damp out rather quickly, however, so they do not produce much of a drum sound. In fact, striking the drum at the center produces quite a dull, thumping sound.

Normal striking technique produces prominent partials with frequencies in the ratios 0.85 : 1 : 1.5 : 1.99 : 2.44 : 2.89. If we ignore the heavily damped fundamental, the

FIGURE 13.11
Sound spectra from a 65-cm timpani tuned to E_3 (a) approximately 0.03 s after striking at the normal point; (b) approximately 1 s later; (c) approximately 0.03 s after striking at the center; (d) approximately 1 s later.

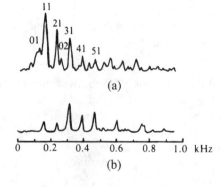

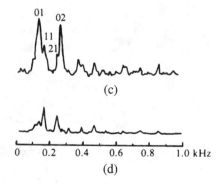

others are nearly in the ratios 1 : 1.5 : 2 : 2.5, a harmonic series built on a nonexistent fundamental an octave below the principal tone. Measurements on timpani of other sizes give similar results (Rossing and Kvistad 1976). It is a little surprising that the pitch of timpani corresponds to the pitch of the principal tone rather than the missing fundamental of the harmonic series, which would be an octave lower. Apparently the strengths and durations of the overtones are insufficient, compared to the principal tone, to establish the harmonic series of the missing fundamental.

13.9 ■ BASS DRUM

Among drums that convey an indefinite pitch or none at all are the bass drum, snare drum, tenor drum, tomtom, conga, and bongo. Many drums of this type have two heads, and each head is given a slightly different tension. Thus, the many inharmonic partials of the two heads produce the indefinite pitch that will blend with music in any key.

Bass drums are typically 50 to 100 cm in diameter. Single-headed, or "gong," drums produce a mellow sound, although two-headed drums with their more indefinite pitch are usually preferred in bands and orchestras. The bass drum is capable of making the loudest sound of all the instruments in the orchestra.

Most drummers tune the *batter*, or beating, head to a greater tension than the *carry*, or vibrating, head; some percussionists suggest that the difference in tension be as great as 75% (giving an interval of about a musical fourth). Quite a different timbre results from setting both heads to the same tension.

In Fig. 13.12 are sound spectra of an 82-cm- (32-in.) diameter drum with the carry head set at a lower tension than the batter head (a) and with both heads at the same tension (b). In both cases, the lowest partial, radiated by the (01) mode in which the two heads vibrate in phase, is the strongest one. In Fig. 13.13(b), the partial at 104 Hz is identified with the higher component in a (01) doublet in which the batter and carry heads move in opposite directions. (Additional restoring force due to compression of the enclosed air raises the frequency of this component, as in the third mode of the guitar shown in Fig. 10.28.)

FIGURE 13.12
Sound spectra of an 82-cm diameter bass drum (a) with the carry head at a lower tension than the batter head; (b) with both heads at the same tension.

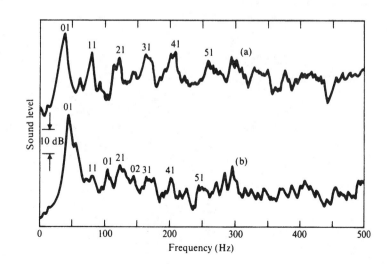

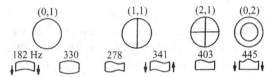

FIGURE 13.13 The six lowest vibrational frequencies of a snare drum. The modal designations (m, n) refer to the membrane modes shown in Fig. 13.9. Mode frequencies are for a typical snare drum. Arrows indicate motion of the drum shell (Fletcher and Rossing 1998).

Frequencies of the (11), (21), (31), (41), and (51) modes have nearly harmonic ratios, as in the timpani, and if their partials were the only ones heard, the bass drum would be expected to have a rather definite pitch. In the frequency range above 200 Hz, however, there are many inharmonic partials that sound louder since the ear is more sensitive to them than to low-frequency sounds. Fletcher and Bassett (1978) found some 160 partials in the frequency range 200 to 1100 Hz.

13.10 ■ SNARE DRUM

The snare drum is a two-headed instrument about 33 to 38 cm (13 to 15 in.) in diameter and 13 to 20 cm (5 to 8 in.) deep. The shell is made from wood, metal, or a synthetic material. Strands of wire or gut are stretched across the lower (snare) head. When the upper (batter) head is struck, the snare head vibrates against the snares. Alternatively, the snares can be moved away from the head to give a totally different sound.

In the snare drum, like the bass drum, there is appreciable coupling between the two heads, especially at low frequency. This coupling may take place acoustically through the enclosed air or mechanically through the shell, and it leads to the formation of mode pairs, as shown in Fig. 13.13. In the first two modes of vibration, the batter and snare heads move in the manner of the (0, 1) mode of an ideal membrane (see Fig. 13.9). In the lower-frequency member of the pair, both heads move in the same direction, and in the higher-frequency member, they move in opposite directions, as in the two-mass vibrator in Fig. 2.7.

A simple two-mass model describes the first two modes reasonably well (Rossing et al. 1992). The batter head is represented by a mass m_b and a spring with a spring constant K_b, the snare head by a mass m_s, and a spring constant K_s; the enclosed air constitutes a third spring with constant K_c connecting the two masses. This system, like the one in Fig. 2.7, has two modes of vibration, whose frequencies are given by

$$f^2 = \tfrac{1}{2}(f_b^2 + f_s^2 + f_{cb}^2 + f_{cs}^2)$$

$$\pm \tfrac{1}{2}\sqrt{[(f_b^2 + f_{cb}^2) - (f_s^2 + f_{cs}^2)]^2 + 4f_{cb}^2 f_{cs}^2},$$

where

$$f_b^2 = \frac{1}{4\pi^2} \frac{K_b}{m_b}, \quad f_s^2 = \frac{1}{4\pi^2} \frac{K_s}{m_s}, \quad f_{cb}^2 = \frac{1}{4\pi^2} \frac{K_c}{m_b}, \quad f_{cs}^2 = \frac{1}{4\pi^2} \frac{K_c}{m_s}.$$

The third and fourth modes in Fig. 13.13, in which the heads move in the manner of the (1, 1) membrane mode, are more difficult to model. In the lower member of the pair, the heads move in opposite directions and air "sloshes" from side to side, so that the mass of the air acts to lower the frequency. In the higher member of the (1, 1) pair, the air moves

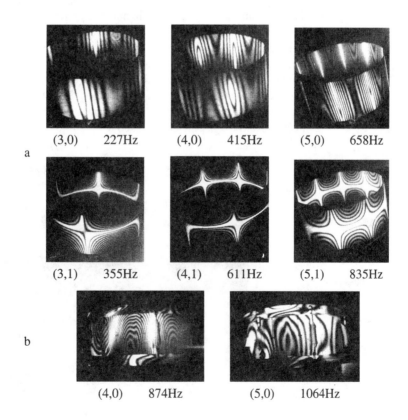

(3,0)   227Hz   (4,0)   415Hz   (5,0)   658Hz

a

(3,1)   355Hz   (4,1)   611Hz   (5,1)   835Hz

b

(4,0)   874Hz   (5,0)   1064Hz

FIGURE 13.14
(a) Holographic interferograms showing six modes of vibration of a snare drum shell without the heads; (b) two modes of a complete drum. (After Rossing et al. 1992.)

a smaller distance and so the frequency is higher. This member results in slight rocking motion of the shell.

Some vibrational modes of a snare drum shell (without the heads) and a complete drum are shown in the holographic interferograms in Fig. 13.14. Although the shell vibrations have a much smaller amplitude than those of the heads, they can influence the sound of the drum, as can be heard by comparing the sound of a snare drum on a drum stand and on a more solid base. Likewise, the snare drum sound depends upon the mass of the shell (but probably not on the material from which it is constructed).

The coupling between the snares and the snare head depends upon the mass and the tension of the snares. At a sufficiently large vibration amplitude in the snare head, properly adjusted snares will leave the head at some point during the vibration cycle and then return to strike it, thus giving the snare drum its characteristic sound. The greater the tension on the snares, the larger the amplitude needed for this to take place.

For the snares to sound at all requires a certain amplitude of the snare head. This critical amplitude increases with the snare tension. The snare tension is optimum when both the head and the snares are moving at maximum speed in opposite directions at the moment of contact. In this case, the impact is the greatest (Rossing et al. 1992).

13.11 ■ OTHER DRUMS

Tomtoms are unsnared drums that are made in a number of sizes, ranging from 20 to 45 cm in diameter. Tomtoms may have either one or two heads; the more indefinite pitch of the two-headed type is usually preferred for orchestral work. A drum set, such as the one shown in Fig. 13.15, usually includes two or three tomtoms of different sizes.

Bongos and *congas* are two popular types of thick-skin drums played with the hands and used extensively in Latin American music. Bongos are typically from 15 to 20 cm in diameter and about 12 cm deep. Conga drums are larger than bongos, having diameters of 25 to 30 cm and thick, tapered shells 60 to 75 cm long (see Fig. 13.16).

FIGURE 13.15
Jazz or trap drum set, including snare drum, pedal-operated bass drum, three tomtoms, two cymbals, and a pedal-operated "hi-hat" (pair of cymbals). (Courtesy of Ludwig Industries.)

FIGURE 13.16
Conga drums.

Many types of drums from Africa, Asia, and other parts of the world are becoming popular in American music. One drum that is particularly interesting acoustically is the *tabla* from north India, a single-headed drum with a closed resonating chamber. The head is tensioned by straps that stretch vertically from top to bottom as shown in Fig. 13.17. Cylindical pieces of wood are inserted between the straps, and by moving these up and down the drum can be finely tuned.

The tabla is loaded at the center by a paste of iron oxide, charcoal, starch, and gum, which hardens but remains flexible. C. V. Raman (who won a Nobel prize for his work in

FIGURE 13.17
Indian drums of the tabla family: *tabla* on the left and *banya* on the right. Note the tensioning straps and the drumhead loaded with black paste at its center.

FIGURE 13.18
Chladni patterns of
six different modes
of vibration in a
tabla all of which
have frequencies
near the third
harmonic. The (02)
and (21) normal
modes are shown in
(a) and (f),
respectively; the
other modes are
combinations of
these two (Raman
1934).

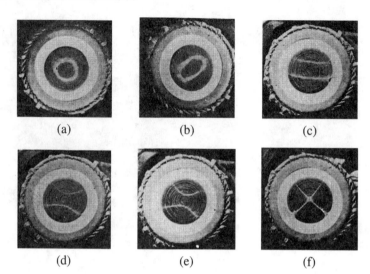

spectroscopy but also wrote several articles on the acoustics of violins and drums) observed that the first four overtones of the tabla are harmonics of the fundamental by virtue of the extra load. Later, he discovered that these five harmonics actually result from nine modes of vibration, several of which have the same frequencies; the (11) mode is the second harmonic, the (21) and (02) modes both vibrate at the third harmonic, the (12) and (31) modes at the fourth harmonic, and the (03), (22), and (41) modes at the fifth harmonic (Raman 1934). Fig. 13.18 shows Chladni patterns of the (02) and (21) modes and several combinations of them which have frequencies near the third harmonic.

The two-headed *mridanga* from south India is similar in acoustic behavior to the tabla. We recorded sound spectra at 35 stages during the process of preparing a mridanga head by loading it in order to illustrate how nine modes in the unloaded head develop into five harmonically related modes in the finished head (Rossing and Sykes 1982).

13.12 ■ VIBRATIONS OF PLATES

Vibrating plates bear much the same relationship to membranes that vibrating bars do to strings. Whereas in strings and membranes the restoring force results from the tension, in bars and plates it results from the stiffness of the solid material. In plates and bars, the overtones tend to be substantially higher than the fundamental. They can vibrate with a variety of boundary conditions, including clamped and free edges.

The vibrations of plates have fascinated physicists for many years. Nearly 200 years ago E. F. F. Chladni published his first book describing his well-known method of sprinkling sand on vibrating plates to make the nodal lines visible. Chladni's lectures throughout Europe attracted many famous persons, including Napoleon. Chladni patterns are used even today to observe the nodal lines of vibrating plates, although some type of electromechanical driver is usually preferred to Chladni's violin bow.

Plates can vibrate in a large number of modes. The modes of a circular plate are often given the labels *m* and *n*, like those of a circular membrane, to designate the number of nodal diameters and nodal circles, respectively. Chladni patterns of a circular plate are shown in Fig. 2.19, and those of the top and back plates of a violin in Fig. 10.17. Chladni observed that the frequencies of the various modes in a plate are nearly proportional to $(m + 2n)^2$, a relationship that has been called *Chladni's law* (Rossing 1982b).

13.13 ■ CYMBALS

Cymbals are among the oldest of musical instruments and have had both religious and military use in a number of cultures. The Turkish cymbals generally used in orchestras and bands are saucer-shaped with a small dome in the center, in contrast to Chinese cymbals, which have a turned-up edge more like a tamtam. Orchestral cymbals are usually between 40 and 55 cm in diameter and are made of bronze. The leading manufacturer of cymbals, the Avedis Zildjian Company, claims that its secret process for treating cymbal alloys was discovered in 1623.

Many different types of cymbals are used in orchestras, marching bands, concert bands, and jazz bands. Orchestra cymbals (see Fig. 2.20(a)) are often designated as French, Viennese, and Germanic in order of increasing thickness. Jazz drummers use cymbals designated by such onomatopoeic terms as crash, ride, swish, splash, ping, and pang. Cymbals range from 20 cm to 75 cm (8 to 30 in.) in diameter.

Given the wide ranges in diameter and thickness, considerable variety of cymbal tone is available to the percussionist. Furthermore, a good cymbal can be made to produce many different tone by using a variety of sticks and striking it at several different places. A large cymbal struck gently near the rim produces a low sound not unlike that of a small tamtam. The fullest sound is obtained by a glancing blow about one-third of the way in from the rim.

The individual modes of vibration of cymbals, shown in Fig. 13.19, are basically those of a circular plate (see Fig. 2.19), but altered by the saucerlike shape of the cymbal. Note that after the (60) mode, the cymbal resonances are combinations of two or more single modes. The mode at 1243 Hz, for example, is a combination of the (13, 0) and (22) modes (see Fig. 13.20(d)). Hologram interferograms of four of these modes are shown in Fig. 13.20.

A cymbal may be excited in many different ways. It may be struck at various points with a wooden stick, a soft beater, or another cymbal. The onset of sound is quite dependent on the manner of excitation. The coupling between vibrational modes in a cymbal is strong, however, so that a large number of partials quickly appear in the spectrum, however it is excited.

Sound spectra of a 40-cm cymbal immediately after striking and after intervals of 0.05, 1.0, and 2.0 s are shown in Fig. 13.21. From this and other similar spectra we can made the following observations:

1. The sound level below about 700 Hz shows a rather rapid decrease during the first 200 ms, after which it decays slowly. This is apparently due to conversion of energy into modes of higher frequency.
2. Several strong peaks in the range 700 to 1000 Hz build up between 10 and 20 ms and then decay.

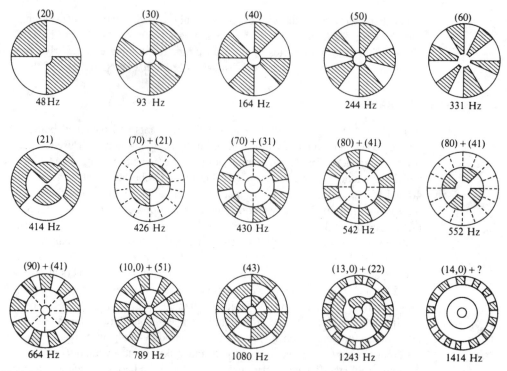

FIGURE 13.19   Modes of vibration of a 38-cm (15-in.) cymbal. The first six modes resemble those of a flat plate, but after that the resonances tend to be combinations of two or more modes (Rossing and Peterson 1982).

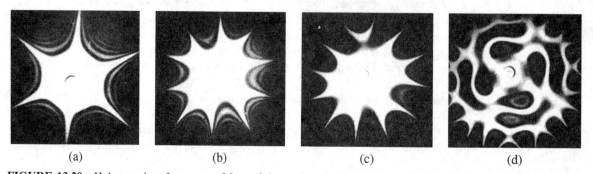

FIGURE 13.20   Hologram interferograms of four of the modes of vibration shown in Fig. 13.17: (a) 30; (b) 50; (c) 60; (d) 13, 0 + 22 (a combination) (Rossing and Peterson 1982).

3. Sound energy in the important range 3 to 5 kHz peaks about 50 to 100 ms after striking.
4. Sound in the range 3 to 5 kHz, which gives the cymbal its "shimmer," is often the most prominent feature from about 1 to 4 s after striking.
5. The low frequencies again dominate the lingering sound, but at a much lower level, so that they are rather inconspicuous.

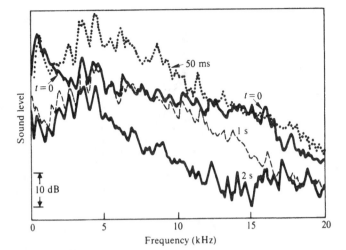

FIGURE 13.21
Sound spectra of a 40-cm cymbal immediately after striking and after intervals of 0.05, 1.0, and 2.0 s (Rossing and Shepherd 1983).

FIGURE 13.22
Phase plots (velocity versus displacement) for a thin crash cymbal center driven at 192 Hz: (a) linear behavior at 0.05 A; (b) harmonics present at 0.15 A; (c) subharmonics (including their harmonics) present at 0.5 A; (d) chaotic vibration at 1 A (Wilbur and Rossing 1997).

Nonlinear Behavior of Cymbals

The conversion of energy from the low-frequency modes that are initially excited when a cymbal is struck into the high-frequency vibrations that are responsible for much of the characteristic cymbal sound embodies some interesting physics. There is considerable evidence that the vibrations exhibit chaotic behavior. The road to chaos appears to follow the following stages: first the generation of harmonics, then the generation of subharmonics, and finally chaotic behavior.

If a flat circular plate is excited sinusoidally, it vibrates in the normal modes of Fig. 2.19 at low amplitudes. Large-amplitude excitation at a frequency near one of these modes leads to the *bifurcation* associated with doubling or tripling of the vibration period. The same behavior is observed for an orchestral cymbal excited sinusoidally at its center, one particular cymbal giving a fivefold increase in period and a subsequent major-chord-like sound based on the fifth subharmonic of the excitation frequency. The fourth and seventh subharmonics have been observed in other cymbals, along with harmonics of these subharmonics (in other words, partials having frequencies $n/4$, $n/5$, or $n/7$ times the excitation frequency).

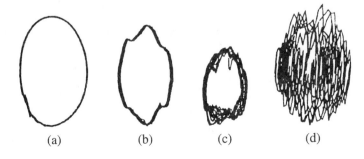

(a)   (b)   (c)   (d)

FIGURE 13.23
Phase plots and
velocity spectra for
a cymbal center
drive at 320 Hz:
(a) harmonic
generation at 0.3 A
drive current;
(b) subharmonics at
0.6 A; (c) chaotic
behavior at 1.4 A
(Wilbur and
Rossing 1997).

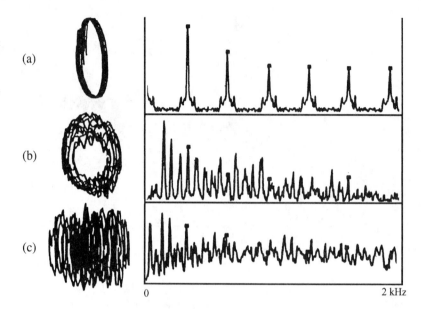

Figure 13.22 shows phase plots (velocity versus displacement near the edge) for a Zildjian thin crash cymbal 41 cm in diameter driven sinusoidally at the center with a shaker. The driving frequency was 192 Hz and the drive amplitude started small and increased in three steps over a 20 : 1 range. Not the successive appearance of harmonics, subharmonics, and chaotic behavior.

Velocity spectra and phase plots are shown in Fig. 13.23 for the same cymbal center driven at 320 Hz. The first set shows harmonics of the driving frequency, whereas the second set at twice the driving current shows subharmonic generation, and the third shows chaotic behavior. In the spectrum of Fig. 13.23, we can see harmonics of one-fifth of the driving frequency, four of which appear between successive harmonics of the driving frequency (marked by small squares). When the drive frequency was raised to 450 Hz, the road to chaos was similar, but the subharmonics are now harmonics of one-seventh of the driving frequency.

Brass, bronze, and steel plates of the same size (and approximately the same thickness) as the cymbal also showed nonlinear behavior leading to chaos. In general, subharmonic generation and chaotic behavior required slightly higher vibration amplitudes in the flat plates than in the cymbals. Subharmonic generation was more difficult to observe in the plates; in general, they tended to move directly to chaotic behavior as the vibration amplitude increased.

A mathematical analysis of cymbal vibrations using nonlinear signal processing methods reveals that there are between three and seven active degrees of freedom and that physical modeling will require a like number of equations. Because of its nonlinear behavior, the cymbal is a difficult but not impossible musical instrument to model for sound synthesis, for example.

FIGURE 13.24
(a) Gong and
tamtam. (b) Typical
shapes of gongs
and tamtams. Note
the deeper rims and
tuning domes of the
gongs.

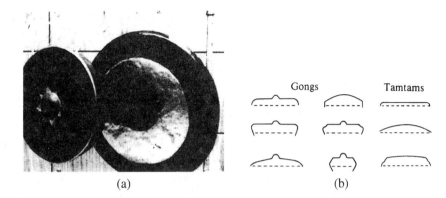

(a)   (b)

13.14 ■ GONGS AND TAMTAMS

Gongs play a very important role in Oriental music, but they enjoy considerable popularity in Western music as well. They are usually cast of bronze with a deep rim and a protruding dome. Gongs used in symphony orchestras usually range from 0.5 to 1 m (20 to 38 in.) in diameter, and are tuned to a definite pitch. When struck near the center with a massive soft mallet, the sound builds up relatively slowly and continues for a long time if the gong is not damped.

Massive gongs are central to every gamelan (ensemble) in Indonesia. Of considerable interest to the acoustician are the gongs used in Chinese opera orchestras, which glide upward or downward in pitch after being struck, due to their special shapes (Rossing and Fletcher 1983).

Tamtams are similar to gongs in appearance, and are often confused with them. The main differences between the two are that tamtams do not have the dome of the gong, their rim is not as deep, and their metal is thinner. Tamtams sound a much less definite pitch than do gongs. In fact, the sound of a tamtam may be described as somewhere between the sounds of a gong and a cymbal. The sound of a large tamtam develops slowly, changing from a sound of low pitch at strike to a collection of high-frequency vibrations, which are described as shimmer. These high-frequency modes fail to develop if the tamtam is not hit hard enough, indicating that the conversion of energy takes place through a nonlinear process. Figure 13.24 shows the shapes of a number of different gongs and tamtams.

13.15 ■ STEELPANS

The Caribbean steelpan is probably the most important new acoustical musical instrument to develop in the twentieth century. In addition to the steelpan being the foremost musical instrument in its home countries, Trinidad and Tobago, steel bands are becoming increasingly popular in Europe, North America, and some Asian countries as well. The modern family of steelpans now covers a five-octave range, and steel bands of today use them to play calypso, popular, jazz, and Western classical music.

The development of steelpans took place in the years following the end of World War II, when the annual celebration of Carnival was resumed with great enthusiasm. Many claims have been made about the invention of the tuned steelpan. Undoubtedly, it resulted from a lot of trial and error on the part of musicians and inventors such as Bertie Marshall, Anthony Williams, and Ellie Mannette. Thousands of 55-gal oil drums left on the beach by the British navy provided ample raw material for experimentation. Although the basic designs have pretty well stabilized, steelpans are still evolving.

Modern steelpans are known by various names, such as tenor (or lead), double second, double tenor, guitar, cello, quadrophonics, and bass. The overlapping ranges of these instruments are shown in Fig. 13.25.

The tenor pan has from 26 to 32 different notes, but each bass pan has only 3 or 4; hence the bass drummer plays on six or more pans in the manner of a timpanist. Different makers still use different designs, although there is a movement in Trinidad and Tobago to standardize designs. Most steelpans are played with a pair of short wood or aluminum sticks wrapped with strips of surgical rubber. The bass pans are hit with a beater consisting of a sponge rubber ball on a stick.

The typical steelpan is constructed from a portion of a 55-gal oil drum. The end of the drum is hammered ("sunk") into a shallow concave well, which forms the playing surface. The depth of the well varies from 10 cm (4 in.) in a bass pan to 19 cm (7.5 in.) in a tenor pan. Part or all of the cylindrical portion of the drum is retained as the skirt, which acts as a baffle, acoustically separating the top and bottom of the playing surface to prevent cancellation of radiated sound; it also radiates considerable sound itself. Skirt lengths vary from about 13 cm (5 in.) for the tenor to 45–70 cm (18–28 in.) in the cello pan. A bass pan retains the entire length of the original drum. Skirt lengths that give the best sound for each pan have been determined by trial and error.

After the end of the oil drum has been hammered into a shallow well, the playing surface is scribed and grooved with a punch to define the individual note areas. The tuner raises each note area up slightly by hammering and then works it up and down to soften the metal. Next the pan is heated over a bonfire, which raises the pitch of each note and appears to harden at least the surface of the metal. The note areas are flattened and then tightened

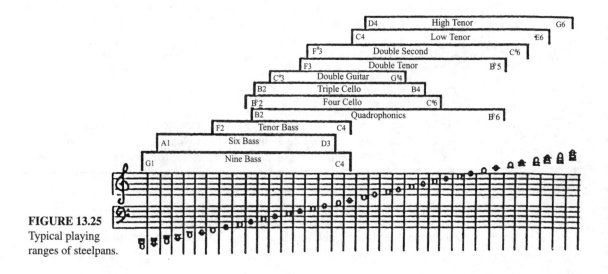

FIGURE 13.25
Typical playing ranges of steelpans.

around the edges by rapid hammer work known as *piening*, which prepares the note areas for tuning.

The actual tuning process begins by lightly tapping the underside of the pan to create a small flat to convex portion of the note area. The grooves between the note areas on the playing surface are retightened with a small hammer, so that when the player strikes the note area, the vibrations are restricted mainly to that note area. The fundamental frequency of a note area can be lowered by glancing blows across the top of the area, and it can be raised by increasing the height at the center or by tapping down at the corners of the area.

A skilled pan-maker also tunes at least one overtone of each note area to a harmonic of the area's fundamental frequency. The first overtone is nearly always the octave, and if the note area is large enough, a second overtone is tuned to the third or fourth harmonic. Tuning the third mode to the musical twelfth (third harmonic) gives the note a more mellow tone, and tuning it to the double octave (fourth harmonic) gives it a bright tone. Tuning the harmonics is more difficult than tuning the fundamental frequency. In general, tapping the underside of the pan along or just outside a boundary of a note area that runs parallel to the nodal line for that mode will lower a harmonic, whereas tapping down or outward on the playing surface just inside this boundary will raise the harmonic frequency.

The layout of the notes on a typical tenor pan is shown in Fig. 13.26. The outer ring has 12 more or less trapezoidal note areas tuned from D_4 (294 Hz) to $C_5^{\#}$ (554 Hz). The middle ring has 12 more or less elliptical note areas tuned from D_5 to $C_6^{\#}$ and the inner ring has four to six near-circular note areas tuned from D_6 to $F_6^{\#}$. Note that the notes in the outer and middle rings are arranged in *circles of fifths*: moving counterclockwise, one goes up a perfect fifth or down a perfect fourth (which is equivalent to going up a fifth and down an octave) on the musical scale. Each note in the middle ring is an octave higher than the corresponding note on the outside ring. The inside ring adds a third octave to five notes (six notes when an optional $F_6^{\#}$ is included). Thus, each note has several harmonics in common with its nearest neighbors, which leads to the strong interaction between notes characteristic of steel pans. (Another note arrangement, sometimes called the *invaders layout*, is also used for tenor pans by some pan-makers.)

Holographic interferograms of the tenor pan driven at several frequencies that excite modes of vibration in the C_5 note area (at 2 o'clock, viewing the pan as a clock face)

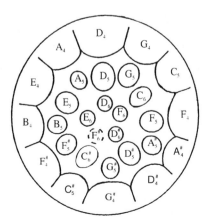

FIGURE 13.26

Layout of note areas on a typical tenor pan.

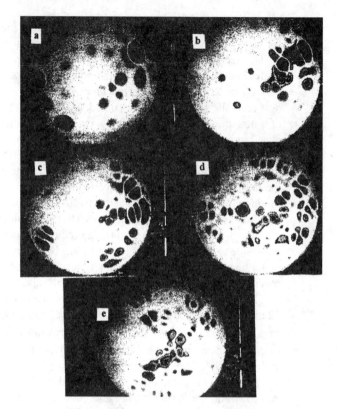

FIGURE 13.27
Vibrations of the
tenor pan at
frequencies that
excite modes in the
C$_5$ note area:
(a) 522 Hz;
(b) 1050 Hz;
(c) 1421 Hz;
(d) 2064 Hz;
(e) 2184 Hz
(Rossing, Hansen,
and Hampton
2000).

are shown in Fig. 13.27. Observe that several other note areas show appreciable vibration, especially at the higher frequencies. At the lowest frequency the active (roughly elliptical) portion of the note area moves in a single phase, while at the frequency of the second harmonic, there is a nodal line parallel to the rim dividing the note area into halves. At 1421 Hz $(2.72f_1)$ there is a radial node; at 2064 Hz $(3.95f_1)$ there are two nodal lines parallel to the rim, and at 2184 Hz there are two nodal lines perpendicular to each other. Besides five modes of the C$_5$ note area, many sympathetic vibrations of other note areas are apparent in Fig. 13.27. The various note areas can be identified by comparing Fig. 13.27 with Fig. 13.26.

Holographic interferograms of several modes observed in the D$_4$, D$_5$, and D$_6$ note areas of the same tenor pan are shown in Fig. 13.28. These are designated by (m, n), where m is the number of circumferential nodal lines and n is the number of radial nodal lines.

13.16 ■ SOUND SPECTRA

The sound spectra of steelpans are rich in harmonic overtones. These appear to have three different physical origins:

1. Radiation from higher modes of vibration of a given note area, tuned harmonically by the tuner;

FIGURE 13.28
Holographic
interferograms of
several modes in
the D_4, D_5, and D_6
note areas of the
tenor pan. Modes
are designated by
(m, n), where m
and n are the
numbers of
circumferential and
radial nodes,
respectively.
Frequency ratios to
the fundamentals
are given (Rossing,
Hansen, and
Hampton 2000).

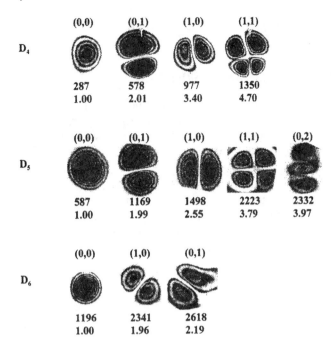

2. Radiation from nearby notes whose frequencies are harmonically related to the struck note;

3. Nonsinusoidal motion of the note area, vibrating at its fundamental frequency.

Sound spectra of two notes on a double-second steelpan are shown in Fig. 13.29. Note that harmonics as high as the ninth harmonic are detected. The first three or four harmonics result from higher modes harmonically tuned, whereas the radiation from harmonically related nearby notes excited by sympathetic vibration could contribute to harmonics such

FIGURE 13.29
Sound spectra from
two notes on a
double-second
steelpan.

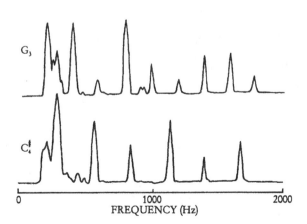

as two, four, six, and possibly eight. The balance of the harmonics are attributed to nonsinusoidal motion of the note area.

13.17 ■ BELLS AND CARILLONS

Bells have been a part of nearly every culture in history. Bells existed in the Near East before 1000 B.C., and a number of Chinese bells from the time of the Shang dynasty (1600–1100 B.C.) are found in museums throughout the world. A set of tuned bells from the fifth century B.C. was recently discovered in the Chinese province of Hubei.

Bells developed as Western musical instruments in the seventeenth century when bell founders discovered how to tune their partials harmonically. The founders in the Low Countries, especially the Hemony brothers (François and Pieter) and Jacob van Eyck, took the lead in tuning bells, and many of their fine bells are found in carillons today.

The carillon also developed in the Low Countries. Chiming bells by pulling ropes attached to the clappers had been practiced for some time before the idea of attaching these

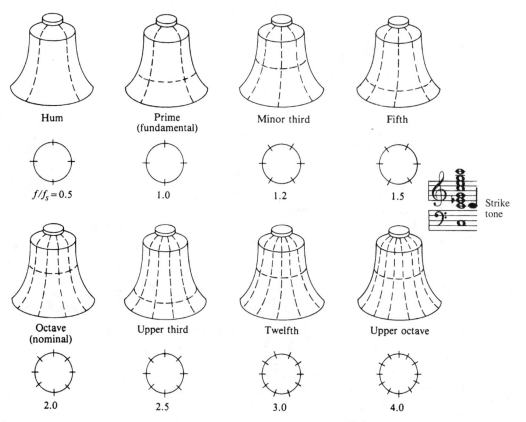

FIGURE 13.30   The first eight vibrational modes of a tuned bell. Dotted lines indicate the approximate locations of nodes. Frequencies relative to the strike note are given. The notes that correspond to these in a C_4 bell are shown on a musical staff. In some bells only five modes are tuned harmonically.

ropes to a keyboard or handclavier occurred to bell ringers in the sixteenth century. Many mechanical improvements during the seventeenth and eighteenth centuries, including the breached wire system and the addition of foot pedals for playing the larger bells, led to development of the modern carillon. Today, the term carillon is reserved for an instrument of 23 (two octaves) or more tuned bells played from a clavier (smaller sets are called *chimes*). The largest carillon in existence is the 74-bell (6-octave) carillon at Riverside Church in New York, whose largest bell is more than 18,000 kg (20 tons).

Perhaps it is stretching the imagination a bit to think of a bell as being a plate, but the general principles of its vibrational behavior are similar. Although the mathematical description of the vibrations of a bell are understandably complex, the principal modes, at least, can be described by specifying the number of circular nodes and meridian nodes. The lowest mode of vibration, (called the *hum* tone), for example, has four meridian nodes, so that alternate quarters of the bell essentially move inward and outward.

Figure 13.30 shows the principal vibrational modes for a carillon bell. The mode called the third is tuned a minor third above the strike note, whereas the upper third is usually a major third above the octave. The *strike note* is determined by the octave, the twelfth, and the upper octave, whose frequencies have the ratios 2 : 3 : 4, just as in chimes (see Section 13.5). Unlike chimes, however, carillon bells have a mode called the prime or fundamental with a frequency at or near the strike note. Careful studies have shown, however, that the pitch of the strike note is determined by the three modes mentioned above rather than by the prime.

Vibrational frequencies of groups 0–X in a church bell with a D_5 strike note are shown in Fig. 13.31. Horizontal lines on the axis at the right indicate relative strengths of several partials in the bell sound (these vary with location, of course). Arrows denote the three partials (octave or nominal, fifth (twelfth), and upper octave) that determine the strike note.

FIGURE 13.31
Vibrational frequencies of groups 0–X in a D_5 church bell. Also shown on the right are the relative strengths at impact of several partials in the bell sound. Arrows denote the three partials in group I that determine the strike note. (From Rossing and Perrin 1987.)

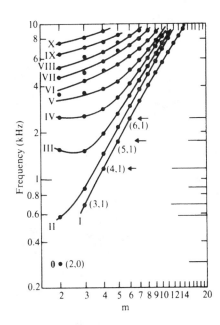

A new type of carillon bell has been developed at the Royal Eijsbouts Bellfoundry in The Netherlands. The new bell replaces the strong minor-third partial with a major-third partial, thus changing the tonal character of the bell sound from minor to major. This requires an entirely new bell profile. The new bell design evolved partly from the use of a technique for structural optimization using finite element methods on a digital computer. This technique allows a designer to make changes in the profile of an existing structure, and then to compute the resulting changes in the vibrational modes (Lehr 1987).

13.18 ■ HANDBELLS

Handbells also date back at least several centuries B.C., although tuned handbells of the present-day type were developed in England in the eighteenth century. One early use of handbells was to provide tower bellringers with a convenient means to practice change ringing. In more recent years, handbell choirs have become popular in schools and churches—some 40,000 choirs are reported in the United States.

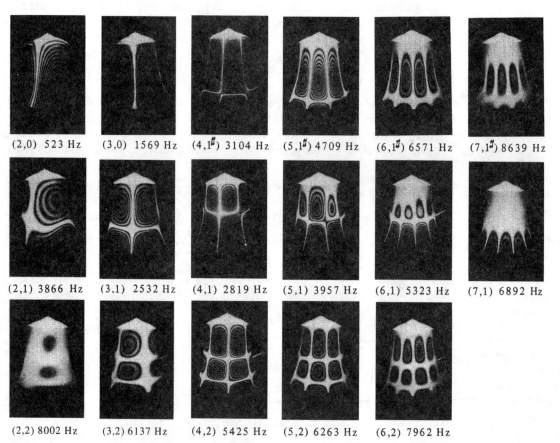

(2,0) 523 Hz	(3,0) 1569 Hz	(4,1$^\sharp$) 3104 Hz	(5,1$^\sharp$) 4709 Hz	(6,1$^\sharp$) 6571 Hz	(7,1$^\sharp$) 8639 Hz
(2,1) 3866 Hz	(3,1) 2532 Hz	(4,1) 2819 Hz	(5,1) 3957 Hz	(6,1) 5323 Hz	(7,1) 6892 Hz
(2,2) 8002 Hz	(3,2) 6137 Hz	(4,2) 5425 Hz	(5,2) 6263 Hz	(6,2) 7962 Hz	

FIGURE 13.32   Time-average hologram interferograms of inextensional modes in a C$_5$ handbell. (From Rossing et al. 1984.)

Although they are cast from the same bronze material and cover roughly the same range of pitch, the sounds of church bells, carillon bells, and handbells have distinctly different timbres. In a handbell, only two modes of vibration are tuned (although there are three harmonic partials in the sound), whereas in a church bell or carillon bell at least five modes are tuned harmonically. A church bell or carillon bell is struck by a heavy metal clapper in order to radiate a sound that can be heard at a great distance, whereas the gentle sound of a handbell requires a relatively soft clapper.

In the so-called English tuning of handbells, followed by most handbell makers in England and the United States, the $(3, 0)$ mode is tuned to three times the frequency of the $(2, 0)$ mode. The fundamental $(2, 0)$ mode radiates a rather strong second harmonic partial, however, so that the sound spectrum has prominent partials at the first three harmonics (Rossing and Sathoff 1980). Some Dutch founders aim at tuning the $(3, 0)$ mode in handbells to 2.4 times the frequency of the fundamental, giving their handbell sound a minor-third character somewhat like a church bell. Such bells are usually thicker and heavier than bells with the English-type tuning.

Hologram interferograms of a number of the modes are shown in Fig. 13.32. The "bull's eyes" locate the antinodes. Note that the upper half of the bell moves very little in the $(7, 1)$ mode; the same is true in $(m, 1)$ modes when $m > 7$.

13.19 ■ SUMMARY

Percussion instruments have experienced a surge of interest, especially in contemporary music. Bars, membranes and plates vibrate in modes that are not harmonically related to the fundamental, and this gives percussion instruments a distinctive timbre. The lower bars of marimbas, xylophones, and vibes have arches cut on the underside in order to tune the first overtone to a harmonic of the fundamental (fourth harmonic, in the case of the marimba and vibes; third harmonic, in the case of the xylophone). These three instruments have a tubular resonator placed under each bar to increase the loudness of the sound. In vibes, these resonators can be opened and closed rapidly to generate a type of vibrato. Chimes have no mode of vibration at the frequency of the strike note, which is a subjective tone.

Drums have membranes of animal skin or synthetic material stretched over some type of air enclosure. The modes of vibration are shifted in frequency away from those of an ideal membrane by the air loading. Timpani convey a strong sense of pitch; the first overtone is tuned to a fifth above the principal tone. Other drums, such as the bass drum, have an indefinite pitch. The Indian tabla has several overtones that are harmonics of the fundamental.

Examples of percussion instruments that are vibrating plates are cymbals, gongs, and tamtams. The sound spectrum of cymbals are characterized by a buildup and subsequent decay of sound in the range 3 to 10 kHz. The spectra of Caribbean steel drums are surprisingly rich in harmonic overtones. Tuned carillon bells and handbells may also be described as platelike. Carillon bells have five to eight modes whose frequencies are harmonics of a fundamental. In a small handbell, only two modes are usually tuned harmonically.

REFERENCES AND SUGGESTED READINGS

Bork, I., and J. Meyer (1982). "Zur klanglichen Bewertung von Xylophonen," *Das Musikinstrument* **31**(8): 1076. English translation in *Percussive Notes* **23**(6): 48 (1985).

Brindle, R. S. (1970). *Contemporary Percussion*. London: Oxford University Press.

Chaigne, A., and V. Doutaut (1997). "Numerical simulations of xylophones. I. Time-domain modeling of the vibrating bars." *J. Acoust. Soc. Am.* **101**: 539.

Fletcher, H., and I. G. Bassett (1978). "Some Experiments with the Bass Drum," *J. Acoust. Soc. Am.* **64**: 1570.

Fletcher, N. H., and T. D. Rossing (1998). *The Physics of Musical Instruments*, 2nd ed. New York: Springer-Verlag. Chapters 18–20.

Flugge, W. (1962). *Statik und Dynamik der Schalen*. New York: Springer-Verlag.

Hardy, H. C., and J. F. Ancell (1963). "Comparison of the Acoustical Performance of Calfskin and Plastic Drumheads," *J. Acoust. Soc. Am.* **33**: 1391.

Lehr, A. (1987). "From Theory to Practice," *Music Perception* **4**: 267.

Raman, C. V. (1934). "The Indian Musical Drums," *Proc. Indian Acad. Sci.* **A1**: 179.

Lord Rayleigh (J. W. Strutt) (1894). *The Theory of Sound*, Vol. 1, 2nd ed. London: Macmillan. Reprinted by Dover, New York, 1945.

Rossing, T. D. (1976). "Acoustics of Percussion Instruments, Part I," *Phys. Teach.* **14**: 546.

Rossing, T. D. (1977). "Acoustics of Percussion Instruments, Part II," *Phys. Teach.* **15**: 278.

Rossing, T. D. (1982a). "The Physics of Kettledrums," *Sci. Am.* **247**(5): 172.

Rossing, T. D. (1982b). "Chladni's Law for Vibrating Plates," *Am. J. Phys.* **50**: 271.

Rossing, T. D. (1984a). *The Acoustics of Bells*. Stroudsburg, PA: Van Nostrand Reinhold.

Rossing, T. D. (1984b). "The Acoustics of Bells," *Am. Scientist* **72**: 440.

Rossing, T. D. (1987). "Acoustical Behavior of a Bass Drum," *J. Acoust. Soc. Am.* **82**: S69.

Rossing, T. D., I. Bork, H. Zhao, and D. Fystrom (1992). "Acoustics of Snare Drums," *J. Acoust. Soc. Am.* **92**: 84.

Rossing, T. D., and N. H. Fletcher (1982). "Acoustics of a Tamtam," *Bull. Australian Acoust. Soc.* **10**(1): 21.

Rossing, T. D., and N. H. Fletcher (1983). "Nonlinear Vibrations in Plates and Gongs," *J. Acoust. Soc. Am.* **73**: 345.

Rossing, T. D., U. J. Hansen, and D. S. Hampton (1996). "Music from Oil Drums: The Acoustics of the Steel Pan," *Physics Today* **49**(3): 24.

Rossing, T. D., U. J. Hansen, and D. S. Hampton (2000). "Vibrational Mode Shapes in Caribbean Steelpans. I. Tenor and Double Second," *J. Acoust. Soc. Am.* **108**: 803.

Rossing, T. D., and G. Kvistad (1976). "Acoustics of Timpani: Preliminary Studies," *The Percussionist* **13**: 90.

Rossing, T. D., and R. Perrin (1987). "Vibrations of Bells," *Applied Acoustics* **20**: 41.

Rossing, T. D., R. Perrin, H. J. Sathoff, and R. W. Peterson (1984). "Vibrational Modes of a Tuned Handbell," *J. Acoust. Soc. Am.* **76**: 1263.

Rossing, T. D., and R. W. Peterson (1982). "Vibrations of Plates, Gongs, and Cymbals," *Percussive Notes* **19**(3): 31.

Rossing, T. D., and D. A. Russell (1990). "Laboratory Observation of Elastic Waves in Solids," *Am. J. Phys.* **58**: 1153.

Rossing, T. D., and H. J. Sathoff (1980). "Modes of Vibration and Sound Radiation from Tuned Handbells," *J. Acoust. Soc. Am.* **68**: 1600.

Rossing, T. D., and R. B. Shepherd (1983). "Acoustics of Cymbals," *Proc. 11th Intl. Congress on Acoustics* (Paris): 329.

Rossing, T. D., and W. A. Sykes (1982). "Acoustics of Indian Drums," *Percussive Notes* **19**(3): 58.

Taylor, H. W. (1964). *The Art and Science of the Timpani*. London: Baker.

Wilbur, C., and T. D. Rossing (1997). "Subharmonic Generation in Cymbals at Large Amplitude," *J. Acoust. Soc. Am.* **101**: 3144 (abstract).

GLOSSARY

harmonics A series of partials with frequencies that are simple multiples of a fundamental frequency. (In a harmonic series, the first harmonic would be the fundamental, the second harmonic the first overtone.)

inharmonic partials Overtones or partials that are not harmonics of the fundamental.

nodes Points or lines that do not move when a body vibrates in one of its modes.

overtones Upper partials or all components of a tone except the fundamental.

radius of gyration A measure of the difficulty of rotating a body of a given mass.

strike note The subjective tone that determines the pitch of a bell or chime; in most tuned bells it corresponds closely to one of the partials, but in chimes it does not.

tension The force applied to the two ends of a string, or around the periphery of a membrane, that provides a restoring force during vibration.

torsional mode An oscillatory motion that involves twisting of the vibrating member.

vibrato Frequency modulation (FM) that may or may not have amplitude modulation (AM) associated with it. Some musicians speak of *intensity vibrato*, *pitch vibrato*, and *timbre vibrato* as separate entities; others understand vibrato to include all three. Sometimes the term **tremolo** is used to describe AM, but this is not recommended because tremolo is used to describe other things, such as a rapid reiteration of a note or even a trill.

Young's modulus The ratio of stress to strain; also called **modulus of elasticity**.

REVIEW QUESTIONS

1. How much does the fundamental bending mode frequency change when the length of a bar is cut in half?

2. Where should a bar be supported in order to best excite its fundamental mode?

3. In a marimba bar, what is the frequency ratio of the first overtone to the fundamental?

4. In a xylophone bar, what is the frequency ratio of the first overtone to the fundamental?

5. Which instrument has motor-driven discs at the tops of the resonators?

6. Which modes determine the strike note of an orchestra chime (tubular bell)?

7. Why is the (0, 1) mode frequency in a kettledrum considerably higher than the corresponding mode in a membrane at the same tension without the kettle?

8. Which of the following drums produce a sound with a definite pitch: timpani, bass drum, snare drum, conga, tabla?

9. Describe the motion of the two heads in the lowest vibrational mode of a snare drum.

10. Explain the delayed sound, or shimmer, in a cymbal.

11. How are the modes of a tabla tuned harmonically?

12. What is the difference between gongs and tamtams?

13. Arrange the following steelpans in order from highest to lowest frequency: cello, guitar, tenor, double tenor, bass, quadrophonics, double second.

14. What are three physical origins of harmonic overtones in steelpans?

15. What is the frequency ratio of the first two vibrational modes in a tuned handbell?

16. Which modes in a carillon bell determine its strike note?

QUESTIONS FOR THOUGHT AND DISCUSSION

1. What is wrong with the statement "The resonators of a marimba prolong the sound"?

2. The lowest mode of vibration of the triangle shown in Fig. 13.8 has four nodes. Make a sketch of the type in Figs. 13.2 or 13.3 to illustrate how it vibrates.

3. Explain why a microphone placed some distance above the center of a kettledrum picks up very little of the principal tone (except, of course, by reflection from the walls, etc.). What modes would be picked up best by a microphone in such a location?

4. Compare the frequencies of the modes of the kettledrum given in Table 13.3 with those given by Rayleigh (Section 13.8).

EXERCISES

1. Compare the ratios of the frequencies of transverse vibrations in the glockenspiel bar in Fig. 13.2 with the theoretical ratios for a thin rectangular bar given in Section 13.1. Can you account for the difference? (*Hint:* Compare Fig. 13.6.)

2. Write an expression for the frequency (f_1) of the lowest mode of vibration of a rectangular bar in terms of its length L and the speed of sound v_L.

3. Write an expression for the frequency ratio of the lowest transverse mode in a bar to the lowest longitudinal mode. Find this ratio for a glockenspiel bar 21.4 cm long and 0.90 cm thick ($K = t/3.46$). Compare this ratio with the ratio of the corresponding values given in Fig. 13.2.

4. What are the actual ratios of the numbers 81, 121, and 169 (discussed in Section 13.5)? How close are they to the ratios 2 : 3 : 4?

5. Using the formulas for longitudinal and transverse vibrations in a bar (Table 13.1), show that the lowest longitudinal and transverse modes will have the same frequency when $L = (9/4)\pi K$. Find the ratio of length to diameter for a bar (rod) of circular cross section having the

frequency for its lowest transverse and longitudinal vibrations.

6. Determine the frequencies of the (41), (51), and (61) modes of the bell in Fig. 13.32. Show that they are nearly in the ratios 2 : 3 : 4, and that they will produce a strike note (virtual pitch) corresponding to D_5 (see Section 7.4). Is there a vibrational mode with this frequency?

7. In large bells, the (61), (71), (81), and (91) modes create a secondary strike note. Determine the frequencies of these modes in Fig. 13.31, and estimate their virtual pitch (see Section 7.4). What note on the scale is this nearest (see Table 9.2)?

8. If a bell does not have perfect circular symmetry (perfection is seldom achieved in practice), one or more modes of vibration will show "doublet" behavior: that is, there will be two modes of vibration with slightly different frequencies. Suppose that such a doublet exists, having components with frequencies of 440 and 442 Hz. At what rate will the bell "warble"? How is it possible to minimize this warble?

EXPERIMENTS FOR HOME, LABORATORY, AND CLASSROOM DEMONSTRATION

Home and Classroom Demonstration

1. *Modes of a rectangular bar*   Sound the first three modes of a steel or aluminum bar (Fig. 13.2 suggests where to hold it and where to strike it for each mode). Note the musical intervals between the modes (the ratio $f_2/f_1 = 2.56$, for example, is slightly less than an octave plus a fourth). Repeat with a bar of hardwood (preferably rosewood) and note the difference.

2. *Modes of a tuned bar*   Sound the first three modes of a tuned marimba, vibraphone, and/or xylophone bar from the lower part of the scale. Note the musical intervals between the modes.

3. *Vibraphone resonators*   Demonstrate the difference in loudness and timbre of a vibraphone (or vibraharp) when the resonators are open and closed. Demonstrate the intensity vibrato obtained by rotating the resonator discs at different rates.

4. *Chimes*   Locate and mark the points where holding a chime firmly will preferentially excite vibrational modes 2 through 6. (They are approximately at 0.5L, 0.36L, 0.28L, 0.22L, and 0.17L, where L is its length.) Then show that modes 4, 5, and 6 are nearly in a harmonic 2 : 3 : 4 fre-

quency ratio, with the audible strike note being one octave below mode 4 and two octaves below mode 6.

5. *Kettledrum*   Contrast the difference in the tone obtained by striking a kettledrum at its center (which emphasizes the symmetric modes) and at the normal playing spot (about one-fourth of the way from edge to center); see Fig. 13.11.

6. *Kettledrum*   Stretch your fingers as far as possible to touch points on the nodal lines of the (2, 1) mode (see Fig. 13.9). Show that the (2, 1) mode is a fifth above the fundamental (1, 1) mode.

7. *Kettledrum*   Show how the pedal, by changing the tension, raises the frequency of all modes nearly proportionately and maintains the tuning of the overtones quite well (but not exactly).

8. *Cymbals*   Strike a cymbal with a wooden drum stick to show how the timbre (especially the way the "aftersound" develops) varies with strike point and the strength of the blow.

9. *Steelpans*   By touching a note area in different places as it is struck, show how the harmonic modes are shaped (see Fig. 13.28, for example).

10. *Chinese opera gongs*   Small Chinese opera gongs show an interesting pitch-glide when struck (Fletcher and Rossing 1998, Section 20.3.1). The larger gong has a downward pitch-glide, whereas the smaller one glides upward.

11. *Handbells*   Several individual modes of vibration can be excited by touching the bell in the vicinity of a node (see Fig. 13.32) and striking it in the vicinity of an antinode. Note the harmonic tuning of the (2, 1) and (3, 1) modes.

12. *Chladni patterns*   See Demonstration 8 in Chapter 2.

Laboratory Experiments

Acoustics of mallet instruments (Experiment 20 in *Acoustics Laboratory Experiments*)

Vibrations of membranes and drums (Experiment 19 in *Acoustics Laboratory Experiments*)

CHAPTER

14

Keyboard Instruments

This chapter features three different types of keyboard instruments. In the piano, strings are set into vibration by striking them with hammers; in the harpsichord, the strings are plucked; and in the pipe organ, sound is produced by blowing air into tuned pipes.

In this chapter you should learn:

- About piano action and how the hammers interact with the strings;
- How the piano soundboard vibrates and radiates sound;
- About clavichords and harpsichords;
- About pipe organs.

14.1 ■ CONSTRUCTION OF THE PIANO

Invented in 1709 by Bartolomeo Cristofori in Florence, the piano has become the most popular and versatile of all instruments. It has a range of more than seven octaves (A_0 to C_8) and a wide dynamic range. Pianos vary in size from small home uprights or spinets to large concert grands.

The main parts of the piano are the keyboard, the action, the strings, the soundboard, and the frame. A simplified diagram of a piano is shown in Fig. 14.1. More than 200 strings extend from the pin block or wrest plank across the bridge to the hitch-pin rail at the far end. When a key is depressed, the damper is raised, and the hammer is "thrown" against the string, setting it into vibration. Vibrations of the string are transmitted to the soundboard by the bridge.

A typical concert grand piano has 243 strings, varying in length from about 2 m at the bass end to about 5 cm at the treble end. Included are 8 single strings wrapped with one or two layers of wire, 5 pairs of strings also wrapped, 7 sets of three wrapped strings, and 68 sets of three unwrapped strings. Smaller pianos may have fewer strings but they play the

FIGURE 14.1
A simplified diagram of the piano. When a key is depressed, the damper is raised, and the hammer is thrown against the string. Vibrations of the string are transmitted to the soundboard by the bridge.

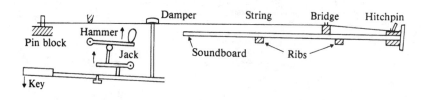

310

FIGURE 14.2
The top view of a
studio grand piano
showing the
cast-iron frame, the
overlapping strings,
hammers, and
dampers. The
cutaway portion at
the treble end
shows the tuning
pin block.
(Courtesy of
Baldwin Piano and
Organ Co.)

same number of notes: 88. A small grand piano with 226 strings is shown in Fig. 14.2. Note that the bass strings overlap the middle strings, which allows them to act nearer the center of the soundboard. The acoustical advantages of wrapped strings and multiple strings for most notes will be discussed in Section 14.2.

The *soundboard* is nearly always made of spruce, about 1 cm thick, with its grain running the length of the piano. Ribs on the underside of the soundboard stiffen it in the cross-grain direction. The soundboard is the main source of radiated sound, just as is the top plate of a violin or cello.

To obtain the desired loudness, piano strings are held at high tensions which may exceed 1000 N (220 lb). The total force of all the strings in a concert grand is over 20 tons! In order to withstand this force and maintain stability of tuning, grand pianos have sturdy frames of cast iron.

The rather complicated action of a grand piano is shown in Fig. 14.3. When a key is pressed down, the damper is raised. At the same time, the capstan sets the whippen into rotation around its pivot. The rotating whippen causes the jack to push against the hammer, knuckle or roller, starting the hammer on its journey toward the string. Just before the hammer strikes the string, the lower end of the jack strikes the jack regulator and rotates away from the knuckle. The freely rotating hammer now strikes the string, rebounds immediately (in order not to damp the string), and falls back to the repetition level. The back check prevents the hammer from bouncing back to strike the string a second time. When the key is lifted slightly, a spring pulls the jack back under the knuckle, so that pressing the key a second time repeats the note. Most upright pianos do not have this feature; thus a note cannot be repeated unless the key returns to its starting position.

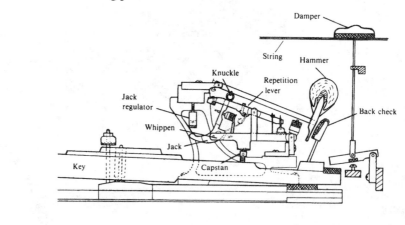

FIGURE 14.3
The action of a grand piano.

The upright piano developed about the middle of the nineteenth century. The nearly rectangular soundboard and the strings in upright pianos are vertical; the hammers travel horizontally. Thus the action is different from that of the grand piano in Fig. 14.3. In full-size upright pianos, which stand 130 to 150 cm ($4\frac{1}{2}$ to 5 ft) in height, the striking mechanism or action is located some distance above the keys and connected to them mechanically by stickers. In studio uprights or console pianos, which stand about 100 to 130 cm ($3\frac{1}{4}$ to $4\frac{1}{4}$ ft) in height, the action is mounted directly over the keys without stickers. In small spinet pianos (less than 100 cm in height), the action is below the keys and drop stickers transmit key motion to the action.

Pianos may have two or three pedals. The right pedal is called the *sustaining* pedal. It raises all the dampers, which allows the struck strings to continue vibrating after the keys are released. The left pedal is some type of expression pedal. In most grand pianos, it shifts the entire action sideways, causing the treble hammers to strike only two of their three strings. This shifting type of pedal is called the *una corda* pedal. In vertical pianos, and a few grands, the left pedal is a *soft* pedal, which moves the hammers closer to the strings, decreasing their travel and thus their striking force.

Many pianos have a third pedal. On most grands and a few uprights, the center pedal is a *sostenuto* pedal, which sustains only those notes which are depressed prior to depressing the pedal, and does not sustain subsequent notes. On a few pianos, the center pedal is a bass sustaining pedal, which lifts only the bass dampers. On a few uprights, the center pedal is a *practice* pedal, which lowers a piece of felt between the hammers and the strings, muffling the tone.

14.2 ■ PIANO STRINGS

The strings are the heart of the piano. They convert some of the kinetic energy of the moving hammers into vibrational energy and pass it on to the bridges and soundboard in a manner that determines the sound quality of the piano.

Piano strings make use of high-strength steel wire. Efficiency of sound production calls for the highest string tension possible, while at the same time minimizing inharmonicity calls for using the smallest string diameter (core diameter in a wrapped string) possible. This results in tensile stresses of around 1000 N/mm^2, which is about half the yield strength of steel wire. For steel with an elastic modulus of 2×10^{11} N/m^2, this results in an elongation of about $\frac{1}{2}$% when the string is under tension. Fortunately, when strings break, it is usually near the keyboard end, so that the broken strings recoil away from the pianist.

An ideal string vibrates in a series of modes that are harmonics of a fundamental (see Section 4.3). Actual strings have some stiffness, which provides a restoring force (in addition to the tension), slightly raising the frequency of all the modes. The additional restoring force is greater in the case of the higher modes because the string makes more bends. Thus the modes are spread apart in frequency and are no longer exact harmonics of a fundamental. In other words, a real string with stiffness is partly stringlike and partly barlike.

The inharmonicity of strings (i.e., the amount by which the actual mode frequencies differ from a harmonic series) is found to vary with the square of the partial number (Fletcher 1964). Thus the second harmonic is shifted four times as much as the fundamental.

The formula may be written

$$f_n = nf_1[1 + (n^2 - 1)A],$$

where f_n = the frequency of the nth harmonic and f_1 = the frequency of the fundamental. For a solid wire without wrapping,

$$A = \frac{\pi^3 r^4 E}{8TL^2},$$

where r = the radius of the string, E = Young's modulus, T = the tension, and L = the length of the string. Thus, the inharmonicity is smallest for long, thin wires under great tension (large L and T, small r).

Because the stiffness of a string increases sharply with its radius (the factor A in the box above increases as r^4), inharmonicity is especially noticeable in the case of the large bass strings. Because wrapped strings are more flexible than are solid strings of the same diameter, the inharmonicity of the bass strings is reduced substantially by the use of wrapped rather than solid strings of the same weight. (The lower strings on a guitar and violin are wrapped rather than solid, for the same reason.)

A small amount of inharmonicity of string partials is considered desirable in pianos. One study of synthesized piano sounds demonstrated the preference of listeners, both musicians and nonmusicians, for tones with inharmonic partials. Tones synthesized with harmonic partials were described as "lacking warmth" and sounding much less like piano tones than those synthesized from slightly inharmonic partials (Fletcher, Blackham, and Stratton 1962). Inharmonicity may also help disguise small tuning errors in the same way that vibrato serves the players of other instruments.

The inharmonicity in strings is the main reason why pianos are "stretch-tuned" (see Section 9.6). If they were not, the upper partials of a note would be slightly sharp with respect to the notes in the upper octaves to which they correspond, and undesirable beats would result when chords are played.

Because pianos are usually tuned by minimizing beats between notes, a stretched scale will automatically be the result. Figure 14.4 shows the deviations from equal temperament that resulted when a spinet piano was aurally tuned by a fine-tuner at a piano factory. Also shown is the *Railsback stretch*, which is an average from 16 different pianos measured by O. L. Railsback. The aural tuning results follow the Railsback curve generally, but with a few deviations that are probably attributable to soundboard resonances. A jury of listeners showed little preference between tuning done electronically according to the Railsback curve and tuning done aurally by the skilled tuner (Martin and Ward 1961).

A large grand piano with long bass strings will show less stretch in the bass than does the small spinet piano in Fig. 14.4. However, the stretch in the upper octaves will not be much different, since the upper strings are not appreciably different in a spinet and a concert grand.

FIGURE 14.4
Deviations from equal temperament in a small piano. (From Martin and Ward 1961. Reprinted by permission of the Amer. Inst. of Physics.)

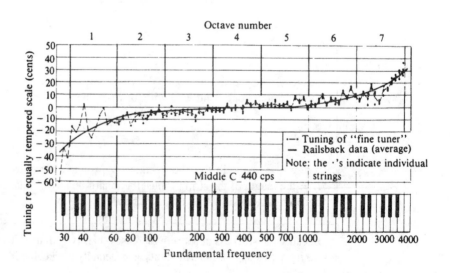

14.3 ■ THE TUNING OF UNISONS

Over most of its playing range, the piano has three strings for each note. Studies have shown that the best piano sound results from tuning these strings one to two cents different from each other (Kirk 1959). If the strings are tuned to exactly the same frequency, the transfer of energy from the strings to the soundboard takes place rapidly, and the decay time of the sound is too small. If the unison strings are tuned too far apart, prominent beats are heard, and what we commonly call a "barroom piano" sound is the result.

When the unison strings are tuned to be a few cents different, the decay curve takes on two different slopes. The hammer sets all three strings into vibration with the same phase, and energy is rapidly transferred to the soundboard initially. Since there are small differences in their frequencies, however, the strings soon get out of phase, and the rate of sound decay slows down, leading to a second slope in the decay curve (*aftersound*). If the frequencies differ by 2 cents, for example, about 400 vibrations would be required for the strings to fall out of phase. The actual coupling between the strings depends, in a somewhat complicated way, on the mechanical properties of the bridge (Weinreich 1977), and the uneven decay of coupled strings is an important characteristic of piano sound. Fine tuning of the unisons is a means of regulating the amount of aftersound. Hammer irregularities can also affect the aftersound, and a skilled piano tuner can probably compensate for these to some extent by adjusting the unisons.

14.4 ■ HAMMER-STRING INTERACTION

The dynamics of the hammer-string interaction has been the subject of considerable research, beginning with Helmholz (1877). The problem drew the attention of a number of Indian researchers, including Nobel laureate C. V. Raman, in the 1920s and 1930s, and it has recently been taken up by Hall (1986, 1987).

When the hammer has less mass than the string, it will most likely be thrown clear of the string by the first reflected pulse. The theoretical spectrum is missing harmonics

FIGURE 14.5
Hammer velocity and hammer-string contact time at various dynamic levels for the C$_4$ note on a grand piano. (Ashkenfelt and Jansson 1988).

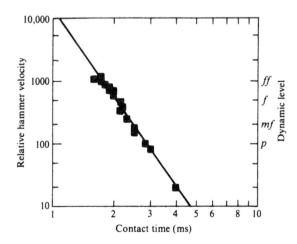

numbered n/β (where β is the fraction of the string length at which the hammer strikes). If the hammer mass is not too small, the spectrum envelope falls off as $1/n$ (6 dB/octave) above a certain mode number. A heavier hammer is less easily stopped and thrown back by the string. It may remain in contact with the string during the arrival of several reflected pulses. Analytical models of hammer behavior are virtually impossible to construct, but computer simulations can be of value.

Hall (1987) has considered the cases of a hard, narrow hammer and a soft, narrow hammer. In the case of the hard hammer, the mode spectrum envelope takes on a slope of -6 dB/octave at high frequencies. For the treble strings, where the hammer mass exceeds the string mass, the spectrum envelope may have a slope as steep as -12 dB/octave at high frequency.

Figure 14.5 shows a rather clear relationship between hammer-string contact time and hammer velocity at different dynamic levels. Striking the key with greater force increases the hammer velocity and decreases the contact time.

14.5 ■ THE SOUNDBOARD

A piano soundboard performs both structural and acoustical functions. Structurally, it opposes the vertical force components of the strings, which are in the range of 10 to 20 N per string, or a total of about 900 to 1800 N (200 to 400 lb). Acoustically, the soundboard is the main radiating member in the piano, transforming some of the mechanical energy of the strings and bridges into acoustical energy.

Nearly all piano soundboards are made by gluing strips of spruce together and then adding ribs at right angles to the grain of the spruce. These ribs are designed to add enough cross-grain stiffness to equal the natural stiffness of the wood along the grain, which is typically about 20 times greater than across the grain. Laminated wood (plywood) soundboards have occasionally been used in low-cost pianos, but these tend to have lower acoustical efficiency and, particularly, less bass response than those of solid spruce.

The unloaded soundboard is generally not a flat panel, but it has a crown of 1 to 2 mm on the side that holds the bridges. When the strings are brought up to tension, the downward force of the bridge causes a slight dip in the crown, and in old pianos the downward bridge force may have permanently distorted the soundboard. Soundboards are often tapered to be thicker near the center and thinner near the edges.

Modern pianos generally have two bridges: a main or treble bridge and a shorter bass bridge. The bridges couple the strings to the soundboard. Piano bridges have an important effect on the tone production. By changing the bridge design, the piano designer can change the loudness, the duration, and the quality of the tone.

The low-frequency vibrational modes of an upright-piano soundboard are shown in Fig. 14.6. In the lowest (0, 0) mode, all vibrating parts of the soundboard move in phase, whereas in the (1, 0) mode, a nodal line divides the soundboard into two parts moving in opposite phase. The (0, 1) mode has a nodal line running lengthwise, and the (2, 0) mode has two transverse nodal lines dividing the soundboard into thirds. The observed mode frequencies lie about midway between those calculated for fixed and simply supported (hinged) edges (Nakamura 1983).

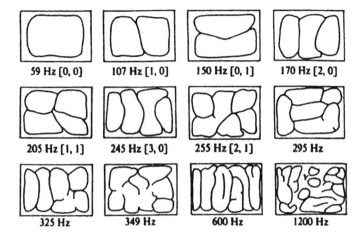

59 Hz [0, 0]   107 Hz [1, 0]   150 Hz [0, 1]   170 Hz [2, 0]

205 Hz [1, 1]   245 Hz [3, 0]   255 Hz [2, 1]   295 Hz

325 Hz   349 Hz   600 Hz   1200 Hz

FIGURE 14.6
Chladni patterns of
an upright-piano
soundboard
(Nakamura 1983).

Grand-piano soundboards have a somewhat more complicated modal structure, starting with the lowest mode at around 50 Hz. Mode shapes for both a 6-ft and a large 9-ft grand are shown in Fletcher and Rossing (1998), where there is further discussion of soundboard design. The best source of piano design, including soundboards, is a series of three papers by Harold Conklin (1996a, b, c).

14.6 ■ PIANO SOUND

Production of sound by a piano is a rather complicated process. The strings are set into vibration by the hammer, and they in turn act on the bridge to set the soundboard into vibration. Vibrational waves travel in many directions on the soundboard, and this leads to a complicated pattern of sound radiation. Eventually, the various components of the sound, which come from different parts of the soundboard, reach our ears; we process them and identify them as coming from a piano.

The various partials in a piano sound build up quite rapidly (typically in 3 ms) and decay at widely different rates. Thus the sound spectrum of the piano is constantly changing with time, as in the case of the percussion instruments, which we discussed in Chapter 13. Many of the clues we use to identify a sound as coming from a piano are contained in the initial interval. This can be demonstrated by playing backward a tape recording of a piano. Now the sounds, which still have the same overall spectrum as before, build up slowly and decay rapidly, and they are more suggestive of a reed organ than a piano (Demo. 29, Houtsma et al. 1987).

Figure 14.7 shows how the spectrum changes over the wide range of the piano (Fletcher, Blackham, and Stratton 1962). In the lowest octave, as many as 45 partials can be detected; at the highest notes, only two or three partials can be detected.

14.7 ■ THE CLAVICHORD

The clavichord, like the piano, depends on struck strings for its sound; but there the similarity ends. A clavichord is a portable keyboard instrument with a soft, delicate sound, well

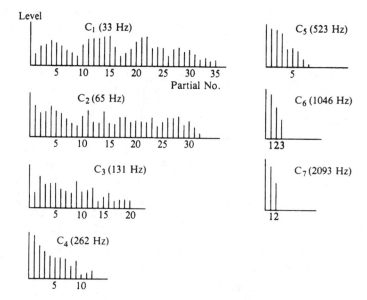

FIGURE 14.7
The spectrum of
high and low piano
notes. (After
Fletcher,
Blackham, and
Stratton 1962.)

suited for small living rooms but not for the concert hall. Its great virtue is its sensitivity; the tone can be varied in loudness and even given a vibrato by varying the force on the key.

The action of a clavichord is shown in Fig. 14.8. A small square piece of brass, called a tangent, is attached to the end of each key. When a key is depressed, the tangent strikes a string (or pair of strings) and causes the portion between the tangent and the bridge to vibrate. A damper prevents the other portion from vibrating and also damps the vibration of the entire string when the tangent is released from the string. The tangent will normally oscillate up and down at a frequency of a few hertz (determined by the mass of the string, the key, and the player's finger; the tension of the string; and the force applied to the key). This up-and-down motion varies the string tension and generates a vibrato. The playing range of the clavichord is typically four octaves, from C_2 to C_6.

The soundboard of a typical clavichord is a nearly rectangular slab of spruce 2 to 3 mm thick. The first two resonances, at about 140 to 330 Hz in one clavichord, are due to coupling between the fundamental modes of the soundboard and the enclosed air, as in a guitar or violin (Thwaites and Fletcher 1981). The next two resonances, around 470 and 570 Hz, are due to coupling between the second modes of the soundboard and the enclosed air. The

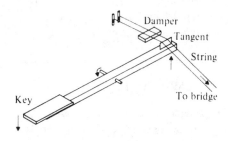

FIGURE 14.8
Clavichord action
(simplified).

clavichord is found to have a fairly uniform sound output except in the lowest octave where it is understandably weak.

14.8 ■ THE HARPSICHORD

Another keyboard instrument that was very popular in the Baroque period is the harpsichord. Like a number of instruments from that period, it has experienced a revival of interest in recent years. Many modern harpsichords are patterned after the best of the early instruments, such as those built by the Ruckers family early in the seventeenth century.

The action of a harpsichord is shown in Fig. 14.9. Pressing the key raises the jack so that the plectrum plucks the string. Plectra were originally carved from birds' quills, but plastic plectra are being used in many modern instruments. The plectrum is attached to a hinged tongue, so that when the jack is lowered, the plectrum contacts the string only briefly before swinging back out of the way. Lowering the jack (by releasing the key) also causes a damper felt to touch the string and damp its vibrations.

The construction of a harpsichord is not unlike that of a grand piano, except that everything is smaller and lighter. The soundboard of a harpsichord is typically 2.5 to 3 mm thick, compared to 10 mm in a typical piano. Strings have about one-third the diameter and about one-tenth the tension of the corresponding piano strings. The thinner strings of the harpsichord have much less inharmonicity (recall from Section 14.2 that inharmonicity is proportional to the fourth power of the radius), even though its bass strings are solid wire.

The string dimensions of harpsichords are usually scaled in some regular fashion from bass to treble. In one modern harpsichord design, for example, the string lengths are proportional to $1/f$ over the top half of the range, with the rate of increase of length decreasing toward the bass end. The bass strings are brass, and their diameters are proportional to $1/f$; the treble strings are steel, and their diameter varies as $(1/f)^{0.3}$. The strings are plucked at a distinct ℓ from the nut, where ℓ is a fraction of the total length L given by $f^{0.6}$ at the treble end. At the lower end $\ell/L = 0.13$. In general, the harmonic content of string vibration increases as the plucking point moves away from the center (Section 10.2), so that the harmonic development of the harpsichord increases in the lower notes (Fletcher 1977a).

In order to vary the loudness and timbre, most harpsichords add one to three additional sets of strings to the original set. The selection of the strings to be played is made by means

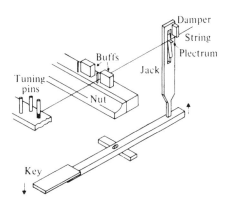

FIGURE 14.9
Harpsichord action
(simplified).

FIGURE 14.10
A Flemish virginal built by Hans Ruckers, 1581. (Reprinted by permission. All rights reserved, The Metropolitan Museum of Art.)

of *stops*. Borrowing from organ terminology, we call a set of strings that play an octave higher a *4-ft* set, and those that play an octave lower *16-ft*; one or more sets of *8-ft* strings play at the pitch of the key pressed. Some harpsichords add a second keyboard, and some of the largest instruments have a pedalboard as well. Couplers between the keyboards provide the harpsichordist with even more tonal options. Many harpsichords have a *buff stop* in which a piece of felt or soft leather is gently pressed against the string near its end to damp it and provide a short, distinctive sound. Sometimes a special set of jacks is provided to pluck the strings very close to the end in order to produce a nasal timbre (*lute stop*).

The *virginal* or *spinet* is a small instrument like a harpsichord with strings running lengthwise in a rectangular case. The keyboard is built into one side of the case, making it a portable tabletop instrument. Virginals were particularly popular in Elizabethan England. An elaborately decorated virginal is shown in Fig. 14.10.

14.9 ■ THE HARP

Although it does not have a keyboard, the harp is related to the harpsichord and other string instruments described in this chapter, so a brief description of it is appropriate here. The modern harp, shown in Fig. 14.11, has 47 strings tuned to the notes of the diatonic scale, plus seven pedals by which each string can be raised or lowered a semitone in order to play the chromatics. With all the pedals in their middle position, the harp plays in the key of C major. Depressing the C-pedal raises all the C's on the harp to $C^\#$. Figure 14.12 illustrates how the pedals raise and lower the pitch of the harp by means of two rotating discs attached to the neck of the instrument, which change the length of the vibrating part of the string.

The strings are attached to a slanted sounding-board at the bottom of the instrument. The nominal range, from C_1 to G_7, is almost as great as that of the piano. The most familiar sound of a harp is a sweeping glissando. The C-strings and F-strings are usually colored differently from the others in order to help the player locate the desired notes.

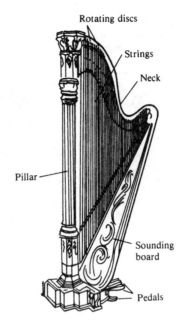

FIGURE 14.11
A harp.

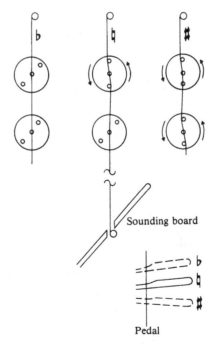

FIGURE 14.12
Mechanism for
tuning the harp.
Depressing one of
the tuning pedals
increases the
tension on all the
strings with that
note name by one
(for naturals) or
two (for sharps)
units.

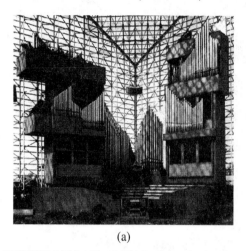

(a)   (b)

FIGURE 14.13   Hazel Wright Organ in the Crystal Cathedral, Garden Grove, California. (a) The chancel organ contains the great organ, swell organ, solo organ, choir organ, positive organ, echo organ, trompetria organs, and pedal organ. (b) The south balcony organ consists of the gallery great organ, string organ, celestial organ, and gallery pedal organ.

14.10 ■ THE PIPE ORGAN: ITS CONSTRUCTION

The pipe organ has been called the "king of musical instruments." No other instrument can match it in size, in range of tone, in loudness, or in complexity. The world's largest organ, in the Convention Hall in Atlantic City, has more than 32,000 pipes of various sizes and shapes. No two pipe organs in the world are exactly alike. A large pipe organ in the Crystal Cathedral in Garden Grove, California is shown in Fig. 14.13. This organ, designed by organist Virgil Fox, has 273 ranks and nearly 16,000 pipes in several locations throughout the church.

The modern pipe organ consists of a large variety of pipes arranged into divisions or organs. Each division is controlled by a separate keyboard or manual, including a pedalboard that is operated by the organist's feet. The pipes in the *swell* division are usually enclosed behind a set of shutters that can be opened or closed (by means of the swell pedal) to change the loudness. The other divisions are not usually enclosed, and their loudness can only be changed by adding or subtracting pipes through the use of *stops*. The principal division of an organ is called the *great organ*, and it usually contains the most stops. Most organs have couplers that allow certain pipes in one division to be controlled from the manual of another.

The windchests contain valves that can be opened to admit air into the pipes. The three main types of windchest actions used today are shown in Fig. 14.14. The oldest type is the *tracker* or mechanical action, in which the keyboards are connected to the chests by rather complicated mechanical linkages. With the second type, the *direct electric*, pressing the key energizes an electromagnet, which in turn opens the valve to allow air to flow into the pipe. The third type is called *electropneumatic*, because electromagnets are used to exhaust air from the bellows, which open the valve into the pipe. Full pneumatic actions, in which the key controls an air valve, are becoming rare.

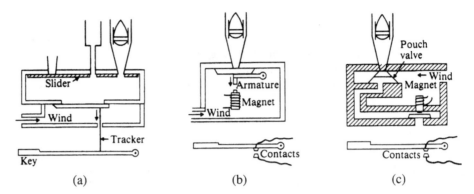

FIGURE 14.14
Windchest actions:
(1) tracker;
(b) direct electric;
(c) electropneu-
matic.

Organ pipes are organized into *ranks* of similar pipes. One rank of pipes will include one pipe for each note (61 in the divisions operated from a keyboard, 32 in the pedal division). Each stop on the organ usually corresponds to one rank of pipes, except in the case of mixture stops, which involve several ranks. On smaller organs, a rank of pipes may be included in more than one stop.

14.11 ■ ORGAN PIPES

There are two basic types of organ pipes: *flue* (labial) pipes and *reed* (lingual) pipes. Flue pipes produce sound by means of a vibrating air jet, in a manner similar to the flute and the recorder (see Section 12.14). Reed pipes use a vibrating brass reed to modulate the air stream.

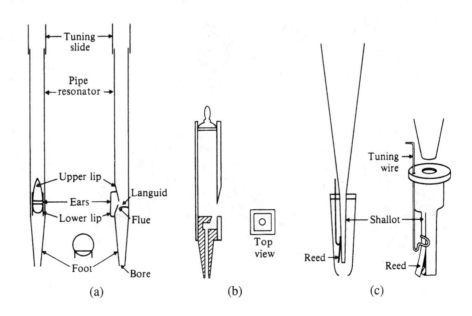

FIGURE 14.15
Organ pipes:
(a) open flue pipe
of metal;
(b) stopped wood
flue pipe; (c) reed
pipe.

The essential parts of a metal flue pipe are shown in Fig. 14.15(a). The air jet passes through a flue or windway, a narrow opening between the languid and the lower lip. Sound is produced when the air jet encounters the upper lip and oscillates back and forth, sometimes blowing into the pipe, sometimes blowing out through the mouth of the pipe. The large flue pipes usually have ears on either side of the mouth to guide the air jet.

Sound Generation in Flue Organ Pipes

When air is admitted at the foot of a flue organ pipe, it flows upward and forms a sheetlike jet as it emerges from the flue (see Fig. 14.15(a)). The jet flows across the mouth of the pipe and strikes the upper lip, where it interacts both with the lip itself and with the air in the pipe resonator. The physics of this interaction is discussed in Chapter 16 of Fletcher and Rossing (1998).

Jets can be described as either *laminar* (arranged in layers or streamlines) or *turbulent* (characterized by eddies or vortices). Jets in organ pipes are almost completely turbulent, which makes it quite difficult to describe them mathematically. Somewhat paradoxically, however, a fully turbulent jet in an organ pipe is more stable than a laminar one, and organ builders go to some lengths to ensure fully turbulent jets by cutting fine nicks along the edge of the languid, for example.

The air flowing during the attack transient in an organ pipe, although quite complicated, can be studied by means of Schlieren photographs. Figure 14.16 shows the

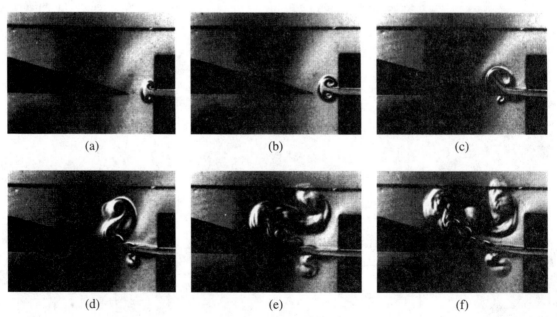

(a)   (b)   (c)

(d)   (e)   (f)

FIGURE 14.16 Schlieren photographs showing air flow at intervals of 0.1 s during the attack when a jet strikes a sharp edge such as the labium of an organ pipe. (Photos furnished by A. Hirschberg.)

jet at intervals of 0.1 s as the jet first strikes the edge, forming vortices. Eventually, this settles down to an acoustically driven jet oscillator.

The stopped wooden flue pipe, shown in Fig. 14.15(b), uses a similar mechanism to produce sound. Since the resonator is now a closed pipe, it need be only half as long as an open pipe in order to sound the same pitch. Wooden pipes usually have a square cross section, and produce a sound with a flutelike quality.

There are three families of flue pipes: (1) diapasons, (2) flutes, and (3) strings. Pipes of the flute family usually have the least overtone content, whereas the bright-sounding strings have the most. String pipes are generally slender cylinders, whereas diapasons are open cylinders of somewhat greater diameter. Flute pipes come in several sizes and shapes, are constructed of either wood or metal, and may be open or closed. Closed pipes sound mainly the odd-numbered harmonics of the fundamental.

The reed pipe, shown in Fig. 14.15(c), has a vibrating reed or tongue, which modulates the flow of air passing through the shallot into the resonator. The reed is pressed against the open side of the shallot by a wire that can be adjusted up and down to tune the vibrating reed. The resonator is usually tuned to about the same frequency as the reed. Reed pipes are rich in harmonics. If the resonator is cylindrical, the odd-numbered harmonics will be favored, as in a clarinet. Conical resonators, however, will reinforce all harmonics, both odd and even.

Figure 14.17 shows several different pipes and the names given to the corresponding organ stops. Resonators of several different shapes are shown, including open and closed cylinders, cones, inverted cones, rectangular cylinders, and closed pipes with chimneys. (The chimney is tuned to boost one of the upper harmonics, around the fifth or sixth.)

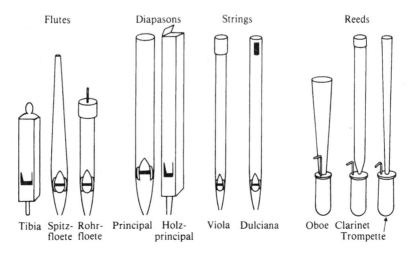

FIGURE 14.17
Organ pipes of various families with the names of the corresponding stops.

Pipe Resonators

Figure 14.18 compares the first four resonance frequencies of open and closed cylindrical, conical, and reverse conical pipes that can be used as organ pipe resonators. Note that the resonance frequencies of cones are essentially the same as those of an open cylinder, although the pressure amplitudes are quite different, as shown in Fig. 12.7. This is not the case with truncated cones and reverse cones, however, where the resonance frequencies are inharmonic except for the closed pipe ($r_2/r_1 = 1$).

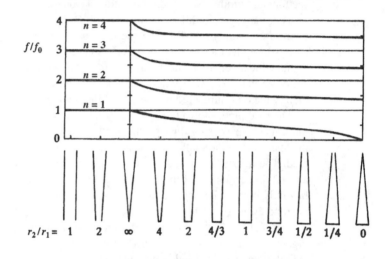

FIGURE 14.18

First four resonance frequencies of open and closed cylindrical and conical pipes.

14.12 ■ PIPE SCALING

The *scale* of a rank of pipes refers to the ratio of diameter to length for the pipe of lowest pitch. Large-scale (large-diameter) pipes tend to have a dominant fundamental and fewer harmonics, whereas small-scale pipes have more harmonics. The reason for this is related to a frequency dependence of the end correction of a pipe. For an open cylindrical pipe, the end correction at the open end is approximately 0.6 times the radius (see Section 4.5), whereas at the mouth of a flue pipe, one adds approximately 2.7 times the radius to calculate the effective length (Strong and Plitnik 1983). The end correction decreases with frequency, so the pipe effectively shortens for the higher partials; thus the pipe resonances are slightly less than an octave apart. However, the spectrum of the sound source (the oscillating jet) has exact harmonics. Thus in a large-scale pipe, only the first few harmonics in the source are reinforced by the natural frequencies of the pipe. For pipes of small scale, such as the strings, the end correction is much smaller to start with; thus the pipe resonances more nearly match the harmonics of the oscillating jet.

Each rank of pipes is usually graduated in diameter according to a fixed relationship for that rank. For large-scale pipes, the diameter may reduce to one-half at the seventeenth note of the scale; for small-scale pipes, the diameter will reduce at a slower rate than this.

TABLE 14.1   Pitch and harmonic number of pipe ranks

Equivalent length (in feet)	32	16	8	4	$2\frac{2}{3}$	2	$1\frac{3}{5}$	$1\frac{1}{3}$	$1\frac{1}{7}$	1
Harmonic number			1	2	3	4	5	6	7	8
Lowest note	C_0	C_1	C_2	C_3	G_3	C_4	E_4	G_4	B^b_4	C_5
"Organ" notation		CCC	CC	C	G	c^1	e^1	g^1	b^{b1}	c^2
Lowest pitch (in Hz)	16	33	65	131	196	261	330	392	466	523

Normally, the mouth width, lip cut-up, and width of the flue opening follow the same scale as the pipe diameter. The mouth width, for example, may vary from 0.6 times the pipe diameter (large-scale flute) to 0.8 times the diameter (string or diapason). Similarly, the cut-up may vary from 0.25 times the mouth width (diapason) to 0.4 times the width (flute); the flue width is typically about 0.03 times the mouth width (Fletcher 1977b).

The pitch of a rank of pipes is expressed as the approximate length of an open pipe having the same pitch as the lowest pipe in that rank. Thus the lowest pipe of an eight-foot rank sounds $C_2(f = 65.4$ Hz), and this is sounded by the lowest key on the manual. The octave from C_4 (middle C) to C_5 then occupies the same central position that it does on the piano keyboard. A four-foot stop plays an octave higher, and a sixteen-foot stop an octave lower. Table 14.1 lists the lowest note and the harmonic number of various ranks.

Mixtures are generally ranks of diapason-scaled pipes designed to augment the normal 8-ft, 4-ft, and 2-ft diapason chorus. They generally have from 3 to as many as 10 ranks, all properly balanced and coupled together. In normal mixtures, only octaves and fifths are used, but the selection of pitches concentrates much of the sound energy over a broad band between 500 and 6000 Hz.

14.13 ■ SOUND RADIATION FROM FLUE PIPES

Because a stopped pipe radiates only from its mouth, the radiation pattern (at least for the fundamental and the lower harmonics, where the mouth is small compared to the wavelength) is nearly isotropic. For the higher harmonics, when the wavelength becomes less than a few times the width of the mouth, the radiation pattern becomes more concentrated

FIGURE 14.19
Calculated
radiation patterns
for three harmonics
of an open flue pipe
standing vertically
(Fletcher and
Rossing 1998).

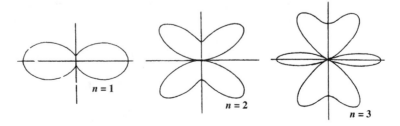

in front of the pipe, with a large angular spread in the vertical plane and a smaller spread in the horizontal plane (see Chapter 17 in Fletcher and Rossing 1998).

An open pipe has two coherent sources, at the mouth and open end, which are in phase for odd harmonics and out of phase for even harmonics. Even though the mouth area is smaller than the area of the open end, the two sources have the same strength, because the power radiated depends only on the total acoustic flow and not on the aperture size, as has been confirmed experimentally by Coltman (1969). Typical radiation patterns for the first three harmonics of an open flue pipe are shown in Fig. 14.19.

14.14 ■ REED PIPES

In reed (lingual) pipes, a curved tongue (the reed) closes against a matching cavity, called the shallot, to modulate the flow of air into the resonator. The vibrating length of the reed is determined by a stiff wire (Stimmkrücke) pressing it against the shallot. Resonators can take on a number of different shapes, including open and closed cylinders and cones. The pipe voicer generally adjusts the reed and the resonator to produce the best sound and then tunes the pipe by adjusting the tuning wire.

Without the resonator attached, the reed frequency varies smoothly with reed length, as would be expected, but attaching the resonator lowers the frequency of the reed and also produces distinct regimes of sound, separated by discontinuities, as the system jumps from one regime to the next one (Rossing et al. 1998).

Reed pipes tend to sound a rather large number of harmonics, which gives them a bright sound, and so they are often used as solo stops on an organ. A model of air flow in reed pipes has been developed by Hirschberg et al. (1990).

Free-reed organ pipes were introduced at the end of the eighteenth century. In contrast to the more common striking reed pipes, the reed does not beat against the shallot but swings freely through a perforated oblong plate of brass similar to those found in reed organs or accordions. Free-reed pipes are characterized by a slower rise time than striking-reed pipes and a "mellow, round" sound (Braasch and Ahrens 1999).

14.15 ■ TUNING AND VOICING ORGAN PIPES

There are several ways of tuning different types of organ pipes. Reed pipes are tuned by moving the tuning wire up and down, thus allowing a longer or shorter length of the reed to vibrate. Flue pipes, on the other hand, are tuned by changing the effective length of the pipe in some way. In the case of a closed pipe, this is accomplished by moving the stopper up or down to change the pipe length. Many open pipes have a tuning sleeve that slides up and down; others have an adjustable slot near the open end. Open pipes without such a tuning device can be tuned by the use of tuning cones. To raise the pitch, the apex of the cone is inserted and tapped to widen the end of the pipe slightly. To lower the pitch, the cup end of the cone is placed over the pipe and tapped to close the end slightly.

One of the most critical of all the steps in organ building is *voicing* the pipes, which means making coarse and fine adjustments in the various parts of the pipe so that it "speaks" properly. Much of the voicing can be done in the organ-builder's shop, but the final voicing or finishing is done after the organ is installed, and it takes into account the acoustics of the room, as well as the acoustics of the organ. One of the objectives of voicing is to achieve a uniformity of loudness and timbre within each rank of pipes. Another is to adjust the initial transient or attack of each pipe.

The attack time varies substantially from one rank of pipes to another and also from large to small pipes within a rank. In general, the number of cycles necessary to build the tone to its steady state remains about the same within a given rank, so the attack time doubles in going one octave down the scale. Reed pipes have shorter attack times than do flue pipes of the same pitch. In many pipes, the attack time is shorter for the upper harmonics than for the fundamental, so the initial sound is higher in harmonic content than the steady state. (Figure 7.15 shows a diapason pipe in which the second harmonic builds up rapidly.) Octave or mixture ranks are often coupled to slow-speaking pedal flue pipes to decrease the attack time. Also important during the attacks are certain characteristic sounds, such as *chiff*.

The main parameters that are adjusted during voicing of flue pipes are (Mercer 1951):

1. *The size of the foot bore.* An increase in bore diameter increases the air flow and produces greater power and greater harmonic content.
2. *The condition of the bore.* Rounding the edges of the bore can change the tone.
3. *Nicking.* Nicks on the languid or the lower lip tend to remove chiff and other extraneous sounds during the attack.
4. *The width of the flue.* This affects both the attack and steady state in a way that is not well understood.
5. *Obstruction near the mouth.* A bar, or "roller beard," placed between the ears is sometimes necessary to prevent a pipe of high pressure or small scale from jumping up an octave in pitch.
6. *The height of the languid.* This controls mainly the articulation or attack; lowering the languid speeds up the attack.
7. *The height of the mouth, or cut-up.* Increasing the cut-up reduces the harmonic content of the tone.
8. *Setting of the upper lip.* Moving the upper lip farther in increases the attack time.
9. *The condition of the upper lip.* Bevelling the upper lip increases the harmonic content of the tone, whereas rounding it decreases harmonic content.

These voicing adjustments tend to be interrelated. For example, increasing the cut-up reduces the harmonic content, whereas increasing the bore diameter produces greater harmonic development and more power. Thus, by combining the two, it is possible to obtain approximately the same tone at increased power.

Laboratory studies (Nolle 1979; Fletcher and Douglas 1980) of the effects of mouth height (cut-up) and mouth position (which relates to adjustments 6 and 8) show that:

1. A high cut-up causes the harmonic content to decrease, the ratio of high to low harmonics to increase, and the overall level to decrease.
2. A low cut-up causes the harmonic content to increase, the ratio of high to low harmonics to decrease, and parasitic oscillations due to the edgetone mechanism (see Section 30.4).

3. Within the optimum operating range, slightly increasing the mouth height favors a "ping" transient, whereas decreasing the height favors a "chiff" transient.

4. A frequency jump to a higher mode is encouraged by a high cut-up or an inward mouth position.

5. Adding ears to a pipe of small scale (diameter to length ratio) increases the range of mouth height and position over which stable oscillation occurs.

Voicing of reed pipes involves adjustments to the reed as well as to the resonator. One of the most important adjustments is in the curvature of the reed. A curved reed does not cut off the flow of air as abruptly as a straight reed does. Also, since tuning is done at the reed, changes in the resonator of a reed pipe will mainly adjust the loudness and timbre of the pipe independently of the pitch.

Many organs have one rank of pipes that is deliberately tuned sharp so that it will produce beats when played with another similar rank. The stop associated with this is called the *voix celeste*.

One or more windchests on an organ are usually equipped with a mechanical device called a *tremolo* or *tremulant*, which causes the air pressure to fluctuate at a regular rate and thus to produce a vibrato.

14.16 ■ SUMMARY

The *piano*, with a playing range of over seven octaves and a wide dynamic range as well, has become the most versatile and popular of all musical instruments. The struck strings transfer vibrational energy through the bridge to the soundboard, which radiates most of the sound. Piano strings have "stretched" partials because of their stiffness; this makes stretch tuning desirable. Most notes on the piano have three strings, and their tuning in relation to each other affects the decay rate as well as the timbre of the note.

The *clavichord*, which also uses struck strings, is a portable instrument with a soft, delicate tone. The strings are struck by tangents, which stay in contact with the strings after striking. The *harpsichord*, which plucks the strings, is experiencing a revival in popularity. It is strung to a much lower tension than the piano, and has a much lighter soundboard. Large instruments have several sets of strings or stops, which may play in different octaves. The strings of a *harp* are plucked by the player rather than by mechanical plectra (as in the harpsichord). Harp strings are tuned to the diatonic scale; to play sharps or flats, the pitch of a string is raised or lowered a semitone by pedals.

The *pipe organ* is the largest of all instruments and has an extremely wide range of tone and dynamics. Its many pipes are organized into ranks and divisions. The main families of pipes (in order of increasing harmonic content) are the flutes, diapasons, strings, and reeds. Flutes, diapasons, and strings use flue pipes with a sounding mechanism similar to that of the flute and the recorder. Reed pipes have a metal reed that modulates the air flow to produce sound. Organ pipe resonators may be cylindrical, conical, or rectangular, with open or closed ends, and made of metal or wood. The pitch of a rank of pipes is denoted by the length of an open pipe sounding the same pitch as the lowest pipe in the rank. Pitch designations range from 1 ft to 32 ft. Voicing of the individual pipes is a very important step in organ building; voicing determines both the initial attack and the steady-state timbre.

REFERENCES AND SUGGESTED READINGS

Ashkenfelt, A., and E. Jansson (1988). "From Touch to String Vibrations—the Initial Course of Piano Tone," Report STL/QPSR 1/88, 31–109, Royal Inst. Technology (KTH), Stockholm.

Blackham, E. D. (1965). "The Physics of the Piano," *Sci. Am.* **99**(6): 88.

Braasch, J., and C. Ahrens (1999). "Attack Transients of Free Reed Organ Pipes," *J. Acoust. Soc. Am.* **105**: 1002 (abstract).

Coltman, J. W. (1969). "Sound Radiation from the Mouth of an Organ Pipe," *J. Acoust. Soc. Am.* **46**: 477.

Conklin, H. A., Jr. (1996a). "Design and Tone in the Mechanoacoustic Piano. Part I. Piano Hammers and Tonal Effects," *J. Acoust. Soc. Am.* **99**: 3286.

Conklin, H. A., Jr. (1996b). "Design and Tone in the Mechanoacoustic Piano. Part II. Piano Structure," *J. Acoust. Soc. Am.* **100**: 695.

Conklin, H. A., Jr. (1996c). "Design and Tone in the Mechanoacoustic Piano. Part III. Piano Strings and Scale Design," *J. Acoust. Soc. Am.* **100**: 1286.

Fletcher, H. (1964). "Normal Vibration Frequencies of a Stiff Piano String," *J. Acoust. Soc. Am.* **36**: 203.

Fletcher, H., E. D. Blackham, and R. Stratton (1962). "Quality of Piano Tones," *J. Acoust. Soc. Am.* **34**: 749.

Fletcher, N. H. (1974). "Nonlinear Interactions in Organ Flue Pipes," *J. Acoust. Soc. Am.* **56**: 645.

Fletcher, N. H. (1977a). "Analysis of the Design and Performance of Harpsichords," *Acustica* **37**: 139.

Fletcher, N. H. (1977b). "Scaling Rules for Organ Flue Pipe Ranks," *Acustica* **37**: 133.

Fletcher, N. H., and L. M. Douglas (1980). "Harmonic Generation in Organ Pipes, Recorders, and Flutes," *J. Acoust. Soc. Am.* **68**: 767.

Fletcher, N. H., and T. D. Rossing (1998). *The Physics of Musical Instruments*, 2nd ed. New York: Springer-Verlag. Chapters 11, 12, 16, and 17.

Fletcher, N. H., and S. Thwaites (1983). "The Physics of Organ Pipes," *Sci. Am.* **248**(1): 94.

Hall, D. E. (1986). "Piano String Excitation in the Case of Small Hammer Mass," *J. Acoust. Soc. Am.* **79**: 141.

Hall, D. E. (1987). "Piano String Excitation II: General Solution for a Hard Narrow Hammer, and III: General Solutions for a Soft Narrow Hammer," *J. Acoust. Soc. Am.* **81**: 535 and 547.

Helmholz, H. L. F. (1877). *On the Sensations of Tone*, 4th ed. Trans., A. J. Ellis. New York: Dover, 1954.

Hirschberg, A., R. W. A. Van de Laar, J. P. Marrou-Maurières, A. P. J. Wijnands, H. J. Dane, S. G. Kruijswijk, and A. J. M. Houtsma (1990). "A Quasi-stationary Model of Air Flow in the Reed Channel of Single-reed Woodwind Instruments," *Acustica* **70**: 146.

Houtsma, A. J. M., T. D. Rossing, and W. M. Wagenaars (1987). *Auditory Demonstrations* (Phillips Compact Disc #1126-061 and text). Woodbury, N.Y.: Acoustical Society of America.

Kirk, R. E. (1959). "Tuning Preferences for Piano Unison Groups," *J. Acoust. Soc. Am.* **31**: 1644.

Klotz, H. (1961). *The Organ Handbook.* St. Louis: Concordia.

Martin, D. W., and W. D. Ward (1961). "Subjective Evaluation of Musical Scale Temperament in Pianos," *J. Acoust. Soc. Am.* **33**: 582.

Mercer, D. M. A. (1951). "The Voicing of Organ Flue Pipes," *J. Acoust. Soc. Am.* **23**: 45.

Nakamura, I. (1983). "The Vibrational Character of the Piano Soundboard," *Proc. 11th ICA*, Paris, 385.

Nolle, A. W. (1979). "Some Voicing Adjustments of Flue Organ Pipes," *J. Acoust. Soc. Am.* **66**: 1612.

Pollard, H. F. (1978). "Loudness of Pipe Organ Sounds," I and II, *Acustica* **41**: 65 and 75.

Rossing, T. D., J. Angster, and A. Miklós (1998). "Reed Vibration and Sound Generation in Lingual Organ Pipes," Paper 2pMU6, 136th meeting, Acoust. Soc. Am., Norfolk, VA, Oct. 1998.

Strong, W. J., and G. R. Plitnik (1983). *Music, Speech and High Fidelity*, 2nd ed. Provo, Utah: Soundprint. Chapters 35, 37, and 38.

Thwaites, S., and N. H. Fletcher (1981). "Some Notes on the Clavichord," *J. Acoust. Soc. Am.* **69**: 1476.

Weinreich, G. (1977). "Coupled Piano Strings," *J. Acoust. Soc. Am.* **62**: 1474.

Weinreich, G. (1979). "The Coupled Motions of Piano Strings," *Sci. Am.* **240**(1): 118.

GLOSSARY

aftersound Second portion of a sound decay having a longer decay time.

chiff A chirplike sound that occurs during attack, especially in flute pipes on an organ.

clavichord A small portable instrument that produces a soft, delicate sound by means of struck strings.

cut-up Height of the mouth opening in an organ pipe; distance from the lower lip to the upper lip.

edgetone The sound produced when an air jet encounters a sharp edge or wedge and oscillates back and forth, first passing on one side, then the other.

flue The narrow windway between the languid and lower lip in a flue pipe.

flue pipe An organ pipe that produces sound by means of a jet of air passing through the flue and striking the upper lip.

harpsichord A keyboard instrument in which strings are plucked by mechanical plectra.

inharmonicity The departure of the frequencies of partials from those of a harmonic series.

jack A device in piano and harpsichord actions that moves up and down when a key is pressed. In the piano, the jack sets the hammer in motion; in the harpsichord, it carries the plectrum.

labial A flue pipe.

laminar flow Fluid flow in which entire layers have the same velocity.

languid A plate that partially blocks an organ pipe and forms one side of the flue.

lingual A reed pipe.

plectrum The small tongue of quill, leather, or plastic that plucks the string of a harpsichord.

soundboard The wooden plate that radiates much of the sound in string instruments.

stretch tuning Tuning octaves slightly larger than a 2 : 1 ratio.

sustaining pedal Right hand pedal of a piano, which raises all the dampers, allowing the strings to continue vibrating after the keys are released.

tangent The small metal square that strikes the string of a clavichord.

tremolo, tremulant A device on an organ that produces a vibrato, usually by varying the air pressure.

turbulent flow Fluid flow characterized by eddies and vortices; the flow velocity tends to vary randomly.

una corda pedal Pedal on grand pianos which shifts the entire action sideways, causing the treble hammers to strike only two of the three unison strings.

virginal A small plucked string instrument in which the strings run parallel to the keyboard.

voicing Adjusting organ pipes to have the desired sound.

voix celeste An organ stop that uses two ranks of pipes with slightly different tunings so that they produce beats.

windchest The important part of the organ that distributes air to selected pipes to make them sound.

REVIEW QUESTIONS

1. What is the approximate range of the piano (in octaves)?

2. Approximately how many strings does a grand piano have?

3. How does the contact time of a piano hammer on a string change with increasing hammer speed?

4. Why are the bass strings of a piano wrapped with wire?

5. What is the main reason for stretch-tuning a piano?

6. What is the effect on string decay time of having three strings for a single note?

7. What is the approximate thickness of a piano soundboard?

8. What mechanical action takes place when the sustaining pedal is depressed?

9. What is the function of a piano bridge?

10. How are the strings in a clavichord excited?

11. How are the strings in a harpsichord excited?

12. What is a virginal?

13. What is the purpose of the pedals on a concert harp?

14. How many pipes does the world's largest pipe organ have?

15. What is a tracker organ?

16. What is a lingual organ pipe?

17. Is the jet in an organ pipe generally laminar or turbulent?

18. What is meant by pipe scaling?

19. Compare the sound power radiated by the mouth and the open end of an organ pipe.

20. Which generally have more harmonics: flue pipes or reed pipes?

QUESTIONS FOR THOUGHT AND DISCUSSION

1. When a chord is played on a piano while the sustaining pedal is depressed, the tone sounds richer than the same chord played without the sustaining pedal. Can you explain why?

2. If a piano key is depressed slowly enough, the hammer fails to contact the string at all. Explain why.

3. Will the two notes shown in Fig. 14.20 sound exactly the same when played on a piano?

4. Can you think of any advantages a tracker action might have over a direct electric action in an organ? any disadvantages?

5. In small pipe organs, the sixteen-foot pipes are almost always stopped wooden pipes. Explain why.

FIGURE 14.20

EXERCISES

1. Suppose that all the strings of a piano were of the same material and also had the same diameter and tension. If the longest string (A_0) were 2 m in length, how long would the highest A-string (A_7) have to be? Is this practical?

2. Show that two unison strings, tuned 2 cents (0.12%) different and initially in phase, will fall out of phase after about 400 vibrations.

3. If a particular string on a harpsichord has one-third the diameter and one-tenth the tension of the corresponding string on a piano, which string will be longer? What will the ratio of lengths be? (See Section 3.2 and 4.3; assume that both strings are steel.)

4. If the diameter of pipes in a particular rank is reduced by a factor of two every seventeen notes, show that the diameter reduces by four over a range of about three octaves.

5. A piano tuner finds that two of the strings tuned to C_4 give about one beat per second when sounded together. What is the ratio of their frequencies? Show that their pitches differ by about 7 cents. (One cent is 1/100 of a semitone and corresponds to a frequency ratio of approximately 1.0006.)

6. By noting the weak partials in the C_1 spectrum in Fig. 14.7, estimate the fraction of the string length β at which the hammer strikes.

7. Calculate the acoustical lengths of open organ pipes tuned to C_0, C_1, and C_2 (frequencies are given in Table 9.2, also Table 14.1). Compare these to the equivalent lengths in Table 14.1.

EXPERIMENTS FOR HOME, LABORATORY, AND CLASSROOM DEMONSTRATION

Home and Classroom Demonstration

1. *Mechanical model of piano action*   Mechanical models of piano actions can sometimes be obtained from piano dealers or manufacturers.

2. *The pedals*   Depress each piano pedal to see and hear what it does.

3. *Trichord*   Damp two of the three strings of one note (piano tuners wedges work the best) and determine the change in the sound.

4. *Piano sound reversed*   Demonstration 29 on the Auditory Demonstrations CD. Piano tones, heard backwards, do not sound like piano tones, even though the spectrum remains unchanged, because the sound of hammer on string comes at the end of a note rather than at the beginning.

5. *Organ pipes*   From an organ builder, obtain as many different organ pipes as possible. These can generally be sounded by blowing on them. Note the different sounds.

6. *Wide scale versus narrow scale*   Compare pipes of wide scale (open flute), medium scale (principal or diapason), and narrow scale (string) having the same pitch.

7. *Closed and open pipes*   Compare the sound spectrum of an open flute pipe to a closed flute pipe of the same pitch.

Laboratory Experiment

Acoustics of organ pipes (Experiment 21 in *Acoustics Laboratory Experiments*)

PART
IV

The Human Voice

It is difficult to overstate the importance of the human voice. Of all the members of the animal kingdom, we alone have the power of articulate speech. Speech is our chief means of communication. In addition, the human voice is our oldest musical instrument.

The human voice and the human ear are very well matched to each other. The ear has its maximum sensitivity in the frequency range from 1000 to 4000 Hz, and that is the range of frequency in which the resonances of the vocal tract occur. These resonances (called formants), we will learn, are the acoustical bases for all the vowel sounds and many of the consonants in speech and singing.

Because speech and singing are such closely related functions of the human voice, it is recommended that all three chapters (15–17) be considered together. It is possible for a singer to begin with Chapter 17, but frequent reference will probably need to be made to Chapters 15 and 16.

CHAPTER

15

Speech Production

Throughout human history, the principal mode of communication has been the spoken word. The systems in the human body that send and receive oral messages are sophisticated in design and complex in function. Our understanding of both hearing and speech has progressed dramatically in recent years, largely because of new techniques for making acoustical as well as physiological measurements. The auditory system and its function were discussed in Chapter 5, and now it is appropriate to devote the same attention to the vocal organs.

In this chapter you should learn:

- How the human vocal organ makes speech sounds;
- How speech sounds are the product of the source, the filter function, and the radiation efficiency;
- About speech articulation by the different parts of the vocal tract;
- About formants as resonances of the vocal tract;
- How the glottis and the vocal tract are studied by speech acousticians.

15.1 ■ THE VOCAL ORGANS

The human vocal organs, as well as a representation of the main acoustical features, are shown in Fig. 15.1. The lungs serve as both a reservoir of air and an energy source. In

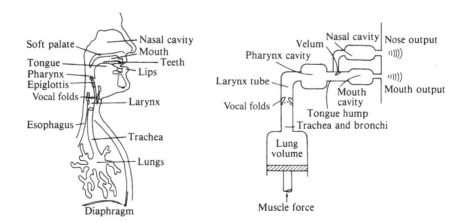

FIGURE 15.1
Human vocal organs and a representation of their main acoustical features. (After Flanagan 1965)

speaking, as in exhaling, air is forced from the lungs through the larynx into the three main cavities of the vocal tract: they pharynx and the nasal and oral cavities. From the nasal and oral cavities, the air exits through the nose and mouth, respectively.

Air can be inhaled or exhaled with little generation of sound if desired. In order to produce speech sounds, the flow of air is interrupted by the vocal cords or by constrictions in the vocal tract (made with the tongue or lips, for example). The sounds from the interrupted flow are appropriately modified by various cavities in the vocal tract and are eventually radiated as speech from the mouth and, in some cases, the nose.

15.2 ■ THE LARYNX AND THE VOCAL FOLDS

The most important sound source in the vocal system is the *larynx*, which contains the *vocal folds* or *vocal cords*. The larynx is constructed mainly of cartilages, several of which are shown in Fig. 15.2. One of the cartilages, the thyroid, forms the projection on the front of the neck known as the Adam's apple.

The vocal folds are not at all like cords or strings, but consist rather of folds of ligament extending from the thyroid cartilage in the front to the arytenoid cartilages at the back. The arytenoid cartilages are movable and control the size of the V-shaped opening between the vocal cords, which is called the *glottis*. Figure 15.3 shows how the arytenoids control the size of the glottis. Normally, the arytenoids are positioned well apart from each other to permit breathing; however, they come together when sound is produced by the vocal folds.

The vocal folds may act on the air stream in several different ways during speech. From a completely closed position in which they cut off the flow of air, they may open suddenly as in a light cough or a glottal stop (such as the glottal "h" that occurs in Cockney English). A glottal stop may also give a hard beginning to a vowel sound, such as the "Idiot!" expressed vehemently. On the other hand, the vocal folds may be completely open for un-

FIGURE 15.2
Various views of the larynx:
(a) back; (b) side.

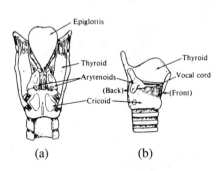

(a)   (b)

FIGURE 15.3
Control of the glottal opening by the arytenoids.

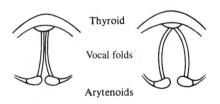

voiced consonants such as "s," "sh," "f," etc. An intermediate position occurs in the "h" sound, where the air stream interacts lightly as it passes between the vocal folds.

The most useful function of the vocal folds, however, is to modulate the air flow by rapidly opening and closing. This rapid vibration produces a buzzing sound from which vowels and voiced consonants are created. These functions of the vocal folds are somewhat analogous to the functions of the lips. The glottal stop corresponds to the action of the lips in a plosive consonant such as p; the light friction of the folds in producing the "h" sound corresponds to the action of the lips in pronouncing "f." The rapid vibration of the vocal folds is similar to the rolling noise made by a child's lips to imitate a motor, or the sound used to indicate coldness ("brrr"), or a trumpet player buzzing his or her lips in a practice exercise. You can feel the vibrations set up by the vocal folds by placing a finger lightly against your Adam's apple. Make sounds "zzzzzz" and "sssss" alternately to turn the vibrations on and off (these are examples of voiced and unvoiced consonants, respectively).

The rate of vibration of the vocal folds is determined primarily by their mass and tension, although the pressure and velocity of the air do contribute in a smaller way. The vocal folds are typically longer and heavier in the adult male than in the female and, therefore, vibrate at a lower frequency (pitch). During normal speech, the vibration rate may vary over a 2 : 1 ratio (one octave), although the range of a singer's voice is more than two octaves. Typical frequencies used in speech are 110 Hz in the male, 220 Hz in the female, and 300 Hz in the child, with wide variations from one individual to another.

Speech scientists describe three different modes in which the vocal folds can vibrate. In the normal mode, they open and close completely during the cycle and generate puffs of air roughly triangular in shape when air flow is plotted against time. In the open phase mode, the folds do not close completely over their entire length, so the air flow does not go to zero. This produces a breathy voice, sometimes used to express shock ("No!") or passion ("I love you"). A third mode, in which a minimum of air passes in short puffs, gives rise to a creaky voice, such as might result if you attempt to talk while lifting a heavy weight. A fourth mode, called *head voice*, or *falsetto*, is normally not used in speech and is discussed in Section 17.6.

The vocal folds are caused to open by air pressure in the trachea, which tends to blow them upward and outward. As the air velocity increases, the pressure decreases between them, and they are pulled back together by the Bernoulli force (see Section 11.4). Ordinarily, however, the restoring force supplied by the muscles exceeds the Bernoulli force. Feedback from the vocal tract has relatively little influence on the vocal fold vibrations (as compared with the close cooperation between the air column in a brass instrument and the player's lips, for example; see Section 11.3).

Another use of the vocal folds is in the production of a whisper. For a quiet whisper, the vocal folds are in much the same position as they are for an "h" sound. For louder whispers, the folds are brought closer so as to interfere more strongly with the flow of air. Say the word "hat" in a loud whisper and note the different rate of air flow during the "h" and the "a" by holding your hand in front of your mouth.

The vocal folds can be observed by placing a small dental mirror far back in the mouth. Using this technique, several investigators have taken high-speed motion pictures (4000 frames/s) of the vocal folds in vibration. Figure 15.4 illustrates this technique, and Fig. 15.5

FIGURE 15.4
Technique for high-speed motion picture photography of the vocal folds. (From Flanagan 1965)

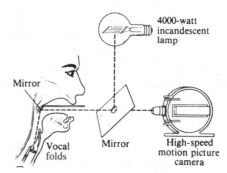

FIGURE 15.5
Successive phases in one cycle of vocal fold vibration. The total elapsed time is approximately 8 ms. (From Flanagan 1965)

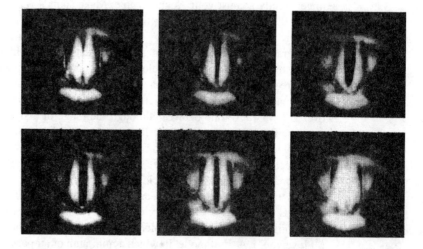

shows one cycle of vocal fold vibration at a frequency of about 125 Hz. Less obtrusive fiber optic probes inserted into the throat allow continuous observation of the vocal folds.

The flow of air through the glottis is roughly (though not exactly) proportional to the area of the glottal opening. For normal vocal effort, the waveform of the air flow is roughly triangular in shape with a duty factor (that is, the ratio of time open to the total period of a vibration) of 30 to 70%, as shown in Fig. 15.6. The sound resulting from this interrupted air flow is characterized as a "buzz" and is rich in overtones. A triangular waveform is composed of harmonics that diminish in amplitude as $1/n^2$ (at a rate of 12 dB/octave), and the sound spectrum of the output of the larynx shows approximately this character for the higher harmonics, as seen in Fig. 15.7.

One might guess that the loud speaking would require greater air pressure from the lungs and a greater amplitude of vocal fold vibration. This is only partly true. Even if we assume the pressure in the trachea to be constant, the pressure in the larynx will fluctuate due to standing waves in the vocal tract. It may be a bit surprising to learn that one of the most important parameters affecting loudness of phonation is the *rate of glottal closure*. Rapid closure introduces higher harmonics in the glottal airflow spectrum, and these harmonics excite resonances of the vocal tract, leading to considerable buildup in sound level.

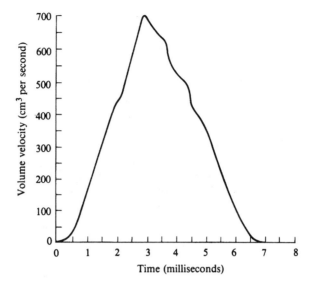

FIGURE 15.6
The variation of air flow in a glottal puff. The curve repeats once every 8 ms (a frequency of 125 Hz).

Although the vocal folds serve as the principal source of sound in speech, other sources are used, especially in the production of unvoiced consonant sounds. Sounds such as "f," "th," "s," "sh," (fricative consonants) and "l" are produced by a turbulent flow of air through a constriction somewhere in the vocal tract. The spectrum of such turbulence is quite broad, with many overtones that are not harmonic. Another source of sound is generated by a sudden release of pressure, such as that used in the plosive consonants p, t, and k. The sounds of consonants will be discussed in Section 15.4.

FIGURE 15.7
A typical waveform of the volume velocity of the glottal output for a fundamental frequency of 125 Hz, and a Fourier spectrum corresponding to this type of waveform. (From Stevens and House 1961.)

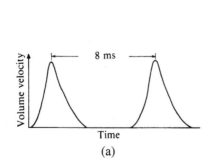

(a)

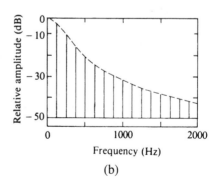

(b)

15.3 ■ THE VOCAL TRACT

The function of the vocal tract is a most remarkable one; it transforms the "buzzes" and "whooshes" from the vocal folds and other sources into the intricate, subtle sounds of speech. This demanding function is accomplished by changes in shape to produce various acoustic resonances. Intensive studies in recent years have produced a great deal of information concerning the details of how this is accomplished. This information is the heart of a branch of science called *acoustical phonetics*.

The vocal tract, as shown in Fig. 15.1, can be considered a single tube extending from the vocal folds to the lips, with a side branch leading to the nasal cavity. The length of the tube is typically about 17 cm, which can be varied slightly by raising or lowering the larynx and by shaping the lips. For the most part, however, the resonances in the vocal tract are tuned by changing its cross-sectional area at one or more points.

The *pharynx* connects the larynx with the oral cavity. It is not easily varied in shape, although its length can be changed slightly by raising or lowering the larynx at one end and the soft palate at the other end. The soft palate also acts as a valve to isolate or connect the nasal cavity to the pharynx. Since food also passes through the pharynx on its way to the esophagus, valves are necessary at the lower end to prevent food from reaching the larynx and to isolate the esophagus acoustically from the vocal tract. The *epiglottis* serves as such a valve, with the "false vocal cords" at the top of the larynx serving as a backup in case some food gets past the epiglottis. The epiglottis, false vocal cords, and vocal folds (cords) are open during normal breathing but closed during swallowing, thus forming a triple barrier to protect the windpipe. The epiglottis and false vocal cords do not appear to play any significant role in the production of speech.

The *nasal cavity* has fixed dimensions and shape, so that it is virtually untunable. In the adult male, the cavity has a length of about 12 cm and a volume on the order of 60 cm^3. The soft palate serves as a valve to control the flow of air from the pharynx into the nasal cavity. If the soft palate is lowered, air and sound waves flow into the nasal cavity and a nasal effect results from resonance within the nasal cavity. If, at the same time, flow through the mouth is blocked, air and sound exit through the nose, and humming results. Nasalized vowel sounds, which are common in French, are made by allowing sound to exit through both the mouth and the nose.

You can observe the soft palate in a mirror as it moves up and down (closing the nasal cavity in the up position). Say "ah" and the soft palate will rise; relax, and it will lower to normal breathing position. A cleft palate is a defect that allows air into the nasal cavity for all sounds, even those that should be entirely oral.

The *oral cavity*, or mouth, is probably the most important single part of the vocal tract because its size and shape can be varied by adjusting the relative positions of the palate, the tongue, the lips, and the teeth. The *tongue* is very flexible; its tip and edges can be moved independently or the entire tongue can move forward, backward, up and down. Movement of the lips, cheeks, and teeth also changes the size, shape, and acoustics of the oral cavity.

The lips control the size and shape of the mouth opening through which sound is radiated. Since the mouth opening is small compared to the wavelength of most components of the radiated sound, the size and shape of the opening are not of particular significance, except as they affect the all-important resonance frequencies of the oral cavity (this will be discussed further in Chapter 17). The mouth radiates more efficiently at higher frequencies where the wavelength approaches the size of the opening. In fact, a rise of 6 dB per octave in radiation efficiency is a good approximation to this effect.

The spectrum envelope of speech sound can be thought of as the product of three components:

$$\text{Speech sound} = \text{source} \times \text{filter function} \times \text{radiation efficiency.}$$

If each of these quantities is expressed in decibels, then the contributions are added rather than multiplied. When the source consists of the vocal folds vibrating in their usual manner, the source function decreases in strength approximately 12 dB per octave (see Fig. 15.7). To this should be added the radiation efficiency of the mouth (which rises approximately 6 dB per octave) giving a net decrease of 6 dB per octave due to the first and last terms in the equation above. It remains to consider the more complicated way in which the filter function of the vocal tract varies with frequency, and that is the subject of Section 15.5.

Inverse Filtering and the Glottogram

Speech sound, we learned, is the product of the glottal flow (source), the vocal tract (filter), and the mouth opening (radiator). It is difficult but not impossible to study these independently. One way to study the glottal flow, for example, is to cancel the filtering effect of the vocal tract by *inverse filtering*. The subject speaks into a mask that measures (by means of a flow resistance and a pressure microphone) the waveform of the air flow. The signal then passes through a set of filters that are adjusted to remove the formants as well as possible. This works quite well, especially for low-pitched phonation and for open vowels. The waveform after filtering is called a *glottogram*, and it gives a fairly accurate representation of the glottal air flow in various modes of phonation.

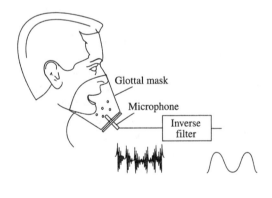

15.4 ■ ARTICULATION OF SPEECH

Before discussing the resonances of the vocal tract, it is appropriate to briefly describe the articulation of English speech sounds, or *phonemes*. In speech structure, one or more phonemes combine to form a syllable, and one or more syllables to form a word. Phonemes can be divided into two groups: vowels and consonants. Vowel sounds are always voiced; that is, they are produced with the vocal folds in vibration. Consonant sounds may be either voiced or unvoiced.

Various speech scientists list from 12 to 21 different vowel sounds used in the English language. This discrepancy in number comes about partly because of a difference of opin-

TABLE 15.1   The vowels of Great American English

Pure vowels						Diphthongs		
ee	heat	/i/	aw	call	/ɔ/	ou	tone	/oʊ/
i	hit	/ɪ/	ů	put	/ʊ/	ei	take	/eɪ/
e	head	/ɛ/	oo	cool	/u/	ai	might	/aɪ/
ae	had	/æ/	ŭ	ton	/ʌ/	au	shout	/aʊ/
uh	the	/ə/	er	bird	/ɜ/	oi	toil	/ɔɪ/
ah	father	/ɑ/				ju	fuse	/ju/

ion as to what constitutes a pure vowel sound rather than a *diphthong* (a combination of two or more vowels into one phoneme). Table 15.1 lists the vowel sounds of Great American, the dialect of English spoken throughout most of western and midwestern United States. Also given are the corresponding symbols from the International Phonetic Alphabet (Denes and Pinson 1973). Figure 15.8 shows the approximate tongue positions for articulating these vowels.

Whereas the vowel sounds are more or less steady for the duration of the phoneme, consonants involve very rapid, sometimes subtle, changes in sound. Thus consonants tend to be more difficult to analyze and to describe acoustically.

Consonants may be classified according to their *manner of articulation* as plosive, fricative, nasal, liquid, and semivowel. The *plosive* or stop consonants (p, b, t, etc.) are produced by blocking the flow of air somewhere in the vocal tract (usually in the mouth) and releasing the pressure rather suddenly. The *fricatives* (f, s, sh, etc.) are made by constricting the air flow to produce turbulence. The *nasals* (m, n, ng) are made by lowering the soft palate to connect the nasal cavity to the pharynx and then blocking the mouth cavity at some point along its length. The *semivowels* or glide consonants (w, y) are produced by keeping the vocal tract briefly in a vowel position and then changing it rapidly to the vowel sound that follows; thus, semivowels are always followed by a vowel. In sounding the *liquids* (r, l), the tip of the tongue is raised and the oral cavity is somewhat constricted.

FIGURE 15.8
Approximate tongue positions for articulating vowels listed in Table 15.1. Numbers 1–8 are the eight cardinal vowels, which serve as a standard of comparison between languages.

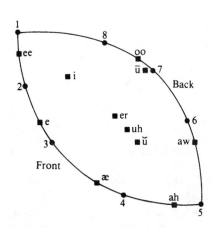

TABLE 15.2   The classification of English consonants

Place of articulation	Plosive		Fricative		Nasal	Semivowel	Liquids
	Unvoiced	Voiced	Unvoiced	Voiced			
Labial (lips)	p	b			m	w	
Labiodental (lips and teeth)			f	v			
Dental (teeth)			th /θ/ (thin)	th /ð/ (then)			
Alveolar (gums)	t	d	s	z	n	y /j/	l, r
Palatal (palate)			sh /ʃ/	zh /ʒ/			
Velar (soft palate)	k	g			ng /ŋ/		
Glottal (glottis)			h				

Phonetic symbols are given where they differ from the English letter.

Consonants are further classified according to their *place of articulation*, primarily the lips, the teeth, the gums, the palate, and the glottis. Terms used by speech scientists to denote place of articulation include *labial* (lips), *dental* (teeth), *alveolar* (gums), *palatal* (palate), *velar* (soft palate), *glottal* (glottis), and *labiodental* (lips and teeth). Finally, consonants are classified as to whether they are *voiced* or *unvoiced*.

Twenty-four consonants of English are thus classified in Table 15.2. Note the seven pairs of voiced/unvoiced consonants. In addition, the pair /tʃ, dʒ/, which refer to the "ch" (church) and "j" (judge) sounds, are sometimes included as separate consonants, although each of them consists of a plosive followed by a fricative (ch ≃ t + sh; j ≃ d + zh). Consonants are more independent of language and dialect than vowels are.

15.5 ■ RESONANCES OF THE VOCAL TRACT: FORMANTS

The *vocal tract* consists of three main sections: the pharynx, the mouth, and the nasal cavity. These can be shaped by movements of other vocal organs, such as the tongue, the lips, and the soft palate (see Fig. 15.1).

Although the pitch and intensity of speech sounds are determined mainly by the vibrations of the vocal folds, the spectrum of these sounds is strongly shaped by the resonances of the vocal tract. It is the character of these resonances that distinguishes one phoneme from another.

The peaks that occur in the sound spectra of the vowels, independent of the pitch, are called *formants*. They appear as envelopes that modify the amplitudes of the various harmonics of the source sound. Each formant corresponds to one or more resonances in the vocal tract. Formant frequencies are virtually independent of the source spectrum.

Figure 15.9 illustrates the effect of formants on the source sound from the larynx. Both the waveform and the spectrum of the source sound are shown along with the waveform and the spectrum of the transmitted speech sound. (Note that in the waveform graphs, the horizontal axis is time; in the spectra, the horizontal axis is frequency.)

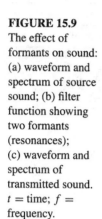

FIGURE 15.9
The effect of
formants on sound:
(a) waveform and
spectrum of source
sound; (b) filter
function showing
two formants
(resonances);
(c) waveform and
spectrum of
transmitted sound.
t = time; f =
frequency.

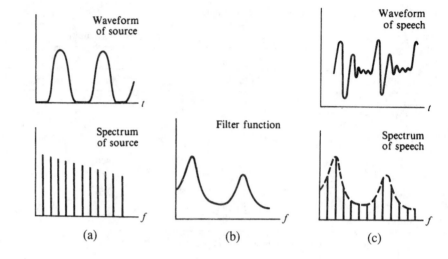

Table 15.3 gives the average formant frequencies for the vowel sounds of men, women, and children. The first nine rows give the average frequencies of the first three formants. The last three rows indicate the relative strengths of the three formants for each vowel. For example, /ɑ/ (ah) has the strongest second formant, only 4 dB weaker than the first formant. For /i/ (ee), on the other hand, the second formant is 20 dB below the first.

TABLE 15.3　Formant frequencies and amplitude of vowels averaged for 76 speakers

Formant frequencies (Hz)		/i/ (ee)	/ɪ/ (i)	/ɛ/ (e)	/æ/ (ae)	/ɑ/ (ah)	/ɔ/ (aw)	/ʊ/ (ủ)	/u/ (oo)	/ʌ/ (u)	/ɜ/ (er)
F_1	M	270	390	530	660	730	570	440	300	640	490
	W	310	430	610	860	850	590	470	370	760	500
	Ch	370	530	690	1010	1030	680	560	430	850	560
F_2	M	2290	1990	1840	1720	1090	840	1020	870	1190	1350
	W	2790	2480	2330	2050	1220	920	1160	950	1400	1640
	Ch	3200	2730	2610	2320	1370	1060	1410	1170	1590	1820
F_3	M	3010	2550	2480	2410	2440	2410	2240	2240	2390	1690
	W	3310	3070	2990	2850	2810	2710	2680	2670	2780	1960
	Ch	3730	3600	3570	3320	3170	3180	3310	3260	3360	2160
Formant amplitudes (dB)		−4	−3	−2	−1	−1	0	−1	−3	−1	−5
		−24	−23	−17	−12	−5	−7	−12	−19	−10	−15
		−28	−27	−24	−22	−28	−34	−34	−43	−27	−20

Source: Peterson and Barney (1952).

15.6 ■ MODELS OF THE VOCAL TRACT

Although the vocal tract, with its many curves and bends, is a rather complex acoustical system, simple models help us to understand the origin of the various formants or resonances.

The simplest acoustic model of the vocal tract is a pipe closed at one end (by the glottis) and open at the other end (lips). Such a pipe has resonances (see Fig. 4.8) given by $f_1 = v/4L$, $f_3 = 3v/4L, \ldots, f_n = nv/4L$ ($n = 1, 3, 5, \ldots$). For a pipe 17 cm long (the typical length of a vocal tract), the resonances occur at approximately 500, 1500, and 2500 Hz, which are surprisingly close to the peaks in the spectrum of the vowel sound /ε/ (typically at 500, 1800, and 2500 Hz).

Suppose we fasten a small loudspeaker to one end of a 17-cm pipe and place a microphone near the open end, as shown in Fig. 15.10. If the loudspeaker were driven by a pure tone (sine wave) of varying frequency, we would note strong resonances at about 500, 1500, and 2500 Hz, as we just discussed. If the loudspeaker were then driven with a sawtooth waveform (or some other waveform with many harmonics) having a frequency of 100 Hz, and the output of the microphone were displayed on a spectrum analyzer, we would see something similar to the spectrum shown in Fig. 15.10(b). The heights of the various harmonics in the source spectrum have now been shaped by the resonances of the pipe. Formants are present at 500, 1500, and 2500 Hz.

In addition to frequency, the parameters *amplitude* and *bandwidth* are used to describe a formant. The amplitude describes the height of a resonance (see the last three rows in Table 15.3), and the bandwidth describes its breadth.

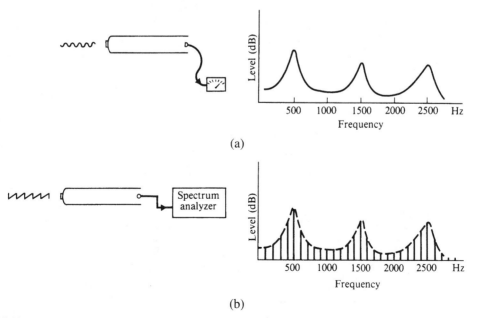

(a)

(b)

FIGURE 15.10   Response of a closed-pipe model of the vocal tract: (a) resonances of a 17-cm pipe excited with a pure tone of varying frequency; (b) spectrum of a 100-Hz sawtooth wave shaped by the resonances (formants) of the pipe.

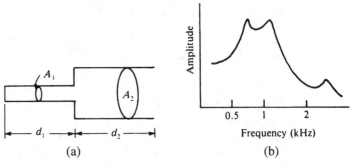

(a)   (b)

FIGURE 15.11   (a) Two-tube approximation of vocal-tract configuration for the vowel /a/ (ah). (b) Approximate form of spectrum envelope of vowel generated by the configuration in (a). (From Stevens 1972.)

Simple models for the vocal tract add to our understanding of how various phonemes might be articulated. Models for the vowel sounds /a/, /i/, and /u/ ("ah," "ee," and "oo") using tubes of two diameters are shown in Figs. 15.11, 15.12, and 15.13 (Stevens 1972).

It is more difficult to construct simple models adequate to describe the articulation of consonant sounds. Stevens (1972) shows, however, that the simple constriction illustrated in Fig. 15.14, if moved to different positions in the vocal tract, can approximate the formants associated with several consonant sounds. The corresponding resonances are shown in Fig. 15.15. For the back portion of the tube, which is essentially closed at both ends, the resonance frequencies are $f_b = nv/2l_b$. For the front portion, the frequencies are $f_f = mv/4l_f$, with $m = 1, 3, 5, \ldots$.

For a given value of the distance l_b from the glottis to the constriction, a vertical line can be drawn and the formant frequencies determined from the intersections with the various curves. Note that for $l_b < 8$ cm, the front cavity (dashed curve in Fig. 15.15) provides the lowest resonance, whereas for $l_b > 10$ cm, the back cavity (solid curve) does so. Near the crossover points, the resonances interact, as represented by the dotted curves. The first formant is not indicated on the graph, but the lowest resonance of the model would be a Helmholtz resonance whose frequency depends on the volume of the back cavity along with the length and cross section of the constriction.

FIGURE 15.12
(a) Two-tube approximation of vocal-tract configuration for the vowel /i/ (ee). (b) Approximate form of spectrum envelope of vowel generated by the configuration in (a). (From Stevens 1972.)

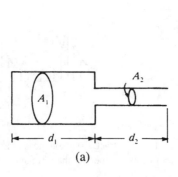

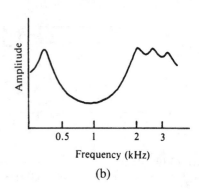

(a)   (b)

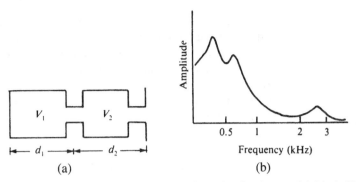

FIGURE 15.13 (a) Approximation of vocal-tract configuration for the vowel /u/ (oo). V_1 and V_2 represent the volumes of the two cavities. (b) Approximate form of spectrum envelope of vowel generated by the configuration in (a). (From Stevens 1972.)

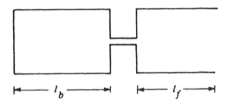

FIGURE 15.14 Idealized model of constricted vocal-tract configuration corresponding to a consonant. The constriction is adjusted to different positions to represent different places of articulation. (From Stevens 1972.)

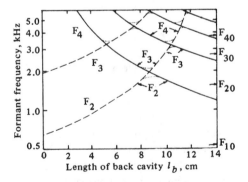

FIGURE 15.15 Relations between natural frequencies and the position of the constriction for the configuration shown in Fig. 15.14. The overall length of the tube is 16 cm and the length of the constriction is 3 cm. The dashed lines represent the lowest two resonances of the front cavity (anterior to the constriction); the solid lines represent the lowest four resonances of the back cavity. The dotted lines near the points of coincidence of two resonances represent the resonant frequencies for the case in which there is a small amount of coupling between front and back cavities. The resonances of a 16-cm tube with no constriction are shown by the arrows at the right. The curves are labeled with the appropriate formant numbers. (From Stevens 1972.)

Constriction positions corresponding to $l_b < 8$ cm represent configurations for uvular and pharyngeal consonants that do not ordinarily occur in English. Rather abruptly moving the constriction from left to right through the F_2–F_3 crossover point ($l_b \approx 8.5$ cm) and at the same time enlarging the constriction toward the vowel configuration resembles the articulation of the velar consonants /g/ or /k/. Similar abrupt shifts near the F_3–F_4 crossover point correspond to the fricatives /s/ and /ʃ/ ("sh"). The labial and labiodental consonants would have $l_b > 13$ cm.

15.7 ■ STUDIES OF THE VOCAL TRACT

FIGURE 15.16
The positions of the vocal organs (based on data from X-ray photographs of the author) and the spectra of the vowel sounds in the middle of the words, *heed*, *hid*, *head*, *had*, *hod*, *hawed*, *hood*, *who'd*. Compare the sounds *hod*, *heed*, and *who'd* with the corresponding two-tube models of the vocal tract in Figs. 15.11, 15.12, and 15.13 (Ladefoged 1962).

A number of speech scientists have made X-ray photographs of the vocal tract during the production of speech sounds. From these photographs, profiles of the vocal tract can be constructed. It is interesting to compare the profiles for the vowel sounds, ɑ, i, and u ("ah," "ee," "oo"), shown in Fig. 15.16, with the simple models shown in Figs. 15.11–15.13.

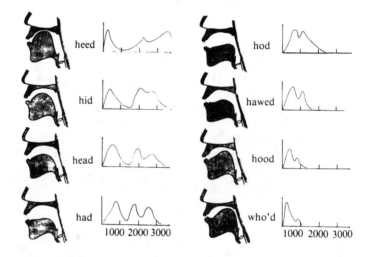

In the case of consonants, the profile of the vocal tract depends to some extent on the vowel sounds that precede and follow the consonant. The location of the constriction (the place of articulation) changes very little, however. Profiles for the stop or plosive consonants are shown in Fig. 15.17. Note that the voiced consonants are nearly identical to the corresponding unvoiced consonants. The only difference between the words "to" and "do," for example, is that the vocal folds vibrate during the /d/ sound, but do not begin vibrating until after the /t/ sound has been articulated.

FIGURE 15.17
Profiles of the vocal tract showing place of articulation of the stop or plosive consonants.

15.8 ■ PROSODIC FEATURES OF SPEECH

Prosodic features are characteristics of speech that convey meaning, emphasis, and emotion without actually changing the phonemes. They include pitch, rhythm, and accent. In English, prosodic features play a secondary role to that of phonemes in the communication of information.

In certain languages, such as Chinese, however, a phoneme can take on several different meanings depending on its "tone." The manner in which the frequency changes with time for the four tones of Mandarin Chinese is shown in Fig. 15.18.

One of the common uses of prosodic features is to change a declarative sentence ("You are going home.") into a question ("You are going home?"). This is done mainly by raising the pitch of the final word. The same sentence could be made imperative by adding stress (increase in both loudness and pitch) to the second word ("You *are* going home!").

Prosodic features tend to indicate the emotional state of the speaker. "Raising one's voice" in anger, for example, increases both loudness and pitch. A state of excitement frequently causes an increase in the rate of speaking. Several attempts have been made to accomplish acoustic "lie detection" by analyzing the prosodic features of recorded speech for evidence of stress.

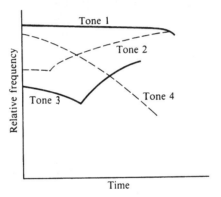

FIGURE 15.18
Frequency changes with time for tones in Mandarin Chinese. (After Luchsinger and Arnold 1965.)

15.9 ■ SUMMARY

The principal parts of the vocal tract are the larynx and vocal cords, pharynx, nasal cavity, oral cavity, tongue, lips, and teeth. Speech sounds originate with the vibrations of the vocal folds or with a constriction of the air flow, are filtered in the vocal tract, and finally are radiated through the lips or nose. Resonances of the vocal tract, called *formants*, determine the vowel sounds, the first and second formant being the most important. Consonants involve rapid changes in sound generated by changing a constriction somewhere in the vocal tract. Consonants can be classified according to their place and manner of articulation.

Simple models of the vocal tract, constructed from tubes of different lengths and diameters, help us understand the acoustical behavior of the vocal tract. Profiles of the vocal tract can be drawn from X-ray photographs. Prosodic features, such as pitch, rhythm, and accent, convey meaning, emphasis, and emotion.

REFERENCES AND SUGGESTED READINGS

David, E. E., Jr., and P. B. Denes, eds. (1972). *Human Communication: A Unified View*. New York: McGraw-Hill.

Denes, P. B., and E. N. Pinson (1973). *The Speech Chain*. New York: Anchor-Doubleday.

Fant, G. (1960). *Acoustic Theory of Speech Production*. The Hague: Mouton.

Flanagan, J. L. (1965). *Speech Analysis, Synthesis and Perception*. New York: Academic.

Ladefoged, P. (1962). *Elements of Acoustic Phonetics*. Chicago: University of Chicago Press.

Lehiste, I., ed. (1967). *Readings in Acoustic Phonetics*. Cambridge, Mass.: MIT Press.

Luchsinger, R., and G. E. Arnold (1965). *Voice-Speech-Language*, Belmont, Calif.: Wadsworth.

Peterson, G. E., and H. L. Barney (1952). "Control Methods Used in a Study of Vowels," *J. Acoust. Soc. Am.* **24**: 175.

Pickett, J. M. (1999). *The Acoustics of Speech Communication*. Needham Heights, Mass.: Allyn and Bacon.

Schouten, J. F. (1962). "On the Perception of Sound and Speech," Congress Report, 4th ICA, Copenhagen, 196.

Stevens, K. N. (1972). "The Quantal Nature of Speech: Evidence from Articulatory-Acoustic Data," in *Human Communication: A Unified View*, p. 51. Eds., E. E. David, Jr., and P. B. Denes. New York: McGraw-Hill.

Stevens, K. N., and A. S. House (1971). "An Acoustical Theory of Vowel Production and Some of Its Implications," *J. Speech & Hearing Research* **4**(4): 75.

Strong, W. J., and G. R. Plitnik (1983). *Music, Speech and High Fidelity*, 2nd ed. Provo, Utah: Soundprint. Chaps. 23–27.

Sundberg, J. (1977). "The Acoustics of the Singing Voice," *Sci. Am.* **236**(3): 82.

Thomas, I. B. (1969). "Perceived Pitch of Whispered Vowels, " *J. Acoust. Soc. Am.* **46**: 468.

GLOSSARY

cardinal vowels Eight vowel sounds that serve as a standard of comparison for the vowels of various languages.

diphthong A combination of two or more vowels into one phoneme.

epiglottis A thin piece of cartilage that protects the glottis during swallowing.

filter A device that allows signals in a certain frequency range to pass and attenuates others.

formants Vocal tract resonances that determine speech sounds.

fricatives Consonants that are formed by constricting air flow in the vocal tract (such as f, v, s, z, th, sh, etc.).

glottis The V-shaped opening between the vocal folds.

larynx The section of the vocal system, composed mainly of cartilage, that contains the vocal folds.

nasals Consonants that make use of resonance of the nasal cavity (m, n, ng).

palate The roof of the mouth.

pharynx Lower part of the vocal tract which connects the mouth to the trachea.

phonemes Individual units of sound that make up speech.

phonetics The study of speech sounds.

plosives Consonants that are produced by suddenly removing a constriction in the vocal tract (p, b, t, d, k, g).

prosodic feature A characteristic of speech, such as pitch, rhythm, and accent, that is used to convey meaning, emphasis, and emotion.

semivowels Consonants for which the vocal tract is formed in a configuration generally used for vowels (w, y).

vocal folds or vocal cords Folds of ligament extending across the larynx that interrupt the flow of air to produce sound.

REVIEW QUESTIONS

1. What are the three main cavities of the vocal tract? Which of these play a role in the production of speech?
2. Describe the way in which the vocal folds vibrate.
3. Describe the role of the vocal folds in producing an "h" sound.
4. The spectrum envelope of speech sound can be thought of as the product of what three components?
5. Describe the spectrum of the glottal source function.
6. Give examples of the following types of consonants: fricative, nasal, liquid, semivowel.
7. Give examples of voiced and unvoiced consonants.
8. Sketch simple two-tube models of the vocal tract configuration for the vowels /ɑ/ and /ɪ/.
9. What voiced and unvoiced consonants are formed with the lips?

10. What voiced and unvoiced consonants are formed with the soft palate?
11. Describe the role of the vocal tract in whistling.
12. What are prosodic features of speech? Give two examples.
13. What are typical vocal fold vibration frequencies in male and female speakers?
14. How is it possible to observe the motion of the vocal folds?
15. How is it possible to observe the shape of the vocal tract for different vowel sounds?
16. How does the glottal waveform change when one speaks louder?

QUESTIONS FOR THOUGHT AND DISCUSSION

1. Discuss the function of each of the principal parts of the vocal tract.
2. If a person partially fills his or her lungs with helium and then speaks, the speech sounds distorted (it is sometimes described as sounding like Donald Duck). Explain this on the basis of formants (the information in Table 3.1 may be helpful). (This type of distortion will be discussed in Chapter 16.)
3. Discuss the acoustics of
   (a) a "hoarse" throat;
   (b) a stuffed nose;
   (c) swollen tonsils.
4. Although the vibrations of the vocal folds are similar to the vibrations of a trumpeter's lips, the control of frequency by the air column through feedback is all but missing in the case of the vocal folds. Can you explain why? (Consider the mass of the vibrating members, the sharpness of the air resonances, and damping in each case.)

EXERCISES

1. Calculate the first three resonances of a tube 11 cm long (the approximate length of a child's vocal tract) open at one end and closed at the other. Compare these to the formant frequencies for /ɛ/ given in Table 15.3.
2. Make a graph of the second formant frequency (vertical axis) versus the first formant frequency (horizontal axis) for the ten vowel sounds given in Table 15.3. Do this for either the average male or female voice. Select a scale for each axis that is appropriate for the data you intend to graph.
3. Take a simple sentence (e.g., "You always give the right answers") and attempt to give it several meanings by changing prosodic features. For each different way of speaking the sentence, indicate the pattern of pitch and loudness used.
4. Express in newtons/meter^2 the maximum and minimum lung pressures used in speech (4 cm and 20 cm of water). Atmospheric pressure (10^5 N/m^2) corresponds to a manometer pressure of about 34 ft of water.
5. Calculate the frequencies of resonance for a tube 16 cm long closed at one end and open at the other, and show that they correspond to F_{10}, F_{20}, F_{30}, in Fig. 15.15.
6. Determine whether there is a "scaling factor" relating male and female vowel formants by the following calculations. Determine the ratios of the female-to-male for-

mant frequencies for the vowels given in Table 15.3. Find the average ratio. Could recording male speech and playing it back 16% faster make it resemble female speech?

7. Suppose a vocal tract 17 cm long were filled with helium ($v = 970$ m/s). What formant frequencies would occur in a neutral tract?

8. Estimate relative male and female vocal tract lengths by:

   (a) Averaging the ratios of a male and female formant frequencies for several vowels;

   (b) Assuming they have about the same ratio as male and female heights.

9. The resonance frequency of a Helmholz resonator (Section 2.3) is

$$f = \frac{v}{2\pi}\sqrt{\frac{A}{lV}},$$

where v is the velocity of sound, A and l are the cross-sectional area and length of the neck, and V is the volume of the resonator. For the model in Fig. 15.14, assume a neck area of 0.6 cm^2, a length of 3 cm, a volume V of 20 cm^3, and a sound velocity of 344 m/s, and calculate the resonance frequency. Compare this to F_{10}, calculated in Problem 5.

EXPERIMENTS FOR HOME, LABORATORY, AND CLASSROOM DEMONSTRATION

Home and Classroom Demonstration

1. *Formant tube*   The experiment illustrated in Fig. 15.10 can be expanded to simulate other vocal tract resonances, as shown in Fig. 15.19. A pure tone is varied over the frequency range of 200–4000 Hz, and the resonances of the tube are observed. Listening to a sawtooth waveform with a fundamental frequency of 100–150 Hz should suggest vowellike sounds as the constriction is moved (b) or tubes of different diameters and lengths are joined together (c).

2. *Movable constriction*   The effect of constricting the air flow at different places in the vocal tract can be simulated by inserting a small nozzle into a short length of pipe, as shown in Fig. 15.20. As the constriction moves up and down the pipe, the sound changes in character, growing louder when the source reaches the position of a pressure maximum for one of the pipe resonances.

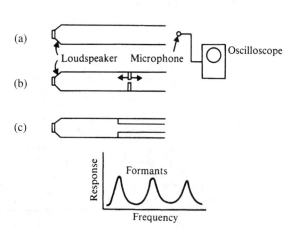

(a) Loudspeaker Microphone Oscilloscope

(b)

(c)

Response Formants Frequency

FIGURE 15.19   Formant tube, which can be used to demonstrate resonances of the vocal tract.

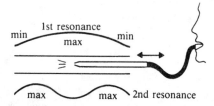

FIGURE 15.20   Demonstration to show the effect of moving a constriction up and down the vocal tract.

3. *Whispered vowels*   During a whisper, the vocal folds produce broadband ("white") noise, which contains a wide range of frequencies and virtually no sensation of pitch. Whispering vowel sounds, however, shapes the vocal tract so that bands of noise near the formant frequencies are emphasized. A rather faint sense of pitch develops, which usually corresponds to the second formant frequency (Thomas 1969).

Professor J. F. Schouten of the Netherlands was well known for his demonstrations of acoustic phenomena as well as his research in hearing in speech. He demonstrated the whispered-vowel phenomenon by whispering the following four lines of vowels to produce the well-known Westminster chime:

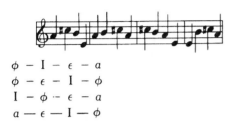

$$\phi - I - \epsilon - a$$
$$\phi - \epsilon - I - \phi$$
$$I - \phi - \epsilon - a$$
$$a - \epsilon - I - \phi$$

The second formant of ϕ (a vowel sound common in Scandinavian and Germanic languages) is around 1760 Hz, which is two octaves above the musical standard $A_4 = 440$ Hz (Schouten 1962).

4. *Single formants* The character of vowel sounds is mainly determined by the first and second formants. The waveform and spectrum /i/ are shown in Fig. 15.21(a). Filtering out one of the formants changes the vowel sound appreciably. The first formant, produced by itself, sounds much like /u/. The second formant by itself has a sharp timbre that produces no phonetic association, since the human voice is unable to produce it singly (Schouten 1962).

5. *Whistling* When one whistles, the vocal tract is excited from the end opposite the vocal folds. The normal whistling range is from about F_5 to F_7 (700 to 2800 Hz), although with practice many people can whistle down to 500 Hz and below. This suggests that the vocal tract acts somewhat like a closed cylindrical pipe (see Fig. 15.10). Whistled sound is very nearly a pure tone, with exceptionally weak overtones. Incidentally, the lowest whistled note for most people is near the pitch of their highest sung note (in falsetto for a male voice); thus, the human vocal system can emit sounds over a total range of about five octaves.

6. *Air flow in various phonemes* Hold a hand directly in front of your mouth and say "what"; note the air flow. Say "ah-h-h," and compare the air flow. Compare pairs of words beginning with voiced and unvoiced plosive consonants (e.g., to/do, pet/bet, kit/bit).

Waveform   Spectral

FIGURE 15.21
(a) Waveform of vowel /i/ and its spectral pattern showing two formants. (b) First formant only gives an /u/-like sound. (c) Second formant only. (After Schouten 1962.)

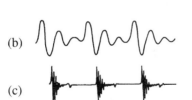
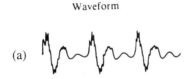
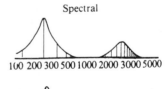
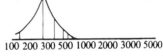
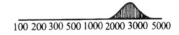

(a)
(b)
(c)

Time   Frequency (Hz)

Laboratory Experiment

Vowel formants (Experiment 22 in *Acoustics Laboratory Experiments*)

16

Speech Recognition, Analysis, and Synthesis

Speech is not just the future of Windows but the future of computing itself.
Bill Gates (quoted by *Business Week*, Feb. 23, 1998)

Our ability to recognize the sounds of language is truly phenomenal. Speech can be followed at rates as high as 400 words per minute. If we assume an average of five phonemes or individual sounds per word, this means recognizing over 30 phonemes per second; even normal conversation requires the recognition of 10 to 15 phonemes per second. In this chapter, we will consider the way in which speech recognition takes place, particularly through certain types of *cues* in the complex speech sounds we hear. Before we consider speech recognition, however, it is appropriate to discuss the acoustical analysis of speech sounds.

In this chapter you should learn:

- About speech spectrograms;
- About recognition of vowels and consonants;
- About filtered and compressed speech;
- About speech recognition and synthesis by computers;
- About speaker identification and voiceprints.

16.1 ■ THE ANALYSIS OF SPEECH

Some speech sounds change rapidly and, therefore, require special techniques for analysis. Graphs of sound level versus time and graphs of sound level versus frequency (sound spectra) are useful but inadequate. It is more useful to display the sound level as a function of both frequency and time. Various techniques have been used for creating such displays.

One way to display three variables is on a three-dimensional graph. When comparing sound level, frequency, and time, this can be approximated by making multiple graphs of sound level versus frequency, each one displaced slightly in time to create perspective. Such a three-dimensional display is illustrated in Fig. 16.1.

Sound spectra have appeared throughout this book. In Section 2.7 we introduced spectra as recipes for describing a complex vibration or sound. In Section 7.10 we discussed how spectrum analysis (or Fourier analysis) is done by spectrum analyzers. An instrument that rapidly analyzes the spectrum of sound is known as a *real-time spectrum analyzer*. Such

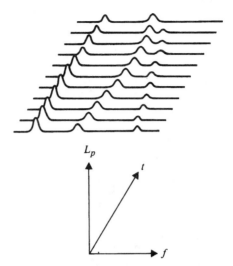

FIGURE 16.1
Three-dimensional
display of sound
level versus
frequency and time.

instruments nowadays are nearly always digital; they generally use a technique called a fast-Fourier transform (FFT) and are referred to as *FFT analyzers*. FFT analyzers are useful in producing three-dimensional ("waterfall") displays such as that shown in Fig. 16.1.

An instrument that is particularly useful for speech analysis is the *sound spectrograph*, originally developed at the Bell Laboratories around 1945. This instrument records a sound-level–frequency–time plot for a brief sample of speech on which the third dimension, sound level, is represented by the degree of blackness in a two-dimensional time-frequency graph.

A modern digital version of the sound spectrograph is shown in Fig. 16.2(b). Filters divide the incoming speech signal into many different frequency bands (from about 50 to 250, depending on the type of analysis being done). The amount of sound power that comes through each filter is measured as a function of time, and the speech spectrograph is printed on the grayscale printer shown at the right. The format printed by the digital sound spectrograph in Fig. 16.2(b) is similar to that of the older analog version (Fig. 16.2(a)), which speech scientists have found so useful over the years.

A speech spectrogram is shown in Fig. 16.3. The horizontal axis is time, and the vertical axis frequency. The vertical striations show the fundamental period of the vocal cord vibrations. Two filter bandwidths are customarily used with the instrument, 45 and 300 Hz. The broader band gives better time resolution at the expense of frequency resolution. In the analog version speech sample must be played many times to record a spectrogram, so playback is at a much higher speed than that used for recording on the magnetic disc.

16.2 ■ THE RECOGNITION OF VOWELS

Although spectrograms of vowel sounds may show four or five formants, the first two or three formants are generally sufficient to identify vowel sounds. On the other hand, experiments have shown that, under some conditions, vowels can be recognized from only the

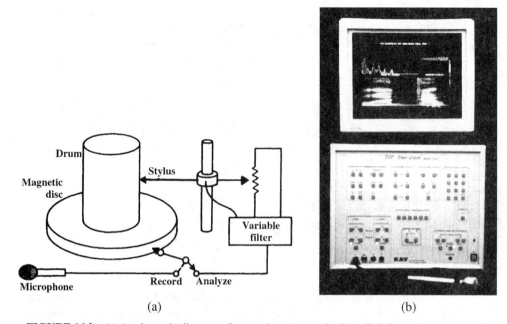

FIGURE 16.2   (a) A schematic diagram of a sound spectrograph; (b) a digital sound spectrograph.

higher formants when the lowest two formants are missing. Thus, in normal speech, there are multiple acoustic cues to aid in the recognition of vowel sounds. Some of these extra cues make it possible to determine vowel sounds even when distortion and interference are present, as the following examples illustrate.

One familiar type of distortion occurs when speech is recorded at one speed and played back at a faster speed, producing what has been called "duck talk." Even when the pitch and all formants are raised by an octave or more (by doubling the speed of a tape recorder, for example), it is possible to understand most of what is being said. Apparently our speech-

FIGURE 16.3

A speech spectrogram. The vertical axis is frequency, and the horizontal axis is time. Sound level is indicated by darkness.

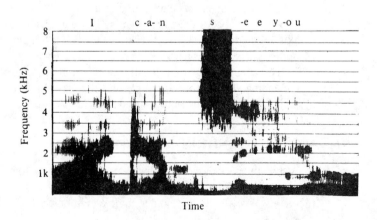

processing system is able to "scale" the entire structure of speech. That is, when we hear speech at a higher pitch, we also look for formants in a higher frequency range. How this is accomplished is not very well understood at the present time.

Another example of distortion caused by formant transposition is *helium speech*. The velocity of sound in pure helium is nearly three times greater than in air (see Table 3.1). If one takes a deep breath of helium, the resonances of the vocal tract (formants) will increase in frequency by some factor, which is typically 1.5 rather than 3, because our exhalation will contain a mixture of helium with nitrogen, carbon dioxide, etc. Speech produced under these conditions sounds quite similar to duck talk, although the nature of the distortion is quite different. Analysis of helium speech indicates that the fundamental pitch is virtually unchanged, because the mixture of gas in the vocal tract has little effect on the vibration frequency of the vocal cords. To understand helium speech, then, we must recognize the raised formants even though the pitch of the vowel sounds corresponds to the normal formants we are accustomed to hearing.

In order to prevent nitrogen narcosis, deep-sea divers breathe mixtures of helium, nitrogen, and oxygen at high pressure, and speech becomes unnatural or even unintelligible. (In Sealab II, for example, the inside pressure was maintained at 6.8 times atmospheric and the gas mixture at 80% helium, 15% nitrogen, and 5% oxygen.) Thus the problem of helium speech has attracted considerable attention, and several experimental speech processors (*formant restorers*) have been developed (Stover 1966).

The vocal tracts of young children are considerably smaller than those of adults; hence, the formant structure is considerably different. The pharynx tends to be proportionately shorter than the oral cavity, so the formant configuration is not scaled to that of an adult. Yet our speech decoder enables us to recognize a vowel spoken by a young child as being the same vowel spoken by an adult. Some of the advantages of multiple speech cues become more apparent in light of these considerations.

16.3 ■ THE RECOGNITION OF CONSONANTS

Unlike vowel sounds, which change slowly, consonant sounds change very rapidly. As we described in Section 15.4, consonants are articulated by constricting or blocking the flow of air somewhere in the vocal tract. The sound cues by which the consonant is recognized often occur in the first few milliseconds after the block is released and air is allowed to flow through the vocal tract.

Figure 16.4 shows a sound spectrogram of a simple phrase, "this is a sound spectrogram." The vowel formants appear as dark horizontal bars, whereas some of the up-and-down movements of these formants signal the consonants. The "s" sound is a burst of noise extending up to 8000 Hz. The fine vertical lines represent the vibrations of the vocal folds.

Experiments with the sound spectrograph have contributed a great deal to our understanding of the recognition of speech sounds, especially of consonants. In some experi-

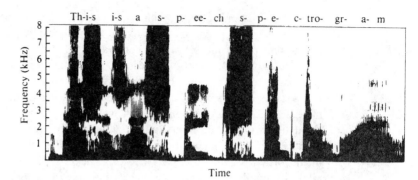

FIGURE 16.4
A sound spectrogram of a spoken phrase. The vertical axis is frequency, and the horizontal axis is time.

ments, certain features of speech are altered or eliminated to determine the intelligibility changes. Another type of experiment, however, generates speechlike sounds artificially. In such artificially synthesized speech, it is relatively easy to adjust separately the acoustic features to determine their effect on speech recognition.

Much early knowledge about the recognition of consonants was obtained with the pattern-playback machine built some years ago at the Haskins Laboratories. This machine works like a speech spectrograph in reverse: when a pattern similar to a spectrogram is fed in, it generates a sound with the designated intensity-frequency-time pattern. Arbitrary patterns may be painted on plastic belts in order to study the effects of varying the features of speech one by one.

A dot presented to the pattern-playback machine produces a "pop" that is like a plosive consonant, but difficult to recognize as any particular consonant unless it is followed by a vowel sound. In experiments by Cooper et al. (1952), listeners were presented with 15-ms noise bursts of varying frequencies, followed by two-formant vowel sounds, as shown in Fig. 16.5. High-frequency bursts were heard as /t/ for all vowels, but bursts at lower frequencies could be heard as either /p/ or /k/, depending on the vowel sound that followed. Bursts were heard as /k/ when they were on a level with or slightly above the second formant of the vowel; otherwise, they were heard as /p/.

Another way to generate plosive consonants on the pattern-playback machine is by a frequency transition in the second formant, which may be upward or downward. Transitions of the second formant of the type shown in Fig. 16.6 will produce the unvoiced plosives /t/, /p/, or /k/, depending on the vowel formants that follow. A remarkable result emerged

FIGURE 16.5
Stimulus patterns for producing /t/, /k/, and /p/ sounds on the pattern playback machine. A single burst of high-frequency noise is heard as /t/. A noise burst at the frequency of the second formant of a following vowel is heard as /k/; a noise burst below the second formant is heard as /p/. (From data in Cooper et al. 1952.)

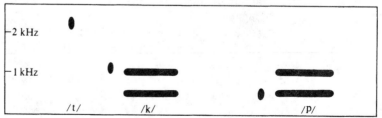

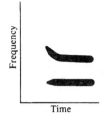

FIGURE 16.6
A formant
transition, which
may produce a /t/,
/p/, or /k/
depending on the
vowel that follows.

from experiments with the pattern-playback machine: All the second-formant transitions perceived as one particular plosive pointed back toward one particular frequency. The transitions in Fig. 16.7, which appear to originate from about 1800 Hz, are all heard as the sound /t/. Similarly, transitions that produce /p/ appear to originate from about 700 Hz, and /k/-producing transitions originate from about 3000 Hz.

The voiced plosives /b/, /d/, and /g/ have associated with their second-formant transition an upward transition in the first formant as well. The first formant is raised from a very low frequency to a level appropriate for the vowel. Figure 16.8 shows patterns that synthesize /b/, /d/, and /g/ sounds before various vowels. Note that although the first formant always moves upward, the second formant can move either upward or downward, depending on the vowel.

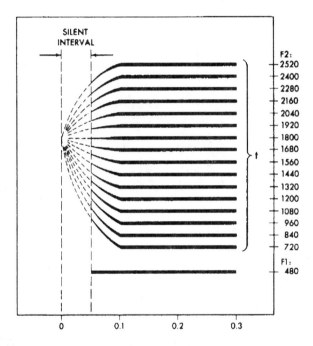

FIGURE 16.7
Second-formant
transitions
perceived as the
same plosive
consonant "t."
(After Delattre,
Liberman, and
Cooper 1955.)

/d/ Sounds

In the case of the consonant /d/, the second formants appear to originate from a "d-locus" at about 1800 Hz; the key to distinguishing the voiced /d/ from the unvoiced /t/, therefore, lies in the cue provided by the first-formant transition. It is interesting to note, however, that patterns extending all the way back to the d-focus, as in Fig. 16.9(a), do not always produce a clear /d/. In order to have a /d/ sound in every case, it is necessary to erase the first part of the transition so that it "points" at the locus but does not actually begin there, as in Fig. 16.9(b). Presumably this is the way we are accustomed to receiving these cues, and major change confuses our speech decoder.

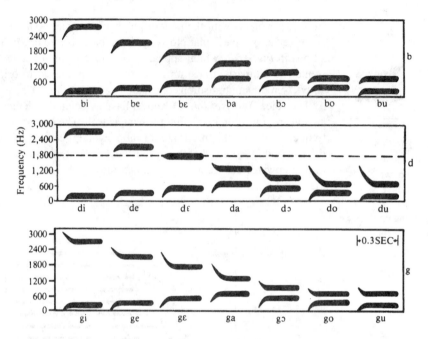

FIGURE 16.8
Spectrographic
patterns sufficient
for the synthesis of
/b/, /d/, and /g/
before vowels. The
dashed line at
1800 Hz shows the
locus for /d/. (From
Delattre, Liberman,
and Cooper 1955.)

For the liquids and semivowels /r/, /ℓ/, /w/, and /j/, the second-formant transition begins at the locus, although the exact character of the transition can vary with context.

Fricative consonants can be distinguished from all other sounds by the hissing noise of the turbulent air stream, which appears as a fuzzy area on speech spectra. We may ask, however, "What are the cues for distinguishing one fricative from another?" Experiments on both natural and synthesized speech have indicated that /s/ and /ʃ/ ("sh") are distinguished

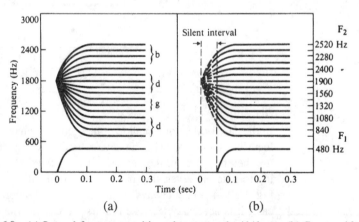

FIGURE 16.9   (a) Second-formant transitions that start at the /d/ locus. (b) Comparable transitions that merely "point" at it, as indicated by the dotted lines. Those of (a) produce syllables beginning with /b/, /d/, or /g/, depending on the frequency level of the formant; those of (b) produce only syllables beginning with /d/. (From Delattre, Liberman, and Cooper 1955.)

from other fricatives by their greater intensities and from each other by their spectra. In the case of /ʃ/, sound energy is concentrated in the 2000- to 3000-Hz range, whereas for /s/, it is above 4000 Hz. The weaker fricatives /f/ and /θ/ ("th") depend on second-formant transitions in the vowel sound that follows to provide clues about the place of articulation.

The *duration* of a sound may provide an important clue for phoneme recognition. In Chapter 15 we pointed out that a fricative appeared to change into a plosive when its duration was shortened. This can be demonstrated by tape recording a normally pronounced word such as see; if the tape is cut to reduce the duration of the initial /s/ from its normal 0.1-s duration to about 0.01 s, the word is heard as "tee."

The effects of third-formant transitions on the perception of consonants are more complicated and less understood than those of second- and first-formant transitions. For example, a third-formant transition provides a clue for the perception of /d/ in "di" but not in "du" (Liberman et al. 1967). Apparently, no simple explanation has been made of this phenomenon, although it may be noted that the second-formant transitions are in opposite directions in the two cases.

16.4 ■ FILTERED SPEECH AND NOISY ENVIRONMENTS

Filters are devices that respond selectively to certain frequencies. The vocal tract acts like a series of filters, each tuned to one of the resonances that we associate with formants; however, electrical filters can be constructed to have a much sharper frequency response than the vocal tract does. Some experiments with electrically filtered speech will be described in this section.

Electrical filters may have high-pass, low-pass, band-pass, or band-reject characteristics (see Section 18.6). A high-pass filter transmits only those frequencies above its cutoff frequency, and a low-pass filter only those frequencies below its cutoff frequency. A band-pass filter has both high and low cutoff frequencies and transmits only frequencies that lie in the band between; a band-reject, or notch, filter rejects only signals between the two cutoff frequencies.

Speech intelligibility is usually measured by *articulation tests* in which a set of words is spoken and a listener or group of listeners is asked to identify them. Articulation tests customarily use lists of specially selected words of one or two syllables. The articulation score, which is the percentage of words correctly identified, will be lower for these isolated test words, of course, than for words used in the normal context of speech.

Articulation scores for filtered speech are shown in Fig. 16.10 for both high-pass and low-pass filtering. The curves are seen to cross at 1800 Hz, where the articulation score for both is about 67%. Normal conversation, therefore, would be completely intelligible by listening only to components above 1800 Hz, or, equally so, by listening only to components below 1800 Hz. It is also possible to achieve an acceptable level of intelligibility for speech after passage through a band-pass filter with a surprisingly narrow passband. The minimum acceptable passband is found to vary with frequency; in the range around 1500 Hz, for example, a 1000-Hz band width is sufficient to give a sentence articulation score of about 90% (Denes and Pinson 1973). Using a narrowband filter ($\frac{1}{3}$-octave bandwidth), intelligibility reached 50% around 2000 Hz but was very low at most frequencies (Chari, Herman, and Danhauer 1977). Needless to say, speech quality deteriorates more

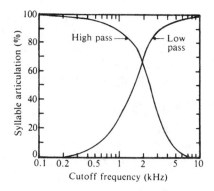

FIGURE 16.10   Intelligibility of filtered speech for different cutoff frequencies of both high-pass and low-pass filters. Note that the two curves cross at about 1800 Hz, where the articulation score is 67% for both types of filter. (After French and Steinberg 1947.)

than does intelligibility with filtering. Filtered speech sounds thin and unpleasant even when it is understandable.

The effects of waveform distortion have also been investigated. *Peak clipping* is a type of distortion that often results from overdriving an audio amplifier, but it is sometimes introduced deliberately into speech communication systems in order to reduce the bandwidth required to carry the speech. Figure 16.11 illustrates moderate and severe peak clipping of a speech waveform. Intelligibility of speech is impaired surprisingly little by peak clipping, although the quality of the speech suffers. Even after severe peak clipping, similar to that shown in Fig. 16.11(c), intelligibility remains at 50 to 90%, depending on the skill of the listener (Licklider and Pollack 1948).

The intelligibility of speech in noisy environments is a timely subject that has been studied at a number of laboratories. The degree of "masking" of speech depends on the intensity and the spectrum of the interfering noise. Using broadband noise or white noise (noise with equal intensity at every audible frequency), the intelligibility of words drops to about 50% when the average intensities of the speech and the noise are about equal. The intelligibility of sentences remains higher, however, because of linguistic and semantic cues. Figure 16.12 indicates how the thresholds of intelligibility and detectability of speech depend on the level of broadband noise.

A tone of lower frequency can mask a tone of higher frequency much more effectively than the converse (see Section 6.10). Thus narrowband noise is most effective in masking speech if its frequency is below the speech band. Potential for speech interference in a

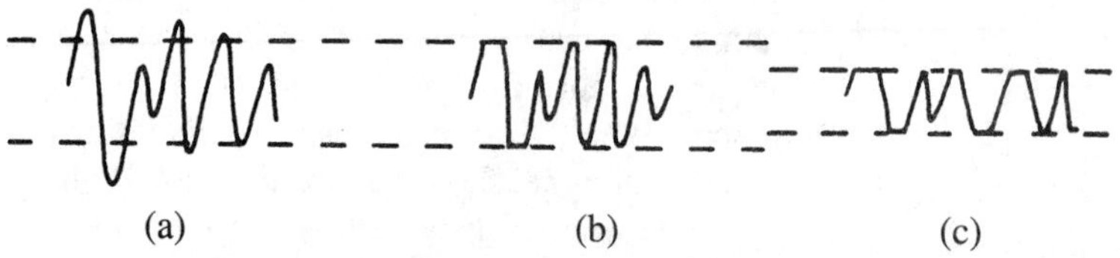

FIGURE 16.11
Peak clipping:
(a) waveform of
original speech;
(b) waveform after
peak clipping;
(c) waveform after
severe peak
clipping.

(a)   (b)   (c)

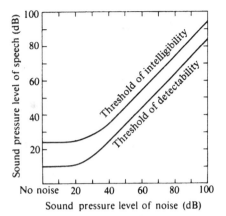

FIGURE 16.12
The thresholds of intelligibility and detectability as functions of the intensity of the masking noise. (From Hawkins and Stevens 1950.)

noisy environment is sometimes expressed by the speech-interference level, which is the average of the noise level in three appropriate frequency bands (see Section 31.5).

An interesting demonstration illustrating one property of speech interference may be performed using *elliptical speech*. As an interfering noise source rises in intensity, one of the first features of speech that is lost is the place of articulation, so "cat" becomes indistinguishable from "tat" and "bed" from "dead," etc. Elliptical speech, in which such substitutions have been made, is difficult to understand under normal conditions, but as the noise level rises, the confusion gradually fades away and linguistic and semantic cues eventually make elliptical speech more understandable.

16.5 ■ THE SYNTHESIS OF SPEECH

"If computers could speak, they could be given many useful tasks. The telephone on one's desk might then serve as a computer terminal, providing automatic access to such things as airline and hotel reservations, selective stock market quotations, inventory reports, medical data, and the balance in one's checking account" (Flanagan 1972a). Providing computers with the ability to speak has been one target of a substantial amount of research on speech synthesis.

Early efforts to imitate speech sounds resulted in various mechanical "talking machines." One such machine, invented in 1791 by Wolfgang von Kempelen of Vienna and later improved by Sir Charles Wheatstone, is shown in Fig. 16.13. A bellows supplies air to a reed, which serves as the main voice source. A leather "vocal tract" is shaped by the fingers of one hand. Consonants, including nasals, are simulated by four constricted passages controlled by the fingers of the other hand.

During his boyhood in Scotland, Alexander Graham Bell had an opportunity to see the Wheatstone reconstruction of von Kempelen's talking machine. Assisted by his father and his brother, he constructed a talking machine of his own, molding the lips, tongue, palate, pharynx, and velum in guttapercha, wood, and rubber. A larynx box of tin had vocal folds made of rubber sheet (Flanagan 1972b).

Modern talking machines use electronic rather than mechanical techniques. Computers have brought about rapid advances in speech analysis. State-of-the-art speech synthesis

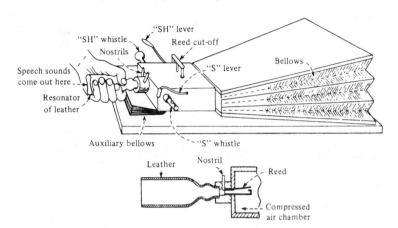

FIGURE 16.13
Wheatstone's
reconstruction of
von Kempelen's
talking machine
(Flanagan 1972b).

generally uses one of two methods: *formant synthesis* by rule or *concatenative synthesis* by computer assembly of speech from pieces of natural speech.

Formant synthesizers employ the source-filter theory of speech production (see Chapter 15, especially Section 15.5). They utilize formant patterns for each phoneme and tracks for F_0 (fundamental frequency) that specify source variations and other aspects of articulation. For example, for the word *saw*, the /s/-phoneme pattern might specify excitation of a filter with a formant at about 5000 Hz, followed by the /aw/ pattern, consisting of three filters with peak frequencies of 600, 1300, and 2500 Hz representing formants F_1, F_2, and F_3 (Bickley, Syrdal, and Schroeter 1999).

Of course there need to be rules for modifying the basic phoneme patterns to create natural connections between them. This is one of the critical problems for good synthesis. Formant synthesizers are efficient because they do not require a large storage capacity and the computational demands are relatively simple. The goal of natural-sounding speech for text-to-speech (TTS) applications continues to drive research on the refinement of formant synthesizers.

Concatenative synthesis methods result in highly intelligible and potentially very natural-sounding speech. Imagine that one recorded all desired words in all desired voices and intonations. This would lead to synthesized speech that sounds very natural, but it would also require a prohibitive amount of storage. Therefore, a compromise has to be made. The voice recordings need to be coded and compressed. There are various ways to do this. A device for coding speech is often called a *vocoder* (short for voice coder), although the term is also applied to a combined speech analyzer and synthesizer.

16.6 ■ SPEECH CODING AND COMPRESSION

Devices such as *channel vocoders* (which transmit information about the output from 16 filters plus another channels for information about unvoiced consonants and fundamental frequency) and *formant vocoders* (which transmit information about the formants themselves) have been successful, but the most popular technique for speech coding has probably been *linear predictive coding* (LPC), which describes a speech waveform in terms of a set of 10 or 12 time-varying parameters derived from analysis of speech samples (Atal and Hanauer 1971). Most "talking chips" in microcomputers make use of LPC, as did the remarkable "Speak and Spell" toy introduced by Texas Instruments in 1978 (Franz and Wiggins 1982). A detailed description of LPC is beyond the scope of this chapter. Other coding techniques include *pitch-synchronous overlap add* (Moulines and Charpentier 1990) and sinusoidal coding techniques (Dutoit 1997).

Bandwidth compression is desirable when speech is to be transmitted over long distances, such as in transoceanic cables, coaxial or optical. The efficiency of a speech-coding method can be expressed in terms of the perceptual quality of the decoded speech versus the required amount of information storage or transmission capacity to move it (the bandwidth). Rate of information is measured in kilobits per second (Kb/s). A speech waveform coded into digital form may require 64 Kb/s to preserve the naturalness and intelligibility of the original speech, although waveform compression techniques can reduce this to 16 Kb/s without noticeable loss of quality. Vocoding techniques, which attempt to preserve perceptual quality rather than waveform accuracy, can often reduce this to 8 Kb/s or even less. Good intelligibility can be obtained as low as 2 Kb/s, although naturalness is lost.

16.7 ■ SPEECH RECOGNITION BY COMPUTERS

Designing a machine that understands language is more difficult than building one that talks. Human listeners have learned to accept a wide range of speech input, including different dialects, accents, voice inflections, and even speech of rather low quality from talking computers. Machines to recognize speech have not yet reached this degree of flexibility, however. Machines that can recognize a limited vocabulary from one speaker will have difficulty recognizing the same words from a different speaker.

Speech recognition may focus either on recognizing individual words or on recognizing connected words in a phrase or sentence. A common strategy for recognizing isolated words is template matching. Templates of appropriate time-varying parameters are created for the words in the desired vocabulary as spoken by selected speakers. These same parameters in a spoken word are then compared to the stored templates, and the closest match is assumed to be the word spoken. Isolated word recognition is practical for such tasks as digit recognition, recognizing simple computer commands, and machine control, but not for general communication.

Continuous speech recognition is much more difficult than isolated word recognition, because it is difficult to recognize the beginning and end of words, syllables, and phonemes. In natural speech, articulatory gestures are made quickly, so that each is modified, to a certain extent, by its neighbors in the spoken sequence. This modification, which can be quite considerable, is known as *coarticulation*. The degree of coarticulation will depend on the rapidity of speech and the mode of speech. Its effect, in some ways, is analogous to the difference between hand-printed letters and handwriting in which the letters are modified as they are connected together.

Much research effort has been devoted to machine recognition of speech, because the potential applications are many. Voice-controlled typewriters and word processors may soon be possible. Voice programming of computers, control of machines, telephone dialing, data entry for materials handling and sorting, financial transactions, etc., while leaving the hands free for other tasks, are of obvious benefit.

16.8 ■ SPEAKER IDENTIFICATION BY SPEECH SPECTROGRAMS: VOICEPRINTS

Can one reliably identify a person by examining the spectrographic patterns of his or her speech? This is a question of considerable legal as well as scientific importance. The Technical Committee on Speech Communication of the Acoustical Society of America asked six distinguished speech scientists to review the matter from a scientific point of view a number of years ago (Bolt et al. 1970). They concluded that "the available results are inadequate to establish the reliability of voice identification by spectrograms." Speech spectrograms, or *voiceprints*, are not analogous to fingerprints, because they do not represent anatomical traits in a direct way. The article by Bolt et al. (1970) does, however, summarize methods used for speaker identification and their validity.

Speech spectrograms portray short-term variations in intensity and frequency in graphical form. Thus they give much useful information about speech articulation. When two persons speak the same word, their articulation is similar but not identical. Thus spectrograms of their speech will show similarities but also differences. However, there are also differences when the same speaker repeats a word, as can be seen in Fig. 16.14.

Our auditory system exhibits an amazing ability to identify speakers, especially if the voices are well known to us, even in the presence of substantial interference. However, wrong identifications are within the experience of all of us. Careful studies, in fact, have shown that listening provides more dependable identification of the speaker than the examination of spectrograms of the same utterances does (Stevens et al. 1968). Work is being done on the design and evaluation of methods for objective voice identification using com-

FIGURE 16.14
Four spectrograms of the spoken word *science*. The vertical scale represents frequency, the horizontal dimension is time, and darkness represents sound level. The two spectrograms at the left are by the same speaker. (Compare similar spectrograms in Bolt et al. 1970.)

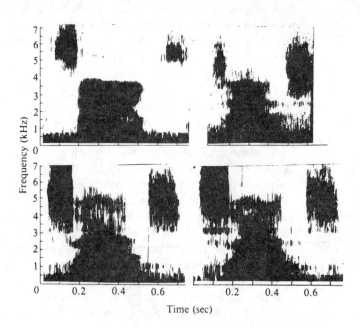

pletely automatic procedures but, at this time at least, they do not inspire great confidence in the use of voiceprints for error-free speaker identification.

16.9 ■ SUMMARY

To analyze speech, it is desirable to display sound level as a function of both frequency and time. This can be done on a three-dimensional graph or by a sound spectrograph.

The first two or three formants are usually sufficient for recognition of vowel sounds even in the presence of distortion or interference. The cues for consonant recognition often depend on the vowel sound that follows. The pattern-playback machine, which generates synthesized speech with specified features, has added much to our knowledge about consonant recognition. Filtering speech and masking speech with noise reduce intelligibility.

It is now possible to build machines that synthesize speech of acceptable quality, and machines that can recognize small vocabularies of words. Other machines can identify a speaker by his or her voiceprint, but not with a high degree of reliability. Future research and development will most likely lead to machines that can speak, understand speech, and even identify a speaker.

REFERENCES AND SUGGESTED READINGS

Atal, B. S., and S. L. Hanauer (1971). "Speech Analysis and Synthesis by Linear Predication of the Speech Wave," *J. Acoust. Soc. Am.* **50**: 637.

Bickley, C., A. Syrdal, and J. Schroeter (1999). "Speech Synthesis," in *The Acoustics of Speech Communication* by J. M. Pickett. Needham Heights, Mass.: Allyn & Bacon.

Bolt, R. H., F. S. Cooper, E. E. David, Jr., P. B. Denes, J. M. Pickett, and K. N. Stevens (1970). "Speaker Identification by Speech Spectrograms: A Scientists' View of Its Reliability for Legal Purposes," *J. Acoust. Soc. Am.* **47**: 369.

Chari, N. C., G. Herman, and J. L. Danhauer (1977). "Perception of One-Third Octave-Band Filtered Speech," *J. Acoust. Soc. Am.* **61**: 576.

Cooper, F. S., P. C. Delattre, A. M. Liberman, J. M. Borst, and L. J. Gerstman (1952). "Some Experiments on the Perception of Synthetic Speech Sounds," *J. Acoust. Soc. Am.* **24**: 597.

David, E. E., Jr., and P. B. Denes (1972). *Human Communication: A Unified View*. New York: McGraw-Hill.

Delattre, P. C., A. M. Liberman, and F. S. Cooper (1955). "Acoustic Loci and Transitional Cues for Consonants," *J. Acoust. Soc. Am.* **27**: 769.

Denes, P. B., and E. N. Pinson (1973). *The Speech Chain*. Garden City, N.Y.: Anchor/Doubleday.

Dutoit, T. (1997). *An Introduction to Text-to-Speech Synthesis*. Dordrecht: Kluwer.

Flanagan, J. L. (1972a). "The Synthesis of Speech," *Sci. Am.* **226**(2): 48.

Flanagan, J. L. (1972b). "Voices of Men and Machines," *J. Acoust. Soc. Am.* **51**: 1375.

Flanagan, J. L., and L. R. Rabiner (1973). *Speech Synthesis*. Stroudsburg, PA: Dowden, Hutchinson & Ross.

Franz, G. A., and R. H. Wiggins (1982). "Design Case History: Speak and Spell Learns to Talk," *IEEE Spectrum* **19**(4): 45.

French, N. R., and J. C. Steinberg (1947). "Factors Governing the Intelligibility of Speech Sounds," *J. Acoust. Soc. Am.* **19**: 90.

Hawkins, J. E., Jr., and S. S. Stevens (1950). "The Masking of Pure Tones and Speech by White Noise," *J. Acoust. Soc. Am.* **22**: 6.

Levinson, S. E., and M. Y. Liberman (1981). "Speech Recognition by Computer," *Sci. Am.* **244**(4): 64.

Liberman, A. M., F. S. Cooper, D. P. Shankweiler, and M. Studdert-Kennedy (1967). "Perception of the Speech Code," *Psych. Rev.* **74**: 431.

Licklider, J. C. R., and I. Pollack (1948). "Effects of Differentiation, Integration, and Infinite Peak Clipping Upon the Intelligibility of Speech," *J. Acoust. Soc. Am.* **20**: 42.

Moulines, E., and F. Charpentier (1990). "Pitch-synchronous Waveform Process Technique for Text-to-Speech Synthesis Using Diphones," *Speech Commun.* **9**, 453–467.

Pickett, J. M. (1999). *The Acoustics of Speech Communication*. Needham Heights, Mass.: Allyn & Bacon.

Rossing, T. D. (1982). *Acoustics Laboratory Experiments*. Copies from the author.

Stevens, K. N., C. E. Williams, J. P. Carbonell, and B. Woods (1968). "Speaker Authentication and Identification: A Comparison of Spectrographic and Auditory Presentation of Speech Material," *J. Acoust. Soc. Am.* **44**: 1596.

Stover, W. R. (1966). "Technique for Correcting Helium Speech Distortion," *J. Acoust. Soc. Am.* **41**: 70.

Strong, W. J., and G. R. Plitnik (1983). *Music, Speech and High-Fidelity*, 2nd ed. Provo, Utah: Soundprint.

GLOSSARY

coarticulation Modification of speech sounds when they are connected to other sounds in a spoken sequence.

concatenative synthesis Uses an inventory of natural speech pieces as building blocks from which an arbitrary utterance can be constructed.

cues Characteristics of speech sounds that help us to recognize them.

formant synthesis Employs the source-filter theory of speech production to synthesize speech.

fricative consonant Consonant that is formed by constructing air flow in the vocal tract (e.g., f, v, s, z, th, sh).

linear predictive coding (LPC) Describing a speech waveform in terms of a set of time-varying parameters derived from speech samples.

masking Obscuring of one sound by another (see Section 6.10).

peak clipping Limiting the amplitude of a waveform so that peaks in the waveform are eliminated; this distorts the waveform.

phonemes Individual units of sound that make up speech.

real-time spectrum analyzer An instrument that rapidly creates a spectrum of a sound.

sound spectrograph An instrument that displays sound level as a function of frequency and time for a brief sample of speech.

spectrogram A graph of sound level versus frequency and time as recorded on a sound spectrograph or similar instrument.

speech synthesis Creating speechlike sounds artificially.

vocoder A combined speech analyzer and synthesizer ("voice coder").

voiceprints Speech spectrograms from which a speaker's identity may be determined.

REVIEW QUESTIONS

1. What is a phoneme?
2. What three variables are plotted by a sound spectrograph?
3. What is a real-time spectrum analyzer?
4. Describe what happens when a speaker inhales helium before speaking.
5. What is a pattern-playback machine?
6. Describe the first and second formants associated with the consonant /t/.
7. Give an example of a consonant whose formants can move in different directions depending upon the vowel that follows.
8. What is a fricative consonant? Give an example.
9. Give an example of a semivowel.
10. What is "elliptical" speech?
11. What is concatenative synthesis?
12. What is linear predictive coding (LPC)?
13. What is a voiceprint?

QUESTIONS FOR THOUGHT AND DISCUSSION

1. When you fail to understand an indistinctly spoken word, is it more apt to be the initial consonant, the final consonant, or the vowel that is not recognized? Try to give reasons for your answer.

2. Discuss why a baritone voice played back at a higher speed than that at which it was recorded does not sound like a soprano.

3. Discuss the acoustics of the frequently heard phrase "he projects his voice." Think of other expressions used to describe good speaking techniques and their possible acoustical basis.

4. Would synthesized speech of high intelligibility but unnatural quality be useful in telecommunication? Would it be acceptable to most users of the telephone?

EXERCISES

1. From the spacing of the small striations in the speech spectrogram shown in Fig. 16.3, estimate the fundamental frequency of the speaker. Can you tell whether the speaker was male or female? (The duration of the spectrogram is 1.9 s.)

2. Recognition of vowels requires frequencies from about 200 to 3000 Hz, whereas recognition of certain consonants requires frequencies up to 8000 Hz. While listening to a radio newscast, quickly turn down the treble tone control, and note which consonants are the most difficult to identify.

3. From Fig. 16.8 estimate the frequency change in the first and second formants during articulation of "di," "da," and "du." Estimate also the time over which these formant shifts take place.

4. Time your own speech rates when speaking normally and when speaking as fast as you can. Then count the number of words in a particular paragraph, and time yourself as you read it at each of these rates.

5. Listen to speech recorded at one rate and then played back at both a faster and a slower rate. Describe the speech you hear (quality, pitch, intelligibility, etc.).

EXPERIMENTS FOR HOME, LABORATORY, AND CLASSROOM DEMONSTRATION

Home and Classroom Demonstration

1. *Vowel formants* Formants of the cardinal vowel sounds spoken by volunteers can be determined by means of a sound spectrograph or a real-time (FFT) spectrum analyzer. If the latter is used, it is best to sum up a number of words or syllables that use the vowel of interest by means of the signal averager.

2. *Changing the fundamental and formant frequencies* Tape record speech at one speed and play it back both faster and slower than recorded. The sounds are still recognizable, because the pitch and the formant frequencies have been changed by the same ratio. The quality has suffered, however.

3. *Changing the fundamental or formant frequencies* If possible, obtain tapes of synthesized speech in which the pitch and formant frequencies are scaled abnormally. Compare to the results obtained in Experiment 2.

4. *Helium speech* Take a deep breath of helium and speak. The resulting speech distortion results from a change in formant frequencies, although the pitch (determined by the vocal-fold vibration frequency) remains the same. *Be careful to replace the oxygen in your lungs as soon as possible.*

5. *Filtered speech* Record a short phrase and play it back through a variety of filters (high-pass, low-pass, octave band-pass, one-third octave band-pass). Compare the relative intelligibilities.

6. *Speech synthesis with a personal computer* A number of speech synthesis programs are available. Compare synthesized vowels with spoken vowels by listening and by real-time spectrum analysis.

7. *Speak and Spell* Texas Instruments' Speak and Spell is an expensive device for demonstrating speech synthesis.

8. *Speaker Identification*   Record spectrograms (or real-time spectra) of the same two sentences or phrases spoken by several volunteers. Can you recognize which ones are spoken by the same person?

9. *Out of Context*   Observe how difficult it is to understand when someone abruptly changes the topic of conversation.

10. *Visual Cues*   The video "Speech Perception" (Acoustical Society of America, Melville, NY, 1997) includes a demonstration showing how one hears a different consonant when listening with the eyes closed (no visual cue) and open.

Laboratory Experiments

Speech sounds: The sound spectrograph (Experiment 23 in *Acoustics Laboratory Experiments*)

Synthesis of vowel sounds (Experiment 24 in *Acoustics Laboratory Experiments*)

Spectra of front vowels (Lab 5 in Pickett 1999)

Spectra of back vowels and diphthongs (Lab 6 in Pickett 1999)

Nasals and glides (Lab 9 in Pickett 1999)

Fricative/stop distinction (Lab 10 in Pickett 1999)

Voiced/unvoiced distinction (Lab 11 in Pickett 1999)

Coarticulation effects (Lab 12 in Pickett 1999)

CHAPTER
17

Singing

It is somewhat ironic that the oldest musical instrument of all, the human voice, is less well understood than the various instruments we discussed in Part III. This is certainly due, in part, to the inaccessibility of its various components within the human body. Studying the human voice might be likened to studying the violin without being allowed to open the case or, at best, to hearing it played from behind an opaque screen with a small hole through which to peek.

In this chapter you should learn:

- About formants in the singing voice;
- About factors influencing the spectra of sung notes;
- About breathing and air flow;
- About registers in singing;
- About the acoustics of choir singing.

The vocal organ, as shown schematically in Fig 15.1, consists of the *lungs*, the *larynx*, the *pharynx*, the *nose*, and the *mouth*. Air from the lungs is forces through the *glottis*, a V-shaped opening between the vocal cords or folds, causing them to vibrate and thus modulate the flow of air through the larynx. The output from the vocal folds is characterized as a buzz (a nearly triangular waveform), rich in harmonics that diminish at a rate of about 12 dB per octave (see Fig. 15.7).

The vocal tract, which consists of the larynx, the pharynx, the oral cavity, and the nasal cavity, acts as a filter-resonator to transform this buzz into musical sound, somewhat in the manner of the tubing of a trumpet or oboe (but without a large amount of feedback to the source). Unlike the horns of the orchestra, however, the vocal tract creates its formants (resonances) mainly by changing its cross-sectional area at various points of articulation along its length. (The vocal tract was discussed in Section 15.3.)

17.1 ■ FORMANTS AND PITCH

In both speech and singing, there is a division of labor, so to speak, between the vocal folds and the vocal tract. The vocal folds or cords control the pitch of the sound, whereas the vocal tract determines the vowel sounds through its formants and also articulates the consonants. The pitch and the formant frequencies are virtually independent of each other in speech, but trained singers (especially sopranos) sometimes tune their vowel formants

FIGURE 17.1
Typical formants of male (♩) and female (♪) speakers represented on a musical staff. (Compare with Table 15.3.)

/i/ee   /u/oo   /I/it   /ʊ/full   /ɜ/er   /ɛ/e   /ɔ/aw   /ʌ/uh   /æ/aa   /ɑ/ah

to match one or more harmonics of the sung pitch. The loudness and timbre of the sung sound depend on both the vocal folds and the vocal tract.

Figure 15.16 shows the vocal tract profiles for 12 English vowels, and Table 15.3 tabulates typical formant frequencies for ten of them as spoken by both male and female voices. In Fig. 17.1, these same data are presented on a musical staff for the reader who is more familiar with this notation. It may be surprising to learn that although female voices are pitched about an octave higher than male voices, the formants usually differ by less than a musical third (less than 25% in frequency).

In Table 15.3 the relative formant amplitudes are given. For most spoken vowels, the second formant is considerably weaker than the first; the "ah" and "aw" sounds have the strongest second formants. We have added dynamic markings (pp, p, mp, mf) to Fig. 17.1 to indicate the relative strengths of the second formant. Although the first and second formants contribute almost equally to vowel sounds (the third formant contributes slightly less), the first formant will usually contribute more to timbre because of its greater amplitude and lower frequency, closer to the fundamental.

Note the position of the formants in Fig. 17.1 in relation to the singing range. In the case of the bass or baritone singer, the fundamental rarely is enhanced by a formant resonance (exceptions are "ee" and "oo" sounds near the top of the singing range). In most cases, the formants enhance higher harmonics of the fundamental; for example, if a bass sings "ah" with a pitch G_2 ($f = 98$ Hz), the first formant gives its greatest boost to the seventh harmonic, the second formant boosts harmonics around the eleventh, and the third formant gives a smaller (but important because of the frequency range in which it lies) boost to the 24th and 25th harmonics and their neighbors. The pitch, of course, remains at G_2, because the overtones are harmonics of this "almost-missing" fundamental. A few people have learned to shape their mouths in such a way that harmonics of a sung pitch can be made audible (see demonstration experiment on Single Formants in Chapter 15).

Another way to present formant frequencies of vowel sounds is shown in Fig. 17.2. Frequencies of the first formant are plotted on the horizontal scale, and those of the second formant on the vertical scale. The egg-shaped regions represent the approximate limits of formant frequencies that the ear will recognize as a given vowel. Note that there is overlap; that is, certain sounds can be interpreted as more than one vowel, depending on the context.

17.2 ■ DIFFERENCES BETWEEN SPOKEN AND SUNG VOWELS

Sung vowels are fundamentally the same as spoken vowels, although singers do make vowel modifications in order to improve the musical tone, especially in their high range.

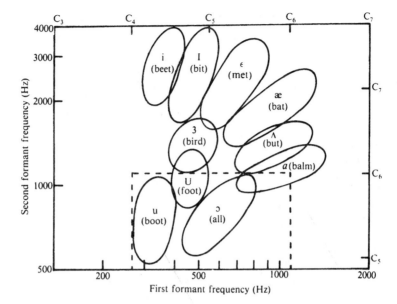

FIGURE 17.2
Frequency of first and second formants for 10 vowels. The dotted line shows the approximate range of a soprano voice. (After Peterson and Barney 1952.)

For example, "ee" is often sung like the umlauted "ü" of the German "für," and the short "e" of bed sounds more like the vowel sound in herd.

Analysis of the individual vowel formants reveals changes that may be substantial. Figure 17.3 shows spectra of the vowel /æ/ (as in bat) spoken and sung by a professional bass-baritone singer. Note that the first formant is virtually unchanged, but the second formant is lower in frequency in the sung vowel. The third and fourth formants remain at about the same pitch but are markedly stronger in the sung vowel.

Four articulatory differences between spoken and sung vowels were noted by Sundberg (1974) as a result of studying X-ray pictures of the vocal tract and photographs of the lip openings. In singing,

1. The larynx is lowered;
2. The jaw opening is larger;

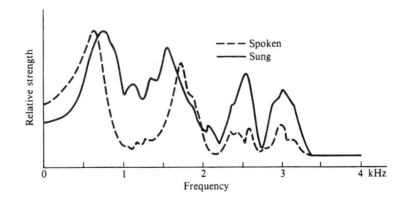

FIGURE 17.3
Spectra of vowel sound /æ/ as spoken and sung by a professional singer.

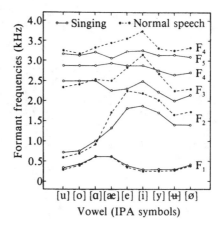

FIGURE 17.4
Formant
frequencies of long
Swedish vowels in
normal male speech
(dashed lines) and
in professional
male singing (solid
lines). (From
Sundberg 1974.)

3. The tongue tip is advanced in the back vowels /u/, /o/, and /ɑ/; and

4. The lips are protruded in the front vowels.

Formant frequencies of nine sung and spoken vowels are shown in Fig. 17.4. Frequencies of the lower two formants do not change very much, although the second formant is slightly lower in singing the front vowels. The third and fourth formants are substantially lower when sung, and a fifth formant is now apparent.

Trained singers, especially male opera singers, show a strong formant somewhere around 2500–3000 Hz. This "singer's formant," which seems to be more or less independent of the particular vowel and the pitch, usually lies between the third and fourth formants and adds brilliance and carrying power to the male singing voice. It is interesting to note that the frequency of this formant is near the resonance frequency of the ear canal, which gives it an additional auditory boost. A formant of 3000 Hz is evident in the spectrograms shown in Fig. 17.5.

Sundberg (1974) attributes the singer's formant to a lowered larynx, which, along with a widened pharynx, forms an additional resonance cavity (about 2 cm long) with a frequency in the range of from 2500 to 3000 Hz. Lowering the larynx also produces the darker vowel sounds favored by most singers. The larynx, which is lowered as much as 30 mm during

FIGURE 17.5
Spectrogram of
vowels /i/, /u/, and
/ɑ/ (ee, oo, ah). The
pitch is E₃
($f = 165$ Hz).
Note the strong
formant at 3 kHz
for all three vowels.
(From van den Berg
and Vennard 1959.)

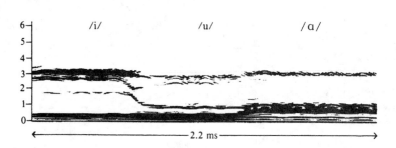

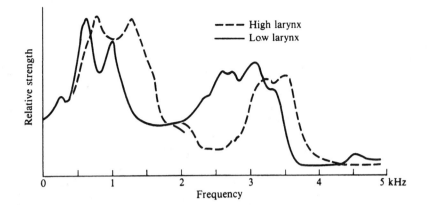

FIGURE 17.6
Spectrum of vowel
sound /ɑ/ sung with
a high and a low
larynx.

swallowing, may be lowered up to 15 mm during singing (Shipp 1977). Untrained singers
tend to raise their larynxes as they raise the pitch.

Figure 17.6 shows spectra of the vowel sound /ɑ/ ("ah") sung with both a high and a
low larynx by a professional bass-baritone singer in our laboratory. The broad resonance
extending from 2500 to 3000 Hz in the spectrum of the low larynx is a blend of the third
vowel formant and the singer's formant.

Because the singer's formant requires a widened pharynx ("open throat"), it is characteristic of good singing in the chest register (see Section 17.4). Professional contraltos
usually have such a formant, but sopranos, who sing mainly in the head register, may not. It
is not usually present in the falsetto voice of the male singer, either. Figure 17.7 shows how

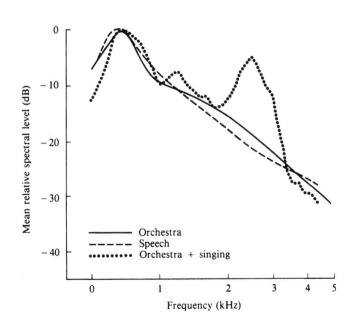

FIGURE 17.7
Idealized average
spectra of normal
speech and
orchestra music.
The dotted curve
shows the average
spectrum of
Jussi Björling
singing with a loud
orchestra
accompaniment.
(From Sundberg
1977a.)

TABLE 17.1   Formant frequencies of basic sung vowels

Formant frequency (Hz)		/i/ (ee)	/I/ (i)	/ɛ/ (e)	/æ/ (aa)	/ɑ/ (ah)	/ɔ/ (aw)	/U/ (ù)	/u/ (oo)	/ʌ/ (u)	/ɝ/ (er)
F_1	M	300	375	530	620	700	610	400	350	500	400
	W	400	475	550	600	700	625	425	400	550	450
F_2	M	1950	1810	1500	1490	1200	1000	720	640	1200	1150
	W	2250	2100	1750	1650	1300	1240	900	800	1300	1350
F_3	M	2750	2500	2500	2250	2600	2600	2500	2550	2675	2500
	W	3300	3450	3250	3000	3250	3250	3375	3250	3250	3050

Source: Appelman (1967).

the singer's formant in the voice of operatic tenor Jussi Björling helped him "cut through" a large orchestra.

It is obvious from Fig. 17.2 that the formant frequencies of different speakers (and singers) may vary rather widely, yet still result in understandable vowel sounds. Furthermore, in certain ranges of singing, the vowel formants change substantially from their normal frequencies. Nevertheless, it is instructive to compare the formant frequencies of typical *sung* vowels given in Table 17.1 to the corresponding formant frequencies of the *spoken* vowels given in Table 15.3.

Formant changes that occur throughout the singing range may be roughly described as the gradual substitution of one vowel sound for another. As the pitch rises, for example, many singers find it convenient to make the following substitutions (Appelman 1967):

Normal range   /i/   /ɛ/   /æ/   /ɑ/   /ɔ/   /u/
High range   /I/   /ɑ/   /ɑ/   /ʌ/   /ʌ/   /U/

Some voice teachers recommend the substitution of a long (closed) vowel for a short (open) vowel as the pitch rises and the converse for a downward crescendo (Haasemann and Jordan 1991). The vowel sounds /I/, /u/, /ʌ/, and /æ/, which do not change substantially with pitch, are termed *stable*.

17.3 ■ FORMANT TUNING BY SOPRANOS

In low voices, the various formants of the local tract emphasize various harmonics of the source sound from the glottis, as we discussed in Section 17.1. Sopranos, however, do much of their singing in a range in which the pitch exceeds the frequency of the first formant. Thus they would not receive the benefit of a boost from formant resonance, and their tones would suffer in quality and loudness. Experienced sopranos have learned how to "tune" their formants over a reasonable range of frequency in order to make a formant coincide with the fundamental or one of the overtones of the note being sung.

For example, a soprano singing /i/ ("ee") at a pitch of F_5 (698 Hz) might find her normal first formant at 310 Hz, more than an octave below the sung pitch. She would receive little support from this formant. However, if she opens her lips somewhat wider than the normal position for speaking /i/, the formant can be pushed up to the vicinity of the sung pitch. Or if she were singing /ɑ/ ("ah") at a pitch of A_4 (440 Hz), she would find her normal

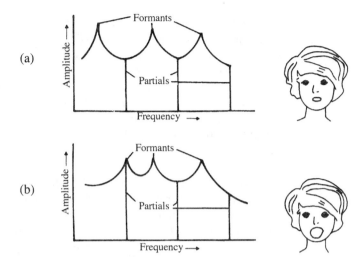

FIGURE 17.8
Formant tuning by wider jaw opening: (a) normal first formant lies below the sung pitch; (b) first formant raised to coincide with sung pitch. (From Sundberg 1977a.)

first formant around 700 Hz, between the fundamental (440 Hz) and the second harmonic (880 Hz) of the sung note. She would probably find it more convenient, in this case, to raise the formant to the vicinity of the second harmonic in order to provide the needed boost. Figure 17.8 shows how formant tuning can be accomplished by increasing the jaw opening to change the shape of the vocal tract.

Figure 17.9, also based on the work of Sundberg, shows the extent to which formant tuning can take place to match one of the harmonics of various sung pitches. The numbered

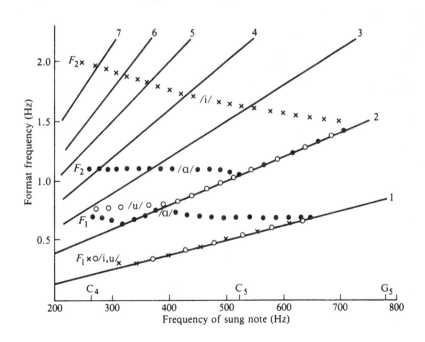

FIGURE 17.9
The tuning of formants to match harmonics of the sung note. F_1 and F_2 are the lowest formants of vowels /i/, /ɑ/, and /u/. The solid lines are the first seven harmonics of the sung note. (After Sundberg 1975.)

lines represent the harmonics of the pitch. Note that formants are usually tuned upward, although downward tuning is also possible.

Formant tuning might be expected to produce objectionable distortion of vowel sounds, but this does not seem to be the case. We are accustomed to recognizing vowels produced at various pitches in the speech of men, women, and children (see Table 15.3) with vocal tracts of different lengths. If the pitch is high, we associate it with relatively high formant frequencies. Recording a vowel sound at one speed and playing it back at another may change it to another vowel sound because the same ratio of f_2/f_1 in the new pitch range is interpreted differently. An "ah" changes to an "oh" when played at half speed (Benade 1976).

Near the top of the soprano range, where formant tuning is particularly marked, it is difficult to distinguish one vowel sound from another. (Try listening to a soprano sing various vowels at a high pitch, and see if you can recognize them.) Composers are quite aware of this difficulty in vowel recognition and generally do not present important text at the top of a soprano's range (if they must, they generally repeat the text at a lower pitch).

17.4 ■ BREATHING AND AIR FLOW

Most singers attach quite a measure of importance to good breathing habits. This may be somewhat of a mystery, however, because the only thing required from the breathing mechanism is that it supply air to the larynx at the desired *subglottal pressure*. Different singers appear to use different muscular strategies to accomplish this. For example, some singers sing with the abdominal wall expanded ("belly out") and some with it contracted ("belly in"). Likewise, different singers use their diaphragms differently during singing. In order to understand the different strategies of breathing, we briefly consider how the human respiratory system operates.

The lungs are spongy, elastic structures; they act somewhat like toy balloons. When inflated, they exert a passive expiratory force that increases with the amount of air inhaled. After maximum inhalation, the pressure is around 2 kPa (equivalent to 10 cm of water gage), which is about one-fiftieth of atmospheric pressure.

FIGURE 17.10
Schematic of the breathing apparatus.
(a) Expansion of the chest cavity by the external intercostals, along with a flattening of the diaphragm, reduces the pressure and causes the lungs to expand and fill with air.
(b) Contraction of the chest cavity by the internal intercostals, along with raising the diaphragm, causes the lungs to contract.

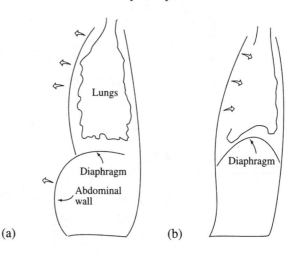

(a)   (b)

Because the lungs have no muscles of their own, breathing is accomplished by changing the size of the chest cavity. There are two basic mechanisms for doing this:

1. Downward movement of the diaphragm to lengthen the chest cavity;
2. Elevation of the ribs to increase the front-to-back thickness of the chest cavity.

Normal quiet breathing is accomplished almost entirely by movement of the diaphragm. During maximum breathing, however, increasing the thickness of the chest cavity may account for up to half of the chest cavity enlargement.

Breathing is handled by two major muscle groups. In one group are the external and internal intercostals that expand and contract the rib cage. The second group, which includes the muscles in the abdominal wall and the diaphragm, changes the abdominal cavity, as shown in Fig. 17.10. The rib cage is an elastic system that is expanded and contracted by the *intercostal muscles* that join the ribs. Muscles can only contract, and they frequently work in opposing pairs (such as the biceps and triceps in the upper arm). The *external* (inspiratory) intercostals function so that a contraction leads to an increase of the rib cage volume, whereas the *internal* (expiratory) intercostals decrease the rib cage volume when they contract. The diaphragm and abdominal muscles also work as an opposing pair, with the diaphragm acting as an inspiratory muscle, and the abdominal muscles are expiratory in nature.

The diaphragm is an important breathing muscle. When relaxed, it assumes a shape like an upside-down bowl; when it contracts, it flattens into a plate, as shown in Fig. 17.10(a). This action also increases the volume of the rib cage, and helps to draw air into the lungs. The diaphragm is a muscle specifically for inhalation. The volume of the abdominal cavity cannot be easily altered, so flattening the diaphragm tends to push the abdominal wall outward. Outward movement of the abdominal wall is, therefore, a visible indication of diaphragm contraction. Conversely, when the abdominal muscle is contracted, the abdominal wall moves inward, and the diaphragm returns to its bowl shape. Thus breathing can be mainly by the use of the intercostal muscles, mainly by the use of the diaphragm, or a combination of both.

The muscular activity required for maintaining the desired subglottic pressure is dependent on the lung volume, because the passive elastic forces of the lungs and the rib cage tend to raise or lower the air pressure inside the lungs, depending on whether the lung volume is greater or less than the *functional residual capacity* (FRC), the volume of air in the lungs at the end of a quiet expiration. When the lungs are filled with a large quantity of air, the passive exhalation force is large, and thus the pressure tends to rise. If this pressure is too high for the intended phonation, it can be reduced by a contraction of the diaphragm and/or the inspiratory intercostals. When the lung volume drops below the FRC due to singing a long phrase, the subglottal pressure will tend to drop, and it can be brought back up to the desired level by contracting the abdominal wall muscle and/or the expiratory intercostals.

Figure 17.11 shows how the total lung capacity is divided into four different volumes:

1. The *tidal volume* is the volume of air moved in and out during normal breathing (about 500 cm^3 in the normal male adult).

FIGURE 17.11
Lung capacity in
the normal young
adult and its
subdivision into
functioning
volumes. The
volume of the male
lung is indicated at
the left, female at
the right.

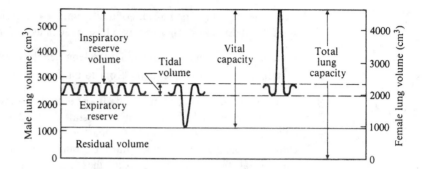

2. The *inspiratory reserve volume* is the volume that can be inspired beyond the normal tidal volume (about 3000 cm^3).

3. The *expiratory reserve volume* is the volume that can be expired by forceful effort at the end of normal tidal expiration (about 1100 cm^3).

4. The *residual volume* is the volume of air that remains in the lung after forceful expiration (about 1200 cm^3).

Corresponding volumes in the female lung average about 20 to 25% less than those given for the male lung.

Vital capacity is the amount of air that can be moved in and out of the lung with maximum effort. The average vital capacity in the young adult male is about 4600 cm^3 and in the young adult female is about 3100 cm^3. Pathological conditions such as tuberculosis, emphysema, chronic asthma, lung cancer, bronchitis, and pleurisy can greatly decrease vital capacity. At a normal breathing rate of 10 breaths per minute, about 5000 cm^3 of air will be moved in and out of the lungs per minute. A young male adult can breathe at a rate as high as 2500 cm^3/s for a short period of strenuous exercise.

17.5 ■ SUBGLOTTAL PRESSURES IN SINGING

Few measurements of subglottal pressure have been published due to the difficulty of measuring it. The most reliable results are obtained by inserting a thin needle into the trachea through the tissues below the cricoid cartilage, and a few singers have submitted themselves to this procedure. There is, however, an indirect method of measurement. When the lips are closed and the glottis is open, the subglottal pressure is equal to the pressure in the mouth cavity. Therefore, the subglottal pressure can normally be determined from the oral pressure during the production of the consonant /p/ (Sundberg 1987).

In order to increase the sound pressure level, it is necessary for the singer to increase the subglottal pressure. This is illustrated in Fig. 17.12, which shows subglottal pressure for a tenor who sang a chromatic scale at piano, mezzoforte, and forte levels. Note that the subglottal pressure increases with phonation frequency as well as with loudness.

During normal quiet breathing, the air pressure in the lungs will be about 100 N/m^2 (1 cm H$_2$O) above and below atmospheric pressure. During maximum expiratory effort

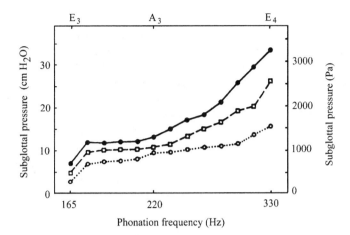

FIGURE 17.12
Subglottal pressure in a tenor who sang a chromatic scale between E$_3$ and E$_4$ at f, mf, and f levels. The pressure increases for increasing level and also with increasing frequency. (After Cleveland and Sundberg 1983.)

with the glottis closed, a pressure of 10,000 N/m^2 is possible in the strong healthy male lung. Fortissimo singing requires a pressure of 3000 to 4000 N/m^2 (compare woodwind blowing pressures given in Fig. 12.22). Figure 17.13 shows the subglottal air pressure (essentially equal to lung pressure) required for different sound levels in the case of four singers.

The relationship between sound level and subglottal pressure, shown in Fig. 17.13, tends to be the same for trained and untrained singers. This is not true of the relationship between sound level and air-flow rate, however. In trained singers, the pressure and rate of air flow tend to increase together with sound level, so a trained singer can sustain a soft tone for a long time. Many untrained singers, however, require a fairly large air-flow rate in order to sing softly, the flow rate reaching a minimum for mf or f dynamic. Flow rates range from about 100 to 400 cm^3/s at different dynamic levels of singing (Bouhuys et al. 1968).

One recent set of experiments (Leanderson, Sundberg, and von Euler 1987) set about determining the role of diaphragm activity during singing (some singers use their diaphragms

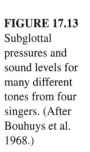

FIGURE 17.13
Subglottal pressures and sound levels for many different tones from four singers. (After Bouhuys et al. 1968.)

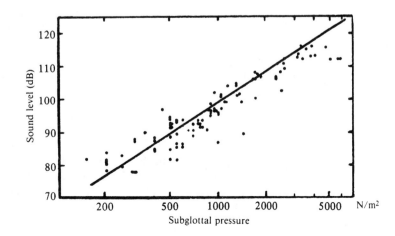

only during inspiration, others contract them during the entire phrase). By simultaneously monitoring pressure across the diaphragm, sound pressure level, and air flow, it was found that the flow rate tended to be higher when the diaphragm was activated, although there were substantial differences in diaphragm use between individual singers. It is probably safe to conclude that use of the diaphragm is not the key to good singing.

Although it would appear that all a singer requires from his or her breathing apparatus is to maintain a stable subglottic pressure, the emphasis put on breathing technique by many singing teachers suggests that the manner of breathing is of some importance.

Even though the way in which the vocal folds vibrate at a given subglottal pressure is determined by the laryngeal musculature, there appear to be some ties between the musculature used for breathing and that used for phonation. The way in which the subglottal pressure is controlled by the respiratory muscle system may generate reflexes that affect the laryngeal muscles.

In normal speech the passive expiratory recoil forces tend to be more important in establishing the desired subglottal pressure, whereas in singing, active muscles appear to be more important (Sundberg 1987). Learning to control these muscles is an important part of learning to sing.

17.6 ■ REGISTERS, VOICES, AND MUSCLES

We discussed the larynx and vocal cords and their functions in speech in Section 15.2. Although their functions are essentially the same in singing, there are a few additional features, such as muscular action, which become important when analyzing the singing voice.

The principal muscles internal to the larynx are the thyroarytenoids, the cricothyroids, and the cricoarytenoids (the names indicate which two cartilages they connect). The *cricoarytenoid muscles* operate the arytenoid cartilages to which the posterior ends of the vocal folds or vocal cords are attached, as shown in Fig. 15.3.

The *cricothyroids* connect the two large cartilages of the larynx, the thyroid, and the cricoid (see Fig. 15.2). They can pull the thyroid forward, with respect to the cricoid, and also downward, closer to it. Both of these actions stretch the vocal folds longitudinally, which is one way of increasing their rate of vibration. The action of the cricothyroid muscles can be observed in two ways: One is to press inward on the Adam's apple while singing a note in midrange. Sudden release of the pressure will cause the pitch to go up. A second experiment consists of placing a finger in the small space between the thyroid and cricoid cartilages while singing. Raising the pitch an octave will force the finger outward as the thyroid is pulled down closer to the cricoid.

The *thyroarytenoids*, also called the vocalis or vocal muscles, form the body of the vocal folds themselves. They extend from the notch of the thyroid to the arytenoid cartilages at the rear and are covered with a membrane that is continuous with the lining of the rest of the larynx. The tension on the vocal folds is a complex balance of forces from all three muscles, and coordination between them is necessary for smooth transition from one pitch to another. In order to hold a steady pitch during a crescendo or diminuendo, these muscles must compensate for the tendency of pitch to rise as the velocity of air flow is increased (due to the Bernoulli force; see Section 11.4).

A good description of the singing voice requires the expression of at least three quantities: fundamental frequency, amplitude, and spectrum (which are closely related, as we know from Chapters 5–7, to the perceived qualities of pitch, loudness, and timbre). The fundamental frequency of the vocal folds is controlled mainly by the laryngeal muscles we have just discussed. The amplitude (loudness) is controlled mainly by the subglottic pressure (which, in turn, is controlled by the respiratory muscles, as discussed in Sections 17.4 and 17.5). The timbre of the singing voice also depends upon the nature of the vocal-fold vibrations and thus depends upon both the laryngeal and the respiratory muscles.

Most descriptions of the way in which these muscles are used in singing include the term *register*. Unfortunately, there is no universally accepted definition of this term. The most common description is that a register is a frequency range in which all tones are perceived as being produced in a similar way and possess a similar voice timbre. According to Holien (1974), "a vocal register is a totally laryngeal event; it consists of a series or a range of consecutive voice frequencies which can be produced with nearly identical phonatory quality." Sundberg's (1987) book on *The Science of the Singing Voice* includes an excellent discussion of registers, and Miller (2000) has written an entire book on the subject.

Much has been said about the various registers used in singing. An idealistic approach is *one register*. The voice, if possible, should produce all the pitches of which it is capable without breaks or radical changes in technique. Some teachers feel that the best way to make this ideal come true is to assume that it *is* true, that it *can* be accomplished.

A more realistic approach is *three registers*. These correspond to the differences in tone caused by different adjustments of the larynx. The registers go by various names, but the most common are *chest*, *middle*, and *head* (in male voices, they are sometimes labeled *chest*, *head*, and *falsetto*). The famous teacher Mathilde Marchesi was obviously a proponent of this approach as she wrote, "I most emphatically maintain that the female voice possesses *three* registers, and not *two*, and I strongly impress upon my pupils this undeniable fact, which, moreover, their own experience teaches them after a few lessons." According to Marchesi (1970), the highest note in the chest register is about E_4 to F_4 for sopranos and F_4 to $F_4^\sharp$ for mezzo-sopranos and contraltos. The highest note in the middle register is about F_5. This is in agreement with the register ranges shown in Fig. 17.14.

The third approach is *two registers*, which considers that every voice has a potential of roughly two octaves of "heavy" mechanism, with about one octave of overlap. This middle octave can be sung in either laryngeal adjustment, and it is possible to combine some of the best qualities of both. Basses and contraltos sing almost exclusively with the heavy mechanism, mixing in just a bit of the light mechanism at the top of their range. However, the light mechanism is never used exclusively except for comic effects. Lyric

FIGURE 17.14
Ranges of three registers (according to Mackworth-Young, 1953). Half notes represent the male voice, quarter notes represent the female voice. Male voices use head register when singing falsetto.

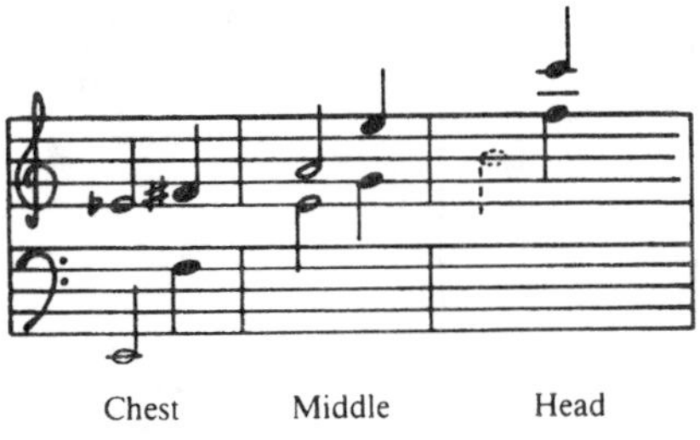

Chest   Middle   Head

and coloratura sopranos, on the other hand, sing with the light mechanism, mixing in just a little of the heavy at the bottom of their range, but never singing in a pure chest voice (Vennard 1967).

Because of the confusion associated with the use of the term register, it is preferable to refer to the two modes of vocal fold vibration as two mechanisms, heavy and light. We will call them *chest* (modal) *voice* and *head* (falsetto) *voice*. The distinguishing feature seems to be in the state of the thyroarytenoid muscles. In the heavy or chest voice, these muscles are active; in the light or head voice, they are virtually passive. An analogy can be drawn between these two modes of vocal fold vibration and the vibrations of the lips of a trumpet player (with active muscles) as opposed to the (passive) vibrations of a clarinet reed.

In the chest voice, the thyroarytenoids or vocalis muscles are active and hence shortened. At the lowest tones, the muscles are relaxed and the vocal folds are thick. Because of their thickness, the glottis closes firmly and remains closed an appreciable part of each cycle of vibration, as it does during speech (see Fig. 15.5).

As the pitch rises in chest voice, the cricothyroid muscles contract and apply tension to the vocal folds. The folds do not elongate rapidly, however, because the thyroarytenoid muscles come into action, and indeed thicken the vocal folds as they do so. At the top notes of the chest voice, the thyroarytenoids of the inexperienced singer sometimes give way to excessive force from the cricothyroids, and the voice "cracks" into an involuntary head tone (Vennard 1967).

In the light mechanism or head voice, the thyroarytenoids offer little resistance to the cricothyroids, which can then apply substantial longitudinal tension to the vocal folds, thus elongating them and making them thin. The vocalis muscles fall to the sides, and the vibration takes place almost entirely in the ligaments with much less amplitude of movement than in the chest voice. The glottis closes only briefly, or not at all, and the resulting sound has fewer harmonics than the chest voice does. According to studies by van den Berg (1968) on isolated larynxes, elongations of 30% are typical, as shown in the graph of stress versus strain in Fig. 17.15. Stress is the force applied per unit area and strain is the percentage by which the length increases.

The manner in which the vocal folds vibrate in the chest voice and in the head voice is shown schematically in Fig. 17.16.

FIGURE 17.15
Stress-strain graph for vocal ligaments under passive tension (similar to head voice). The horizontal axis represents applied stress of force. (From van den Berg 1968.)

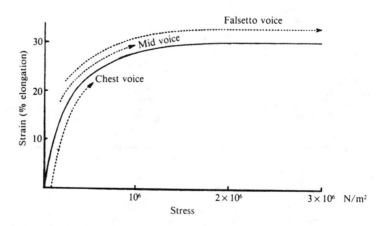

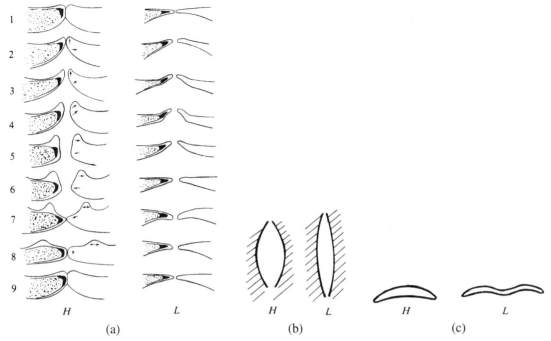

FIGURE 17.16 Schematic diagrams of vocal folds vibrating: (a) side view (from Titze 1973); (b) top view; (c) edge view. In each diagram *H* denotes the heavy mechanism (chest voice) and *L* the light mechanism (head voice).

When vibrating in the light mechanism, the vocal folds are up to 30% longer, are appreciably thinner, and have a smaller effective mass. The folds do not ordinarily close completely during any part of the cycle (compare the *open-phase* speech mode described in Section 15.2). This results in fewer harmonics of the fundamental and also in less effi-

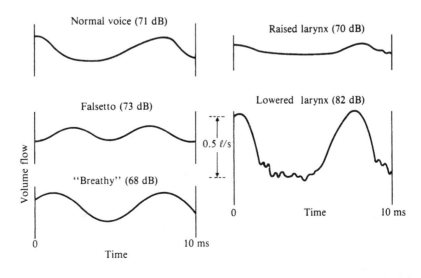

FIGURE 17.17
Waveforms of
glottal air flow
during various
modes of singing.
(After Sundberg
1978.)

cient conversion of breath power into sound power. The waveforms of the glottal air flow during speech and their spectra of overtones were shown in Figs. 15.6 and 15.7. Waveforms of glottal air flow during various modes of singing are shown in Fig. 17.17.

Two other more unusual registers deserve to be mentioned as well. One is the *Strohbass* register used by male voices to produce very low bass notes, such as those required in some Russian choir music. This is also called the *vocal fry* register, because it makes use of a loose glottal closure that is termed vocal fry by many speech therapists. Although used mainly to extend the voice below the singer's normal range, it is also used to help train the low notes in the chest or modal register (McKinney 1994).

Some singers denote a special "whistle," or flageolet, register at the top of the female vocal range. Although early writers suggested that the vocal folds actually puckered like lips to form a whistle, it is probably more correct to say the flutelike sound results from air passage through a small opening between the arytenoid cartilages.

17.7 ■ OTHER FACTORS INFLUENCING THE SPECTRA OF SUNG NOTES

It is characteristic of nearly all musical instruments that raising the dynamic level increases the levels of the higher harmonics more rapidly than that of the fundamental (see, for example. Figs. 11.13(c) and 12.16). The same effect is observed in singing, as can be seen in Fig. 17.18. In loud singing a greater fraction of the total sound energy appears in the higher harmonics as compared to soft singing.

The reason for this gain in energy in the higher harmonics can be seen by comparing the glottal air-flow waveforms (*glottograms*) in Fig. 17.19. As the loudness of phonation is increased, the rate of closure of the glottis (indicated by the slopes of the heavy lines drawn along the trailing edges of the waveforms) increases. Fourier analysis shows that waveforms with rapid rates of rise or fall have spectra rich in harmonics (compare Fig. 7.11).

Normally, a male voice tends to have weaker fundamental and stronger harmonics than a female voice, as shown in Fig. 17.20(a). Also, a male voice singing falsetto has a stronger fundamental and weaker harmonics than when singing the same note in the modal register, as shown in Fig. 17.20(b). In Fig. 17.20 the vertical axis shows only the deviation from

FIGURE 17.18
Sound pressure level in the first four harmonics at different total sound pressure levels. In soft phonation, the fundamental dominates, but the higher harmonics take on increasing importance as the loudness increases. (From Sundberg 1987.)

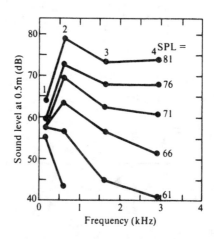

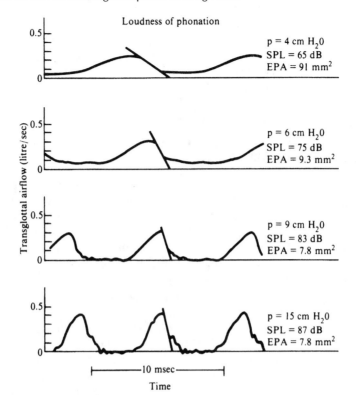

FIGURE 17.19
Glottal waveforms
for four different
levels of phonation.
The rate of glottal
closure increases as
the phonation level
increases. (From
Sundberg 1987.)

FIGURE 17.20
(a) Relative
strengths of
harmonics in male
and female voices.
(b) Relative
strengths of
harmonics in a
male voice in the
modal and falsetto
registers. In both
cases the vertical
axis shows the
deviation from the
overall decrease of
12 dB/octave that
characterizes voice
source. (From
Sundberg 1987.)

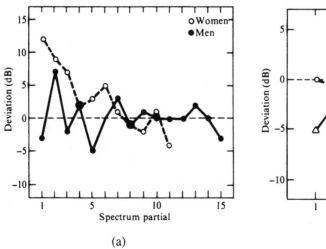

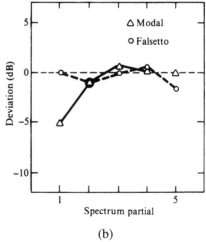

(a)

(b)

the overall decrease of 12 dB/octave that characterizes the voice source in both speech and singing.

A certain number of tones can be sung using either chest or head voice. At a given frequency, the chest voice source offers greater resistance to glottal flow than the head voice. Thus, it is possible to increase the subglottal pressure and thereby achieve greater loudness. Furthermore, the chest voice has greater harmonic content, which makes it possible to sound louder at low fundamental frequencies.

17.8 ■ CHOIR SINGING

Choral singing and solo singing are two distinctly different modes of musical performance, making different demands on the singers. Most research on the acoustics of singing has been directed at solo singing, and so less is known about the voice use in choir singing.

The first author had the opportunity of participating in some experiments at the Royal Institute of Technology (KTH) in Stockholm, which compared identical passages sung by experienced singers in solo and choir modes. A number of differences were noted, in both male and female singers.

Male singers tended to employ a more prominent singer's formant in the solo mode, as can be seen in Fig. 17.21, while the fundamental is emphasized more in the choir mode, as might be expected (Rossing, Sundberg, and Ternström 1986). It appeared that this was accomplished through adjustments in both articulation (adjustment of formant frequencies) and phonation (change in the glottal waveform).

Female singers also tend to produce more energy in the range 2 to 4 kHz in the solo mode, as shown in Fig. 17.22, although different subjects appear to differ substantially in spectral characteristics in this frequency range. It is more difficult to obtain accurate glottal waveforms from female singers, so it is difficult to distinguish changes in articulation from voice source changes. The extent of vibrato appeared to be greater in the solo mode (Rossing, Sundberg, and Ternström 1987).

Other studies by the Stockholm group have observed the degree of unison and accuracy of intervals in choir singing, both under normal conditions and when singers were deprived of feedback from other singers. In a good amateur choir, the standard deviation in the notes

FIGURE 17.21
Average spectrum envelopes for a male singer who sang a phrase as a solo singer and as a choral singer. In the latter case his lowest partials are somewhat stronger and his singer's formant is slightly weaker. (From Rossing, Sundberg, and Ternström 1986.)

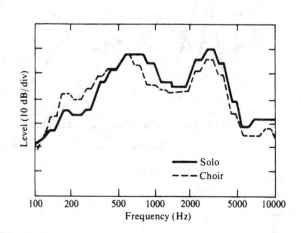

FIGURE 17.22
Average spectrum
envelopes for a
female singer who
sang the same
phrase at two
different sound
levels as a solo
singer and as a
choral singer.
Choral singing
gives slightly
weaker high
partials. (From
Rossing et al.
1987.)

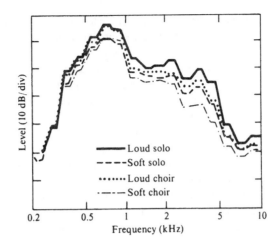

sung by members of the bass section was found to be 16 cents (Ternström and Sundberg 1988). In another experiment, singers were asked to sing a note at constant sound level and with the same pitch, as reference tones presented at different levels. When the reference tone was the vowel /a/, they were able to do this quite well, but when the vowel was /u/, the singers sang about 25 cents sharp with the softest reference tone and about 45 cents flat with the loudest reference tone. Apparently the relatively low number of harmonics in the /u/ tone (due to the low frequencies of the first and second formants) was the cause of this (Sundberg 1987).

17.9 ■ POPULAR SINGING AND OTHER STYLES

Because there are more popular ("pop") singers than classical singers in contemporary society and their media exposure is considerably greater, more young singers emulate the stars of popular entertainment than of opera and classical music. A wide variety of singing styles exist in nonclassical music, and relatively few of them have been studied scientifically. We will, however, attempt a brief discussion of a few of them.

There are a few easily recognized features of most popular singing styles. First of all, the texts of songs play a very important role in the total effect, more so than in most classical singing. The texts are often witty or carry an emotional message; it is important that they be understandable, even on first hearing. Therefore, the singer is permitted less of the vowel modification used by classical singers to enhance the instrumental beauty of the vocal line.

Second, a high value is put on naturalness of the sound, even at the expense of beauty. The use of some vocal techniques cultivated in classical singing are avoided in order not to make the voice sound well trained. The dark, or covered, voice seems particularly offensive to the impression of naturalness. On the other hand, unevenness and certain features of individual voices are readily tolerated (Schutte and Miller 1993).

Third, the performer is generally considered more important than the composition, and so the song is freely changed in order to show off the singer's voice to best advantage.

However, this can easily be carried too far, as in some of the modifications made to the national anthem by some popular singers.

We have already discussed how registers overlap and how many singers can sing a number of notes in two different registers. *Belting* describes a manner of loud singing used by female popular singers to extend the chest register above its normal range. It is characterized by a raised larynx and matching the first formant with the second harmonic on open (high F_1) vowels.

Schutte and Miller (1993) compare the spectra of the same vowel sung by a versatile mezzo soprano in classical, popular, and belt styles. In the classical style, the F_1 and F_2 formants are lower, making the first two harmonics about equal in amplitude (giving the sound a darker quality), whereas in the popular (also called "legit," or Broadway) style, the formants are higher, so that the first harmonic dominates. In the belt style, F_1 and F_2 are raised so much that the second and third harmonics are 27 dB above the first harmonic, resulting in a loud, bright, "edgy" sound.

Professional male country singers were found by Stone, Cleveland, and Sundberg (1999) to demonstrate characteristics different from those found in classical singers. The inspiratory and expiratory patterns of breathing, as well as the voice source properties and formant frequencies, were found to be quite distinctive in country singers. Whereas the classically trained singers exhibit different characteristics in their singing and speaking voices, country singers showed similar features, including a "speaker's formant" (a prominent F_4) in both. On the other hand the "singer's formant" found in classical singers, was generally missing. This probably reflects the emphasis on text in country singing as well as the fact that they use electronic voice amplification to be heard over an orchestra, and hence do not require a singer's formant to be heard.

Ordinarily male singers do not deliberately tune their vocal tract resonances (formants) to the vocal-fold vibration frequency (as a soprano does). If a male singer unknowingly does so, he may be in for a surprise. The strong acoustic feedback from the vocal tract to the larynx disrupts the national motion of the vocal folds and a *voice break* occurs (Sundberg 1981). The situation is similar to the "wolf note" on a string instrument due to strong interaction between string and body resonances.

By tuning their vocal tract resonances to a *harmonic* of the vocal-fold vibration frequency, however, Tibetan monks produce a very interesting sound. Using this technique while singing a very low note, a single monk can accentuate certain harmonics and produce what sounds like a chord (Smith, Stevens, and Tomlinson 1967). At a higher fundamental frequency, the tuned harmonic may sound more like a whistle. The listener hears a tonal vowel with a pitch corresponding to the fundamental frequency of the voice, accompanied by the whistling of the tuned harmonic. The whistle tracks the fundamental tone.

This technique, sometimes called harmonic singing or throat singing, has been used by David Hykes, who has developed a remarkable control of frequency and sharpness of vocal tract resonance. On a single sustained vowel, he is able to slide the resonance up and down the harmonic series, causing individual harmonics to "pop out" in succession. Alternatively, he can hold a steady whistle frequency while moving the fundamental up and down in pitch. From a perceptual point of view, the difference between emphasizing a single harmonic this way, as shown in Fig. 17.23(a), and the more familiar emphasis of several harmonics by vowel formants, as shown in Fig. 17.23(b), is interesting. The fifth

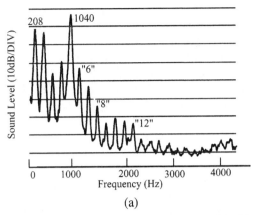
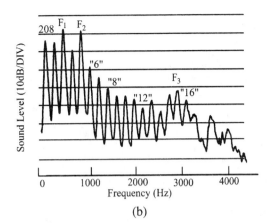

FIGURE 17.23 The spectra of the sung vowel /a/ with fundamental frequency of 208 Hz. (a) A single harmonic (fifth) stands out in harmonic chant. (b) Normal vowel formant emphasizes several harmonics that blend together. (After Hartmann 1997.)

harmonic in Fig. 17.23(a) is considerably stronger than the other harmonics and thus it pops out and is clearly heard (Hartmann 1997).

17.10 ■ SUMMARY

In singing, as in speaking, the vocal folds act as a source of sound, which is filtered by the vocal tract. The resonances (formants) of the vocal tract determine the vowel sounds as well as the timbre of the sung tone. Sung vowels and their formants are slightly different than spoken vowels, one of the most important differences being the appearance of a *singer's formant* around 2500–3000 Hz. One of the important results of vocal training is to learn to lower the larynx and open the pharynx to create this extra formant. Sopranos often sing at pitches above their normal formants, and therefore must "tune" these formants if they are to reinforce the sung notes.

There appear to be two mechanisms for singing: In one, the vocalis muscles are active; in the other, they tend to be passive. They can be referred to as *heavy mechanism* (chest voice) and *light mechanism* (head voice). During normal breathing, about 500 cm^3 of air is moved per breath. In a trained singer, both air flow rate and pressure increase with sound level.

In loud singing, a greater fraction of the total sound energy appears in the higher harmonics, partly due to the higher rate of closure of the glottis. Singers tend to concentrate more energy in the range 2 to 4 kHz in solo singing, whereas they emphasize the fundamental more in choir singing.

REFERENCES AND SUGGESTED READINGS

Appelman, D. R. (1967). *The Science of Vocal Pedagogy.* Bloomington, Ind.: Indiana University Press.

Bartholomew, W. T. (1940). "The Paradox of Voice Teaching," *J. Acoust. Soc. Am.* **11**: 446.

Benade, A. H. (1976). *Fundamentals of Musical Acoustics.* New York: Oxford. (See Chapter 19.)

Bloothooft, G., and R. Plomp (1984, 1985, 1986). "Spectral Analysis of Sung Vowels. I, II, and III," *J. Acoust. Soc. Am.* **75**: 1259; **77**: 1580; **79**: 852.

Bjorklund, A. (1961). "Analyses of Soprano Voices," *J. Acoust. Soc. Am.* **33**: 575.

Bouhuys, A., J. Mead, D. F. Proctor, and K. N. Stevens (1968). "Pressure-Flow Events During Singing," *Annals N. Y. Acad. Sci.* **155**: 165.

Cleveland, T., and J. Sundberg (1983). "Acoustic Analysis of Three Male Voices of Different Quality," in *Proc. Stockholm Music Acoustics Conference (SMAC 83),* ed. A. Askenfelt, S. Felicetti, E. Jansson, and J. Sundberg. Stockholm: Royal Swedish Acad. of Music.

Haasemann, F., and J. M. Jordan (1991). *Group Vocal Technique.* Chapel Hill, N.C.: Hinshaw Music.

Hartmann, W. H. (1997). *Signals, Sound, and Sensation.* Woodbury, N.Y.: American Inst. Physics. 124.

Hollien, H. (1974). "On Vocal Registers," *J. Phonetics* **2**: 125–143.

Large, J. (1972). "Towards an Integrated Physiologic-Acoustic Theory of Vocal Registers," *Nat. Assn. Teachers of Singing, Bull.* **28**: 18, 30.

Leanderson, R., J. Sundberg, and C. von Euler (1987). "Role of Diaphragmatic Activity During Singing: A Study of Transdiaphragmatic Pressures," *J. Appl. Physiol.* **62**: 259.

Mackworth-Young, G. (1953). *What Happens in Singing.* London: Neame.

Marchesi, M. (1970). *Bel Canto: A Theoretical and Practical Vocal Method.* (Dover reproduction of original undated publication by Enoch & Sons, London.)

McKinney, J. C. (1994). *The Diagnosis and Correction of Vocal Faults.* Nashville, Tenn.: Genevox Music Group.

Miller, D. G. (2000). *Registers in Singing.* Roden, Netherlands: author copyright.

Peterson, G. E., and H. L. Barney (1952). "Control Methods Used in a Study of the Vowels," *J. Acoust. Soc. Am.* **24**: 104.

Rossing, T. D., J. Sundberg, and S. Ternström (1986). "Acoustic Comparison of Voice Use in Solo and Choir Singing," *J. Acoust. Soc. Am.* **79**: 1975.

Rossing, T. D., J. Sundberg, and S. Ternström (1987). "Acoustic Comparison of Soprano Solo and Choir Singing," *J. Acoust. Soc. Am.* **82**: 830.

Schutte, H. K., and D. G. Miller (1993). "Belting and Pop, Non-Classical Approaches to the Female Middle Voice: Some Preliminary Considerations," *J. Voice* **7**: 329–334.

Seymour, J. (1972). "Acoustic Analysis of Singing Voices, Parts I, II, III," *Acustica* **27**: 203, 209, 218.

Shipp, T. (1977). "Vertical Laryngeal Position in Singing," *J. Research in Singing* **1**: 16. (abstract in *J. Acoust. Soc. Am.* **58**: S95.)

Smith, H., K. N. Stevens, and R. S. Tomlinson (1967). "On an Unusual Mode of Chanting by Certain Tibetan Lamas," *J. Acoust. Soc. Am.* **41**: 1262–1264.

Stone, R. E., T. Cleveland, and J. Sundberg (1999). "Formant Frequencies in Country Singers' Speech and Song," *J. Voice* **13**: 161–167.

Strong, W. J., and G. R. Plitnik (1977). *Music, Speech and High Fidelity,* Provo, Utah: Brigham Young University Press. (See Chapter 6B.)

Sundberg, J. (1974). "Articulatory Interpretation of the 'Singing formant,' " *J. Acoust. Soc. Am.* **55**: 838.

Sundberg, J. (1975). "Formant Technique in a Professional Female Singer," *Acustica* **32**: 8.

Sundberg, J. (1977a). "The Acoustics of the Singing Voice," *Sci. Am.* **236**(3).

Sundberg, J. (1977b). "Singing and Timbre," in *Music Room and Acoustics.* Stockholm: Royal Academy of Music.

Sundberg, J. (1978). "Waveform and Spectrum of the Glottal Voice Source," Report STL-QPSR 2–3, 35. Stockholm: Speech Transmission Lab., Royal Institute of Technology.

Sundberg, J. (1981). "Formants and Fundamental Frequency Control in Singing. An Experimental Study of Coupling Between Vocal Tract and Voice Source," *Acustica* **49**: 47–54.

Sundberg, J. (1987). *The Science of the Singing Voice.* DeKalb, IL: Northern Illinois University Press.

Ternström, S., and J. Sundberg (1988). "Intonation Precision of Choir Singers," *J. Acoust. Soc. Am.* **84**: 59.

Titze, I. R. (1973). "The Human Vocal Cords: A Mathematical Model, Part I," *Phonetica* **28**: 129. (See also Part II, *Phonetica* **29**: 1.)

van den Berg, Jw. (1968). "Register Problems," *Ann. New York Academy of Sciences* **155**: 129.

van den Berg, Jw. and W. Vennard (1959). "Toward an Objective Vocabulary for Voice Pedagogy," *N. A. T. S. Bull.,* Feb. 1959.

Vennard, W. (1967). *Singing: The Mechanism and the Technic.* New York: Carl Fischer.

GLOSSARY

belting A manner of loud singing used by female popular singers to extend their chest register above its normal range.

chest (modal) voice Mode of singing associated with a heavy mechanism or active vocalis muscles.

cricoarytenoids The muscles of the larynx that help to apply tension to the vocal folds.

cricoid Lower cartilage of the larynx.

cricothyroids The muscles of the larynx that determine the relative position of cricoid and thyroid cartilages and thus affect vocal fold tension.

diaphragm The dome-shaped muscle that forms a floor for the chest cavity.

external (inspiratory) intercostals Intercostal muscles used to breathe air into the lungs.

flow rate The volume of air that flows past a point and is measured per second.

formant A resonance of the vocal tract.

functional reserve capacity (FRC) Volume of air in the lungs at the end of a quiet expiration.

harmonic singing Tuning vocal tract resonances to a harmonic of the vocal-fold vibration frequency to produce single or multiple tones.

head voice Mode of singing associated with a light mechanism, passive vocalis muscle, and elongated, thin vocal folds.

heavy mechanism Mode of vocal fold vibration in which the vocalis muscles are active, and the vocal folds or cords are thick.

intercostal muscles Muscles joining the ribs that are used for breathing.

internal (expiratory) intercostals Intercostal muscles used to breathe air out from the lungs.

larynx The source of sound for speaking or singing.

light mechanism Mode of vocal fold vibration in which the vocalis muscles are relaxed, and the vocal folds elongated and thin.

middle register A combination of light and heavy mechanism that lies between the chest and head registers.

pharynx The lower part of the vocal tract connecting the larynx and the oral cavity.

singer's formant A resonance around 2500 to 3000 Hz in male (and low female) voices that adds brilliance to the tone.

speaker's formant A prominent fourth formant in the speaking and singing voice of country singers.

Strohbass (vocal fry) register Register used for very low bass notes; makes use of a loose glottal closure termed *vocal fry*.

subglottal pressure Amount by which the air pressure in the lungs exceeds atmospheric pressure.

thyroarytenoids (vocalis muscles) The muscles that form part of the vocal folds.

thyroid The upper cartilage of the larynx.

tidal volume The volume of air moved in and out of the lungs during a normal breath.

vital capacity The volume of air that can be moved in and out of the lungs with maximum effort during inhalation and exhalation.

vocal fry Loose glottal closure that allows air to bubble through with a frying-pan sound.

vocal tract The tube connecting the larynx to the mouth consisting of the pharynx and the oral cavity.

vocalis muscle The thyroarytenoid muscle.

whistle register Very high register in the female voice in which the arytenoids form a whistle.

REVIEW QUESTIONS

1. What are the four main parts of the vocal organ?

2. In singing, the vocal tract is tuned by changing its length. (T or F)

3. What vowel has a low F_1 and a high F_2?

4. A soprano can match first formants of what two vowels with her fundamental frequency?

5. What vowel has the highest F_1?

6. What is a *singer's formant*?

7. How is a singer's formant formed?

8. Why does a soprano tune the first formant to match the fundamental frequency?

9. What is meant by subglottal pressure?

10. How is the subglottal pressure measured?

11. How much air is moved in and out of the lungs in normal breathing?

12. Doubling the subglottal pressure produces about how much change in the sound level?

13. What is the approximate range of air flow during singing?

14. Describe the vocal folds when singing in *chest* voice.

15. What is vocal fry? For what type of singing is it employed?

16. How does glottal closure rate change as the phonation level increases?

17. What are two differences in the male singing voice in solo and choir singing?

18. What are two differences in the female singing voice in solo and choir singing?

19. What is *belting*?

20. Describe the technique used by a Tibetan monk to sing what sounds like a chord.

QUESTIONS FOR THOUGHT AND DISCUSSION

1. Try to sing as many notes as possible in both chest and head registers. Can you sing in both registers? How much overlap is there in your voice?

2. Is a stress of 10^6 N/m^2 (See Fig. 17.15) a large stress? What is the breaking stress of a piece of cotton cord? nylon thread?

3. Normal speaking is done in chest voice. Is it possible to speak in a head voice? Is speech intelligibility affected?

4. Place either a cardboard tube, a length of pipe, or your cupped hands around your lips to extend the vocal tract and lower the formant frequencies. Describe the tone produced. What is often called a dark, or covered, tone is produced by extending the vocal tract at the lower end. Is this equivalent to what you have done?

EXERCISES

1. Find the frequencies that correspond to the three singing registers designated in Fig. 17.14.

2. What harmonics of G$_2$ ($f = 98$ Hz) are enhanced by the formants of /i/? of /u/?

3. Compare the first three formant frequencies in Fig. 17.4 to those in Table 17.1 for the sung vowels /u/, /ɑ/, and /i/.

4. Find the lengths of closed pipes that would resonate at 2500 and at 3000 Hz. Are these reasonable lengths for the cavity formed by the (closed) glottis and the (open) pharynx?

5. The power (in watts) used to move air in or out of the lungs is equal to the pressure (in N/m^2) multiplied by the flow rate (in m^3/s). Find the power for:

   (a) Quiet breathing ($p = 100$ N/m^2, flow rate = 100 cm^3/s);

   (b) Soft singing ($p = 1000$ N/m^2, flow rate = 100 cm^3/s);

   (c) Loud singing ($p = 4000$ N/m^2, flow rate = 400 cm^3/s).

6. According to Fig. 17.13, a pressure of 4000 N/m^2 will produce a sound level of about 120 dB.

   (a) Find the intensity and sound pressure that correspond to this sound level (see Chapter 6).

   (b) Compare the sound pressure at the mouth to the steady subglottal air pressure.

   (c) Assuming a mouth opening of 20 cm^2, calculate the total radiated sound power.

   (d) What portion of the total power calculated in Exercise 5 is converted into sound? (*Answer:* About 0.1%.)

EXPERIMENTS FOR HOME, LABORATORY, AND CLASSROOM DEMONSTRATION

Home and Classroom Demonstration

1. *Waveforms of vowel sounds*   By connecting a microphone to an oscilloscope, display the waveforms for different vowel sounds. A male voice singing "oo" in falsetto at E$_4$ (near the first formant frequency) produces nearly a sine wave with few overtones, for example. Singing "ee" at the same frequency adds small wiggles due to the upper harmonics (mainly the sixth and seventh), which are near the second formant. Finally singing "ah" in a normal chest voice at about

that same pitch can produce a tone in which the second and third harmonics exceed the fundamental because of the high first-formant frequency.

2. *Darkened vowel sound*   Produce something akin to dark, or covered, vowel sounds by singing with a short length of tubing surrounding your lips. This lengthens your vocal tract and lowers the formant frequencies. (Of course, in actual covered singing the length of additional tubing is at the other end of the vocal tract, in the vicinity of the vocal folds). This works the best when your mouth opening is made large, as in singing "ah" or "ee." (Strong and Plitnik 1997).

3. *Formant frequencies*   Use an FFT analyzer to record spectra of as many vowels as possible. Try to determine the formant frequencies and compare them to the frequencies for spoken vowels in Table 15.3 and in Fig. 17.2. Compare the same vowel spoken and sung.

4. *Identifying vowels at high tessitura*   Have a trained soprano sing various vowel sounds near the top of her singing range and try to identify each vowel sound. (It may be difficult to do out of context if the fundamental frequency is higher in frequency that the first formant.)

5. *Scaled formants*   Record a series of sung vowels on tape and play them back at different speeds. Note that some of the vowel sounds change to others ("ah" changes to "oh" at half-speed, for example).

6. *Different ways of breathing*   Attempt to breathe with your ribs alone, keeping your abdominal wall stationary (try to "freeze" it in the belly-in and belly-out positions as well as in the normal position). Breathe by keeping your ribs as stationary as possible (in both the raised and lowered positions) and moving your abdominal wall to activate your diaphragm.

7. *Vibrato rate*   Display the waveform of a good singer on an oscilloscope so that you can see the vibrato rate. Ask the singer to increase and decrease the vibrato rate and amplitude.

8. *Voice source analyzed by inverse filtering*   This videotape from the Royal Institute of Technology in Stockholm includes several find demonstration experiments on the voice source in singing as well as speaking.

Laboratory Experiments

The singing voice (Experiment 25 in *Acoustics Laboratory Experiments*)

PART
V

Electroacoustics

Electroacoustics deals with the conversion of electrical energy into acoustical energy or the reverse. The discipline that deals with practical applications of electroacoustics is sometimes called audio engineering, and it generally combines elements of physics, music, and electrical engineering. Chapters 18–22 present the basic principles of electroacoustics or audio engineering.

Chapter 18 covers the principles of electronic circuits in a rather elementary way. It may provide a welcome review for students who have already taken a course in physics or electrical engineering. Music students and others who are not familiar with electrical circuits may wish to read further in one or more of the reference texts listed at the end of the chapter. It is worth taking the time to carefully answer all of the questions and work through all of the exercises or problems. Chapters 19 and 20 discuss loudspeakers, microphones, amplifiers, and tuners, the basic building blocks of audio and electroacoustics systems.

Chapter 21 is an introduction to digital computers and digital techniques. Again, some students who have had previous experience with digital computers will find it a review, but others may have to proceed more slowly. In addition to providing an introduction to digital techniques that will be applied in the discussion of sound recording, this chapter also serves as an introduction to Part VII on Electronic Music Technology. Finally, Chapter 22 discusses sound recording, using both analog and digital techniques.

CHAPTER

18

Electronic Circuits

The world of *audio* deals with acoustic energy that has been translated into electrical signals and then modified by electronic circuits. Electronic circuits are made up of components with special electrical properties. These components may be passive devices (which do not require a power source separate from their controls) or active devices (which do require a power source separate from their controls).

Before learning about electronic circuits, we ought to consider a few basic principles of electricity. Although we use electricity every day in countless ways, it is curious how few people learn to understand what electricity is or how it actually performs the useful tasks we call on it to do.

Electricity is everywhere in nature. All the common, everyday materials that we handle are held together by electrical forces. These forces are created by positive and negative charges coming from the protons and electrons in atoms. The light that reaches our eyes consists of electric (as well as magnetic) waves, as do the waves that bring us heat from the sun. When we use the term electricity in everyday language, however, we usually mean *electrical energy* or an *electric current*, and that is the way in which we will use the term electricity here.

In this chapter you should learn:

- About direct current (dc) with voltages, currents, and resistance;
- About electrical energy and power;
- About alternating current (ac) with inductors and capacitors;
- How low-pass, high-pass, band-pass, and notch filters are made;
- About vacuum tubes, diodes, and transistors;
- About the characteristics of amplifiers;
- About oscillators and signal generators;
- About transformers and power supplies.

18.1 ■ DIRECT CURRENT

An electric current is the rate of flow of electric charge (charge per unit time passing a point in the circuit) and is capable of carrying energy from one place to another. An electric current in a wire consists of a cloud of negatively charged electrons moving through the wire, but in other conducting media (for example, the liquid in a battery), the current may be carried by charged atoms. A direct current (dc) is an electrical current that flows in only one direction. An early model of electric current, proposed by Benjamin Franklin and others, was that of a charged fluid (practical electricians to this day sometimes refer to electricity as "juice"), and it may be helpful to our understanding of electricity to draw analogies between the flow of electricity in wires and the flow of water in pipes.

The electrical circuit in Fig. 18.1(b) and the water circuit in Fig. 18.1(a) have many similarities. In the water circuit, a pump raises the water pressure, which causes the water to flow around the circuit. It could be made to perform some work (like turning a paddle wheel) as it drops to the lower reservoir. In the electrical circuit, the battery raises the potential (voltage) in one part of the circuit so that an electrical current will flow. The current "drops" through the light bulb and some of its energy is converted to light and heat. The circuit diagram in Fig. 18.1(c) could be used to represent either the water circuit or the electrical circuit; V represents the *voltage* supplied by the battery or the pressure supplied by the pump, and R represents the *resistance* to the flow of water or electric current on the part of the small pipe or the light bulb.

Clearly, the amount of current or water that flows depends on the size of V (electrical voltage or water pressure) and R (resistance). Large V and small R leads to a large current. In fact, the relationship between current, voltage, and resistance (referred to as *Ohm's law*) is one of the important laws of nature. Using the symbol I for current, it is written

$$I = \frac{V}{R}. \tag{18.1}$$

To use Ohm's law, one needs a consistent set of electrical units. In the international system (SI), V is expressed in *volts*, R in *ohms*, and I in *amperes* (frequently abbreviated amps). The preferred symbols for the units are V for volts, Ω for ohms, and A for amperes. (Note that italic V stands for the potential (voltage), and Roman V stands for the unit volts. For I and R, the quantity and the unit are different symbols.)

If the circuit contains only a single source of voltage (potential) and a single resistance, as in Fig. 18.2, the use of Ohm's law to calculate the current is straightforward. In Fig. 18.2, the current is $I = \frac{V}{R} = \frac{3}{6} = 0.5$ A.

If the circuit contains two or more voltage sources in series, these can be added to give the total source voltage.* In a two-cell flashlight, two 1.5-V batteries are connected in

FIGURE 18.1
(a) A water circuit in which a pump raises the water pressure, resulting in a current flow. (b) An electrical circuit in which a battery raises the electrical potential (voltage) so that electric current can flow through the light bulb. (c) A circuit diagram that could represent either of the systems illustrated in (a) and (b).

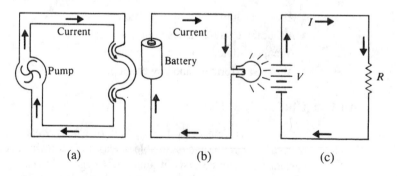

(a)   (b)   (c)

*If the sources are batteries, they must be connected in the same direction or polarity in order for the total source voltage to be the sum of the individual voltages. If they are connected in opposition, their voltages subtract, as one discovers by inserting one of the batteries into a flashlight in the reverse direction.

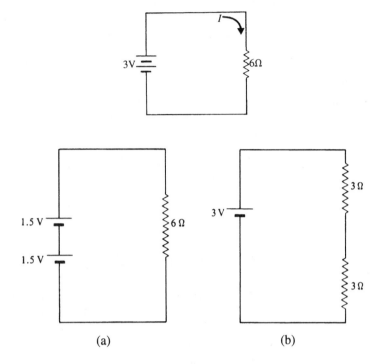

FIGURE 18.2
An electrical circuit
with a single
voltage source and
a single resistance.

FIGURE 18.3
(a) An electrical
circuit with two
voltage sources in
series. (b) A circuit
with two
resistances in
series. In both of
these circuits, the
current is the same
as that in the circuit
shown in Fig. 18.2.

series to provide a source of 3 V, as shown in Fig. 18.3(a). If the resistance of the bulb were 6 Ω, the current would be 0.5 A, as in the circuit in Fig. 18.2.

Similarly, if two resistances R_1 and R_2 are connected in series, the total resistance is their sum. This can be represented by the equation

$$R = R_1 + R_2. \tag{18.2}$$

Figure 18.3(b) shows a circuit with two 3 Ω resistances in series in which the current is 0.5 A, the same as in the circuits shown in Figs. 18.2 and 18.3(a). Note that the total voltage (3 V) supplied by the battery is equally divided between the two resistances, so that a voltmeter connected across either one of them would read 1.5 V. Voltages across series elements add together, whether they are the source voltages of two or more batteries connected in series or the voltage drops across two resistances in series.

Examples Involving the Use of Ohm's Law to Calculate Current and Voltage

In the first circuit shown in the figure, a 6-V battery causes an electric current to flow through a 3 Ω resistor (which could be a small light bulb). The current that flows is easily calculated to be $I = \frac{V}{R} = \frac{6}{3} = 2$ A.

In the second circuit, two resistors connected in series impede the flow of electrical current. Thus, the total resistance through which current must flow is $R = 10 + 14 = 24$ Ω. The current will be $I = \frac{V}{R} = \frac{12}{24} = 0.5$ A. If we wish, we could use Ohm's law in the form $V = IR$ to calculate the voltage that appears across ei-

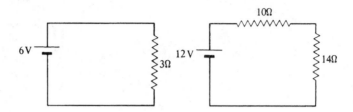

ther one of the resistors alone. For example, the voltage across the 14- Ω resistor is $V_{14} = IR = (0.5)(14) = 7$ V. Clearly the voltage across the other resistor would be the total voltage less V_{14}:

$$V_{10} = V - V_{14} = 12 - 7 = 5 \text{ V.}$$

The same answer could be arrived at by use of Ohm's law:

$$V_{10} = IR = (0.5)(10) = 5 \text{ V.}$$

An arrangement of two resistors in series such as this is often called a *voltage divider*, because the entire voltage is divided between the two resistors. Frequently the point of division is made variable, as shown in the accompanying figure. If a voltage V is applied across the resistor R, a fraction $\frac{R_1}{R}$ appears across the variable part R_1. The output voltage, therefore, is $\frac{VR_1}{R}$. A voltage divider is used for the volume control in a radio or amplifier. Such a device is also called a *potentiometer*, or simply a *pot*.

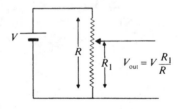

Another way to connect circuit elements is in parallel, as shown in Fig. 18.4. Connecting two identical voltage sources in parallel does not increase the total voltage, but it does increase the available current, because each source supplies half the current in the circuit, as shown in Fig. 18.4(a). Similarly, connecting two resistances in parallel provides two

FIGURE 18.4
(a) A circuit with two voltage sources in parallel. (b) A circuit with two resistances in parallel. In both circuits, the total current is $I = I_1 + I_2$.

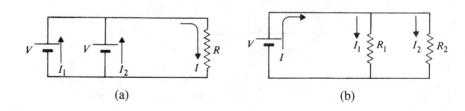

(a)   (b)

alternate paths for the current, so the total current is the sum of I_1 and I_2, the currents through R_1 and R_2, respectively.

A formula for calculating the total resistance of two or more resistances in parallel can be derived by using Ohm's law. In the circuit shown in Fig. 18.4(b), the voltage across each resistor is V, since they are both connected directly to the voltage source. Thus the currents in the two resistances are $I_1 = \frac{V}{R_1}$ and $I_2 = \frac{V}{R_2}$. The total current is

$$I = I_1 + I_2 = \frac{V}{R_1} + \frac{V}{R_2}.$$

But we know that if R is the total resistance in the circuit, $I = \frac{V}{R}$. Thus if each term in the equation is divided by V, we obtain an expression for R:

$$\frac{1}{R} = \frac{1}{R_1} + \frac{1}{R_2}. \tag{18.3}$$

For example, if $R_1 = R_2 = 10\ \Omega$, then

$$\frac{1}{R} = \frac{1}{10} + \frac{1}{10} = \frac{2}{10},$$

so

$$R = \frac{10}{2} = 5\ \Omega.$$

The total resistance R of two resistances in parallel will always be less than either individual resistance.

The formula can be extended to any number of resistances in parallel:

$$\frac{1}{R} = \frac{1}{R_1} + \frac{1}{R_2} + \frac{1}{R_3} + \cdots + \frac{1}{R_n}. \tag{18.4}$$

18.2 ■ ELECTRICAL ENERGY AND POWER

Work, energy, and power were defined in Chapter 1; we recall that power is the rate at which work or energy is delivered (energy per unit of time delivered). Electrical power is expressed in watts and electrical energy in joules, just as mechanical power and energy are (one joule = one watt-second). However, when expressing electrical power, the kilowatt-hour (kWh), a much larger unit of energy, is frequently used (1 kWh = 3.6×10^6 J; see Section 1.10).

When you pay your electric bill, you pay for the total amount of *electrical energy* consumed in all the circuits in your house. The energy depends on the voltage, the average current, and the total time during which the current flows. The pump, in this case, is the power-generating station, which maintains a constant voltage (usually 120 V) throughout your house. A kilowatt-hour would light a 100-W bulb for 10 h. In comparison, a clarinetist would have to blow his or her instrument for 20,000 h (more than 2 years) without stopping in order to generate a kilowatt-hour of acoustic energy.

> ### Calculating Electric Power and Energy
>
> Electric *power* is the product of current and voltage: $\mathcal{P} = IV$. Power is expressed in *watts* when I is in *amperes* and V is in *volts*. A 60-W light bulb is designed to draw a current of 0.5 A. Thus $\mathcal{P} = 0.5 \times 120 = 60$ W. Power is the rate at which energy is used; it is measured in watts and kilowatts (1 kW = 1000 W). Energy is the product of power and time. Thus watt-second, watt-hour, and kilowatt-hour are units used to measure energy.

18.3 ■ ALTERNATING CURRENT

Alternating current (ac) is electrical current that reverses its direction many times each second. Alternating current is generated in a microphone, and alternating current drives the voice coil in a loudspeaker. Nearly all electrical power in the United States is ac with a frequency of 60 cycles per second (60 *hertz*, abbreviated Hz). This means that 120 times each second, the current in a light bulb stops momentarily and reverses its direction. Ordinarily you are not aware of this interruption, but if you look near the end of a fluorescent lamp, you can sometimes observe a flicker.

FIGURE 18.5
(a) Simple alternating current from a 60-Hz power line.
(b) Complex alternating current from an audio amplifier.

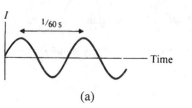

(a)

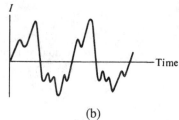

(b)

Figure 18.5(a) shows how the current varies with time for a simple alternating current with frequency of 60 Hz, which is the type supplied by power companies in the United States. The waveform closely resembles that of a simple harmonic oscillator, shown in Fig. 2.2. Figure 18.5(b) shows alternating current with a more complex waveform such as one might observe in the output of an audio amplifier.

FIGURE 18.6
Alternating current with an effective 120 V has a voltage that actually varies from −170 V to 170 V. It is equivalent in power to a 120-V direct current.

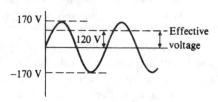

Ohm's law holds for ac circuits as well as for dc (direct current) circuits. The same is true for formulas used to calculate power and energy, with one caveat: In the case of ac, we use *effective* voltage, current, and power. An effective ac voltage of 120 V (such as we are supplied by the power company) means that the voltage rapidly varies from -170 to $+170$ V, as shown in Fig. 18.6. Rarely, however, do we need to concern ourselves with this fact when using ac electricity, because the power available from ac with an effective voltage of 120 V is the same as would be available from a 120-V dc.

In electronic circuits that use ac, we commonly encounter *capacitors* and *inductors*. They are circuit elements whose impedance to the flow of electricity varies with frequency. The electrical *impedance Z* is defined as the ratio of voltage to current: $Z = \frac{V}{I}$. Impedance in an ac circuit is analogous to resistance in a dc circuit. Like resistance, it is expressed in ohms.

A typical example of an inductor is a coil of wire wound around a core of magnetic material. The buildup of magnetism in the core prevents the current from changing as rapidly as it would otherwise; thus, an inductor impedes the flow of ac even if the resistance of its wire is small. In an inductor, the impedance increases with frequency, so it is more difficult for high-frequency current to flow, as shown in Fig. 18.7(a).

A simple capacitor can be made from two parallel plates separated by air. As positive charge is added to one plate and negative charge is added to the other, an electric field is created between the two plates. The capacitance of the plates (its capacity to store charge) increases as the area of the plates increases or as the distance between the plates decreases.

In a capacitor, the impedance decreases as the frequency increases, so that it is easier for high-frequency current to flow, as shown in Fig. 18.7(b). In electrical circuits, we denote inductance by the symbol L and measure it in *henries* (abbreviated H). Capacitance is denoted by C and measured in *farads* (abbreviated F).

The various electrical units are appropriately named after scientists of the eighteenth and nineteenth centuries: Georg Ohm, Andre Amperé, Alessandro Volta, James Watt, James Joule, Joseph Henry, and Michael Faraday.

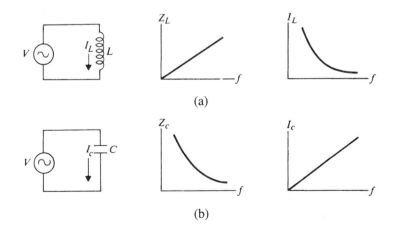

FIGURE 18.7
Dependence of impedance Z and current I on frequency f (a) for an inductor; (b) for a capacitor.

The formulas that relate V, I, Z, L, and C are as follows:

$$\text{For an inductor:} \quad Z_L = 2\pi f L, \qquad I_L = \frac{V}{Z_L} = \frac{V}{2\pi f L}; \quad \text{and}$$

$$\text{For a capacitance:} \quad Z_C = \frac{1}{2\pi f C}, \qquad I_C = \frac{V}{Z_C} = 2\pi f C V,$$

where V = the potential in volts (V), Z = the impedance magnitude in ohms (Ω), I = the current in amperes (A), L = the inductance in henries (H), f = the frequency in hertz (Hz), and C = the capacitance in farads (F). *Note:* In electronic circuits, capacitance is almost always expressed in μF (microfarads) or pF (picofarads), because a farad is a very large amount of capacitance: $1\mu\,F = 10^{-6}$ F and 1 pF $= 10^{-12}$ F.

Examples Involving Alternating Current

1. In Fig. 18.7(a), find the impedance and the current when $L = 5$ H, $f = 60$ Hz, and $V = 120$ V.

   Solution

   $$Z_L = 2\pi f L = 2(3.14)(60)(5) = 1885\ \Omega;$$

   $$I = \frac{V}{Z} = \frac{120}{1885} = 0.064\ \text{A} = 64\ \text{mA}.$$

2. In Fig. 18.7(b), find the impedance and the current when $C = 5\ \mu$F, $f = 200$ Hz, and $V = 50$ V.

   Solution

   $$Z_C = \frac{1}{2\pi f C} = \frac{1}{2(3.14)(200)(5 \times 10^{-6})} = 159\ \Omega;$$

   $$I = \frac{V}{Z} = \frac{50}{159} = 0.312\ \text{A}.$$

18.4 ■ TRANSFORMERS

A *transformer* consists of a pair of closely coupled inductors. Usually a transformer is constructed by winding two coils of wire on a core of iron or other magnetic material. The coil or winding that is conducted to the voltage source (input) is called the *primary*; the winding that is connected to the load (output) is called the *secondary*. The circuit diagram used to symbolize a transformer is shown in Fig. 18.8.

Transformers, by their nature, function only when the current in the primary is changing. Consequently, they are ac devices. One main function of a transformer is to "transform" the primary voltage up or down into the secondary voltage. If the number of turns of wire in the secondary is less than in the primary, then the secondary voltage will be less than the primary voltage. The converse of this is also true. In fact, the primary-to-secondary voltage ratio is exactly equal to the primary-to-secondary winding turns ratio:

FIGURE 18.8
A transformer
connected to an ac
source of voltage.
The inductor
connected to the
voltage source is
called the primary.
The inductor
connected to the
load resistor is
called the
secondary.

$$\frac{V_p}{V_s} = \frac{N_p}{N_s} \quad \text{or} \quad V_s = V_p \frac{N_s}{N_p}. \tag{18.5}$$

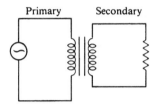

Primary   Secondary

If, for example, the number of turns in the primary is 300 for a given transformer and the number of turns in the secondary is 100, then the secondary voltage will be 40 V when the primary voltage is 120 V:

$$V_s = 120 \frac{100}{300} = 40 \text{ V}.$$

Note that the transformer does not supply any power, so that even though primary and secondary voltages may differ, equal amounts of power are found in the primary and secondary for an ideal transformer: $P_p = P_s$. Most transformers (except for superconducting transformers) are not ideal and will have losses.

18.5 ■ ELECTRICAL RESONANCE

Mechanical vibrators, such as pendulums, mass-spring systems, open and closed pipes, and Helmholtz resonators, were discussed in Chapter 2. Each of these vibrators has its own natural frequency and, if acted on by a driving force at this frequency, resonance occurs (see Chapter 4).

FIGURE 18.9
(a) A simple circuit
with inductance L
and capacitance C
in series.
(b) Response of this
circuit to a driving
voltage; when
$f = f_0$ resonance
occurs and the
current I reaches
its maximum value.

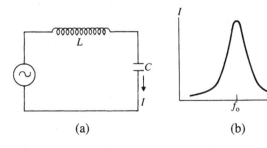

(a)   (b)

It is possible to have resonance in electrical circuits as well. The most common example of an oscillatory circuit is one with inductance and capacitance, such as the simple circuit shown in Fig. 18.9. The resonance frequency of this circuit is

$$f_0 = \frac{1}{2\pi\sqrt{LC}}. \tag{18.6}$$

As the frequency of the source is varied, the current shows a maximum at the resonance frequency f_0, which is analogous to the maximum amplitude at resonance in Fig. 4.2. The maximum current and the line width in the electrical circuit are determined by the internal resistance of the inductor, the capacitor, and the driving source.

A slightly different kind of electrical circuit, which also shows resonance, is the one shown in Fig. 18.10, in which an inductor and a capacitor are connected in parallel. Now as the frequency of the driving source is increased, the current in the inductor decreases but the current in the capacitor increases (see Fig. 18.7). At the resonance frequency f_0, the two currents are equal, but, for reasons that we will not discuss, they are in opposite phase (that is, the currents I_L and I_C at any instant are in opposite directions). Thus the net current from the source drops to a minimum value (which would be zero except for internal resistance in the inductor and capacitor). A graph of current versus frequency (see Fig. 18.10(b)) is a sort of "upside-down resonance curve," but if we prefer the more familiar line shape, we can make a graph of impedance Z versus frequency, as in Fig. 18.10(c).

FIGURE 18.10

(a) A circuit with inductance L and capacitance C in parallel. (b) The total current reaches a minimum value at the resonance frequency f_0. (c) Impedance Z reaches a maximum at f_0.

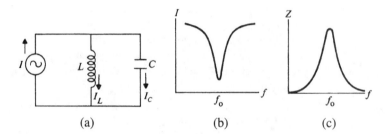

(a)   (b)   (c)

A very common application for a circuit of this type is the tuning circuit in a radio or television receiver, shown in Fig. 18.11. In this case, the tuning capacitor C is variable, so the resonance frequency can be changed in order to select different stations. The AM broadcast band extends from 550 kHz to 1600 kHz, and stations are assigned to frequencies

FIGURE 18.11

Tuning circuit in a radio or television receiver uses a circuit with a variable frequency of resonance.

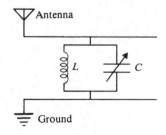

that differ by 10 kHz, so the resonance linewidth should be somewhat less than 10 kHz. (The FM band is somewhat less crowded and greater linewidths can be used to allow for higher fidelity.)

Examples Involving Resonance of Electrical Circuits

1. Find the resonance frequency of a 100-mH inductor and a 0.1-μF capacitor connected in series.

   Solution

   $$f = \frac{1}{2\pi\sqrt{LC}} = \frac{1}{2(3.14)\sqrt{(10^{-1})(10^{-7})}}$$

   $$= \frac{10^4}{2(3.14)} = 1592 \text{ Hz}$$

2. If $L = 0.2$ mH and $C = 200$ pF in a radio tuning circuit, to what frequency is the radio tuned?

   $$f = \frac{1}{2\pi\sqrt{LC}} = \frac{1}{2(3.14)\sqrt{(2 \times 10^{-4})(2 \times 10^{-10})}} = 7.96 \times 10^5 \text{ Hz}$$

   $$= 796 \text{ kHz.}$$

3. Show that the currents I_C and I_L in Fig. 18.10 are equal when

   $$f = \frac{1}{2\pi\sqrt{LC}}$$

   Solution Use the formulas in the box in Section 18.3 and set $I_C = I_L$:

   $$I_C = 2\pi f C V = I_L = \frac{V}{2\pi f L}.$$

   Multiply each side of the equation by f and divide by $2\pi C V$:

   $$f^2 = \frac{1}{(2\pi)^2 LC} \quad \text{or} \quad f = \frac{1}{2\pi\sqrt{LC}}.$$

18.6 ■ FILTERS

It is frequently desirable to filter out signals of high or low frequency or to alter the balance between them. This can be accomplished by using a passive frequency-selective circuit called a *filter*. An example of a filter is the treble tone control circuit in an audio amplifier which attenuates high-frequency signals by the desired amount. Such a filter is called a *low-pass* filter, because signals of low frequency pass through without attenuation. Other types of filters are called high-pass, band-pass, and band-reject.

A *high-pass* filter passes signals of high frequency but attenuates those of low frequency. A *band-pass* filter passes signals within a certain frequency band but attenuates all others. The reverse of a band-pass filter is a *band-reject*, or *notch*, filter, which attenuates only signals within a certain frequency band and passes all others. The *cutoff frequency* of a high-pass or low-pass filter is the frequency at which the response has dropped to 71% of its maximum value. Band-pass and band-reject filters are described by a resonance frequency at the center of the frequency band they filter or by the cutoffs of the pairs of filters comprising them. Characteristics of the four basic filter types are shown in Fig. 18.12.

Modern electronic instruments frequently make use of rather sophisticated filter networks. However, rather basic filters can be constructed from simple combinations of a

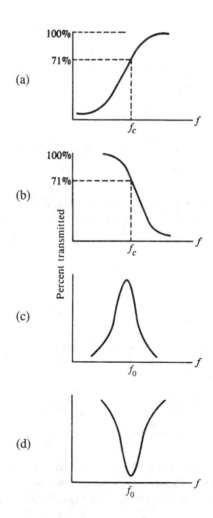

FIGURE 18.12
Characteristics of basic filter types: (a) high-pass; (b) low-pass; (c) band-pass; (d) band-reject. The cutoff frequency is f_c and the resonance frequency is f_0.

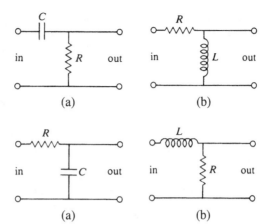

FIGURE 18.13
High-pass filters.

FIGURE 18.14
Low-pass filters.

resistor with a capacitor or an inductor, whose impedances, you will recall, depend on frequency.

Either of the networks shown in Fig. 18.13 can act as a high-pass filter. In the first, the low frequencies are attenuated by the series capacitor, whereas high frequencies are allowed to pass. In the second network, the inductor acts as a shunt (*short circuit*), reducing low-frequency signals, whereas high frequencies pass on through the filter.

Low-pass filters are just the reverse of high-pass filters. In the first example in Fig. 18.14, high frequencies are shunted by the capacitor; in the second example, frequencies above the cutoff are attenuated by the inductor in series.

The cutoff frequency of a basic *RC* high-pass filter is

$$f_c = \frac{1}{2\pi RC}. \tag{18.7}$$

Expressing *C* in farads and *R* in ohms gives f_c in hertz. The cutoff frequency of a basic *RL* high-pass or low-pass filter is

$$f_c = \frac{R}{2\pi L}; \tag{18.8}$$

inductance *L* is expressed in henries.

Practical filters in electronic instruments frequently combine two or more of the simple networks we have just discussed. A bandpass filter, for example, can be constructed by combining a high-pass filter and a low-pass filter. (The high-pass filter, of course, must have a slightly lower cutoff frequency than the low-pass filter does.) Another way to make a bandpass filter, however, is by using a resonance circuit, as shown in Fig. 18.15. The

FIGURE 18.15
A bandpass filter, which utilizes a parallel combination of an inductor and a capacitor.

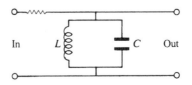

resonance frequency of the parallel LC circuit is

$$f_0 = \frac{1}{2\pi\sqrt{LC}},$$ (18.9)

and this becomes the center frequency of the pass band.

18.7 ■ ACTIVE DEVICES

Circuit elements discussed so far in this chapter (resistors, inductors, and capacitors) are passive devices; that is, they do not contain a power source. They have the ability to reduce the power of a signal but not to increase it. We will now introduce active circuit elements that use an external power source to control and amplify signals. Active devices fall into two classes, vacuum tubes and semiconductor devices.

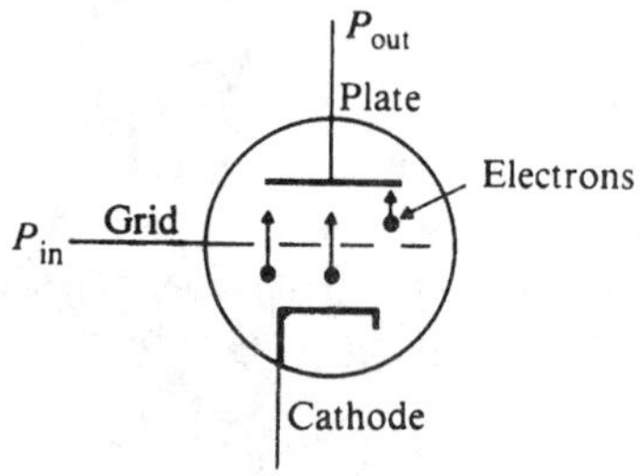

FIGURE 18.16
A vacuum tube that may be used to amplify electrical signals.

At the beginning of the twentieth century, development of the *electronic valve* opened the door to electronic circuits. The ability to control the flow of current in a circuit transformed the electrical world from power production to global communications. The vacuum tube diode allowed current to flow in one direction but not the other. The triode (shown in Fig. 18.16) has the ability to amplify currents. A hot cathode emits electrons, a plate collects them, and one or more grids control their rate of flow. A small power, P_{in}, supplied to the grid controls a larger power, P_{out}, in the plate; hence, amplification takes place. Armed with these two electronic valves (and modifications that followed), radio, radar, television, and music synthesis were developed. Vacuum tubes have been replaced by semiconductor devices in most applications; however, they are still used in musical electronics.

FIGURE 18.17
A diode consisting of a p-region, an n-region, and a junction between them. The symbol for the diode indicates the direction of easy current flow.

The properties of conductors (which allow a flow of current) and insulators (which resist the flow of current) have been known for many years. Research into the properties of semiconductor materials, such as silicon (with electrical properties between those of

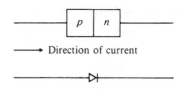

FIGURE 18.18
An *npn* junction
transistor. The flow
of electrons from
emitter to collector
is controlled by
applying an
electrical signal to
the base.

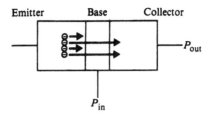

conductors and insulators) led to the development of new electronic valves called diodes
and transistors.

Semiconductor diodes (Fig. 18.17) consist of a tiny silicon crystal with two regions of
carefully selected impurities: a *p*-region with a deficiency of electrons and an *n*-region
with an excess of electrons. Current flows easily from the *p*-region to the *n*-region but not
in reverse. This one-way device is used as a detector in radio circuits and a rectifier in
power supplies.

An npn transistor is divided into three regions (Fig. 18.18): two *n*-regions (called the
emitter and collector) separated by a *p*-region (called the base). A flow of a substantial
electron current between the emitter and the collector is controlled by a relatively small
current applied to the base, thereby becoming a current amplifier. Figure 18.19 compares
a transistor amplifier to a water amplifier. In both cases, a small amount of power, P_{in},
controls a larger amount of power, P_{out}, giving a power gain of P_{out}/P_{in}. The arrow on the
transistor symbol shows the direction of the positive current flow (even though electrons
are actually flowing in the opposite direction). The output power actually comes from the
battery; however it is controlled by the power into the base.

FIGURE 18.19
(a) A water
amplifier circuit in
which a small
amount of power
P_{in} applied to a
faucet controls a
relatively large
amount of power
P_{out} in the water
from a reservoir.
(b) A transistor
amplifier circuit in
which a small
amount of power
P_{in} applied to the
base controls the
flow of a larger
power P_{out} in the
collector circuit.

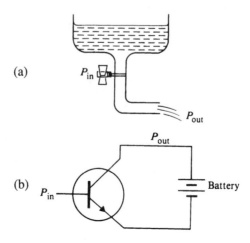

FIGURE 18.20
Amplifier symbol
used in simplified
circuit diagrams.

P_{in} ———————▷——————— P_{out}

18.8 ■ AMPLIFIERS

Often in simplified electronic circuit diagrams, an amplifier is merely denoted by a simple triangular symbol with input and output, as shown in Fig. 18.20. The active element may be either a vacuum tube or a transistor. One of the most significant parameters in describing an amplifier is its *voltage gain*, that is, the ratio of the output voltage to the input voltage. If this ratio remains constant, the amplifier is said to have a *linear* characteristic. (See Fig. 18.21.)

Two other important parameters of an amplifier are its current gain and its power gain. *Current gain* is the ratio of output current to input current, and *power gain* is the ratio of output power to input power. Since electrical power is the product of current times voltage, it is not difficult to see that the power gain will be the current gain times the voltage gain.

An amplifier might be termed a *voltage amplifier* or a *power amplifier*, depending on whether it is designed primarily to have a large voltage gain or a large power gain. An example of a voltage amplifier is a preamplifier designed for use with a microphone. The amplifier that drives a loudspeaker, on the other hand, is a power amplifier. A power amplifier has a low output impedance so that it can deliver a large output current.

Audio amplifiers nearly always incorporate several individual stages, some of which are voltage amplifiers (the early stages) and some of which are power amplifiers (the output stage and the stage that precedes it).

An ideal amplifier is one in which the output faithfully resembles the input but at a higher level. No amplifier is capable of performing this task perfectly. Real amplifiers invariably produce some distortion, although in high-quality amplifiers, the distortion may be small enough to be unnoticeable.

We will discuss two types of distortion: limited frequency response and nonlinear gain. An audio amplifier should have a "flat" frequency response (that is, gain that is independent of frequency) over the audible range of at least 20 to 20,000 Hz. Nearly all high-quality amplifiers meet this criterion. In low-quality amplifiers, however, low-frequency "roll-off"

FIGURE 18.21
A high-fidelity
amplifier has a
linear characteristic
(that is, the graph of
output versus input
is a straight line).

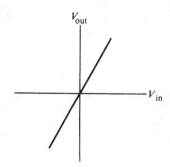

FIGURE 18.22
Gain versus frequency characteristic of an amplifier showing roll-off in gain at high frequency (B) and low frequency (A). The bandwidth is measured between the 3-dB points.

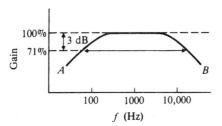

(at point A in Fig. 18.22) may be caused by coupling capacitors that are too small and therefore present too much impedance to the passage of low-frequency signals from one stage to the next. High-frequency roll-off (at point B in the figure) could be caused by stray capacitance in the circuit or by transistors that have insufficient gain at high frequency.

The frequency response of an amplifier is often described by the bandwidth between the "3-dB points," that is, the frequencies at which the voltage gain has dropped to 71% of its midfrequency value. At these points, the gain is 3 dB below its midfrequency value.

In most amplifiers, there is a range of input signal V_{in} for which the output faithfully resembles the input. For larger signals, however, the output may be distorted, as shown in Fig. 18.23. Supplying an amplifier with an input signal that takes it outside its linear region is referred to as "overdriving" the amplifier. This results in "clipping" the top and bottom of the signal. The distorted output frequently contains harmonics that were not present in the input signal.

FIGURE 18.23
A distorted output may result if an amplifier is driven outside its linear region. The distorted output has a greater harmonic content than the input.

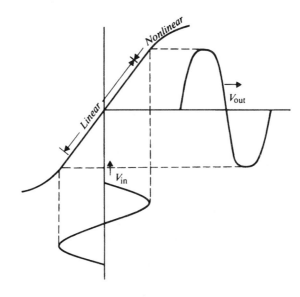

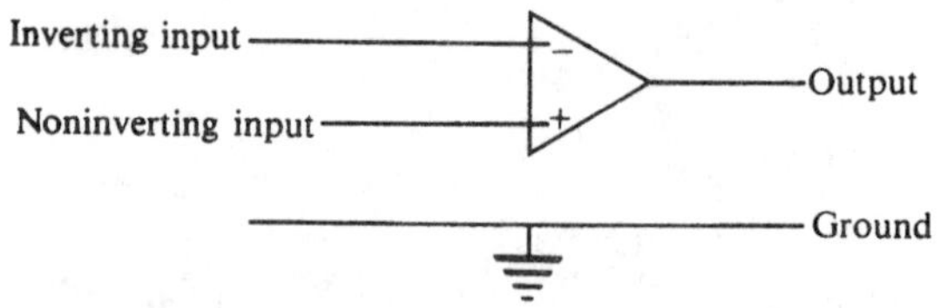

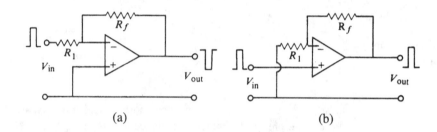

FIGURE 18.24
A schematic for an operational amplifier (op amp).

FIGURE 18.25
An operational amplifier with negative feedback:
(a) inverting amplifier;
(b) noninverting amplifier.

18.9 ■ OPERATIONAL AMPLIFIERS

Integrated circuit technology has made operational amplifiers, or *op amps*, so common that mention should be made of them in this chapter.

An operational amplifier is a high-gain amplifier that actually consists of several transistors, resistors, and other components fabricated on a single semiconductor chip. In the simplified schematic in Fig. 18.24, two inputs are shown: One is called *inverting* ($-$) and one is called *noninverting* ($+$). A signal applied to the noninverting input produces an output signal with the same phase as the input, whereas a signal applied to the inverting input produces an output of opposite phase.

A resistor R_f is usually connected from the output to the inverting input, as shown in Fig. 18.25, to provide negative feedback. Negative feedback increases the frequency response and reduces distortion at the expense of a reduced gain. The input signal may be applied to either the inverting or the noninverting input, as shown.

The voltage gain is determined by the ratio of the resistances R_f/R_1. In the inverting amplifier, the voltage gain is $A = R_f/R_1$, whereas in the noninverting amplifier, it is $A = (R_f/R_1) + 1$.

18.10 ■ OSCILLATORS AND FUNCTION GENERATORS

The opposite of negative feedback is positive feedback. Negative feedback decreases the gain of an amplifier but makes it more stable; positive feedback *increases* the gain but may make it *unstable*. An amplifier that becomes sufficiently unstable generates signals of its own.

Amplifier instability is not necessarily bad. An amplifier that is unstable at one frequency, only, can be useful as an *oscillator*, or *generator*. This selective instability can be accomplished by means of positive feedback through a band-pass filter. This is, in fact, how most audio generators operate. The band-pass filter can be made tunable so that the generator can be tuned to the desired frequency (see Fig. 18.26).

FIGURE 18.26
Positive feedback
through a
band-pass filter
causes oscillation at
a single frequency
determined by the
filter.

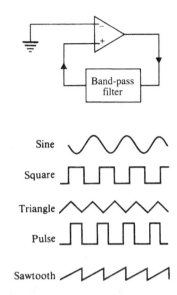

FIGURE 18.27
Waveforms
available from a
function generator.

Audio generators or oscillators used in the laboratory and studio are usually tunable over a wide frequency range (e.g., 20 Hz to 1 MHz). There is frequently a *range*, or *multiplier*, switch, which changes the frequency by factors of 10, plus a continuous frequency control for selecting the desired frequency within the range. Audio generators often furnish square waves as well as sine waves (pure tones), and an output amplitude control is usually included.

Popular in the laboratory are function generators, which generate a variety of electrical waveforms or functions (see Fig. 18.27). Sine waves, square waves, and triangle waves are nearly always available, and other waveforms such as the pulse and the sawtooth may be available as well. The frequency of many function generators can be controlled electrically; that is, the frequency can be changed from that indicated on the frequency dial by applying an electrical control voltage.

18.11 ■ POWER SUPPLIES

Nearly all electronic circuits require a source of direct current. This can be supplied by batteries, as it is in portable equipment, but a more economical source of power is the ac power line. By using a transformer and diodes, it is possible to change 120-V ac power to dc power at any voltage desired. This can be done with the simple rectifier circuit shown in Fig. 18.28.

An ac of 120 V is supplied to the primary of the transformer. By selecting the appropriate ratio N_p/N_s, the desired ac voltage V_s will appear at the secondary. This ac is then *rectified*, or changed, to a pulsating dc by means of the diode, which permits current to flow in one direction only (see Section 18.7). The waveform of the pulsating dc voltage V_R across the load resistor is shown in Fig. 18.28(b).

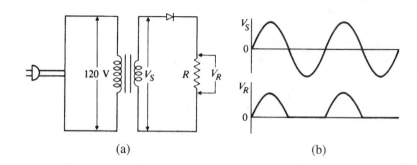

(a)   (b)

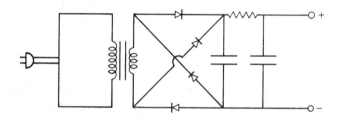

A more practical power-supply circuit, shown in Fig. 18.29, works on the same principle as the simple rectifier circuit shown in Fig. 18.28, but by using additional diodes and
capacitors, the pulsations in the dc voltage are smoothed out so that a steady dc is available.

18.12 ■ SUMMARY

An electric current is the flow of electric charge; an electric current can carry energy from
one place to another. An analogy can be drawn between an electrical circuit with a battery
and a water circuit with a pump. Alternating current (ac), like direct current (dc), conveys
energy or power, but it reverses its direction of flow periodically. Inductors and capacitors
are important elements in ac circuits, and they can be described by an impedance of opposition to the flow or current. An electrical circuit with inductance and capacitance has a
resonance frequency at which the current reaches a maximum (series circuit) or a minimum
(parallel circuit).

A filter is a frequency-selective circuit that may have a high-pass, low-pass, band-pass,
or band-reject characteristic. Band-pass filters may incorporate circuits that have a resonance at the desired frequency.

A diode allows current to flow easily in one direction but resists its flow in the reverse
direction. A transformer can change the voltage of an alternating current. A power supply
generally uses a transformer and one or more diodes to produce dc power from ac power.

An amplifier is a device in which a small amount of input power controls a larger amount of output power. Electrical amplifiers accomplish this by using a transistor, vacuum tube, or similar device as a valve.

Distortion can be reduced in an amplifier by negative feedback. Positive feedback, on the other hand, causes an amplifier to become unstable and eventually to oscillate. Audio generators or oscillators often are amplifiers with positive feedback at a selected frequency. Function generators provide a variety of waveforms and often have provision for voltage control of frequency.

REFERENCES AND SUGGESTED READINGS

Horowitz, P., and W. Hill (1989). *The Art of Electronics*, 2nd ed. New York: Cambridge University Press.

Meade, R. L. (1994). *Foundations of Electronics*. New York: Delmar Publishers.

Mimms, F. M. (1983). *Getting Started in Electronics*. Radio Shack.

White, D., and R. Doering (1997). *Laboratory Manual for Electrical Engineering Uncovered*. Upper Saddle River, N.J.: Prentice Hall.

GLOSSARY

ac (alternating current) Electric current that reverses its direction of flow several times each second (120 times each second, in the case of electrical power distributed in the United States having a frequency of 60 Hz).

amplifier A device in which a small amount of input power controls a larger amount of output power.

base, collector, emitter The three regions of a transistor; in the most common usage, the input signal is applied to the base and the output is taken from the collector.

capacitor A device that stores energy by creating an electrical field between two conductors; ac can flow through a capacitor but dc cannot.

current The flow of electrical charge. Current is measured in amperes.

current gain The ratio of output current to input current.

dc (direct current) Electric current that flows in one direction only, such as that supplied by a battery.

diode A device that allows easy current flow in one direction only.

distortion A measure of the difference between the output and input signals in an amplifier.

energy The ability to do work. Various units are used to measure energy, such as joules, BTUs, calories, or foot-pounds.

filter An electrical circuit that passes alternating currents of some frequencies and attenuates others. Basic filter types are high-pass, low-pass, band-pass, and band-reject.

function generator An audio generator that provides several different waveforms or functions at the desired frequency.

impedance A measure of the opposition to the flow of electric current by a circuit element such as a resistor, capacitor, or inductor. Impedance is measured in ohms.

inductor A device that stores energy by creating a magnetic field, usually within a coil of wire.

Ohm's law A fundamental law that relates electric current I, voltage V, and resistance R; written $I = \frac{V}{R}$, or $V = IR$.

operational amplifier (op amp) A high-gain amplifier with a large amount of negative feedback and high input impedance.

potential, or **voltage** A measure of the electrical force, or pressure, that causes a current to flow; typically supplied by a generator or a battery and measured in volts.

power The rate at which energy is supplied or that rate at which work is done. It is measured in watts; one watt equals one joule per second.

power gain The ratio of output power to input power.

rectifier A diode that is used to change ac into pulsating dc.

resistor A device that converts electrical energy into heat.

resonance The natural frequency of a system, at which its response to a mechanical or electrical force reaches a maximum.

transformer A device that changes ac at one voltage into ac at a higher or lower voltage.

transistor A solid-state amplifying device consisting of a crystal of germanium or silicon with carefully selected impurities.

vacuum tube An amplifying device in which electrons from a hot cathode flow through one or more grids before reaching a plate.

voltage amplifier An amplifier designed mainly for voltage gain.

voltage gain The ratio of output voltage to input voltage.

REVIEW QUESTIONS

1. What is the difference between direct current and alternating current?

2. What actually moves in an electric current?

3. As one increases the resistance in a circuit, what must be done with the voltage to keep the current the same?

4. When one adds resistors in series, is the resultant resistance higher or lower than the individual resistance?

5. When one adds resistors in parallel, is the resultant resistance higher or lower than the individual resistance?

6. To increase the electrical power, does one increase the voltage or the current?

7. What is the difference between an inductor and a capacitor in the method it uses to store energy?

8. Which passes current easiest at high frequencies, a capacitor or an inductor?

9. What is the purpose of a transformer in a power supply?

10. If the inductance is increased in a resonant circuit is the frequency increased or decreased?

11. Which type of filter blocks higher frequencies?

12. What is the difference between a semiconductor diode and a transistor?

13. Why does an amplifier reduce its gain at high and low frequencies?

14. What is the cause of clipping in an amplifier?

15. What is an op amp?

16. How does an oscillator differ from an amplifier?

QUESTIONS FOR THOUGHT AND DISCUSSION

1. Explain why a fluorescent lamp flashes 120 times per second (why not 60 times, for example?). Also, why does it remain lit between flashes?

2. Sketch a simple circuit for a battery charger using a diode as a rectifier.

3. The circuit shown below might serve as a treble tone control in an amplifier.

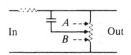

Explain how it works. Will more treble sound be heard in position A or in position B?

4. List several advantages of using transistors rather than vacuum tubes. Can you think of any possible advantages of tubes?

5. Is it possible to produce a nearly sinusoidal wave (single frequency) by filtering a square wave? Would you use a high-pass or a low-pass filter? (See Section 7.10, Fig. 7.12 in particular.)

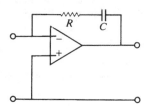

6. (a) Will the negative feedback in this amplifier be greater at high or at low frequency? Explain.

   (b) Will the voltage gain be greater at high or low frequency? Explain.

EXERCISES

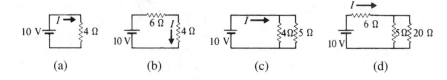

(a)   (b)   (c)   (d)

1. In each of the above circuits, find the current I.

2. Using the figure, determine how much power is consumed if

(a) $R = 20 \ \Omega$;

(b) $R = 40 \ \Omega$.

3. Find the resonance frequency of each circuit shown here.

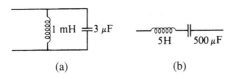

(a)   (b)

4. The accompanying circuit represents a flashlight.

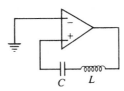

(a) How much resistance should the bulb have in order for the current to be 2 A?

(b) How much power will be supplied to the bulb?

5. What is the total resistance of

(a) Resistances of 5 Ω, 6 Ω, and 15 Ω in parallel?

(b) Resistances of 20 Ω and 5 Ω in parallel?

6. Show when resistances R_1 and R_2 are connected in parallel, the total resistance is

$$\frac{R_1 R_2}{R_1 + R_2}.$$

Test this out with the resistances given in Exercise 5(b).

7. A 5-V source with an internal resistance $r = 10 \ \Omega$ delivers power to a 5- Ω load. How much power is delivered to the load? (*Hint:* Find the current and then use Ohm's law to find the voltage across the load. The product of these two quantities will be the power delivered to the load.) Repeat for a "matched" load, $R = 10 \ \Omega$, and a load of $R = 15 \ \Omega$.

8. A transformer is to furnish an ac of 6 V from a 120-V power source.

(a) What should the turns ratio be?

(b) If the current in the load is 2 A, what is the current in the primary?

(c) Verify that power in the primary and the secondary are the same by multiplying current and voltage in the primary and the secondary.

9. Find the cutoff frequency of each circuit shown and tell whether it is a high-pass or low-pass filter.

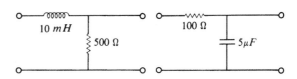

10. (a) Find the voltage gain of the amplifier shown.

(b) Is it an inverting or a noninverting amplifier?

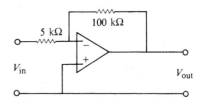

11. A certain amplifier has a voltage gain of 100 and a current gain of 50. What is its power gain? Using the same formula as that used for sound power (see Section 6.4), express the power gain in dB.

12. Two amplifier stages, each having a power gain of 500, are cascaded, with the output of one connected to the input of the other.

(a) Show that the combined amplifier has a power gain of 250,000.

(b) Show that the total gain in dB is the sum of the individual amplifiers.

13. A series LC circuit acts as a band-pass filter with

$$f_0 = \frac{1}{2\pi \sqrt{LC}}.$$

(a) Explain how the operational amplifier circuit shown here functions as an oscillator.

(b) At what frequency will it oscillate if $L = 5$ mH and $C = 10 \ \mu F$?

14. An oscillator usually has several ranges. If the capacitor C in the circuit shown can be varied from 100 to 500 pF, what frequency range can be tuned with each value of L?

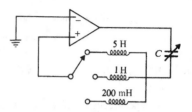

EXPERIMENTS FOR HOME, LABORATORY, AND CLASSROOM DEMONSTRATIONS

Home and Classroom Demonstrations

1. *Static electricity*   Hold one lead of a small neon lamp in your hands as you drag your feet across the carpet. Touch the other end to a metal doorknob and watch the lamp light up.

2. *Electrified plastic and glass*   Place small bits of paper on the table. Run a plastic comb through your hair then move the comb close to the bits of paper. They will be attracted to the comb. (This works best on a dry day.) Electrons are transferred from your hair to the comb, leaving it negatively charged which attracts the bits of paper. You can also rub a glass rod with a silk cloth which removes electrons from the glass, leaving it positively charged. Positive charges will also pick up the bits of paper.

3. *Opposite and like charges*   Hang small pieces of foam plastic from threads so they are spaced about an inch apart. Without being charged, the plastic will hang straight. Place an electrified comb (see above) between the two pieces. Negative charge will be transferred to both pieces of plastic, making them repel (move away from) each other, because like charges repel. Place the comb near one piece of plastic and the charged glass rod near the other. In this case negative charge is transferred to one piece and positive charge is transferred to the other. In this case they will move closer together, because opposite charges attract.

4. *Water analogy*   Demonstrate the analogy of water flow to electric current with the following experiment. Fill a plastic $\frac{1}{2}$-gal milk bottle with water. The height of the water represents the voltage. Drill a small hole at the bottom of the bottle and see how long it takes to fill up a $\frac{1}{2}$-c measure. The hole represents a resistance and the water flow represents the current. With the same hole, fill the bottle half full and see how long it takes to fill the $\frac{1}{2}$ c. Because the height (voltage) is reduced by one-half, the flow (current) will also be reduced, so it will take twice as long to fill up the measure. Next enlarge the hole (decrease the resistance) and repeat the experiment. See if you can prove Ohm's law by using water.

5. *Basic series parallel circuit*   Make up a basic series parallel circuit (shown here) on a display board using lamp sockets and a standard wall switch. Compare the effect of connecting R_1 and R_2 (series) with that of connecting R_2 and R_3 (parallel, with R_1 shorted out). Use various sizes of lightbulbs in different positions to predict which will be the brightest. Using all three lamps, and the switch to predict what will happen when some are in parallel and others are in series.

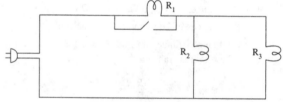

6. *Diode current control*   The switch on a flashlight breaks and connects the circuit, allowing current to flow (lighting up the lightbulb) or not (turning off the lightbulb). With the switch in the *off* position, connect a wire across the switch so current bypasses the switch and turns the light on. Try the same thing with a diode, turn it around, and try it again. When the diode is in the forward position, current flows and the light turns on. In the reverse position current does not flow, and the light remains off.

Laboratory Experiments

Audio tests and measurements (Experiment 27 in *Acoustics Laboratory Experiments*)

Amplifier performance 1 (Experiment 28 in *Acoustics Laboratory Experiments*)

Amplifier performance 2 (Experiment 29 in *Acoustics Laboratory Experiments*)

Series and parallel components (Lab 2 in *Laboratory Manual for Electrical Engineering Uncovered*)

RC filters (Lab 13 in *Laboratory Manual for Electrical Engineering Uncovered*)

Resonant filters (Lab 14 in *Laboratory Manual for Electrical Engineering Uncovered*)

Loudspeakers

It is virtually impossible to amplify sound waves (that is, to add energy to them as sound waves). Electrical signals, on the other hand, are relatively easy to amplify. Thus, a practical system for amplifying sound includes input and output *transducers*, together with the electronic amplifier. The input and output transducers are usually a microphone and a loudspeaker, respectively. A microphone converts a sound signal to an electrical signal, and a loudspeaker converts the amplified and modified electrical signal to a much louder sound, as shown in Fig. 19.1.

Similarly, to record music it is necessary to convert sound signals to electrical signals, which can be amplified in order to drive the recording equipment, as in Fig. 19.2. The playback system includes an amplifier and a loudspeaker to convert electrical signals into sound. Microphones and signal modifying equipment will be discussed in Chapter 20, while recording equipment will be dealt with in Chapter 21.

In this chapter you should learn:

- How loudspeakers are constructed;
- About the characteristics of loudspeaker enclosures;
- About designing loudspeaker enclosures;
- About multispeaker systems;
- About loudspeaker efficiency and distortion;
- About noise-canceling headphones.

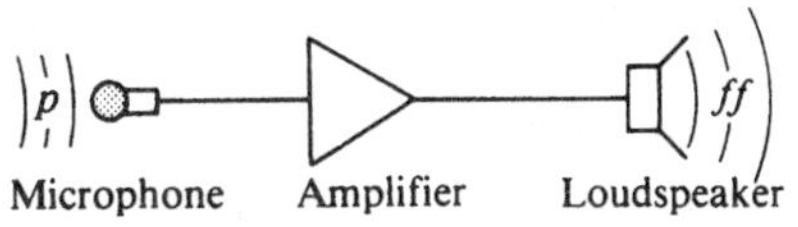

FIGURE 19.1
Public address system consisting of a microphone, an amplifier, and a loudspeaker. Sound is converted to an electrical signal, which is amplified and used to generate a louder sound.

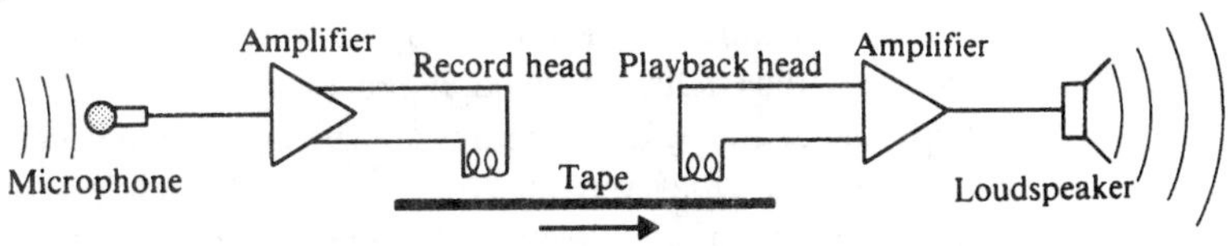

FIGURE 19.2
Tape recording system. To the system shown in Fig. 19.1 are added record and playback heads and a second amplifier.

423

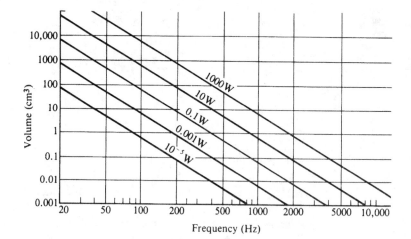

FIGURE 19.3
Volume of air moved by a loudspeaker as a function of sound power and frequency.

19.1 ■ LOUDSPEAKERS AS TRANSDUCERS

A loudspeaker is a transducer that converts electrical energy into acoustic energy. Actually, this conversion takes place in two steps: The electrical signal causes mechanical motion of the speaker cone or diaphragm, which in turn causes the pressure waves in the air that we call sound.

How much air do we actually need to move? That depends on the sound power and the frequency, as shown in Fig. 19.3. It is clear from this graph that loudspeakers for low-frequency reproduction (*woofers*) must be considerably larger than high-frequency transducers (*tweeters*).

There are two principal types of loudspeakers: those in which the vibrating surface (cone or diaphragm) radiates sound directly into the air, and those in which a horn is interposed between the diaphragm and the air. Although *horn* loudspeakers have a higher efficiency, most home loudspeaker systems rely on *direct radiator* loudspeakers, at least at low frequencies.

19.2 ■ STRUCTURE OF DYNAMIC LOUDSPEAKERS

A *dynamic* loudspeaker is an electromagnetic transducer consisting of a voice coil, a cone diaphragm, and a magnet structure with a small gap in which the voice coil moves (see Fig. 19.4). The magnet structure is designed to provide a strong magnetic field across the gap, so that when a current flows in the voice coil it will experience a force. Although early dynamic speakers used electromagnets, modern loudspeakers have high-efficiency permanent magnets, thus eliminating the need to supply a current to the electromagnets.

An alternating electric current in the voice coil causes it to move in and out in the magnetic field, thus radiating sound. In order to move a larger quantity of air and thereby radiate more sound, a cone diaphragm is attached to the voice coil and moves in and out with the coil. In general, the larger the cone the better the radiation at low frequencies because of the greater volume of air moved by the cone. On the other hand, a large cone has a large mass,

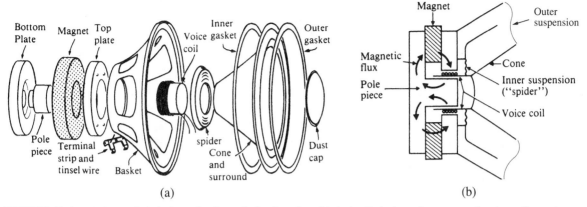

FIGURE 19.4 (a) An exploded view of a dynamic loudspeaker. (b) A detailed view of magnet and voice coil structures; note the path of magnetic flux. (Part (a) courtesy of Speakerlab, Inc., Seattle, Washington.)

and it is difficult to move at high frequencies. Thus it is difficult (although not impossible) to build a single loudspeaker that radiates efficiently at both high and low frequencies. Many schemes have been devised to overcome this difficulty, but the most common solution is to use two or more speakers with an electrical or mechanical crossover to separate low-frequency and high-frequency signals, as will be discussed in Section 19.10.

Several different magnet designs are used in loudspeakers. Three of the most common types are illustrated in Fig. 19.5. Each one consists of a cast magnet of a special magnetic alloy or a ceramic material. In general, the greater the magnetic flux, the greater the efficiency of the speaker. However, efficiency is not the only criterion in magnet design. For high-fidelity reproduction of sound, the magnetic field in the gap must be uniform throughout the maximum distance that the voice coil moves.

The magnet is usually one of the most costly components in loudspeaker manufacture, so there is a tendency to make the magnets as small as practical. High-quality speakers, however, require substantial magnet systems. Some manufacturers of loudspeakers customarily specify the magnet weight in the loudspeaker specifications, whereas other manufacturers specify the magnetic flux across the gap.

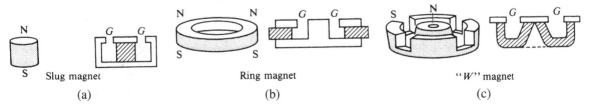

FIGURE 19.5 Three types of permanent magnets commonly used in loudspeakers. At the right are shown the complete magnet systems, consisting of the magnet itself (shaded) plus a yoke of soft iron to complete the flux path and provide a circular gap in which the voice coil moves (indicated by G).

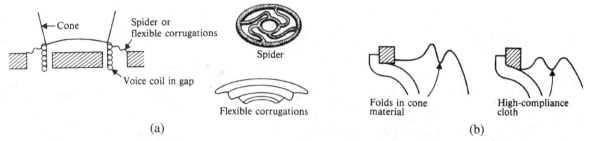

FIGURE 19.6   (a) The inner suspension of a loudspeaker consists of a flexible spider or cloth corrugations, which keep the voice coil centered in the magnet gap. (b) The outer suspension may consist of high-compliance cloth or soft folds in the cone material itself.

An important consideration in loudspeaker design is the suspension of the cone and the voice coil. In addition to supporting the cone, the suspension system must keep the voice coil accurately centered in the magnet gap, provide springlike action to bring the coil back to its equilibrium position with no input signal, and also to provide some degree of mechanical damping. This is quite an order!

The *compliance* of a speaker indicates how flexible its suspension is. Compliance is expressed in units of distance divided by force (m/N or cm/dyne). Most suspension systems use a flexible inner suspension to keep the voice coil centered in the magnet gap plus an outer suspension to support the cone. The inner suspension may be a plastic or thin metal "spider" with long, thin legs (see Fig. 19.6(a)) or a plastic-impregnated corrugated cloth. The outer suspension may consist of soft folds in the cone material itself, or a high-compliance cloth may be used to attach the cone to the frame (see Fig. 19.6(b)). The outer suspension is usually designed to provide edge damping to inhibit standing waves of short wavelength in the cone.

19.3 ■ ELECTROMAGNETIC DRIVERS

Loudspeakers, like electric motors, illustrate the most basic principles of electromagnetism: the interaction of electric currents, magnetic fields, and mechanical forces (or motion). These principles, in fact, underlie most of modern industrial technology, and so it is important to understand them.

In 1820, Hans Christian Oersted discovered that electric currents affect a magnetic compass, and therefore that magnetism is related to electricity. This discovery led to the observation that electric currents are the source of magnetism. In an electromagnet the electric current flows in a wire; in a permanent magnet a sort of electric current flows within the atoms of the magnetic material. In both cases, we can imagine a *magnetic field* as a sort of map showing the direction that a compass needle would point at every point in the field. It is helpful to draw a magnetic field vector **B** that shows both the direction and the strength of the magnetic field. Often **B** is called the *magnetic B-field* or simply the magnetic field.

The *B*-field, since it points in the direction of a compass needle, will point from the north pole of a horseshoe magnet to its south pole, as shown in Fig. 19.7(a). Around a

FIGURE 19.7
Magnetic B-field in
or near: (a) a
horseshoe
permanent magnet;
(b) a
current-carrying
wire; (c) a
current-carrying
coil and a
cylindrical
permanent magnet.

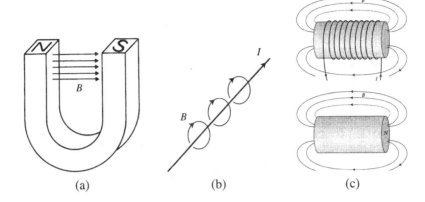

(a)   (b)   (c)

current carrying wire, the B-field is circular, as shown in Fig. 19.7(b). In or near a current carrying coil or near a cylindrical permanent magnet, the B-field makes the path shown in 19.7(c). You may wish to sketch the B-field for each of the magnets in Fig. 19.5.

Force on a Current-Carrying Wire in a Magnetic Field

A current carrying wire in a magnetic field experiences a force, as shown in Fig. 19.8(a). The force is in a direction perpendicular to both the current and the B-field, as shown. Is it upward or downward? One can use the fingers of the right hand to determine this: If the forefinger points in the direction of the current and the middle finger in the direction of the B-field, then the thumb will indicate the direction of the force, as shown in Fig. 19.8(b).

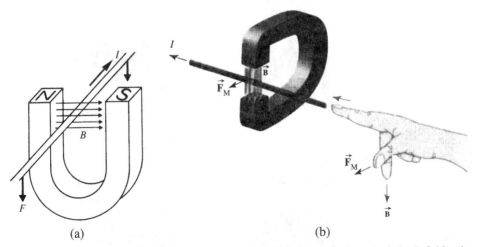

(a)   (b)

FIGURE 19.8 (a) Force on a current-carrying wire in a magnetic field is perpendicular to both the B-field and to the current. (b) Right-hand rule for determining the direction of the force.

In a loudspeaker driver the alternating current is constantly reversing its direction, so the force reverses its direction at the same frequency.

Force on a Loudspeaker Drive Coil

In a loudspeaker, the current-carrying wire is in the form of a cylindrical coil and the magnetic B-field is radial across the magnet gap, so the driving force is along the axis of the speaker, alternating in direction as the current alternates. The driving force is proportional to BLI, where B is the strength of the B-field, L is the total length of wire in the magnetic field, and I is the current. To maximize the driving force, for a given current, it is clear that both B and L should be as large as possible. Maximizing B is done by using as powerful a magnet as possible and minimizing the width of the gap (G in Fig. 19.5). Maximizing L could be accomplished by winding as many turns of wire in the voice coil as possible. More turns of wire means greater resistance, however, and more ohmic (I^2R) power loss in the coil. Some high-efficiency voice coils are wound with flat ribbon or wire with a square, rather than round, cross section in order to maximize L while minimizing the gap width.

Two different gap/coil geometries are used in loudspeaker drivers, as shown in Fig. 19.9. The overhung coil in Fig. 19.9(a) is the most common. The distance x_{max} represents the distance the coil can move in one direction and maintain a constant number of turns in the gap. x_{max} tends to be smaller in the underhung coil shown in Fig. 19.9(b). In order for the force to be proportional to the current as the coil moves (that is, to maintain a linear relationship between F and I), the number of turns in the gap should remain constant. The overhung arrangement generally allows the largest voice coil excursion.

Voltage Induced in a Wire Moving in a Magnetic Field

When a wire moves in a magnetic field, a voltage is induced in the wire. Faraday's law of induction states that the voltage induced in a moving wire is proportional to the length of the wire, the B-field, and the speed of the wire. This transducer principle is used in

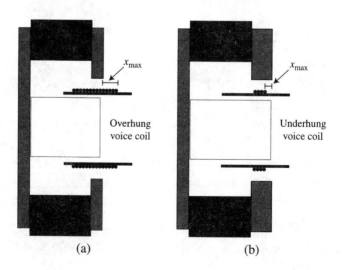

FIGURE 19.9
Overhung (a) and underhung (b) coil geometries.

Overhung voice coil

Underhung voice coil

(a) (b)

electrical generators (in automobiles, in power plants, etc.), and it is also used in dynamic microphones.

When a loudspeaker voice coil moves in a magnetic field, it not only works as a motor but it works as a generator as well. A voltage is induced in the moving voice coil that opposes the voltage applied to the coil by the audio amplifier. This so-called "back voltage" opposes the flow of current. It is most apparent at or near the resonance frequency of the loudspeaker, as we shall see.

19.4 ■ MECHANICAL AND ELECTRICAL CHARACTERISTICS

The dynamic loudspeaker is a rather complicated vibrator; however, a good starting place for understanding it is to model it as a simple mass-spring system (see Chapters 2 and 4). The mass of the system is the combined mass of the voice coil, the cone, and the air that moves with the cone. The spring constant of the system is determined by the inner and outer suspension elements. Normally, one speaks of the compliance C of a loudspeaker (flexibility of the suspension), which is the reciprocal of the spring constant K (stiffness of the suspension):

$$K = 1/C \tag{19.1}$$

A very compliant loudspeaker will move easily and have a low restoring force. The resonance frequency of a loudspeaker driver (sometimes referred to as the "cone resonance") is:

$$f_d = \frac{1}{2\pi}\sqrt{\frac{K}{m}} = \frac{1}{2\pi\sqrt{Cm}} \tag{19.2}$$

When a loudspeaker is driven by an oscillating current, it acts as a resonant system and has a characteristic Q (Section 4.1), defining the sharpness and height of the resonance peak. The Q of a resonant system expresses the ratio of the stored energy to the energy dissipated during each cycle of oscillation. If the damping is small, the Q is large. For a mass-spring system, the Q is given by:

$$Q = 2\pi f_d \frac{m}{R} = \frac{1}{R}\sqrt{\frac{m}{C}} \tag{19.3}$$

If the loudspeaker were driven with a force of constant amplitude and vibrated in a vacuum, it would have a response curve much like the simple mass-spring oscillator in Fig. 4.2. It doesn't, of course. The sea of air in which the cone moves contributes both mass and resistance. There are contributions to damping from both electrical and magnetic effects. In order to consider mechanical and electrical effects together, it is often wise to draw an equivalent electrical circuit. Such a circuit can be kept reasonably simple by including only the most important parameters or it can be made more exact and complicated by including small effects as well as large ones.

The electrical impedance of a loudspeaker can be measured by driving it with a constant current (an audio generator with a large resistance in series, for example) and measuring the voltage across the speaker terminals with a voltmeter. (Electrical impedance is voltage divided by current, as discussed in Section 18.3). At its resonance frequency, the voice coil

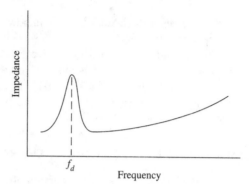

FIGURE 19.10
Impedance of a
loudspeaker as a
function of
frequency.

moves at maximum amplitude in the magnetic field, and therefore generates a maximum
back voltage (see Section 19.3). Above the resonance frequency, the impedance (voltage)
goes through a minimum, and then slowly increases with frequency due to the inductance
of the voice coil ($Z_L = 2\pi f L$), as shown in Fig. 19.10. The minimum impedance is close
to the "nominal" impedance of the speaker, generally 8 ohms.

19.5 ■ BAFFLES AND ENCLOSURES

In order for a loudspeaker to radiate sound efficiently, especially at low frequency, it must
be mounted on some sort of baffle. To understand why a baffle is important, note that
when the loudspeaker cone moves forward, a wave of increased pressure is generated at
the front, but a wave of decreased pressure is generated on the rear side of the cone. If
there is no baffle, there will be a short-circuit path around the edge of the cone from front
to back, as shown in Fig. 19.11. When this path is less than a quarter wavelength, the
radiation efficiency drops off rapidly. Thus the baffle must be large enough to eliminate
paths shorter than a quarter wavelength at the lowest desired frequency.

FIGURE 19.11
(a) Without a baffle
there is a
short-circuit air
path from the front
side of the cone to
the back side. (b) A
baffle eliminates
this path or greatly
increases the path
length.

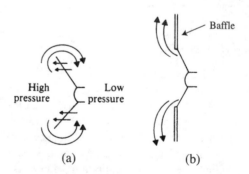

Two simple experiments demonstrate the important effect of a baffle on sound radiation. If a tuning fork is set into vibration and slowly passed through a slot in a
sheet of paper, as shown in Fig. 19.12(b), the loudness increases noticeably as each

tine passes through the slot. The paper, acting as a crude baffle, partially blocks the short-circuit path from one side of the vibrating tine to the other.

In the second experiment, a small speaker (two or three inches in diameter) with no baffle radiates very little sound at low frequency. Music thus sounds distorted. Placing the same speaker against a piece of plywood with an appropriate hole greatly increases the radiation of sound at low frequency, as illustrated in Fig. 19.12(d).

FIGURE 19.12
Two experiments that demonstrate the effect of a baffle: (a) tuning fork with no baffle; (b) tuning fork with one tine in a slot cut in a sheet of paper; (c) small loudspeaker with no baffle; (d) same loudspeaker with a plywood baffle.

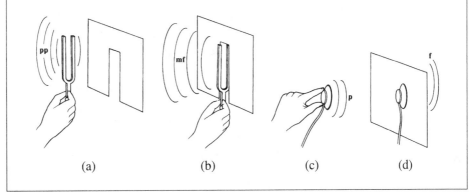

(a)   (b)   (c)   (d)

Mounting the loudspeaker in the wall of a large enclosure or room (effectively an "infinite" baffle) eliminates all short-circuit paths from back to front, but no use is made of the sound radiated from the back of the cone. Greater efficiency is obtained if some use is made of this back wave, especially at low frequency where it is difficult to achieve efficient radiation of sound. Therefore, many modern loudspeaker systems have a type of speaker enclosure that not only serves as a baffle but also as an acoustic circuit to transmit the back wave to the front of the enclosure in the proper phase to reinforce the sound radiated from the front of the cone. Three common types are the bass-reflex, drone-cone, and rear-horn enclosures.

19.6 ■ AIR SUSPENSION LOW-FREQUENCY LOUDSPEAKERS

The closed-box is the simplest of all loudspeaker enclosures. If the enclosure is made so large that the compliance of the enclosed air is greater than the compliance of the driver suspension, the system is essentially an infinite baffle system. If the compliance of the enclosed air is considerably less than that of the speaker driver, however, the system is called an *acoustic-suspension* or *air-suspension* system. In such a system, shown in Fig. 19.13, air in the box, compressed by the inward motion of the speaker cone, furnishes more of the restoring force than the inner and outer suspension of the cone.

Acoustics pioneer Harry Olson and his associates patented the air-suspension design in 1949, and it was made popular in the 1950s by Edgar Villchur and Henry Kloss, who founded AR (Acoustic Research). Drivers for use in air-suspension speakers generally have a low free-air resonance frequency, a relatively high cone mass, and a long voice coil. Furthermore, the driver should have a fairly high Q_d of 0.3 or greater. For best results,

FIGURE 19.13
An air-suspension
speaker system
consists of a
high-compliance
speaker driver
mounted in an
airtight box.

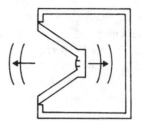

the ratio of resonance frequency to Q_d (sometimes called efficiency-bandwidth product) should be 50 or less (Dickason 2000).

Air-suspension systems are designed to operate with large excursions of the voice coil and the cone. This allows a relatively small speaker cone to radiate bass tones satisfactorily, even though the radiation of low frequencies requires the movement of a relatively large volume of air (see Fig. 19.3). Air-suspension loudspeakers tend to have a low efficiency for conversion of electrical power to sound power, but if space is an important consideration, they make it possible to obtain good performance from small speakers.

To predict a speaker's low-frequency performance in enclosures of various volume, the resonance frequency f_d, the Q-value Q_d, and the compliance C_d of the driver must be known. (Sometimes speaker manufacturers give the compliance as an equivalent air volume V_d which would have a compliance equal to that of the driver. This is given by $V_d = \rho c^2 C_d$, where ρ is the air density and c is the speed of sound.) Various formulas have been derived to predict the resonance frequency and the Q-value in the closed box, as well as the so-called cutoff frequency f_3 at which the bass response is down 3 dB. Examples of these formulas and how they apply to real loudspeakers are given in Appendix A11.

19.7 ■ EQUIVALENT ELECTRICAL CIRCUIT FOR A LOUDSPEAKER

The dynamic loudspeaker, like most mechanical vibrators, can be represented by an equivalent RLC circuit (see Section 18.3), which includes a voltage source plus resistive, inductive, and capacitive elements. In such a circuit, compliance is represented by a capacitor, mass by an inductor, and damping by a resistor. The enclosure, likewise, can be represented by additional circuit elements.

A simple circuit that is widely used is the one in Fig. 19.14 by Australian engineer Richard Small (1972), which includes the effect of a closed-box or air suspension enclosure. The driving voltage is proportional to the output voltage of the source e_g, the B-field,

FIGURE 19.14
Equivalent circuit
representing a
closed-box speaker
system. Each
circuit element
represents an
acoustic function of
the driver or the
enclosure. (Small
1972.)

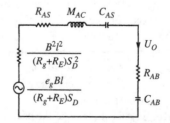

and the voice coil length L (see Section 19.3), while the current in the circuit represents the volume velocity U_o. R_{AS} represents the driver suspension losses, while R_{AB} represents the internal energy absorption. C_{AS} represents the compliance of the driver suspension, while C_{AB} represents the compliance of the air in the enclosure. M_{AC} represents the mass of the speaker cone and voice coil assembly plus the air load.

Applying the circuit analysis you learned in Chapter 18, it is quite clear that the circuit in Fig. 19.14 is a series RLC circuit, and that it has a resonance frequency determined by M_{AC}, C_{AS}, and C_{AB}. The system Q is determined by a combination of all the circuit elements. The system Q, in fact, will do much to determine the acoustical character of the loudspeaker system. $Q = 1$ (critical damping) will give the best transient response; $Q = 1/\sqrt{2} = 0.71$ (Butterworth response) will give the maximum flat response with minimum cutoff; $Q > 0.71$ (Chebyshev response) will give the greatest power handling response with somewhat degraded transient response (Dickason 2000). When Q becomes as large as 1, a "booming" bass results. Typical amplitude responses for different values of Q are shown in Fig. 19.15. The rolloff below f_3 (cutoff) in a closed-box system is essentially 12 dB/octave.

The choice of Q and other parameters is partly a matter of taste. For listening to classical piano music, for example, good transient response is important, while for some types of pop music, a strong bass is more important.

19.8 ■ VENTED BOXES: BASS-REFLEX ENCLOSURES

A *bass-reflex enclosure* has a port or a duct through which the back wave is radiated to the front. The enclosure is a sort of Helmholtz resonator with a piston driver at the closed end, as shown in Fig. 19.16(c). A mechanical analog is shown in Fig. 19.16(d). As the speaker cone C is driven in and out by the voice coil, the air in the enclosure is compressed in

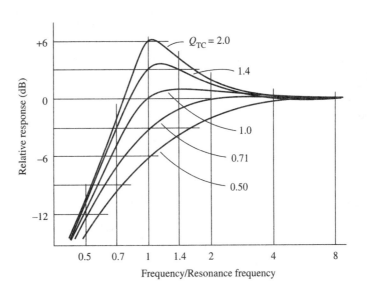

FIGURE 19.15
Relative amplitude response versus frequency for a closed-box loudspeaker system (after Small 1972).

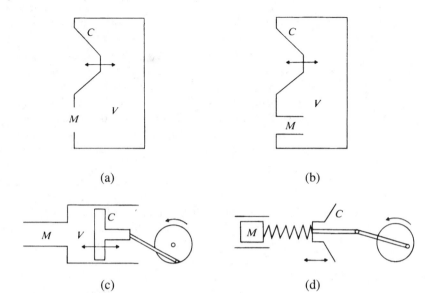

FIGURE 19.16
Bass-reflex speaker enclosures (a) with a port; (b) with a duct; (c) represented as a Helmholtz resonator driven by a piston; (d) a mechanical analog.

the same manner as the spring in Fig. 19.16(d). The mass of the air in the port or duct is represented by mass M.

All four systems shown in Fig. 19.16 behave in the manner described in Section 2.4. At frequencies below resonance, the mass M moves in phase with the driver C (c.f. Fig. 2.7(a)), which in (a) and (b) means that sound radiated from the port or duct will be out of phase with that radiated from the speaker (in phase with the back side of the speaker means out of phase with the front side). At frequencies above resonance, however, the mass in each case moves out of phase with the driver (c.f. Fig. 2.7(b)), which in (a) and (b) means that sound radiated from the port or duct will be in phase with radiation from the front side of the speaker cone. Although the area of the port or duct is smaller than the speaker, the amplitude of motion may be considerably greater near resonance, so that the sound radiated from the port exceeds that radiated from the speaker cone. This is the "reflex action" from which the enclosure derives its name.

FIGURE 19.17
Output from a loudspeaker in a bass-reflex enclosure compared with the same speaker in an infinite baffle and a closed box. f_0 is the resonance frequency of the loudspeaker and also the bass-reflex enclosure in this design.

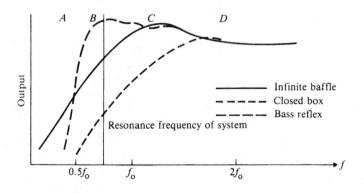

Figure 19.17 shows the output from a loudspeaker in a bass-reflex cabinet compared to the same loudspeaker in an infinite baffle and in a closed box. The bass-reflex cabinet and the loudspeaker have the same resonance frequency f_0. Note that the response at frequencies between $0.5 f_0$ and f_0 is increased by the bass-reflex enclosure at the expense of a sharper cutoff below $0.5 f_0$.

The acoustic behavior of the bass-reflex enclosure can be understood by comparing four different frequency ranges in Fig. 19.17. In range A, the sound radiated from the port or duct is out of phase with that radiated by the loudspeaker, which results in cancellation. In range B, just below resonance, sound radiation from the port exceeds that from the speaker cone, and so even though the two are out of phase, the total radiation is increased. In range C, above resonance, the two are in phase, and radiation is enhanced, but this enhancement grows small in range D, well above resonance. Note that the resonance frequency of the loudspeaker in the bass-reflex enclosure is different from the resonance frequency f_0 of the same speaker when it is not enclosed.

At low frequency, radiation from the vent contributes substantially to the sound output of the system. It does so, however, by increasing the acoustic load on the rear of the cone, reducing cone motion and thus the directly radiated sound from the speaker. In this sense, the bass-reflex principle subtracts as much as it adds.

Compared to closed-box systems, vented enclosures have several advantages:

1. Lower cone excursion near resonance results in higher power-handling capacity and lower modulation distortion;
2. Lower cutoff frequency using the same driver;
3. A higher efficiency (theoretically +3 dB) for the same enclosure volume.

Bass reflex enclosures are more difficult to align than closed-box systems, however.

The original patent for a bass-reflex enclosure was granted to A. C. Thuras in 1932. The theory is described and equivalent circuits developed in classic books by Olson (1947) and Beranek (1954) and in a paper by Thiele (1961). Because of the many circuit elements, there are many combinations of circuit values which will produce the desired cutoff frequency, Q-value and other acoustic parameters. Each combination that produces a useful frequency response is called an *alignment*. Thiele (1961), for example, gives 28 useful alignments.

In some of the alignments (notably the so-called Butterworth alignments), the enclosure is tuned to the speaker cone resonance frequency, as in the simple prototype used in Fig. 19.16 to illustrate the basic principle. In other alignments the two may be quite different. If the enclosure is large, it may be tuned considerably below the speaker resonance. Design of a bass-reflex or vented-box enclosure can be quite challenging, although simplified formulas and procedures are given in a number of popular books and articles. Examples of these formulas and how they apply to real loudspeakers are given in Appendix A11.

Measuring the electrical impedance of a speaker in a bass-reflex enclosure, using the technique described in Section 19.4, will generally give an impedance versus frequency curve with two resonance peaks rather than one, as shown in Fig. 19.18. Resistor R should be 200 Ω or more so that the current supplied to the loudspeaker does not change appre-

FIGURE 19.18
(a) Circuit for
determining the
resonance
frequency of a
speaker. (b) Voltage
across the speaker
(input impedance)
when unenclosed
and when mounted
in a bass-reflex
enclosure.

ciably with speaker impedance. The voltage across the speaker terminals will then be proportional to impedance. The electrical impedance of the speaker should show two peaks, as shown in Fig. 19.18(b), one peak slightly above and one slightly below the free speaker cone resonance.

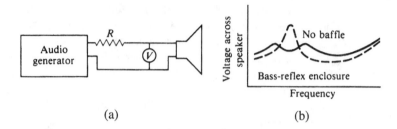

(a)   (b)

Drone Cones

A speaker enclosure with a passive speaker, or *drone cone*, is similar in principle to a bass-reflex enclosure (see Fig. 19.19). The drone cone, which is a loudspeaker cone without a voice coil or a magnet, is driven by the back wave from the main speaker. Like the air in the port or duct of a bass-reflex speaker, it radiates a substantial part of the total sound near the resonance frequency of the enclosure. The drone cone is usually either the same size as the main speaker or slightly larger.

Because the drone cone has a substantial amount of mass, the volume of the cabinet can be modest in size and yet achieve a low resonance frequency. An additional advantage over the bass-reflex enclosure with a port is that large air velocities in the latter can lead to turbulence, whereas the larger area of the drone cone reduces the air velocity at high power levels.

Rear Horn and Acoustic Labyrinth Enclosures

An acoustic horn is a very effective acoustic radiator. (The horn loudspeaker will be discussed in Section 19.9.) It is possible to use a horn to radiate the back wave from a cone loudspeaker, as shown in Figs. 19.20(a) and (b). The cone loudspeaker in each of these arrangements radiates directly from the front side of its cone, but the back side drives a modified horn that radiates indirectly.

The acoustic labyrinth in Fig. 19.20(c) incorporates an open pipe that is approximately half a wavelength in length for low frequency sound, so that sound from the back of the speaker cone arrives at the front in phase with the direct-radiated sound.

FIGURE 19.19
A drone-cone
enclosure
substitutes a
speaker without a
voice coil for the
port or duct of a
conventional
bass-reflex
enclosure.

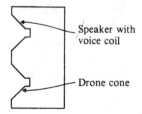

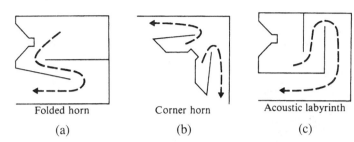

FIGURE 19.20
Three arrangements
for rear loading of a
cone speaker with a
horn or labyrinth.

| Folded horn | Corner horn | Acoustic labyrinth |
| (a) | (b) | (c) |

19.9 ■ HORN LOUDSPEAKERS

A *horn* loudspeaker consists of an electrically driven diaphragm and a voice coil coupled acoustically to a horn. The acoustic horn transforms high pressure at the throat to lower pressure distributed over the large mouth of the horn. The acoustic horn is the acoustical analog of an electrical step-down transformer, as shown in Fig. 19.21.

A well-designed horn loudspeaker is a very efficient radiator of sound, with efficiency reaching as high as 40 to 50% compared to a typical 3 to 5% for a cone-type loudspeaker. Therefore, it is widely used in public address systems (especially outdoors) and theater sound systems, where a large amount of power is desired. A horn capable of radiating low frequencies becomes large and cumbersome, however, unless a folded-horn arrangement such as the one shown in Fig. 19.22 is used.

One type of low-frequency folded horn, first designed in 1941 by Paul Klipsch, is still in use today. It is called the Klipschorn and is shown in Fig. 19.23. Designed to stand in the corner of a room, it uses the walls of the room as an important part of the horn. The efficiency (of converting electrical energy to acoustical energy) is about 30% over a frequency range of 50 to 350 Hz (Klipsch 1941).

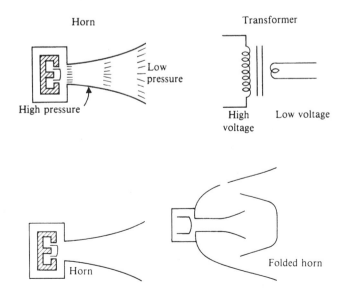

FIGURE 19.21
An acoustical horn
is analogous to an
electrical
transformer.

FIGURE 19.22
Straight horn and
folded horn.

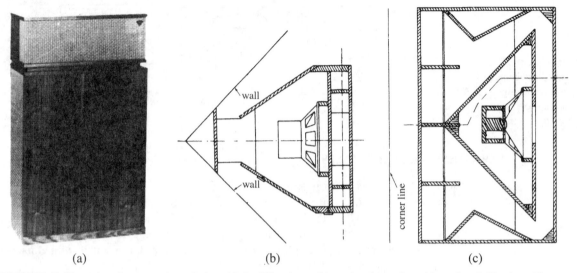

(a)   (b)   (c)

FIGURE 19.23 A low-frequency horn design: (a) the Klipschorn; (b) sectional top view; (c) sectional side view. (Courtesy of Klipsch and Associates.)

19.10 ■ MULTISPEAKER SYSTEMS

Because of the difficulty of obtaining high performance at both high and low frequencies from the same speaker, nearly all high-fidelity loudspeaker systems have two or more speakers. A typical speaker system will consist of a large-cone low-frequency *woofer*, a small high-frequency *tweeter*, and an electrical crossover network that feeds low-frequency signals to the woofer and high-frequency signals to the tweeter, as shown in Fig. 19.24(a). More elaborate systems add a midrange speaker and a second crossover network to direct a band of frequencies to this speaker, as shown in Fig. 19.24(b). Crossover frequencies vary widely among different manufacturers.

A woofer usually has a large voice coil, a deep cone, and a fairly soft suspension in order to have a low frequency of resonance. A tweeter, on the other hand, has a small lightweight voice coil and a relatively stiff suspension. Horn tweeters have become very popular, because *if properly oriented*, they provide good angular dispersion of high-frequency sounds

FIGURE 19.24
Two-way and three-way loudspeaker systems along with typical division of frequencies by the crossover networks. Crossover frequencies vary among manufacturers.

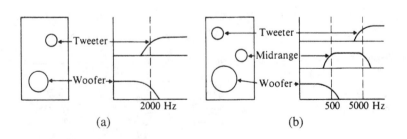

(a)   (b)

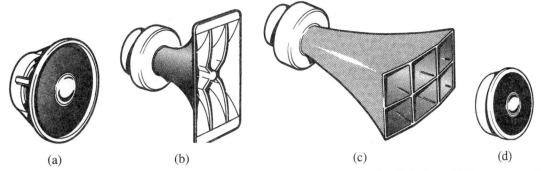

FIGURE 19.25 Several types of tweeters: (a) cone type; (b) diffraction horn; (c) multicellular horn; (d) dome tweeter. Types (b) and (c) include special horns for good angular dispersion of high-frequency sound.

that tend to be quite directional. A dome tweeter substitutes a lightweight diaphragm (typically aluminum) for the cone in order to decrease the mass that moves with the voice coil, and thus to improve the response at high frequency. Dome tweeters are popular as "super tweeters" in multiple-speaker systems. Several types of tweeters are illustrated in Fig. 19.25.

Several manufacturers supply coaxial two-way and even three-way (triaxial) speaker systems with two or three different speakers mounted on the same frame. This is a convenience in designing the enclosure. Figure 19.26 illustrates multispeaker systems, including two-way, three-way (see Fig. 19.24), and coaxial three-way. The coaxial system shown has cones attached to the woofer voice coil; other triaxial systems have three independent voice coils.

The crossover network may consist of simple high-pass and low-pass filters, as shown in Fig. 19.27. The purpose of the crossover network is to send high frequencies to the tweeter and low frequencies to the woofer. If a separate midrange speaker is included, it may be fed through a band-pass filter. If the midrange speaker is a second cone attached to the voice coil of the woofer, the crossover will be mechanical rather than electrical.

FIGURE 19.26
Three examples of multispeaker systems: (a) a two-way speaker system commonly used as a studio monitor; (b) a three-way speaker used for stereo music; (c) a coaxial speaker with two drivers on the same axis. (Courtesy of ElectroVoice.)

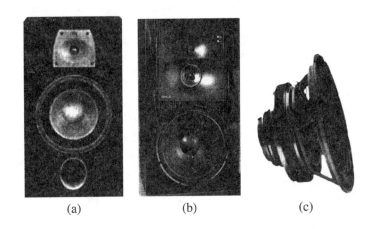

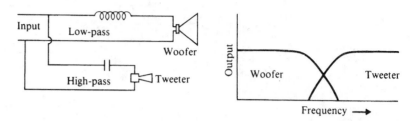

FIGURE 19.27
Crossover network consisting of high-pass and low-pass filters.

A recent trend in music listening and home theater is to extend the bass range by adding an extra-large woofer called a subwoofer. Because of its size, the subwoofer is placed in its own enclosure and often has its own amplifier. Low frequencies are nondirectional, so a single subwoofer can be placed anywhere in the listening room.

19.11 ■ OTHER LOUDSPEAKER TYPES

New speaker designs are constantly being marketed. Some of them are variations of more familiar designs; others are quite different. We mention only a few of them.

Electrostatic Speakers   The electrostatic speaker, shown in Fig. 19.28, is similar in construction to the condenser microphone. A perforated screen serves as one electrode, and a metal film on a plastic diaphragm forms the other. The plastic diaphragm carries a polarizing charge that is attracted or repelled by the charge in the screen.

Electrostatic speakers have smooth response at high frequency, but they are inefficient at low frequency due to the difficulty in obtaining a large excursion of the diaphragm. They have the added disadvantage of a high input impedance, which makes them somewhat difficult to drive efficiently with a transistor amplifier.

Planar Speakers   Several planar- or flat-speaker systems, both of magnetic and electrostatic type, have appeared, but they tend to be inefficient and have not become very popular.

Cylindrical Radiators   Several loudspeakers designed to radiate equally in all directions have appeared on the market. In some cases, the loudspeaker is pointed toward a conical reflector. The Ohm F speaker features sideways radiation from the outside of a vertical cone. Cylindrical radiators can be placed almost anywhere in the room, but interference effects can distort the sound field in the room when they are used as components in a stereophonic system.

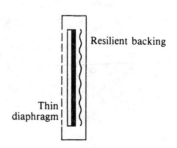

Resilient backing

Thin diaphragm

FIGURE 19.28
An electrostatic speaker.

Air-motion Transformer   This speaker uses a flexible pleated diaphragm to which strips of conductive aluminum have been bonded. Motion of the aluminum strips in a magnetic field causes the pleats to open and close and thus to pump air in and out.

Ribbon Speakers   Like a ribbon microphone, a ribbon loudspeaker consists of a metal foil in a magnetic field. The ribbon consists of an aluminum or copper conductor pattern bonded to a Mylar film (or some other nonconducting material). The length and thickness of the conductor determines the impedance and power-handling capability of the loudspeaker. When the audio signal passes through the conductor (which is located in a magnetic field), the ribbon moves, producing the audible sound wave. Ribbon speakers are bidirectional with sound emitted from the back as well as the front, resulting in sound cancellation, especially at lower frequencies. In spite of the weak bass, ribbon speakers are known for their clarity at mid- and high frequencies.

19.12 ■ RECENT TRENDS IN LOUDSPEAKERS

Over the past 20 years the fundamentals of loudspeaker systems have remained the same; however, new technologies and new applications for sound have made a big difference in the design of loudspeakers. The traditional material for making loudspeaker cones has been paper. The use of plastics (such as polypropylene), which are more uniform in construction and more rigid, results in smoother response curves, improved sensitivity, and better damping. Several improvements have been made to the voice coil. These include (1) edge-winding the voice coil with flat wire, which allows more turns of the coil in the same area, (2) having a double-voice coil to direct two signals into a single woofer, and (3) using Ferrofluid (a viscous substance with suspended magnetic particles) in the magnetic gap to conduct heat away from the tweeter voice coil and control output at resonance through increased damping. Enclosures are often tall and narrow, placing smaller drivers at ear level rather than larger drivers on the floor. More attention is paid to alignment of the voice coils of the tweeter and woofer so that sound arrives at the same time. There is more emphasis in a larger dynamic range to accommodate CDs. Extending bass and treble is important for stereo imaging and clarity of sound. More detail is being paid to modeling the loudspeaker and crossover design. Finally, there is more attention to loudspeaker placement within a listening room. Most of the changes, rather than being fundamental, stem from new technologies and listening tastes.

19.13 ■ LOUDSPEAKER EFFICIENCY

The purpose of a loudspeaker is to convert electrical energy into mechanical energy and to radiate it as acoustic energy. Unfortunately, only a small part of the electrical energy is converted to sound; in most home loudspeaker systems, 90 to 99% of it is wasted as heat. The percentage of electrical power radiated as sound determines the *efficiency* of the loudspeaker. In spite of the fact that it is now possible to buy high-fidelity amplifiers that deliver enormous amounts of electrical power, loudspeaker efficiency remains an important consideration.

Loudspeaker manufacturers tend not to be very informative about speaker efficiency. Almost never are efficiencies stated as simple ratios or percentages. The term "power-handling capability" is quite meaningless; it merely tells how much electrical power a speaker system can absorb without damage. Obviously, a low-efficiency loudspeaker system will be able to "handle" a larger amount of power (and convert it to heat) than can a comparable system with a high efficiency. But loudspeakers are not intended to be space heaters.

The efficiency of a loudspeaker depends on many design parameters, such as magnet strength, cone area, type of enclosure, resistance of the voice coil, and mass of cone and coil. It may change with position of the loudspeaker in the room. The efficiency increases with the square of the magnetic field strength and also with the square of the cone diameter, so these tend to be the most important parameters. The magnetic field strength depends on the weight of the magnets (sometimes stated), but also on other parameters, such as magnet materials and air gap size. Horn loudspeakers are much more efficient than cone speakers, but low-frequency horns are considered too large for most home sound systems. Air-suspension speaker systems tend to have very low efficiencies but are popular because of their compactness.

Some loudspeaker manufacturers state the sound level that will be obtained per watt of input on the speaker axis at 1 m (or some other distance) from the speaker. This depends on both the efficiency of the loudspeaker and its directivity (how much of its total power is radiated straight ahead). The directivity factor of a loudspeaker varies markedly with frequency, so the on-axis sound level will also vary and, to be meaningful, it should be stated for both high and low frequencies. The on-axis sound level should be expressed as, for example, 89 dB at 1 m with 1-W input at 500 Hz. It certainly would be a great help to the intelligent buyer of high-fidelity sound systems if manufacturers would specify the efficiency (preferably in percent) of their loudspeaker systems at several different frequencies!

19.14 ■ LOUDSPEAKER DISTORTION

The trend in recent years has been toward the compact use of low-efficiency speaker systems. In order to produce bass sound from a compact speaker system, long-throw air-suspension drivers are used. Radiation of low-frequency sound requires moving a substan-

FIGURE 19.29
Distortion as a function of output sound pressure level for three loudspeakers: (1) 8-in.-diameter cone; (2) 12-in.-diameter cone; (3) large horn. Total distortion is largely intermodulation (IM) from a mixture of the 41-Hz and 350-Hz test signals. (From Klipsch 1972.)

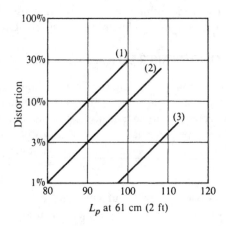

tial quantity of air (see Fig. 19.3); this can be done by a speaker cone of small diameter moving through a large amplitude or by a large-diameter cone moving through a much smaller amplitude. As a general rule, the distortion will be less in the latter case, as the curves in Fig. 19.29 illustrate. These curves should not be regarded as descriptive of all speakers of those sizes, of course. The amount of distortion will depend on many design parameters, such as magnet size, voice-coil design, suspension, etc. Certainly there are speaker systems with eight-inch and twelve-inch drivers with lower distortion. Nevertheless, minimizing cone travel by using large-diameter high-efficiency drivers tends to reduce distortion as well as to conserve amplifier power. There is also distortion introduced by the travel of the voice coil exceeding the uniform magnetic field. Those who love to play their music loud often hear this type of distortion.

One source of distortion in loudspeakers, which cannot be eliminated even by careful design, is intermodulation due to the Doppler effect (see Section 3.7). If the loudspeaker cone is making large excursions at low frequency, it will act as a moving source for high-frequency sound. The amount of distortion depends on the amplitude of the low-frequency excursions as well as the frequencies of the two interacting sounds (Klipsch 1972).

The crossover frequency in two-way speaker systems is typically at 2000–4000 Hz; even in three-way systems, the woofer often receives signals up to 1000 Hz. Thus, many pairs of frequencies can interact to produce a wide assortment of intermodulation distortion products resulting in a "muddy" sound at high levels of power. Fortunately, average power levels are quite low, and noticeable distortion due to the Doppler effect occurs mostly during loud peaks.

In the future, hopefully, accurate specifications of the efficiency, directivity, distortion, and frequency response of loudspeakers will be made available to prospective users.

19.15 ■ EARPHONES (HEADPHONES)

Earphones have been used in electronic communications for many years, but more recently they have become popular for the individual listening to stereophonic music. Several different types of earphones or headphones are in use, including *dynamic (moving coil)*, *electrostatic*, *piezoelectric*, and the *magnetic (metal-diaphragm)* type used in telephone receivers. Dynamic, electrostatic, and piezoelectric earphones are shown in Fig. 19.30.

Dynamic earphones resemble small loudspeakers in their construction and operation. The diaphragm is analogous to the cone of the loudspeaker but usually has no inner suspension or spider. The electrostatic earphone with its light plastic diaphragm is analogous to a condenser microphone. Some electrostatic earphones have a permanently charged electret diaphragm as do electret-condenser microphones. Piezoelectric earphones are similar to crystal (piezoelectric) microphones (see Chapter 20).

Earphones differ in the way they couple sound to the ear. *Circumaural* earphones have a liquid- or foam-filled doughnut-shaped pad that makes a nearly airtight seal against the head. Circumaural earphones have high efficiency at low frequencies, because the moving diaphragm produces pressure changes directly in the ear canal. Efficiency is lost if the seal is not tight, however.

Open-air, or *supra-aural*, earphones do not depend on an airtight seal against the wearer's head. The diaphragm is spaced away from the ear by a foam pad that leaks air in

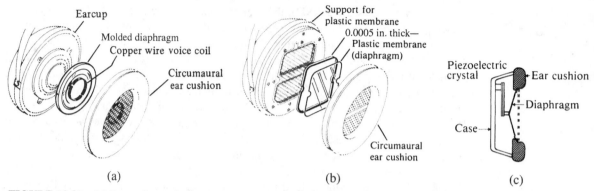

FIGURE 19.30   (a) Dynamic earphone. (b) Electrostatic earphone. (c) Piezo-electric earphone. (Parts (a) and (b) courtesy of Koss SAC Training Lesson No. 9.)

a controlled way. Open-air earphones are lighter in weight than the sealed type, but tend to be somewhat less efficient at low frequency.

Recently miniature dynamic earphones that extend into the ear like a stethoscope have become available. Their advantage is the fact that they are very lightweight, one popular model having a mass of only 33 g.

Listening to stereophonic music with earphones is quite a different experience from listening to a normal room environment.

Two new developments in headphone technology are noise-canceling headphones and multichannel headphones. The normal home listening environment is reasonably quiet, so normal headphones work well. There are environments (such as on an airplane) that are so noisy it is difficult to hear, even with headphones. In 1989 Bose developed a special noise-canceling headphone for use by airline pilots. A built-in microphone and amplifier listen to the ambient noise and feed the noise to a computer chip, which reverses the phase and then feeds the inverted noise back to the headphones. The negative-feedback noise mixes with the ambient noise resulting in noise cancellation, reducing the amplitude of the noise. This technology is called active-noise cancellation. Commercial versions of noise-canceling headphones are offered by several companies to be used by airline passengers to listen to music and movies.

Dolby, who developed surround-sound technology, also developed the multichannel headphone technology. Using special signal processing they were able to make a two-channel headphone sound like a surround-sound system. In 1999 Sony commercialized the technology as 5.1 channel headphones. Using surround decoding chips, Dolby Digital, Dolby Surround, or DTS can be emulated (see Section 20.10).

19.16 ■ SUMMARY

A loudspeaker converts electrical power into sound. Most loudspeakers use cone-type direct radiators, but a wide variety of loudspeaker enclosures are in use. In an air-suspension system, the air in the sealed enclosure provides most of the restoring force for the speaker cone. In a bass-reflex speaker, the back wave from the speaker cone causes a substantial amount of low-frequency sound to be radiated from a duct or port. Horn loudspeakers have very high efficiencies, but low-frequency horns are very large. Although speaker efficiency is a very important consideration, it is rarely specified in the case of systems designed for home use.

Bidirectional ribbon speakers used as tweeters offer extra clarity. Advanced materials have improved loudspeaker characteristics. Specialized headphones include noise-canceling headphones and multichannel headphones.

REFERENCES AND SUGGESTED READINGS

Ballou, G. (1987). *Handbook for Sound Engineers*. Indianapolis: Howard Sams.

Beranek, L. L. (1954). *Acoustics*. New York: McGraw-Hill. Reprinted by Acoust. Soc. Am., 1986.

Cohen, A. B. (1968). *Hi-Fi Loudspeakers and Enclosures*, 2nd ed. New York: Hayden.

Dickason, V. (2000). *The Loudspeaker Design Cookbook*. Peterborough, NH: Audio Amateur Press.

Institute of High Fidelity (1974). *Guide to High Fidelity*. Indianapolis: Howard Sams.

Klipsch, P. W. (1972). "Modulation Distortion in Loudspeakers: Part III," *J. Audio Eng. Soc.* **20**: 827.

Olson, H. F. (1947). *Elements of Acoustical Engineering*, 2nd ed. Princeton, NJ: Van Nostrand.

Rossing, T. D. (1980). "Physics and Psychophysics of High-Fidelity Sound, Part III," *Physics Teach.* **18**: 426.

Small, R. H. (1972). "Closed-Box Loudspeaker Systems—Part I: Analysis," *J. Audio Eng. Soc.* **20**: 798.

Small, R. H. (1973). "Vented-Box Loudspeaker Systems," *J. Audio Eng. Soc.* **21**: 363, 438, 549, and 635.

Thiele, A. N. (1961). "Loudspeakers in Vented Boxes," *Proc. I.R.E. (Australia)* **22**: 487. (Reprinted in *J. Audio Eng. Soc.* **19**: 382, 471 (1971).)

Weems, D. B. (1996). *Designing, Building, and Testing Your Own Speaker System*, 4th ed. New York: McGraw-Hill.

GLOSSARY

air-suspension speaker A loudspeaker mounted in the front of an airtight box so that the pressure of the enclosed air furnishes a major part of the force that restores the speaker cone to its equilibrium position.

baffle An arrangement that reduces interference between sound radiated from the front and rear of a speaker by increasing the path length from front to back.

base-reflex enclosure A speaker enclosure in which the back wave from the speaker is radiated through a port or duct in the front.

compliance A measure of flexibility; it is expressed in units of distance divided by force (m/N or cm/dyne).

crossover network A network designed to send high frequencies to the tweeter and low frequencies to the woofer. The crossover frequency is the approximate point of division between high and low frequencies.

diffraction horn A horn that has one broad dimension and one narrow one, designed to spread sound (by means of diffraction) in the direction of the narrow dimension.

drone cone A passive loudspeaker that has no voice coil or magnet.

electromagnetic induction Generation of an electrical voltage in a wire that moves across a magnetic field.

horn loudspeaker A system that uses an acoustic horn to enhance sound radiation from a moving diaphragm.

impedance The ratio of voltage to current. The relevant current, in the case of source impedance or output impedance, is the current that the device can deliver; in the case of input impedance, it is the current that the device draws from the source.

infinite baffle A large baffle or an enclosure that prevents interference between sound radiated from the front and back of the speaker cone.

magnetic field (*B*-field) A map of the magnetic force around a current-carrying wire or magnet.

magnetic flux A magnet "current" that flows from a north pole to a south pole. The product of *B*-field and area.

magnetic induction An electrical voltage appears on a conductor that moves in a magnetic field.

piezoelectric crystal A crystal that generates an electric voltage when it is bent or otherwise distorted in shape or, conversely, distorts in response to a voltage.

Q factor Ratio of stored energy to the energy dissipated during each cycle of oscillation. Measure of the sharpness of a resonance.

subwoofer A loudspeaker designed to produce extra-low-frequency sound, below the woofer.

transducer A device that converts one form of energy into another.

tweeter A loudspeaker designed to produce high-frequency sound.

woofer A loudspeaker designed to produce low-frequency sound.

REVIEW QUESTIONS

1. What is the best method to amplify sound waves?
2. What is the function of a transducer?
3. Which type of loudspeaker uses a coil of wire in a magnetic field to produce motion?
4. How does the compliance of a loudspeaker relate to the spring constant of a spring?
5. Why is a large loudspeaker cone preferred for radiating low-frequency sound?
6. Why is a small loudspeaker cone preferred for radiating high-frequency sound?
7. Why does the electrical impedance of a loudspeaker show a peak at its resonance frequency?
8. What is an air-suspension speaker system?
9. What is a bass-reflex speaker system?
10. What is the purpose of a baffle?
11. What effect does increasing the Q of a loudspeaker have on the frequency response curve?
12. What is the effect of decreasing the volume of an airtight enclosure in an air-suspension system?
13. How can you lower the resonance frequency in a bass-reflex enclosure without changing the volume?
14. What is the advantage of a horn loudspeaker over a cone-type loudspeaker?
15. Which driver in a loudspeaker system requires the most electrical power?
16. What is an electrostatic speaker?
17. What is a subwoofer?
18. How do noise-canceling headphones work?

QUESTIONS FOR THOUGHT AND DISCUSSION

1. What are the advantages and disadvantages of air-suspension loudspeakers?
2. Why does a horn loudspeaker have a higher efficiency than a cone loudspeaker?
3. If the baffle in which a loudspeaker is mounted is insufficient in size, which will suffer most: the bass response or the treble response? Why?
4. Why does a woofer have a large-diameter cone? Why does a tweeter have a small cone?
5. How can you get a good bass response from tiny loudspeakers?
6. How would your taste for music (classical music versus rock music) help determine the Q of the loudspeaker system best suited to your type of music?

EXERCISES

1. Compare the sound power level of a loudspeaker with an efficiency of 1% to one with an efficiency of 10% supplied with the same electrical power.
2. Compare the cone areas of speakers having diameters of 20 cm, 30 cm, and 38 cm (8 in., 12 in., and 15 in.).
3. A loudspeaker 20 cm in diameter is mounted at the center of a 1-m square baffle board.

    (a) Determine the path length from the center of the front side to the center of the back side of the speaker.

    (b) At what frequency will this path be equal to one-half wavelength of sound?

4. (a) From the graph shown in Fig. 19.3, determine the volume of air that must be moved in order to generate a sound power of 0.1 W at 100 Hz.

(b) How far must a speaker cone move in order to move this volume of air if the cone has a diameter of 8 in.?

5. For satisfactory operation down to a cutoff frequency f_c, (a) a horn loudspeaker should have a mouth diameter at least one-fourth of the wavelength at f_c, and (b) its diameter should double no more rapidly than every one-ninth of the wavelength at f_c. Determine the mouth diameter and the total length of a horn loudspeaker with $f_c = 100$ Hz if the throat diameter is 5.4 cm.

6. If a loudspeaker can produce a sound pressure level of 92 dB for 1 W of input power at a distance of 1 m, estimate its efficiency in converting electrical power to acoustic power. (*Hint*: Assume that the sound is uniformly radiated into a hemisphere so that the sound level at 1 m is 8 dB less than the sound power level; see Section 6.2.)

7. How large a volume of air must be moved by a loudspeaker cone radiating 0.1 W of acoustical power at 50 Hz? How far would the cone of a 12-in. loudspeaker (actual cone diameter is 25 cm) have to move to displace this volume?

8. A certain loudspeaker has a compliance of 8.7×10^{-4} m/N and a mass (cone plus voice coil) of 71 g. Estimate its resonance frequency. (See Section 2.1; the compliance is the reciprocal of stiffness or spring constant: $C = 1/K$.)

9. Express the compliance of the loudspeaker in Exercise 8 in cm/dyne (1 dyne $= 10^{-5}$ N). If you placed the speaker face up and added a 100-g mass, how far would the speaker deflect? Is this a practical method for measuring compliance?

EXPERIMENTS FOR HOME, LABORATORY, AND CLASSROOM DEMONSTRATIONS

Home and Classroom Demonstrations

1. *Loudspeaker polarity*   Try reversing the leads connected to a pair of stereo speakers and observing the result. Of course, this causes a reverse polarity, but can you hear it? Compare mono versus stereo and high versus low frequencies.

2. *Look under the hood*   Remove the cover off your loudspeakers. Measure the diameter of the drivers and estimate the air moved. Look for any crossover circuits available.

3. *Loudspeaker as a microphone*   Hook up a speaker to the inputs of an amplifier and run the output to a second speaker. After showing that speaker A can be a microphone and speaker B a speaker, reverse the leads and do it the other way around.

4. *Resonance of a bass-reflex loudspeaker*   Place a lighted candle at the opening of the port of a bass-reflex loudspeaker. Apply a variable-frequency sine wave at a fairly high level to the loudspeaker. Enclosure resonance is easily seen.

5. *Styrofoam cup loudspeaker*   Wrap about 50 turns of fine wire into a flat coil, and glue this to the bottom of a Styrofoam cup. Connect it to the output of an audio amplifier, and play music through the amplifier. When a ceramic magnet is brought up close to the coil, you should hear a sound emanating from the cup.

6. *Effect of baffling a speaker*   Connect a small loudspeaker (5–10 cm. in diameter) to an audio amplifier, and play music through it. Then hold it tightly against a board with a circular hole in it (which matches the diameter of the speaker). You should hear a much stronger sound, especially at low frequency where the "acoustic short circuit" is especially noticeable without the baffle.

7. *Effect of baffling a tuning fork*   Cut a slot in a sheet of paper just large enough to pass the tines of a tuning fork through. Set the tuning fork into vibration, and slowly move it through the slot. You should notice a marked increase in the sound as each tine passes through the slot.

8. *Loudspeaker cone resonance*   Supply the loudspeaker cone a nearly constant current by connecting a large ($\sim$1000-Ω) resistor in series with an audio amplifier. Connect a digital voltmeter across the loudspeaker and slowly sweep the frequency. A maximum should occur at the frequency of the cone resonance. The cone resonance can be lowered by adding mass.

9. *Loudspeaker in a cabinet*   Demonstration 8 can be repeated with a loudspeaker cone in a bass-reflex cabinet.

Laboratory Experiments

Loudspeaker performance: Efficiency, frequency response, and directivity (Experiment 30 in *Acoustics Laboratory Experiments*)

Loudspeaker performance: Distortion, impulse response, and phase response (Experiment 31 in *Acoustics Laboratory Experiments*)

20

Microphones, Amplifiers, and Tuners

Chapter 19 focused on loudspeakers, which are output transducers. We now turn our attention to input transducers (microphones), which convert acoustic waves into electrical signals. Microphones are essential components in sound reinforcement and sound-recording systems. Once an electrical signal is obtained from a microphone, the signal can be amplified, modified, mixed with other signals, and then returned to listeners (sound-reinforcement system) or stored for future playback (sound-recording systems).

A home entertainment system selects previously recorded sounds from several media (compact disc, tape, etc.) along with broadcast sound (radio), which is amplified and delivered to two or more loudspeakers. A tuner is the component in a home entertainment system that receives broadcast signals (tunes them in) and passes them on to the amplifier. Often the tuner is built into the same box with the amplifier.

This chapter discusses microphones, amplifiers, and tuners used in sound recording, sound reinforcement, and home entertainment.

In this chapter you will learn:

- About the physical characteristics of microphones;
- About the electrical characteristics of microphones;
- About microphone sensitivity;
- About the characteristics of audio amplifiers;
- About distortion in audio amplifiers;
- About the properties of AM/FM tuners;
- About surround-sound receivers.

20.1 ■ MICROPHONES AS TRANSDUCERS

A microphone is a transducer that produces an electrical signal when actuated by sound waves. Microphones may be designed to respond to variations in air pressure due to the sound wave or to variations in particle velocity as the sound wave propagates. Most microphones in common use are *pressure* microphones.

A pressure microphone has a thin diaphragm, which moves back and forth with the rapid pressure changes in a sound wave. The diaphragm is connected to some type of electrical generator, which may be a piezoelectric crystal (crystal microphone), a moving coil (dynamic or magnetic microphone), or a variable capacitor (condenser microphone). Each of these microphone types will be discussed briefly.

One can also classify microphones according to their ability to pick up sound arriving from different directions. Microphones may be omnidirectional (that is, nearly the same sensitivity in all directions), unidirectional (that is, considerably more sensitive in one direction), or bidirectional in their response pattern. The directionality of a microphone is usually determined by the acoustical design of its housing.

Crystal Microphones

Crystal microphones (see Fig. 20.1) use piezoelectric crystals that generate a voltage when deformed by a mechanical pressure. Rochelle salt crystals generate a large electric signal and are used in some inexpensive microphones. Better-quality crystal microphones generally use a ceramic material, however, which is much less sensitive to humidity, temperature, and mechanical shock. Ceramics commonly used include barium titanate, lead titanate, and lead zirconate. The piezoelectric effect is described in the following box.

Crystal microphones have a comparatively large electrical output voltage, which makes them convenient for use in portable sound equipment and tape recorders. On the other hand, the high-frequency response of most crystal microphones is less "flat" than that of condenser microphones due in part to the greater mass that must move. Therefore, they are not often used in the professional recording of sound.

The *piezoelectric* (pronounced "pah-ee-zo-electric") effect was discovered in 1880 by the brothers Pierre and Jacques Curie, who observed that certain crystals acquire electric charge when they are compressed or distorted. The amount of charge that appears is proportional to the distortion and disappears when the force is removed. If leads are attached to the crystal at the appropriate places, an electrical output signal is obtained.

The piezoelectric effect is reversible; if an electrical voltage is applied to a piezoelectric crystal, it changes its shape. This has led to the use of piezoelectric crystals as ultrasonic generators and even as loudspeakers for high-frequency sound. Quartz, tourmaline, topaz, and Rochelle salt are examples of natural piezoelectric crystals.

FIGURE 20.1

A crystal microphone. Sound pressure on the diaphragm causes a deformation of the crystal, generating an electrical signal.

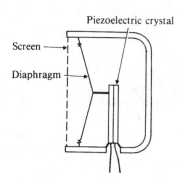

FIGURE 20.2
A dynamic (moving coil) microphone. Sound pressure on the diaphragm causes the voice coil to move in a magnetic field.

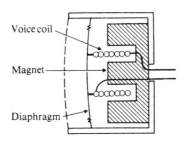

Voice coil

Magnet

Diaphragm

Dynamic Microphones

In a *dynamic* or magnetic microphone, an electrical signal is generated by the motion of a conductor in a magnetic field. This is an example of *electromagnetic induction*, or the dynamo principle, discussed in Section 19.3. In the most common type of dynamic microphone, sound pressure on a diaphragm causes the attached voice coil to move in the field of a magnet, as shown in Fig. 20.2. Movement of the voice coil thus generates an output voltage.

The electrical voltage generated by a dynamic microphone is quite small, but the impedance of the voice coil is also small. Because of the low source impedance, the power generated by a dynamic microphone is not necessarily small, even though the voltage output is small. The low source impedance is a distinct advantage when the microphone is located a substantial distance from the amplifier, as will be discussed later (see Section 20.3). If a higher output voltage is desired, a transformer can be used, preferably at the amplifier end of the transmission line. It is essential that the mass of the moving coil be kept very small in order to yield good response at high frequency.

A dynamic microphone is similar in principle to a small loudspeaker, and, indeed, a loudspeaker will serve as a low-quality microphone. This is frequently done in two-way intercommunication units. The mass of a loudspeaker cone plus voice coil is too large to give satisfactory response at high frequency, however, and so its fidelity as a microphone is generally low.

FIGURE 20.3
A simple capacitor (condenser) consisting of two conducting plates connected to a battery. One plate acquires a positive charge, the other a negative charge. The amount of charge changes as the spacing between the plates changes.

Dynamic microphones are rugged, have a broad frequency response, and are able to withstand the high sound levels that occur in popular music. Thus they are widely used in both live performance and recording.

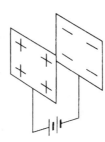

Condenser Microphones

When two metal plates that face each other are connected to a battery, they acquire and store electric charge (see Fig. 20.3). Such an arrangement of charged plates is called a *capacitor*, or *condenser*. The amount of electric charge that can be stored in a condenser depends on the size of the plates and on their spacing. Thus, if one of the plates moves, an electric current will flow in the circuit as charge arrives and departs from the plate. This is the principle of the condenser, or electrostatic, microphone.

In a condenser microphone, one plate is usually the thin movable diaphragm, and the other plate is the fixed backing plate, as shown in Fig. 20.4. As the diaphragm is moved by the pressure of a sound wave, a small current flows in the circuit.

A condenser microphone has a very high source impedance, and for this reason, a preamplifier is usually incorporated into the microphone itself. The diaphragm can be made very light; therefore, the condenser microphone is capable of excellent response at high frequencies. A major disadvantage is the need for a relatively high voltage source to maintain electrical charge (bias) on the plates of the capacitor.

Electret-Condenser Microphones

A type of microphone that retains most of the advantages of the condenser microphone but eliminates the need for a high-voltage bias supply is the *foil-electret*, or *electret-condenser*, microphone, shown in Fig. 20.5. The diaphragm consists of a plastic foil less than one-thousandth of an inch thick with an even thinner layer of metal attached. The foil has been given a permanent electrical charge by a combination of heat and high voltage or by electron bombardment during manufacture.

FIGURE 20.4
A condenser microphone. The diaphragm constitutes one plate of a capacitor and it moves with sound pressure.

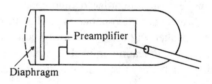

FIGURE 20.5
An electret condenser microphone. A thin metalized plastic diaphragm is tightly stretched across a perforated backing plate. The holes in the back plate couple to an air cavity.

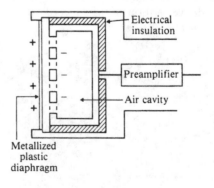

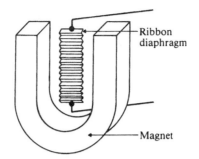

FIGURE 20.6
A ribbon
microphone. A
lightweight ribbon
diaphragm moves
in a magnetic field,
thus generating an
electrical signal.

Electret-condenser microphones often incorporate a built-in preamplifier that is usually a field-effect transistor. Because of their high input impedance, field-effect transistors are a better match for condenser microphones than the junction transistors we discussed in Chapter 18 (see Appendix A.9).

Ribbon Microphones

A *ribbon* microphone is a magnetic microphone in which the lightweight ribbon diaphragm is also the moving conductor (see Fig. 20.6). The ribbon responds to the acoustic velocity (the speed at which air particles are moving) rather than the pressure of the sound wave. A ribbon microphone responds readily to sound waves arriving from the front or back but is insensitive to sound arriving from the sides. Thus, it has a bidirectional response.

Ribbon microphones were popular in the early days of radio broadcasting, because their bidirectional nature allowed performers to stand on opposite sides facing each other. They are sensitive to moving air currents, however, and therefore are not satisfactory for outdoor use. Their use is now mainly restricted to special situations, such as a vocal soloist who wants to sing very close to the microphone and minimize the danger of "popping" when plosive consonants (p, t, d, etc.) are sung.

20.2 ■ MICROPHONE PICKUP PATTERNS

The physical construction of a microphone determines its directivity, or pickup pattern. We saw an example in the ribbon microphone, which is bidirectional. Another common pattern is the omnidirectional (omni) microphone, which is sensitive to sound coming from all directions. Omnidirectional microphones are often used for recording and stage plays to pick up sound from performers spread across a wider area.

In many applications, it is desirable to have a microphone with maximum sensitivity in one direction only (unidirectional). One such application is in sound systems in which acoustic feedback may lead to oscillation if the microphone picks up sound from the loud-speaker.

The most popular unidirectional microphone is the *cardioid* microphone, which has a heart-shaped pickup pattern (see Fig. 20.7b). A cardioid microphone can be made by adding a hole in the rear of the case of a sealed omnidirectional microphone. When a sound wave arrives from the rear, the pressure maxima and minima reach both front and

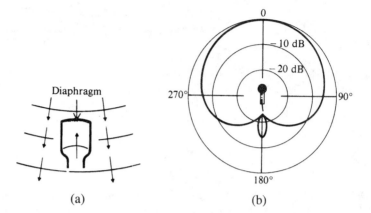

FIGURE 20.7
A simple cardioid
microphone (a) and
its pickup
pattern (b).

back of the diaphragm at the same time, and its deflection is thus minimal. A sound wave arriving from the front produces normal deflection, however.

It is relatively difficult to obtain a cardioid response pattern over a wide range of frequencies; thus, many cardioid microphones have an uneven frequency response, especially a roll-off at bass frequencies. Off-axis response, in particular, will be colored by a variation in the directional response with frequency. This is offset, however, by the unidirectional characteristic of the cardioid microphone, which reduces the pickup of background noise.

For even greater directivity, line microphones or reflector microphones can be used. The line microphone shown in Fig. 20.8(a) (sometimes called a "shotgun" microphone) has a tube with side holes through which sound can enter. Sound waves approaching from nearly head-on will leak into the tube at each side hole, building up into a traveling wave within. Sound waves from other directions, however, will enter the tube in varying phases through the side holes, and these contributions will tend to cancel each other.

A reflector microphone consists of a microphone element at the focus of a parabolic reflector, as shown in Fig. 20.8(b). Parabolic reflectors are also used in spotlights and tele-

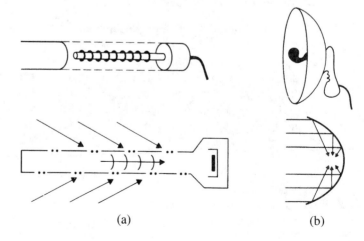

FIGURE 20.8
Directional
microphones:
(a) line
microphone;
(b) reflector
microphone.

scopes; they are shaped so that all the sound (or light) arriving from straight ahead is reflected toward the focus. A reflector microphone has even greater directivity than a line microphone does.

20.3 ■ MICROPHONE IMPEDANCE

Microphones are generally classified as having *high impedance* or *low impedance* (referring to the electrical impedance of the output of the microphone). Crystal, condenser, and electret-condenser microphones have high output impedances from 50,000 ohms and up. Dynamic microphones, on the other hand, have low impedances, typically from 50 to 600 Ω.

If a high-impedance microphone is used with a long microphone cable, high-frequency signals will be attenuated (a 20-ft cable, for example, may result in a 6-dB loss at 10,000 Hz). This loss may be prevented by using a line-matching transformer or by locating a preamplifier near the microphone. Most commercial condenser microphones include a built-in transformer, resulting in a low output impedance. Low-impedance microphones can be used with long cables without much loss of signal at high frequency. Low-impedance (600-Ω) cables are frequency used in sound systems.

20.4 ■ MICROPHONE SENSITIVITY

There are basically two ways to express the sensitivity of a microphone: in terms of voltage output or in terms of power output for a given sound pressure. Sensitivity is measured by placing the microphone in a 94-dB sound field (sound pressure = 1 Pa) and measuring the open-circuit voltage on the output. The voltage sensitivity, S_V, is expressed in decibels (dBV):

$$S_V(\text{dBV}) = 20 \log V/p, \tag{20.1}$$

where V is the open-circuit output voltage and p is the sound pressure in pascals. (Note that 0 dBV = 1V/1 Pa). A sound field of 74 dB (1 microbar = 0.1 Pa) was used in the past to measure microphone sensitivity. If you encounter an old sensitivity rating (given in volts/microbar), add 20 dB to determine the actual voltage sensitivity.

Output-power sensitivity, S_P (in dBm, where m means that 0 dB is at 1 mW), is used in professional sound reinforcement. It is calculated from the voltage sensitivity and the microphone impedance:

$$S_P(\text{dBm}) = S_V - 10 \log Z + 24, \tag{20.2}$$

where Z is the electrical output impedance of the microphone. An alternative power sensitivity rating G_m, recommended by the Electrical Industries Association (EIA), gives the output power level for a 0-dB sound field as

$$G_m(\text{dBm}) = S_P - 94. \tag{20.3}$$

Most microphone manufacturers provide specifications for the open-circuit voltage sensitivity and the output impedance. Some manufacturers include the output-power sensitivity, whereas only a few manufacturers give the EIA sensitivity. To compare the sensitivities of two microphones, sensitivity ratings found in the specification may be converted to open-circuit voltage sensitivities and then compared.

Of primary interest to the microphone user is the output voltage that can be expected at a given sound level. The open-circuit output voltage is expressed by

$$20 \log V = S_V + L_P - 94, \tag{20.4}$$

where V is the open-circuit voltage, S_V is the open-circuit voltage sensitivity, and L_P is the sound pressure level. The following example illustrates how to compute output voltage.

EXAMPLE 20.1   A microphone has a sensitivity listed as 1.2 mV/Pa and an output impedance of 200 Ω.

(a) What is the open-circuit voltage sensitivity in dBV?
(b) What is the power sensitivity in dBm?
(c) What is the EIA sensitivity in dBm?

Solution

(a) $S_V(\text{dBV}) = 20 \log V/p = 20 \log(1.2\text{mV}/1 \text{ Pa}) = -58$ dBV
(b) $S_P(\text{dBm}) = S_V - 10 \log Z + 24 = -58 - 10 \log 200 + 24 = -57$ dBm
(c) $G_m(\text{dBm}) = S_P - 94 = -57 - 94 = -151$ dBm

EXAMPLE 20.2   A microphone has a sensitivity listed as -56.6 dB (0 dB $= 1$ mW/Pa) and a 600-Ω output impedance.

(a) What is the voltage sensitivity in dBV?
(b) What is the open-circuit voltage with a singer singing at 102 dB?

Solution

(a) $S_V = S_P + 10 \log Z - 24 = -56.6 + 10 \log 600 - 24 = -52.8$ dBV
(b) $20 \log V = S_V + L_P - 94 = -52.8 + 102 - 94 = -44.8$ dB
$\qquad V = 10^{-44.8/20} = 5.75$ mV

20.5 ■ AMPLIFIERS

Home entertainment amplifiers increase the signals of CD players, tape players, and tuners from line level (from around half a volt to several volts) to a level appropriate for loudspeakers. Even though the phonograph is becoming obsolete, modern amplifiers still include a preamplifier that boosts the low-level phonograph signal (a millivolt or two) to line level. An amplifier may be packaged as a separate component or integrated with a tuner into a receiver.

The amplifier has controls to select the desired program source, adjust the desired signal level ("volume"), and adjust the tonal balance between high, middle, and low frequencies. The tonal balance is usually adjusted by means of bass and treble controls, although some amplifiers provide equalizers which adjust the gain in several frequency bands individually. The action of typical treble and bass tone controls is illustrated by Fig. 20.9.

Many amplifiers have a loudness control. With the loudness switch turned on, the frequency response of the amplifier will automatically change with the volume level. At high level, the response will be essentially flat, but as the volume level is reduced, the bass and treble are increased to compensate for the reduced sensitivity of the ear to low-level sounds of high and especially low frequency.

The power amplifier section is designed to supply substantial amounts of audio power to the loudspeakers with as little distortion as possible. Because large amounts of power are involved, efficiency becomes an important consideration. The large output transistors are supplied DC power from the power supply; some of it is converted into audio output, the rest into waste heat. Because this heat must be carried away by heat "sinks," designers try to minimize its production.

One way to minimize waste heat in the output transistors is to have them conduct electricity only when an audio signal is present, and then only as much as necessary. A circuit so designed is called a class-B amplifier, as opposed to a class-A amplifier in which the transistors conduct nearly all the time. Unfortunately, a class-B amplifier generates substantially more distortion than does a class-A amplifier. The solution arrived at in most power amplifiers is to operate in an intermediate condition known as *class-AB*.

Waste heat is usually carried away from the output transistors by attaching them to large aluminum or copper heat sinks, which are good conductors of heat. The heat sinks in turn are cooled by air motion, or convection. For this reason, it is important to comply with the manufacturer's instructions regarding the free circulation of air past the output circuitry. Transistors, which normally last for years, can be destroyed in a wink by overheating or overloading electrically!

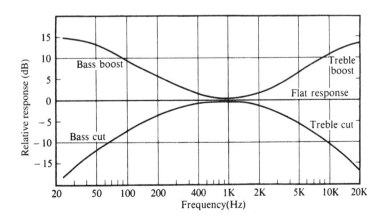

FIGURE 20.9
Action of typical bass and treble controls.

20.6 ■ DISTORTION

Distortion refers to any signal not in the original recording or program source that appears in the reproduced sound. Distortion can be generated in any component in the audio system. Once generated, it will most likely be reproduced by the remaining components along with the desired sound.

The three types of distortion that are most significant in a high-fidelity system are harmonic distortion, intermodulation distortion, and transient distortion. Harmonic and intermodulation distortion usually result from a nonlinear transfer characteristic in some component. Transient distortion results from insufficient damping.

Harmonic distortion results in the creation of harmonics of the fundamental frequency that are not present in the original signal. "Clipping" of waveform peaks, which results in substantial harmonic distortion, can occur because of overloading some component. Small amounts of harmonic distortion often escape notice, because musical tones already contain several harmonics. Harmonic distortion is expressed as a percentage of the total signal.

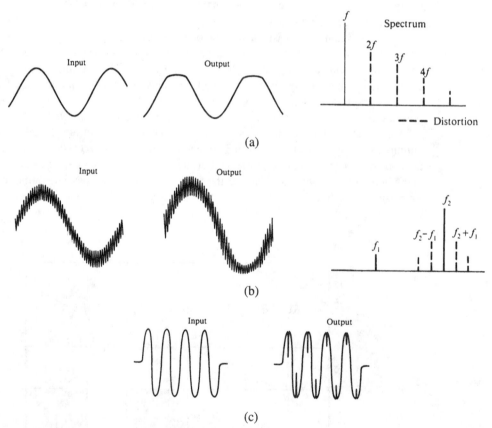

FIGURE 20.10 Three types of distortion: (a) harmonic; (b) intermodulation; (c) transient. In each case, input and output waveforms are shown, and spectra of (a) and (b) are shown as well.

Intermodulation distortion may result when tones of two or more different frequencies are reproduced at the same time (almost always the case in music). A nonlinear characteristic in some component results in the creation of sum and difference frequencies. These new frequencies are harmonically and musically unrelated to the desired tones, and are therefore more noticeable and objectionable than harmonic distortion products. Intermodulation distortion is also expressed as a percentage of the total signal.

Transient distortion occurs when some component cannot respond quickly enough to a rapidly changing signal. In a loudspeaker, it may occur because of insufficient damping of a mechanical resonance. The loudspeaker depends on the amplifier to damp its motion, and an amplifier with a damping factor of at least 20 is desirable. (The *damping factor* expresses the ratio of speaker impedance to "internal resistance" in the amplifier output.)

Transient intermodulation distortion (TIM) can occur in an amplifier because time delay in the feedback loop causes momentary overloading of some circuit element. Thus the percussive sound of a cymbal is distorted, although the steady sound of a violin is not.

Figure 20.10 illustrates the three types of distortion just discussed. If the input signal is a pure tone of frequency f, harmonic distortion will create distortion products with frequencies $2f$, $3f$, etc. Intermodulation distortion may appear as amplitude modulation (AM) or frequency modulation (FM). AM distortion, shown in Fig. 20.10(b), produces distortion products at the sum and difference frequencies. (FM distortion, not shown, produces additional frequencies as well.) One type of transient distortion is shown in Fig. 20.10(c).

20.7 ■ AMPLIFIER POWER AND DISTORTION

The most familiar specification of amplifiers is the output power; it is also one of the most misleading. In an electronics catalog from the early 1970s, one can find an assortment of expressions such as music power, IHF power, peak power, or rms power. In 1974, the Federal Trade Commission (FTC) ruled that the continuous average power per channel, the power frequency response, and the total harmonic distortion at full rated power must be stated "clearly, conspicuously, and more prominently than other representations." Furthermore, the test conditions for measuring power were clearly defined, including a preconditioning at one-third of rated output (the heating of the output transistors is usually greater at one-third of rated output than at full power). A few manufacturers still use the incorrect term rms power to refer to continuous average power, however.

How important is power? It is not easy to answer this question, because room conditions, loudspeaker efficiencies, and listening habits differ widely. A typical average listening level of 65 dB in a room may require from 0.01 to 0.03 mW of acoustic power. Even with a low-efficiency (1%) loudspeaker, this requires only 0.001 to 0.003 W of electrical power. However, at an orchestra concert, you would hear peaks to 105 dB or even more. To reproduce such peaks, 40 dB above the average level, would require a 10,000-fold increase in power! This would require 10 to 30 W of power from the amplifier.

Table 20.1 is a rough guide to the amplifier power requirements recommended in a publication by quality equipment manufacturers (Institute of High Fidelity 1974). The figures appear to pertain to living rooms with large amounts of absorption. In most dormitory rooms, 10 W per channel even into low-efficiency speakers will reproduce peaks well in excess of 100 dB.

TABLE 20.1 A rough guide to power-amplifier requirements for various speaker efficiencies and room sizes

Amplifier power (continuous watts per channel)	Highest sound pressure level (in dB) possible for a room of the indicated volume (in ft^3)								
	Low-efficiency systems			Medium-efficiency systems			High-efficiency systems		
	2000 ft^3	3000 ft^3	4000 ft^3	2000 ft^3	3000 ft^3	4000 ft^3	2000 ft^3	3000 ft^3	4000 ft^3
10	94 dB	92 dB	91 dB	97 dB	95 dB	93 dB	102 dB	101 dB	100 dB
20	97 dB	95 dB	94 dB	100 dB	98 dB	96 dB	105 dB	104 dB	103 dB
35	99.5 dB	97.5 dB	96.5 dB	102.5 dB	101.5 dB	98 dB	107 dB	106 dB	105 dB
50	101 dB	99 dB	98 dB	104 dB	102 dB	100 dB	109 dB	108 dB	107 dB
75	103 dB	101 dB	100 dB	105 dB	103.5 dB	101.5 dB	110.5 dB	109.5 dB	108.5 dB
100	104 dB	102 dB	101 dB	107 dB	105 dB	103 dB	112 dB	111 dB	110 dB
125	105 dB	103 dB	102 dB	108 dB	106 dB	104 dB	113 dB	112 dB	111 dB

Notes: Ninety to ninety-five decibels of typical sound pressure level would be about as loud as what you would hear when sitting in the balcony seats at a live concert. One hundred decibels is fairly loud, about what you could expect at midorchestra seating in a concert hall. One hundred ten decibels is reached by an orchestra playing full blast, with the listener in the front rows of the hall. One hundred twenty-six decibels is considered to be the threshold of physical pain. (From the Institute of High Fidelity, © 1974. Reprinted by permission.)

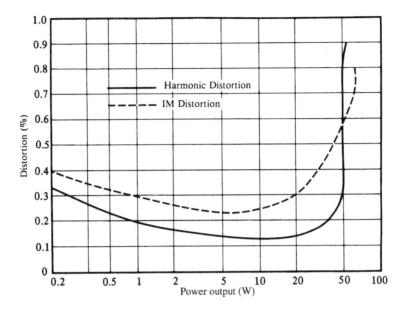

FIGURE 20.11
Typical amplifier distortion curves. (From the Institute of High Fidelity, © 1974. Reprinted by permission of the Electronic Industries Association.)

Salespeople have been known to recommend the purchase of high-power amplifiers because the large amount of reserve power will supposedly assure less distortion. This is not normally the case. In most modern amplifiers, harmonic and intermodulation distortion remain low until one approaches the rated output power. Rarely is the amplifier the main source of distortion in a high-fidelity system, in any case.

The harmonic and intermodulation distortion of a typical amplifier are shown in Fig. 20.11. Intermodulation is generally measured by the application of two signals with widely spaced frequencies (e.g., 60 Hz and 7000 Hz) in a 4:1 ratio of amplitudes. In both cases, distortion is expressed as a percent of the output signal.

20.8 ■ AM/FM TUNERS

Radio stations are assigned a frequency in one of two bands: 540–1600 kHz (5.4–16×10^5 Hz) and 88–108 MHz (8.8–10.8×10^7 Hz). Broadcasts in the upper band use *frequency modulation* (FM), whereas broadcasts in the lower band use *amplitude modulation* (AM). At the lower frequency of the AM band, signals can be transmitted over long distances, but because the AM band is crowded, stations are allowed to transmit audio frequencies up to 5000 Hz only. This severely limits the fidelity of sound reproduction. Furthermore, AM signals are subject to interference from electrical noise, both natural (atmospheric electricity) and human-made (car ignition, motors, etc.). On the other hand, FM stations are allowed to transmit a much greater range of audio frequency, but the high-frequency waves in the FM band are normally limited in range to "line of sight" or slightly more. Within metropolitan areas, however, FM radio has become the major source of high-fidelity program material.

Figure 20.12 illustrates the difference between amplitude modulation and frequency modulation. In both cases, audio information is superimposed on a radio carrier wave of

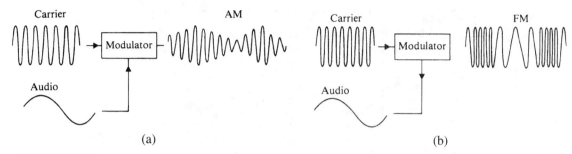

FIGURE 20.12   A schematic diagram showing (a) amplitude modulation (AM); (b) frequency modulation (FM).

high frequency. In AM broadcasting, the amplitude is varied (modulated) in accordance with the audio signal; in FM broadcasting, it is the frequency that varies.

The AM band, which has a total frequency range of 1060 kHz, is divided into broadcast channels that are 10 kHz wide, which is wide enough for the 5000-Hz audio signal plus channel separation, so it does not interfere with adjacent channels. On the FM band, on the other hand, with a total frequency range of 20 MHz, channels can be 200 kHz wide. This bandwidth, 20 times greater than that of AM stations, allows not only a wider range of audio frequency but a wider dynamic range as well; furthermore, it permits the broadcasting of stereophonic sound.

The tuner selects the AM or FM signal from the desired station, amplifies it, and then *demodulates*, or detects, it (that is, extracts the audio signal). (The technical details of how this is accomplished will not be described.) Many modern AM/FM tuners are finely engineered instruments, capable of quality sound reproduction. Furthermore, the performance specifications are much more accurate and complete than for other components in an audio system. A few of the more important specifications are as follows.

1. *Sensitivity* describes the ability of the tuner to pick up weak signals. The sensitivity is usually expressed as the electrical signal (in microvolts) at the antenna, which will result in an audio signal 30 dB greater than the noise level or, more recently, in dBf for 50 dB quieting.

2. *Signal-to-noise (S/N) ratio* expresses the ratio of audio signal to noise for a strong input signal. As the input signal increases, both noise and distortion usually decrease, as shown in Fig. 20.13. The S/N ratio for stereo reception is generally somewhat less than for mono reception.

3. *Total harmonic distortion (THD)*, also shown in Fig. 20.13, compares (in percent) the harmonic distortion generated in the tuner to the undistorted signal.

4. *Selectivity* measures the ability of the tuner to reject stations on nearby channels. It is usually expressed in decibels of attenuation for a station two channels (400 kHz) away.

5. *Capture ratio* expresses the ability of a receiver to reject the weaker of two signals on the same channel. The capture ratio (in decibels) tells how much smaller the one signal must be to be rejected; a capture ratio below 3 dB is desirable.

6. *Stereo separation*, the separation between stereo channels, should be at least 30 dB at all audio frequencies.

FIGURE 20.13
Typical noise and
distortion in an FM
tuner. (From the
Institute of High
Fidelity, © 1974.
Reprinted by
permission of the
Electronic
Industries
Association.)

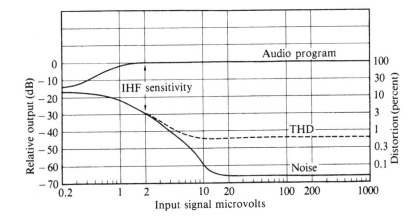

Other tuner specifications may include *image rejection, i-f rejection* and *spurious-response rejection*, all of which describe the ability to reject some type of undesired signal.

20.9 ■ STEREO BROADCASTING

Much of the program material of FM stations these days is broadcast in stereo. However, it is broadcast in a compatible format, so that it can be received on either stereophonic or monophonic tuners. This is done by mixing the left (L) and right (R) signals together to create sum and difference signals, which are then broadcast.

The sum (L + R) is broadcast as a conventional FM signal and is received by both monophonic and stereophonic receivers. At the same time, a difference signal (L − R) is broadcast at a higher frequency. A monophonic FM radio, of course, receives (L − R) along with the regular audio signal, but it is not heard because of its high frequency. A stereophonic receiver, on the other hand, reconstructs the desired (L) and (R) signals from the (L + R) and (L − R) signals received (Fig. 20.14). Because the signal-to-noise ratio of a tuner is usually greater when operated in the mono mode, it is sometimes desirable to use this mode when listening to a weak or distant FM station.

20.10 ■ AV RECEIVERS: SURROUND SOUND

The most recent advances in home entertainment systems combine music and video in the home theater. A key component of the home theater system is the AV (audio-video)

FIGURE 20.14
Composition of a
stereo broadcast
signal.

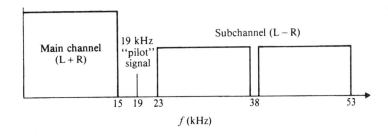

receiver. The AV receiver combines, in a single box, a tuner, a power amplifier, and a surround-sound decoder. It is the surround-sound decoder that allows home theater.

Development of Surround Sound

Many of us think of surround sound (using multiple speaker channels in an audio system) as a new technology. Actually, it has been used in the motion picture industry for many years. When Disney studios released *Fantasia* in 1940, they used a new technology called Fantasound, which included three front audio channels located on the screen plus two surround channels located on the sides of the theater. The movies *Oklahoma* (1955) and *Sound of Music* (1965) used a six-channel system with five screen channels and one surround channel.

Stereophonic sound came into home listening as a simplification of theater sound. The physical nature of LP records made it difficult to have more than two channels. Attempts at quadraphonic sound in the late 1960s and early 1970s failed, largely because of competing formats and resistance to placing four large speakers in the home. The stereo format dominates the home market, which has expanded to include FM radio, many tape formats, and CD audio.

Meanwhile, the movie industry expanded surround sound by using Dolby Stereo (an amplitude-phase matrix derived from quadraphonic sound technology) to place four channels of sound onto an optical sound track that had space for only two tracks. Three channels were used at the screen in addition to one surround channel. The requirement for extra power in the bass with movies such as *Star Wars* and *Close Encounters of the Third Kind* led to dedicated subwoofers in the system. Finally, with *Superman I* the surround sound was split into two separate channels, left and right. In 1987 the Society of Motion Picture and Television Engineers (SMPTE) developed a standard for the minimum number of channels needed to create the desired sensations in the theater. This standard is called 5.1, which means five amplified channels are needed (front left, front center, front right, surround left, and surround right) as well as a subwoofer channel with its own amplifier. New digital formats, such as DVD, have taken surround sound from the theater to home entertainment.

Surround-Sound Technology

A typical home theater system is shown in Fig. 20.15, which consists of an AV receiver connected to various input sources and a set of loudspeakers. Audio inputs include analog devices such as phono, tape deck, and VCR as well as digital devices such as CD, DVD, and digital TV. The output meets the 5.1 surround-sound specification by giving five separate amplified outputs plus a separate subwoofer output that is not amplified. Typical subwoofers have a built-in amplifier.

As surround sound migrated from the cinema to home theater, several encoding schemes combined up to four signals into two channels. One of the early schemes (called *Hafler stereo*) passed the stereo signals (L and R) to the left and right speakers and the different signal (L − R) to the surround speakers, thus synthesizing the ambient sound. Including delay in the surround speakers resulted in a greater "feel" of a real concert hall. In 1982 Dolby Surround Sound was released for the home theater market, which added a center

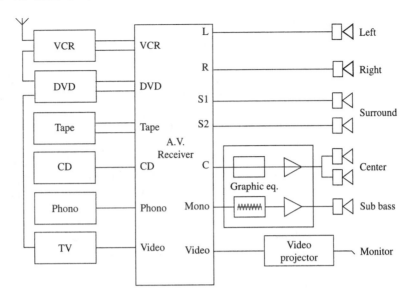

FIGURE 20.15
Typical home
theater system.

channel (L + R). The encoding is a 4-2-4 scheme, in which four original channels (left, right, center, and surround) are encoded into two stereo tracks and then decoded back into the four original channels. The sound intended for the surround channel is recorded on top of the left and right channels but out of phase with respect to each other. The surround channel is recreated by taking the difference (L−R) and then adding delay. Dolby Surround ProLogic is a common method used to record movies onto VHS tape. Sound localization is improved by actively sensing the dominant direction of sound and increasing the gain to loudspeakers in that direction. This allows the sitting area to be expanded while preserving the surround effect.

Dolby Digital was introduced to meet the new standards being developed for digital television, digital video discs, and direct broadcast satellite. Dolby Digital has a full 5.1 implementation, where signals for five channels and a dedicated subwoofer are stored separately rather than combined onto two channels through a matrix. Other multichannel surround sound systems are available. Digital Theater Systems (DTS), commonly found on laser discs, is used for mixing and reproducing live performances. It is capable of 5.1 surround sound but has less compression than Dolby Digital. Sony Dynamic Digital Sound (SDDS) adds two extra screen channels for a 7.1 capacity. Lucasfilm combined with Dolby Labs to develop TDX Surround EX, which extends Dolby Digital to 7.1 by adding rear surround speakers.

Because many surround-sound formats are available (although they may be incompatible), it is important to determine which formats are decoded when you select an AV receiver. It is possible to find receivers that will decode Dolby ProLogic, THD, and Dolby Digital.

20.11 ■ SUMMARY

A microphone produces an electrical signal when actuated by sound waves; most microphones respond to variations in sound pressure. There are many different microphone types, including crystal, dynamic, condenser, electret-condenser, and ribbon. The most popular directional characteristics are omnidirectional and cardioid (unidirectional). Important microphone specifications include sensitivity and output impedance. High-impedance microphones require transformers or in situ preamplifiers if long microphone cables are to be used.

Radio stations transmit in two frequency bands; the greater bandwidth allocated to stations in the FM band makes high-fidelity stereophonic sound transmission possible. The performance of FM tuners and receivers is well described by rather elaborate specifications.

Integrated amplifiers include a preamplifier and a control section as well as a power amplifier. Power amplifiers produce large audio signals with very small amounts of noise and distortion. Distortion is of three main types: harmonic, intermodulation, and transient. Distortion in loudspeakers is more prevalent than in amplifiers, but is seldom in the manufacturer's specifications. High-efficiency, large-cone speakers tend to have less distortion at high power levels than do low-efficiency, long-throw speaker systems.

Home theater systems use six channels which include left, center, and right loudspeakers in front, two surround-sound speakers on the side, and a subwoofer with a separate amplifier.

REFERENCES AND SUGGESTED READINGS

Alldrin, L. (1997). *The Home Studio Guide to Microphones*. Hal Leonard.

Bartlett, B. (1991). *Stereo Microphone Techniques*. Woburn, Mass.: Focal Press.

Carr, J. J. (1999). *Exploring Solid-State Amplifiers*. Indianapolis: Howard W Sams & Co.

Harley, R. (1997). *Home Theater for Everyone*. Acapella Pub.

Holman, T. (2000). *5.1 Channel Surround Sound—Up and Running*. Focal Press.

Hood, J. L. (1997). *Valve and Transistor Audio Amplifiers*. Butterworth-Heinemann.

Huber, D. M., and P. Williams (1998). *Professional Microphone Techniques*. Hal Leonard.

Kamichik, S. (1999). *Designing Power Amplifiers*. PROMPT Publications.

Self, D. (1996) *Audio Power Amplifier Design Book*. Butterworth-Heinemann.

Slone, R. (2000). *High-Power Audio Amplifier Guidebook with Projects*. New York: McGraw-Hill.

White, P. (2000). *Music Technology: Basic Microphones*. Sanctuary Pub Ltd.

GLOSSARY

amplitude modulation (AM) The method of radio broadcasting in which the amplitude of a carrier wave is determined by the audio signal. The AM band extends from 540 to 1600 kHz.

AV receiver A home entertainment component including a tuner, amplifier and surround sound decoder.

carrier wave A high-frequency electromagnetic wave capable of being modulated to transmit a signal of lower frequency.

class AB A power amplifier circuit in which current flows in each transistor for slightly more than one-half of the cycle.

complementary symmetry An *npn* and a *pnp* transistor that otherwise have identical electrical characteristics; such pairs of transistors are used as audio power amplifiers.

condenser microphone A microphone in which the diaphragm serves as one plate of a small capacitor or condenser. As the diaphragm moves, the electrical charge on the condenser varies.

distortion Signals that appear in the output of a sound reproduction system that were not present in the original input signal.

dynamic microphone A microphone that generates an electrical voltage by the movement of a coil of wire in a magnetic field.

electret-condenser microphone A condenser microphone that has an electrified foil between the plates of the capacitor, thus eliminating the need for a polarizing voltage as required in an air-dielectric condenser microphone.

electromagnetic induction Generation of an electrical voltage in a wire that moves across a magnetic field.

frequency modulation (FM) The method of radio broadcasting in which the frequency of a carrier wave is modulated (altered slightly) by the audio signal. The FM band, which extends from 88 to 108 MHz, allows stations sufficient bandwidth to transmit high-fidelity stereophonic sound.

harmonic distortion The creation of harmonics (frequency multiples) of the original signal by some type of nonlinearity in the system (the most common cause is overdriving some component).

impedance The ratio of voltage to current. In the case of source impedance or output impedance, it is the current that the device can deliver; in the case of input impedance, it is the current that the device draws from the source.

intermodulation (IM) distortion The creation of sum and difference frequencies from signals of two different frequencies.

piezoelectric crystal A crystal that generates an electric voltage when it is bent or otherwise distorted in shape.

push-pull An arrangement used in power amplifiers; a positive voltage in the input causes the current to rise in one transistor, whereas a negative voltage does the same for another transistor. Push-pull amplifiers usually have an output transformer but do not require complementary transistors.

sensitivity (microphone) Voltage or power generated in a microphone at a given sound pressure level.

sidebands Sum and difference frequencies created in the modulation process.

surround sound A multichannel loudspeaker system used in home theaters.

transducer A device that converts one form of energy into another; in this chapter, we discussed the conversion from acoustic energy to electrical energy (microphone).

transient distortion Overshoot, or other unprogrammed response, that results from the inability of some component to follow a very rapid change in signal.

velocity microphone A microphone that responds to particle velocity rather than to sound pressure.

REVIEW QUESTIONS

1. Which type of microphone is very inexpensive and produces a large electrical signal?

2. Which type of microphone is sensitive to acoustic velocity rather than pressure?

3. Which type of microphone requires a high-voltage bias?

4. Which type of microphones use electromagnetic induction?

5. Which type of microphone pattern might best be used to pick up a quarterback's signals at a football game?

6. What is the advantage of using a low-impedance microphone over using a high-impedance microphone?

7. Which audio component requires the use of an audio preamplifier?

8. Which class of power amplifier is most efficient with least distortion?

9. Which type of distortion is produced by clipping the waveform peaks?

10. Which type of modulation is most commonly used for stereo broadcasts?

11. Which surround-sound decoding scheme uses the 5.1 standard?

QUESTIONS FOR THOUGHT AND DISCUSSION

1. What are the advantages and disadvantages of using microphones of low impedance as compared to those of high impedance?

2. Describe the directional pattern of a cardioid microphone. For what applications are cardioid microphones superior?

3. Would it be possible to broadcast stereophonic programs on AM radio? Why?

4. Why is intermodulation distortion generally more objectionable to the ear than an equal amount of harmonic distortion?

5. Why would Dolby Digital give more realistic sound than Dolby ProLogic?

EXERCISES

1. Compare the voltage outputs from two microphones: Microphone A has a sensitivity −60 dB re 1 V per 1 Pa and microphone B has a sensitivity of −66 dB compared to the same reference.

2. Calculate the actual voltage output of microphone A at a sound pressure of 1 Pa. To what sound pressure level does this correspond (see Chapter 6)?

3. If two 8-Ω speakers are connected to the same output terminals of the amplifier, what net impedance will be "seen" by the amplifier?

4. An amplifier has a damping factor of 30 when used with an 8-Ω speaker. What will the damping factor be if used with a 16-Ω speaker?

5. An 8-Ω loudspeaker is capable of handling 30 W of audio power. How much current will flow when it is receiving 30 W of power? What size of fuse or circuit breaker should be used to protect it from overload?

6. What is the wavelength at the center of the FM band? (See Exercise 1 in Chapter 3.) How long should the horizontal loop of an FM antenna be, if it is to be one-half of a wavelength (in order to be resonant)?

EXPERIMENTS FOR HOME, LABORATORY, AND CLASSROOM DEMONSTRATIONS

Home and Classroom Demonstrations

1. *Microphone styles* Acquire different types of microphones (including dynamic, condenser, and electret microphones) in various configurations to show the variety of microphones available in the market. Each microphone should be discussed in terms of methods of transduction, functional design, acoustical and electrical characteristics, and application in the studio. As an alternative, a catalogue (or the Internet) containing a variety of microphones could be used to select a microphone for a specific application based on its characteristics.

2. *Proximity effect* Connect a cardioid microphone to an amplifier and loudspeaker. Demonstrate the increased bass response as the source (talker or singer) gets closer to the microphone. Discuss why pop singers seem to swallow their microphones. Repeat the experiment with an omnidirectional (omni) microphone and note the difference.

3. *Sensitivity measurements* Connect a pink noise source to an amplifier and loudspeaker. Adjust the volume control to read 90 dB on a sound level meter at a given distance from the source. Measure the voltage (with a high-impedance voltmeter) on the output of a microphone at the position of the sound level meter.

4. *Phantom power* Connect a condenser microphone to a mixer with switchable phantom power. Turn the phantom power off and determine the output level as you sing into the microphone. Turn the phantom power on and note the difference.

5. *Preamp clipping* Apply a sine wave to the input of a preamplifier and view both the input and the output signals on an oscilloscope. Note that the output signal is also a sine wave, but with an increased amplitude. Determine the gain of the amplifier by comparing the output amplitude with the input amplitude. Increase the amplitude of the signal generator until the output waveform is no longer a sine wave. The new waveform is produced by clipping the signal from overloading the input. To hear the harmonic distortion from clipping, connect a power amplifier and loudspeaker to the output of the preamp. You should hear the change of timbre of the tone as soon as it starts clipping.

6. *Series and parallel connection of loudspeakers* Connect an 8-ohm loudspeaker to the output of an amplifier and listen to some piece of music. Without changing the settings on the amplifier attach two 8-ohm loudspeakers in series to the output and listen to the same music. Now attach the two 8-ohm loudspeakers in parallel and listen again. Compare what you heard in the three cases.

Laboratory Experiments

Audio tests and measurements: an introduction (Experiment 27 in *Acoustics Laboratory Experiments*)

Amplifier performance: frequency response and harmonic distortion (Experiment 28 in *Acoustics Laboratory Experiments*)

Amplifier performance: hum, noise, and intermodulation distortion (Experiment 29 in *Acoustics Laboratory Experiments*)

Microphones (Experiment 32 in *Acoustics Laboratory Experiments*)

CHAPTER
21

Digital Computers and Techniques

Electronics technology, as we have seen, has been extremely useful in the field of music. Music has been analyzed, synthesized, recorded, transmitted from place to place, and conceptually augmented by the advent of the vacuum tube, the transistor and the integrated circuit. Furthermore, the nonlinear switching mode of transistor operation together with advances in integrated circuit fabrication techniques has led to an unprecedented revolution in digital electronics. Computers that once filled large buildings (and cost substantial fortunes) are now thoroughly outpaced by consumer products accessible to nearly everyone. The advantages of digital processing techniques for musical sound are so useful and ubiquitous that we consider them in this chapter in considerable detail.

In this chapter, you should learn:

- About binary data and how numbers are represented in computers;
- About the basic organization of digital computers and computer programming;
- About digital signals, including sampling and quantization, aliasing, dithering, and linear and nonlinear encoding.

21.1 ■ WHAT IS A COMPUTER?

Calculating with numbers has little to do with many modern computer applications. Increasingly, computers allow people to communicate in useful ways, both with each other and with information repositories of various kinds, such as online libraries and databases. Computers are useful as intermediary controllers of complex devices ranging from kitchen appliances to robotic space vehicles. Computers help not only to analyze complex information, such as the structures of interstellar space and human DNA, but also to visualize and understand them as well. Computers allow us to withdraw cash from bank accounts at 4:00 A.M., to identify the best driving route to someone's house, and to do a host of other things for which the calculation of numbers per se plays a relatively minor part.

21.2 ■ BINARY DATA

Computers spend most of their time *processing information* rather than calculating numbers. The unit of information—according to classical information theory—is the *bit*, which stands for *binary digit*. Bits can be used to represent virtually anything that can be measured, from the temperature on Mars to the shape of the letters that spell out the text of this book to the voltages coming out of microphones standing in front of an orchestra playing Beethoven's *Fifth Symphony*.

A bit of information is properly thought of as "standing for" one of two possibilities. The world is so full of dichotomies that there are many general ways to name these two possible states: yes or no, true or false, yin or yang, up or down, right or left, etc. As the name indicates, however, it is common practice to name the two possible bit states 1 and 0, in accordance with binary (i.e., base 2) numbers.

If 1 bit can be used to distinguish between two possibilities, 2 bits can be used to distinguish among four possibilities, namely, 00, 01, 10, and 11. The number of possibilities doubles each time we add a bit (e.g., for 3 bits we have all the possibilities for 2 bits proceeded by 0 or all those possibilities preceded by 1). It is easy to see that, in general, N bits may be used to distinguish among 2^N possibilities. Conversely, information theory rests on the principle that one choice out of S equally likely possibilities represents $E = \log_2 S$ bits of information (note that E—which stands for *entropy*—is not necessarily an integer and that *bits of information* are not the same as *bits of memory*.

Computers use electronic circuits to store information in binary form. A transistor operating in its nonlinear (switching) mode, for example, can be turned either on or off—one of these states might be associated with 1, the other with 0 (it doesn't matter which information state is associated with which electrical state).

It requires more than simply turning a transistor on or off to *store* a bit of information, however. In order to be stored, a circuit must have two *stable* states, that is, it must remain in one state or the other once it is put there, just as a light switch remains on or off until an external force changes it. We can, for example, interconnect two NOR gates to form a simple circuit called a flip-flop that can store 1 bit of information (see Fig. 21.1).

A flip-flop made out of transistorized logic gates can store information only so long as electrical power is supplied to the circuit. A flip-flop is, therefore, an example of a *volatile memory*, because the contents of that memory are lost whenever the power supply to the circuit is turned off. Although volatility is an arguable disadvantage, transistor-based (also called *solid-state*) memory can generally be changed (*written*) or retrieved (*read*) at very high rates of speed. The speed of solid-state memory depends primarily on how quickly the

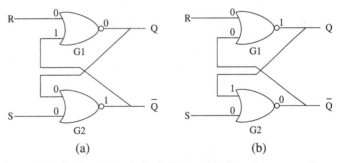

(a)   (b)

FIGURE 21.1   An S-R flip-flop: If both the S (Set) and R (Reset) inputs are 0, two NOR gates* interconnected as shown can be in either state A or state B. The output of the flip-flop is labeled Q (the complemented output is also available). Temporarily holding input S to 1 causes the flip-flop to enter state B ("set," or Q = 1); temporarily holding input R to 1 causes the flip-flop to enter state A ("reset," or Q = 0). Holding both R and S to 1, simultaneously setting and resetting the flip-flop, would generate an indeterminate output. The S-R flip-flop is thus bistable and can be used as a one-bit memory. (*NOR gates give a 0 as an output if either input is a 1.)

transistors can be switched on and off—times that are typically measured in nanoseconds (10^{-9} s) or picoseconds (10^{-12} s).

Another form of data storage might use a small region of magnetizable material (such as magnetic oxide) to store data. Magnetizing the material in one direction would represent 1; in the other direction, 0. Magnetic storage has the advantage of being *nonvolatile*, because the magnetic state of the material is retained after the power supply to the circuit is switched off. The disadvantage of magnetic memory is that it typically is one or two orders of magnitude slower than solid-state memory.

Computer memory can be volatile and fast or nonvolatile and relatively slower. It is common for many bits to be read or written at one time. The standard unit of digital memory (as opposed to the unit of information) is a *byte*, which is simply an ordered collection of 8 bits. Because 1 byte clearly represents 1 out of $2^8 = 256$ possibilities, it might be used to store a simple character of text (such as an upper- or lower-case letter in the English alphabet, a decimal digit (i.e., 0–9), a punctuation symbol, space, tab, backspace, etc.) A standard way of associating 8-bit codes with simple text symbols is called ASCII (for American Standard Code for Information Interchange). The ASCII code for the letter *A*, for example, is 0100 0001, for *a* is 0110 0001, for ? is 0011 1111, etc. Although ASCII provides enough possibilities for simple English text, the increasing need to represent characters in many languages has driven the development of UNICODE, which is similar to ASCII, except that text is represented by a series of 16-bit codes. UNICODE can represent far more accented and umlauted characters and alphabets, *glyphs* that occur in languages such as Chinese, Japanese, and Korean, as well as multilingual documents.

Computers often process information in units other than bytes; for example, there is fairly wide agreement that a group of 4 bits (half a byte) is called a *nibble*. Computers also process words, which might consist of 2, 4, 8, or 16 bytes (depending on the computer) and occasionally other odd-sized "chunks" of data.

21.3 ■ NUMBER REPRESENTATIONS

When computers represent numerical information, the bits in a computer byte or word have a special interpretation. In the simplest form, the bits may simply be interpreted from left to right as a simple binary number. Recalling that any number is a "sum of products" of digits multiplied by base-dependent digit *weights*, we see that the ASCII bit code for *A* can be interpreted as the following (decimal) number:

$$(0 \times 2^7) + (1 \times 2^6) + (0 \times 2^5) + (0 \times 2^4) + (0 \times 2^3) + (0 \times 2^2)$$
$$+ (0 \times 2^1) + (1 \times 2^0)$$
$$= (0 \times 128) + (1 \times 64) + (0 \times 32) + (0 \times 16) + (0 \times 8)$$
$$+ (0 \times 4) + (0 \times 2) + (1 \times 1)$$
$$= 65_{10}.$$

The computer does not know whether the byte code 0100 0001 represents the ASCII code for *A* or the decimal number 65—it is entirely up to programmers to keep track of the proper interpretation of each byte of data in a computer's memory. Fortunately for programmers, sophisticated computer software such as compilers, interpreters, assemblers,

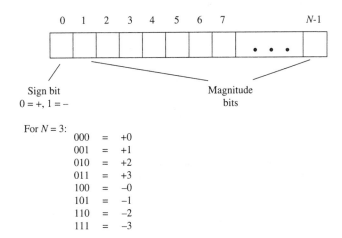

FIGURE 21.2
In sign-magnitude
notation, the first
bit of an N-bit
word signifies the
sign (typically 0 for
plus and 1 for
minus). The
remainder of the
bits represent the
magnitude of the
number. Note that
two forms of zero
are possible.

and debuggers can be used to help keep track of the myriad of such details involved in most computer programs.

In order to represent more than 1 out of 256 possible numbers, *words* of memory can be used. A 2-byte word can represent, for example, one of $2^{16} = 65{,}536$ possible values. Interpreting the bits directly as before, we see that the values represented are 0 through 65,535. Such values are often called *unsigned*, because only nonnegative values are included.

There are two common ways of accommodating negative numbers. One is to designate the first bit as a *sign bit*; the remaining bits can then be used to represent the magnitude of the number (see Fig. 21.2). Assuming we again use 2-byte words for numbers, the *sign-magnitude* approach to representing numbers allows for numbers from $-32{,}767$ to $+32{,}767$. A minor problem with this approach is that it allows two different representations for zero: $+0$ and -0—it is necessary to arrange for the bit patterns 0000 0000... and 1000 0000... to be considered to be numerically equal to each other.

Another common approach is the *2's complement* convention (see Fig. 21.3). For positive numbers, 2's complement notation looks exactly like sign-magnitude. A negative number is formed by adding 1 to the *complement* of a number (the complement is a number formed by inverting all the bits). In 4-bit nibbles, 0001 represents $+1$, whereas -1 is repre-

FIGURE 21.3
In 2's complement
notation, negative
numbers are
formed by adding 1
to the complement
of the positive
value. Carries from
the most significant
bit are ignored.
Note that the first
bit still indicates
the sign of the
number.

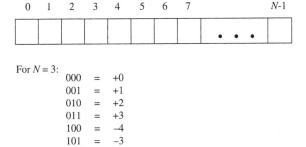

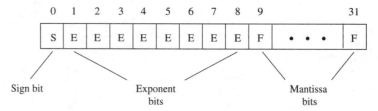

FIGURE 21.4 In the widely used IEEE convention for floating point numbers, bit 0 is a sign bit, bits 1–8 are exponent bits, and bits 9–31 are "fraction" bits. In most situations, the exponent x is equal to $E - 127$, where E is the (unsigned) number formed by the set of E bits above. Mantissa m is equal to "1.F," where F is the fraction formed by the set of F bits shown above.

sented by $1110 + 0001 = 1111$ (*carry* bits out of the leftmost part of the nibble are ignored). The 2's complement technique for 16 bits thus represents the numbers $-32{,}768$ through $+32{,}767$. It has the advantage of a unique representation for zero and the disadvantage (usually minor) of a slightly asymmetric allowance for negative and positive numbers.

So far we have considered only integer numbers, i.e., numbers in which the *binary point* (the binary equivalent of the decimal point) is located just to the right of the rightmost bit. It is possible to locate the binary point at any position, including to the left of the leftmost bit, in which case the 16 bits we have been discussing represent fractional values less than 1. Interpreted in this way, the ASCII code for A corresponds to the number

$$(0 \times 2^{-1}) + (1 \times 2^{-2}) + (0 \times 2^{-3}) + (0 \times 2^{-4}) + (0 \times 2^{-5}) + (0 \times 2^{-6})$$

$$+ (0 \times 2^{-7}) + (1 \times 2^{-8})$$

$$= \left(1 \times \frac{1}{4}\right) + \left(1 \times \frac{1}{256}\right) = \frac{65}{256} = 0.25390625.$$

The concept of binary fractions is readily extended to binary representations for numbers expressed in *scientific notation*, such as 1.6×10^{19}. A typical 4-byte *floating-point* value uses the 1 byte to represent an "exponent" plus 3 bytes to represent a fractional "mantissa" (see Fig. 21.4).

Binary representation can lead to some surprising consequences. Just as the simple ratio $\frac{1}{3}$ has no exact representation as a decimal number (it is usually written $0.33\ldots$), some values that have exact decimal representations do not have exact binary representations. The rational value $\frac{1}{10}$, for example, easily written as 0.1 in decimal, unfortunately has not exact representation as a binary fraction. Just as adding $0.33\ldots$ to itself 3 times in decimal gives a result of $0.99\ldots$, adding the (inexact) representation of $\frac{1}{10}$ to itself 10 times in binary also yields $0.99\ldots$, not 1.

21.4 ■ ORGANIZATION OF A COMPUTER

A typical digital computer basically consists of the following (see Fig. 21.5):

- A fast, volatile *working memory* (random-access memory, or RAM)
- *Nonvolatile memory* (such as read-only memory, or ROM, one or more hard disk drives, or HDDs, diskette and/or tape drives, etc.)

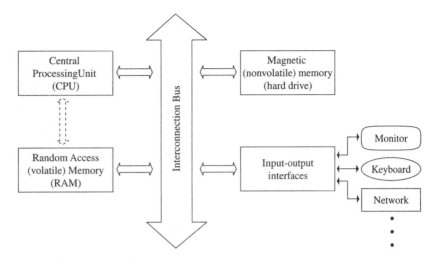

FIGURE 21.5
Basic organization
of a typical digital
computer.

- One or more *central processing units*, or CPUs
- Provision for various types of *input-output*, or I-O (keyboard, mouse, display, loudspeakers, printers, network interfaces, modems, etc.)
- Various control, interconnection, and *support elements*, such as bus connectors, ports, power supplies, etc.

Computers generally run under the control of a permanently resident program called an *operating system*. Commonly used operating systems include various versions of Windows, UNIX, MAC OS, etc. In response to actions taken by the computer user, the operating system loads necessary programs into its working memory from its nonvolatile memory, which is organized by the operating system into files containing both programs and data. Once in the working memory, programs can execute rapidly, often performing millions of operations each second. Under control of the selected program, the central processing unit accesses data in the memory or coming in from input-output devices, performs logical, symbolic, or numerical operations according to the instructions in the program and the values of the data, and transmits new data back to the memory and/or input-output devices. Operating systems allow computers seemingly to execute more than one program simultaneously (*multithreading* and/or *multitasking*), perhaps using more than one central processor simultaneously (*multiprocessing*).

Programmers use computer-aided design tools such as assemblers, compilers, interpreters, and debuggers to develop the programs that control the operations of a computer. Large, complex computer programs are often called *applications*. A detailed account of computer design and computer programming is beyond the scope of this book; however, many excellent references are available on these subjects.

21.5 ■ THE CPU

The central processing unit (CPU) is the heart and brain of the computer. It includes the arithmetic/logic unit (ALU) plus control and timing circuitry and registers. In a large mainframe computer, the CPU may include many circuit boards filled with components; in a microcomputer, on the other hand, the entire CPU is contained in one integrated circuit called a *microprocessor*. A *microcontroller* is a specialized microprocessor with memory and I/O in a single integrated circuit used to control devices, such as a synthesizer or mixer. The microprocessor uses a clock signal to keep track of its operations. The clock speed gives a rough indication of how many instructions can be executed in each second. As a consequence of the oft-quoted *Moore's law*, which states that the number of transistors on an integrated circuit will double every 12 to 18 months, clock speeds have doubled every 2 or 3 years ever since integrated circuits were invented (around 1960).

21.6 ■ MEMORY

Computer memories can be classified in several ways. One is according to whether it can retain its contents when the power is turned off. A memory that can do so is called *nonvolatile* memory; one that loses its contents is called *volatile*. Another way of classifying memories is according to addressability and read/write capabilities. A *random-access* memory (RAM) is one in which data can be read from and written into any location with minimum delay. A *read-only* memory (ROM) also offers random access to any location but only to read its contents. Data in a *serial* or *sequential access* memory, on the other hand, are available only at certain times, and the access time may depend upon its storage location. Yet another way of classifying memory is according to the storage medium and the way in which it is used. Most common computer memories are semiconductor, magnetic, and optical stage.

1. *Main memory.* The main memory in a computer is used to store the data and the programs that are currently being executed by the CPU. The main memory must have a fast access time, and therefore it is almost always in a semiconductor random access memory.

2. *ROM.* Nearly every computer includes nonvolatile read-only memory. ROM is used to store programs that interface the computer with input/output devices (such as the BIOS or basic I/O system), a program to test the computer system prior to operation, and a "bootstrap" program that tells the computer what to do when it is first turned on. Erasable/programmable memory (EPROM) can be erased by ultraviolet light and then reprogrammed. Electrically erasable memory (EEPROM) can be erased and programmed while still in the computer.

3. *RAM.* Random-access memory is located in semiconductor integrated circuits (ICs). RAM may be either static or dynamic. Static memory cells (SRAM) hold information for as long as power is supplied, whereas dynamic memory cells (DRAM) must be frequently *refreshed* to maintain their contents. DRAM is slower than SRAM, but it stores large amounts of data in a small space; hence, it is used for the main memory of the computer. SRAM is commonly used for *cache* memory (a small, fast "scratch-pad" memory accessed directly by the CPU) because of its speed.

4. *Mass memory.* Mass memories are intended to store large amounts of data inexpensively. The most commonly used devices are magnetic hard discs (storing many gigabytes, terabytes, or petabytes of data) and removable diskettes (storing at most a few

megabytes). In both of these devices, the data are stored as the magnetic remnant state in a small spot on the magnetic oxide coating, as in magnetic tape recording. For removable discs with larger storage capacity than diskettes, there are *zip* and *jaz* discs, which store up to a few gigabytes.

21.7 ■ INPUT AND OUTPUT (I/O)

A computer can calculate at high speeds and store large amounts of data. However, if you are not able to access the data, these capabilities are of little use. The purpose of I/O is to enter data into the computer and get data back out. A common interface to the computer consists of a keyboard for entering data and a monitor for looking at the data. Additionally, there are usually a mouse for selecting commands or graphical input, one or more serial ports (often with one or more attached modems) for serial communications, a parallel printer port, and a sound card for sound and sound-related I/O.

On the surface, a keyboard is a simple device: push down a key and a character is displayed on the monitor. Actually, a lot goes on under the hood. Inside the keyboard is a microcontroller, which is itself a small computer. The microcontroller scans all the keys of the keyboard and detects when a key is pressed or released. When a key is pressed, a special character (the keycode) is sent to the microprocessor. When the key is released, a different keycode is sent. The microcontroller can also send a series of the same keycode when a key is held down for some time. This is called the typematic feature of the keyboard. Once the keycode is received by the microprocessor, it is converted into an ASCII character and stored in a small buffer (a temporary storage location in RAM) until the program retrieves it for use.

Monitors are similar to television sets in that they display information visually. Unlike television, a monitor is organized as an array of picture elements, or *pixels*. Pixels are arranged in rows and columns across a computer display. The numbers of rows and columns per inch (or mm) define a display's *resolution*. These pixels can display characters or graphics. The simplest mode of display is monochromatic and turns the pixel either on or off. In color monitors, each pixel can represent a variety of colors. If a pixel only needs to be turned on or off, it can be represented by a single bit of memory. If N bits of memory are used instead, each pixel can (theoretically) represent 2^N colors. The images displayed on the monitor are stored in a special location in memory called video RAM. The image is updated by the microprocessor by writing new data to the video RAM.

A mouse is an input device used to interact with the monitor. It consists of a small trackball that rotates as it is moved. There are two pairs of optical or mechanical sensors that detect movement in the horizontal and vertical directions. A cursor on the monitor screen is updated as the trackball moves. In addition to the trackball the mouse has one, two, or three pushbuttons used to select a location on the screen. Selection or activation is performed by a single- or double-click of the button or by holding the button down while dragging the cursor across an area of the screen.

Data are communicated to external devices, such as a printer or a modem, using either parallel or serial communications. In parallel communications eight separate wires are used, one for each bit sent. In serial communications bits are sent one at a time over a single wire. Parallel communication is faster and uses shorter cables for devices, such as printers, located close to the computer. Serial communication allows data to be sent to other de-

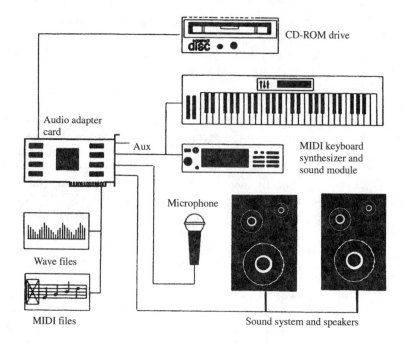

FIGURE 21.6
The hardware
components of a
multimedia system.

vices or over telephone lines using a modem. Many computers have direct connections to bit-serial computer networks, such as Ethernet, which require specialized cards to convert information to network formats.

Sound cards expand the capability of a computer to include sound I/O. A typical sound card offers several multimedia capabilities (Fig. 21.6). It includes controller circuitry for a CD-ROM drive, which can play audio CDs (Chapter 22) or CD-ROM software. A musical instrument digital interface (MIDI) port allows the computer to control music keyboards and synthesizers (Chapter 28). The sound card contains a digital synthesizer for playing back MIDI files. A microphone can be connected to a (possibly present) sampler on the sound card, which generates wave files (a format for storing sound) for future playback. Finally, the sound card has a mixer and a small amplifier for externally powered loudspeakers.

21.8 ■ COMPUTER BUSES

A bus is a common group of wires used to interconnect the memory and I/O with the CPU. The CPU places an address (location of data) on the address bus and control commands (read data from memory, write data to I/O, etc.) on the control bus. Data are transferred on the data bus between the CPU and memory or the CPU and I/O. Normally the CPU, memory, and I/O are placed on a printed circuit board (called a motherboard), so the address, data, and control buses are also on the motherboard. There is often a need to add resources beyond those contained on the motherboard, such as increased memory or a scanner. These resources are made available with expansion boards, which are connected to one or more computer buses on the motherboard.

When IBM released the original personal computer in 1982, it included several connectors (called the I/O channel by IBM) for attaching expansion boards. The I/O channel had an 8-bit data bus and a 20-bit address bus. When the IBM AT was released (about 1986), the data bus grew to 16 bits and the address bus, to 24 bits. The AT bus added a small connector in line with the I/O channel connector so old 8-bit boards or new 16-bit boards could be used. At this time, many companies were making IBM-compatible computers. IBM then introduced a new PS/2 series of computers developing the proprietary microchannel bus, which was not compatible with the AT bus. Meanwhile, other computer manufacturers got together to standardize the AT bus, calling it the Industry Standard Architecture (ISA) bus. When the 80386 microprocessor was used in computers (in the early 1990s), it required a 32-bit data bus, so the Extended ISA architecture (EISA) bus was developed. The ISA bus and EISA bus operate at 8 MHz, which is much slower than microprocessor speeds.

Two more recent buses have been developed to increase I/O speed for plug-in cards: (1) the peripheral component interconnect (PCI), allowing much higher speed than the older EISA bus; and (2) the accelerated graphics port (AGP), which operates at the microprocessor clock frequencies for video cards. Two more recent high-speed serial buses are: (1) the Universal Serial Bus (USB), with speeds up to 12 Mbps and (2) IEEE 1394 (also called FireWire, or I-Link), operating at transfer rates up to 400 Mbps.

21.9 ■ COMPUTER PROGRAMMING

Computers are interesting because they can be programmed; that is, their function is determined by sets of instructions (or commands) written by programmers. Strictly speaking, the computer understands only commands written in machine language, which uses a suitable numerical code of 1s and 0s for every operation. However, machine language is unnecessarily detailed for most computer programmers.

Assembly language is an example of a low-level language (having the ability to directly access internal registers within the CPU) consisting of a set of mnemonics and directives that help programmers write machine language. An assembler is a computer program that translates each assembly-language instruction into a machine-language instruction. For example, MOV CL,BL is an example of an assembly-language instruction that moves a byte of data from a register called BL to a register called CL. The equivalent machine-language instruction might be 10001010 11001011. Assembly language is specific to the microprocessor being used to execute the program; therefore, an assembly program written for an Intel microprocessor would not run on a Motorola microprocessor.

A number of high-level languages (such as BASIC, FORTRAN, PASCAL, and "C") have been developed since the 1950s to simplify programming. Before any of these user languages can be employed, a special program called a compiler, or an interpreter, must translate the program from user language to the correct machine language for the particular computer. This allows a programmer to learn one high-level language and to write programs for many different microprocessors, allowing the compiler to do the translation.

Programming methodology continues to grow and change. Programming languages such as BASIC solve problems with a step-by-step approach. Another method of approaching problems is object-oriented programming (OOP). Computer languages such as Smalltalk, Visual Basic, C++, and Java define a set of objects, characteristics the objects manifest, and actions the objects can perform. An example of an object is a window displayed on the monitor. It has characteristics of color and position and can take the action of closing when a mouse is clicked on the EXIT button. Object-oriented program languages are commonly used when programming with graphic user interfaces (GUIs) such as Windows and Macintosh operating systems.

Several music software applications are available for home musicians using either a PC or a Macintosh computer. One important software application is called a *sequencer*. A sequencer acts like a multitrack tape recorder. In theory, a full orchestra score can be created one track at a time and then played back all together. Once the score is entered, the tempo can be changed or the piece transposed into other keys. Notation software (such as the commercial Finale program) acts like a word processor for music. Besides the musical notes, all the symbols normally seen on musical scores (and many seldom seen) can be entered and properly formatted for publication, as well as synthesized using sound-card capabilities. A patch editor/librarian allows you to create and store a multitude of sounds that can be used by your synthesizers to play back music. There are also music composition and theory applications. With the multitude of high-quality music software available, the computer has become an essential tool for musicians.

21.10 ■ DIGITAL SIGNALS

We have seen how transducers are used to convert sound signals into their electrical analogs and vice versa: Various types of microphones transduce sound signals into analogous electrical signals, and various types of headphones and loudspeakers transduce analog electrical signals into sound signals. In their analog electrical form, sound signals can be transmitted from place to place, recorded, processed, analyzed, synthesized, amplified, and reproduced.

The advantages of the analog electrical representation of sound are evident: electrical signals travel much faster than the speed of sound, allowing long-distance conversations to occur in real time with very little delay. Because of the relationship of electricity and magnetism, analog electrical signals are easily interchanged with analog magnetic signals that are very useful for recording and reproduction. Analog electrical signals can be modified in various useful ways, such as amplification and filtering. As we will see in Chapter 27, analog electrical signals can be synthesized in ways that may or may not depend on naturally occurring sounds.

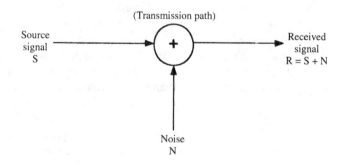

FIGURE 21.7
Basic model of noise contamination for all transmitted signals.

But analog electrical signals also have inherent disadvantages. Foremost among these is the fact that spurious electrical signals resulting either from random sources or from electrical processing devices themselves can contaminate analog signals (see Fig. 21.7). The problem with such noise contamination is that, because the frequencies in the (undesired) noise signals largely overlap the frequencies in the (desired) analog signals, they are virtually impossible to separate. Thus, although noise-reduction circuits can reduce such noises as tape hiss in magnetic sound recordings (see Chapter 22), no amount of clever processing can ever eliminate such noise contamination entirely.

Digital signals, on the other hand, can be made so nearly impervious to noise (under normal circumstances) that they have revolutionized audio technology and replaced analog signals in a wide variety of applications (see Fig. 21.8). The reason is that a digital signal is an electrical signal that varies between only two predetermined values, such as 0 V and +5 V. One of these values (either) represents a binary 1 value; the other represents a binary 0. Suppose, in a digital transmission, the transmitter sends a signal of +5 V, representing a binary 1. The receiver of that signal will receive the transmitted signal, plus any noise that has contaminated the signal along the transmission path. As long as the receiver is able to tell that the signal being transmitted represents a 1 and not a 0, the noise is irrelevant, because the receiver can generate an entirely new, untarnished +5-V signal based on its detection of the transmitted signal. Of course, the receiver could become confused if the noise is great enough to entirely obliterate the transmitted signal. But, unlike the analog receiver, which outputs the received signal plus any unsuppressed noise that contaminated it during transmission (and there always is some), the digital receiver essentially reconstructs a new, essentially "perfect" replica of the transmitted signal, thereby eliminating transmission noise altogether. A properly made copy of a digital signal not only exhibits high fidelity to the original, it is *identical* to it.

FIGURE 21.8
When an analog signal is contaminated by noise (upper figure), the noise generally becomes inseparable from the signal. When a digital signal is contaminated by noise (lower figure), it is usually possible to reconstruct the uncontaminated source signal perfectly. Thus a digital copy of a digital original can be lossless.

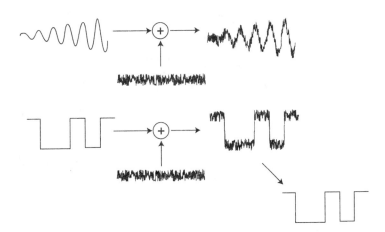

21.11 ■ PULSE-CODE MODULATION

One of the most widely used forms of digital signal involves a particular type of encoding called *pulse-code modulation*, or PCM. In this context, the terms *pulse* refers to the 1 value against a background of 0 values (or vice versa). In PCM, therefore, the pattern of 1s and 0s is varied according to the shape of the analog electrical signal. PCM is not the only type of digital encoding of a signal: others include pulse-amplitude modulation (PAM) and pulse-width modulation (PWM), in which the amplitude or width (in time) of the pulse is varied instead of the pattern of pulses.

In order to create a PCM-encoded *digital signal*, an analog electrical signal is first passed through an analog *low-pass filter*, which (ideally) removes all frequency components above a certain critical frequency called the *Nyquist frequency* (frequencies below this value are unaltered). The *bandlimited* result is then applied to the input of a device called an analog-to-digital converter, or ADC. The ADC essentially does two things: it measures—or *samples*—the (bandlimited) analog signal at successive instants in time, and it *encodes*—or *quantizes*—each sample into a corresponding binary code, or number. The resulting *sequence of binary numbers* is known as the (PCM-encoded) digital signal (see Fig. 21.9).

FIGURE 21.9
The digital audio chain: the analog signal is digitized by removing all frequencies above half the sampling rate, then sampling and encoding (quantizing) the sample values. The resulting digital signal can be transmitted, stored, and processed in a variety of ways. The digital-to-analog converter produces a signal that needs to be "smoothed" before it is amplified for playback through loudspeakers.

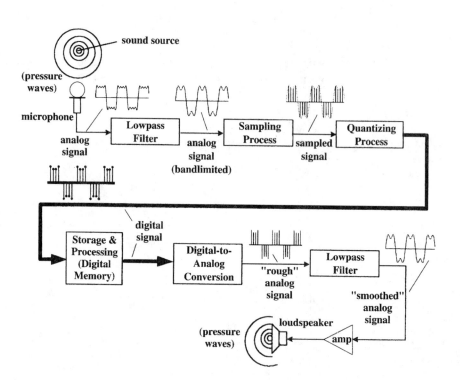

21.12 ■ THE SAMPLING THEOREM

The low-pass filter and sampling process operate according to the *sampling theorem* or *Nyquist theorem*, which states that at least R samples per second are needed to adequately represent a signal containing frequencies up to $R/2$ hertz. R hertz is the sampling frequency, and $R/2$ hertz is called the *Nyquist frequency*, after Harry Nyquist, the originator of the sampling theorem. Each "sample" of the analog signal is a (theoretically perfect) measurement of its amplitude at a given instant in time (this is represented by a signal's exact height, at a given instant in time, above or below the horizontal axis in a waveform plot). According to the sampling theorem, to represent frequencies up to 20,000 Hz, at least 40,000 samples per second must be used. In practice it is usually necessary to set R to a value higher than this minimum owing mainly to limitations of the band limiting low-pass filter.

21.13 ■ QUANTIZATION

The quantizing process consists of assigning an N-bit binary number to the value of the instantaneous amplitude sample. Suppose an analog electrical signal representing a sound always lies within an amplitude range of ± 5 V (see Fig. 21.10). If the quantizing process had only $N = 1$ bit available for encoding, then it could at best distinguish between two voltage ranges, for example, positive or negative. In that case, any voltage of 0 through 5 V might be encoded as (say) a 0, whereas any voltage greater than -5 V and less than 0 V might be encoded with a 1. If $N = 2$ bits were available, then four amplitude regions could be distinguished between ± 5 V. If we designate two codes to correspond to the maximum and minimum voltages (a common practice), the other two codes then correspond to two evenly spaced voltages in between.

21.14 ■ QUANTIZATION ERROR

When the analog signal is reconstructed from its digital samples, each numerical code is assigned a particular voltage level. In the $N = 2$-bit case just discussed, the 11 code might

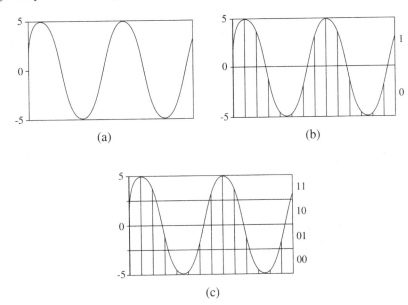

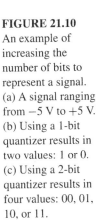

FIGURE 21.10
An example of increasing the number of bits to represent a signal. (a) A signal ranging from -5 V to $+5$ V. (b) Using a 1-bit quantizer results in two values: 1 or 0. (c) Using a 2-bit quantizer results in four values: 00, 01, 10, or 11.

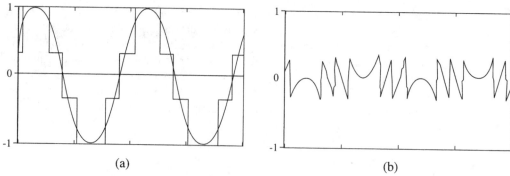

FIGURE 21.11   When the digitized signal is reconstructed into an analog signal, quantization error causes the reconstructed signal to resemble a staircase waveform (a). We can think of this staircase signal as the sum of the original analog signal (solid curve in (a)) and a random "quantization error" signal (shown inverted in (b)).

represent $+5$ V, the 10 code, $+\frac{5}{3}$ V, the 01 code, $-\frac{5}{3}$ V, and the 00 code, -5 V. Because any amplitude sample with an original value from 0 to $+\frac{10}{3}$ V would be represented during signal reconstruction as $+\frac{5}{3}$ V, we see that there is a maximum error of $\pm\frac{5}{3}$ V for any of the codes. This error—which is limited in size to one-half the size of a quantization region—is called *quantization error* and is inherent in the quantizing process.

The bad news is that quantization error is inherent in digital signals, just as tape hiss is inherent in analog recording. The good news is that each time we add 1 bit to the digital sample size, we halve the size of the maximum quantization error. Using the same coding convention as in Fig. 21.10, for a signal ranging in amplitude from -5 to $+5$ V (a total amplitude range of 10 V), an $N = 16$-bit ADC makes a quantization error of only $\pm 5/(2^N - 1) = \pm\frac{5}{65535} \approx \pm 7.629 \times 10^{-5} = \pm 76.29$ μV. If we assume that this error is random (i.e., that it has no discernible pattern or predictability), then quantization error can be treated as random (i.e., *white*) noise added to a digital signal by the encoding process (see Fig. 21.11). If the peak amplitude of the (largest) signal is A (i.e., the amplitude range is $\pm A$), the peak amplitude of the quantization error "noise" is always $A/(2^N - 1)$. The signal-to-quantization error noise ratio (SQNR) is then approximately $6N$ decibels, where N is the number of bits used to encode each sample.

A 16-bit ADC therefore has a best-case SQNR of about 96 dB, meaning that the noise is about 96 dB below the level of the maximum amplitude signal. When the signal is smaller, however, the level of the noise does not decrease (just as tape hiss does not decrease when low-amplitude sounds are recorded). The worst-case SQNR therefore occurs for signals with the smallest amplitudes: those that toggle only the least significant bit (lsb) of the ADC (Fig. 21.11 shows the relative sizes of the signal and quantization error noise for a signal that toggles only two least significant bits).

21.15 ■ ALIASING

As previously mentioned, the sampling rate R must be at least twice that of the highest frequency contained in the sampled analog signal. Because analog signals may in general contain *any* frequency (even ones that are inaudible), it is necessary to pass the analog

FIGURE 21.12
A frequency sampled more than twice per period is said to be "oversampled" (a). A frequency sampled exactly twice per period is said to be "critically sampled" (b). A frequency sampled less than twice per period is said to be "undersampled" (c). Notice that the samples in the undersampled case correspond to a different ("aliased") frequency (dashed line) according to Equation (21.1).

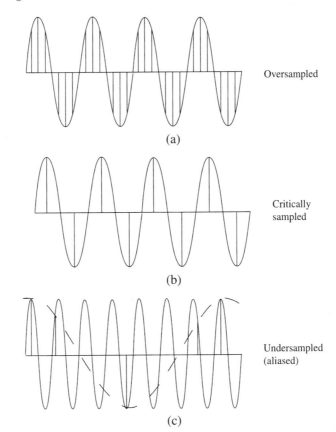

(a) Oversampled

(b) Critically sampled

(c) Undersampled (aliased)

signal through a low-pass filter to remove any frequencies above half the sampling rate before it is sampled. If this is not done properly, an extremely objectionable form of signal distortion called *aliasing* (or *foldover*) can occur (see Fig. 21.12).

Aliasing misrepresents frequencies higher than the Nyquist frequency at a different value lower than the Nyquist frequency according to the formula

$$f_{\text{alias}} = f - \text{int}\left(\frac{f}{R} + \frac{1}{2}\right)R, \tag{21.1}$$

where f_{alias} is the aliased frequency, f is the frequency in the analog signal, and int( ) means *the integer part of*. If the magnitude of f is less than $\frac{R}{2}$, then $f_{\text{alias}} = f$ in the preceding formula. But if f is greater than $\frac{R}{2}$, it is "reflected" back so that it lies within the range $-\frac{R}{2} < f < +\frac{R}{2}$ (see Fig. 21.13). For example, if $R = 10{,}000$ Hz, the Nyquist rate is $\frac{R}{2} = 5000$ Hz. Sampling any frequency up to 5000 Hz in this situation does not present a problem. But if we attempt to sample 6000 Hz, Eq. (21.1) shows that the resulting digital signal will be indistinguishable from 4000 Hz (because 6000 Hz exceeds the Nyquist rate by 1000 Hz, it is aliased 1000 Hz *below* the Nyquist rate).

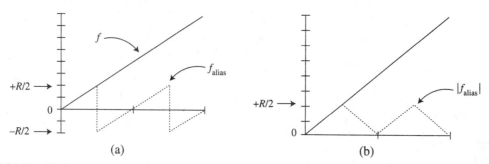

FIGURE 21.13 (a) shows how increasing frequency, f, is mapped into aliased frequency in Equation (21.1). Because negative frequencies usually sound just like positive frequencies, (b) depicts how aliased frequencies are typically perceived by the ear.

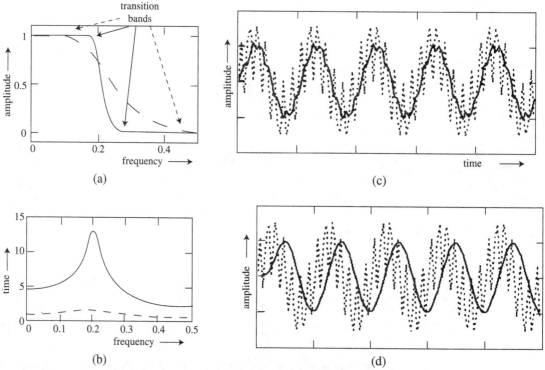

FIGURE 21.14 Comparing the phase distortion in gradual versus sharp cutoff lowpass filters. As the transition from passband to stopband is narrowed, the phase distortion (and hence the group delay) increases dramatically. (a) Gradual (wide transition band, dashed) and sharp (narrow transition band, solid) lowpass filter amplitude responses. (b) Gradual (dashed) and sharp (solid) lowpass filter group delay responses. Note large delays around the cutoff frequency. (c) Input (dotted) and output (solid) signals for the gradual cutoff lowpass filter. Note that not all high frequencies in the input are removed, and that time delay is relatively small. (d) Input (dotted) and output (solid) signals for the sharp cutoff lowpass filter. Note that high frequencies in the input are removed, and that time delay is significant.

Assuming there is no particular relation between the frequencies in the signal and the sampling rate, aliasing can result in strong frequency components falling at essentially random places, causing beats, heterodyning, and other kinds of highly significant distortion. Thus, it is important to ensure that all frequencies above half the sampling rate are removed before the sampling process is carried out. Note that this is true even if the removed frequencies are above the human hearing range, because aliasing such frequencies would fold them down into the audible range.

21.16 ■ PHASE DISTORTION

Another important source of distortion in digital signals is the analog low-pass filter used to band limit the signal. Any low-pass filter that attempts to have a very flat passband, a very narrow transition band, and a very deep stopband will necessarily introduce significant phase distortion around the region of the transition band (see Fig. 21.14). The phase response of a filter determines how much delay is introduced into the signal at various frequencies (such delay is measured in terms of *phase delay* or *group delay*). Phase distortion can be significant when some frequencies are delayed by perceptibly different amounts than others. Whether phase distortion is audible depends greatly on the signal and the listener, but there are many cases in which such distortion is highly objectionable to some listeners. Because it is impossible to reduce the phase distortion while keeping a narrow transition band, it is generally desirable to relax the transition band so that it transits from pass- to stopband more gradually. To avoid distorting audible frequencies, the preferred technique is usually to sample the signal at a faster rate.

21.17 ■ OVERSAMPLING

Oversampling (see Fig. 21.15) has two main advantages. First, for a signal that is oversampled by a factor L, the level of the quantization error noise is reduced by a factor of L. Sampling twice as fast as necessary (i.e., $L = 2$) has approximately the same benefit as adding 1 bit to the encoder, thereby increasing the SQNR by about 6 dB. Every time we double the sampling rate, we can therefore remove 1 bit from the encoder with no ill effects. This is because the quantization noise is uniformly spread out over the entire frequency range up to half the (now higher) sampling rate, so half of it may be eliminated by filtering in the digital domain using the techniques of digital signal processing that we are about to discuss. A digital low-pass filter can have a much lower phase-distortion characteristic than an analog filter (including no phase distortion at all).

The second advantage of oversampling is that it reduces the complexity of the analog low-pass filter, which can now make a more gradual transition from passband to stopband, generally making it simpler and less costly. Its lower performance requirements also greatly reduce the amount of phase distortion introduced into the signal before the sampling process.

The most extreme form of oversampling reduces the quantization requirement for the ADC all the way to a single bit. Such ADCs are called *delta-sigma modulators* and use extremely high sampling rates. As shown in Fig. 21.15, once the signal is oversampled, it is digitally filtered and then *downsampled*, i.e., reduced to a sampling rate necessary for

FIGURE 21.15
An oversampling
analog-to-digital
converter first
increases the target
sampling rate, R,
by a factor of L. A
digital lowpass
filter removes
unnecessary high
frequencies from
the digital signal.
The digital lowpass
filter computes its
output to a
precision of
$N + \log 2(L)$ bits.
An L-to-1
downsampler
retains every Lth
sample of the
processed digital
signal, resulting in
a sampling rate
of R.

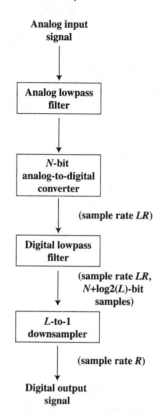

representing audio signals (once it is properly filtered, we can reduce the sampling rate of a digital signal by a factor of two merely by discarding every other sample, for example).

21.18 ■ DITHERING

Because quantization effects normally become more noticeable at low-signal amplitudes, it becomes important to ensure that the sampling process and the signal are truly decorrelated, i.e., that there is no detectable pattern in the quantization errors for small signals. Although it might sound somewhat counterintuitive, an effective way to ensure decorrelation between the signal and the sampling process is to add a small amount of (random) noise to the analog signal just before it is sampled (see Fig. 21.16). This process, called *dithering*, can ensure that the quantization errors are truly random, which has the effect of spreading out their effects over all frequencies. This is far preferable to concentrating these effects at one or a few frequencies, which can result in objectionable buzzes or other artifacts that can be clearly audible when signal amplitudes are low, such as near the end of a reverberant tail just before it dies away to silence.

It is even possible to "shape" the dithering noise so that quantization effects are gathered to extremely high frequencies, where they are likely to be altogether inaudible.

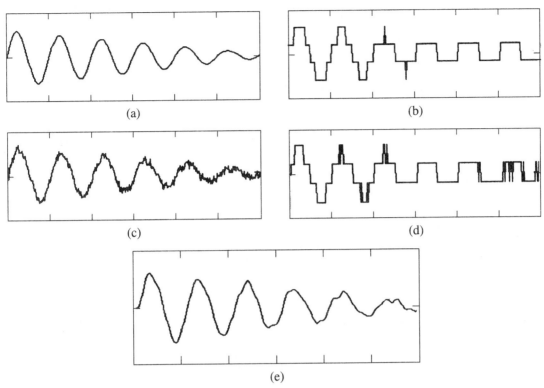

(a)

(b)

(c)

(d)

(e)

FIGURE 21.16 Signal (a) is a decaying sinusoid. Signal (b) shows how the quantization process distorts signal (a) into a square wave as it becomes very small. Signal (c) is the same as signal (a) with "dithering" noise added. Signal (d) shows that when signal (c) is quantized, it retains information about very small amplitudes as it rocks back and forth between its two smallest levels. The advantages of dithering are especially evident in signal (e), which is a lowpass filtered version of signal (d).

21.19 ■ COMPANDING

In addition to dithering, it is possible to combat quantization error noise by using nonuniform amplitude encoding. It is easiest to understand nonuniform encoding by considering the size of a *quantization region*, which is the range of analog voltages that are all mapped into the same bit code. As we saw previously, for "linear" PCM, each of the 2^N regions of quantization for N bits has the same size, and the maximum quantization error is everywhere equal to half of the size of the quantization region. Normally we think of the difference between the analog voltage and the voltage at the center of each quantization region as the quantization error.

It is possible, however, to make the quantization regions near 0 V smaller at the cost of making the quantization regions larger when they are far from zero (see Fig. 21.17). This is, in effect, an *instantaneous* compression/expansion scheme (also known as a *companding* scheme), in which the signal is compressed during encoding and later expanded during decoding. It has the effect of making the quantization error smaller for low-amplitude signals at the cost of making the error larger for high-amplitude signals. Instead of having

FIGURE 21.17
Nonlinear
quantization
generally narrows
the quantization
levels that are close
to zero, while
widening those far
from zero. This
reduces the size of
quantization errors
for small signals at
the cost of
increasing the
errors for large
signals. However,
large signals tend to
mask the errors.

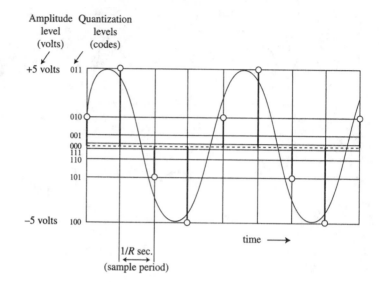

a constant level of quantization error noise, nonuniform encoding tends to maintain the quantization error noise at a certain number of decibels below the level of the signal. Non-linear encoding then relies on the greater loudness of high-amplitude signals to *mask* the greater quantization error noise introduced by coarser encoding.

There are so many ways to adjust the quantization regions that it is necessary to have international standards for nonlinear PCM. Two such standards currently exist: μ-law, which is used in the United States and Japan, and A-law, which is used throughout Europe. μ-law encoding is accomplished according to the formula

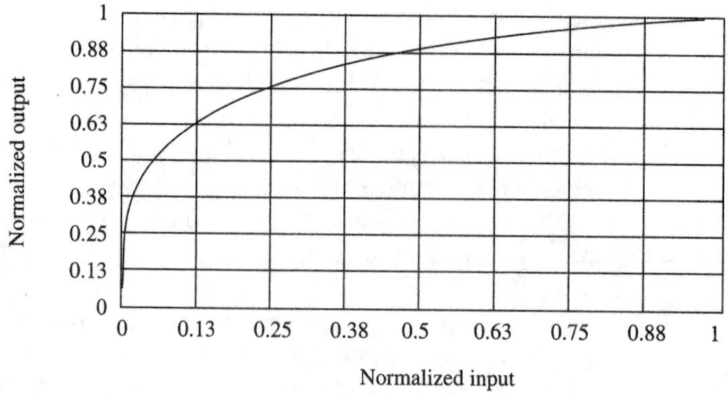

FIGURE 21.18
The μ-law
compression curve
defined by
Equation (21.2)
with μ set to 255.
Input values (at the
bottom) are
mapped to output
values (at the left).
This nonlinear
mapping
compresses the
quantization levels
near zero and
expands those far
from zero.

$$F(x) = \mathrm{sgn}(x)\frac{\ln(1 + \mu|x|)}{\ln(1 + \mu)}, \qquad (21.2)$$

where x is the (linear) sample value (limited to 13 bits), $F(x)$ is the nonlinearly encoded sample value, $\text{sgn}(x)$ is 1 for positive x and 0 for negative x, $\ln(x)$ is the natural logarithm of x (to the base $e = 1.618\ldots$), and $\mu = 255$ (this is, therefore, also called μ-255 encoding; see Fig. 21.18). It compresses 13 linearly encoded bits into 8 nonlinearly encoded bits (A-law encoding uses a slightly different technique and achieves nearly the same performance).

The principal advantage of nonlinear encoding is that it can achieve almost 13-bit performance for a memory cost of only 8 bits per sample, which is a considerable saving. That is, the *dynamic range* of the signal for 8-bit μ-law (the range between the loudest and softest representable signals) is about the same as for 13-bit linear PCM (about $13 \times 6 = 78$ dB).

A disadvantage of nonlinear encoding is that samples must be decoded (i.e., converted to linear quantization) if they are to be processed (two nonlinearly encoded digital signals cannot be mixed by simply adding them together, for example). Such nonlinear coding and decoding is typically accomplished using a software or hardware coder-decoder, also known as a *codec*.

21.20 ■ SUMMARY

Computers play a significant role in music. Computers consist of a central processing unit (CPU), various forms of memory for storing programs and data, input/output (I/O) devices for entering and retrieving data, and groups of wires called buses for communicating between the CPU and memory or I/O. Music applications can be programmed in assembly language, a high-level language, or an object-related language. An audio signal can be sampled, converted into digital information, modified by a computer, and then converted back to an audio signal. When sampling a signal, the sampling rate must be at least twice the lowest frequency in the signal being sampled or aliasing (unwanted frequencies) occur.

REFERENCES AND SUGGESTED READINGS

Chamberlain, H. (1987). *Musical Applications of Microprocessors*, 2d ed. Indianapolis: Hayden.

Moore, F. R. (1990). *Elements of Computer Music*. Upper Saddle River, N. J.: Prentice Hall.

Pohlmann, K. C. (1989). *Principles of Digital Audio*, 2d ed.: H. R. Sams.

Rosch, W. C. (1994). *Hardware Bible*, 3d ed.: H. R. Sams.

Steiglitz, K. (1996). *A Digital Signal Processing Primer with Applications to Digital Audio and Computer Music*. Addison-Wesley.

Strawn, J., ed., with contrib. by J. W. Gordon, F. R. Moore, J. A. Moorer, T. L. Petersen, J. O. Smith, and J. Strawn. (1993). *Digital Audio Signal Processing: An Anthology*. A-R Editions.

Strawn, J., ed. (1988). *Digital Audio Engineering*. A-R Editions.

Zoelzer, U. (1997). *Digital Audio Signal Processing*. New York: John Wiley.

GLOSSARY

aliasing (foldover) In a digital signal, the misrepresentation of frequencies above half the sampling rate at incorrect frequencies below half the sampling rate.

analog-to-digital converter (ADC) A circuit that converts numbers from an analog to a digital representation.

arithmetic/logic unit (ALU) A circuit within the central processing unit that performs arithmetic and logic operations.

ASCII American Standard Code for Information Interchange, which assigns 8-bit codes to commonly used symbols on an English computer keyboard.

assembler A computer program that translates symbolic written codes into internal computer (machine) instructions on a more-or-less one-to-one basis.

bandlimited signal An analog signal that has no frequencies higher than a known given frequency.

binary number A number in base 2 representation; a 1 or a 0.

binary point In base 2 (binary) arithmetic, a period symbol used to denote the boundary between whole units (to its left) and fractional units (to its right).

bit A binary digit, used in the sense of a minimum unit of computer memory as well as a small, two-valued unit of information.

bus A group of wires connecting the CPU with memory and I/O.

byte An ordered collection of 8 bits of memory.

central processing unit (CPU) The portion of a computer that performs arithmetic, logic, and program sequencing operations.

codec A hardware or software device capable of both coding and decoding digital signals.

compander A device capable of both compressing and expanding the dynamic range of a signal.

compiler A computer program that translates statements written in a high-level computer language (such as C or Pascal) into either low-level symbolic assembler code or directly into internal computer (machine) instructions.

debugger A computer program designed to assist a programmer in diagnosing and correcting the operation of another computer program.

delta-sigma modulator An analog-to-digital (digital-to-analog) conversion system that operates by encoding in 1-bit bit measurements at a very high sampling rate.

digital-to-analog converter (DAC) A circuit that converts numbers from a digital to an analog representation.

dither A low-amplitude noise added to an analog signal to ensure that low-amplitude signals are decorrelated from the sampling process, thus improving their distortion and noise characteristics.

downsampling The process of lowering the sampling rate of a digital signal by discarding samples: a $1/M$ downsampler would retain 1 out of every M samples, for example, of a suitably filtered digital signal.

DRAM (dynamic random access memory) A type of volatile semiconductor memory that stores large amounts of data in a small space. It must be frequently refreshed.

EPROM (erasable/programmable read-only memory) Read-only memory that can be erased by ultra-violet light and then reprogrammed. **EEPROM (electrically erasable/programmable read-only memory)** can be erased and programmed while still in the computer.

flip-flop A digital circuit capable of acting as a memory for a single bit of information by remaining in one state until it is actively placed in a second state.

floating point A means of representing numbers using one set of bits for an exponent x and another set for a mantissa m, interpreted as m times b to the power of x (b is typically 2).

high-level language A computer language that can be used to program a wide variety of computers rather than accessing registers of a specific CPU.

input-output (I/O) The portion of a computer that deals with transferring information into or out of the computer's memory.

interpreter A computer program that translates statements in a computer language into machine instructions that are carried out one by one as soon as each statement is complete.

low-level language A computer language, such as assembly, designed to access the individual registers of a CPU.

memory, nonvolatile A type of computer memory that does not disappear when the power is turned off, such as a disk or tape memory.

memory, volatile A type of computer memory that disappears more or less immediately when the power is turned off, such as random-access memory (RAM).

memory, working The (typically volatile) memory in a computer used temporarily while a program is running.

microcontroller A single integrated-circuit chip containing a CPU, memory, and I/O used for small control applications.

microprocessor The central processing unit of a computer constructed on a single integrated circuit chip.

MIDI (musical instrument digital interface) An interface that allows a synthesizer or other electronic instrument to be controlled by a computer or another electronic instrument (see Section 29.8).

multiprocessing The simultaneous use of more than one processor (CPU) to run one or more programs.

multitasking A computer technique of rapidly switching from one task to another in order to simulate the execution of many tasks at once.

nibble An ordered collection of 4 bits of computer memory.

NOR gate (not-or gate) NOT-OR circuit, which inputs 0 if either (or both) inputs are 1 (i.e., the output is the opposite of an OR gate).

Nyquist frequency Half the sampling frequency. Frequencies above the Nyquist frequency, if not filtered out before sampling, will appear at other frequencies less than the Nyquist rate (aliasing or foldover).

operating system A "master" computer program residing permanently in a computer that controls the execution of other programs, the overall organization of information into files and folders, the control of input-output devices, etc.

oversampling (1) Increasing the sampling rate by some factor L (see upsampling); (2) Any frequency that is sampled more than twice per period. Any frequency that is sampled less than twice per period is said to be **undersampled**.

pixel A picture element on a monitor.

pulse-code modulation (PCM) A particular type of digital signal that encodes a corresponding analog signal as an ordered collection of evenly spaced samples of its amplitude, with each sample linearly encoded as a certain number of bits.

quantizing The process of assigning a discrete digital value to a theoretically continuous analog amplitude.

quantization error The difference between the quantized value and the "true" value of a sample.

random-access memory (RAM) A computer memory in which any storage location can be addressed for read or write operations.

read-only memory (ROM) A computer memory that stores information permanently.

sample A measurement of the instantaneous value of the amplitude of an analog signal.

sampling (Nyquist) theorem A theorem stating that a bandlimited signal containing frequencies up to F hertz must be sampled at least $2F$ times per second in order to avoid aliasing.

sign-magnitude The technique of representing negative integers using 1 bit to represent the sign and the rest of the bits to represent the magnitude of a number.

SQNR (signal-to-quantization error noise ratio) Ratio of the amplitude of the signal to the amplitude of the quantization error noise, expressed in dB.

SRAM (static random access memory) A type of memory that holds information as long as power is applied. It is generally faster than DRAM and does not require refreshing.

2's complement The technique of representing negative integers as 1 plus the bit-complement of the corresponding positive integer.

UNICODE A coding standard that assigns 16-bit codes to commonly used symbols used in many languages.

unsigned An ordered collection of bits that represents a numerical magnitude.

upsampling The process of increasing the sampling rate of a digital signal by inserting samples: a factor-L upsampler would insert $L-1$ zeros between each sample of a digital signal (this signal would typically then be digitally filtered).

word An ordered collection of bits, typically 2, 4, or 8 bytes in length.

REVIEW QUESTIONS

1. What is a bit? A byte? A nibble? A word?

2. Why is solid-state memory usually volatile?

3. What does ASCII stand for?

4. What is the main advantage of UNICODE?

5. When are sign-magnitude and 2's complement representations the same, and when are they different?

6. How could 0.375 be represented in binary?

7. Why do computers need an operating system?

8. What is the primary advantage of a digital signal over an analog signal?

9. What does PCM stand for?

10. What is the sampling theorem?

11. What is the analog equivalent of quantization error?

12. What is the SQNR of 24-bit linear PCM?

13. If the sampling rate is 44,100 Hz, at what frequency will 22,000 Hz appear? 23,000 Hz?

14. What component of analog-to-digital conversion introduces phase distortion?

15. What are the advantage(s) and disadvantage(s) of oversampling?

16. How does adding noise to an analog signal improve its digital representation?

17. What are the advantage(s) and disadvantage(s) of non-linear amplitude encoding?

18. What minimum sampling rate could be used to capture a harmonic at 10 kHz?

19. What is the theoretical dynamic range of a 16-bit sample?

20. What is the difference between an ADC and a DAC?

21. What is the difference between a serial bus and a parallel bus?

22. What is the difference between a low-level programming language and a high-level programming language?

QUESTIONS FOR THOUGHT AND DISCUSSION

1. How might a computer system compensate for the fact that some values (such as 0.1 base 10) do not have exact binary representations?

2. A digital microphone is one that directly produces a digital signal. Similarly, a digital loudspeaker is one that directly accepts a digital signal. How might such devices operate, and what would be their advantages over their analog equivalents?

3. What advantages might pulse-amplitude or pulse-width modulation have over PCM, if any?

4. Does aliasing have any potential usefulness?

5. In the dithered signals shown in Fig. 21.16, how is information about the waveform represented? Why does the filtered version of the dithered digital signal conform better to the original?

6. A hand-held computer, like a large computer, uses binary numbers internally. Why is it unnecessary to understand binary numbers in order to use it?

7. What are the advantages of programming in a high-level language such as C rather than in machine language?

8. If an audio signal is sampled using an ADC and then converted back to analog using a DAC, would there by any difference between the original signal and the regenerated signal?

EXERCISES

1. Write the binary representations of the following decimal numbers.
   (a) 12
   (b) 22
   (c) 36

2. Write the decimal representations of the following binary numbers.
   (a) 1101001
   (b) 100110
   (c) 1010101

3. Try to multiply binary numbers 101 times 110, using the same procedure that you use to multiply ordinary decimal numbers. Check your answer by multiplying the corresponding decimal numbers.

4. If the ASCII code for the letter A is 0100 0001, what would be the codes for B, C, and so on?

5. According to Fig. 21.4, what would be the 32-bit floating-point representation of 0.5 (base 10)?

6. If the sampling rate is 30 kHz, what is the highest frequency of audio signal that can be successfully coded in a digital representation? A frequency of 25 kHz in the original signal will give rise to what foldover frequency in the coded signal? How can foldover be prevented?

7. How many different levels can be represented by a 13-bit binary code? What is the possible dynamic range for this signal?

8. Up to 6 billion bits of binary data stored on a compact disc are read out in 74 min. What is the maximum bit rate per second?

9. What would be the approximate signal-to-quantization error-noise (SQNR) ratio of a linear PCM signal quantized to 24 bits?

EXPERIMENTS FOR HOME, LABORATORY, AND CLASSROOM DEMONSTRATION

Home and Classroom Demonstration

1. *Digital computers*   Open up a typical home computer and identify all the components you can.

2. *Sampling theorem*   Demonstrate aliasing by having one person or object rotate in place in front of a classroom. Students hold their eyes closed until they momentarily open them when the instructor says "peek." Vary the rotation rate (frequency) and peek frequency (sampling rate) to illustrate oversampling, critical sampling, and undersampling.

3. *Quantization error (and dithering)*   Use a magnifying glass or other method to examine a raster-based computer display closely. By displaying diagonal lines of various types, visually demonstrate the effects of quantization error. (If it is available, demonstrate software that minimizes such visual effects by dithering diagonal lines.)

4. *Companding*   Use a digital audio recording processing package such as Pro Tools to compress the dynamic range of an example of music with a significant dynamic range, such as percussion music. Demonstrate and/or discuss the effect of such compression on such music as Ravel's *Bolero* (subtitled: *Crescendo for Orchestra*).

5. *Companding*   Listen to a Dolby-B encoded analog tape recording with and without the decoding and discuss how such differences are reflected in μ-law encoded digital samples.

6. *Convolution and digital filtering*   Write a computer program that demonstrates convolution by creating a digital filter with an impulse response consisting of the tune for *Frere Jaques*. Verify the impulse response by listening to it. Demonstrate convolution (and superposition) by exciting the filter with impulses spaced four beats apart, thus creating the round.

7. *Sound cards*   Examine two or more sound cards, identifying as many components as you can.

Laboratory Experiment

SDK-88: The SDK-88 microprocessor trainer allows students to examine how a microcomputer works and to program in machine language.

CHAPTER
22

Sound Recording

Although sound recording began more than 100 years ago, spectacular advances in recording technology in recent years have had a profound effect on our musical culture. Millions of people enjoy high-fidelity reproduction of music in their own homes. The production and distribution of CDs, tapes, and DVDs has become a major industry.

There are several ways to classify sound recording. The most obvious is probably as *analog* or *digital*. Another is according to the physical principle used. The four principal methods used in recording have been mechanical recording, magnetic recording, optical recording, and direct electronic recording. We discuss each of these methods, although we concentrate mainly on digital recording and playback using optical methods.

In this chapter, you should learn:

- How sound recording developed;
- About mechanical recording on phonograph records;
- About magnetic tape recording, analog and digital;
- About magnetic recording on hard disks;
- About electronic recording in solid-state memories.

22.1 ■ THE ORIGINS OF SOUND RECORDING

The 1870s saw the invention of the telephone (1876), the phonograph (1877), and the electric lamp (1878). Thomas Edison, who had been trying to build a device to record the dots and dashes of the telegraph, developed a machine that could record the human voice on a tinfoil-wrapped cylinder turned with a hand crank (see Fig. 22.1). Sound entered a mouthpiece to a diaphragm connected to a needle. The needle moved up and down, creating "hill-and-dale" grooves in the tinfoil. A second stylus rode up and down on the

FIGURE 22.1
Edison's first tinfoil phonograph, 1877.

impressions on the foil and drove the diaphragm of a primitive loudspeaker. Edison called his talking machine a *phonograph.*

For the first demonstration, Edison recited "Mary had a little lamb" into the mouthpiece. Needless to say, the recording was low-fidelity, but it was a first. The phonograph was an instant sensation. Edison brought it to the White House to demonstrate it to President Hayes. In 1878 Edison obtained a patent for the phonograph and created the Edison Speaking Phonograph Company to oversee the manufacture and exhibition of the talking machines (Beyer 1999).

Other inventors worked on recording sound. Alexander Graham Bell used his Volta Prize money (which he won for inventing the telephone) to establish an acoustics laboratory. In this laboratory his cousin Chichester Bell and Charles Tainter, a scientist and instrument maker, set about to improve on the phonograph. The Bell-Tainter *graphophone,* introduced in 1877, used cylinders of wax instead of tinfoil and also added an electric motor to turn the cylinder at a steadier rate. Stimulated by this competition, Edison replaced his tinfoil with paraffin-coated paper. Later on, a gold coating was vaporized onto it, thus making a negative from which other cylinders could be imprinted.

In 1891 Emile Berliner invented the *gramophone,* which introduced two major improvements. He used a flat disk rather than a cylinder, and he replaced hill-and-dale recording with a back-and-forth motion of the stylus in the plane of the disc. Berliner's gramophone was generally regarded as being superior to Edison's phonograph. Elridge Johnson improved the gramophone with a spring motor in 1897 and founded the Victor Talking Machine Co. with the famous "little nipper" dog as trademark. The Grammy trophies that are awarded for excellence in sound recording each year, incidentally, are a replica of an early Berlin gramophone.

A major breakthrough came in 1925 with the invention of the vacuum-tube amplifier. Being able to amplify the sound prior to recording greatly improved the sound quality. Now you could listen to your favorite music in your own living room from a shellac record spinning at 78 rpm. Still, 78-rpm records were noisy and had limited playing time. Long-playing (LP) records were introduced in 1948. These records turned at a slower speed ($33\frac{1}{3}$ or 45 rpm) and had a narrower groove, resulting in a much longer playing time (up to 30 min). They were made of vinyl plastic, which greatly reduced the surface noise of the old shellac records.

22.2 ■ STEREOPHONIC DISK RECORDING

Stereophonic recording requires two independent channels of recorded information. On stereo disks, this is accomplished by recording the right and left channels along two sides of the same groove, as shown in Fig. 22.2. The sides of the groove are at right angles to each other, so that motion of the stylus in the direction marked L does not affect its motion in the direction marked R.

An important consideration in the recording and reproduction of sound is the compatibility of different types of records with different kinds of playback equipment. Monophonic and stereophonic disk records are compatible as follows:

1. When a two-channel stereo record is played on a monophonic reproducing system, the output signal will be a combination of both recorded channels. This is because variation in either side of the groove moves the stylus laterally.

2. When a single-channel monophonic record is played on a stereophonic reproducing system, lateral motion of the stylus will induce the same voltage in both stereo pickup coils, so the monophonic sound will appear in both channels.

FIGURE 22.2
(a) Two channels of stereo information, L and R, recorded on two sides of groove. (b) A simplified diagram of moving-magnet stereo pickup. The motion of the stylus and magnet in direction R induces a voltage in coil R but almost no voltage in coil L.

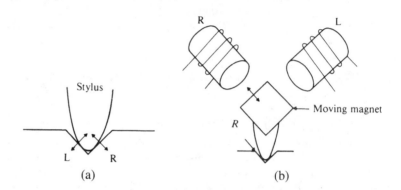

22.3 ■ MAGNETIC TAPE RECORDING

Modern recording tape consists of a thin coating of magnetic particles on a plastic base. Iron oxide (Fe_2O_3 and Fe_3O_4) powders are still the most widely used, although chromium dioxide (CrO_2) and metal powders have recently become very popular because of the improved signal-to-noise ratio that can be achieved through their use.

The magnetic coating on the tape is made up of very tiny particles (typically 0.5 μm in length) held in a plastic binder. Particles this small have desirable *single-domain* properties, which means that their magnetization remains stably oriented in one of two opposite directions. (Thus, a recorded signal remains on the tape until it is erased.)

The magnetic coating may be regarded as having a large number of tiny bar magnets in random alignment in the unmagnetized state. When a magnetic field is applied by means of the record head, the magnets align themselves in the direction of the field, as shown in Fig. 22.3. This model should not be carried to the point of thinking that the magnetic particles move or rotate, however. They do not. The motion involved in alignment of the magnets is inside the atoms themselves.

The magnetic properties of recording tape can be represented by a *hysteresis loop* of the type shown in Fig. 22.4. The magnetization M is plotted vertically and the magnetic field H is plotted horizontally. As the field H increases from the unmagnetized state O, the magnetization M increases (as the magnets align with the field) until magnetic *saturation* is reached at point S. When the field is now reduced to zero, the magnetization does not return to zero, but rather to the magnetic *remanence* M_r. In order to reduce the magnetization to zero, a field must be applied in the reverse direction. This reverse field H_c is called the *coercivity*. Further increase in the reverse field saturates the magnetization in this direction, indicated by $-S$ on the hysteresis curve. Note that the hysteresis loop is traversed in one direction only.

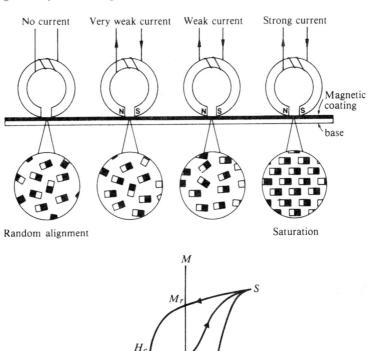

FIGURE 22.3
Magnetic domains within the oxide coating shown as tiny bar magnets. As the magnetic field increases, the magnets align along the field.

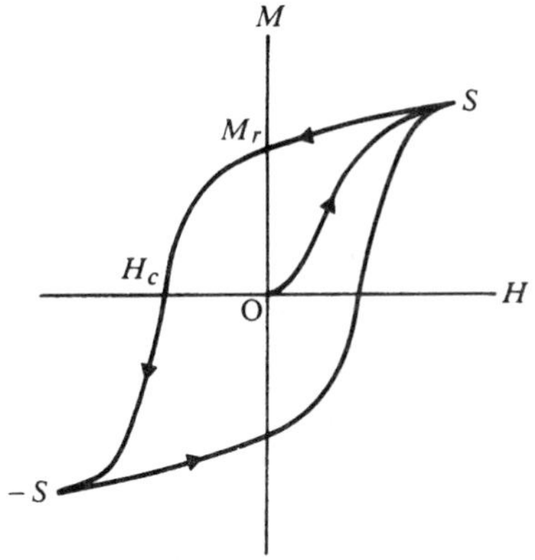

FIGURE 22.4
Magnetic hysteresis loop. (See the text for an explanation of the symbols.)

Different magnetic oxides have different magnetic properties. For example, CrO_2 has greater magnetization than Fe_2O_3, and thus it is capable of a greater signal-to-noise ratio (which is particularly important in cassette recorders). However, it also has a larger coercivity, which means that the record bias must be increased. Most recorders can adjust to either type of material by changing the bias to the recording head. Hysteresis loops of CrO_2 and Fe_2O_3 are compared in Fig. 22.5. Bias will be discussed further in Section 22.5.

In a typical tape recorder, the tape passes three recording heads in succession. First, the *erase head* applies a rapidly oscillating magnetic field that erases the old information (demagnetizes the tape); then the *record head* magnetizes the tape in the desired pattern; and finally a *playback head* reads the recorded pattern and generates an output voltage as the tape passes. Many tape recorders use a single head for the record and playback functions, but this is slightly less convenient, because the tape cannot be monitored during recording.

The basic principles of magnetic tape recording are simple: The record head generates an alternating magnetic field across its gap, and this field leaves a pattern to the magnetic remanence of the tape passing by, as illustrated in Figs. 22.6 and 22.7. When the tape

FIGURE 22.5
Magnetic hysteresis loops for chromium dioxide and one type of iron oxide commonly used in recording tape. Note that CrO_2 has a higher remanent magnetization M_r and a higher coercivity H_c.

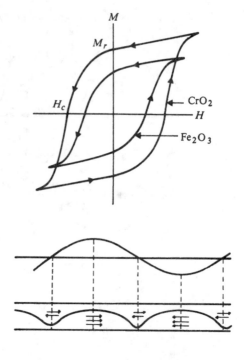

FIGURE 22.6
Magnetization of a magnetic tape on which a pure tone has been recorded. Arrows represent the magnetization.

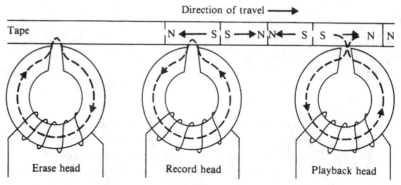

FIGURE 22.7
Arrangement of erase, record, and playback heads in a magnetic tape recorder.

arrives at the gap in the recording head, it sets up a magnetic "disturbance" that induces an electrical signal in the coil windings. Although the basic principles are simple enough, there are several practical considerations.

22.4 ■ TAPE SPEED AND FREQUENCY RESPONSE

Analog magnetic tape recorders usually operate at one of the following speeds: 15, $7\frac{1}{2}$, $3\frac{3}{4}$, or $1\frac{7}{8}$ inches per second (in/s); the corresponding metric speeds are 38, 19, 9.5, and 4.8 cm/s, respectively. The choice of speed is a compromise between tape economy and performance. Professional recording is often done at 15 in/s, whereas cassette recorders generally use $1\frac{7}{8}$ in/s.

The high-frequency limit usually occurs when the wavelength of the recorded signal approaches the size of the gap in the playback head. A simple calculation will indicate the frequency at which this might be expected.

FIGURE 22.8
High-frequency cutoff occurs when a wavelength of recorded signal approaches the head gap size. (a) The wavelength is twice the gap size; magnetic flux travels through the head. (b) The wavelength equals the gap size; no flux reaches the coil windings.

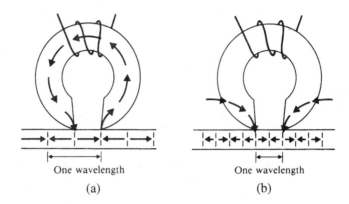

One wavelength
(a)

One wavelength
(b)

The recorded wavelength for a pure tone is the tape velocity divided by the frequency of the tone. Thus, for a 10-kHz tone recorded at $3\frac{3}{4}$ in/s,

$$\text{wavelength} = \frac{3.75 \text{ in/s}}{10,000 \text{ Hz}} = 3.75 \times 10^{-4} \text{ in} = 9.5\mu\text{m}.$$

This is only a few times larger than the gap size in a typical playback head and is, therefore, near the upper limit that can be reproduced with fidelity.

The graph of relative output as a function of frequency for different tape speeds (see Fig. 22.9) shows the frequency characteristic for a tape without equalization. Note that the

FIGURE 22.9
The frequency characteristic for an unequalized tape at four different tape speeds.

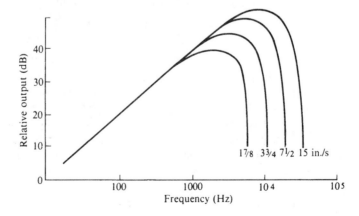

FIGURE 22.10

FIGURE 22.10
The recorded
magnetization as a
function of signal
current for varying
amounts of
high-frequency
bias. With no bias
(dashed curve) the
distortion is large.
(After Westmijze
1953.)

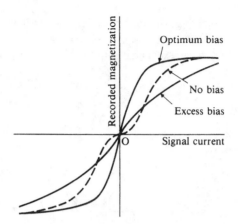

output rises with frequency (since the rate of flux change through the head increases) until the wavelength of the recorded signal on the tape approaches the gap width and then falls off rapidly.

22.5 ■ BIAS AND EQUALIZATION

In order for the magnetization to be a linear function of the recording signal, it is necessary to use a magnetic *bias*. Although older analog tape recorders occasionally used a constant field as bias (and even a constant erase field), high-frequency erase and bias currents are used almost exclusively in modern equipment. A high-frequency current of 40 to 150 kHz is supplied to the erase head and also superimposed on the signal at the record head.

High-frequency bias has two very important advantages over DC bias: (1) the recorded magnetization is a linear function of the sound signal over a larger dynamic range, and (2) when the sound signal dips to zero, the tape remains demagnetized and therefore generates less noise in the playback head. The relationship between the current in the recording head and the recorded magnetization in the tape is shown in Fig. 22.10 (compare with the inner curve OS in Fig. 22.4). The bias amplitude is selected to make use of the linear portion of the curve. A physical description of high-frequency bias is given in the following box.

High-frequency bias can be understood by referring to the waveforms shown in Fig. 22.11. The current supplied to the recording head (c) is the sum of the bias (a) plus the sound signal (b); the bias current must be substantially larger than the signal current. During the time that the tape is next to the recording gap, the bias field is large enough to drive the magnetization rapidly around the hysteresis loop from one magnetic state to the other. As the tape moves away from the gap, however, the field diminishes, and the magnetization traverses a steadily diminishing "minor" hysteresis loop (e). In the first case shown, the magnetization diminishes toward zero; in the second case, the excursions are unequal, and a remanent magnetization results as the tape leaves the field of the gap.

FIGURE 22.11
The use of
high-frequency bias
in tape recording:
(a) the
high-frequency bias
current (typically
40 to 150 kHz);
(b) the sound signal
current; (c) the total
current supplied to
the recording head;
(d) the field "seen"
by the tape as it
leaves the record
gap; (e) the
magnetic states of
the tape as it leaves
the gap field
(Rossing 1980).

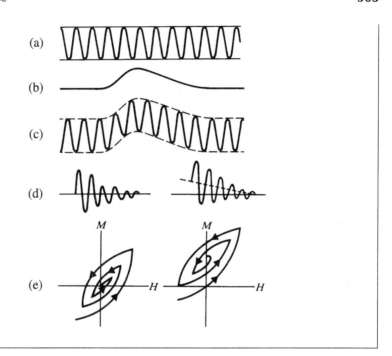

The equalization of response in tape recording is a more complex problem than in disk recording. The voltage induced in the playback coils is proportional to the *rate of change* of magnetic flux through the playback head. Thus, for a given amplitude of variation of the tape magnetization, the output signal increases linearly with the frequency of the recorded signal and the tape speed (the response rises with frequency at the rate of 6 dB per octave). There are other complicating factors, however (see Fig. 22.9, for example), so that it is not enough to merely attenuate the high frequencies by 6 dB/octave. In fact, it is common practice to provide equalization during both recording and playback in order to reduce tape noise. Nearly all tape recorders provide for different equalization networks for different tape speeds.

22.6 ■ TAPE NOISE

Unrecorded tape will produce a hiss when it is played back. This is due to microscopic random variation in the magnetization M. If the tape has been carefully erased with a high-frequency field, the average magnetization at any point will be zero, but M may vary in small regions the size of the gap. Erasing with the steady field of a permanent magnet (called D.C. erase), used on a few inexpensive tape recorders, results in a much higher level of tape noise than does high frequency erase. Some tapes have a lower level of tape noise than others.

Quality tape recorders are capable of recording signals that will play back 50 dB or more above the level of the tape noise. An important tape recorder specification is its *signal-to-noise ratio*. Unfortunately, there are several different ways of measuring this ratio (e.g.,

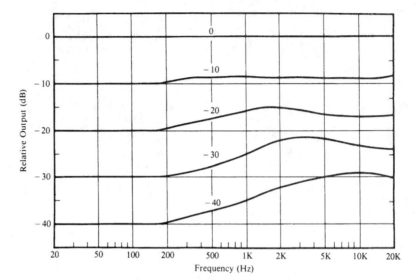

FIGURE 22.12
Dolby encoder
characteristics.
Low-level
high-frequency
signals are boosted
during recording
and attenuated
during playback.

weighted or unweighted signal, how much distortion is allowed in the signal, etc.), so that a comparison is difficult.

Dolby noise-reduction systems have become very popular lately, especially in cassette tape recorders. These systems change the equalization with the level of the recorded signal. In the Dolby B-system, low-level high-frequency signals are boosted 10 dB when recorded and attenuated by the same amount during playback. This has the effect of reducing tape noise during soft passages, and yet loud sounds will not overdrive the tape (the extra noise reduction is not needed during loud passages). The Dolby C-system provides about 20 dB of noise reduction at high frequency by using two compressors in series (Dolby 1983). Figure 22.12 shows the preemphasis used in the Dolby B-system encoder.

The Dolby A-system, slightly more sophisticated than the B-system, is widely used in professional recording. Dolby SR (spectral recording) is also intended for professional use as a replacement for Dolby A. It uses a sliding-band filter technique, which makes a 24-dB noise reduction for frequencies between 800 Hz and 6 kHz and a 16-dB reduction below 800 Hz. Dolby C and Dolby S (a simplified version of Dolby SR) are optimized for cassette recording. Other noise-reduction systems in use include the DBX system and the Philips DNL system, all of which improve the signal-to-noise ratio by at least 5 to 10 dB, which is especially important in low-speed cassette recorders.

22.7 ■ DIGITAL TAPE RECORDING

For many years magnetic tape recording has been a standard method for recording analog data. Although the principles involved in recording digital and analog data on magnetic tape are the same, there are some important differences:

1. The bandwidth required for digital recording is at least 30 times greater than that required for analog recording;

2. Only two values (1 and 0) are used in digital recording, so linearity is not necessary;

3. Bits rather than waveforms are recorded, so the data bits in one channel can be recorded on several tracks, or several channels can be recorded on one track by means of time sharing.

Analog tape recorders nearly always have stationary heads. In general, the maximum frequency that can be recorded depends on the tape speed and the head gap size (see Section 22.4). With tape speeds ranging from 4.8 to 38 cm/s ($1\frac{7}{8}$ to 15 in/s), it is possible to record frequencies across the entire range of audibility.

With a sampling rate of 44.1 kHz, digital recording of 16-bit words requires a bit rate of $44,100 \times 16 = 705,600$ bits per second. Allowing for error detection, synchronization, etc., pulse frequencies of 1 MHz and higher will be recorded. Achieving the required high-frequency response in digital recorders is possible by using very high tape speeds or by using rotating heads.

1. *Rotating heads.* Rotating heads were developed for videotape recording where bandwidths of over 4 MHz are required. Two heads are attached to a cylindrical drum which rotates opposite to the direction of tape motion, as shown in Fig. 22.13. The tape is wrapped partway around the drum at an angle, so that the tracks of recorded information run diagonally across the tape. The two heads record alternate tracks.

To use a video recorder for digital audio recording, the audio signal must be processed to conform to the video signal format. Synchronization pulses appropriate for the television format used in the video recorder (NTSC in the USA and Japan) are added. Video recorders use frequency modulation (FM), so changes in frequency represent ones and zeros. An appropriate block of audio data goes into each video frame.

Using videotape recorders for digital audio offers an inexpensive way to achieve the high-frequency response required. The additional electronic instrumentation required is fairly simple, because an FM modulator is already included in the VTR.

2. *Stationary heads.* Although the use of videotape recorders offers a low-cost method for digital audio tape recording, it makes tape editing and synchronous multi-track recording quite difficult. For this reason, most professional recording has been done using recorders with stationary heads.

One of the first successful digital tape recorders, developed around 1976 by Thomas Stockham and colleagues at Soundstream, used a 42.5-kHz sampling rate and 16-bit quantization. It was used to master many early digital recordings. Around 1977, the 3M Company introduced a 32-track recorder with a sampling rate of 50 kHz.

FIGURE 22.13
Rotating head tape recorder for video or digital audio recording.

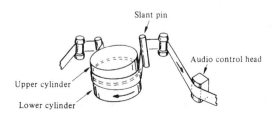

22.8 ■ DIGITAL AUDIOTAPE RECORDERS FOR HOME USE

In 1985, representatives of a large number of manufacturers of DATs agreed on tentative specifications both for stationary-head records (S-DAT) and rotating-head recorders (R-DAT). A year later, a final technical standard for R-DAT was published (Feldman 1987).

R-DAT uses a rotating head design and helical scan quite similar to that used in video-tape recorders. The head rotates at 2000 rpm and the tape passes it at an angle of 6°23′. Unlike videotape, however, the tape is wrapped around only $\frac{1}{4}$ of the head drum to mini-mize tape wear. Setting the two heads at slightly different angles reduces crosstalk between tracks and allows a denser track spacing than with standard videotape recorders. Tape width is 3.81 mm (0.15 in) and the tape speed is 8.15 mm/s (0.32 in/s). Cassettes of dimensions $73 \times 54 \times 10.5$ mm ($2\frac{7}{8} \times 2\frac{1}{8} \times \frac{13}{32}$ in), slightly smaller than audio cassettes, as shown in Fig. 22.14, allow a playing time of 2 h.

The S-DAT format records 20 data tracks side by side, and these densely spaced tracks are read by magneto-resistive (MR) heads similar to those used in computer disks. Al-though stationary-head recorders offer some advantages over rotary-head recorders, the S-DAT format was no competition for the R-DAT format, and S-DAT recorders quickly disappeared from the commercial scene. Accordingly, R-DATs came to be known simply as DATs.

The music industry was worried that DATs would be used to copy CDs. For this reason, the earliest DATs would record at a sampling rate of only 48 kHz rather than at the sam-pling rate of 44.1 kHz used in CDs (although they would play back tapes recorded at either sampling rate). Nowadays, most DATs record at either 44.1-kHz or 48-kHz sampling rates. Most DATs incorporate a serial copy-management system (SCMS) for copyright protec-tion. Decks that follow the SCMS standard are locked out of digital copying by a code known as ID6 so that copyrighted material can be copied only once. DATs in professional use are free of SCMS, however, and the definition of a professional machine apparently lies in the type of digital input signal it will accept.

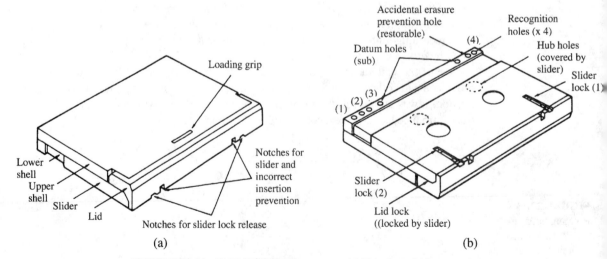

FIGURE 22.14   DAT (RDAT) tape cassette. (a) Top view; (b) bottom view.

It is well to point out the difference between copying ("dubbing") a tape by connecting the analog output of a tape player to the analog input of a tape recorder, and copying by connecting the digital output of a tape player to the digital input of a DAT. The latter type of copying might be called "cloning," because it results in a nearly exact copy. A tape can be digitally copied (cloned) many times with little or no loss in fidelity. Large-scale digital copying or cloning is what the music recording industry wants to prevent.

In 1992 Philips introduced its digital compact cassette (DCC) system, which recorded eight 185-μm-wide tracts in each direction on a 3.8-mm tape similar to that used in analog tape cassettes. In fact, the idea was to have a recorder that would play analog cassettes as well as record and play digital cassettes. The DCC system used an audio compression system called PASC to reduce the required bit rate by about a factor of 4. Like S-DAT recorders and computer disk drives, the DCC recorders used magneto-resistive reading heads. Like the S-DAT, the digital compact cassette system was not a commercial success. This goes to show how quickly a good idea can fade away in the audio market.

22.9 ■ DIGITAL MULTITRACK RECORDING

In professional recording studios many audio tracks are recorded and then mixed down to two tracks for stereo. Professional reel-to-reel recorders using the digital audio stationary head (DASH) format allow digital multitrack recording along with cut-splice and punch-in/punch-out editing (adding sound by pushing a switch). Many DASH machines use half-inch tape, and they can record up to 48 digital tracks. The high cost of DASH recorders limits their use to large studios, however. Two popular recorders, designed for professional use but affordable for small home recording studios, are the Alesis ADAT and the Tascam DA-88. Both these machines record on tape videocassettes and have special provisions for synchronizing the multiple tracks.

The ADAT digital recording system was released by Alesis Corporation in 1991. Since then, it has become a leader in digital-tape multitrack recording. The ADAT records eight tracks of 16-bit linear audio on a standard S-VHS videocassette. Using a VCR tape transport similar to the R-DAT, the ADAT operates at a tape speed of $3\frac{3}{4}$ in./s for more than 1 h of recording time. Using proprietary synchronization, up to 16 ADATs can be linked together for 128 tracks of recording. SMPTE code and MIDI code synchronization is also recognized. The ADAT Multichannel Optical Interface allows one ADAT to be connected with another to transfer digital data rather than converting it back to audio with a DAC and resampling with an ADC. In 1995 Alesis announced the ADAT-XT recorder, which increased tape transport speed by a factor of four, allowed 18-bit 128-times oversampled recording and 20-bit 8-times oversampled playback, and had many control and editing features. This was followed in 1998 by the ADAT Type II (XT-20), which uses 20-bit sampling and a computer interface for editing.

In 1993 Tascam Corporation introduced the DA-88, which provides an alternative format for digital multitrack recording. The DA-88, also an 8-track, 16-bit recorder, uses a special transport mechanism for the smaller Hi-8 videotape rather than the S-VHS tape, allowing up to 108 min of music to be recorded. The DA-98 provides 18-bit recording, whereas the high-resolution version, the DA-98HR, allows switching between 16-bit and 24-bit recording.

Digital tape recorders are easy to use for recording continuous music (such as a concert) but more difficult for recording, editing, and mixdown in the studio. The recent use of computer hard disks for recording provides the flexibility needed in the studio.

22.10 ■ HARD-DISK RECORDING

Hard disks are commonly used for data storage in computers. Because digital sound is just another form of data, it is possible to store tracks of sampled sounds on a hard drive. Hard disks consist of a set of aluminum platters rotating on a spindle with movable read/write heads all enclosed in a box (Fig. 22.15). The platters are coated with a magnetic material similar to magnetic tapes. At first, iron oxides were used for the coating; however, a newer thin-film magnetic medium is currently used to allow a higher density of data storage than tape. The read/write heads move in and out like a phonograph arm to the location of the data on the disk. Unlike a phonograph record, which stores sound as a long continuous spiral, hard drives store data in concentric rings (called tracks) divided into wedge-shaped sectors. This means that sound samples must be broken up into sector-sized blocks before being stored on the disk. Because music is usually recorded as a continuous stream, a record buffer of random-access memory (RAM) is used to hold the sampled data until it is allocated to segments on the disk. Likewise, on playback a replay buffer is used to hold the samples from the disk until they are reconstructed into a continuous sound stream.

Hard disks have a great advantage over tape recorders for editing sound. Previous editing methods included physically cutting and splicing the tape, which was especially difficult for DATs, where tracks are recorded diagonally on the tape instead of vertically. More modern methods include selecting blocks of data from recorded tracks and then recording edited data over the top of old data. This is destructive editing, because the old data no longer exist. Because sound is stored in multiple segments on a hard disk, it is easy to retrieve data to be edited and store the new data in new sectors. Also, it is possible to randomly access any data instead of trying to find it on the tape. For this reason, hard-disk editing has largely replaced tape editing in many sound studios. Digital audio workstations combine hard-disk recording with powerful editing tools.

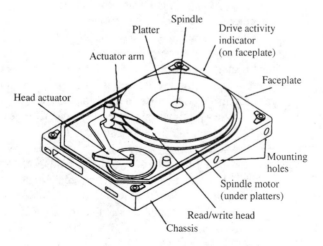

FIGURE 22.15
Hard-disk drive
showing parts.

22.11 ■ COMPACT DISC DIGITAL AUDIO

Perhaps the most important development in high-fidelity sound reproduction in recent years is the Compact Disc Digital Audio system. Only 12 cm in diameter, a compact disc can store more than 6 billion bits of binary data to be read out by a laser. This is equivalent to 782 megabytes, or more than the capacity of 1500 half-megabyte floppy disks. Over 275,000 pages of text, each holding 2000 characters, could be stored on a compact disc for display on a television monitor.

Used for digital audio, a compact disc stores 74 min of digitally encoded music. This music can be reproduced with very high fidelity over the full audible range of 20 to 20,000 Hz. The dynamic range and the signal-to-noise ratios can both exceed 90 dB, and the sound is virtually unaffected by dust, scratches, and fingerprints on the disc. Unlike most other digital recording media, compact discs can be replicated in large quantities from a master disc.

Recorded information is stored in pits impressed into the plastic surface of the disc, which is then coated with a thin layer of aluminum to reflect the laser beam (Fig. 22.16). Pits are about 0.5 μm wide and 0.11 μm deep, arranged in a spiral track similar to the spiral groove in a phonograph record, but much narrower. The track spacing on a compact disc is about 1.6 μm.

The track on a compact disc, which spirals from the inside out, is about 3 mi in length. The track of pits is recorded and read at a constant 1.25 m/s, so the rotation rate of the disc must change from about 8 to 3.5 rev/s as the spiral diameter changes. Each pit edge represents a binary 1, whereas flat areas within or between the pits are read as binary 0s.

The laser beam, applied from below the compact disc, passes through a transparent layer 1.2 mm thick and focuses on the aluminum coating, as shown in Fig. 22.17. The spot size of the laser on the transparent layer is 0.8 mm, but at the signal surface where the pits are recorded, its diameter is only 1.7 μm. Thus, any dust or scratch smaller than 0.5 mm will not cause a readout error because it is out of focus. Larger blemishes are handled by error-correcting codes.

The optical pickup used to read a compact disc is shown in Fig. 22.18. A semiconductor laser emits a beam of infrared light (790-nm wavelength) that is eventually focused to a

FIGURE 22.16
Tracks of pits recorded on a compact disc. Also shown is the focused laser beam used to read the recorded information.

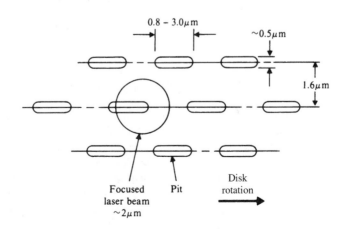

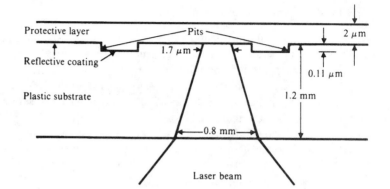

FIGURE 22.17
Cross section of a compact disc (not to scale).

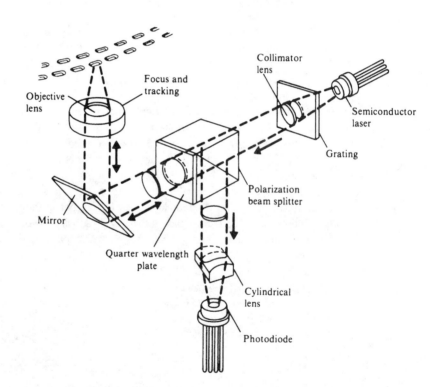

FIGURE 22.18
Optical pickup in a compact disc player. Coherent light from a semiconductor laser is focused on one recorded track on the disc. The reflected light is directed to a photodiode.

tiny spot 1.7 μm in diameter. The reflected beam is directed to a photodiode that generates electrical signal to be amplified and decoded.

Included in the sophisticated optical pickup are a diffraction grating, a polarization beam splitter, a quarter-wavelength plate, and several lenses. A semiconductor laser employs an aluminum-gallium-arsenide (AlGaAs) pn junction, similar to that in a light-emitting diode (LED). The diffraction grating creates two secondary beams that are used for tracking the primary beam.

When the laser beam strikes a land area between two pits, it is almost totally reflected. When it strikes a pit (which appears like a bump from the reading side), whose 0.11-μm height is roughly one-quarter wavelength of the laser light, the part of the beam reflected from the pit cancels the part reflected from the land, so little or no light returns. The reflected light of varying intensity is directed on the photodiode; a change in intensity will eventually be interpreted as a binary 1 and unchanged intensity as a 0.

The Compact Disc Digital Audio system, a joint development of Philips (in The Netherlands) and Sony (in Japan), is a very sophisticated system. An elaborate error correction system, called *cross-leave Reed-Solomon code* (CIRC), was developed to deal effectively with both random errors, caused by inaccurate cutting, and burst errors, due to dirt and scratches on the disc. Packing density is maximized by rearranging the data bits, using a channel code called 8-to-14 modulation (EFM).

Compact disc players also include sophisticated servomechanisms for keeping the laser beam centered on the track and for keeping it focused exactly on the reflecting surface within the disc. Decoding the recorded information and digital-to-analog conversion in most players incorporates digital filtering plus a technique called oversampling (see the box).

Although many of the technical details of the Compact Disc Digital Audio system are beyond the scope of this book, we briefly describe three innovative features: autofocusing, EFM coding, and oversampling.

1. *Autofocus.*   To distinguish between pits and land areas, the laser beam must stay focused within about 0.5 μm of the reflecting surface. The flat surface of the disc, however, may have deviations as large as 0.5 mm, 1000 times greater. Thus

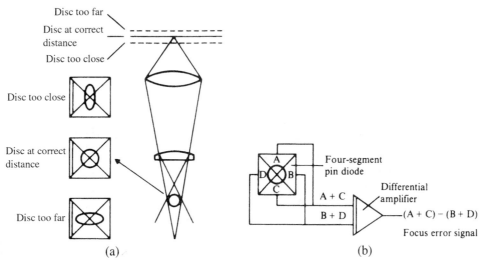

(a)   (b)

FIGURE 22.19   A square pin diode photodetector, divided into four identical triangular segments, and a difference amplifier control a focus servo that moves a cylindrical lens up and down to correct for disk surface irregularities. (Miyaoka, ©1984 by IEEE).

the objective lens must refocus rapidly as the surface deviates during the rotation of the disc. A servodriven autofocus system that uses a four-quadrant photodiode and a differential amplifier to control the servo disc makes the rapid refocus possible.

The cylindrical lens, shown just above the photodiode in Fig. 22.18, projects a circular laser spot on the photodiode if the reflecting surface of the disc is exactly at the focus of the objective lens but an elliptical spot if it is above or below the focus, as shown in Fig. 22.19. The four-quadrant diode provides electrical signals proportional to the light intensity on each of its four quadrants, and these are combined to form the error focus signal, as shown in Fig. 22.19. The objective lens is attached to a movable coil in a magnetic field, similar to that which drives a loudspeaker cone.

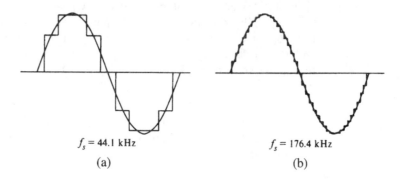

FIGURE 22.20

A sinusoidal signal at 4.41 kHz sampled with a sampling rate F_s of 44.1 kHz (a) and with a frequency four times higher (b). In (b) the staircase curve approximates more closely to the analog waveform, and the high frequencies present in the staircase signal are more easily filtered out.

$f_s = 44.1$ kHz

(a)

$f_s = 176.4$ kHz

(b)

2. EFM coding.   Compact discs use a modulation or channel code called 8-to-14 modulation (EFM). The EFM is a code in which 8 bits of data are represented by 14 channel bits. This may seem wasteful, but by requiring that every one in channel bits (corresponding to a transition) be separated by 2 to 10 channel zeros, EFM allows bit sizes to be less than the diameter of the laser beam that reads them. Thus EFM actually results in a greater bit density. In the compact disc system, a 1.7-μm-diameter laser spot reads pits whose length varies incrementally from 0.833 to 3.054 μm.

It turns out that of the 16,384 (2^{14}) possible 14-bit patterns, only 267 of them satisfy the specified requirement of having two to 10 binary 0s in succession, but this is more than enough to code the 256 (2^8) possible 8-bit blocks. Blocks of 8 bits are translated into blocks of 14 bits using a conversion table (Miyaoka 1984).

3. Oversampling.   After decoding and error correction, the digital signal exists as a series of 16-bit words. Each word represents the numerical value of the acoustical signal, as sampled 44,100 times per second. The digital-to-analog converter (DAC) generates an electrical voltage of the appropriate magnitude for each word and holds it constant until the next word arrives. The resulting "staircase" curve resembles the original signal waveform plus a high-frequency signal with spectral components at 44.1 kHz and its harmonics, as shown in Fig. 22.20(a). These high-frequency components have to be removed by a low-pass filter.

To avoid the use of sharp cutoff analog filters and to improve the signal-to-noise ratio, most manufacturers have introduced digital filters with oversampling, as shown

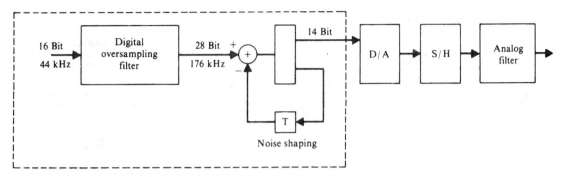

FIGURE 22.21   Digital filter with oversampling. (Reproduced with permission of the publisher, Howard W. Sams & Co., Indianapolis, *Principles of Digital Audio* by Kenneth C. Pohlmann, copyright 1985.)

in Fig. 22.21. Rather than suppress high-frequency components after the signal has been converted back to analog form, a digital oversampling filter is added before the DAC. In this filter, each 16-bit word is multiplied four times by different 12-bit co-efficients, the resulting 28-bit words are averaged together and output at a rate either twice (88.2 kHz) or four times (176.4 kHz) that of the original sampling frequency. This has the net effect of raising the high-frequency component in the staircase to 88.2 kHz or 176.4 kHz (as shown in Fig. 22.21(b)), which is more easily removed by filtering. Oversampling is discussed in Section 21.17.

22.12 ■ RECORDABLE CDS

Originally, the CD was a playback-only medium. However, development of the *write-once-read-many* (WORM) technology now allows musicians to create their own master CDs. Such CDs are generally called CD-R discs. Then, fully erasable CDs (CD-RW) appeared on the scene.

Recordable CD-R discs can be recorded on, or "burned," once and played back many times. They generally have an organic dye (such as phthalocyanine) located in concentric rings (not a spiral) between a polycarbonate substrate and a reflective metal layer (Fig. 22.22(a)). Pits are created when the organic dye is heated with a powerful 6 to 8-mW laser (rather than the 1- to 2-mW lasers used for playback), causing the dye to decompose and change its reflection. A thin layer of silver or gold is coated over the dye to protect it. Some of the dye/coating combinations appear green, blue, or yellow. An alternative method is to use two metals that form an alloy when heated, resulting in different reflection properties. As can be seen in Fig. 22.22(c) and (d), the pits created by burning a CD-R disc are very similar to those of a standard CD. In any case, they can be played by CD-Audio players or CD-ROM drives. An earlier method of recording by having the laser actually burn a tiny hole in the coating is no longer used, but the term *burning* a CD still remains.

Rerecordable CDs (CD-RW) may be created using either an optical phase change or a magneto-optical technology. An optical phase-change medium consists of a polycarbonate substrate on which a stack (usually five layers) is deposited. The recording layer is sand-

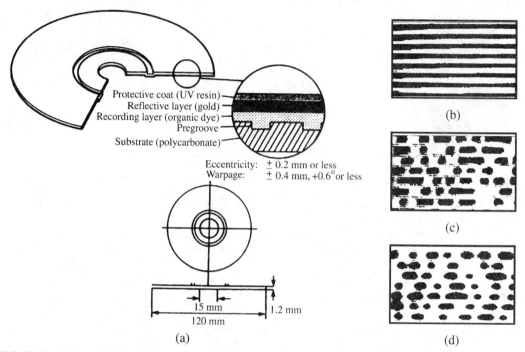

FIGURE 22.22   (a) Specification for blank recordable compact disc (CD-R), showing disc composition and recording area dimensions. Microscopic view of CD-R (b) before and (c) after recording, with (d) standard recorded CD for comparison.

wiched between dielectric layers that draw excess heat from the phase-change layer during the writing process. The phase-change layer is a crystalline compound made up of silver, indium, antimony, and tellurium. When this rather exotic material is heated to a certain temperature and cooled, it becomes crystalline, but if it is heated to a higher temperature, it becomes amorphous (noncrystalline) when cooled. The crystalline areas reflect the laser light better than the amorphous areas, which tend to absorb the light.

Magneto-optical technology (Schouhamer Immink and Braat 1984), similar to that used in recordable minidiscs, is described in the following section.

22.13 ■ MINIDISCS

Just as the CD replaced phonograph records, the MiniDisc was designed to replace the tape cassette. It records and stores up to 74 min of music (roughly the same as a CD) on a miniature 64-mm-diameter disc housed in a 7-cm by 6.75-cm by 0.5-cm cassette. It can store up to 160 MB of recorded music using 16-bit, 44.1-kHz audio. It uses an audio compression algorithm called adaptive transform acoustic coding (ATRAC). ATRAC performs a Fourier transform (see Sec. 2.7) and divides the sound into 52 frequency bands. The content of each band is compared with the threshold of hearing (see Chapter 4) to determine which signals to throw away, because they would not be audible. A 5:1 compression is achieved by this method (see Appendix A.8).

MiniDiscs are of two types. Polycarbonate optical discs, similar to CDs, are used for prerecorded music, whereas magneto-optical (MO) discs are used for home recording. The MD optical pickup has the remarkable ability to read both the prerecorded and the MO discs. For prerecorded discs, the pickup reads the amount of light reflected from the pits and land areas, as in a CD pickup. For MO discs, on the other hand, the pickup reads the rotation of the plane of polarization by the data tracks in the magnetic recording surface.

Because of its compact size, the MD system is especially suitable for portable use, as in Walkman-type players and car audio systems. Thus, a buffer memory chip has been added to hold approximately 10 s of playback data. If the pickup is jarred from its position on the disc, the memory will continue to supply data while the pickup repositions itself, thus making it less susceptible to skipping and mistracking than an ordinary CD player. To fill the buffer memory, the MD pickup reads data from the disc at 1.4 megabits/s, nearly five times the rate at which the decoder uses that data.

One of the features of the MD is complete random-access capability; the digital addresses for each selection or track are recorded, as in CDs. To achieve random access in a recordable disc (whose final recorded configuration is unknown at the time the disc is manufactured), recordable discs are provided with a circumferential microgroove, or pregroove, for tracking and spindle-servo control during recording and playback. Addressed information is recorded at intervals of 13.3 ms by zigzags in the pregroove. Therefore, the disc has all the addresses already notched along the groove before recording occurs. At the beginning of the disc (the inner circumference) is a user table of contents area, where track numbers and addresses are recorded in order to speed editing.

Magneto-Optical Recording

Recording on MO media combines some of the advantages of magnetic recording (rapid recording, unlimited reversibility) with those of optical readout (high bit density, contactless read and write). Recording is done thermomagnetically, and playback makes use of the MO Kerr effect. The magnetic medium is generally a thin magnetic layer of amorphous (noncrystalline) rare-earth transition-metal alloy, such as terbium-iron-cobalt, sandwiched between two layers of a dielectric, such as silicon nitride. The easy axis of magnetization of the magnetic layer is perpendicular to the recording surface, and the magnetic layer has a very high coercivity (see Section 22.3), so that its remanent magnetization will point into or out from the surface. Stable magnetic domains as small as 100 nm have been observed by means of Lorentz microscopy (Hansen and Heitmann 1989).

Recording is done thermomagnetically by simultaneously heating a small spot with the laser and applying a magnetic field. The coercivity decreases with increasing temperature and goes to zero at the Curie temperature. The MO layer is heated by momentarily increasing the power of the playback laser while a magnetic field is applied by a recording head on the opposite side of the disc. As the spot cools, its magnetization remains in the direction of the field from the recording.

Playback makes use of the Kerr MO effect, the rotation of the plane of polarization of light when it is reflected from the magnetized surface. The plane of polarization is rotated one direction if the MO layer is magnetized upward and in the other direction if it is magnetized downward, as shown in Fig. 22.23. (The rotation at the magnetic surface is only about 0.2°, although this is increased to about 1° by the dielectric layers.) The

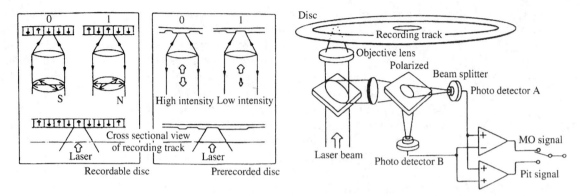

FIGURE 22.23 MiniDisc optical pickup can read either prerecorded disc (with pits) or recordable disc (with magneto-optical bits).

reflected light passes through a polarization beam splitter to two photodiodes arranged so that diode A receives more light if the Kerr rotation is in one direction and diode B receives more if it is in the other direction. The electrical signals from the diodes are subtracted to determine whether a 1 or a 0 is read.

Note that the optical playback system is also used to read prerecorded discs. In this case, the amount of light reflected depends on whether or not a pit exists on the recording surface. If no pit exists, a high proportion of the light is reflected back through the beam splitter and analyzer into the photodiodes; if a pit exists, less light reaches the photodiodes due to interference. In this case, the electrical signals from the photodiodes are added to determine whether a 1 or 0 is read.

22.14 ■ DVD

DVD, which once stood for digital video disc or digital versatile disc, represents a new generation of optical disc-storage technology. DVD aims to encompass home moves, music, computer storage, and other data-storage applications in a single digital format. DVD may become the most successful consumer electronic product of all time.

The storage capacity of a DVD is phenomenal. Although it looks very much like a CD, it can store up to 17 GB (gigabytes), or 26 times more than a CD. This is possible because it uses narrower tracks (0.74 μm versus 1.6 μm) and a laser of shorter wavelength (640 nm green versus 780 nm infrared) and records in one or two layers and on one or two sides of the disc. The physical formats include DVD-ROM (read only), DVD-R (record once), and DVD-RW (record and erase), just as in the case of CDs, and the application formats include DVD-video (often called just DVD) and DVD-audio.

A single-layer, single-sided DVD-video carries 2 h of video, whereas a two-layer, double-sided DVD can hold 8 h of high-quality video (or 30 h of VHS quality video). Up to eight tracks of digital audio (for multiple languages, etc.), each with as many as eight channels, can accompany the video. Up to nine camera angles, which the viewer can select during playback, can be included. Special-effects playback include freeze, step, slow, fast, and scan. Some players include such additional features as six-channel analog sound, reverse single-frame stepping, reverse play, and digital zoom. DVD-video is usually encoded from digital studio master tapes to MPEG-2 format.

22.15 ■ DVD-AUDIO

When DVD was released in 1996, there was no DVD-audio format. The DVD-audio specification (announced in 1999) allocated the video portion of the DVD to audio. Table 22.1 compares the specifications of DVD-audio with those of an audio CD. The DVD-audio provides up to six channels of audio in the 5.1 surround-sound format (or many other possible formats, living up to its name of versatile), with a 96-kHz sampling rate of 24-bit samples, giving a frequency response of 48 kHz and a dynamic range of 144 dB. The two-channel (stereo) format increases the frequency response to 96 kHz. The most practical advantages of DVD-audio are an extended playing time for CD-quality music (8 h for single-layer DVDs) and the surround-sound option for the listening environment. Because the sampling rates are higher than for other DVDs, DVD-audio cannot be played back on common DVD players. At this writing, there are DVD players that play only video, some that play only audio, and some that play both types. Interestingly enough, most DVD players will also play CDs.

TABLE 22.1   Specification comparisons of DVD-audio with audio CD

Specification	CD	DVD-Audio
Disc capacity	650 MB	4.7 GB—single layer
		8.5 GB—dual layer
		17 GB—double-sided
Audio channels	2 (stereo)	Up to 6 (5.1)
Sampling rate	44.1 kHz	44.1, 48, 88.2, 96, 176.4, or
		192 kHz
Frequency response	5–20 kHz	0–96 kHz (max)
Sample size	16 bits	12, 16, 20, or 24 bits
Dynamic range	96 dB	144 dB
Maximum data rates	1.4 Mbps	9.6 Mbps

22.16 ■ SOLID-STATE MEMORY RECORDING

All the sound recorders mentioned so far require many mechanical parts to move the tape across recording heads or move recording heads to locations on a disc. It is possible to create a recording system with no moving parts. In 1995, the Nagra Ares-C, said to be the first tapeless digital recorder, was introduced. This recorder samples data at frequencies from 16 to 48 kHz and stores the data on Flash RAM cards. With data compression, it records up to 3.15 h on a 192-MB card. A palm-size recorder, the Ares-P, which also uses Flash RAM cards, is shown in Fig. 22.24.

Flash memory is a type of solid-state memory, which is used, for example, to hold control code in computers and to remember presets in audio radios. It is so named because the microchip is organized so that a section of memory cells can be erased in a single action, or "flash."

FIGURE 22.24
Nagra Ares-P
palm-size tapeless
solid-state recorder.

Another example of solid-state recording is the ChipCorder record and playback chip, which can store up to 16 min of sound. The ChipCorder has been used in hundreds of products, including telephone answering machines, cellular phones, alarms, pagers, toys, and even in greeting cards. It uses a multilevel storage in which more than 250 voltage levels are stored in each memory cell. This means that sound in either analog or digital form can be stored.

The IS22C040 sound-storage chip from Integrated Silicon Solution, Inc., is a one-time programmable chip that stores and plays back sound. It provides 8-kHz sampling for 32 s or 6-kHz sampling for 40 s. Sound is stored in 8 or 32 selectable storage segments so several short songs can be stored on a single chip. Solid-state recording of sound is new, and one can expect many new products on the market before the ink in this book is dry.

22.17 ■ SUMMARY

Our technology today offers several alternatives for recording and playing back sound. Phonographs, after close to a century of use, are now becoming antiques. Recording analog signals on magnetic tape is still alive with cassette tapes, but it is fading. Both these recording technologies have been largely replaced with the compact disc (CD), which is inexpensive and easy to use and gives much higher fidelity than phonograph disks or tapes. Serious recording uses digital technology, both two-track stereo (DAT) and multitrack (ADAT). Studios most often use hard-disk recording for advanced editing capabilities. The computer world has led to optical recording with CD-R and MO technology. Finally, there is also the choice of recording directly to memory or digital chips. Changes in the past few years, including DVD-audio, forecast changes in the future as well.

REFERENCES AND SUGGESTED READINGS

Alkin, G. (1996). *Sound Recording and Reproduction.* Woburn, Mass.: Focal Press.

Beyer, R. T. (1999). *Sounds of Our Times.* New York: Springer-Verlag, Chapter 5.

Camras, M. (1987). *Magnetic Recording Handbook.* New York: Van Nostrand Reinhold.

Daniel, E. D., C. D. Mee, and M. H. Clark (1999). *Magnetic Recording: The First 100 Years.* New York: IEEE Press.

Feldman (1987). "Four DAT Recorders," *Audio* **71**(7): 36.

Hansen, P., and H. Heitmann (1989). "Media for erasable magnetooptic recording," *IEEE Trans. Magnetics* **25**: 4390.

Miyaoka (1984). "Digital Audio Is Compact and Rugged," *IEEE Spectrum* **21**(3): 35.

Pohlman (1985). *Principles of Digital Audio Technology*. Indianapolis: H. W. Sams.

Rossing, T. D. (1981). "Anhysteretic Magnetization and Tape Recorder Bias," *Am. J. Phys.* **49**: 655.

Rossing, T. D. (1987). "The Compact Disc Digital Audio System," *Phys. Teach.* **25**: 556.

Rossing, T. D. (1994). "Digital Techniques for Sound Recording," *Phys. Teach.* **32**: 458.

Rumsey, F. (1994). *Sound and Recording*. Woburn, Mass.: Focal Press.

Schouhamer Immink, K. A., and J. J. M. Braat (1984). "Experiments Toward an Erasable Compact Disc Digital Audio System," *J. Audio Eng. Soc.* **32**: 138.

Watkinson, J. (1997). *Audio for Television*. Woburn, Mass.: Focal Press.

Westmijze, W. K. (1953). "Studies on Magnetic Recording. III. The Recording Process," *Philips Res. Rep.* **8**: 245.

GLOSSARY

ADAT Digital multitrack recording system that uses S-VHS video cassettes.

bias That which is added to the desired signal to produce a composite. In the case of magnetic tape recording, for example, the bias is generally a constant magnetic field that oscillates at a high frequency.

CD-R A recordable compact disc that can be recorded once but played back many times.

CD-RW A recordable and erasable compact disc on which you can read and write.

coercivity The magnetic field that must be applied in the reverse direction to reduce the magnetization to zero.

Compact Disc Digital Audio (CD) An efficient means for digital sound recording that uses an optical pickup to read the recorded information.

DASH A digital multitrack recorder with stationary heads used in professional studios.

demagnetization Erasing a magnetically recorded signal by applying a rapidly varying magnetic field of large amplitude; in a demagnetized tape, the tiny magnetic "domains" are oriented in nearly random directions.

dither Low-level noise added to the signal to reduce the effect of quantization error.

Dolby system A widely used noise-reduction system that boosts low-level high-frequency signals when recorded and reduces them in playback.

domain (magnetic) A tiny region in which the atomic magnets point in the same direction.

8-to-14 modulation (DFM) A code used to represent blocks of eight numbers on a compact disc.

equalization Boosting some frequencies and attenuating others to meet some prescribed recipe. Equalization is usually applied during both recording and playback.

gramophone Flat-disk recording system invented by Emile Berliner in 1891.

hysteresis (magnetic) The effect of past history on the magnetic state of a material; when a magnetic field is applied momentarily and removed, the material does not revert to its original state. Magnetic hysteresis may be described by a hysteresis loop, a graph of magnetization versus applied magnetic field.

interleaving A means for rearranging recorded data to minimize the loss of information due to playback errors.

Kerr magneto-optical effect Rotation of the plane of polarization of light upon reflection from a magnetic material. The direction of rotation indicates the direction of the magnetization.

magneto-resistive reading head Tape-reading head with a material whose resistance changes in a magnetic field. A magneto-resistive head senses magnetic flux (field), whereas the more common inductive head senses rate of change of flux (field). Magneto-resistive heads can be made very thin.

MiniDisc (MD) A miniature version of a Compact Disc for digital sound recording.

oversampling A method for increasing the rate of digital samples to the DAC in order to avoid the need for an analog filter with a sharp cutoff.

polarized light light waves that vibrate in one plane only.

pulse-code modulation A means for representing a sequence of binary numbers by a series of electrical pulses.

punch in/punch out Editing a recording by substituting a new section.

quarter wavelength plate device that changes plane polarized light to circularly polarized light.

R-DAT Rotary-head digital audio tape recorder.

remanence (magnetic) The net magnetization that remains after the magnetic material has been saturated and the field has been removed.

saturation (magnetic) State of maximum magnetization when the magnetic domains are aligned as well as possible.

S-DAT Stationary-head digital audio tape recorder.

serial copy management system (SCMS) Device that prevents making ("cloning") multiple copies with a digital recorder.

signal-to-noise ratio The ratio (usually expressed in dB) of the average recorded signal to the background noise.

REVIEW QUESTIONS

1. What were three major inventions during the 1870s?
2. What is magnetic hysteresis?
3. Why is it important to have high coercivity in a magnetic recording material?
4. What is the advantage of a CrO_2 coating on magnetic tape over iron oxide?
5. What is the relationship between tape speed and frequency response?
6. What is the purpose of bias on a magnetic tape?
7. How does the Dolby noise-reduction system work?
8. What is the main advantage of digital recording over analog recording?
9. Why are rotating heads used for digital tape recording?
10. How many tracks can be recorded simultaneously using the ADAT system?
11. Under what conditions is the hard-disk recorder most useful?
12. How does a CD player read data from the CD?
13. How does the CD-R record data to a disc?
14. How does a MiniDisc player prevent skipping and mistracking when the player is jarred?
15. What is the Kerr magneto-optical effect? How is it used in sound recording?
16. Why can't a DVD-audio disc be played on some DVD players?

QUESTIONS FOR THOUGHT AND DISCUSSION

1. Why was Berliner's back-and-forth stylus superior to Edison's hill-and-dale recording? (*Hint:* Where does dust settle in a phonograph groove?)
2. Why is the signal-to-noise ratio in a cassette recorder usually lower than it is in a reel-to-reel recorder?
3. Why does a high-frequency erase field result in a lower level of noise than does a unidirectional (DC) erase field (used in some old tape recorders)?
4. Why does a high tape speed give better high-frequency response?
5. Why are digital tape recordings virtually free from wow and flutter?
6. What is the purpose of the low-pass filter used in the playback of digital recordings?
7. Why can't a magneto-optically recorded CD-RW disc be used in an ordinary CD player? Why was the digital compact cassette (DCC) system not a commercial success?
8. Why does the rotational speed of a CD change from 8 to 3.5 rev/s?

EXERCISES

1. What is the wavelength of an 8000-Hz signal recorded on cassette tape moving at $1\frac{7}{8}$ in/s?
2. What is the wavelength of a 100-kHz bias field recorded on tape moving at a speed of $7\frac{1}{2}$ in/s?
3. Express each of the following dimensions in a CD as numbers of wavelengths of the laser light: pit width, pit depth, diameter of the focused laser spot.
4. Up to 6 billion bits of binary data stored on a compact disc are read out in 74 min. What is the maximum bit rate per second?
5. Show that using a laser with a wavelength of 635 nm (green) increases the storage capacity of an optical disc (such as DVD) by a factor of 1.5 over that of the 790-

nm (red) laser used in CDs. How much could the storage capacity be increased by using a 430-nm (blue) laser? A 350-nm (ultraviolet) laser?

6. Show that a 16-bit representation of signals allows a signal-to-noise ratio of 96 dB. What is the signal-to-noise ratio for a 24-bit representation?

EXPERIMENTS FOR HOME, LABORATORY, AND CLASSROOM DEMONSTRATION

Home and Classroom Demonstration

1. *Surface noise on phonograph records* Compare the surface noise on a vinyl LP record with that of an older 78-rpm phonograph record.

2. *Stereo phono cartridge* If the two sets of coils in a stereo magnetic pickup are connected in series, the pickup will respond only to vertical or to horizontal motion of the stylus, depending upon the polarity of the series connection. Play recorded material using only one coil; then connect the two coils series-aiding and series-opposing. The demonstration is most effective with a monophonic recording.

3. *Building an Edison-type phonograph* An Edison phonograph can be built with simple materials. See: http://freeweb.pdq.net/headstrong/phono.htm.

4. *Magnetization patterns on magnetic tape* Use a *tape viewer* (which has tiny magnetic particles suspended in a solution) to view the magnetization patterns on a tape, both in the erased state and with music recorded on it.

5. *Tape recorder bias* If possible, disconnect the bias from the record head of a tape recorder and note the result. Then bring a permanent magnet into the vicinity of the record head to produce a DC bias. Reconnect the high-frequency bias. This demonstration is most effective if the recorder has a separate playback head so that the recorded material can be monitored immediately after recording.

6. *Diffraction from a compact disc* Hold a penlight or other point source of light so that it illuminates a compact disc at different angles. Note the spectrum you see at several different angles. Which color is reflected through the greatest angle? (The CD, on which concentric circles of pits are spaced at 1.6 μm (1.6×10^{-6} m), is acting as a reflection grating.)

7. *Greeting card with ChipCorder* Obtain a "talking" greeting card. Record and playback a message (approximately 10 s).

Laboratory Experiments

Magnetic tape recording (Experiment 33 in *Acoustics Laboratory Experiments*).

Diffraction from tracks of a compact disc (Experiment 5.3 in *Light Science* by T. D. Rossing and C. J. Chiaverina. New York: Springer-Verlag, 1999).

PART VI

The Acoustics of Rooms

In Chapter 3, we defined sound waves as longitudinal waves that travel in a solid, liquid, or gas. The most common path from source to receiver is through air, but not necessarily a *direct* path through the air. In a room, most of the sound waves that reach the listener's ear have been reflected by one or more surfaces of the room or by objects within the room. In a typical room, sound waves undergo dozens of reflections before they become inaudible.

It is not surprising, therefore, that the acoustical properties of the room play an important role in determining the nature of the sound heard by a listener. Performers may not be able to change the acoustics of a concert hall, but they can (consciously or unconsciously) adapt their performance to the particular hall, so that listeners receive the optimum quality of sound. To do this, it is important to understand some of the principles of room acoustics.

Chapter 24 discusses systems for the electronic reinforcement of sound. Sound systems may be necessary to provide an adequate level of sound in a very large auditorium, or they may be designed to compensate partially for some acoustical defect (such as inadequate or excessive reverberation). Chapter 25 discusses the properties of small rooms, home listening rooms, and recording studios.

CHAPTER

23

Auditorium Acoustics

The subject of concert hall acoustics is almost certain to provoke a lively discussion by both performers and serious listeners. Musicians recognize the importance of the concert hall in communication between performer and listener. The opening of a new concert hall invokes a flurry of reviews, opinions, and criticisms of its acoustical qualities. Amateur and professional critics try to compare a piece of music performed in the new hall with how they remember it sounding in other halls on other occasions (by other performers perhaps). Opinions of new halls tend to polarize toward the extremes of very good or very bad. In spite of the extensive research on the acoustics of concert halls, the opinion still prevails in some circles that their acoustic design is "black magic." This chapter will show you that there are good scientific principles that can be applied in designing concert halls, churches, and classrooms.

In this chapter you should learn:

- About direct, early, and reverberant sound;
- About reverberation time and sound absorption;
- About criteria for good acoustics;
- How concert halls are rated.

23.1 ■ SOUND PROPAGATION OUTDOORS AND INDOORS

Before discussing the acoustics of auditoriums, concert halls, and other large rooms, let us briefly review how sound propagates in various environments.

An environment in which the sound pressure is proportional to $1/r$ (where r is the distance from the source) is called a *free field*. When a source of sound is small enough to be considered a point source and is located outdoors away from reflecting objects, a free field results. Sound waves travel away from the source in all directions, the wave fronts having the shape of spheres, as shown in Fig. 6.1. The sound pressure in a free field is proportional to $1/r$, as shown in Fig. 23.1 (r is the distance from the source). Thus

FIGURE 23.1
The way in which sound pressure p and sound pressure level L_p decrease with distance r in a free field.

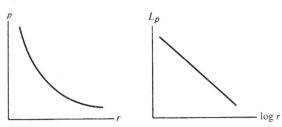

525

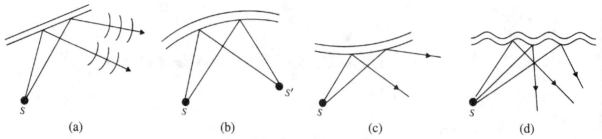

| | | | |
| (a) | (b) | (c) | (d) |

FIGURE 23.2   Reflection of sound by various surfaces: (a) flat surface acts like a mirror; (b) concave surface concentrates sound in the region S'; (c) convex surface scatters sound; (d) rough surface leads to diffuse reflection.

the pressure p is halved when the distance r doubles. From the definition of the decibel, given in Chapter 6, we see that the sound pressure level decreases 6 dB each time the distance r is doubled. Free-field conditions rarely occur indoors, except in reflection-free *anechoic* rooms. (Anechoic means *echo-free*; this is generally achieved by covering the walls, ceiling, and floor with wedges of sound-absorbing material.)

Indoors, sound travels only short distances before encountering walls and other obstacles. These obstacles reflect and absorb sound in ways that largely determine the acoustic properties of the room. Figure 23.2 illustrates the reflection of sound by flat, curved, and rough surfaces (*rough* in this case means that irregularities have dimensions comparable to the sound wavelength).

23.2 ■ DIRECT, EARLY, AND REVERBERANT SOUND

Let us examine how sound travels from the source to the listener in an auditorium. Sound waves travel at 344 m/s (1130 ft/s); the direct sound, therefore, may reach the listener after a time of anywhere from 0.02 to 0.2 s (20 to 200 ms), depending on the distance from source to listener. A short time later, the same sound will reach the listener from various reflecting surfaces, mainly the walls and ceiling. In Fig. 23.3, these reflections are shown

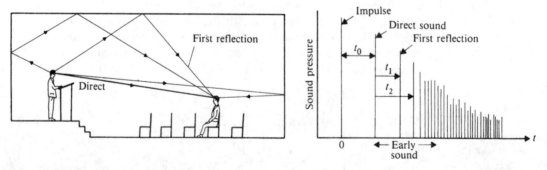

FIGURE 23.3   Paths of direct and reflected sound from source to listener with corresponding time delays for a sound impulse. (From *Music, Acoustics, and Architecture.* © 1988 Leo Beranek. Used with author's permission.)

arriving with various time delays t_1, t_2, t_3, etc. The first group of reflections, reaching the listener within about 50 to 80 ms of the direct sound, is often called the *early* sound.

After the first group of reflections, the reflected sounds arrive thick and fast from all directions. These reflections become smaller and closer together, merging after a time into what is called *reverberant* sound. If the source emits a continuous sound, the reverberant sound builds up until it reaches an equilibrium level. When the sound stops, the sound pressure level decreases at a more or less constant rate until it reaches inaudibility. In the case of the impulsive sound shown in Fig. 23.3, the decay of the reverberant sound begins immediately, and there is no equilibrium level. This is illustrated by the decay curve of sound pressure shown in Fig. 23.4(a) for a 400-m^3 classroom occupied by 50 students. The sound-pressure level decay is shown for the same room when occupied and unoccupied. Note that in the occupied room the sound dies out slightly faster due to absorption by the occupants (Jesse 1980).

A reasonably simple but accurate analysis of the acoustics of an auditorium can be obtained from a careful study and comparison of *direct*, *early*, and *reverberant* sound. In

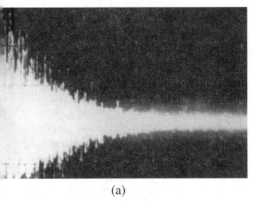

(a)

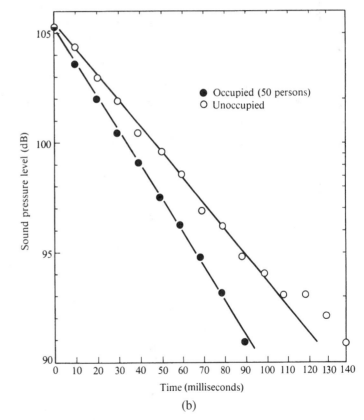

(b)

FIGURE 23.4 Sound decay curves in a 400-m^3 classroom: (a) sound pressure as a function of time (room unoccupied); (b) sound pressure level as a function of time for the same room occupied and unoccupied. An exponential decay of sound pressure corresponds to a linear decay of sound pressure level. (From Jesse 1980.)

Section 23.7 we will consider the criteria for good concert hall acoustics and how these criteria relate to direct, early, and reverberant sound. In the intervening sections, the character of these three types of sound will be discussed in more detail.

23.3 ■ DIRECT AND EARLY SOUND: THE PRECEDENCE EFFECT

Some sound sources heard in auditoriums and concert halls are nondirectional (that is, they radiate essentially the same intensity of sound in all directions). Thus the level of *direct* sound depends only on the distance from the source, decreasing by 6 dB for each doubling of the distance. Other sources are quite directional (see Fig. 10.32, for example). Brass instruments toward the upper end of their playing range (where the size of the bell is comparable to the sound wavelength) are also quite directional.

Our auditory system has an uncanny ability to determine the direction of a sound source, even in the presence of many distracting sounds. The manner in which we localize the source of the direct sound was discussed in Section 5.5. For sounds of low frequency, localization depends mainly on the observation of a very slight difference in the time of arrival (or the phase of steady sounds) at our two ears. For sounds of high frequency (above about 1000 Hz), the difference in sound level at our two ears, due to the shadow cast by our head, provides the main clue.

The arrival of the reflected sound from many directions adds complications. When an orator articulates a phoneme or a musician attacks a note, our ears are provided with several reflections that closely follow the direct sound. The spectrum and time envelope of these reflected sounds will be more or less identical to those of the direct sound and if they arrive within about 50 to 80 ms of the direct sound, the ear does not hear them as separate sounds. Rather, they tend to reinforce the direct sound, a fact that is especially important to listeners located quite a distance from the source. For rapidly varying sound, such as speech, the limit is probably around 50 ms, but for more slowly varying music, the limit is more like 80 ms.

Most remarkably, however, the auditory processor continues to deduce the direction of the source from the first sound reaching the ears, which it interprets as following the direct path. This remarkable ability of our auditory system, called the *precedence effect*,* has considerable significance in our perception of stereophonic and surround sound. The source is perceived to be in the direction from which the first sound arrives provided that (1) successive sounds arrive within 35 ms, (2) the successive sounds have spectra and time envelopes reasonably similar to the first sound, and (3) the successive sounds are not too much louder than the first.

As a result of studying 76 of the world's leading concert and opera halls, Beranek (1996) concluded that a concert hall is considered "intimate" if the delay time between direct and first reflected sound is less than 20 ms. If the auditorium has the traditional rectangular shape, this first reflection for most listeners will come from the nearest side wall, although listeners located near the center may receive their first reflection from the ceiling. In some large concert halls, a portion of the audience will be too far removed from both ceiling and side walls to receive early reflections within the desirable time interval; in those cases, reflecting surfaces of some type are often suspended from the ceiling.

*Sometimes it is called the *Haas effect*, because of a paper by H. Haas in *Acustica* 1: 49 (1951), which describes it in some detail. Similar effects had apparently been noted by others, including Joseph Henry, as early as 1856.

More recent studies have shown, however, that early reflections from side walls are not equivalent to early reflections from the ceiling or from an overhead reflector. One study showed a high preference for concert halls with ceilings sufficiently high that the first lateral reflection reaches the listener before the first overhead reflection (West 1966). Others have found that if the total energy from lateral reflections is greater than the energy from overhead reflections, the hall takes on a desirable "spatial responsiveness" or "spatial impression" (Barron 1971; Jordan 1975).

23.4 ■ REVERBERANT SOUND

The characteristic of auditoriums familiar to most people is the reverberation time. Although the behavior of reverberant sound is too complicated to be described by a single number, the reverberation time at midfrequency (usually 500 to 2000 Hz) does give quite a fair indication of the "liveness" of the auditorium or concert hall.

Instead of the impulsive sound source used in Figs. 23.3 and 23.4, let us switch on a steady source, leave it on for a time interval T, and then switch it off. The growth and decay of the reverberant sound are shown in Fig. 23.5.

The rate of growth, the rate of decay, and the reverberant sound level can be deduced by considering the sound energy. The source supplies the energy, which is stored in the air space of the auditorium and eventually absorbed by the walls, the ceiling, the objects within the auditorium, and, under certain conditions, the air itself. The *reverberant level* is reached when the rate at which energy is supplied by the source (that is, the source power) is equal to the rate at which sound is absorbed.

Although the reverberant sound, like the early sound, reinforces the direct sound and adds to the overall loudness (which is an important consideration in a large auditorium), too greater a level of reverberant sound may result in a loss of clarity. In a good concert hall the direct sound should be substantially louder than the background sound at all locations. In large auditoriums, this may call for electronic reinforcement of the direct sound.

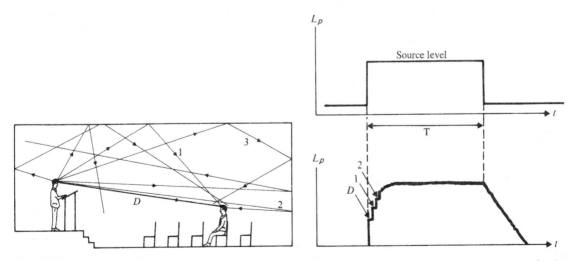

FIGURE 23.5 Growth and decay of reverberant sound in a room: D represents direct sound; 1, 2, 3, etc., are early reflections.

(In Chapter 24, electronic reinforcement of both direct and reverberant sound will be discussed.)

When the sound source is a full-size symphony orchestra, most listeners find the level of direct sound to be optimum when they are seated about 20 m (60 ft) from the source (Beranek 1962). This distance would place one at roughly the center of many of the world's best concert halls.

In principle, it is easy to determine the theoretical reverberation time of a room. The sound energy stored in the room depends on the power of the source and the volume of the room; the rate at which that energy is absorbed depends on the area of all surfaces and objects in the room and their absorption coefficients. In a bare room, where all surfaces absorb the same fraction of the sound that reaches them, the reverberation time is thus proportional to the ratio of volume to surface (Sabine 1922):

$$RT = K \frac{\text{volume}}{\text{area}}. \tag{23.1}$$

In general, large rooms have longer reverberation times than do small rooms.

If the sound energy were uniformly distributed throughout the room, its decay would follow a curve like that shown in Fig. 23.6(a), which is called an exponential curve. The logarithm of such a curve is a straight line, and so the decay of sound pressure level is a straight line, as shown in Fig. 23.6(b). It is customary to define the reverberation time (abbreviated RT or T_{60}), as the time required for the sound level to decrease by 60 dB. Often it is not convenient to measure the time for the entire 60-dB decay; assuming that the decay is a straight line, however, one can double the time required for a 30-dB decay.

Decay curves similar to those shown in Fig. 23.6(c) and (d) are often observed in auditoriums. In Fig. 23.6(c), there are two different reverberation times, one describing the initial and one the final portion of the decay. This may indicate an insufficient distribution of sound and can lead to a feeling of "dryness" in a hall even though the final reverberation time falls within acceptable limits. Spikes on the decay curve of the type shown in Fig. 23.6(d) result from the storage of sound energy in the form of room resonances.

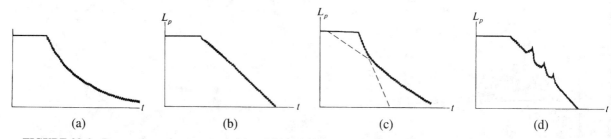

(a)   (b)   (c)   (d)

FIGURE 23.6 Decay of reverberant sound: (a) and (b) are decay curves of sound pressure and sound level in a somewhat idealized room with uniform energy distribution; (c) and (d) illustrate curves often observed for auditoriums; (c) shows different initial and final reverberation times; (d) shows peaks due to prominent standing waves or room resonances.

23.5 ■ CALCULATION OF THE REVERBERATION TIME

How long does it take to empty a bathtub of water? It depends, no doubt, upon the volume of the tub and the size of the drain. If V is the volume of water and A is the area of the drain, we might expect that the time T required to drain the tub would be approximately $T = kV/A$, where k is a constant. Now think about what happens when a room is filled with reverberant sound that is absorbed as the sound dies away (see Eq. (23.1)).

Let us consider a hypothetical room of volume V with hard walls that absorb very little sound but with a window of area A through which sound can escape to the outside. The reverberation time now depends on the ratio of the room volume to the area of the absorbing window. When V is in cubic meters (m^3) and A in square meters (m^2), the constant K in Eq. (23.1) equals 0.16 s/m, so the formula for reverberation becomes

$$RT = 0.161 \frac{V}{A} \qquad (V \text{ in m}^3; A \text{ in m}^2). \qquad (23.2)$$

If room dimensions are given in feet, the formula may be written

$$RT = 0.049 \frac{V}{A} \qquad (V \text{ in ft}^3; A \text{ in ft}^2). \qquad (23.3)$$

In a real room, we can compare the absorbing power of each surface to that of the window in our hypothetical room. The window is assumed to "absorb" all the sound that reaches it; thus we assign it an absorption coefficient $a = 1$. A surface that absorbs half of the sound power will have an absorption coefficient $a = 0.5$. Two square meters of this material have an absorption equivalent to 1 m^2 of open window. A surface having an area S and an absorption coefficient a thus has an absorption

TABLE 23.1 Absorption coefficients for various materials

| Material | Frequency (Hz) | | | | | |
	125	250	500	1000	2000	4000
Concrete block, unpainted	0.36	0.44	0.31	0.29	0.39	0.25
Concrete block, painted	0.10	0.05	0.06	0.07	0.09	0.08
Glass, window	0.35	0.25	0.18	0.12	0.07	0.04
Plaster on lath	0.14	0.10	0.06	0.05	0.04	0.03
Plywood paneling	0.28	0.22	0.17	0.09	0.10	0.11
Drapery, lightweight	0.03	0.04	0.11	0.17	0.24	0.35
Drapery, heavyweight	0.14	0.35	0.55	0.72	0.70	0.65
Terrazzo floor	0.01	0.01	0.02	0.02	0.02	0.02
Wood floor	0.15	0.11	0.10	0.07	0.06	0.07
Carpet, on concrete	0.02	0.06	0.14	0.37	0.60	0.65
Carpet, on pad	0.08	0.24	0.57	0.69	0.71	0.73
Acoustical tile, suspended	0.76	0.93	0.83	0.99	0.99	0.94
Acoustical tile, on concrete	0.14	0.20	0.76	0.79	0.58	0.37
Gypsum board, $\frac{1}{2}$ in.	0.29	0.10	0.05	0.04	0.07	0.09

$$A = Sa \qquad (23.4)$$

equivalent to an open window of area A.

The total absorption in the room is found by adding up the contributions from each surface exposed to the reverberant sound:

$$A = S_1 a_1 + S_2 a_2 + S_3 a_3 + \cdots. \qquad (23.5)$$

The absorption is sometimes expressed in sabins or metric sabins, units named in honor of Wallace Sabine, a pioneer in the study of room acoustics. One sabin is the absorption of one square foot of open window, and one metric sabin is the absorption of one square meter of open window. It is less confusing, however, to express absorption A in square meters (or square feet). Table 23.1 gives the absorption coefficients a for a number of materials at different frequencies.

EXAMPLE 23.1   Let us calculate the reverberation time for a room 20 m × 15 m and 8 m high. The walls are painted concrete block, the ceiling plaster on lath, the floor carpet on concrete. We neglect all furnishings and do the calculation at $f = 500$ Hz:

Walls:	$A_1 = (2 \times 15 \times 8 + 2 \times 20 \times 8)(0.06) = 34$;
Ceiling:	$A_2 = (15 \times 20)(0.06) = 18$;
Floor:	$A_3 = (15 \times 20)(0.14) = 42$;
	$A = A_1 + A_2 + A_3 = 94 \text{ m}^2$;
	$V = 20 \times 15 \times 8 = 2400 \text{ m}^3$;
	$RT = 0.161\, V/A = (0.161)(2400)/(94) = 4.1 \text{ s.}$

Neglecting the furnishings, as in Example 23.1, leads to an unrealistically long reverberation time. In a typical room, the furnishings will contribute a substantial fraction of the total absorption. People are also good absorbers of sound, each member of the audience contributing 0.2 to 0.6 m^2 (metric sabins) of absorption, depending on frequency. The values of sound absorption of various types of seats, occupied and unoccupied, are given in Table 23.2.

23.6 ■ AIR ABSORPTION

In a large auditorium, the air itself contributes a substantial amount to the absorption of sound at high frequencies. The absorption of air depends on the temperature and relative humidity, and an additional term, mV, proportional to the volume should be added to the absorption A. The constant m is given in the last two lines of Table 23.2. The reverberation time for a large auditorium is

$$RT = 0.161 \frac{V}{A + mV} \qquad \text{(large auditorium)}. \qquad (23.6)$$

TABLE 23.2   Sound absorption by people and seats, and air absorption

Material	125	250	500	Frequency (Hz) 1000	2000	4000	8000	Unit*
Wood or metal seats, unoccupied	0.014	0.018	0.020	0.036	0.035	0.028		m^2
Upholstered seats, unoccupied	0.13	0.26	0.39	0.46	0.43	0.41		m^2
Audience in upholstered seats	0.27	0.40	0.56	0.65	0.64	0.56		m^2
Air absorption (per m^3):								
20°C, 30% RH	—	—	—	—	0.012	0.038	0.136	m^{-1}
20°C, 50% RH	—	—	—	—	0.010	0.024	0.086	m^{-1}

Note: Values of sound absorption are given in m^2; to convert to ft^2, multiply by 10.8. Values of air absorption are given in m^{-1}; to convert to ft^{-1}, divide by 3.3.

EXAMPLE 23.2   Let us add 200 upholstered seats to the room used in Example 23.1. We assume half of them occupied and calculate the reverberation time at 2000 Hz. We also consider the absorption of the air at 20°C, 30% relative humidity.

Walls: $A_1 = (2 \times 15 \times 8 + 2 \times 20 \times 8)(0.09) = 50;$

Ceiling: $A_2 = (15 \times 20)(0.04) = 12;$

Floor: $A_3 = (15 \times 20)(0.60) = 180;$

Empty seats: $A_4 = (100)(0.43) = 43;$

Occupied seats: $A_5 = (100)(0.64) = 64;$

Air absorption: $mV = (0.012)(2400) = 29;$

$$A = 50 + 12 + 180 + 43 + 64 = 349 \ m^2;$$

$$V = 20 \times 15 \times 8 = 2400 \ m^3;$$

$$RT = 0.161 \frac{2400}{349 + 29} = 1.02 \ s.$$

Note that in Example 23.2, nearly one-third of the total absorption results from the seats and the audience. This is not unusual behavior at this frequency. The formulas we have used assume that the reverberant sound is distributed uniformly throughout the room, which is not always the case. Near large absorbers, the sound energy may be less than at other locations within the room; more complicated formulas take this and other factors into account. Nevertheless, the reverberation times calculated from the formulas given in this chapter will be reasonably close to the measured values.

23.7 ■ CRITERIA FOR GOOD ACOUSTICS

From our discussion of reverberation, the interdependence of reverberant sound level and reverberation time should be clear. Increasing the absorption decreases both the reverberant *level* and the reverberation *time*. The optimum reverberation time is thus a compromise between clarity (requiring a short reverberation time), loudness (requiring a high reverberant level), and liveness (requiring a long reverberation time). The optimum reverberation time will depend on the size of the auditorium and the use for which it is designed. An auditorium intended primarily for speech should have a shorter reverberation time than one intended for music. Figure 23.7 indicates reverberation times considered desirable for auditoriums of various sizes and functions.

The acoustic requirements for concert halls, opera houses, lecture halls, theaters, and churches are quite different, but there are a number of common requirements that must be met:

1. *Adequate loudness.* Everyone must be able to hear the speaker or performer. The room should not be too large or have excessive absorption.
2. *Uniformity.* Listeners in all parts of the room should hear as nearly the same sound as possible. There must be a sufficient number of sound diffusing surfaces to avoid "dead" spots. All sections of an orchestra should blend together in a balanced way.
3. *Clarity.* There must be sufficient absorbing surfaces that the reverberant sound does not mask following sounds.
4. *Reverberance, or liveness.* The listener should feel bathed in sound from all sides, but at the same time be able to localize the sound source. Clarity and liveness may be partly contradictory.
5. *Freedom from echoes.* Reflected sound should arrive early enough to reinforce the direct sound but not be perceived as a separate echo.

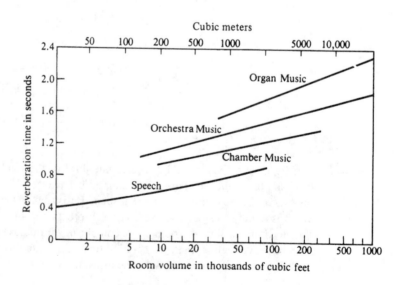

FIGURE 23.7
Desirable reverberation times for auditoriums of various sizes and for various functions.

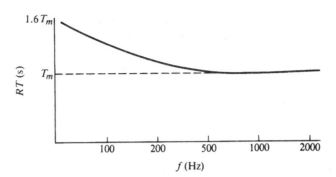

FIGURE 23.8
Variation of
reverberation time
with frequency in a
good concert hall.

6. *Low level of background noise.* The noise from heating and ventilating systems and from external sources should be kept very low.

A feeling of liveness or reverberance is especially important at low frequency to give support to bass notes. Fortunately, many building materials have lower absorption coefficients at low frequencies. Figure 23.8 shows how the reverberation time may vary with frequency in a good concert hall.

Recent work has stressed the importance of having sufficient reflected sound arriving from the sides. Such lateral reflections arriving with time delays of from 25 to 80 ms add to the feeling of spaciousness, whereas overhead reflections during the same period add mainly to the early sound (Barron 1971; Reichardt et al. 1975).

23.8 ■ CONCERT HALLS

Contrary to a common opinion, the scientific principles involved in acoustical design of concert halls are well understood. In spite of that, some concert halls, both old and new, suffer from acoustical problems. Many of the deficiencies result from the most common problem of all: poor communication among the architects, the acoustical designers, the building contractors, and the musicians who will use the hall.

Music and acoustics grew up quite independently of each other, and thus they developed different vocabularies to describe their important concepts. In his book on concert and opera halls, Beranek (1996) attempts to bridge the language gap by defining a number of terms in a way that they can be understood by both musicians and acousticians. Some of them are as follows:

1. *Intimacy, or presence.* A hall is said to have *acoustical intimacy* if music played in it gives the impression of being played in a small hall.

2. *Reverberation, or liveness. Reverberation* refers to sound that persists in a room after a tone is suddenly stopped. *Liveness* is related primarily to the reverberation times at frequencies above 350 Hz. (A hall can sound live and still be deficient in bass.)

3. *Spaciousness: apparent width (ASW).* A concert hall has one of the attributes of spaciousness if the music appears to emanate from a source wider than the visual width of the source.

4. *Spaciousness: listener envelopment (LEV).* Listener envelopment is highest when the reverberant sound appears to come from all directions.

5. *Clarity.* Clarity is the degree to which the discrete sounds in a musical performance stand apart from one another.

6. *Warmth.* Warmth is defined as liveness of the bass or fullness of bass tones (75 to 350 Hz) relative to that of the midfrequency tones (350 to 1400 Hz). The term *dark* is also applied to a hall with a strong bass.

7. *Loudness.* Loudness is largely related to the size and shape of the hall and to its reverberation.

8. *Balance.* Some of the ingredients that combine to give good balance are acoustical; some are musical. Good balance is promoted by good stage design.

9. *Blend.* Mixing of the sounds from the various instruments of the orchestra depends partly on sound-reflecting surfaces close to the stage.

10. *Ensemble.* The ability of the performers to play well together also depends partly on the sound-reflecting surfaces close to the stage.

Important Criteria for Concert Halls

Table 23.3 gives acoustical data for several concert halls. The time delay t_1 is given for selected locations near the center of the main floor and the middle of the balcony. Reverberation times RT in seconds are given at frequencies of 125, 500, and 2000 Hz. The two halls in London are nearly at the extremes of acceptability. The Royal Festival Hall in London is considered "dry," excellent for chamber music and music from the Baroque period. The Royal Albert Hall is not very popular with musicians; the type of orchestral music that can be performed there is limited by its long reverberation, which also adds excitement to certain types of music, such as Tchaikovsky's *1812 Overture.* The Royal Festival Hall has been improved by the addition of "assisted resonance," which will be discussed in Chapter 24.

TABLE 23.3   Acoustical characteristics of concert halls

	Year built	Volume (m^3)	Floor area (m^2)	Number of seats	t_1 (ms) Floor	t_1 (ms) Balc.	RT (s) 125	RT (s) 500	RT (s) 2000 Hz
Symphony Hall, Boston	1900	18,740	1550	2630	15	7	2.2	1.8	1.7
Orchestra Hall, Chicago	1905	15,170	1855	2580	40	24	—	1.3	—
Severence Hall, Cleveland	1930	15,700	1395	1890	20	13	—	1.7	1.6
Carnegie Hall, New York	1891	24,250	1985	2760	23	16	1.8	1.8	1.6
Opera House, San Francisco	1932	21,800	2165	3250	51	30	—	1.7	—
Arie Crown Theatre, Chicago	1961	36,500	3265	5080	36	14	2.2	1.7	1.4
Royal Festival Hall, London	1951	22,000	2145	3000	34	14	1.4	1.5	1.4
Royal Albert Hall, London	1871	86,600	3715	6080	65	70	3.4	2.6	2.2
Concertgebouw, Amsterdam	1887	18,700	1285	2200	21	9	2.2	2.1	1.8
Kennedy Center, Washington	1971	19,800	1220	2760	—	—	2.5	2.2	1.9

Source: After Beranek (1962) with additions.

In addition to the criteria for good acoustics in Section 23.7 (adequate loudness, uniformity, clarity, fullness, freedom from echoes, and low level of background noise) and reverberation, there are some additional criteria that are important in concert halls.

Spatial impression A sufficient portion of the early sound should arrive from the side (from side wall reflections). In recent work, two different aspects of spatial impression have been identified. One is called the Auditory Source Width (ASW) which refers to the impression that the music appears to emanate from a source wider than the visual width of the source. A second aspect, called Listener Envelopment (LEV), refers to the impression that the reverberant sound appears to come from all directions.

Early decay time The initial rate of decay of reverberant sound is more important than the total reverberation time. A rapid initial decay is interpreted by the ear as meaning that the reverberation time is short.

Things to be avoided in auditorium design include the following:

1. *Echoes.* An echo is a strong reflected sound that is sufficiently delayed (over 50 ms, usually) from the direct sound that it can be heard as a separate entity rather than as a continuation of the original sound. When echoes are heard in an auditorium, a likely culprit is the rear wall.
2. *Flutter echoes.* Flutter echoes are a series of echoes that occur in rapid succession; they usually result from reflections between two parallel surfaces that are highly reflective.
3. *Sound focusing.* Focusing of sound can be caused by reflection from large concave surfaces (see Fig. 23.2). Certain sounds will be heard too loudly near the focus of a curved surface.
4. *Sound shadows.* Under balconies at the rear of the auditorium, there may be insufficient early sound, because most of the reflections from the side walls and ceiling do not reach this area even though they are in a direct line of sight to the performer and therefore receive the direct sound.
5. *Background noise.* This will be discussed in Section 23.9.

A study of 22 European concert halls by Schroeder, Gottlob, and Siebrasse (1974) showed the following:

1. The greater the early decay time, the greater the preference for the hall, up to a reverberation time (determined from the early decay time) of 2 s. Above 2 s, the preference for the hall decreased with increasing reverberation time.
2. Narrow halls were generally preferred to wide ones.
3. Considerable preference was shown for halls having a high *binaural dissimilarity*; that is, listeners preferred dissimilarity of sound at their two ears, such as might result from a high degree of asymmetric sound diffusion.
4. Halls with less *definition* were preferred. Definition represents the ratio of energy in the first 50 ms to the total energy.

Size and Shape of Concert Halls

Among the world's concert halls generally considered to have excellent acoustics are several halls that are essentially rectangular in shape, such as the Musikvereinsaal in Vienna, Symphony Hall in Boston, and the Conertgebouw in Amsterdam. One reason for their favorable rating is that the strong, early reflections from the sidewalls, which reach every listener, lead to a sense of spaciousness. This has given rise to a preference for the rectangular "shoebox" design. Many of the classic concert halls that are considered to have good acoustics have seating capacities of less than 2000 seats, which also avoids many acoustical problems, such as lack of loudness and intimacy and low spatial impression.

For financial reasons, modern concert halls generally have seating capacities somewhat greater than the classic halls. If a rectangular hall is made too large, listeners in the rear of the hall do not have a good view of the performers and experience a rather poor direct-to-reverberant sound ratio. When visual considerations suggest a wide hall and acoustical considerations indicate a narrow hall, compromises must be made. Lateral reflecting surfaces have been provided by several innovative schemes. Sometimes the hall is made asymmetric in order to give such surfaces, as in Segerstrom Hall in Costa Mesa, California. Sometimes the audience is seated on all sides of the stage, as in the Berlin Philharmonic.

Many modern concert halls have been provided with features that can be varied to "tune" the hall. These have included movable reflecting panels and absorbing surfaces that can be raised or lowered (see also Section 23.11). The success of this arrangement depends upon obtaining a consensus from musicians and listeners, not always an easy thing to do.

23.9 ■ BACKGROUND NOISE

Nothing is more disappointing than to have a performance of music spoiled by background noise from an air-conditioning system or from nearby traffic. Although this does not often happen in major concert halls, it is all too common in churches, school auditoriums, and smaller concert halls where a qualified acoustician has not been consulted.

Background noise can be of internal or external origin. A common source of internal noise is an improperly designed ventilating system. Low-frequency noise from the ventilating equipment itself may be carried through ducts into the auditorium, and a broad band of noise is often generated by air flow in the ducts and grills. In addition, low-frequency vibrations from machinery may be transmitted by the building structure into an auditorium. Other noise originates from noisy doors, inadequate isolation from corridor noise, and so on.

External noise is somewhat more difficult to eliminate. Depending on the hall's location, external noise may come from traffic on a nearby street, aircraft overhead, etc. Possible solutions include placing the auditorium at the center of a building complex, constructing heavy walls, making certain no windows face the offending street, and so forth.

A family of noise criteria curves is shown in Fig. 23.9. The NC curves specify the maximum noise level permissible in each octave band for a particular NC rating. For example, if architectural specifications call for NC-20 or below, the sound pressures in all octave bands must be below those shown in Fig. 23.9. A concert hall should meet at least the NC-20 curve and preferably the NC-15 standard.

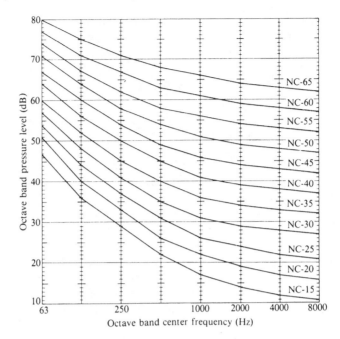

FIGURE 23.9
Noise criteria (NC) curves. (From *Noise and Vibration Control*, revised. © 1988 Leo Beranek. Used with author's permission.)

23.10 ■ AVERY FISHER HALL: A CASE STUDY

One of the most famous concert halls in the world is Avery Fisher Hall, originally called Philharmonic Hall, in New York (see Fig. 23.10); its history illustrates the tribulations of acoustic design. Since its opening in 1962, it has undergone several renovations, ending with a complete rebuilding in 1976. An apparent success at last, its metamorphosis has taught us a great deal about auditorium acoustics.

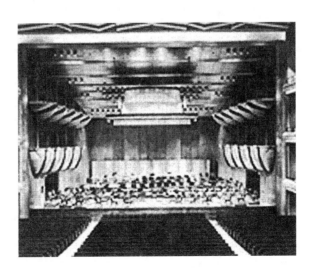

FIGURE 23.10
Avery Fisher Hall at Lincoln Center in New York.

Philharmonic Hall was designed by architect Max Abramovitz after an extensive research study by acoustician Leo Beranek of the world's concert halls. Acoustical expertise was furnished by members of Beranek's firm (Bolt Beranek and Newman) as well as by Hope Bagenal of England. The hall was intended to rank with the best in the world.

When the hall opened in 1962, most musicians and listeners were disappointed. The major defects were (1) weak bass; (2) a lack of liveness; (3) echoes from the rear wall; (4) inadequate sound diffusion; (5) poor hearing conditions for the musicians on stage (see Schroeder et al. 1966). Scientists from the Bell Telephone Laboratories were called on to evaluate the acoustics, and a distinguished committee of acoustical consultants headed by V. O. Knudsen was asked to recommend improvements. Changes made during the summers of 1963, 1964, 1965, 1969, and 1972, costing more than $2 million, improved the hall, but criticisms continued. In 1975, the hall was completely redesigned by architect Philip Johnson and acoustician Cyril Harris. The reconstruction, following somewhat the proven lines of Boston's Symphony Hall and the more recent Kennedy Center (Washington) and Orchestra Hall (Minneapolis), cost more than $5 million.

What went wrong with the original Philharmonic Hall, the product of so much careful planning? Beranek (1975) maintains that the final plans were expanded and modified without his consent or that of the orchestra. To enlarge the seating capacity from 2400 to 2600, the side walls were spread out into a more "modern" fan shape. The adjustable ceiling was eliminated for reasons of economy, as were some irregularities on the side walls, which would have acted as diffusers of sound.

The designers of the hall placed a great emphasis on acoustical intimacy, and suspended 136 panels ("clouds") from the ceiling to provide some of the early reflections that would come from the side walls in a smaller hall. It is now known that early sounds arriving from above and from the sides are not nearly so equivalent as was thought. The designers apparently counted rather heavily on last-minute adjustments during "tuning week" to optimize several parameters.

Several of the problems in the hall resulted from insufficient diffusion of sound. This resulted in a decay curve with two slopes. Although the total reverberation time was about right, the initial slope was too steep (early decay time too short), thus giving the hall the impression of dryness, or lack of reverberance. The story behind the redesign of Avery Fisher Hall (renamed in honor of its chief benefactor) is told in a dramatic account by Bliven (1976).

23.11 ■ VARIABLE ACOUSTICS

Quite a number of attempts have been made to design auditoriums with variable acoustics. They have met with varying success. Usually it is the reverberation time that is varied by changing the amount of absorption in the room. One technique is the use of wall panels that can be rotated, one side with a high absorption coefficient and one with a low absorption coefficient. Another technique is the use of extended or retracted absorbing panels or blankets to change the absorption. In a large hall, the amount of absorption that must be added or subtracted in order to produce a noticeable change is very large.

Another technique for varying the acoustics, which will be discussed in Chapter 24, is to enhance the reverberation and/or the direct sound by electronic means. There are distinct limits to the range of variability, however, and the baseline from which these variations take place should be one of good initial acoustical design.

23.12 ■ CHURCHES

Churches and synagogues are not primarily concert halls, and yet they share the same requirements for good acoustics discussed in Section 23.7. Figure 23.7 illustrated how different the optimum reverberation time is for speech and organ music. Obviously some compromise is called for in church design.

Most of the old cathedrals in Europe have long reverberation times, and the spoken word was generally not as important as it is in most contemporary worship. (Was that cause or effect?) In many old churches (especially in northern Europe), the pulpit is strategically placed in the center of the congregation and has a large canopy to reflect the preacher's voice. One reads stories about preachers with "booming" voices who learned to speak so that they could be understood in a rather unfriendly acoustical environment.

It is important that the background-noise level in churches be kept as low as in concert halls. Heating, ventilating, and air-conditioning (HVAC) systems must be designed with great care. Churches on a noisy street must include adequate insulation from external noise. Architects who design small churches may not always understand the importance of including a qualified acoustical consultant in the planning.

Electronic reinforcement of sound, which will be discussed in Chapter 24, is often necessary in larger churches, although in the authors' opinion it is too frequently overdone. Nothing is more distracting than a poorly designed sound-reinforcement system, and we have heard many! Smaller churches should be able to function without electronic reinforcement (except, perhaps, sound reinforcement for the hard of hearing).

23.13 ■ CLASSROOMS

The need for good acoustics in classrooms is simple: students must be able to understand the teacher and each other. Providing adequate speech intelligibility is a matter of controlling three types of classroom noise: *reverberation; heating ventilation and air-conditioning noise; and noise from outside the classroom.*

Studies have shown that reverberation times in a quiet classroom should be $\frac{1}{2}$ s or less in order to avoid speech interference. In classrooms, as in all spaces, reverberation increases as the room volume increases and is reduced if more sound-absorbent surfaces are introduced. Special consideration should be given to the speech intelligibility range, 500 to 4000 Hz.

Deciding on the maximum acceptable noise level of classroom HVAC systems is critical. According to the American National Standards Institute (ANSI), the range of values for lecture halls and classrooms should be NC-25 to NC-30. A teacher using a normal voice will produce a sound level of about 46 dB at the ears of a student 30 ft away. NC-30 corresponds roughly to a background noise level of about 36 dB, which should produce the 10-dB difference in speech and noise level required for speech intelligibility. (Note that the ANSI standard is based on speech intelligibility, whereas some other higher suggested maximum levels are based on "annoyance." However, common sense says that it is not enough that students are not annoyed by HVAC noise; they must be able to understand the teacher as well.) Control of HVAC noise is discussed in Chapter 32.

Setting criteria for exterior noise is somewhat more complex. Traffic noise tends to be constant, and the criteria for HVAC noise are appropriate. Requirements for intermittent noise, such as from aircraft flyovers, can be slightly less stringent. Speech interference levels (SIL) of SIL-50 are generally acceptable; these are roughly equivalent to NC-50. Lower levels are clearly desirable if learning is to be maximized.

It is rather ironic that only recently has much attention been paid to classroom acoustics. The Department of Education, the Acoustical Society of America, and other professional organizations have organized several conferences on this important topic, and we hope that increased attention will be paid to it in the near future. We urge readers of this book to become activists in their own communities. Students deserve the opportunity to learn in quiet classrooms!

23.14 ■ SUMMARY

In a free field, sound level decreases by 6 dB for each doubling of the distance. Inside rooms, however, in which sound reflects from many surfaces, the sound level is greater than the free-field level. In analyzing the acoustics of an auditorium, attention should be given to the direct, early, and reverberant sounds and the proportion of the early sound arriving from the sides. The ability of our ears to locate a source by analyzing the direct sound and virtually ignoring the early sound is called the precedence effect. Reverberant sound builds up and decays at a rate characterized by the reverberation time RT, which depends on the volume of the room and the absorption of all the surfaces in it.

Criteria for good concert hall acoustics (e.g., intimacy, liveness, warmth, clarity, etc.) can be related to measurable parameters. Optimum values for some parameters, such as reverberation time, depend on the size and use of the auditorium. Background noise, due to internal and external sources, should be kept at a minimum.

Churches share the same requirements for good acoustics. Reverberation times must accommodate church music as well as the spoken word. HVAC systems must be designed with great care. Students in classrooms must be able to understand the teacher at all times, and this requires the control of reverberation, HVAC noise, and external noise.

REFERENCES AND SUGGESTED READINGS

Barron, M. (1971). "The Subjective Effects of First Reflections in Concert Halls—The Need for Lateral Reflections," *J. Sound Vibr.* **15**: 475.

Beranek, L. L. (1962). *Music, Acoustics and Architecture.* New York: Wiley.

Beranek, L. L. (1971). *Noise and Vibration Control.* New York: McGraw-Hill.

Beranek, L. L. (1975). "Changing Role of the Expert," *J. Acoust. Soc. Am.* **58**: 547.

Beranek, L. L. (1996). *Concert and Opera Halls: How They Sound.* Melville, N.Y.: Acoustical Society of America.

Bliven, B., Jr. (1976). "Annals of Architecture," *New Yorker* (Nov. 8): 51.

Cremer, L., H. A. Müller, and T. J. Schultz (1982). *Principles and Applications of Room Acoustics,* vols 1 and 2. London: Applied Science.

Doelle, L. L. (1972). *Environmental Acoustics.* New York: McGraw-Hill.

Jesse, K. E. (1980). "A Classroom Acoustical Absorption Experiment," *The Physics Teacher* **18**: 41.

Jordan, V. L. (1975). "Auditoria Acoustics: Developments in Recent Years," *Applied Acoustics* **8**: 217.

Knudsen, V. O. (1963). "Architectural Acoustics," *Sci. Am.* **209**(5): 78.

Knudsen, V. O., and C. M. Harris (1950). *Acoustical Designing in Architecture*. New York: J. Wiley. Reprinted by The Acoustical Society of America, 1978.

Marshall, A. H., D. Gottlob, and H. Altruz (1978). "Architectural Conditions Preferred for Ensemble," *J. Acoust. Soc. Am.* **64**: 1437.

Reichardt, W., O. Abdel Alim, and W. Schmidt (1975). "Definition and Basis of Making an Objective Evaluation Between Useful and Useless Clarity Defining Musical Performances," *Acustica* **32**: 126.

Sabine, W. C. (1922). *Collected Papers on Acoustics.* Reprinted by Dover, New York, 1964.

Schroeder, M. R. (1979). "Binaural Dissimilarity and Optimum Ceilings for Concert Halls: More Lateral Sound Diffusion," *J. Acoust. Soc. Am.* **65**: 958.

Schroeder, M. R., B. S. Atal, G. M. Sessler, and J. E. West (1966). "Acoustical Measurements in Philharmonic Hall," *J. Acoust. Soc. Am.* **40**: 434.

Schroeder, M. R., D. Gottlob, and K. F. Siebrasse (1974). "Comparative Study of European Concert Halls: Correlation of Subjective Preferences with Geometric and Acoustic Parameters," *J. Acoust. Soc. Am.* **56**: 1195.

West, J. E. (1966). "Possible Subjective Significance of the Ratio of Height to Width of Concert Halls," *J. Acoust. Soc. Am.* **40**: 1245.

GLOSSARY

anechoic Echo-free; a term applied to a specially designed room with highly absorbing walls.

direct sound Sound that reaches the listener without being reflected.

early sound Sound that reaches the listener within a short time (about 50 ms) after the direct sound.

free field A reflection-free environment, such as exists outdoors or in an anechoic room, in which sound pressure varies inversely with distance ($p \propto 1/r$).

HVAC Heating, ventilating, and air-conditioning system.

noise criteria (NC) curves A family of curves defining levels of room noise in several octave bands.

precedence effect The ability of the ear to determine the direction of a sound source from the direct sound without being confused by the early sound that follows.

reverberant sound Sound that builds up and decays gradually and can be "stored" in a room for an appreciable time.

reverberation time The time required for the stored or reverberant sound to decrease by 60 dB.

sabin, metric sabin Units for measuring absorption of sound; the sabin is equivalent to one square foot of open window, the metric sabin to one square meter.

REVIEW QUESTIONS

1. How does the sound pressure change with distance from the source in a free field?

2. In what two environments does a free field occur?

3. Express the speed of sound in m/s and in ft/s.

4. Why does the direct sound arrive first at the listener's ear?

5. What is the precedence effect?

6. A hall is considered intimate if the first reflection arrives within how long after the direct sound?

7. Why do large rooms generally have longer reverberation times than small rooms?

8. The reverberation time is the time for the reverberant sound level to decrease by how much?

9. Why is a longer reverberation time desired for organ music than for speech?

10. NC curves are used to express what?

11. Why is sound focusing by curved surfaces usually undesirable in a concert hall?

12. How low should the background noise level be in a concert hall?

13. What is meant by a "shoebox" design?

14. Give an example of a successful concert hall having a shoebox design.

15. What are three sources of noise in a typical classroom?

16. What is the maximum noise level that will allow speech intelligibility in a typical classroom?

17. What is a desirable reverberation time in a classroom?

QUESTIONS FOR THOUGHT AND DISCUSSION

1. Why does the use of cushioned seats help to make the reverberation time of an auditorium independent of audience size?

2. Which is easier to correct, a reverberation time that is too long or one that is too short?

3. What are desirable reverberation times for speech and for orchestral music in an auditorium with $V = 1000 \text{ m}^3$?

4. An auditorium is thought to have excessive reverberation, especially at low frequency. It is proposed that the ceiling be covered with acoustic tile to reduce this. What do you think of this solution? Is there a better one?

EXERCISES

1. Compare the absorption of 100 m^2 of plastered wall with that of 100 m^2 of carpeted floor at

   (a) 125 Hz;

   (b) 2000 Hz.

2. An auditorium has dimensions 40 m × 20 m and a ceiling height of 15 m. The front and back walls are covered with plywood paneling; the side walls and ceiling are plaster. The floor is wood. There are 1100 wooden seats. Estimate the reverberation time (500 Hz) when

   (a) The hall is empty;

   (b) Half the seats are filled;

   (c) All the seats are occupied.

3. Estimate the time delay t_1 of the first reflected sound for a person seated near the center of the auditorium described in Exercise 2. Does the first reflection arrive from the side or from overhead?

4. If the ceiling in this auditorium were covered with acoustical tile, by how much would the reverberation time be decreased?

5. Specify reasonable values for the reverberation times at 100, 200, 500, and 1000 Hz for a 2000-m^3 concert hall to be used primarily for orchestral music.

6. If two hard parallel walls are spaced 30 m apart, calculate the repetition rate of the flutter echo that might result. What efforts might be made to prevent its occurrence?

7. Repeat a historic experiment done by Joseph Henry more than 130 years ago. Clap your hands periodically as you move away from a large, flat wall. Determine how far away you have to be in order to distinguish the echo from the original sound. Divide twice this distance by the speed of sound to obtain the "limit of perceptibility," as Henry called it [see *J. Acoust. Soc. Am.* **61** 250 (1977)]. Compare your result to that given in Section 23.3.

8. Find the reverberation time at 8000 Hz for a very live room having a volume of 1000 m^3 when the temperature is 20°C and the relative humidity is 30%. Assume that absorption by the walls is negligibly small. Would your answer be different if $V = 100 \text{ m}^3$ instead?

9. Show that in a free field, a graph of L_p versus $\log r$ is a straight line having a slope of −20 (see Fig. 23.1).

EXPERIMENTS FOR HOME, LABORATORY, AND CLASSROOM DEMONSTRATION

Home and Classroom Demonstration Experiments

1. *Resolving time of echoes*   Clap your hands as you move away from a large, flat wall on a building. Note the minimum distance away from the wall at which you can distinguish the echo from the original sound. Find the echo-resolving time of your ears by dividing the distance the sound traveled by the speed of sound (see *The Physics Teacher* **16**: 600 (1978)).

2. *Sound decay curves*   Using a digital oscilloscope or photographing the trace of a regular oscilloscope (as in Fig. 23.4), record sound decay curves in various rooms and try to determine the reverberation time of the room.

3. *Change in sound level with distance*   Measure the sound pressure level as a function of distance from a noise source and make a graph similar to Fig. 23.1(b) (or Fig. 24.1). Determine the extent of the free field.

4. *Effect of echoes*   Play Demonstration 35 on the *Auditory Demonstrations* CD. In an anechoic room, then in a conference room, and finally in a very reverberant space, you hear a hammer striking a brick followed by an old Scottish prayer. Playing these sounds backward focuses attention on the echoes that occur.

5. *Effect of concert halls*   Play the several orchestral selections as recorded in an anechoic room and then add the simulated reverberation equivalent to three classic concert halls (Denon CD 8723).

6. *Effect of concert halls*   Play a reverberation-free Mozart recording, adding simulated reverberation equivalent to 11 European concert halls (Demonstration record accompanying *Music Perception in Concert Halls* by M. R. Schroeder, Publication 26, Royal Swedish Academy of Music, Stockholm, 1979.

7. *Maximum and minimum sound levels in a large room*   Carry a sound-level meter around a large room and determine the maximum and minimum sound levels in the sound field of a broadband noise source.

8. *Standing waves in a large room*   Carry a sound-level meter around a large room and determine the maximum and minimum sound levels in the sound field of a 1000-Hz tone. What is the difference in maximum and minimum levels? How far apart are the maxima? Repeat with a 200-Hz tone.

9. *Noise level*   Determine the noise level in a room using both A-weighting and C-weighting on your sound-level meter.

10. *Noise criteria rating*   Measure the background noise level in octave bands and make a graph to determine the noise level (see Fig. 23.9). What is the lowest NC curve that all your readings will fit under?

Laboratory Experiments

Reverberation time (Experiment 34 in *Acoustics Laboratory Experiments*)

Reflection and absorption of sound (Experiment 36 in *Acoustics Laboratory Experiments*)

24

Electronic Reinforcement of Sound

A small- to medium-sized room with good acoustical design should not require electronic reinforcement for either speech or music, and an unnecessary sound system is an annoying distraction. Nevertheless, in many large auditoriums, sound systems are necessary to obtain adequate loudness and good distribution of sound. This is especially true in athletic arenas or other enclosures with high levels of background noise. The sound system should be integrated into both the visual and acoustical design of the room. A sound reinforcement system is optimum when most listeners are hardly aware of its presence.

In this chapter you should learn:

- How sound systems can reinforce the direct, early, or reverberant sound fields;
- About the design of sound systems;
- About loudspeaker directivity;
- About acoustic feedback and equalization;
- About outdoor sound systems.

Electronic reinforcement of either the direct, early, or reverberant sound (see Section 23.2) is possible, although it is generally the direct sound that most needs to be reinforced. Reinforcement of the reverberant sound in dry auditoriums, however, can increase the reverberation time and create a feeling of liveness. If excessive reverberation or background noise interferes with clarity of speech, selective reinforcement of the direct sound over a selected range of frequency can improve the situation. It should be remembered, however, that electronic reinforcement of the direct sound may increase the reverberant level as well.

Sections 24.1–24.3, along with the example on p. 554, discuss sound sources and sound fields in a more quantitative way than is found throughout the rest of the chapter. This is necessary for a full understanding of the subject. However, it is possible to skim these sections and still gain some appreciation for the electronic reinforcement of sound.

24.1 ■ SOUND SOURCES IN A ROOM

For the purposes of this discussion, we characterize sound sources by their *power* and *directivity*. Both of these parameters will vary with frequency, of course. A source tends to be more directional at high frequencies where the wavelength of sound is comparable to the dimensions of the source.

The power of a source is expressed in watts. The sound power level L_w compares the power W of a source to the reference power $W_0 = 10^{-12}$ W (see Eq. 6.1). The average source power of a person speaking at a conversational level is about 10^{-5} W ($L_w = 70$ dB). (It has been said that it would take 50 speakers 5 years to generate enough energy to boil a cup of tea.)

The directivity factor Q is defined as the ratio of the sound intensity at a distance r in front of a source to the sound intensity averaged over all directions. A source that radiates equally in all directions (a spherical source) has a directivity factor $Q = 1$. A hemispherical source has $Q = 2$; a simple source in a corner (which radiates into one-quarter of a sphere) has $Q = 4$; a source at a three-surface corner (in the corner of a room at the floor or ceiling, for example) has $Q = 8$.

The sound pressure level L_p depends on the power and the directivity of the source, its distance, and the strength of the reflected sound. In a free field (away from reflecting surfaces), the sound pressure level at a distance r meters from the source is

$$L_p = L_w + 10 \log \frac{Q}{4\pi r^2}. \tag{24.1}$$

24.2 ■ SOUND FIELDS

We speak of the distribution of sound in space as a sound field. The character of the field due to a sound source varies with the distance from the source and the acoustic environment. Figure 24.1 illustrates how sound level varies with distance from the source in a room. When the distance is small compared to the dimensions of the source, the sound level varies with location, because some parts of the source may radiate more strongly than others. This part of the sound field is called the *near field*, and it is crosshatched in Fig. 24.1.

Farther away from the source, the sound pressure varies as $1/r$; this is called the *free field*. In the free field, the sound level decreases by 6 dB each time the distance r is doubled. As the distance r increases, the contribution from reflected sound takes on increasing importance, and we enter the *reverberant field*. The sound level in a large room eventually reaches the reverberant level at which it no longer decreases with increasing distance.

The reverberant level depends on the absorption of various surfaces in the room as well as on the power of the source. If the total absorption A is given in square meters (see Section 23.5), the reverberant sound level is $L_p(\text{reverb.}) = L_w + 10 \log 4/A$. The sound

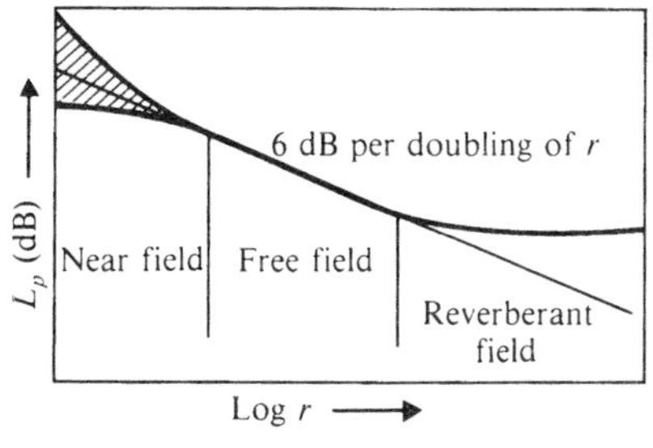

FIGURE 24.1
Variation of sound pressure level L_p with distance from the source r in a typical large room. In the free-field region, intensity varies as $1/r^2$, so L_p decreases by 6 dB each time r is doubled. In the near field, L_p depends on the geometry of the source. (Compare Fig. 23.1.)

level due to both direct and reverberant sound is

$$L_p = L_w + 10 \log \left(\frac{Q}{4\pi r^2} + \frac{4}{A} \right). \tag{24.2}$$

Let us consider a hypothetical room.

EXAMPLE 24.1   Suppose an auditorium has dimensions $20 \times 30 \times 10$ m and an average absorption coefficient $\bar{a} = 0.15$. A speaker standing in the corner radiates 10^{-5} W of acoustical power. Thus, $A = 0.15 \times 2(20 \times 30 + 20 \times 10 + 30 \times 10) = 330$ m^2. For a corner source, $Q = 4$. At a distance of 5 m from the source,

$$L_p = 10 \log 10^7 + 10 \log \left(\frac{4}{4\pi(25)} + \frac{4}{330} \right) = 70 - 16 = 54 \text{ dB}.$$

The reverberant level in this rather live room will be

$$L_p = 10 \log 10^7 + 10 \log 4/330 = 70 - 18 = 52 \text{ dB}.$$

Provided the room noise is not too great, it should be possible to hear the speaker without sound reinforcement. On the other hand, a listener seated 30 m away, who receives only 35 dB of direct sound, 17 dB below the reverberant level, will have trouble understanding the speaker. Reinforcement of the direct sound would improve the quality of sound this listener hears.

Studies made in the Netherlands on the intelligibility of speech have led to the following formulas for percentage articulation loss for consonants (percentage of sounds identified incorrectly) in a reverberant room (Peutz 1971):

$$\% \text{ AL} = \frac{200r^2(\text{RT})^2}{V} + k, \tag{24.3}$$

where RT = the reverberation time (s), r = the distance from the source (m), V = the volume (m^3), and k = the constant for each listener that indicates listening ability (1.5% for the best listener to 12.5% for the poorest). As the distance from the source increases, a distance D is reached beyond which the articulation loss remains constant at a value

$$\% \text{ AL} = 9RT + k. \tag{24.4}$$

This critical distance D is given by

$$D = 0.20\sqrt{V/\text{RT}}. \tag{24.5}$$

For skilled speakers and listeners, a % AL of 25 to 30 as calculated from these formulas may be acceptable, because speech includes a fair amount of redundancy (that is, it is possible to "guess" a few lost consonants on the basis of the text). A better strategy, however, is to reduce % AL to 15 or less in order to accommodate the "average" speaker and listener.

EXAMPLE 24.2   Determine the % AL in the room described in Example 24.1, where

$$RT = 0.161V/A = 0.161(6000)/330 = 2.9 \text{ s}.$$

Assume $k = 7$ for a typical listener.

$$\% \text{ AL} = \frac{200(2.9)^2 r^2}{6000} + 7 = 0.28r^2 + 7.$$

Thus, % AL = 15 for $r = 5.4$ m.

$$D = 0.20\sqrt{6000/2.9} = 9.1 \text{ m}.$$

$$\% \text{ AL} = 9RT + a = 9(2.9) + 7 = 33 \qquad \text{for } r > D.$$

The % AL is excessive in much of the room because of the reverberation. The direct sound should probably be reinforced.

Equations (24.3) and (24.4) for articulation loss assumed nondirectional sources ($Q = 1$). When a directional source is used, Davis and Davis (1975) recommend dividing Eq. (24.3) by the directivity factor Q, because the ratio of direct sound to reverberant sound increases by that factor. Also D increases by the factor Q. Thus,

$$\% \text{ AL} = \frac{200r^2 T^2}{QV} + a; \tag{24.3a}$$

$$D = 0.20\sqrt{QV/T}. \tag{24.5a}$$

If the main problem in a room is background noise, the intelligibility can obviously be improved with electronic reinforcement of sound. The sound level should be raised at least 25 dB above the background noise for all listeners, if possible. However, if the problem is with a poor ratio of direct sound to reverberant sound (such as one finds in old cathedrals, for example), the sound system must use speakers with high directivity factors Q. Otherwise, the reverberant sound, as well as the direct sound, will be reinforced. Speech clarity can be improved by reinforcement of only midrange frequencies (500–4000 Hz), which carry most of the speech intelligence.

24.3 ■ POWER CONSIDERATIONS

Equation (24.2) can be used to determine the power capacity of a sound system to be used for sound reinforcement, because it works for any source. What must be determined are the desired levels of direct sound and reverberant sound. The direct-sound level is given by

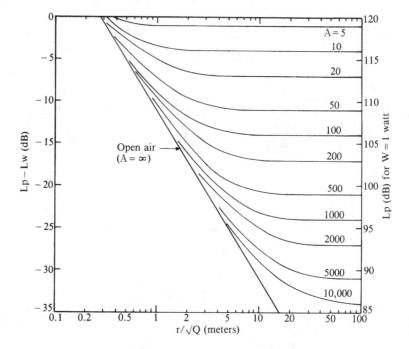

FIGURE 24.2
Chart for
determining sound
pressure level L_p
in a room, where r
is the distance from
the source, Q is the
directivity of the
source, W is its
power, L_w is its
sound power level,
and A is the total
absorption (in m^2).

Eq. (24.1), and the reverberant sound is represented by Eq. (24.6):

$$L_p \text{ (direct)} = L_w + 10 \log Q/4\pi r^2, \tag{24.1}$$

$$L_p \text{ (reverb)} = L_w + 10 \log 4/A. \tag{24.6}$$

[Note that the combined direct and reverberant sound, given by Eq. (24.2), is not simply the sum of L_p (direct) and L_p (reverb).] In these equations, A and r^2 are in square meters; if they are expressed in square feet, add 10 dB to each equation.*

For speech, a sound pressure level of 65 to 70 dB will be adequate, provided that it is at least 25 dB above the noise level in the room. Music, however, spans a wide dynamic range, and peak levels of 90 to 100 dB or more may be required. Systems used to amplify musical instruments in rock groups deliver 110 to 120 dB peaks, although usually not without some distortion.

Figure 24.2 is a useful chart for determining the acoustic power needed in a room. Each curve represents a single value for the total absorption A. The vertical axis at the right gives the sound pressure level for a source power W of 1 W, and the vertical axis on the left can be used to determine the sound pressure level with a source of any sound power level.

EXAMPLE 24.3   Determine the acoustical power necessary to reach a level of 100 dB in a room with a total absorption $A = 400$ m^2. At a large distance ($r/\sqrt{Q} > 10$ m), $L_p - L_w \simeq$

*Because 1 m^2 = 10.76 ft^2, $10 \log 10.76 = 10.32$ dB is a more precise figure; however, adding 10 dB is usually accurate enough.

-21 dB (see the chart). Thus, $L_w = L_p + 21 = 121$ dB, which corresponds to $W = 1.26$ W of acoustical power that must be delivered by the sound reinforcement system.

24.4 ■ LOUDSPEAKER PLACEMENT

One of the most important considerations in designing a sound system is the placement of loudspeakers in order to give the best coverage over the listening area. Most systems use a large single source or a distribution of small sources throughout the room. In most auditoriums, single-source systems are preferred because they preserve best the spatial pattern of the sound field.

A single source generally consists of a cluster of loudspeakers with directivity factors Q selected to give the best coverage of the audience. The preferred location of a single source is on the centerline of the room, near the front, over the speaker's head. Vertical displacement of the source is not particularly distracting, because of our inability to localize sound in a vertical plane. The loudspeakers should be aimed toward listeners at the rear of the auditorium, as shown in Fig. 24.3(a).

Another arrangement, which provides satisfactory coverage in a long room with a low ceiling, is the distributed-speaker system, shown in Fig. 24.3(b). Each unit mounted in the ceiling covers 60° to 90°. If the room is long, it is important to have an electronic time delay for the rear speakers; otherwise, the direct sound arrives after the sound from the loudspeakers, so that it appears to be a distracting echo. Some time-delay systems use digital delay networks; others use magnetic tape or disc. Loudspeakers should not be placed along the side walls of an auditorium, where their crossfire will cause the listener to hear sound from several loudspeakers at the same time.

Another loudspeaker arrangement to be avoided is the common practice of putting one speaker on each side of the stage area or front wall, as shown in Fig. 24.4. Listeners seated at points A or B hear sound from one of the speakers before they hear the direct sound. If this arrangement of speakers is necessary, sound to both loudspeakers should be delayed electronically so the direct sound is heard first. However, the single loudspeaker shown in Fig. 24.3(a) is much to be preferred.

24.5 ■ LOUDSPEAKER DIRECTIVITY

Three types of loudspeaker arrays are commonly used in sound systems: (1) cone radiators, (2) line or column radiators, and (3) horn radiators. An ordinary cone-type loudspeaker has

FIGURE 24.3   (a) Central loudspeaker system in a large auditorium. (b) Distributed loudspeaker system in a long, low room.

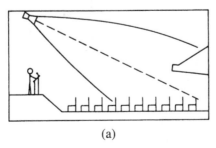

(a)

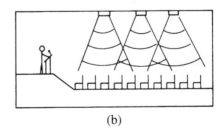

(b)

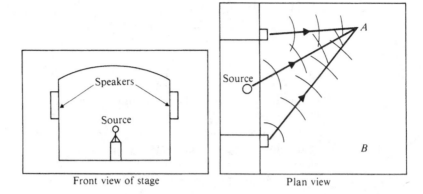

FIGURE 24.4
Unsatisfactory
arrangement of
loudspeakers.
Listeners on side *A*
hear the sound from
one speaker before
the direct sound.

Front view of stage   Plan view

a directivity factor Q that depends on frequency. When the wavelength of sound is much larger than the cone size, the radiation is quite uniform in the forward hemisphere; even directly backwards, the sound level is only 10 or 15 dB less than in the forward direction, depending on the design of the speaker housing. The directivity factor Q of the speaker will be about 2. At higher frequencies, the radiation is mainly in the forward direction, and Q increases to 20 or more depending on speaker size and wavelength. The radiation pattern and Q of a typical 8-in. loudspeaker are shown in Fig. 24.5. The polar radiation pattern indicates the relative sound level at each angle compared to the level at the same distance directly in front of the speaker.

Combining two or more elements changes the directivity. A vertical column of speakers is equivalent to increasing the speaker size in the vertical direction but not the horizontal. Thus the beam spreads out much less in the vertical direction than in the horizontal. Line or column radiators are sometimes used in sound systems because they increase Q while maintaining the broad distribution of a single speaker in the horizontal direction.

Horn loudspeakers nearly always have greater efficiency in converting electrical energy to acoustical energy than do cone-type loudspeakers. They also can be designed to have greater directivity. Thus they are well suited for use in sound systems. Their main disadvantages lie in their larger size and the difficulty in achieving good low-frequency response.

FIGURE 24.5
Radiation pattern
and directivity
factor Q for a
typical 8-in.
cone-type
loudspeaker. (After
Davis and Davis
1975).

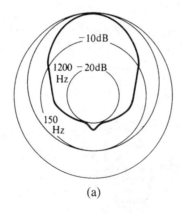

(a)

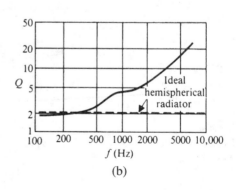

(b)

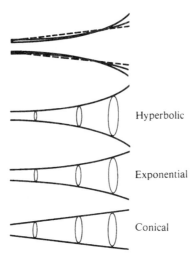

FIGURE 24.6 A comparison of the rates of expansion of conical, exponential, and hyperbolic horns.

To achieve smooth response down to 100 Hz, for example, the horn mouth should have an area of at least 8 ft^2. Many systems use horn loudspeakers at middle and high frequencies, and cone-type loudspeakers at low frequency.

Horns are called conical, exponential, or hyperbolic according to the way in which their area expands with distance from the driver. Figure 24.6 compares the rates of expansion. The two most common horn designs in use are the *multicellular horn* and the *radial/sectoral horn*. The radial/sectoral horn has straight sides on two boundaries and curved sides on the other two boundaries, as shown in Fig. 24.7(a). The multicellular horn consists of several exponential horns with axes passing through a common point. At the lower frequencies, the entire unit radiates as one horn, but at high frequencies, each horn radiates its own narrowing beam. Figure 24.7(b) shows a multicellular horn.

Constant directivity horns, although they are larger in size than exponential or multicellular horns, are designed to maintain their dispersion pattern over a wide-frequency band and are thus quite popular in sound reinforcement systems.

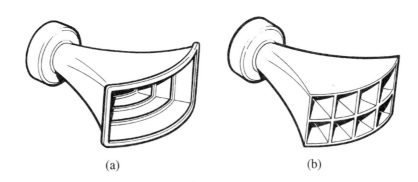

FIGURE 24.7 (a) A radial/sectoral horn. (b) A multicellular horn.

(a)　　　　　　　　(b)

FIGURE 24.8
Acoustic feedback
from loudspeaker
to microphone to
amplifier can cause
a sound system to
become unstable
and go into
oscillation.

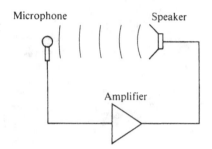

24.6 ■ ACOUSTIC FEEDBACK

You have probably been in an auditorium when the sound system begins to howl or squeal loudly. This phenomenon results from *acoustic feedback*, an example of positive feedback, which, as we noted in Chapter 18, can cause an amplifier to act as an oscillator. Acoustic feedback occurs when the microphone picks up sound from the loudspeaker and sends it to the amplifier to be reamplified. If the electrical gain is greater than the acoustical loss, the signal continues to build up and the system goes into oscillation. This process is illustrated in Fig. 24.8. If the gain is not quite large enough to send the system into oscillation, acoustic feedback may still cause speech to sound "tinny" due to a long decay time at certain frequencies.

In most auditoriums, acoustic feedback limits the amount of gain that can be achieved from an amplifying system. Microphones will always be in the reverberant sound field, of course, and sometimes they receive a fairly large amount of direct sound from the loudspeakers as well. By turning down the amplifier gain, one will find that greater sound levels can be tolerated without acoustic feedback, but the sensitivity to the desired sound will decrease as well.

The useful gain without feedback oscillation can be increased by using microphones of high directivity (see Chapter 20) and placing them as far from the loudspeakers as possible and well off the axis of directional loudspeakers. Since feedback oscillation tends to occur at the frequencies of prominent room resonances (peaks in the reverberant sound field), room equalization will raise the oscillation level substantially.

Consider the case illustrated in Fig. 24.9. We define three distances, and also factors $G(\theta)$ to express the off-axis output of the loudspeaker and $F(\phi)$ to describe the directivity of the microphone. Thus, r = the distance from the loudspeaker to the

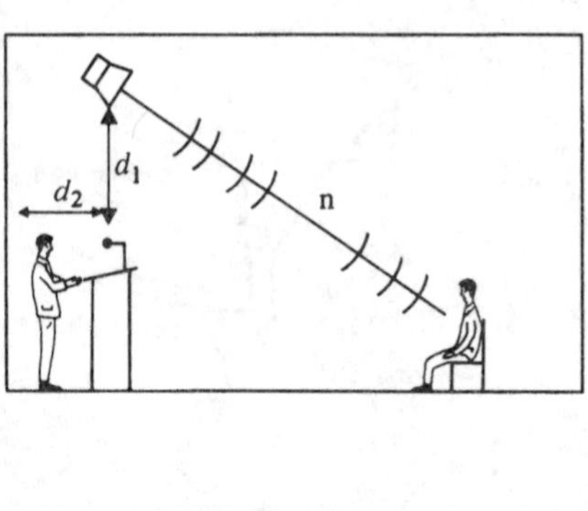

FIGURE 24.9
Parameters of
sound systems with
single microphone
and loudspeaker.

listener, d_1 = the distance from the microphone, d_2 = the distance from the source to the microphone, $G(\theta)$ = the sound level at angle θ compared to the sound level on the loudspeaker axis at the same distance, and $F(\phi)$ = the directivity of the microphone. The sound level at a listener (on the axis of the loudspeaker) due to direct sound from the loudspeaker will be

$$L_p(1) = L_w + 10\log\frac{Q}{4\pi r^2}.$$

The sound fed back to the microphone directly from the loudspeaker and from the reverberant field in the room will be

$$L_p(m) = L_w + 10\log\left(\frac{Q}{4\pi d_1^2} + \frac{4}{A}\right) - G(\theta) - F(\phi).$$

The sound picked up by the microphone from a nondirectional source with power level $L_w(s)$ will be

$$L_p(s) = L_w(s) + 10\log\frac{1}{4\pi d_2^2}.$$

It is up to the sound engineer to design the system such that $L_p(1)$ will be sufficiently large to provide clarity to the listener, yet $L_p(m)$ will be small enough to prevent feedback oscillation. The directivity factors $G(\theta)$ and $F(\phi)$, as well as the Q of the loudspeaker, are within the engineer's control. Obviously if d_2 (source-to-microphone distance) can be kept small, the task is much easier.

24.7 ■ EQUALIZATION

One of the most important steps in the installation of a sound system is the adjustment of its frequency response to complement that of the room in which it is installed. This is called *equalization*; unfortunately, it is a step that is too often neglected. Equalization serves two important functions: (1) it results in a more natural sound by compensating for predominant room resonances, and (2) by suppressing acoustic feedback oscillation, it permits the sound system to be operated at a higher level.

Pipe-organ builders have long recognized the importance of room resonances in churches and auditoriums, and they regulate or voice individual pipes accordingly. It is unthinkable for an organ builder not to make this final adjustment to the room, in spite of the labor involved. Many expensive sound systems, however, are installed without any provision for equalizing their response to the auditorium in which they will be used.

Very crude equalization can be obtained by means of the bass and treble tone controls, which boost or cut the low and high frequencies. This is usually not sufficient to accomplish the goals of equalization, however. Some type of selective filtering is generally necessary. The most economical type of filter set has eight to ten band-pass filters (see Chapter 18) with pass bands one octave wide, each of which controls one octave of sound. More flexi-

FIGURE 24.10
A filter set for equalization of a sound with 36 $\frac{1}{3}$-octave filters. (Courtesy of United Recording Electronics Industries.)

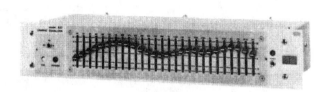

bility is obtained by using $\frac{1}{3}$-octave filters, but this adds to the cost, because three times as many filters are required. A filter set with $\frac{1}{3}$-octave filters is shown in Fig. 24.10.

Figure 24.11 illustrates how equalization can increase the usable gain of a sound system. The sound system before equalization will generally show peaks in gain at certain frequencies. These correspond to room resonances, enhanced perhaps by the sound system itself (the character of this enhancement depends on the placement of the microphones and loudspeaker). As the amplifier gain is increased, acoustic feedback causes the system to oscillate at the highest of these peaks.

FIGURE 24.11
Frequency response of a typical auditorium with a sound system before and after equalization. The solid curve indicates the sound level in each $\frac{1}{3}$-octave band of frequency for an input with equal level in each band (pink noise). (After Davis and Davis 1975).

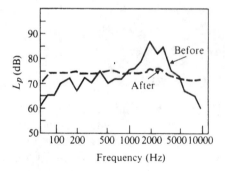

After the system is equalized, the response curve may look like the dashed line in Fig. 24.11. It is clear that the level may now be safely raised several decibels without danger of oscillation due to feedback. Sometimes tunable band-reject filters with narrow bands ($\frac{1}{6}$-octave or $\frac{1}{10}$-octave) are used to reduce prominent resonances. Such a filter set is shown in Fig. 24.12.

Another useful device for suppressing acoustic feedback oscillation in a sound system is a *frequency shifter*, which shifts the frequencies in the microphone signal by an inaudible amount, typically from 3 to 5 Hz. This small shift in frequency is often enough to avoid feedback due to room resonances that are narrow but prominent.

FIGURE 24.12
A feedback
suppressor that
includes four
tunable band-reject
filters with $\frac{1}{6}$-octave
bandwidths. Each
filter can be tuned
to the frequency of
a room or sound
system resonance.
(Courtesy of United
Recording
Electronics
Industries.)

24.8 ■ TIME DELAY

A sound system with loudspeakers at several positions in a large auditorium generally requires some type of time delay for best results. As we pointed out in Section 23.3, sound that arrives up to 50 ms after the direct sound (provided, of course, that it is similar in spectrum and time envelope) will reinforce the direct sound and yet preserve the apparent direction of the sound source. Time delay is especially important in the case of supplementary speakers positioned in problem areas, such as underneath a balcony.

The best time-delay systems use digital time delay. In a digital system, the audio waveform is sampled at regular intervals, and the samples are converted to digital numbers by a device called an *analog-to-digital converter* (ADC). The resulting series of numbers is circulated in a digital memory and read out after the desired delay time has passed. The numbers are then converted back to voltages in order to reconstruct the original audio waveform. This conversion of numbers back to voltages is done by a *digital-to-analog converter* (DAC).

24.9 ■ ENHANCEMENT OF REVERBERATION

The fact that some halls are used for varied purposes makes the adjustment of reverberation time desirable. Maximum clarity of speech demands a short reverberation time; a pipe organ sounds best in a reverberant room (see Fig. 23.7). Some control of reverberation time is possible by using movable wall panels, etc., but these techniques are generally expensive. A solution that is becoming increasingly popular is the use of "assisted resonance" or the electronic enhancement of reverberation.

One method that has been used for reverberation enhancement places a loudspeaker and a microphone in a *reverberation chamber*, a small room of 300 to 3000 m³ with highly reflecting walls, ceiling, and floor. By amplifying the reverberant sound in this chamber and feeding it back into the main auditorium through loudspeakers, the reverberation time may be increased. The reverberation chamber usually has nonparallel walls to increase the number of resonances, but some type of equalization is necessary to smooth out the frequency response curve of the chamber.

A simple means of providing reverberation, used in early electronic organs and music synthesizers, is a spring with a transducer at each end. The transducers are usually piezoelectric crystals or magnetic pickups similar to phonograph pickups. One transducer generates a sound wave, which propagates down the spring, stimulating a second transducer at the other end. The sound wave travels back and forth many times on the spring before dying out, and this approximates the decay of reverberant sound. The effect tends to resemble the decay of sound in a one-dimensional room or tunnel, however; units with several springs of varying length minimize this effect.

A pleasing two-dimensional effect is obtained by placing a number of transducers on a thin plate (Kuhl plate) or foil. By careful design and provision for damping the plate, a pleasing effect can be obtained. A reverberation unit of high quality using a gold foil is available commercially.

Electromechanical reverberation systems, such as those just described, have largely been replaced by digital reverberators using digital delay lines. A digital delay line consists of a series of connected registers; a signal that enters the delay line will exit a certain time later. A simple reverberator can be constructed from a delay line by including a feedback connection with some attenuation to prevent overflow. Practical reverberators often combine several simple reverberators having different reverberation times plus an all-pass reverberator to obtain natural sounding reverberation (Schroeder 1962).

Some digital reverberators use digital signal processing (DSP) to model the familiar spring (one-dimensional), plate (two-dimensional), or chamber (three-dimensional) electromechanical reverberators. The reverberance obtained from each of these three types of electromechanical reverberators was different, and the user is thus given some choice in timbre. Digital techniques for reverberation are discussed in Section 4.4 of Moore (1990).

One of the most elegant systems for enhancing the reverberation of a large concert hall is the *assisted resonance* system developed by acoustician P. H. Parkin and his associates for the Royal Festival Hall in London. The Royal Festival Hall, opened in 1951, had some desirable acoustical characteristics but was found to be deficient in low-frequency reverberation. The system developed for this hall consists of 172 independent channels, each with its own tuned resonator spaced over the frequency range 58 to 700 Hz. At the lower frequencies, each channel uses a microphone placed inside a tuned Helmholtz resonator; above 300 Hz, tuned pipes are used. The gain of each channel is adjustable individually to obtain satisfactory reverberation at present and perhaps in the future in the event that changes are made in the hall. Figure 24.13 shows the results of tests made using music as a source.

FIGURE 24.13
Reverberation time in the Royal Festival Hall with and without assisted resonance. (From Parkin and Morgan 1970.)

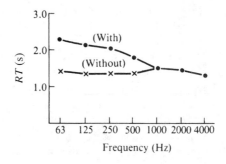

In order to ensure acceptance of the system, the reverberation time was increased gradually over several concert seasons. As Parkin and Morgan (1970) describe it, "the policy has been to keep discussions to a minimum, because of the passions likely to be aroused in some breasts by the thought of loudspeakers in the RFH." After conducting two concerts with the Berlin Philharmonic Orchestra, Herbert von Karajan commented that the acoustics of the Royal Festival Hall are probably now among the finest in the world. He and several members of the orchestra were interested in what had been done to the hall since they had last performed in it two years earlier. Work on the system has been aimed at slightly increasing the reverberation time in the full auditorium to the same value as the empty auditorium, so that artists can rehearse and perform under nearly identical conditions.

24.10 ■ MICROPHONE PLACEMENT AND MIXING

Microphones are generally placed in the direct field of the speaker or performer, so that each time the distance from the source is doubled, the microphone output is reduced by 6 dB. This means that doubling the distance reduces the available gain before feedback by 6 dB, but it also reduces the effect of microphone-to-source variations, which is important if the speaker wishes to be able to move slightly without producing a change in level.

The microphone-to-source distance also has an effect on the timbre of the sound, especially if the microphone has a cardioid pickup pattern (see Section 20.2). As the distance decreases, the proximity effect tends to create a "bassy" sound. Closing in on the microphone can increase breath noise and pop noise from plosive consonants.

When a microphone is placed on a large surface, such as a floor, a gain of nearly 6 dB can be realized. However, if the microphone is a small distance away from a large surface, such as a floor, cancellation of certain frequencies can occur due to interference between the direct and reflected sound waves, as shown in Fig. 24.14. The canceled frequency can be determined by simple geometry. If the microphone were 3 m from the source and both were 1.5 m above the floor, for example, the path difference would be 1.23 m, and the canceled frequency would be about 140 Hz.

FIGURE 24.14
An example of cancellations caused by reflections from the floor.

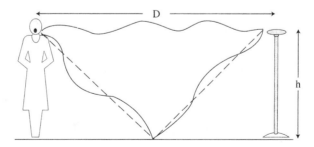

The direct path length from the source to the microphone is D (Fig. 24.14). The reflected wave travels a distance given by $2[(D/2)^2 + h^2]^{1/2}$. The path difference is $2[(D/2)^2 + h^2]^{1/2} - D$. Cancellation occurs when this difference is $\lambda/2, 3\lambda/2, 5\lambda/2,$

etc., because then the direct and reflected waves arrive at the microphone in opposite phase. This series of cancellation frequencies is an example of *comb filtering*.

Most sound-reinforcement systems have more than one microphone. As the number of open microphones increases, the potential for acoustic feedback increases as well. Thus, it is a good idea to minimize the number by turning off unused microphones. This can be done by a sound system operator, but it can also be done by having a sound-operated switch in each channel or by means of an automatic microphone mixer. An automatic mixer tries to attenuate each channel by an amount equal to the difference between that channel and the sum of all channel levels.

Although most amplifiers used for sound reinforcement have level controls for the various inputs as well as a master gain control, many sound-reinforcement systems include an audio control console similar to those used in recording studios. These may be either analog or digital. If the mixing and signal processing is done in the digital domain, the inputs for microphones and other analog sources must have high-quality ADCs. Digital control consoles have memories in which all the parameters (gain of each microphone preamplifier, equalization, etc.) can be stored, and these can easily be programmed to change in time during an event.

24.11 ■ REINFORCEMENT FOR THE HEARING IMPAIRED

Many people with hearing loss are able to function well in face-to-face situations but are somewhat lost in noisy or reverberant settings. Even people who wear hearing aids have problems in reverberant rooms. Speech intelligibility can be increased by providing a means to enhance the sound at the listener's ear. This can be done with one of four types of wireless transmission-receiver systems: magnetic induction, FM broadcasting, AM broadcasting, or infrared light.

A magnetic induction, or loop, system employs a large loop of wire wrapped around the seating area to set up a magnetic field that can be picked up by a hearing aid with a T-coil designed for magnetic coupling of the earpiece of a telephone (about 60% of hearing aids in the United States have these). Portable receivers can be supplied to users who do not have such a hearing aid.

FM and AM broadcast systems and infrared systems require special receivers. FM receivers produce a signal of high quality, and the Federal Communications Commission (FCC) has reserved a band of frequencies (72 to 76 MHz) for their use. AM broadcast systems operate in the regular broadcast band or below (10 to 1600 kHz). Infrared systems are widely used in professional theaters and concert halls, because they are easily confined to the main auditorium. However, they don't function very well in a brightly lighted room.

24.12 ■ OUTDOOR SOUND SYSTEMS

The discussion in this chapter has focused on sound systems for use indoors in auditoriums and large rooms. For purposes of comparison, it is interesting to consider an outdoor system such as that in Fig. 24.15. In the past few years there have been a number of rock music

FIGURE 24.15
Large outdoor
sound system for
musical
entertainment
event. Line array
loudspeaker
systems suspended
from support truss
left and right of
stage canopy.
(Photo courtesy
JBL Professional.)

concerts held outdoors with thousands in attendance. Portable sound systems generating kilowatts of acoustic power have been set up to provide coverage to large areas having high levels of background noise.

Clearly, sound power takes on substantially more importance outdoors, where there is no reverberant sound field for reinforcement. Thus, large power amplifiers and arrays of loudspeakers with high efficiency are needed.

An example of a large outdoor system is that used to "broadcast" the concerts-in-the-park series to audiences as large as 100,000 people in New York's Central Park. The system, described by Rosner and King (1977), is designed to cover an area of about 80,000 m^2 and to provide peak levels of 105 dB at 45 m and 99 dB at 90 m from the stage. The frequency response is flat to ± 3 dB over the range 63 to 1000 Hz and has a downward slope of 2 dB per octave above 1000 Hz. The loudspeakers are built into two 40-ft portable towers. Each tower has 8 woofers in 4 bass horn cabinets, 6 midrange radial horns, and 12 tweeters. Sixteen 80-W amplifiers in each tower supply the electrical power. The system includes a mechanical reverberation unit to provide liveness for the outdoor environment.

24.13 ■ SUMMARY

Electronic enhancement of either direct, early, or reverberant sound is possible, although it is generally the direct sound that must be reinforced. Sound sources are characterized by their power (in watts or decibels) and their directivity, which can be expressed by a factor Q. At various distances from the source, the sound field in a room is characterized as the near field, the free field, or the reverberant field. The clarity of speech, described by a percentage of articulation loss (% AL) of consonants, can be improved by using a well-designed sound system. Loudspeaker design and placement are important factors in the performance of a sound system. Horn loudspeakers usually have the highest efficiency and directivity.

Acoustic feedback can cause oscillation to occur in a sound system at a frequency determined by the room resonances or the distance from the loudspeakers to the microphones. This annoyance can be prevented by equalizing the system to smooth out peaks in the gain.

A sound system that includes loudspeakers at several locations in a large auditorium may require time delay. Digital time-delay systems provide the best performance but are expensive. Enhancement of reverberation can improve the sound of music in a dry concert hall.

REFERENCES AND SUGGESTED READINGS

Ballou, G. M. (1991). *Handbook for Sound Engineers: The New Audio Cyclopedia*. Woburn, Mass.: Focal Press.

Beranek, L. L. (1954). *Acoustics*. New York: McGraw-Hill. Reprinted by the Acoustical Society of America, 1986.

Boner, C. P., and C. R. Boner (1968). "Sound-Reinforcing System Design," *Appl. Acoust.* **1**: 115.

Davis, D., and C. Davis (1975). *Sound System Engineering*. Indianapolis: Howard Sams.

Everest, F. A. (1978). *The Complete Handbook of Public Address Sound Systems*. Blue Ridge Summit, Penn.: TAB Books.

Klepper, D. L. (1970). "Sound Systems in Reverberant Rooms for Worship," *J. Audio Eng. Soc.* **18**(4): 391.

Moore, F. R. (1990). *Elements of Computer Music*. Upper Saddle River, N.J.: Prentice Hall.

Parkin, P. H., and K. Morgan (1970). " 'Assisted Resonance' in the Royal Festival Hall, London: 1965–1969," *J. Acoust. Soc. Am.* **48**: 1025.

Peutz, V. M. A. (1971). "Articulation Loss of Consonants as Criterion for Speech Transmission in a Room," *J. Audio Eng. Soc.* **19**: 915.

Rosner, A., and L. S. King (1977). "New Mobile Sound Reinforcement System for the Metropolitan Opera/New York Philharmonic Orchestra Park Concerts," *J. Audio Eng. Soc.* **25**: 566.

Schroeder, M. R. (1962). "Natural Sounding Artificial Reverberation," *J. Audio Eng. Soc.* **10**: 219.

GLOSSARY

acoustic feedback Sound from a loudspeaker picked up by a microphone (either in the direct field or the reverberant field) and reamplified.

all-pass filter A filter with a constant ("flat") gain at all frequencies but with frequency-dependent phase shift.

analog-to-digital converter (ADC) A circuit that converts (analog) voltages to a digital or numerical representation.

articulation loss of consonants (% AL) A measure of speech intelligibility; the percentage of consonants heard incorrectly (strongly influenced by noise or excessive reverberation).

comb filter A filter with amplitude minima regularly spaced in frequency.

digital-to-analog converter (DAC) A circuit that generates a voltage proportional to a digital number.

equalization Changing the gain of a sound system at certain frequencies to compensate for room resonances and other peaks in the response curve.

free field That part of the sound field where the sound level decreases by 6 dB for each doubling of distance.

L_p Sound pressure level $= 10 \log p/p_0$ (see Chapter 6).

L_w Sound power level $= 10 \log W/W_0$ (see Chapter 6).

near field That part of the sound field where the sound level varies from point to point because of the radiation pattern of the source.

Q **(directivity factor)** Comparison of the sound power radiated directly ahead of a sound source to that radiated in all directions.

reverberant field That part of the sound field in which sound level is independent of distance from the source.

room constant A quantity that describes the absorption in a room; it is slightly greater than the total absorption A used to calculate reverberation time (see Section 23.5).

REVIEW QUESTIONS

1. How is the directivity factor defined?

2. What is the directivity factor Q for a source that radiates equally well in all directions?

3. How does the sound pressure vary with distance from the source in a free field?

4. Why is a single loudspeaker cluster preferred in most auditoriums?

5. Under what conditions should you use loudspeakers with high directivity?

6. How does the efficiency of horn loudspeakers compare to that of cone loudspeakers?

7. What is a constant directivity horn?

8. What causes acoustic feedback in a sound system?

9. What are two important benefits of equalization?

10. How is equalization accomplished?

11. Why is time delay needed in some sound reinforcement systems?

12. Describe two methods for enhancing reverberation electronically.

13. What are the advantages in speaking close to the microphone in a large auditorium? The disadvantages?

14. What are four types of wireless systems that can be used to provide sound reinforcement for the hearing impaired?

15. Why must an outdoor sound system deliver substantially more power than an indoor system?

QUESTIONS FOR THOUGHT AND DISCUSSION

1. Draw a cross section (to scale) of the auditorium in Example 24.1 on p. 548 (30 m long, 10 m high). Indicate the correct placement of a single-source loudspeaker cluster, and measure the angle of the beam that would give coverage to the rear two-thirds of the auditorium.

2. What are the advantages and the disadvantages of using horn loudspeakers? Why are they not often used in home hi-fi systems?

3. Explain the advantages and disadvantages of using directional microphones (e.g., cardioid type) in auditorium sound systems.

4. Give examples of rooms (on or off campus) in which each of the following loudspeaker arrangements would be expected to give the best results: (a) a single cluster of loudspeakers front and center; (b) clusters of loudspeakers on each side near the front; (c) multiple loudspeakers placed in the ceiling throughout the room.

5. If you were asked to design a sound system for a church, what acoustical measurements would you make before designing the system?

EXERCISES

1. At what frequency does the wavelength of sound equal the diameter of a 15-in. woofer? a 2-in. tweeter?

2. Assuming the outdoor sound system described in Section 24.12 were to convert 2.5 kW of electrical power to acoustic power with an average efficiency of 20% and to radiate it into a 90° angle ($Q = 4$), determine the sound level at 90 m. (First determine L_w for the source.) Compare this to the design level at this distance.

3. What are the upper and lower frequencies passed by an octave-band filter tuned to 500 Hz? (In other words, what frequencies are one-half octave above and below 500 Hz?)

4. If we assume the record and playback heads are spaced 1 cm apart, what tape speed is required to give a delay time of 20 ms? Is this practical?

5. If a microphone is 5 m in front of a loudspeaker, at what frequency might oscillation take place due to acoustic feedback? Could it also occur at harmonics of this frequency? Explain.

6. Check the statement in Section 24.1 concerning the 50 speakers and the cup of tea. (It takes about 1050 J to raise the temperature of a $\frac{1}{4}$-L cup of water 1°C; a joule is equal to a watt-second.)

7. A sound source having $Q = 2$ and a sound power $L_w = 80$ dB radiates into a room with a total absorption $A = 500$ m^2.

   (a) Using Fig. 24.2, find the sound pressure level L_p at distances of 5 m and 15 m from the source.

   (b) Make the same computations using Eq. (24.2).

EXPERIMENTS FOR HOME, LABORATORY, AND CLASSROOM DEMONSTRATION

Home and Classroom Demonstration Experiments

1. *Sound field*   With a sound-level meter, measure the sound pressure level as a function of distance from a broadband noise source and make a graph similar to Fig. 24.1 (or Fig. 23.1(b)). Determine the extent of the free field.

2. *Equalization*   If the lecture hall has a sound system with an equalizer (or if one can be borrowed), show how equalization affects the amplified sound in a room. Using a pink noise source (equal power in each octave band), measure the sound level in each octave band, and equalize these levels. (A real-time octave band or $\frac{1}{3}$-octave band analyzer is ideal for this.)

3. *Loudspeaker directivity*   Loudspeaker directivity can be demonstrated (rather crudely) by rotating a loudspeaker so that its axis "sweeps" across the classroom. (It is best to use octave bands or $\frac{1}{3}$-octave bands of noise; sine waves give rise to standing waves, which confuse the intended demonstration.)

4. *Acoustic feedback*   Demonstrate acoustic feedback by *carefully* placing a microphone a few meters in front of a loudspeaker. Estimate the time delay (distance/speed) and compare its reciprocal to the frequency at which feedback first occurs.

Laboratory Experiments

Filtered sound (Experiment 37 in *Acoustics Laboratory Experiments*)

Demonstration experiments 1–4, when done quantitatively, will serve as laboratory experiments.

25

Small Rooms, Home Listening Rooms, and Recording Studios

Chapter 19 discussed loudspeakers, Chapter 20 discussed microphones, amplifiers, and tuners, and Chapter 22 discussed systems for recording and playing back recorded sound. These are the basic components needed for recording and reproducing high-fidelity sound. However, there is more to high-fidelity sound than merely assembling high-quality components. Although it is certainly important to have components with low distortion and noise, ample power-handling capacity, and so on, the importance of the acoustical properties of the listening room are often underestimated. Just as in the case of live performance in a concert hall, the room has much to do with the quality of the sound we hear. In fact the listening room is the weakest link in many high-fidelity sound systems.

In this chapter, you should learn:

■ About the acoustics of small rooms and how the acoustics differ from those of large rooms;

■ How to predict the resonance frequencies in small rooms and their importance;

■ About sound images from multiple sources;

■ About the sound field in listening rooms;

■ About the use of sound diffusors and absorbers;

■ How to design a home theatre;

■ About the acoustic requirements for recording studios and control rooms.

25.1 ■ ACOUSTICS OF SMALL ROOMS

In Section 23.2, we defined three sound fields: direct sound, early reflected sound, and reverberant sound. Much of the acoustical character of an auditorium or concert hall depends on the balance and the time relationships between these three types of sound. The first reflected sound typically reaches a listener from 10 to 30 ms after the direct sound. Reverberation times of 1 to 3 s are common.

In a small room, such as a home listening room, quite a different time relationship exists. Walls and ceiling are so close to the listener that many reflections arrive within a few milliseconds after the direct sound. It is, therefore, not important to distinguish between early reflected sound and reverberant sound. Achieving "intimacy" (which depends on a short time delay between direct and first reflected sound—see Section 23.7) is no problem at all in a small room. In fact, the inherent intimacy in a small room makes it difficult to simulate the acoustics of a larger space.

Home listening rooms seldom sound reverberant even though the level of the reverberant sound may exceed the level of the direct sound ($4/A > Q/4\pi r^2$ in Eq. 24.2) except for listeners seated very close to the source. That is because the reverberation time of a small room is generally short, typically less than half a second for a home listening room. To produce a concert-hall sound, some reverberant sound must be included on the recording or some type of time-delay device ("room expander") used.

According to the Sabine formula (Eq. 23.1), the reverberation time is proportional to the room volume V and inversely proportional to the absorption A. Thus the reverberation time of a small room that has limited volume can only be increased by making the absorption small. This tends to produce a "shower room" type of reverberation, however, because the individual resonances of the room become prominent. The individual resonances of a small room with little absorption are similar to those of a rectangular box, as described shortly.

Consider the resonances of a rectangular box with dimensions L, W, and H. The frequencies of the various resonances are

$$f_{lmn} = \frac{v}{2}\sqrt{\left(\frac{l}{L}\right)^2 + \left(\frac{m}{W}\right)^2 + \left(\frac{n}{H}\right)^2},\qquad(25.1)$$

where v is the speed of sound and l, m, and n are integers 0, 1, 2, ...

Modes in which two of the integers l, m, and n are zero are called *axial* modes, because they consist of standing waves propagating back and forth parallel to two pairs of room boundaries and being reflected by the third pair. Modes in which only one of the integers l, m, and n is zero are called *tangential*, and they consist of waves reflecting off two pairs of surfaces (like balls on a pool table). Modes in which none of the integers l, m, and n are zero are called *oblique* modes. In general, axial modes store more acoustic energy because the sound waves travel farther between reflections.

Contours of equal sound pressure for the (2, 0, 0) axial mode and the (3, 2, 0) tangential mode are shown in Fig. 25.1. Note the pressure maxima along the boundaries and the pressure maxima that occur in the corners. Pressure maxima occur in the corners for all room modes, which suggests corners as good locations for sound absorbers. Placing a loudspeaker or microphone near a corner maximizes the response, but only at such low frequencies that the distance to the corner is smaller than a quarter wavelength or so.

EXAMPLE 25.1   Find the lowest resonances of a box with dimensions of 2.0, 1.0, and 0.8 m. The lowest resonance occurs when $l = 0$, $m = 0$, and $n = 1$ (because L is the longest dimension); we label this the 100 mode, and its frequency is

$$f_{100} = \frac{344}{2}\sqrt{\left(\frac{0}{2.0}\right)^2 + \left(\frac{0}{1.0}\right)^2 + \left(\frac{1}{0.8}\right)^2} = 86 \text{ Hz}.$$

The next few resonances are found by letting l, m, and n be other small integers. The results are:

$$
\begin{array}{lll}
f_{100} = 86 \text{ Hz}, & f_{010} = 172 \text{ Hz}, & f_{001} = 215 \text{ Hz} \\
f_{200} = 172 \text{ Hz}, & f_{110} = 192 \text{ Hz}, & f_{101} = 232 \text{ Hz} \\
f_{111} = 289 \text{ Hz}, & f_{020} = 344 \text{ Hz}, & f_{002} = 430 \text{ Hz}
\end{array}
$$

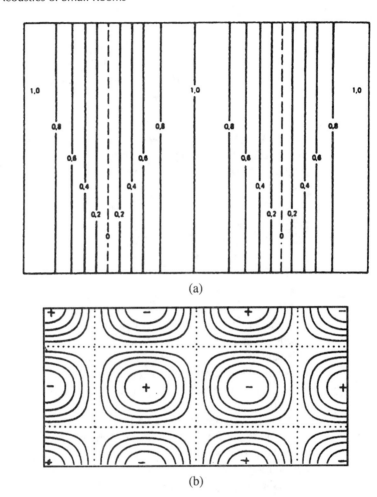

FIGURE 25.1
Contours of equal
sound pressure in a
rectangular room.
(a) (2, 0, 0) axial
mode; (b) (3, 2, 0)
tangential mode.
Dashed lines
represent nodes.

Suppose that you stand in a shower room with hard, smooth walls that reflect sound well. The dimensions may not be too different from those of the box in Example 25.1. As you sing up and down the scale, you may hear strong resonances near the calculated frequencies. The resonances are close enough together in frequency to give a "bathroom baritone," full-bodied, resonant sound.

Unfortunately, dormitory rooms may have acoustical behavior that is more like the shower room just described than like a living room amply endowed with carpet, upholstered furniture, and other sound absorbers. Many purchasers of high-fidelity sound systems have experienced disappointment that the sound they hear at home is colored by room resonances that didn't exist in the dealer's listening room. For this reason, some dealers allow potential buyers to audition high-fidelity systems at home.

The point of equality between the direct and the reverberant field can be estimated from the graph shown in Fig. 24.2. First the total absorption A is calculated; then the point at which the appropriate A-curve would intersect the direct-field ($A = \infty$) curve, if extended horizontally to the left, gives the value of $r/\sqrt{Q}$ at which the two fields contribute equally. If the directivity factor Q of the source is known, the distance r can be determined.

EXAMPLE 25.2   A home listening room has total absorption $A = 20$ m^2. The source, a loudspeaker placed along one wall, has a directivity factor ranging from $Q = 2$ at low frequency to $Q = 10$ at high frequency. At what distance r_c will the direct and reverberant sound fields be equal in level?

Solution   The $A = 20$ curve in Fig 24.2, if extended horizontally to the left, would intersect the direct-field line at $r/\sqrt{Q} = 0.5$. Thus at low frequency, $r_c = (0.5)\sqrt{2} = 0.7$ m; at high frequency, $r_c = (0.5)\sqrt{10} = 1.6$ m.

Most listeners will be at least 2 m from the loudspeakers and hence will be well into the reverberant field for low-frequency sound. At higher frequencies, the total absorption A will probably increase, as will the directivity factor Q of the loudspeaker, so the point of equality between the two fields moves to a greater distance. A distance of 3 to 4 m from the source may be required to get well into the reverberant field. Nevertheless, even in a "dead" room, a substantial portion of the sound is reverberant sound that arrives from all directions.

An important difference between the reverberant fields of large and small rooms is that in a large room, the reverberant sound decays over an appreciable time interval, whereas in a small room the buildup and decay times are short. This provides an important auditory clue to room size.

As the dimensions of the room increase, the number of resonances within the audible range multiplies rapidly. In a large room there are so many resonances that the room appears to take on a fairly smooth frequency response, although there may be an emphasis or a deficiency in some particular range of frequency. Not so in a small room with hard walls, however; room resonances may give emphasis to certain frequency ranges, and may result in considerable coloration of sound from a high-fidelity system.

25.2 ■ ROOM PROPORTIONS

From Eq. (25.1) and Example 25.1, it is easy to see that the frequency distribution of the modes of a room is determined by its dimensions. In a cube, for example, three axial modes would have the same frequencies for each value of the integer (e.g., $f_{100} = f_{010} = f_{001}$). In a listening room such concentration of modes at certain frequencies would be undesirable, because it would emphasize these frequencies too much and provide too little resonance at frequencies in between. Distribution of mode frequencies for a cube and a rectangle (with dimensions in the ratios 1 : 2 : 3) having equal volume are shown in Fig. 25.2. Note that

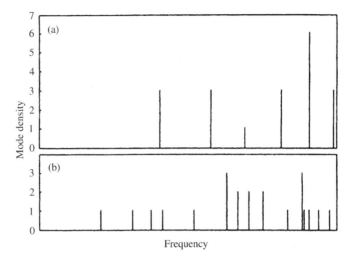

FIGURE 25.2
Distribution of
mode frequencies
in a cube and a
rectangle (with
dimensions in the
ratios 1 : 2 : 3)
having equal
volume (Fletcher
and Rossing 1998).

the cube has a very "peaky" response with many coincident modes, whereas the 1 : 2 : 3 room has a more even spread of resonances.

Even better mode distribution can be obtained by avoiding integer multiples entirely. A rectangular room with dimensions in the ratio 5 : 3 : 2 has a smooth response, because it avoids overlapping modes, as do other nonintegral ratios. The *golden ratio* 1.618 : 1 : 0.618 ($\sqrt{5} + 1 : 1 : \sqrt{5} - 1$) also gives a very smooth response. Rooms with oblique walls do also, but rectangular rooms are generally preferred for other reasons.

As frequency increases, the number of modes greatly increases. The number of modes with frequencies within the range 0 to an upper limit f is given approximately by the expression

$$N_f \approx \frac{4\pi}{3} V \left(\frac{f}{v} \right)^3 + \frac{\pi}{4} S \left(\frac{f}{v} \right)^2 + \frac{L'}{8} \frac{f}{v}. \tag{25.2}$$

In this expression, V is the volume of the room, S is its area, $L' = 4(L + W + H)$ is the sum of the lengths of all the edges of the room, and v is the speed of sound (Kuttruff 1997).

25.3 ■ SOUND IMAGES FROM MULTIPLE SOURCES

In order to better understand stereophonic and surround sound, as well as room acoustics, let us consider various ways in which our auditory system perceives sounds from multiple sources. First we consider monophonic two-speaker arrangements (that is, two loudspeakers receiving signals from the same source). Recall that at low frequencies the main clue to the direction of a source is the difference in the arrival times of the sound at our two ears, whereas at high frequencies it is the interaural intensity difference that dominates (see Section 5.5).

In Fig. 25.3 the listener is located on the median plane equidistant from both speakers. If both speakers receive the same signal at the same strength, an "image" will be created at location A on the median plane, as shown in Fig. 25.3(a). If the signal strengths at the

FIGURE 25.3
Location of sound
images from two
sources S_L and S_R
with the same
program material:
(a) identical
sources; (b) same
signal at different
levels; (c) one
source delayed by
increasing the
distance from
source to listener.

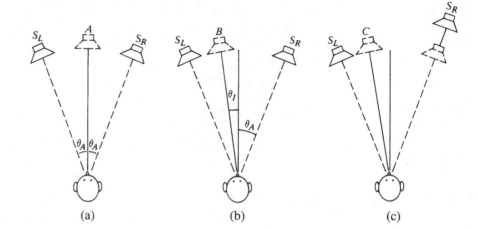

two speakers are different, however, the image will shift toward the speaker receiving the stronger signal, as indicated by B in Fig. 25.3(b).

> The angle of the image θ_I with respect to the median plane can be calculated from the following equation:
>
> $$\frac{\sin \theta_I}{\sin \theta_A} = \frac{p_L - p_R}{p_L + p_R},\qquad(25.3)$$
>
> where θ_A is the angle of each speaker with the midplane and p_L and p_R are the signal strengths (that is, the sound pressures at the listening point due to sound from the two speakers).

If one of the speakers is moved to a greater distance from the listener than the other, the sound image moves toward the nearer source (position C). If S_R is farther away by $\frac{1}{3}$ m or more (so that the sound delay is more than 1 ms), the image coincides with S_L. This is another example of the precedence effect (see Section 23.3). However, if the sound received from S_R is made greater than that from S_L, the image moves back toward the median plane. Thus, it is possible to trade off amplitude for time delay within certain limits.

> The extent to which this trade-off works and the effective *trading ratios* (the difference in arrival time divided by the equivalent difference in level) have not been completely established, although they have been the subjects of recent experiments. The following results are representative of several experiments:
>
> 1. The trading ratio is frequency-dependent, as would be expected from the fact (see Section 5.5) that localization of the sound source at low frequency is mainly by comparison of the arrival times (or phases) of the sound at our two ears, whereas intensity differences dominate at high frequency.

2. The trade-off is not complete. Although at low and mid frequencies a large fraction of the image shift due to a change in distance from the source can be compensated for by a change in level, the image (position C in Fig. 25.3(c)) cannot be restored completely to the midplane (Gilliom and Sorkin 1972).

3. There is disagreement among the results of different experiments. At 200 Hz, for example, trading ratios ranging from 60 to 147 μs/dB are reported; at 500 Hz, the range is all the way from 10 to 200 μs/dB.

4. At high frequency, where the ear is insensitive to phase, no trade-off is noted for steady tones. In one experiment, localization of a 2400-Hz tone was found to be entirely based on interaural level. However, when the 2400-Hz tone was modulated at 200 Hz, the trade-off was comparable to that obtained with a 200-Hz tone by itself (Young and Carhart 1974).

You may wish to perform some simple experiments with your own stereo system on the roles of time and intensity in sound localization. By switching the amplifier to its monophonic mode, apply the same signal from an audio generator to both loudspeakers. While seated at equal distances from the two speakers, adjust the balance control until the sound appears to originate from an image midway between the speakers. Now move closer to one speaker and try to restore the image to the midplane by manipulating the balance control. Try the experiment with tones of high and low frequency, music, and speech.

Figure 25.4 illustrates time/intensity trading and also the approximate range of time and intensity differences over which the precedence effect applies. The horizontal axis gives the time by which a pulse to the speaker on the left (L) is delayed with respect to the speaker on the right (R). The vertical axis is the amount by which the sound level due to speaker L exceeds that from speaker R. The ascending curve at the left end indicates the approximate combination of time and intensity differences that will center the source image between the speakers. Note that when $L_L - L_R$ exceeds about 15 dB, it is impossible to compensate completely with a time delay. Also, when the time delay exceeds about 1 ms, the precedence effect (Section 23.3) defeats time/intensity trading. A listener seated only

FIGURE 25.4
Range of time delay and intensity difference over which time-intensity trading takes place and also the limits of applicability of the precedence effect. (After Madsen 1970.)

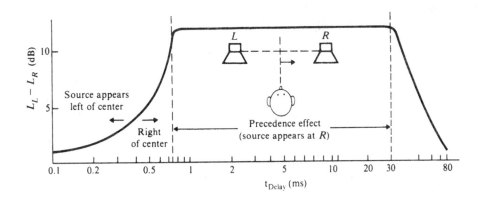

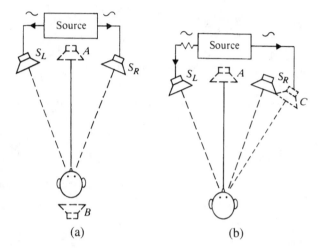

FIGURE 25.5
Location of sound
images from two
out-of-phase
sources. (After
Rossing 1981.)

about 0.33 m (1 ft) closer to the speaker on the right would experience such a 1-ms time delay.*

Now consider what happens when the phase of the signal to one of the speakers is reversed (sometimes hi-fi sets are inadvertently installed this way!). Provided that the listener is located on the median plane, the image appears to move from position *A* to a position inside or in back of the listener's head (position *B* in Fig. 25.5(a)). If the signal level at one of the speakers is reduced sufficiently, however, the sound image may shift to position *C* beyond the other speaker.

In the examples thus far considered, the same signal has been presented to both source speakers, although with different intensities and phases. If the spectra of the two sources are different, the image appears to be broadened. In particular, if one speaker is given a signal with a high-frequency emphasis and the other a signal with a low-frequency emphasis, as shown in Fig. 25.6(a), both speakers will appear to deliver the full spectrum and the apparent size of the image is broad. Furthermore, the listener can shift away from the median position without losing the effect. If, however, the frequency crossover is abrupt, as shown in Fig. 25.6(b), there will be a noticeable difference in timbre between the two sources (Gardner 1973).

Other differences between two sources, other than spectrum shape, are found to broaden the sound image. One effective way to achieve image broadening is to add reverberation to one source but not to the other.

In summary, three important properties of sound from multiple speakers are (1) the degree of *fusion* into a single image, (2) the *broadening* of the fused image, and (3) the *displacement* of the image in space. These are strongly influenced by differences in *level*, *spectrum*, *phase*, *time of arrival*, and *reverberant sound* between the two sources.

You can perform a simple experiment that illustrates sound localization in a "live" room (with little absorption). Let the source be a steady tone (provided by an audio

---

*This has the unfortunate consequence that the stereophonic effect works best in a limited area called the *sweet spot*. Home theater systems generally add a center loudspeaker for speech, as we discuss in Section 25.9.

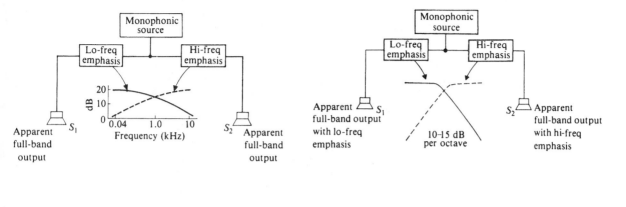

FIGURE 25.6   Broadening of sound images by complementary frequency emphasis at two sources. (a) Broad crossover range (approximately eight octaves) results in an apparent full-band output from both sources. (b) Narrow crossover range results in different source timbre. (From Gardner 1973.)

generator and loudspeaker) and move around the room. You will find it very diffi-cult to locate the source. When you are near a maximum, for example, the apparent direction of the source may change with a small movement of your head. But now substitute a percussive sound source for the steady one; have a friend clap his or her hands or strike a bell. If you close your eyes and point to the source as it moves around the room, you will make few errors. This is a good illustration of the prece-dence effect; you derive the clues needed to localize the source from the direct sound, even though the reflected sound, which follows in a few milliseconds, may be much louder.

25.4 ■ WHAT IS HIGH-FIDELITY SOUND?

Before we describe the various types of sound-reproducing systems, let us attempt to define high-fidelity sound.

To achieve realism in reproduced sound, five conditions should be satisfied:

1. The frequency range of the reproduced sound should be sufficient to retain all the audi-ble components in the source sound, and the sound spectrum of the reproduced sound should be identical to that of the source.
2. The reproduced sound should be free of distortion and noise.
3. The reproduced sound should have loudness and dynamic range comparable to the orig-inal sound.
4. The spatial sound pattern of the original sound should be reproduced.

5. The reverberation characteristics (in space and time) of the original sound should be preserved in the reproduced sound (Rossing 1979).

No sound-reproducing system is able to satisfy all five of these criteria completely. The extent to which a given system is able to satisfy them determines its fidelity. A high-fidelity system should satisfy them to the degree desired by the discerning listener.

High-fidelity components were discussed in Chapters 19–20. Most quality components manufactured by reputable companies are capable of fulfilling the first two criteria reasonably well (assuming proper installation and optimum adjustment). The most common exceptions are the noise and distortion one encounters in phonograph records and record players and the distortion that may be present in loudspeakers (for example, harmonic distortion, intermodulation distortion, and transient distortion). The distortion in most modern amplifiers is acceptably low when they are operated within their rated output.

Many high-fidelity systems are capable of reproducing sound at the level heard in a concert hall, as required to fulfill the third criterion. However, it is doubtful whether such a level would be pleasing in a small room. Sound levels in a concert hall reach 100 dB and more, whereas in a home listening room, music played at a level that reaches 85 dB will sound loud. The dynamic range in a room is limited by the tolerable top level and by the threshold that can be heard above the background noise, which may be about 25 to 30 dB in a home listening room and 30 to 35 dB in a concert hall (see Section 23.9).

Figure 25.7 gives a general idea of the loudness and dynamic range at which music may be heard in rooms of various sizes. Attempts to reproduce the full dynamic range of a concert hall in a small listening room would not create a very pleasing effect. The amount of sound power needed to achieve various sound levels is discussed in Section 24.3.

Criteria 4 and 5 also are requirements of the system rather than the individual components and often tend to be overlooked by high-fidelity sound enthusiasts (*audiophiles*). Attempts to improve the ambience or spatial characteristics of reproduced sound have led to the development of stereophonic and surround-sound systems, direct-reflecting speaker systems, room expanders, stereophonic spreaders and shifters, etc.

FIGURE 25.7
Top level, threshold level, and dynamic range for music in rooms of various sizes.

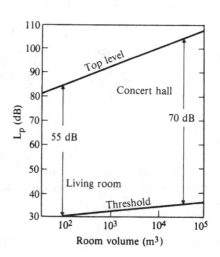

25.5 ■ SINGLE- AND MULTICHANNEL SOUND-REPRODUCING SYSTEMS

Monophonic system   A monophonic system consists of one microphone, one amplifier, and a single loudspeaker in the listening room. Ambience is provided by the acoustic characteristics of the listening room. The performance of a high-quality monophonic system in a room with good acoustics should not be underestimated. Not many good monophonic systems exist, however. Figure 25.8(a) illustrates monophonic sound.

Monaural system   A monaural system differs from a monophonic system in that sound is fed to only one ear of the listener. This type of sound reproduction, used in the telephone and in some psychoacoustics experiments, would definitely not be called high-fidelity. Figure 25.8(b) illustrates monaural sound.

Binaural system   A binaural system, which uses two microphones to feed two earphones, is capable of satisfying all five criteria for sound realism. The microphones are usually placed in a "dummy head" to reproduce the directionality of the human auditory system at all frequencies. One disadvantage of binaural reproduction is the failure of the sound to change as the head of the listener is moved; this robs the listener of some important clues about the spatial characteristics of the sound and tends to give the impression that the sound source is located inside the head of the listener. Partly because of this problem, few binaural recordings have been made available commercially. Advanced systems for virtual reality applications can actually compensate for head movements using head-tracking devices.

Listening to stereophonic music with stereo earphones produces a greatly exaggerated stereo effect that is interesting but not realistic; the source image usually appears to be inside or above the head. This is not true binaural reproduction however, because the recording microphones were probably not positioned in a pattern resembling that of the auditory system. A binaural system is illustrated in Fig. 25.8(c).

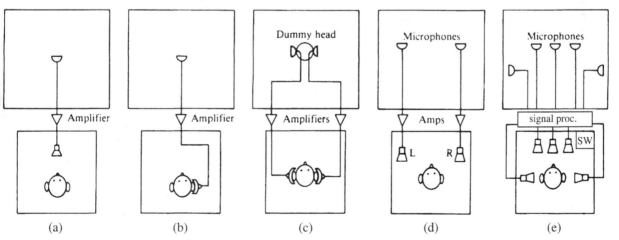

FIGURE 25.8   (a) Monophonic sound reproduction: A single microphone feeds a single loudspeaker in the listening room. (b) Monaural sound reproduction: Sound is fed to one ear only by means of an earphone. (c) Binaural sound reproduction: Two microphones, placed in a dummy head, feed two earphones. (d) Stereophonic sound reproduction: Two microphones feed two loudspeakers in the listening room. (e) Surround sound: Multiple microphones feed multiple loudspeakers.

Stereophonic system   Sound picked up by two microphones is fed to two loudspeakers in the listening room. This popular system, shown in Fig. 25.8(d), will be discussed in Section 25.6.

Surround-sound system   Sound picked up by multiple microphones feeds multiple loudspeakers in the listening room. This system, shown in Fig. 25.8(e) will be discussed in Section 25.10.

25.6 ■ STEREOPHONIC SOUND

The most popular, and perhaps the most successful, system for reproducing sound with a spatial dimension is the stereophonic system. In the basic stereophonic (or stereo) system, sound from the source is picked up by two microphones, recorded or transmitted in two separate channels and reproduced by two loudspeakers. In theory, there are many ways to do this. In practice, several different arrangements of microphones and loudspeakers have been used with varying degrees of success.

Although experiments with stereophonic sound took place in the 1930s at the Bell Telephone Laboratories in the United States and at Electric and Musical Industries Ltd. (EMI) in England, stereophonic sound recording did not develop commercially until the mid-1950s. The Bell Labs experiments are described in a series of papers by Fletcher et al. (1934), whereas the EMI work by Alan Blumlein and his colleagues resulted in a series of patents (Lipshitz 1986). Early stereophonic recordings of the Philadelphia Orchestra (conducted by Leopold Stokowski), made by Bell Labs engineers in 1931–32, are of remarkably high quality.

Blumlein's preferred technique for recording stereophonic sound was to use a pair of bidirectional (figure-of-eight) microphones with their axes of maximum response at an angle of 90° to each other. This arrangement, still used by many recording engineers, is referred to as a stereosonic system, or a *Blumlein pair*.

Stereo listening tests were performed at a convention of the Audio Engineering Society several years ago, using a musical selection recorded in a concert hall with six different microphone arrangements used for stereophonic recording. In the listening tests, 34 listeners were seated in a favorable stereophonic listening area, and 30 listeners were seated in unfavorable positions. Although the listening position had little influence on some parameters such as intimacy, dynamic range, and brilliance, it was found that impressions of liveness, warmth, and source width were very much dependent on listening position (Ceoen 1972). A recording system that used a pair of cardioid microphones spaced 17 cm apart with their axes of maximum response at an angle of 110° was judged to give the best results overall.

Many different microphone arrangements have been used for stereophonic recording. They may be classified as *coincident pairs*, *near-coincident pairs*, *spaced pairs*, or *baffled pairs*. The following are examples of systems that have been rather widely used:

1. *XY system.*   A coincident pair of cardioid microphones with their axes of maximum response at an angle of 135°.

2. *Stereosonic system, or Blumlein pair.*   A coincident pair of velocity microphones with their axes of maximum response at an angle of 90°.

3. *MS (midside) system.*   A coincident pair consisting of a forward-pointing cardioid microphone and a sideways-pointing velocity microphone. The sum and difference of the signals from these two microphones are recorded as the right and left stereo channels. (This system is also described in Blumlein's early patents.)

4. *ORTF system.*   A nearly coincident system consisting of two cardioid microphones spaced 17 cm apart with their axes of maximum response at an angle of 110°. (ORTF is the French broadcasting system.)

5. *NOS system.*   A pair of cardioid microphones spaced 30 cm apart with their axes of maximum response at an angle of 90°. (NOS is the Dutch broadcasting system.)

6. *A-B, or spaced pair, system.*   A pair of microphones spaced several feet apart. The microphones can have any pickup pattern but omnidirectional seems to be the most popular. This technique, widely used by amateurs, tends to give an exaggerated stereo effect if the spacing is too great.

7. *OSS, or baffled, pair.*   Two omnidirectional microphones are separated a few inches by a baffle in between them. The baffle is often a hard disk with an absorbent foam (as in the Jecklin disk).

The specific arrangement of microphones is always a matter of taste. For some people, coincident microphone techniques sound dry or analytical—too "correct." More of a feeling of spaciousness can be created by introducing some signal delay between the two channels or by using spaced pairs. In recording studios, each instrument often has its own (or even more than one) microphone, and the signals are mixed together to create a stereophonic signal. A comprehensive discussion of microphone techniques is given by Streicher and Everest (1998).

One of the important criteria for realism in reproduced sound is that the spatial sound pattern of the original sound should be reproduced. Stereophonic sound systems accomplish this to a much greater degree than do monophonic systems, providing the listener is seated in a favorable stereophonic listening area (sometimes referred to as a sweet spot). What is a "favorable" listening area? It means being fairly close to the median plane and at a distance such that the two speakers are separated by an angle of about 40 to 90°. This frequently presents difficulties in arranging the home living room for good stereo.

In a rectangular room, the best location for the speakers is usually in the corners of the end wall. Corner placement provides good room coverage and also enhances the radiation of low-frequency sound. Figure 25.9 shows three loudspeaker arrangements in a room having dimensions in the ratio 3 : 2. Note that the most favorable stereo listening area in Fig. 25.9(a) is substantially larger than those in Fig. 25.9(b) and (c). The favorable listening area may be moved to one side or the other of the median plane by changing the balance control of the amplifier to give one speaker a greater gain. There are practical limits to this, however, because only the relative loudness of the speakers, and not the relative phase, is changed.

Although the limiting angles (40° to 90°) suggested here are somewhat arbitrary, it is well known that the spatial effect of stereophonic sound disappears outside a certain range

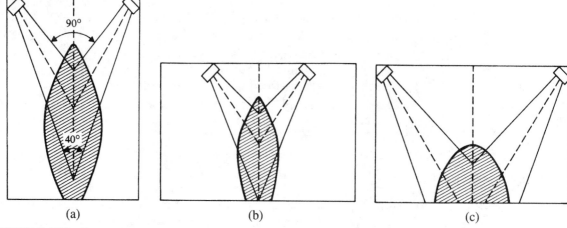

FIGURE 25.9   Favorable stereo listening areas for three different loudspeaker arrangements in a rectangular room with dimension in the ratio 3 : 2. (After Rossing 1981.)

of angle. If the angle is too narrow, the source appears to be monophonic. If the angle is too great, the listener hears two distinct sources with a distracting "hole" in the center.

If you are installing a stereophonic sound system in a home listening room, considerable experimentation with speaker placement is recommended. The results vary somewhat from room to room due to the influence of absorbing and reflecting surfaces in the sound field.

25.7 ■ THE SOUND FIELD IN LISTENING ROOMS

Stereophonic sound-reproducing systems can be quite successful in satisfying four of the five criteria for realism in reproduced sound, including the reproduction of the spatial sound pattern, when the loudspeakers and the listeners are in favorable locations. However, most listening rooms do not begin to produce the ambience experienced in a concert hall.

We can walk into a strange room blindfolded and rather quickly and accurately estimate its size by listening to the sound in the room. We do not have to make any special effort to do this; our auditory processor has been programmed by experience. Presumably, when reflected sounds follow the direct sound with little delay, an auditory impression of smallness is created. Conversely, when the reflections arrive with a distribution in time and space that is characteristic of a large concert hall, a feeling of spaciousness is created. Various attempts have been made to introduce some of the features of concert hall sound into home listening rooms. Placing additional speakers in the room creates some feeling of spaciousness, because the sound arrives from several directions.

In the room described in Example 25.2 (see Section 25.1), nearly all listeners would find themselves well into the reverberant field for sounds of low frequency, but for sounds of high frequency, the effect would depend on where they were seated. This situation is amplified by the fact that the absorption in most listening rooms is greater at high frequency. Although the direction of a source is determined by the direct sound (precedence effect), the tonal balance appears to be derived from the total sound reaching the ear. This strongly suggests that the frequency response of a loudspeaker should be described by the total *power* radiated at each frequency rather than the sound pressure level on its axis (which depends on the directivity factor Q at that frequency).

In one experiment, the spectral distributions of sound from a high-quality loudspeaker, placed at 22 locations in 8 listening rooms, were measured (Allison and Berkovitz 1972). When these measurements were compared to similar curves made at favorable seat locations in concert halls, it was found that in the important midband range (250–2500 Hz), the curves were similar. Below 250 Hz, however, the sound levels of the combined loudspeaker-room systems were considerably below those in the concert halls. This loss of low-frequency sound in the smaller rooms can be attributed to inadequate stiffness of walls, windows, etc., which causes them to absorb low-frequency sound by vibrating sympathetically. At high frequency, on the other hand, the average sound level in listening rooms was higher than in concert halls, relative to the low frequencies. This experiment suggests that to emulate the tonal balance heard in a concert hall, the tone controls of a high-fidelity system should be adjusted for bass boost and treble cut, rather than for flat response.

Another factor to be considered is the loading effect of nearby walls on a loudspeaker radiating sound of long wavelength. Placing a loudspeaker in a corner of a room can increase its total radiated power at low frequency by as much as 9 dB. What happens, in effect, is that the walls and floor of the room form a sort of pyramidal horn that increases the efficiency of radiation (see Section 19.9). The walls also cause an unevenness in frequency response, however, which can be minimized by placing the loudspeaker so that the distances from the woofer cone to each of the nearby reflecting surfaces differ by at least a factor of two (Allison 1979).

25.8 ■ SOUND DIFFUSORS

The sound we hear in a listening room is a combination of the direct sound and reflections from many surfaces in the room. In a large auditorium, there is substantial delay between the direct and reflected sound (see Section 23.2). Reverberant sound comes from a complex mixture of reflections coming from many different surfaces at different times. In small rooms the sound is often absorbed before there is a uniform mixture. The purpose of sound diffusors is to provide this mixture of reflected sounds, even in a small room.

As we learned in Section 23.1, sound waves incident on a surface will be absorbed or reflected (or a combination of both). Flat walls and concave surfaces tend to direct the sound, whereas convex or rough surfaces scatter the sound in several directions. Geometric shapes attached to room surfaces help to scatter and diffuse the sound. Rather than reflecting sound in a single direction, as a flat surface does (Fig. 23.2(a)), triangular shapes, rectangular protrusions, and semicylindrical surfaces—as shown in Fig. 25.10—scatter sound in many directions, resulting in a diffuse sound field, even in a small room.

In Section 3.10 we learned that waves passing through a narrow opening spread out in all directions due to diffraction. Their wavelength, compared to the size of the opening, determines how much they spread. With two or more openings, waves spread out from each opening, causing constructive and destructive interference, depending on the phase of the waves. This is the principle of the diffraction grating used to study light; many slits spread light into its various frequencies (colors).

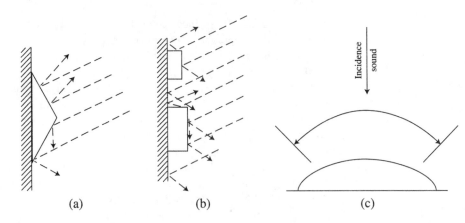

FIGURE 25.10
Diffusion by geometric shapes: (a) triangular shape; (b) rectangular protrusions; (c) semicylindrical surface.

(a)   (b)   (c)

In 1975, acoustician Manfred Schroeder suggested using an acoustical grating to create an effective sound diffusor. The theory of phase-grating sound diffusors is based on number theory; Schroeder used a scheme called maximum-length sequences, a stream of fixed-length digital 1s and 0s with some interesting statistical properties. Schroeder bent a piece of sheet metal into such a pattern of 1s and 0s to confirm his theory of acoustical scattering from such an object. (The piece of metal is in the museum of the University of Göttingen, where he worked.)

Since then, *quadratic-residue diffusors* and *primitive-root diffusors* have been developed. Figure 25.11 shows a cross section of a phase-grating sound diffusor consisting of a structure with a sequence of wells to scatter sound within a certain frequency band. The sequence is repeated along the diffusors. The maximum depth of the wells determines the effective low-frequency limit of the diffusors. The well depth should be $1\frac{1}{2}$ times the wavelength at the lowest frequency. The highest frequency is determined by the well width, which is half a wavelength at the highest frequency. The actual sequence of wells is determined by number theory, which is beyond the scope of this book. Figure 25.12 shows the ability of a phase-grating sound diffusor to scatter sound in all directions. A variety of practical phase-grating diffusion panels are available commercially from RPG Diffusor, Inc.

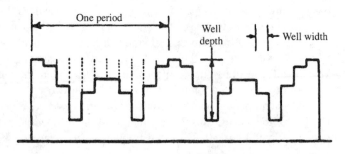

FIGURE 25.11
Cross section of a grating (quadratic residue) sound diffusor.

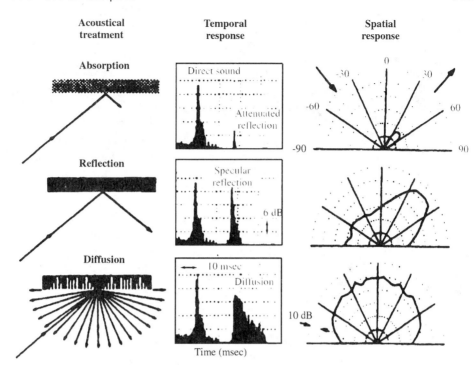

FIGURE 25.12
A comparison of
sound incident on
three surfaces.

25.9 ■ SOUND ABSORPTION

Sound absorption in a room depends mainly upon the surface area of the walls, ceiling, and floor and on the nature of the surfaces, as we learned in Section 23.5. The reverberation time depends upon the volume to surface ratio. In a small room, this ratio is usually rather small, and so sounds decay rather rapidly. (Indeed, we notice a clear difference in the sound when a carpet or a sofa is removed.) Furthermore, a small listening room generally has furniture whose upholstery adds a generous amount of absorption. On the other hand, recorded and broadcast music generally carries the reverberation signature of the concert hall or studio in which it was recorded, and so the listening room should be almost free of reverberation in order not to distort this.

Porous materials, such as drapery, carpets, glass fiber, and acoustical tile, convert acoustic energy to heat energy as the vibrating air particles interact with the tiny fibers in the absorber. Porous materials absorb very well at high frequency (see Table 23.1). Panels of wood, glass, gypsum board, and even plaster on lath, on the other hand, vibrate in response to sound waves of long wavelength and thus absorb rather well at low frequency but have very little absorption at high frequency. This is convenient for the acoustician or architect, because it allows "tuning" of the room by careful choice of absorbing materials.

A third type of absorber that depends upon the principle of the Helmholtz resonator (see Section 4.7) can provide absorption over a selected frequency band. A Helmholtz resonator absorbs sound energy at or near its resonance frequency. Some old churches in Scandinavia had clay pots imbedded in the walls to act as Helmholtz resonators (Brüel 1951). A more practical way to achieve this type of absorption, however, is to cover a

FIGURE 25.13
A *bass trap* sound
absorber.

cavity with a perforated panel. The resonance frequency is given approximately by

$$f = 200t\sqrt{\frac{p}{d}}, \tag{25.4}$$

where p is the perforation percentage (hole area/panel area $\times$ 100), t is the effective hole length with the correction factor (panel thickness + (0.8) $\times$ hole diameter), and d is the depth of the air space (inches).

Home listening rooms are frequently used for other purposes as well. A living room that is to be used for the performance of chamber music, for example, should not have as much absorption as a room for listening to recorded music. Thus, multipurpose use presents a dilemma. It is possible to partially vary the absorption of a room by making changes, such as drawing the drapes, for example. Other times, portable absorbers can be added or taken away. One type of absorber, shown in Fig. 25.13, is often called a *bass trap* because of its absorbing ability at low frequency. The trap uses cavity resonance with alternating layers of porous material and air to absorb frequencies having a quarter wavelength equal to the depth of the trap. For example, a trap 1 m deep would absorb a frequency of about 84 Hz. Such absorbers are most effective when placed in the corners of a room, where all modes have pressure maxima.

25.10 ■ HOME THEATERS

For many people a home theater means purchasing a large-screen TV and a surround-sound system and installing them in their favorite family room. Surround-sound systems were discussed in Section 20.10. In this section, we will consider how application of acoustic design principles can lead to a quality home theater which better simulates a movie theater.

One consideration in home theater design is the size and shape of the room itself. Typically, a home theater will be larger than the average high-fidelity listening room but small enough to be considered a small room acoustically. As we learned in Section 25.2, there are certain room-dimension ratios that favor the distribution of room-mode frequencies.

The next consideration in home theater design is loudspeaker placement. The 5.1 surround-sound standard uses five loudspeakers (left, right, center, left surround, and right surround) plus a subwoofer. (The center speaker is optional in some recordings but important in motion pictures.) In Section 25.3 we learned that two properly phased loudspeakers could create a central image in a limited area (the sweet spot). In home theaters this is impractical because there is often more than one listener. This is why home theater systems usually have a center loudspeaker that primarily reproduces the speech dialogue, which is intended to be centrally localized. (Compare this with the use of a single loudspeaker for speech reinforcement in auditoriums in Section 24.4.) Movie theaters use a central speaker behind the screen for the same reason.

The left and right surround loudspeakers can be placed as a dipole surround (Fig. 25.14(a)) or a matching surround (Fig. 25.14(b)). In the United States, the most common loudspeaker arrangement follows the THX dipole-surround format, where the surround speakers are located on the side walls to the left and right of the listener. In the matching-surround setup, the listener sits in the center of a circle with all five loudspeakers being equidistant. The left and right speakers are 60° apart, with the center speaker mid-

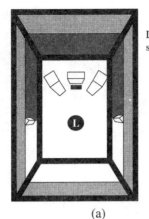
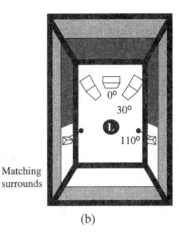

FIGURE 25.14
Surround-sound
loudspeaker
placement patterns:
(1) dipole
surrounds;
(b) matching
surrounds.

(a)   (b)

way between them. The left and right surround speakers are 140° apart and 110° from the
forward position, as seen in Fig. 25.14(b).

The next task is to control early reflections. In a large auditorium early reflections enhance the sound and give a feeling of spaciousness; if they arrive early enough they give
a feeling of intimacy. In a small room, however, early reflections interact with the direct
sound, resulting in a "hilly" (rather than flat) response that has peaks and valleys in the
spectrum. The spatial effect in surround-sound systems is achieved mainly with signal processing rather than room acoustics, and so early reflections are reduced.

25.11 ■ SOUND-RECORDING STUDIOS

Sound-recording studios vary widely in size and design. Studios as small as 300 m^3
(1000 ft^3) may be used to record soloists or small ensembles. A small chamber-music
studio with a volume of about 1000 m^3 (35,000 ft^3) would accommodate a small orchestra, choir, or instrumental ensembles, whereas a large music studio, such as the one in
Fig. 25.15, may have a volume of 2000 m^3 (70,000 ft^3) or more. The studio must be large
enough so that the musicians feel comfortable, but the path length of sound reflections to
the microphones must be kept small.

Unlike in home theaters, reverberation time is carefully controlled in recording studios.
As seen in Fig. 23.7, the desired reverberation time for a studio depends upon the studio
volume and the style of music being recorded. Reverberation times in a studio are usually
shorter than those found in the concert hall. In a chamber-music studio the reverberation
time is typically 0.9 to 1.2 s, whereas a larger studio could vary between 1.2 and 2.4 s.
Panels with variable absorption can be used to alter the reverberation times in studios to
adapt to different recording venues. One example is a rotatable panel that is flat on one side
with absorptive material, the other side being convex and treated with hardboard to make
it reflective. Reverberation time is changed by rotating the reflective side out to increase
reverberation time or rotating the absorptive side out to decrease it (see Section 23.11).

Creating diffuse sound by spreading out room resonance frequencies and scattering
reflections is even more important in recording studios than in home theaters. To reduce the

FIGURE 25.15
Large recording
studio.

risk of flutter echoes and to distribute room resonance frequencies more evenly, recording studios are often built in irregular shapes rather than with parallel walls. The wide variety of studio shapes is left up to the imagination of the designer. Diffusor panels are also used in combination with absorption to control reflections in the studio. Placement of diffusors and absorbers will depend upon the shape of the room. One important parameter is the initial time delay (ITD), which is the time difference between the direct sound and the first reflected sound reaching the microphone. This helps determine the intimacy of the music. This parameter is controlled by placement of the performer and the microphone and the surface on which the reflection occurs.

A key element in studio design is noise isolation. The ambient noise must be kept as low as possible. The first step in noise control takes place before the studio is built; a site-noise survey characterizes the noise in the environment. From this, a specification for noise isolation is written to give construction details of the walls, doors, and windows so that it is quiet inside the studio. Details about environmental noise are found in Chapter 30 and noise-control ideas are given in Chapter 32. Along with controlling outside noise, it is essential to control noise from the heating and air conditioning systems of the studio itself.

There are circumstances that call for sound isolation within the recording studio. The overall ensemble sound is usually desired from orchestras and choirs; therefore, microphones are placed a distance from the musical group so that the sounds blend before they get to the microphones. For solos and smaller ensembles, close miking is often used to record individual sounds of the performers to be mixed later in postproduction. In this case it is undesirable for the sound of one performer to be picked up by the microphone of another performer. The *rule of three* is often used to make sure the distance from a musician to any other microphone is at least three times the distance to his or her primary microphone. In some cases (such as in rock music), extra isolation is required. This is especially true for a drum set, which requires its own isolated room called a drum cage. The purpose of the drum cage is to reduce the high sound levels typical of a drum set at the position of the other instruments. A vocal booth is also provided to isolate vocalists from instrumentalists.

25.12 ■ CONTROL ROOMS

Control rooms are usually associated with recording studios. They are treated separately here because they differ in their acoustical needs. A control room contains a mixing console and recording equipment with space for a mixing engineer or mixer (the person responsible for mixing sounds from the studio onto recorded tracks) and possibly other observers. Monitor loudspeakers deliver the sound from the studio (or previously recorded tracks) to the mixing engineer. An observation window separating the control room from the studio allows the mixer to know exactly what is happening in the studio. The acoustic requirement for the control room is to be absolutely neutral so that there is no coloration of sound between what is delivered by the monitors and what is received at the mixer's ears.

The primary consideration for the observation window is transmission loss, the reduction of sound between the studio and the control room (see Section 32.5). The transmission loss of a single-pane window increases about 3 dB with every doubling of the window mass. The mass can be increased by making a thicker window or by laminating the glass (glass being the most common material used for observation windows). Plastic is also used, but the density of plastic is half that of glass, so the window would have to be twice as thick for the same transmission loss. To increase the transmission loss, a second pane is commonly used, with an air gap between the two panes. Besides the added transmission loss for the second pane, there is a 3-dB increase of transmission loss for each doubling of the distance between the two panes of glass.

As with home theaters, controlling early reflections is of major importance in control-room design. The effect of early reflections can be seen in Fig. 25.16. The sound energy received at the ear consists of direct sound followed by early reflected sound delayed in time (Fig. 25.16(a)). As the direct and early reflected sound interfere, a hole in the frequency response occurs, coloring the sound (Fig. 25.16(b)). We have adapted to this coloration caused by early reflections when listening to home hi-fi, considering it part of the ambiance. In home theaters, however, the ambiance is provided by signal processing in the system and should not be modified. It is even more important to control early reflections in a control room so that the sound the mixer hears is the actual sound coming from the recording studio. This includes the ITD (discussed with recording studios), which determines the intimacy of the studio. Being a smaller room, the ITD of the control room (1–5 ms) is shorter than the ITD of the studio, obscuring the sound from the studio. Elimi-

FIGURE 25.16
Comb filtering:
(a) time delay
between direct
sound and first
reflection;
(b) degradation in
response curve
from interference.

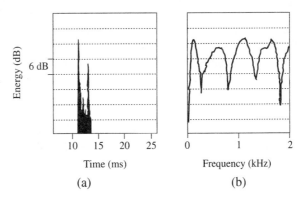

FIGURE 25.17
Creating a
reflection-free zone
by shaping
reflective surfaces
to direct sound
away from the
mixer.

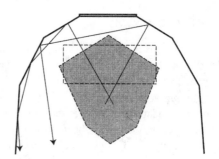

nating early reflections from the front end of the control room lengthens the control room ITD (10–20 ms) because the first reflection now comes from the rear of the room. With the control room ITD longer than the studio ITD, the mixer can hear the sound from the studio more clearly.

The need to control early reflections led to the live-end dead-end (LEDE) concept of control-room design. The idea was to make the front half of the control room a reflection-free zone by adding massive absorption, whereas the back half was very diffusive. The mixing engineer would turn up the gain on the system to compensate for losses due to absorption. A reflection-free zone can also be created by shaping reflective surfaces (Fig. 25.17) to direct the sound toward the rear of the control room away from the mixer. This reduces the need for extra gain to compensate for absorptive losses because there is no dead end.

Double-shell control rooms are common, consisting of a heavy outer shell and a lighter inner shell. The outer shell (often made out of concrete) is used for noise isolation and control of low-frequency modes. The inner shell can be made of lighter materials. It is shaped to control early reflections and provide diffusion.

25.13 ■ SUMMARY

High-fidelity sound reproduces much of the spectrum, dynamic range, and spatial characteristics of the original sound with minimal addition of distortion and noise. Important considerations in understanding the sound field in a listening room include (1) room resonances, (2) the relationship of reverberant to direct sound, (3) sound images from multiple sound sources, and (4) absorption and diffusion from walls and furniture within the room.

Home theaters generally use surround-sound technology with five loudspeakers plus a subwoofer to create a theaterlike environment. Control of early reflections and diffusion of sound are important considerations in maintaining the proper acoustic environment. Sound-recording studios must be carefully isolated from neighboring noise sources. Control rooms are carefully designed with a reflection-free zone in the front half of the room and a reflecting-diffusing zone in the rear of the room.

REFERENCES AND SUGGESTED READINGS

Allison, R. F. (1979). "Influence of Listening Rooms on Loudspeaker Systems," *Audio* **63**(8): 37.

Allison, R. F., and R. Berkovitz (1972). "The Sound Field in Home Listening Rooms," *J. Audio Eng. Soc.* **20**: 459.

Ballou, G. (1998). *Handbook for Sound Engineers*. Indianapolis: Howard W. Sams.

Benson, K. B. (1988). *Audio Engineering Handbook*, New York: McGraw-Hill, Chapter 13.

Brüel, P. V. (1951). *Sound Insulation and Room Acoustics*. London: Chapman and Hall.

Ceoen, C. (1972). "Comparative Stereo Listening Tests," *J. Audio Eng. Soc.* **20**: 19.

Davis, D., and C. Davis (1987). *Sound System Engineering*, 2nd ed., Indianapolis: Howard W. Sams, Chapter 9.

Egan, M. D. (1988). *Architectural Acoustics*. New York: McGraw Hill.

Everest, F. A. (1989). *The Master Handbook of Acoustics*. Blue Ridge Summit, Penn.: TAB Books.

Everest, F. A. (1997). *Sound Studio Construction on a Budget*. New York: McGraw-Hill.

Fletcher, H., et al. (1934). Symposium on Wire Transmission of Symphonic Music and Its Reproduction in Auditory Perspective. *Bell System Tech. J.* **13**: 239.

Fletcher, N. H., and T. D. Rossing (1998). *Physics of Musical Instruments*, 2nd ed. New York: Springer-Verlag.

Gardner, M. B. (1973). "Some Single- and Multiple-Source Localization Effects," *J. Audio Eng. Soc.* **21**: 430.

Gilliom, J. D., and R. D. Sorkin (1972). "Discrimination of Aural Time and Intensity," *J. Acoust. Soc. Am.* **52**: 1635.

Hirsch, J. (1979). "Audio Listening-Room Expanders," *Popular Electronics* **15**(2): 41.

Holman, T. (2000). *5.1 Surround Sound: Up and Running*. Boston: Focal Press.

Kuttruff, H. (1997). "Sound in Enclosures," in *Encyclopedia of Acoustics*, Vol. 3, ed. M. J. Crocker. New York: Wiley.

Lipshitz, S. P. (1986). "Stereo Microphone Techniques . . . Are the Purists Wrong?" *J. Audio Eng. Soc.* **34**: 716.

Madsen, E. R. (1970). "The Disclosure of Hidden Information in Sound Recording," (Lecture given May 1970).

Møller, H. (1974). "Relevant Loudspeaker Tests in Studios, in HiFi Dealers' Demo Rooms, in the Home, etc.,—Using $\frac{1}{3}$-Octave, Pink-Weighted, Random Noise," 47th Audio Eng. Soc. Convention, Copenhagen. Reprinted as B & K Application Note 17-197 (B & K Instruments, Marlborough, Mass.).

Olson, H. F. (1969). "Home Entertainment: Audio 1988," *J. Audio Eng. Soc.* **17**: 390.

Olson, H. F. (1972). *Modern Sound Reproduction*. New York: McGraw-Hill.

Rossing, T. D. (1979). "Physics and Psychophysics of High-Fidelity Sound, Part I," *Physics Teach.* **17**: 563.

Rossing, T. D. (1981). "Physics and Psychophysics of High-Fidelity Sound, Part IV," *Physics Teach.* **19**: 293.

Schroeder, M. R. (1975). "Diffuse Sound Reflections by Maximum-Length Sequence," *J. Acoust. Soc. Am.* **57**: 149.

Schroeder, M. R. (1975). "Models of Hearing," *Proc. IEEE* **63**: 1332.

Schroeder, M. R., and B. S. Atal (1963). "Computer Simulation of Sound Transmission in Rooms," *IEEE Int. Conv. Record*, Pt. 7: 150. See also *Am. J. Phys.* **41**: 461 (1963).

Streicher, R., and F. A. Everest (1998). *The New Stereo Soundbook*. Pasadena: Audio Engineering Associates.

Young, L. L., Jr., and R. Carhart (1974). "Time-Intensity Trading Functions for Pure Tones and a High-Frequency AM Signal," *J. Acoust. Soc. Am.* **56**: 605.

GLOSSARY

ambience Spaciousness; the degree to which sound appears to come from many directions.

bass traps A tuned absorber used to absorb low frequencies in a room.

binaural Sound reproduction using two microphones (usually in a "dummy" head) feeding two headphones, so that the listener hears the sound he or she would have heard at the recording location.

cardioid microphone A microphone with a heart-shaped directivity pattern designed to pick up sound in one direction preferentially.

comb filters Filters that reject regularly spaced frequencies through interference.

control room The important part of a recording studio in which the mixing, processing, and recording equipment, and listening monitors are located.

diffusion The mixing of sound to make it more homogeneous by scattering off irregular objects.

high-fidelity sound Sound that reproduces much of the spectrum, dynamic range, and spatial characteristics of the original sound and adds minimal distortion and noise.

home theater A room designed and equipped for viewing movies as well as listening to music.

initial time delay (ITD) The time difference between the arrival of the direct sound and the first reflected sound at the ear.

live-end, dead-end (LEDE) A control-room design philosophy that makes the front half of a control room absorptive and the back half diffusive.

monaural Sound reproduction using one microphone to feed a single headphone, such as is used in telephone communication.

monophonic Sound reproduction using one microphone to feed one or more loudspeakers with one signal.

pink noise Random noise that has the same power in each octave or $\frac{1}{3}$-octave band.

quadraphonic Sound reproduction using four microphones to feed four loudspeakers; usually two are in front of the listener and two are behind or to the sides.

reverberant sound Sound that reaches the listener after a large number of reflections; as one moves away from a sound source, the sound level reaches a steady value called the reverberant level.

stereophonic Sound reproduction using two microphones to feed two loudspeakers.

surround sound A system using multiple channels of sound and multiple loudspeakers to envelop the listener and create a feeling of spaciousness.

sweet spot The listening area in a room where the best spatial imaging takes place.

transmission curve A frequency-dependent curve showing how well sound is transmitted through a barrier.

REVIEW QUESTIONS

1. What is the difference between sound fields in a large room and a small room?
2. What is the effect of room resonances on sound quality?
3. How can you determine the acoustic size of a room?
4. What is the precedence effect?
5. If I am equidistant from two loudspeakers, each producing the same sound, where will the image of the sound be?
6. If the left loudspeaker is moved farther away, how will the image of the sound move?
7. What is high-fidelity sound?
8. What happens to a sound wave as it hits a reflector? an absorber? a diffusor?
9. Name three materials used for high- and midfrequency absorbers.
10. What is the difference in how a bass trap and a diaphragmatic absorber reduce low-frequency sounds?
11. How can a Helmholtz resonator absorb sound?
12. What is the effect of early reflection in home theater listening?
13. How is a drum cage used in a recording studio?
14. Does slanting one pane of glass in a two-pane observation window reduce the sound transmitted between the studio and control room?
15. How is it possible to avoid massive absorption in the dead-end portion of a control room?

QUESTIONS FOR THOUGHT AND DISCUSSION

1. In binaural sound reproduction, why are the two microphones placed in a dummy head rather than merely given that same separation in air?
2. Comment on the practice of connecting a third speaker between the two stereo channels to eliminate the hole in the middle.
3. Is the restriction on listener location in the room likely to be a serious deterrent to the acceptance of quadraphonic sound?
4. Is it possible to have realistic stereophonic sound in an automobile? What about surround sound?
5. What is the difference between a binaural system and a stereophonic system for recording and reproducing sound?
6. Which geometric shape would be best for diffusing sound?
7. Would an airspace behind a foam absorber help as much as an airspace behind an acoustic tile?
8. Which gives the best listening environment, the dipole surround or matching surround? Why?
9. Why would you want the reverberation time to be less when recording an ensemble in a studio than when listening to the same ensemble in a concert hall?
10. What are the pros and cons of a double-shell control room versus a single-shell control room?

EXERCISES

1. Calculate the frequencies of the first three resonances of a room with dimensions 5 m × 10 m × 2.5 m. Do they have any significance acoustically?

2. Suppose that two loudspeakers, each 30° from the median plane, carry the same program material, but the loudspeaker on the left has twice the signal strength of the one on the right. Describe the location of the image. (Calculate $\sin \theta_I$, and look in a set of tables or use a pocket calculator to determine the angle θ_I corresponding to your value of $\sin \theta_I$; $\sin 30° = \frac{1}{2}$.)

3. Draw a diagram of a room with dimensions in the ratio 4 : 3 and indicate the favorite listening areas for two different placements of a pair of speakers in a stereo system.

4. From the chart given in Fig. 24.2, determine the distance from a loudspeaker with $Q = 5$ in a room with $A = 500$ at which direct and reverberant sound are equal in level. (*Hint:* $L_p - L_w$ should be 3 dB above the reverberant level.)

5. The sketch below represents the room described in Exercise 1. Indicate the location of the maximum sound pressure level when the room is excited in each of its first three resonances.

6. Suppose the time/amplitude trading ratio is found to be 100 μs/dB. If one speaker is 0.2 m farther away than a second, how much greater must its sound power level be in order that the sound image will appear to be on the median plane?

EXPERIMENTS FOR HOME, LABORATORY, AND CLASSROOM DEMONSTRATION

Home and Classroom Demonstrations

1. *Stereophonic music on an oscilloscope*   One of the best ways to present stereophonic music visually is to display the two channels on the X and Y axes of an oscilloscope. Switch the amplifier to mono to obtain a linear trace. Switch to stereo for a two-dimensional display. Note that the correlation between channels varies widely in different stereo recordings. Old monophonic recordings reprocessed for stereo are easily recognized.

2. *Combining direct and reflected sound: comb filtering*   Place a microphone 17.2 cm from a wall and direct white noise from a loudspeaker toward the wall. Observe the spectrum of the sound picked up by the microphone. The reflected signal will reach the microphone 1.0 ms after the direct sound, which results in minima in the observed spectra at 500 Hz, 1500 Hz, 2500 Hz, etc. Changing the distance from the microphone to the wall changes the comb frequencies.

3. *Maximum and minimum sound levels in a small room*   Carry a sound-level meter around a small room and determine the maximum and minimum sound levels in the sound field of a broadband noise source.

4. *Standing waves in a small room*   Carry a sound-level meter around a small room and determine the maximum and minimum sound levels in the sound field of a 1000-Hz tone. What is the difference in maximum and minimum levels? How far apart are the maxima? Repeat with a 200-Hz tone.

5. *Time-intensity trading*   Trade-off between amplitude and time delay can be illustrated with the arrangement in Fig. 25.3(c). Try to determine the limits of time-intensity trading.

6. *Differences in sound recording methods*   Several CDs that illustrate the sound quality obtained from different microphone arrangements are available. For example, tracks 35–44 on Denon PG6006 present music by Mozart and Bruckner recorded using *single-point miking*, with auxiliary microphones, and with multimicrophone methods.

Laboratory Experiments

Acoustical modes of a box (Experiment 35 in *Acoustics Laboratory Experiments*)

Reflection and absorption of sound: the impedance tube (Lab 36 in *Acoustics Laboratory Experiments*)

PART
VII

Electronic Music Technology

For most people, the vast majority of musical experiences are now electronically mediated by amplifiers, filters, loudspeakers, and recording devices, even when the instruments heard are of the traditional "acoustic" variety. Increasingly, the music being heard is fully or partly *synthesized* in ways that may or may not be apparent to the listener. Just as movie special effects now can be seamlessly incorporated into natural images, electronic music technology has progressed to the point where its users can decide whether to make an "artificial" sound or not. This capability is due to several key developments over the past century or so, starting with *sound recording* and *transmission*, progressing through the development of *analog* and *digital electronics*, the advent of *vacuum tubes*, *transistors* and *integrated circuits*, and on to *computers*, *digital signal synthesis*, *analysis* and *processing*, and the *Internet*.

Modern capabilities also rest on many developments specific to electronic music, including *electronic musical instruments*, *musique concrète* and *elektronische musik*, *voltage-controlled synthesizers*, *computer music*, *digital music synthesizers* and *samplers*, *digital audio signal processing*, *music synthesizers* and *samplers*, and *perception-based methods for compressing sound signals* for efficient storage and transmission over almost any type of communications channel.

Chapters 26–29 describe some of the principles of electronic music technology with emphasis on both the concepts and the techniques. The reader should have an understanding of basic electronic circuits and digital techniques as presented in Chapters 18 and 21; a reading of all five chapters in Part 5 is recommended.

CHAPTER
26

Electronic Music Technology and its Origins

Electronic music technology allows its users to synthesize sounds that go well beyond those of traditional musical instruments as well as to imitate them in increasingly transparent ways. It allows recorded sounds to capture not only the tone qualities but also the spatial characteristics of the original sound sources and their reverberant environments. By accounting for human perception, electronic music technology allows the best possible sound quality to be transmitted or stored using the smallest possible amount of information. The word *technology* comes from the Greek *technē*, meaning art or skill, plus *logos*, meaning word. The Greeks also have the word *technologia*, meaning the systematic treatment of an art. Music technology in general is, therefore, "what can be said about the art or skill of music." We can thus define *music technology* as *the sum total of what we know how to do in order to make music*, such as how to make musical instruments and rooms in which music may be suitably heard. *Electronic music technology* is that part of music technology based primarily on the field of electronics and, especially, digital electronics.

This basis has led (so far) to the beginnings of artificial musical intelligence, the synthesis of new musical sounds, and the accurate and flexible re-creation of traditional musical sounds by artificial means. It increasingly includes the incorporation of arbitrarily specifiable sound into computer-based media, such as virtual reality, movies, games, and radio-, TV-, and web-based telecommunications. Not only do today's computers and related devices talk and sing and play music, they increasingly listen to us doing the same things, and in increasingly interactive and naturalistic ways. It seems clear that as we use a computer keyboard and mouse less and less, our interactions with machines are becoming more like our interactions with each other. As machines become more adept at experiencing humans on their own terms, they also become more adept at helping us communicate more effectively with each other. The Internet is already connecting people with each other and with their most profound desiderata (for better or worse) in unprecedented ways. Such technological capabilities are already changing human culture in far-reaching ways.

Technology is clearly an expression of our own imaginings and desires as well as our know-how, with all the consequent social and ethical implications. Therefore, it is helpful to view technology both in terms of its content and its history. The relevant technical details (i.e., its content) describe the current state of technological power. We must keep in mind, however, that little changes more rapidly than the current state of technology, especially computer-related (i.e., "high") technology. History tells us much of what we know about what we really intended all along to do with technological power, given the chance. Fortunately, music technology has a rich and extensive history, and very little of it appears to be as troubling as some other types of technological capability, such as that of warfare. We begin our overview of music technology with a brief overview at some of its conceptual

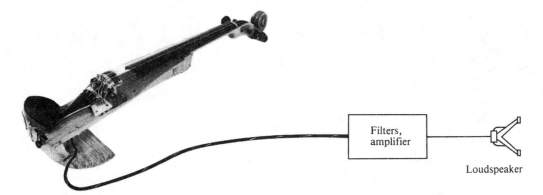

FIGURE 26.1   Electronic violin with piezoelectric pickups and filters replacing the resonances of the (missing) violin body. (From Mathews and Kohut 1973).

beginnings. As we saw in Chapter 9, the ancient Greeks—especially the Pythagoreans—laid important foundations for the relationship of music and mathematics. Knowledge of such relationships—combined with knowledge of acoustics and electronics—allows the construction of many interesting types of musical instruments, such as the electronic violin shown in Figure 26.1.

In this chapter, you should learn:

- About the major components of contemporary music technology;
- About some of the ancient precursors to contemporary music technology;
- About early technology for recording sound;
- How understanding electricity and electronics enables contemporary music technology;
- About electroacoustic music developments early in the twentieth century;
- How quantum physics and transistors form the basis for contemporary music technology.

26.1 ■ CONTEMPORARY MUSIC TECHNOLOGY

Contemporary music technology is based on concepts that have been honed over millenia but only recently have come within the practical purview of technologists. Sound has always been somewhat mysterious to many people, giving rise to questions such as the one noted at the beginning of this book about trees falling in forests. A more general version of this question might be, If a tree falls in the forest where no one can hear, is there still a tree? Or a forest? Or anything that falls? The dichotomous nature of human perception and reality has become more profound, rather than less, as times goes by. Most scientists today ascribe to the view that there is an objective reality that we necessarily perceive subjectively. Science is in many ways the most successful human attempt yet to arrive at a point of view that is independent of any particular subjective viewpoint, with the hope that it is the same for all observers rather than only those who accept certain subjective premises.

Contemporary technology is based—insofar as possible—on contemporary science. If we wish to transmit sound from place to place using the minimum information content (i.e., bandwidth), we might start by asking, Which parts of sounds do people perceive, and which do they not? Obviously, if we can avoid transmitting the parts of a sound that simply do not matter to people who listen to them, we can reduce what needs to be transmitted accordingly. How do we determine what matters and what does not? The best scientific evidence so far is found in the principles of masking (see Section 6.10), which says that certain sounds will be masked (i.e., not subjectively perceived) in the presence of others. Because the relationship between the "subjective" and "objective" characteristics of sound is called *psychoacoustics*, the best available predictor of what a person will actually hear in a given sound is called a *psychoacoustic model* of that sound. Such models are fundamental to the operation of current audio compression standards such as MUSICAM, MP3, and AC-3. These models are, however, a subject of intensive research and will undoubtedly change rapidly as the technology of digital sound representation continues to evolve.

Contemporary technology is not limited by contemporary science, however. Whenever a scientific basis is either lacking altogether or is established only marginally, technologists are free to use their own intuition. In other words, technology is at least partly subjective. For example, some of the most famous problems of science and mathematics are widely believed to behave in certain ways, although no one has yet been able to prove it conclusively. For many years, for example, no one could prove that four colors were enough to color an arbitrary map in such a way that no two countries with a nonzero-length border would have the same color (the famous four-color problem). Today, extremely complicated computer-generated "proofs" of this theorem exist, even though no human could possibly comprehend them due to their length. So the question remains: Are four colors enough? Most mapmakers have for centuries realized that to buy only four colors of ink is enough, even though they cannot prove it. Another example is Fermat's last theorem. The details of this simple, yet famous (and unproved), mathematical conjecture are left to the interested reader to investigate.

In technology, a good hunch can be as acceptable as a supported scientific result or a proved mathematical conjecture. If no counterexample can be found (as in the case of the four-color theorem or Fermat's last theorem), a technologist is likely to proceed as if these things were established.

The tradition of musical thought forms the basis for contemporary musical technology. Though few useful records are available (one of the most important developments in communications technology—the printing press—was invented by Johannes Gutenberg only in fifteenth–century Germany), we do know a few things about ancient precursors to contemporary musical technology.

26.2 ■ ANCIENT PRECURSORS

The earliest documented *automated* musical instruments are attributed to a group of scholars called the *Banū Mūsā* in ninth-century Baghdad (Farmer 1929). The *Banū Mūsā*, which literally means "the sons of Moses," were named Muhammad, Ahmad, and Al-Hasan. They were actually the sons of ibn Shākir, who was one of the founders of what we now call algebra. The *Banū Mūsā* were among the most celebrated scholars of their day, whose favorite sciences were reportedly geometry, mechanics, music, and astronomy.

One musical work by the *Banū Mūsā* that has survived is a treatise on automatic musical instruments—including a hydraulic organ—entitled *Al-ālat illatītuzammir binafsihā (The Instrument That Plays By Itself)*. One instrument they described apparently incorporated a pinned cylinder (such as those found in modern music boxes) playing flute pipes.

Working with the *Banū Mūsā* sometime around A.D. 830, Abu Ja'far Muhammad ibn Mūsa al-Khwarismi wrote a book entitled *Kitab al-jabr wa'l-muqabala (Rules of Restoration and Reduction)*. This book laid out the basics of algebra (the word *algebra* is taken from the title of the book and we derive the term *algorithm* from the name of the author).

Another notable precursor to contemporary music technology can be found in the work of Francis Bacon. Shortly before his death in 1627, the noted English philosopher and essayist described the essence of contemporary sound technology, at least three centuries before most of it was invented. In 1624 he wrote what is for all intents and purposes the first science fiction novel, entitled *The New Atlantis*, which describes imaginary technological developments of many types, including the following prescient passage.

> We have also our sound-houses, where we practice and demonstrate all sounds, and their generation. We have harmonies which you do not, of quarter-sounds, and lesser slides of sounds. Divers instruments of music likewise to you unknown, some sweeter than any you have; together with bells and rings that are dainty and sweet. We represent small sounds as great and deep; likewise great sounds extenuate and sharp; we make diverse tremblings and warblings of sounds, which in their original are entire. We represent and imitate all articulate sounds and letters, and the voices and notes of beasts and birds. We have certain helps which set to the ear do further the hearing greatly. We have also divers strange and artificial echoes, reflecting the voice many times, and as it were tossing: it and some that give back the voice louder than it came; some shriller, and some deeper; yea, some rendering the voice differing in the letters or articulate sound from that they receive. We have also means to convey sounds in trunks and pipes, in strange lines and distances.

In contemporary parlance, we might have used words like sound synthesis for Bacon's "sound generation," microtones instead of Bacon's "lesser slides of sounds." We might also have used terms such as audio signal processing, hearing aids, artificial reverberation, and sound spatialization, not to mention telephone and radio, instead of Bacon's equivalents. And if ever sound has been conveyed over "strange lines and distances," streaming compressed audio over the Internet would have to qualify. These are examples of how, before anything can be invented, someone has to imagine it.

26.3 ■ EARLY SOUND RECORDING

The true beginning of the contemporary era of music technology is probably 1877, when both Thomas Alva Edison and Emile Berliner invented devices that could record sound. For the first time, sound could be captured and heard at a time different from its original production, fundamentally changing the relationship of people to sound and especially to music. Edison's initial photograph was realized with a moving strip of paper tape coated with paraffin wax. An indentation was impressed on the wax by a vibrating needle attached to a thin diaphragm. Edison started the Edison Speaking Phonograph Company in 1878. He later refined his recording mechanism to include a continuously grooved, revolving metal cylinder wrapped in tinfoil. Berliner's initial device, invented virtually simultaneously with

Edison's, used both cylinders and discs for sound recording. Edison's and Berliner's original sound recording systems were purely mechanical—electrons had yet to be discovered. It is interesting to speculate on how sound recording technology might have developed in the absence of electronics, although this is not the way things happened.

In 1880, Chichester Bell and Charles Tainter devised and patented several means for transmitting and recording sound. Bell and Tainter's device recorded on wax-coated cylinders rather than tinfoil. By the late 1880s, Bell and Tainter had formed the American Graphophone Company. By the early 1890s, Berliner became the first person to market prerecorded stamped disc records for his gramophone, which later became the Victrola, manufactured by the Victor Talking Machine Company.

As is often the case, a clear understanding of technology can matter little in the absence of an equally clear insight into the desires and needs of people. There were no books on mass media in the early days of sound recording, and it was unclear just what people wanted to hear on prerecorded cylinders and disks. Edison reasoned that most people really only wanted to hear good music, so he saved considerable expense by recording little-known (but excellent) musicians who at first remained anonymous on his company's recordings. Berliner reasoned that people would pay more—or at least more readily—to hear famous musicians perform than equally competent but little-known ones. Berliner's recordings of such famous musicians as operatic tenor Enrico Caruso and virtuoso violinist Fritz Kreisler won the day. Even though Edison was first to market sound recordings and his recordings were of superior technical quality, Berliner was first to recognize how "star power" holds sway over people's pocketbooks. Berliner's company therefore captured the greater market share despite Edison's technological precedence and superiority (Norman 1998).

26.4 ■ UNDERSTANDING ELECTRICITY

Electricity was poorly understood at the time (by contemporary standards), and electronics was still science fiction. Edison had observed, for example, that an electrical current could be made to flow through a vacuum if one metal contact was heated near another that was positively charged. He even patented this "Edison effect" in 1883, but he was never able to explain it, nor did he ever make practical use of it.

It was not until 1897 that Joseph John Thompson confirmed the existence of lightweight, negatively charged particles called electrons (for which he later received the newly established Nobel Prize).

Guglielmo Marconi patented the radio in 1898. In the same year, Thaddeus Cahill invented the Dynamophone (also known as the Telharmonium; see Fig. 26.2), a machine that produced music using *dynamos*, similar in concept to the tone-wheel generators later used in the Hammond organ (discussed in Section 26.6). The Dynamophone is reported to have weighed more than 200 tons. It was designed to transmit sound over telephone wires, but the telephone wires of the day were unfortunately too delicate for the powerful signals. (Because electronic amplifiers had not yet been invented, the electromechanical tone generators had to supply enough electrical power to drive the loudspeakers directly.) *McClure's Magazine* crowed "democracy in music ... the musician uses keys and stops to build up voices of flute or clarinet, as the artists uses his brushes for mixing color to obtain a certain hue ... it may revolutionize our musical art ...."

FIGURE 26.2
Thaddeus Cahill's
Dynamophone, also
known as the
Telharmonium, ca.
1898.

26.5 ■ THE AGE OF ELECTRONICS

In 1906 Lee De Forest patented the *Audion*, which exploited the Edison effect (now understood) by controlling the flow of electrons from a heated metal *cathode* to a nearby positively charged metal plate, or *anode*. These elements were encapsulated in a sealed glass case from which air had been removed (and/or replaced with inert gases), thus creating a *vacuum tube*. To control the flow of electrons, De Forest used a fine *grid* of wires positioned between cathode and anode. In effect, small changes in the electric field of the grid leveraged much larger changes in the current flowing to the plate, thus creating a basic electronic amplifier. Even though the vacuum tube itself had been invented a year earlier by J. Ambrose Fleming in England, it was De Forest's addition of the control grid that gave the vacuum tube its great versatility and usefulness. One of De Forest's first applications of the Audion was the wireless transmission of sounds generated by the Telharmonium (1907).

The first radio broadcast was in New York City in 1910, and the first radio station was built there in 1920. In 1915, De Forest patented the use of his Audion as an oscillator by feeding back some of its output signal to its input. His patent was entitled *Electrical Means for Producing Musical Notes*. Thereafter, controllable oscillators provided a basic source of artificial signals for electronic sound synthesis. The birth of musical electronics is thus essentially the same as the birth of electronics itself.

26.6 ■ EARLY ELECTRONIC MUSIC

In 1919, Lev (Leon) Theremin invented the *Aetherophone* in Russia (later called the *Theremin* or *Thereminovox*; see Fig. 26.3). The instrument uses two vacuum-tube oscillators operating at frequencies substantially higher than audible. One frequency is fixed, and the performer varies the other. The two frequencies are mixed (added) so that they

FIGURE 26.3
While playing the Theremin, one hand controls the pitch while the other controls the amplitude of the sound coming from the triangular speaker (ca. 1920).

beat with each other (see Section 8.4). This beating, or heterodyning, signal is amplified and applied to a loudspeaker. Two antennas atop and aside the box detect the proximity of the player's hands. One hand varies the capacitance in the feedback circuit of the variable-frequency oscillator, causing a shift in the pitch of the output signal. The other hand controls loudness by varying the capacitance in a carefully balanced amplifier circuit. Near the hands of an expert performer (such as the legendary Clara Rockmore), the Theremin can produce gliding, ethereal tones reminiscent of an eerie female singing voice. This instrument has been used by many composers and was notably featured in Miklos Rózsa's Oscar-winning music for the 1945 Alfred Hitchcock movie *Spellbound*, starring Ingrid Bergman and Gregory Peck and featuring a bizarre dream sequence designed by Salvador Dali.

In 1922, Ottorino Respighi called for a phonograph recording of nightingales singing in the score for his orchestral tone poem *Pini di Roma (Pines of Rome)*. This represents the first incorporation of recorded sound into a major composition of music intended for live performance.

In 1928, Maurice Martenot built the *Ondes Martenot* (first called the *Ondes Musicales*; *ondes* in French means "waves"). The instrument used technology similar to the Theremin, except a movable electrode was used to produce variations in electrical capacitance, which in turn varied the frequency of one or more vacuum-tube oscillators. Performers wore a metal ring that passed over the keyboard. Many French composers, including Olivier Messiaen and Edgar Varèse, have used this instrument. Other instruments, such as Friedrich Trautwein's *Trautonium*, found favor in Germany with important composers such as Paul Hindemith, Richard Strauss, and Edgar Varèse.

The Theremin, Ondes Martenot, and Trautonium were important advances in music technology but required a great deal of skill to play, especially to play well. Despite their incorporation into many significant works by major composers, they never found a permanent place in most symphony orchestras, nor could they be constructed inexpensively enough for home use. For most people, therefore, such instruments were relegated to the realm of musical curiosities.

26.7 ■ ELECTRICAL ORGANS

Laurens Hammond used magnetic tone-wheel generators (essentially small dynamos such as those used by Cahill) in 1935 as the basis for his electromechanical organ. The *Hammond organ* uses a set of many-sided steel tone wheels rotating close to magnetized rods (see Fig. 26.4). This action induces an electrical voltage in a coil of wire wrapped around the end of each rod. The frequency and waveform of the resulting signal depends on the number of sides on the tone wheel and their shape, as well as its rotation speed. Tone wheels can have a hundred or more sides and may be operated in opposing directions to generate higher frequencies. One advantage of this system is that it remains permanently in tune, because the same motor drives all the tone wheels at the same speed. Another advantage is that the entire system can be conveniently retuned by increasing or decreasing the rotation speed. Because the waveform produced by each tone wheel is generally complex, the strengths of each of its harmonic components can also be adjusted by means of a system of drawbars, allowing the performer to choose the relative amplitude of each harmonic.

The player of a traditional pipe organ controls tone color for each keyboard by operating switches called *stops*. Each stop typically activates a set—or *rank*—of pipes. Different ranks have different tone colors. Because each stop could be either on or off, the player could theoretically select any of 2^N tone colors for each keyboard, where N is the number of stops available for that keyboard. Of course, not all the combinations would sound different, because some of the ranks are typically very soft, whereas others are very loud, and loud ranks tend to mask soft ones.

Some Hammond organs included a set of nine *drawbars* for each keyboard, allowing the performer to set the relative strengths of the first seven harmonics, plus a so-called subfundamental (an octave below the fundamental) and a subthird (an octave below the third; see Fig. 26.5). Pulling an individual drawbar out increases the volume of the associated sound component. Pushing it in all the way (i.e., setting it to zero) essentially silences the associated component. Because each drawbar has 9 settings (0 through 8), the player can theoretically select any of $9^9 = 387,420,489$ tone colors for each 9-drawbar keyboard. As with pipe-organ stops, not all combinations of drawbar settings are useful.

Notably, Hammond also introduced the *Leslie* speaker, which uses a rotating reflector facing a loudspeaker to achieve a vibrato effect (see Fig. 26.6). Hammond and others also used electromechanical units based on springs in order to achieve a simple but effective

FIGURE 26.4

An octagonal steel tone wheel rotating close to a magnetized rod induces a current in a pickup coil (late 1930s).

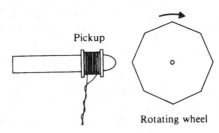

Pickup

Rotating wheel

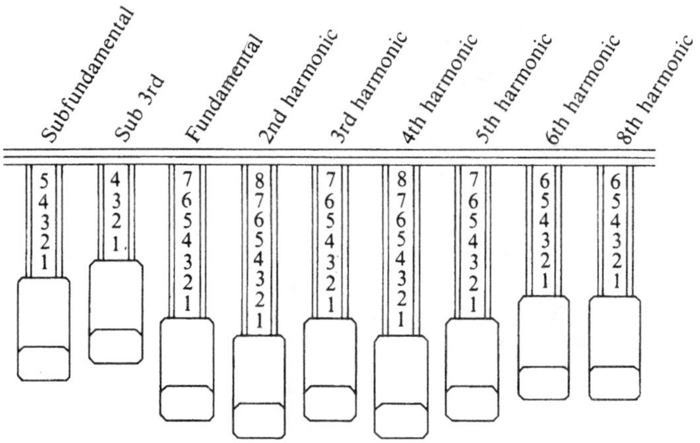

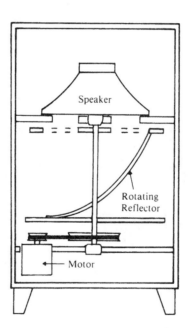

FIGURE 26.5
A separate set of Hammond drawbars allows the performer to select the relative proportions of partials in the tone for each keyboard (late 1930s).

FIGURE 26.6
The reflector in a Leslie speaker could rotate at variable speeds, using the Doppler effect to produce vibrato for the Hammond organ (late 1930s).

kind of reverberation effect (Fig. 26.7). Sound introduced at one end of the spring travels to the other, where it is reflected with some frequency-dependent loss. It then travels in the other direction, reflects again, and so on. Small mechanical fluctuations in the spring ten-

FIGURE 26.7
A simple spring reverberation unit. More than one unit was typically used (late 1930s).

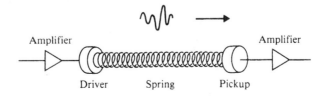

sion also tend to vary the time delay through the spring in a frequency-dependent manner. The result is a crude but effective simulation of the effect of sound bouncing around the inner surfaces of a reverberant room.

In addition to electromechanical systems such as the one used by Hammond, many other types of electronic organs have been designed and built. Most use electronic tone generators consisting of one or more oscillators.

Some organs use the principle of *additive synthesis*, wherein only simple, basic waveforms such as sinusoids are combined to produce complex waveforms for musical tones. Others use *subtractive synthesis*, which starts with harmonic-rich sawtooth or other complex waveforms that are passed through a system of filters to adjust the strengths of various harmonics (additive and subtractive synthesis are discussed further in Chapter 29).

One obvious problem that must be solved by the designer of an electronic organ is the substantial amount of electronic circuitry needed to produce every pitch, loudness, and tone color desired. One of the simplest ways to reduce the amount of circuitry needed (and also the cost of the resulting instrument) is to use a single master oscillator operating at a relatively high frequency, such as a few megahertz. The signal produced by this oscillator goes through a network of 12 *frequency dividers*, each of which produces one of the 12 pitch classes needed for an equal-tempered keyboard. The various octaves of these 12 pitches are obtained by dividing these frequencies further by successive powers of 2 (see Fig. 26.8). The advantages of this approach are low cost and permanent tuning, but the output of a frequency divider is much more limited in scope than the output of one or more oscillators.

One system, for example, uses a master oscillator frequency of 3872 kHz; the master oscillator is crystal-controlled for great stability in frequency. A series of divider circuits generate the 12 frequencies of the upper octave indicated in Table 26.1.

Another approach is to use one oscillator for each pitch and frequency dividers to obtain all lower octaves of the generated pitches. This system has the advantage that the individual pitch classes can be adjusted, allowing it to play in Pythagorean or just tuning, if desired. But to avoid the problems that result from hearing only various combinations of the frequency-divider output, it is necessary to use one or more oscillators for each and every note the organ can produce. Even small, advanced vacuum tubes require a fair amount

FIGURE 26.8
In an eight-stage frequency divider, a single oscillator furnishes a square wave in the top octave and eight bistable circuits generate that same pitch in eight lower octaves. Note that the use of bistable circuits limits the outputs to square waves.

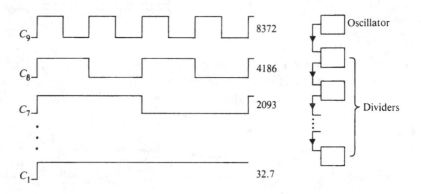

TABLE 26.1 Divisors used to generate the upper octave from the master oscillator at 3872 kHz

Note	Divisor	True Scale Frequency (Hz)	Digital Frequency (Hz)	Error (cents)
C^8	925	4186.01	4186.0	0
B^-	980	3951.07	3951.1	0
$A^{\#7}$	1038	3729.31	3730.4	+0.5
A^7	1100	3520.00	3520.0	0
$G^{\#7}$	1165	3322.44	3323.5	+0.55
G^7	1235	3135.96	3235.3	−0.3
$F^{\#7}$	1306	2959.96	2960.3	+0.1
F^7	1386	2793.83	2793.6	0.14
E^7	1468	2637.02	2637.0	0
$D^{\#7}$	1555	2489.02	2490.0	+0.5
D^7	1648	2349.32	2349.7	+0.2
$C^{\#7}$	1746	2217.46	2217.8	+0.2

Source: Douglas (1976)

of power and produce a great deal of heat. A large electronic organ requiring 400 to 500 oscillators would be quite expensive, large, and unreliable. (If you think of a vacuum tube as a highly educated lightbulb, how long do you think a panel of 400 to 500 lightbulbs would glow before at least one burned out?)

Current electronic organs and pianos generally use digital circuitry and techniques such as those described in Chapter 29. More likely than not, the internal workings of a recently-designed electronic organ or piano will resemble that of a digital synthesizer or sampler, even though the "look and feel" of the instrument (i.e., its "user interface") may seem quite traditional.

26.8 ■ MODERN ELECTRONICS

In 1947 three physicists named Bardeen, Shockley, and Brattain working at Bell Laboratories in New Jersey invented the *transistor*. Instead of controlling the flow of electrons through a vacuum, the transistor controls the flow of electrons through solid crystals of special materials called *semiconductors*. Understanding how transistors work is a part of *solid-state physics*, which in turn relies heavily on *quantum mechanics* (see Appendix A.9).

If De Forest's addition of the grid to the vacuum tube created the field of electronics, the invention of the transistor surely revolutionized it. Just as a metallic grid can be used to control the flow of electrons from heated cathode to anode in a vacuum tube, a *base* made of semiconductor material can be made to control the flow of current from *emitter* to *collector* in a simple transistor (see Fig. 26.9). Unlike the tube, however, the transistor requires no vacuum in order to operate, it does not need to heat up in order to operate, it uses only a tiny amount of power, and it can be made almost vanishingly small. Furthermore, if one transistor can be fabricated on a small piece of semiconductor, so can many interconnected

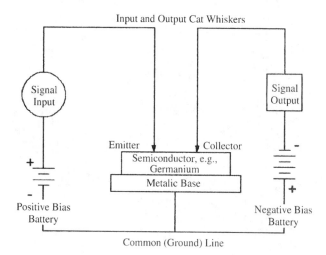

Input and Output Cat Whiskers

Signal Input

Signal Output

Emitter   Collector

Semiconductor, e.g., Germanium

Metalic Base

Positive Bias Battery

Negative Bias Battery

Common (Ground) Line

FIGURE 26.9
When a small signal flows from emitter to collector "cat's whiskers" in a simple transistor (the first kind of transistor produced), the larger output current is controlled by the (smaller) signal input. (After Rockett 1948).

ones, which is the basis for the *integrated circuit* (see Fig. 26.10). Transistors multiplied the amount of electronic circuitry that could be placed in an area the size of a thumbnail by millions (see Fig. 26.11).

Transistors can also be operated in different ways, allowing them to be used either as amplifiers or as switches. Transistor amplifiers form the basis for *analog* electronics;

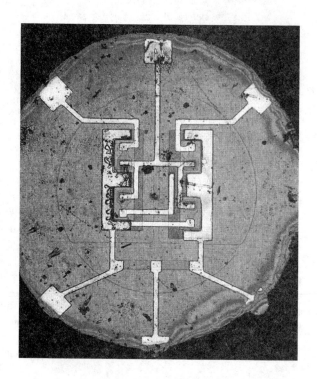

FIGURE 26.10
A dual flip-flop integrated circuit constructed by the Fairchild Camera and Instrument Corp., ca. 1961. The white lines are aluminum interconnects and the four conical shapes around the center are transistors. The device diameter is .06 inch (1.524 mm). (From Augarten, 1984).

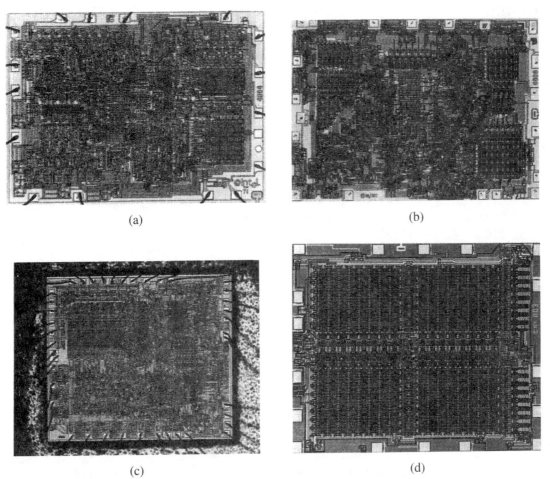

(a)

(b)

(c)

(d)

FIGURE 26.11 Four historic integrated circuits from Intel: (1) 4004 microprocessor (1970); (b) an 8008 microprocessor (1972); (c) an 8080 microprocessor (1974); and (d) an 1103 1k-bit memory chip (1970). (From Augarten 1984.)

switching transistors form the basis for *digital* electronics. In a transistor amplifier, the control element (base) directly, or *linearly*, controls the output current in the circuit. In a transistor switch, the control element merely turns the output current on and off. Switching transistors operate in their *nonlinear* mode and form the basis for circuits that implement *binary logic*, which is based on two values, TRUE and FALSE. Logically speaking, it doesn't matter what electrical quantities are used to represent true and false values, so long as there are two reliably distinguishable levels.

Operating in their logic mode, transistor can be fabricated and combined to form *logic gates*, such as AND and NOT. A two-input AND gate, for example, has an output that is TRUE when both of its inputs are TRUE. If either or both input is FALSE, the output is FALSE (see Fig. 26.12). A NOT circuit simply outputs TRUE if its input is FALSE

(input) A — NOT >o— B
(Output)

A	B
0	1
1	0

(inputs) A / B — AND — C
(Output)

A	B	C
0	0	0
0	1	0
1	0	0
1	1	1

(inputs) A / B — OR — C
(Output)

A	B	C
0	0	0
0	1	1
1	0	1
1	1	1

(inputs) A / B — NAND >o— C
(Output)

A	B	C
0	0	1
0	1	1
1	0	1
1	1	0

and FALSE if its input is TRUE. Combining the AND and OR gates produces a NAND (from NOT-AND) gate. A NAND gate has an output that is the logical inverse of the AND gate. Remarkably, only NAND gates and their interconnections are needed—in principle at least—to construct an entire digital computer, including all memory, arithmetic, and logic functions. Jack Kilby, working at Texas Instruments, fabricated the first integrated circuit in 1958. In 1965, Gordon Moore, cofounder of the Intel Corporation, penned his famous *Moore's law*, which states that the number of transistors per square inch in integrated circuits will double every year for the foreseeable future (this pace has slowed a bit: the current definition of Moore's law replaced 1 year with 18 mo.) In 1971, Marcian Hoff of the Intel Corporation fabricated the first complete digital central processing unit (CPU) on a single chip of silicon about the size of a fingernail.

26.9 ■ MUSIC TECHNOLOGY TODAY

Given the development of electrical and electronic technologies alongside the longstanding relationships among music, mathematics, and acoustics, the contemporary musician enjoys a plethora of means by which music can be created, modified, and distributed. Electronic music systems normally involve *analog* electronics, if only because loudspeakers and sound reinforcement equipment are inherently analog in nature, and without loudspeaker systems, electronic music would have no voice. Since the early 1980s, much electronic equipment has undergone a gradual transformation into digital form; *digital* electronics have certain inherent advantages when it comes to precision and repeatability of results. We will investigate these two basic realms of music technology in the next two chapters.

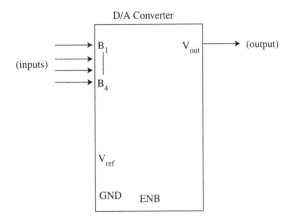

FIGURE 26.13   Symbol for an integrated circuit that performs digital-to-analog (D/A) conversion. The input is a 4-bit binary number, B_1 through B_4. The chip divides the reference voltage, V_{ref} successively by two, and uses transistor switches to turn each successive voltage on and off according to the values of the input bits. The output voltage is then just the sum of these switched voltages according to $V_{out} = B_1 \times (V_{ref}/2) + B_2 \times (V_{ref}/4) + B_3 \times (V_{ref}/8) + B_4 \times (V_{ref}/16)$.

Music technology tends to include both analog and digital components operating in an integrated and coordinated fashion. This also means that key elements of contemporary music technology are the converters that transform signals from the analog to the digital domain and vice versa: so-called analog-to-digital and digital-to-analog converters (see Fig. 26.13).

26.10 ■ SUMMARY

Music technology has a long history spanning ancient Greece, the Middle East during the Middle Ages, and Europe during the Renaissance. Modern music technology dates from 1877, when Edison and Berliner invented the first sound recording devices. The first specifically electrical musical instrument was Cahill's Dynamophone (1897). Advances in scientific understanding of electricity led to De Forest's invention of the Audion (1906), which ushered in the age of electronics. Using the vacuum tube as an oscillator allowed such important early advances in electronic music as the Theremin (1919) and the Ondes Martenot (1928). The Hammond organ (1935) introduced such advances as the tone-wheel generator, drawbar system, and Leslie speaker. The invention of the transistor (1947) allowed electronic devices to be miniaturized. Integrated circuits (1960) allowed small electronic devices to grow exponentially in complexity, eventually allowing entire digital processors to be made on a single chip (1971). By operating transistors in both analog and digital modes, contemporary music technology can exploit virtually all the advances made in electronic technology.

REFERENCES AND SUGGESTED READINGS

Augarten, S. (1984). *Bit by Bit: an Illustrated History of Computers*. New York: Ticknor & Fields.

Bacon, Frances (1624). *The New Atlantis*, Kessinger (paperback), 1997.

Douglas, A. (1976). *The Electronic Musical Instrument Manual*, 6th ed. Blue Ridge Summit, PA: TAB Books.

Farmer, H. G. (1929). *A History of Arabian Music to the XIIth Century*. London: Luzac & Co., Ltd., 1967.

Mathews, M. V., and J. Kohut (1973). "Electronic Simulation of Violin Resonances," *J. Acoust. Soc. Amer.* **53**:1620.

Moore, F. R. (1996). "Dreams of Computer Music—Then and Now," *Computer Music Journal* **20**(1).

Norman, D. A. (1998). *The Invisible Computer*. Cambridge, MA: The MIT Press.

Rhea T. (1989). "Clara Rockmore: The Art of the Theremin," *Computer Music Journal* **13**(1).

Rockett, F. H. (1948). "The Transistor." *Sci. Am. 181* (Sept. 1948). (Reprinted in *Revolutions of Science, Sci. Am,* 1999)

Welch, W. L. and L. B. S. Burt (1994). *From Tinfoil to Stereo: The Acoustic Years of the Recording Industry, 1877–1929*. Gainesville: University Press of Florida.

GLOSSARY

additive synthesis Sound synthesis based on adding together many simple waveforms, such as sinewaves at various frequencies, amplitudes, and phase offsets.

algorithm Any step-by-step instructions that solve a particular problem, such as the algorithm for performing long division.

analog electronics Electronic circuits in which the active elements (vacuum tubes and/or transistors) operate in their linear mode, i.e., as amplifiers.

analog-to-digital converter (ADC) A circuit that converts information—typically a waveform representing a sound—from analog representation (such as voltage versus time) to digital form (such as a sequence of binary numbers representing the waveform).

anode The positive (+) terminal in an electrical device such as a battery or vacuum tube.

Audion De Forest's first three-element vacuum tube (also known as a *triode*), consisting of a cathode, an anode, and a grid in between.

binary logic Any system of logic in which all properly formed statements are either TRUE or FALSE.

binary number A base 2 number in which the digits are limited to either zero (0) or one (1).

bit A contraction of the words *binary digit*; one bit is a single 1 or 0 and is either a basic unit of digital memory or a basic unit of information.

cathode The negative (−) terminal in an electrical device such as a battery or vacuum tube.

digital electronics Electronic circuits in which the active elements (vacuum tubes and/or transistors) operate in their nonlinear mode, i.e., as switches that are turned either on or off.

digital-to-analog converter (DAC) A circuit that converts information—typically a waveform representing a sound—from digital representation (such as a sequence of binary numbers representing the waveform) to analog form (such as voltage versus time).

drawbar A multiposition control on a Hammond organ that selects the amount of a particular harmonic frequency (or other sound component) that will be used to synthesize a tone.

dynamo A device that generates electrical energy by moving a wire in the presence of a magnetic field, or by moving a magnet near a wire.

frequency divider A circuit that accepts a single frequency as an input and outputs multiple signals at frequencies that are integer divisors of the input frequency.

grid A fine mesh of wires in a vacuum tube that controls the flow of electrons from cathode and anode.

logic gate A digital electronic circuit that has one or more binary (i.e., two-valued) inputs and one or more binary outputs. Output values are determined by logical operations such as AND, OR, and NOT.

oscillator A circuit that outputs a signal with a specific waveform at a controllable frequency, amplitude, and phase.

subtractive synthesis Sound synthesis based on the controlled attenuation or removal (by filtering) of components from a multicomponent waveform, such as a sawtooth or triangular waveform.

transistor A device fabricated from semiconductor material (such as silicon) in which a base element controls the flow of current from an emitter element to a collector element.

vacuum tube A device consisting of a glass shell from which air is removed (or replaced by inert gases) in which electrons flow from a cathode to an anode, possibly controlled by a grid or other elements.

REVIEW QUESTIONS

1. What does the word technology mean?
2. How are music, technology, and human history related?
3. What is the relation between technology and science?
4. Who were the *Banū Mūsā*?
5. Where do the words *algebra* and *algorithm* come from?
6. When was the first work of science fiction written?
7. Who invented the first sound-recording devices?
8. Why did the early Victrola become more popular than the phonograph?
9. When was the electron discovered?
10. In what year were the radio and Dynamophone first announced?
11. What is the significance of the Audion?
12. Who invented the vacuum tube?
13. How does an oscillator operate?
14. What was the first source of artificial signals in electronic music synthesizers?

15. How was the Theremin played?
16. In what piece of music was recorded sound first employed?
17. What sound-generating principle did Thomas Cahill's invention share with Laurens Hammond's?
18. Which drawbars on a Hammond organ do not control harmonics?
19. How does a spring reverberator operate?
20. How many oscillators are used in a large electronic organ?
21. What are the advantages of transistors over vacuum tubes?
22. What is an integrated circuit?
23. Name three kinds of logic gates and describe their function.
24. What is Moore's law?
25. What is the difference between analog and digital circuits?

QUESTIONS FOR THOUGHT AND DISCUSSION

1. Is it possible to study subjective phenomena objectively? Is there a role for subjectivity in science?

2. How might sound recording technology have developed in the absence of electronics?

3. What are some possible advantages of tone generation by addition? By subtraction?

4. Why does the Theremin use the difference tone between two high-frequency oscillators as a tone generator rather than, say, using the capacitance of the hand to control the frequency of an audio-frequency oscillator directly?

5. Many reverberation units use two or more springs with different lengths to spread out the resonances. Can you explain how this works?

6. On a Hammond organ, what are the functions of the sub-fundamental and subthird drawbars?

7. How many parallels to contemporary audio technology terms can you find in the quote from Bacon's *New Atlantis* (for example, lesser slides of sounds = microtones)?

8. How many possible two-input logic functions exist? List them and give logic names (AND, OR, etc.) to as many as you can.

EXERCISES

1. Show that magnetic tone wheels with 196 and 185 teeth, rotating at the same speed, generate tones approximately one semitone different in pitch.

2. Express 3872 kHZ in Hz and MHz.

3. How many ranks of pipes would a keyboard on a traditional pipe organ need to offer the player as many combinations of settings as Hammond drawbars (ignore the question of how many settings would be useless)?

4. Show how a two-input OR function (output is FALSE only if both inputs are FALSE; output is TRUE otherwise) can be made by interconnecting only NAND gates. Is your answer the only possible (correct) answer?

5. Suppose that a small loudspeaker is mounted on a photograph turntable so that its center moves in a circle of 10-cm radius at $33\frac{1}{3}$ rev/min.
   (a) What is its speed as it moves around the circle?
   (b) What is he maximum shift in frequency (f/f_s) due to the Doppler effect?
   (c) To what musical interval does this correspond?
   (d) Would the pitch shift be noticeable to a listener?

EXPERIMENTS FOR HOME, LABORATORY, AND CLASSROOM DEMONSTRATION

Home and Classroom Demonstrations

1. *Music box operation*   Examine the pinned disk from a music box. Try to determine what music it plays from a visual examination. Suggest ways of manufacturing such objects from written and/or performed music.

2. *Mechanical sound-recording and transmission*   Stretch a piece of aluminum foil across the opening of a paper tube or similar object and talk into it (or have someone else talk into it). Try to estimate the amplitude of the movements of the foil. Estimate the mass per unit area of the foil and, from this, estimate the force it generates when vibrating. Connect two such devices together with a long piece of string and demonstrate an improved "tin-can telephone."

3. *Electromechanical sound generation*   Construct a small dynamo by using a wire wrapped around a magnetized rod, near which a steel regular polygon rotates. Amplify the current generated in the wire and experiment with different shapes of steel polygons, noting the various pitches and tone qualities that can be produced.

4. *Point-contact junctions*   Construct a "foxhole radio" from a piece of no. 22 wire, a blue or rusty razor blade, a coat hanger, a paper tube from a roll of toilet paper, a safety pin, and an earphone. Discuss the similarities of its operation with the original point-contact transistor.

5. *Theremin*   Build or buy a theremin. Theremins and theremin kits are available from Big Briar in Asheville, North Carolina, a company founded by electronic music pioneer Robert Moog. Circuit diagrams for building theremins are available on the Internet. Theremin music is demonstrated on a video "Theremin, an Electronic Odyssey" (Orion Home Video, 1995).

Laboratory Experiments

With an oscilloscope, observe the waveforms available from various electronic instruments, such as the theremin and early electrical organs. What is their frequency range?

CHAPTER
27

Analog Electronic Music

As we saw in the last chapter, an understanding of electricity and electronics technology, combined with an understanding of the scientific principles of acoustics and music, allows the creation of modern music technology in its many forms. Advances in technology usually create opportunities to do things that were not previously possible (which is why they are called advances). What actually gets done, however, depends not only on technological opportunities but also on the concepts in the minds of those who exploit them. Instead of having the fixed physical structures of traditional musical instruments, electronic instruments can synthesize or capture virtually any sound. Musical concepts in the minds of the founders of electronic music forever changed notions of what musical instruments and music itself are, or could be, giving rise to musical explorations of entirely new kinds, much to the delight (and chagrin!) of listeners ever since. The initial implementation of electronic music technology in analog form provided both the crucible and the conceptual basis for the more powerful digital forms to come later. In this chapter we will examine these basic concepts from the standpoint of how electronics technology impacted fundamental concepts of music and musical instruments starting around the middle of the twentieth century.

In this chapter, you should learn:

- About the basic elements of electronic music;
- About the conceptual and technological basis of *musique concrète*;
- About the conceptual and technological basis of *electronische musik*;
- About analog synthesizers;
- About digital control of analog synthesizers.

27.1 ■ THE ELEMENTS OF ELECTRONIC MUSIC

Electronic music can be described in many ways, but from a technological standpoint it is useful to define a set of basic functional elements from which most electronic music techniques can be discussed and understood. Figure 27.1 shows such a set of basic elements, many of which have been described elsewhere in this book.

Microphones

In electronic music terms, a *microphone* is really a form of sensor, but it is so important that it deserves a special category of its own. A microphone transforms sound pressure

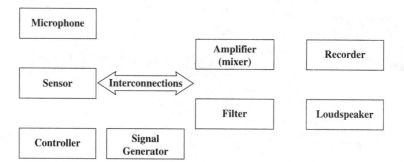

FIGURE 27.1
The basic elements
of electronic music.

variations at a particular point in space into analogous variations in some electrical quantity such as voltage (see Chapter 20). As we shall see, it is possible to create electronic music in which microphones play anywhere from an essential role to no role at all, just as it is possible to create visual art that may or may not involve the use of a camera.

Sensors

The *sensor* element refers to transducers, such as to knobs, dials, switches, keys, pedals, and other controls that might be manipulated by a musician in order to convey musical information. They generally sense the position, velocity, or acceleration of physical manipulations or in some cases sound waves, heartbeats, electrical currents in the brain, temperature or altitude changes in the atmosphere, etc. Traditional musical instruments obviously involve a variety of ways of "sensing" the actions of a performer in ways that affect the sounds produced.

Controllers

A *controller* might be a sensor as well, if the output of a sensor is used to control the actions or signals used by other elementary devices. More sophisticated controllers are possible, however, from pinned rotating cylinders such as those described by the *Banū Mūsā* to stored program computers executing sophisticated algorithms for musical decision making.

Signal Generators

A *signal generator* is an artificial source of signals that might be used for control purposes or as an oscillatory waveform that represents one or more (or all) components of a sound. It is a fundamental alternative to the microphone in the sense that it is a possible source of sound signals. Electronic music systems might, of course, contain both microphones and signal generators.

Interconnections

In order to utilize more than one of the elements described here, it is necessary to interconnect them in various ways. The way in which a given set of elements is interconnected can completely change the significance of any given element. Because the number of elements in a given electronic music system may be large, important strategies have been devised to allow flexible and even dynamic ways of achieving interconnections.

Amplifiers and Mixers

Amplifiers are useful both as sources of electrical power that drive electronic signals along interconnection paths and for driving loudspeakers. Amplifiers and related devices are essential parts of many electronic devices that deal with sound signals. Amplifiers are also essential components of *mixers*, which are essentially amplifiers with multiple, individually controllable inputs.

Filters

A *filter* element in a restricted sense is a device that modifies the amplitude or phase of a signal in a frequency-dependent manner. More generally, filters can modify or process signals in any manner whatsoever. A linear filter, for example, outputs only frequencies that are present at its input. A nonlinear filter, however, can output frequencies that are not present at its input. Filters can be used not only to modify signals but to measure or analyze them in various ways.

Recorders

A *recorder* is any device that stores a signal in such a way that it can be later retrieved. A recorder is a form of memory.

Loudspeakers

A *loudspeaker* is a transducer that renders the acoustic information in a signal as vibrations in the atmosphere that can be perceived as sound by one or more listeners (see Chapter 19). It can be anything from a simple earphone to a coordinated array of transducers intended to generate a coherent pressure wave field containing spatial information about complex sounds.

27.2 ■ *MUSIQUE CONCRÈTE*: THE CONCEPTUAL BASIS

In 1948, Pierre Schaeffer defined a new kind of music by assembling a succession of recorded sounds, the first case consisting of a rhythmic succession of locomotive steam and wheel sounds punctuated by steam whistles. He later wrote in his *A la Recherche d'une Musique Concrète (Musique Concrète Research* 1952):

> This determination to compose with materials taken from an existing collection of experimental sounds, I name *musique concrète* to mark well the place in which we find ourselves, no longer dependent upon preconceived sound abstractions, but now using fragments of sound existing concretely as sound objects defined and whole ....

The fundamental idea of *musique concrète* is thus to replace the (abstract) sounds of traditional musical instruments with the (concrete) sound of nature. The first example was Schaeffer's musical composition *Etude aux Chemins de Fer (Railroad Study)*. There are a few notable examples of the incorporation of natural sounds into music before 1948, such as Ottorino Respighi's incorporation of recorded nightingales in his orchestral piece

Pines of Rome (1924) and George Antheil's use of airplane engines on stage in his *Ballet Mécanique* (1926). Before the advance of more sophisticated sound-recording technology, however, such examples were limited to more-or-less verbatim quotations of natural sounds.

The earliest experiments of Pierre Schaeffer and his colleague Pierre Henry occurred at the *Radiodiffusion Française* (French Radio) in Paris. Sounds were at first recorded onto a rotating disk directly with a vibrating lathe. Such recordings could be edited only by playing back multiple disks at the same time and switching among them with a mixer. Although editing sounds in this manner is possible, it certainly is not quick or easy. Prospects for *musique concrète* were furthered greatly with the advance of magnetic recording. Magnetic tape recorders, in particular, allowed sounds to be edited simply by cutting the tape with a razor blade and splicing the cut pieces back together in various ways. This is, in fact, the way that much electronic music was assembled until the advance of digital sound editing and mixing in the 1980s.

Magnetic tape recording allowed recorded sounds not only to be quoted literally but also to be repeated, or looped, simply by splicing the end of a piece of magnetic tape back to its beginning, forming a *tape loop*. Looping tape-recorded sounds in this manner could, for example, impart to them a rhythmic quality that they may not have possessed in their original form. Not only could any sound be repeated, but its tape playback speed could be changed, thus accelerating or decelerating the rhythm (with corresponding changes in pitch). By playing the tape backward, the sound could also be reversed in time, yielding an entirely new perspective on what is otherwise a natural sound. Looping techniques in modern samplers owe both their name and origin to such tape-based techniques.

Pierre Schaeffer followed his *Railroad Study* with other examples of *musique concrète*, including *Etude aux Tourniquets* (sounds of toy spinning tops and percussion instruments), *Etude Violette*, *Etude Noire* and *Diapason Concertino* (piano sounds), and *Etude Pathétique* (sounds of saucepans, canal boats, voice, harmonica, and piano). Recordings of traditional musical instruments are sometimes used in *musique concrète*, but these instruments are generally serving in their capacity as sources of sound objects as well as occasional references to traditional music.

27.3 ■ *MUSIQUE CONCRÈTE*: THE TECHNOLOGICAL BASIS

In terms of the elements of electronic music described in Fig. 27.1, we see the four key elements of *musique concrète*: the microphone, the recorder, the mixer, and the loudspeaker. Of course, these elements need to be interconnected, amplifiers (and possibly filters) are needed for the speaker(s), and so on, but these are incidental to the concept of *musique*

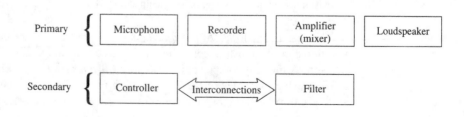

FIGURE 27.2
The basic elements of *musique concrète*.

concrète (see Fig. 27.2). Other issues include the need to play back the recorded sounds in various ways. Opportunities for different kinds of playback vary according to the details of the recorder. Disk recorders might allow random starting points and variable speeds of playback. Tape recorders allow variable speeds, directions, and looping of playback, though they do not allow random starting points as well as disk recordings. Tape recordings, however, can be copied, cut into various pieces, and reassembled in various ways, effectively allowing flexible temporal control of playback.

27.4 ■ *ELEKTRONISCHE MUSIK*: THE CONCEPTUAL BASIS

As *musique concrète* was being developed in France, a group of scientists, engineers, and musicians in Germany began working on another extension of traditional music that became known as *elektronische musik*, or electronic music. Just as *musique concréte* was based on modified and reordered recordings of natural sounds, *elektronische musik* was based on the notion that virtually any sound could theoretically be constructed by adding together appropriate mixtures of sinusoids and other artificially synthesized sounds.

The initial impetus for *elektronische musik* came in 1948 when Homer Dudley, of Bell Telephone Laboratories in Murray Hill, New Jersey, visited Werner Meyer-Eppler, then the director of the Institute of Phonetics at the University of Bonn. Dudley was renowned for his work on a device called the *vocoder* (for *voice coder*), which used filters to both analyze and synthesize the sounds of human speech. The vocoder was the basis for a similar device—also due to Dudley—featured at the 1939 World's Fair in New York City. Called the *voder* (or *voice operation demonstrator*), a highly skilled operator "performed" speech by pressing on keys while simultaneously operating pedals, knee-operated sliders, and a variety of other control devices. Fair visitors were amazed that the operator could make Dudley's voder say understandable things in languages she didn't speak and even to sing in limited ways.

The result of these meetings were demonstrations and collaborations resulting in a series of lectures given in Darmstadt by Meyer-Eppler and Robert Beyer during the summer of 1950 entitled "The World of Sound of Electronic Music." As a result of this and related activity, the technical staff at *Westdeutscher Rundfunk* (WDR, for West German Radio) in Cologne committed resources to the idea of establishing a studio to investigate the musical ramifications of such ideas. The first musical result of the Cologne studio for *elektronische musik* was a composition completed in 1952 called *Musica su Due Dimensioni* for flute, percussion, and electronically generated tape by composer Bruno Maderna, who worked in collaboration with Meyer-Eppler.

The first director of the Cologne studio for *elektronische musik* was a composer named Herbert Eimert, who was very much a proponent of a compositional technique called *serialism*. In the early 1920s the Viennese composer, music theorist, and painter Arnold Schoenberg published his method of composing music with 12 tones. Schoenberg showed that basing an entire musical composition on a fixed sequence of the 12 different equal-tempered pitch classes can lead to a conceptual unification of musical materials not unlike the traditional system, which relates all pitches to a single key-note, called tonality. Because it was an alternative to the traditional system of tonality, Schoenberg's approach has sometimes been called *atonality*, though this is a linguistic misnomer.

Schoenberg had two very influential students named Alban Berg and Anton Webern. Webern, in particular, extended Schoenberg's ordering principles beyond pitch rows to fixed sequences for virtually every musical characteristic, including loudness (musical dynamic), duration (rhythmic value), articulation (*staccato, legato, pizzicato,* etc.), and tone color (*timbre*). This total-ordering approach to music composition became known as serialism and attracted many influential composers in Europe (such as Boulez, Stockhausen, Nono, Dallapiccola, and Ligeti) and America (such as Stravinsky, Babbitt, Carter, Wuorinen, Davidovsky, and Martino).

In effect, the amplitudes, frequencies, and phases of sinusoids (as well as any other sound parameters that composers could think in terms of) became grist not only for the mill of *elektronische musik* but for serialist composers as well. Eimert (1959) wrote: "It is certain that no means of musical control could have been established over the electronic material had it not been for the revolutionary thought of Anton Webern."

27.5 ■ *ELEKTRONISCHE MUSIK*: THE TECHNOLOGICAL BASIS

In terms of the elements of electronic music described in Fig. 27.1, the key elements of *elektronische musik* were the signal generator, the recorder, the mixer, and the loudspeaker. As in the case of *musique concrète*, these elements need to be interconnected, amplifiers (and possibly filters) are needed for the speaker(s), and so on (see Fig. 27.3). In some ways, and especially because of equipment limitations, the need to mix several sounds together and the need to play back the recorded sounds in various ways are essential to the practice of *elektronische musik* if not necessarily its theory. The equipment complement at the original studio in Cologne, for example, consisted of a sine-wave generator, a white-noise generator, two electronic keyboard devices, a mixer, a four-track tape recorder, and a monophonic tape recorder. Composers worked by recording a sinusoid on one channel of tape, then another (possibly with a different loudness contour, or "envelope") on a second channel, then a third, and then a fourth. The duration of the sound was determined by cutting the tape, which traveled at 15 or 30 in/s. These might be mixed together and rerecorded on the monophonic recorder, then transferred back on channel one of the four-track recorder, etc. When all the bits and pieces of a complete composition were prepared, they could all be spliced together into a final result.

Although theoretically sound, *elektronische musik* was extremely tedious to produce in this manner. The advantages of constructing sounds on such a microscopic level were often offset by the difficulty of the procedure, the opportunities for errors, and limitations caused by the buildup of tape hiss and other noises during multilayered mixing operations.

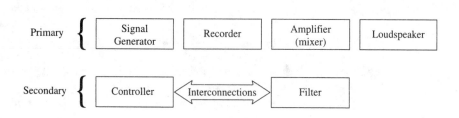

FIGURE 27.3
The basic elements
of *elektronische
musik.*

27.6 ■ TAPE MUSIC

It soon became apparent to practitioners of both *musique concrète* and *elektronische musik* that the philosophical differences in sound source (tape-recorded natural sounds versus electronic signal generators) were less important than the manipulative opportunities offered by the new techniques of electronic music. Karlheinz Stockhausen, one of the principle proponents of *elektronische musik*, as a result of his work in Schaeffer's studio in 1952–1953, later observed: "Wherein lies the difference between instrumental sounds, between any audible events: violin, piano, the vowel a, the consonant sh, the wind?" One result of such observations was Stockhausen's landmark composition *Gesang der Jünglinge*, which is based on sounds recorded from the speech of a young boy blended with electronically produced tones. The result is experienced by the listener as a kind of sound-continuum, in which the natural and the artificial together provide a sound palette much richer than either does alone.

Other composers in other places soon began to carry out experiments using electronic music. Notably, Toshiro Mayazumi, who had worked with Schaeffer in Paris, and others started an electronic music studio in 1953 at NHK (Nippon Houso Kyokai, or Japan Broadcasting Corporation) in Tokyo. This effort resulted in Mayazumi's *Etude I* in 1955 as well as later works by composers Minao Shibata, Makoto Maroi, Toshi Ichiyanagi, and Joji Yuasa.

In 1952, composers Vladimer Ussachevsky and Otto Luening presented works of their own at a concert organized by famed conductor Leopold Stokowski at the Museum of Modern Art in New York City. This concert of electronic music was the first of its kind in the United States, and it led to a demonstration and interview on the *NBC Today* show by Ussachevsky and Luening. As a result of the pioneering work of Ussachevsky and Luening, the Rockefeller Foundation funded the first studio for electronic music in the United States at Columbia University in 1955.

Joining forces with composer Milton Babbitt of Princeton University, Ussachevsky and Luening founded the Columbia-Princeton Electronic Music Center. This center comprised several tape studios and, notably, the first significant electronic music synthesizer.

27.7 ■ EARLY ELECTRONIC MUSIC SYNTHESIZERS

As electronic music studios grew in size and number, several problems became increasingly evident. The equipment they contained tended to be very expensive and required a great deal of technical knowledge to use and maintain. The early electronic music studios were similar in their requirements to radio stations and recording studios, where they were therefore often located. Institutional-sized budgets and technical staffs were needed to keep them going.

As the amount of equipment increases, the problem of interconnections increases combinatorially. Straightforward patch cords soon become tangled forests when more than a few devices are interconnected, especially when many sound channels are needed simultaneously. Finally, when many devices are used, each typically requires from a few to many parameters to be set, such as the frequency, waveshape, and amplitude of each signal generator. It is frequently desirable to make changes in interconnections and parameters in a time-dependent manner, depending on the requirements of the musical composition being created.

By integrating and coordinating the operation of most or all the elements of electronic music (see Fig. 27.1), synthesizers address these problems of size, cost, and flexible, dynamic interconnections and parameter control.

27.8 ■ THE MARK II ELECTRONIC MUSIC SYNTHESIZER

In 1955, the RCA Electronic Music Synthesizer (see Fig. 27.4) was designed and built by Harry Olson and Herbert Belar at RCA's Sarnoff Laboratories in Princeton, New Jersey. A more advanced version—the Mark II—was installed in 1959 at the 125th Street studio of the Columbia-Princeton Electronic Music Center in New York at a cost of $10,000, funded by a grant from the Rockefeller Foundation. Its principle users were composers, who described it in the following terms.

> (Charles Wourinen) ... a 750-vacuum-tube affair in which information was encoded by a fiendish combination of 4-bit binary switches, banks on the walls and console of this nearly room-sized machine, and two particularly clever paper drives, each of which would encompass two channels of information. Holes were punched in the paper that then passed over a metal roller to which contact was made by a set of brushes. The brushes were arranged so that they would lie over the holes that passed beneath, making continuous contact. Time was represented on the machine as the number of holes at a certain rate of the paper drive. Pitch

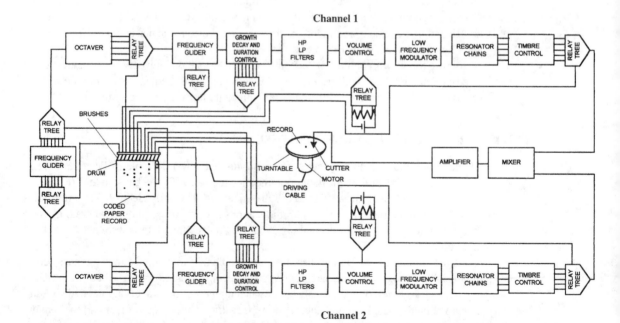

FIGURE 27.4   Schematic of the RCA electronic music synthesizer designed by Olson and Belar (ca. 1955).

and other information were represented in binary numbers by a combination of preset switches and banks of holes in the paper (Boulanger 1984).

(Milton Babbitt) The machine was extremely difficult to operate. First of all, it had a paper drive, and getting the paper through the machine and punching the holes was difficult. We were punching in binary. The machine was totally zero, nothing predetermined, and any number we punched could refer to any dimension of the machine. There was an immense number of analog oscillators, but the analog sound equipment was constantly causing problems. I couldn't think of anything that you couldn't get, but other composers gave up—it was a matter of patience .... There were many people who would look at this machine and say, "It's a computer." But it never computed anything. It was basically just a complex switching device to an enormous and complicated analog studio hooked into a tape machine. And yet for me it was so wonderful because I could specify something and hear it instantly (Chadabe 1997).

Although the RCA synthesizers are more of historical than technological interest today, they introduced several important concepts into electronic music. First of all, the variable-speed paper tape drives—mechanically problematic though they were—separated control of the synthesizer from the physical limitations of human performers: no longer were the actions of the synthesizer limited by ten fingers, two feet, two hands, etc. Neither did the synthesizer produce ancillary sounds such as bow scratches, key clicks, or valves rattling. A musical trill consisting of precisely 32 (or 33 or 34) alternations per second could readily be obtained, and rhythms too complex for human performance could be precisely realized. This allowed some of the first methodical investigations into aspects of human perception that had been hitherto inaccessible. For example, what is the just noticeable difference (jnd) of trill speed? How many layers of complex polyrhythms can be distinguished using similar or contrasting tone qualities?

Finally, synthesizers can provide for at least a certain measure of real-time control, allowing them to be played like traditional musical instruments.

27.9 ■ VOLTAGE CONTROL

The next important technological advance in electronic music synthesizers was the concept of *voltage control*. Robert Moog, working near New York City, and Donald Buchla, working in San Francisco, built successful voltage-controlled synthesizers in the mid-1960s (see Fig. 27.5). Since they were based on transistorized modules, the Moog and Buchla synthesizers could be made much smaller, lighter, and cheaper than the vacuum-tube-based RCA synthesizers. This allowed voltage-controlled synthesizers to be successfully developed as consumer products.

The principal of voltage control is straightforward: *applying a voltage to a control input has the same effect as manually turning a knob.* Any of the elements of electronic music can, in principle, be voltage-controlled as well as manually controlled. For example, a voltage-controlled oscillator (VCO) is one that generates one or more waveforms with a frequency that is set by either a manually turned knob, a control voltage, or some combination of both. The control voltage can be the output of an electronic circuit such as an electronic keyboard, a knob-adjusted voltage source (such as a battery attached to a potentiometer), another oscillator, or any (mixed) combination of such voltages (see Fig. 27.6).

FIGURE 27.5   The Mini Moog Sonic Six (1974–1979). It included a 49-key keyboard and two VCO modules with sawtooth, triangular, and rectangular (pulse/square) outputs. Knobs adjusted such things as the width of the rectangular waveforms, portamento rates, and attack and decay times for the envelope generators. It also contained a VCF and two LFOs (low-frequency oscillators). It could produce as many as two notes at once.

FIGURE 27.6
A generic voltage-controlled oscillator (VCO) module. The frequency of the output signal is controlled by the sum of the control input voltage(s) and the setting of the frequency control knob. Other knobs set the output waveshape, the width of the pulse waveform, and the output signal amplitude.

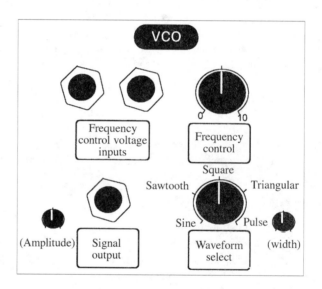

Similarly, a voltage-controlled amplifier (VCA) is one whose gain is controlled by either a knob or a control voltage, a voltage-controlled band-pass filter (VCF) is one whose center frequency and bandwidth are both controlled by knobs or control voltages, and so on.

27.10 ■ MODULES AND PATCHES

A voltage-controlled synthesizer typically consists of several modules such as VCOs, VCAs, and VCFs, all of which are interconnectable. The strategy used for such interconnections may rely on cables or switches of various kinds—in any case, a given inter-

FIGURE 27.7
A forest of patch cords interconnect the modules of a Moog synthesizer.

connection among modules is called a *patch* (no doubt named for the lightweight cables called *patch cords* designed specifically for such interconnections). Complex patches often required users to color-code their patch cords (see Fig. 27.7).

To combat the "forest of patch cords" problem, some analog synthesizers employ innovative interconnection methods, ranging from switches to slider pots to matrix pins (see Fig. 27.8). Although internal programming of interconnections tends to restrict the number of possible interconnections that can be made at one time, it also reduces programming time and improves reliability, making the synthesizer more practical for a live performance.

For example, by patching the output of one VCO module into the frequency control input of another, we can modulate the frequency of the second oscillator with the signal of the first (see Fig. 27.9). A more useful patch might add a modulating waveform to a source of constant voltage: the constant voltage would then set the center frequency, around which frequency modulation occurs. This type of frequency modulation (FM) could be used to produce a vibrato effect, with the vibrato rate and depth being determined by the frequency and amplitude of the modulating VCO.

Similarly, the signal output of one VCO may be added to the control voltage output of a keyboard in order to obtain a vibrato effect. The frequency and amplitude of the modulating VCO would then control the rate and depth of the vibrato (see Fig. 27.10).

Such patches may be combined in various ways. Figure 27.11 shows a patch that simultaneously produces both frequency and amplitude modulation using two VCOs and a VCA.

Another important module on most voltage-controlled synthesizers is the *ADSR envelope generator*. The ADSR (for attack-decay-sustain-release) envelope generator typically produces a control voltage that is intended to be applied to the control input of a VCA. Typical ADSR modules combine four knobs with four control inputs for setting each of the four segments of the envelope signal (see Fig. 27.12). In addition, a "trigger" input detects when a control voltage exceeds a preset value, thus initiating—or triggering—the

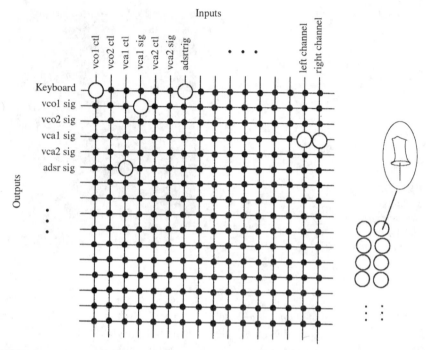

FIGURE 27.8 A pin matrix can be shown to connect an arbitrary number of available outputs (rows) to an arbitrary collection of signal and control inputs (columns). Placing a patch pin at a matrix intersection established electrical connection between the output row and input column. The patch shown connects the output of a keyboard to both a VCO and to the trigger input of an ADSR envelope generator. When a key is pressed, the triggered envelope uses a VCA to modulate the amplitude of the signal produced by the VCO. The result is placed into both output channels.

FIGURE 27.9
The signal output of a VCO 1 is patched into the frequency control input of VCO 2. Assuming both are set for sine waveforms, and that the frequency and amplitude of VCO 1 are small, the waveforms produced will resemble those shown.

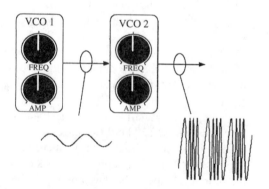

envelope function. The envelope proceeds through its attack and decay phases and remains at its sustain level until the trigger input goes below the preset value, thus initiating the release phase. The ADSR inputs typically control the respective speeds of the attack, decay, and release phases, plus the level of the sustain phase. Some ADSR modules also include a *retrigger* input that restarts the envelope signal from the beginning of the attack phase regardless of which phase it is currently in. Such envelope retriggering is somewhat akin in purpose to the repetition lever in a grand piano action (see Section 14.1)—it allows notes to be rapidly restarted.

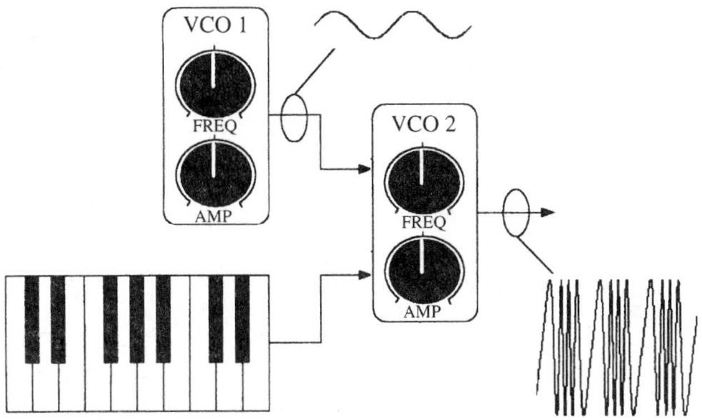

FIGURE 27.10
A vibrato patch: the frequency and amplitude of VCO 1 control the rate and depth of a vibrato around the pitch determined by the keyboard. Note that both control inputs and the frequency knob setting of VCO 2 are added together, further allowing this knob to be used as a pitch bend.

Voltage-controlled filters (VCFs) may be low-pass, high-pass, band-pass, or band-reject filters with parameters such as cutoff frequency, center frequency, and bandwidth controlled by knobs and/or control voltages. The principle use of a VCF is to control the *timbre* of a sound, which is often a complex waveform (such as a sawtooth waveform) from a VCO. A simple (and very effective) patch involving a sawtooth VCO, an ADSR envelope generator, and a VCA may be used to produce what on many modern synthe-

FIGURE 27.11
An AM/FM patch: VCO 1 controls the rate and depth of both the vibrato imparted by VCO 2 and the tremolo imparted by the VCA. The keyboard is patched into a second frequency-control input of VCO 2.

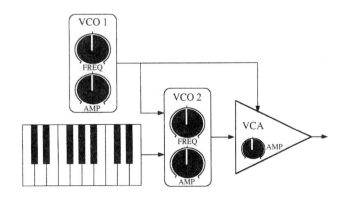

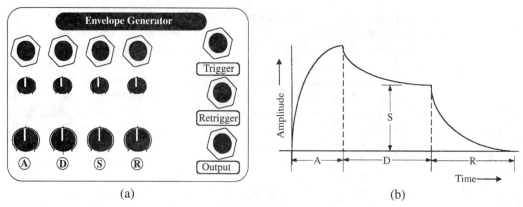

(a)   (b)

FIGURE 27.12   A typical ADSR envelope module. A new envelop is generated when the trigger input goes above a preset value. It passes through the attack (A) and decay (D) phases, then holds the sustain (S) level until the trigger goes below the preset level, which starts the final release (R). Knobs and control voltages determine how much time the A, D, and R segments take as well as the S level. Each control input has an attenuator; a retrigger input causes the envelope to immediately start again at the beginning of the attack phase.

sizers is labeled *analog brass* because it mimics an early analog synthesizer imitating the sound of traditional brass instruments (see Fig. 27.13). This patch is based on the finding of computer music pioneer Jean-Claude Risset that the bandwidth of a brasslike tone is proportional to its amplitude. It is the basis for a tone quality found commonly on later synthesizers and called analog brass.

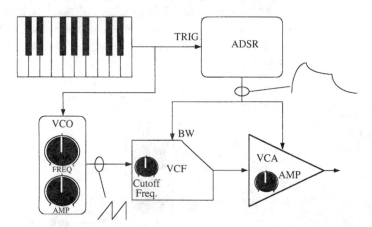

FIGURE 27.13   A patch for analog brass. A harmonic-rich (sawtooth) waveform is passed through both a filter (VCF) and an amplifier (VCA). Using voltage control, the bandwidth (BW) of the filter and the gain of the amplifier are both made to track the same ADSR envelope, causing the bandwidth of the output to shrink in proportion to the amplitude envelope, imparting a brass-like quality to the sound.

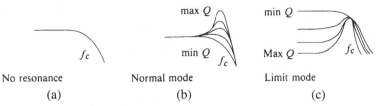

No resonance   Normal mode   Limit mode

(a)   (b)   (c)

FIGURE 27.14   Characteristics of a voltage-controlled low-pass filter with cutoff frequency f_c, without resonance (a), with resonance (b), and with resonance in *limit mode* (c). Because these are filter responses, the horizontal axis is frequency and the vertical axis is amplitude.

Some VCFs include a feature that adjusts the resonance, or Q factor, of the filter at or near its cutoff frequency (Q is essentially center, or cutoff, frequency divided by bandwidth: thus a high-Q filter has relatively narrow bandwidth). In some VCFs, increasing Q increases the gain of the filter at its resonance frequency, whereas in others, the resonance gain remains the same and the gain at other frequencies is proportionally decreased, which is sometimes known as *limit mode* (see Fig. 27.14). A VCF with a high Q can "ring," i.e., it can oscillate on its own, sometimes with no input signal at all.

27.11 ■ CONTROL VOLTAGE CHARACTERISTICS

Depending on the synthesizer, control voltages may or may not be interchangeable with waveform signals. In addition, the modules may respond to control voltages either linearly or exponentially (see Fig. 27.15). These variations of control methods represent the choices of the synthesizer designer, and there are advantages and disadvantages to whichever choices are made.

The principal advantage of exponential control voltages is that the responses of VCOs and VCAs follow Fechner's hypothesis (see Section 5.6) that sensation is proportional to the logarithm of stimulus. With exponential response, a VCO, for example, responds with a certain number of semitones per volt. Thus, equal increases in control voltage result in

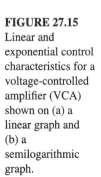

FIGURE 27.15
Linear and exponential control characteristics for a voltage-controlled amplifier (VCA) shown on (a) a linear graph and (b) a semilogarithmic graph.

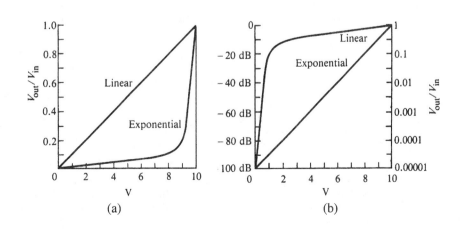

equal ratios of frequency increases, i.e., equal musical intervals. Thus, if a VCO increases its frequency by one octave per volt, adding $\frac{1}{12}$ V will increase the pitch by a semitone, and adding 1 V to a control signal will transpose all pitches up by one octave. Such an exponential control voltage is sometimes more convenient for musical applications because it relates directly to musical pitch.

Similarly, equal increases in control voltage result in equal ratios of increase in amplifier gain for an exponentially responding VCA, i.e., equal increases on a decibel scale. Such a VCA might vary its gain by 10 dB/V, for example, which readily translates into 1 dB per tenth of a volt, etc.

Because waveform signals are inherently linear, the primary disadvantage of exponential control voltage response is that signals and control voltages typically may not be mixed. Adding the output of a low-frequency VCO with a sinusoidal waveform and amplitude of $\frac{1}{12}$ V to a constant voltage will produce a vibrato of ± 1 semitone if the controlled VCO responds exponentially at 1 V per octave. If true frequency modulation is desired, it may be necessary to pass one or more signals through a conversion module designed to translate between linear and exponential scales.

27.12 ■ OTHER MODULES

In addition to those already mentioned, voltage-controlled synthesizers may contain other modules that perform other functions. Most synthesizers contain noise sources such as white-noise or pink-noise modules. A *white-noise generator* produces a randomly fluctuating signal in which all frequencies are present in more or less equal proportion. By analogy with white light, then, which contains equal proportions of all visible frequencies of light, a white-noise generator produces a sound that is rather high pitched due to the uneven response of the ear to different frequencies. Especially at low amplitudes, white noise sounds to the ear like the inverse of the equal-loudness curves (see Fig. 6.4).

A *pink-noise generator* produces a randomly fluctuating signal that sounds more like all frequencies to the ear. This is typically achieved by filtering white noise so that low frequencies are emphasized. Instead of having equal energy at all frequencies, as does white noise, pink noise typically has equal energy per octave. Because octaves occupy increasing amounts of frequency bandwidth as they go up in pitch, the energy density (per unit frequency) of pink noise thus decreases with increasing frequency.

Many synthesizers contain a *reverberation module* similar to the spring unit described in Chapter 26. An *envelope-follower module* measures the average amplitude of a signal and generates a control voltage that is proportional to it. Thus the gain of a VCA (or the frequency of a VCO, or the bandwidth of a VCF) can be made to "follow" the amplitude envelope of an arbitrary signal. Such a control signal can then impart some of the quality of the input sound to virtually any other sound (this is the source of many "talking" effects that can be superimposed on almost any sound).

A particularly interesting module called a *ring modulator* multiplies two signals together to form its output signal. A *modulator* is a circuit that modifies something about one signal in response to another. An *amplitude modulator*, for instance, uses one signal to control—or modulate—the amplitude of another signal (we will also discuss *frequency modulation* in some detail later on). An amplifier, for example, multiplies its input signal by a positive value called the *gain* of the amplifier. Because the signal can generally be either positive or negative but the gain is always positive, we can think of an amplifier as a *two-quadrant* amplitude modulator, because it must accommodate only the following two cases of multiplication.

1. Input signal is +, gain is +, output is +.
2. Input signal is −, gain is +, output is −.

Because a ring modulator multiplies two *arbitrary* signals, either one of which can be positive or negative at any time, it must accomplish *four-quadrant* amplitude modulation according to the following:

1. Input 1 is+, input 2 is +, output is +.
2. Input 1 is +, input 2 is −, output is −.
3. Input 1 is −, input 2 is −, output is +.
4. Input 1 is −, input 2 is +, output is −.

The ring modulator gets its name from the fact that the analog circuit that accomplishes four-quadrant multiplication is traditionally drawn with its four multiplying parts lying on a circle.

Multiplying one signal by another in this manner produces many interesting results. Essentially, the output signal consists of frequencies located at all possible sums and differences of frequencies taken from the two input signals. As a simple example, if input 1 is a complex waveform containing frequencies at 100 Hz and 200 Hz, and input 2 is a sinusoid at 10 Hz, the output would consist of a signal containing the four frequencies 100 ± 10 Hz and 200 ± 10 Hz. More complex inputs can produce dazzlingly complex outputs that generally sound distorted in ways ranging from pleasant to disturbing.

As explained in Section 8.4, alternating constructive and destructive interference between sinusoids at 100 Hz and 102 Hz will give rise to a beat frequency of 2 Hz. Figure 8.6 also shows that the frequency of the resulting waveform is the average of the two frequencies, in this case, 101 Hz. The beating waveform can be thought of either as resulting from adding the 100 Hz and 102 Hz sinusoids or from multiplying the 101 Hz sinusoid by one of 1 Hz (which way we hear it is determined by the size of the frequency difference.

Another way to understand the equivalence of adding and multiplying sinusoids is to observe that this is not just a physical or perceptual phenomenon but an inherent property of the sinusoids themselves. We learned in Section 2.1 that sinusoids represent basic vibration patterns called simple harmonic motion. A sinusoid always consists of just one frequency—any other vibration pattern is the result of multiple frequencies.

A sinusoid is any curve that is identical in shape to the mathematical sine or cosine functions (the sine and cosine functions themselves have the same basic shape, only different starting points). It is possible to demonstrate many interesting things about sinusoids by using their basic mathematical properties (see Appendix A.10). By combining the basic definitions of sine and cosine with the Pythagorean theorem, for example, we can show that

$$\cos A + \cos B = 2 \cos\left(\frac{A+B}{2}\right) \cos\left(\frac{A-B}{2}\right) \tag{27.1}$$

where A and B are any angles. Such relationships are called trigonometric *identities*, because they show two different ways to express the same thing. If we let $A = 2\pi f_A t$ and $B = 2\pi f_B t$, $\cos A$ and $\cos B$ can equally well represent two sinusoids at frequencies f_A and f_B. The left side of Eq. (27.1) then represents the sum of the two sinusoids. The right side correspondingly represents one sinusoid at the average frequency, $(A+B)/2$, multiplied by another at one-half of the difference frequency, $(A-B)/2$. Beats occur at twice the latter frequency because a sinusoidal waveform passes through both its positive and negative peaks during each cycle, resulting in two places of maximum amplitude in the product waveform. There is also a factor of 2 multiplying the right side of the equation because the peak amplitude of the left side is twice that of a single sinusoid whenever both peaks add in the same direction (up or down).

Another trigonometric identity that says the same thing in a different way is

$$\cos A \cos B = \tfrac{1}{2}[\cos(A-B) + \cos(A+B)] \tag{27.2}$$

27.13 ■ OTHER CONTROLLERS

In addition to the many types of voltage-controlled modules found in analog synthesizers, Donald Buchla invented an analog automation device called a *sequencer* that allowed a series of control voltages to be specified and then played back in sequence. Initially an attempt to reduce the labor involved in splicing tapes, the sequencer allowed virtually any melody, loudness sequence, or other series of sounds to be played back automatically. The main restriction was that the length of the sequence could not exceed the number of stages in the sequencer (originally 8 or 16 in a Buchla synthesizer). Clever application of the sequencer makes it possible to automate the playing of a nonrepeating series of sounds as well (listen, for example, to composer Morton Subotnick's landmark composition *Silver Apples of the Moon*, which was realized on a sequencer-equipped Buchla synthesizer in 1967).

27.14 ■ DIGITAL CONTROL AND THE *GROOVE* SYSTEM

As analog voltage-controlled synthesizers grew in capabilities and complexity (see Fig. 27.16), the problems of rapidly manipulating the interconnections of increasing numbers of signal and control modules grew as well. By the late 1960s, digital computers had advanced to the point where they could be used to generate some or all of the signals needed to control an analog synthesizer, which gave rise to the *hybrid synthesizer*. In a hybrid synthesizer all the waveforms were typically analog, but some or all of the control signals are generated using a digital computer.

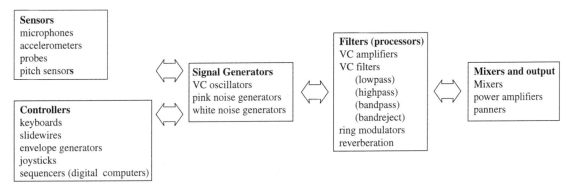

FIGURE 27.16 The "assembly line" for sound in an analog electronic music synthesizer. Compare the modules to the basic elements of electronic music mentioned in Fig. 27.1.

FIGURE 27.17 Schematic diagram of the GROOVE system, ca. 1969. In response to user inputs, including interactive controls and a computer program, the computer updates digitally stored time functions, a bank of DACs, and SPDT relays at a rate determined by an interrupt oscillator. A user-supplied patch bay determines how these signals affect an extensive bank of analog synthesis equipment.

Max Mathews and F. R. Moore created such a system in the late 1960s at Bell Laboratories. It was called *GROOVE*, for *G*enerated *R*ealtime *O*perations *O*n *V*oltage-controlled *E*quipment. It consisted of a Honeywell DDP-224 computer attached to a bank of 14 DACs, as well as a bank of computer-controlled relays and a variety of analog synthesizer modules (see Fig. 27.17). Some of the analog modules were commercially produced by Moog, ARP, and other companies; many were custom designs. An interchangeable "patch bay" allowed users to rapidly reconfigure interconnections between the DACs and the analog modules, as well as among the modules themselves. Each SPDT (single-pole, double-throw) relay appeared on the patch bay as well, allowing some of the interconnections to be controlled directly and dynamically by the computer (which ones depended on how a user configured the patch panel in use).

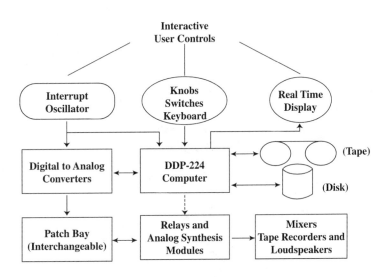

As a state-of-the-art, late 1960s-vintage computer, the DDP-224 consisted of a random-access memory (RAM), a central processing unit (CPU), a (digital) magnetic tape drive, and a magnetic disk drive in one room and various control devices in another. All electronics were based on *discrete* (i.e., physically separate) germanium transistors; integrated circuits were still in the process of being invented. The machine had 32,768 words of magnetic core memory (each word contained 24 bits; bytes were not yet standard as memory units). A core memory module containing 4K words occupied a box about the size of a medium-sized refrigerator. The removable (i.e., interchangeable) disk packs held about 1 megabyte of data each, and the disk drive was about the size of a washing machine. Connected to the computer were a vector graphics display, an IBM typewriter for text input, several knobs and switches whose positions could be sensed by the computer, and a control panel consisting of blinking lights showing the current state of all machine registers. Any bit in any register could be set or reset by pressing on the corresponding control panel light, thus defining the starting point for low-level programming.

The GROOVE software system consisted of a set of programs written in FORTRAN and assembly language that ran on the DDP-224 computer in real time, the speed of which was determined by an electronic oscillator the computer could sense through *interrupt* signals. Each time an interrupt signal occurred (about 100–200 times a second), the computer would execute a *foreground* program that necessarily had to accomplish everything in less than about 5 or 10 ms. Whenever the foreground program was not running, a *background* program stored in the computer memory would run.

The background task of GROOVE was to update the display viewed by the user, which showed time marching from left to right across the screen. The vertical axis was divided into several regions, like staves in a musical score. Each region displayed the graph of one function of time, whose significance was determined by the user. One function of time might control the frequency of a VCO, another the gain of a VCA, another the bandwidth of a VCF, etc., depending on the patch used. Values of each time function changed at the interrupt oscillator rate—about 100 times per second (this rate could be manually adjusted by the user).

The foreground program of GROOVE sensed the positions of all inputs, such as knobs, keyboards, joysticks, etc., and executed the user-supplied program that determined how these data were used to determine the outputs to the DACs (which in turn controlled the analog synthesis modules). In addition, the user-supplied program determined how the displayed functions of time were updated according to the input data and various signal-processing tactics the user wished to employ, such as combining keyboard data with a knob to create a pitch bend effect. Previously stored functions of time were read from disk and optionally updated and restored, allowing both a nonbinding rehearsal and a permanent update mode for later replay.

Many composers from Mathews and Moore to Pierre Boulez, Laurie Spiegel, Emmanuel Ghent, and Vladimer Ussachevsky experimented with the GROOVE system, which provided one of the most control flexible systems for electronic music ever devised. But, as the DDP-224 computer exemplified, electronics technology had begun to change from analog to digital.

27.15 ■ SUMMARY

Electronics has transformed music into a very general art of sound. Electronic music itself grew out of advances in electronics and sound-recording technology combined with acoustical principles and new musical concepts such as serialism. One major school of thought was *musique concrète*, which was based on creating a new kind of music that relied on combining and manipulating natural sounds instead of traditional musical instruments. A second major school of thought was *elektronische musik*, in which arbitrarily complex sounds could be synthesized from simple, basic components. Both these approaches allowed exploration of sounds that had never before been accessible to musicians and listeners. Work in both of these increasingly nondistinct areas was originally hampered by available technology, which made the manipulation of sounds extremely expensive and labor intensive. Early experiments in electronic music nevertheless provided an augmented conceptual basis for later developments in electronic music that were based on increasingly inexpensive and powerful solid state electronics (i.e., transistors). The first primary result was the analog synthesizer, which capitalized on modular design principles, the principle of voltage control, innovative interconnection schemes such as patch cords and pin matrices, increasingly sophisticated control devices such as ADSR envelope generators and sequencers, and, ultimately, control by a digital computer.

REFERENCES AND SUGGESTED READINGS

Boulanger, R. (1984). "Interview with Roger Reynolds, Joji Yuasa and Charles Wourinen," *Computer Music Journal* **8**(1).

Chadabe, J. (1997). *Electric Sound: The Past and Promise of Electronic Music.* Upper Saddle River, N.J.: Prentice Hall.

Dudley, H. (1939). "The Vocoder," *Bell Labs Record* **17**.

Eimert, H. (1959). "What is Electronic Music?" in *Die Reihe* **1**, Bryn Mawr, P.A.: Theodore Presser.

Hartmann, W. M. (1975). "The Electronic Music Synthesizer and the Physics of Music," *Am. J. of Physics* **43**.

Howe, H., Jr. (1975). *Electronic Music Synthesis.* New York: Norton.

Mathews, M. V., and F. R. Moore (1970). "GROOVE—A Program to Compose, Store, and Edit Functions of Time," *Commun. ACM* **13**(12).

Olson, H. F., and H. Belar (1955). "Electronic Music Synthesizer," *J. Acoust. Soc. Amer.* **27**.

Schaeffer, P. (1952). *A la Recherche d'une Musique Concrète.* Paris: Editions du Seuil.

Stockhausen, K. (1961). "Two Lectures" in *Die Reihe* **5**. Bryn Mawr, P.A.: Theodore Presser.

GLOSSARY

ADSR envelope generator A, D, S, and R refer to the parameters of an envelope: A = attack time; D = (initial) decay time; S = sustain level; R = (final) release time.

amplitude envelope The manner in which peak or average amplitude varies with time; the envelope is comprised of the attack and decay of a tone, among other things.

elektronische musik A type of music based on the control of many electronically generated signals that together form a musical result.

envelope follower A circuit that generates a control voltage that duplicates the amplitude envelope of a signal taken from an acoustical musical instrument or other sound source.

exponential A function or characteristic that varies as e^x. In a typical exponential control characteristic, an increase of 1 V doubles some parameter such as frequency.

hybrid synthesizer Use of a digital computer to control the analog circuitry in a synthesizer; also a synthesizer that combines analog and digital generators or modules.

linear A function or characteristic that varies in direct proportion to a control signal. In a typical linear control characteristic, an increase of 1 V adds a fixed amount to some parameter such as frequency.

modulator A device that controls (modulates) a characteristic of one signal (amplitude, frequency, etc.) with another signal.

musique concrète A type of music based on manipulating the "concrete" sounds of nature (rather than the "abstract" sounds of traditional musical instruments).

panning Applying a set of output signals to a series of loudspeakers sequentially so that the sound seems to move in space.

patch (Verb) to interconnect; or (noun) a set of interconnections that causes a synthesizer to produce certain types of sounds.

pink noise Low-pass-filtered random noise for which the energy contained in each octave band is the same.

pitch bend A control that lowers or raises an otherwise fixed pitch (such as one specified by a note on a keyboard).

Q A filter parameter that specifies the sharpness of a resonance (technically, ratio of energy stored to energy dissipated per cycle).

ring modulator A circuit that multiplies two input signals, thereby forming an output signal containing the sums and differences of all pairs of frequencies present in the two input signals.

sequencer A circuit that switches through a predetermined sequence of control signals.

synthesizer An instrument that creates complex sounds by generating, altering, and combining various electrical waveforms, typically by means of voltage-controlled modules.

VCA (voltage-controlled amplifier) An amplifier whose gain varies linearly or exponentially in proportion to a control voltage.

VCF (voltage-controlled filter) A filter whose cutoff frequency (or center frequency and bandwidth) varies linearly or exponentially in proportion to one or more control voltages.

VCO (voltage-controlled oscillator) An oscillator whose frequency varies linearly or exponentially in proportion to one or more control voltages.

REVIEW QUESTIONS

1. What is the basic concept of *musique concrète*?
2. How were sounds first recorded in *musique concrète*?
3. Name three advantages of magnetic tape recording for sound manipulation.
4. What was the fundamental concept of *elektronische musik*?
5. What is a vocoder?
6. How were the original concepts of *elektronische musik* linked to *serialism*?
7. Why were *musique concrète* and *elektronische musik* originally so tedious to produce?
8. Where was the first major electronic music synthesizer located?
9. How did the RCA Mark II avoid the limitations of human performance?

10. What is the basic principle of voltage control?
11. What is a VCO? A VCA? A VCF? An LFO? An ADSR generator?
12. What is a patch?
13. What is a sequencer?
14. How does frequency modulation work?
15. How does one synthesize analog brass?
16. What are the advantages and disadvantages of exponential control voltages?
17. Which sound has a higher pitch, white noise or pink noise? Why?
18. Describe the operation of a ring modulator.
19. What is a hybrid synthesizer?
20. What did functions of time control in the GROOVE system?

QUESTIONS FOR THOUGHT AND DISCUSSION

1. What are the principal advantages and disadvantages of *musique concrète* and *elektronische musik*?
2. In what ways do synthesizers replace traditional musical instruments or performers?
3. What kinds of music, if any, can be made electronically that cannot be made with traditional instruments?

4. About how much of the music heard today is produced electronically?
5. What is the purpose of the limit mode shown in Fig. 27.14?
6. What effect does turning the "width" knob have on the resulting sound for the VCO shown in Fig. 27.6?

EXERCISES

1. The control characteristic of a certain VCO is stated as 1 V per octave.

   (a) Is this a linear or exponential characteristic?

   (b) If a control voltage of 3 V causes it to oscillate at $C_3(f = 131$ Hz), what voltage will be required to give $G_3(f = 196$ Hz)?

2. If one input to a ring modulator contained two frequency components at $f_1 = 200$ Hz and $f_2 = 350$ Hz and the other input was a sinusoid at $f = 50$ Hz, what frequencies would be present at the output? What would you expect the pitch of the resulting sound to be?

3. Make a drawing similar to Fig. 27.13 showing what patch is represented by the pin matrix shown in Fig. 27.8.

4. Make a drawing (to scale) of an ADSR envelope with a maximum amplitude of 2 V and the following parameters: $A = 50$ ms, $D = 20$ ms, $S = 1.5$ V, $R = 100$ ms.

EXPERIMENTS FOR HOME, LABORATORY, AND CLASSROOM DEMONSTRATIONS

Home and Classroom Demonstrations

1. *Musique concrète*   Using a portable tape recorder collect a variety of sounds from your environment. Using a home or laboratory computer, edit and combine these sounds into an interesting collage of arbitrary length.

2. *Elektronische Musik*   Tape record sounds produced by electronic oscillators and noise generators, filtering them before or after recording as desired. Using a home or laboratory computer, edit and combine these sounds into an interesting collage of arbitrary length.

3. *Electronic music*   Using your local library and/or other sources, find as many recorded examples of electronic music made before about 1955 as you can. Also find as many examples made within the past five years as you can. Listen to these examples and compare their features.

4. *Analog music synthesizers*   Obtain the oldest working analog music synthesizer you can and experiment with its capabilities.

Laboratory Experiments

Electronic Music Synthesizers (Experiment 38 in *Acoustics Laboratory Experiments*).

Digital Audio Signal Processing

All audio signals may be recorded and/or transmitted from place to place. Electrical analog audio signals can be of extremely high fidelity to their mechanical counterparts, but they are constantly plagued by the tendency to become contaminated by noise. It is, therefore, important to carefully shield high fidelity analog audio signals from myriad sources of unwanted electromagnetic noise. In most cases, the best way to make a signal nearly impervious to noise is to convert it to digital form. Digital signals are essentially immune to noise contamination unless the noise becomes so powerful that it obliterates the signal altogether.

Just as analog signals can be amplified, mixed, and filtered, digital signals can be processed. Instead of using electronic circuit elements such as vacuum tubes or inductors directly, digital signals are processed with numerical operations such as addition, subtraction, multiplication, and division. Of course, the digital circuits that perform these numerical operations are still made of transistors and capacitors, but these circuit elements are generally operating in their nonlinear—or digital—mode under control of some sort of program. And just as more complex processors for analog signals—such as filters for analog signals having frequency-dependent gain and phase characteristics—can be made by combining simpler processing elements such as resistors and capacitors, digital filters can be made by combining elementary numerical operations such as adding and multiplying.

In this chapter you should learn:

- About the time and frequency domains;
- About digital signal transforms;
- About convolution;
- About digital filtering.

28.1 ■ DIGITAL AUDIO-SIGNAL PROCESSING

A comprehensive discussion of digital signal processing would go well beyond the scope of this book. The basic ideas, however, involve understanding relationships between the time (waveform) and frequency (spectrum) representations of signals and methods for modifying the signal in either of these two domains.

28.2 ■ DIGITAL SIGNAL PROCESSING AND MATHEMATICS

Because a digital signal is essentially a list of numbers representing the shape of a corresponding waveform, it should come as no surprise that digital signal processing involves processing numbers—i.e., it is inherently mathematical in nature. Just as capacitors and inductors affect analog electrical signals in frequency-dependent ways, equivalent mathematical operations affect digital signals in corresponding ways.

For example, we can amplify a digital signal merely by multiplying all its values by any value greater than 1. Stated mathematically, we might say that a digital output signal, $y(n)$, is equal to a digital input signal, $x(n)$, multiplied by a constant gain factor G, where G is greater than 1, i.e., $y(n) = G \times x(n)$, $G > 1$, and $n = 0, 1, 2, \ldots$. Similarly, we can mix two digital signals in many cases merely by adding them together, number by number, forming a "sum" signal. The mathematical way of saying the same thing might read $y(n) = x_1(n) + x_2(n)$.

Other mathematical operations yield many interesting and useful results. For example, if we replace each value of a digital signal by the average of itself and the previous value, we smooth out (slightly) any rapid variations in the signal. Because rapid variations are associated with high frequencies, such an averaging operation attenuates the high frequencies in that signal, whereas low frequencies tend to be much less affected. The averaging operation therefore implements a simple low-pass filter. The mathematical version of this filter is very simple: $y(n) = [x(n) + x(n-1)]/2$. This demonstrates how mathematics is *the* essential tool of digital signal processing.

Before you become concerned that the mathematics will become impenetrable for the average reader, remember that computers can only add, subtract, multiply, and divide (though they can do such operations very rapidly). Digital signal processing, therefore, cannot rely on any more complicated mathematical operations than simple arithmetic. If you do not use mathematics on a regular basis, however, you may find it useful at this point to review the mathematical representation of sound waveforms, especially the sine and cosine functions of trigonometry (see the Appendix A.10).

28.3 ■ THE TIME DOMAIN

We saw (in Chapter 21) that a digital signal is an ordered set of sample values representing the amplitude of a waveform at a sequence of equally spaced instants in time. If we think of an analog waveform as a mathematical function of time $f(t)$, where t is time, then the equivalent digital signal sampled at a rate of R samples per second can be represented as $x(nT)$, where $T = 1/R$ and $n = 0, 1, \ldots$, is called the *sample index*. It is common practice

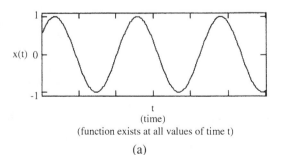

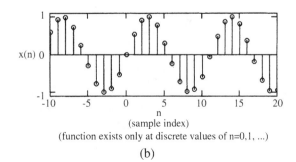

(a) (b)

FIGURE 28.1   Representation of a continuous waveform (a) in the samples time domain (b).

in discussions of digital signals to omit explicit reference to the sampling rate (because *all* digital signals have *some* sampling rate) and to speak of a digital signal simply as $x(n)$, where $n = 0, 1, \ldots,$ as before (see Fig. 28.1). Thus, if a function of time is described as $f(t) = A \sin(2\pi f t + \phi)$, where A is peak amplitude, f is frequency, and ϕ is the phase offset at $t = 0$, then we simply substitute nT for t to obtain the equivalent digital function. This results in the expression $x(nT) = A \sin(2\pi f nT + \phi)$, or some simply, $x(n) = A \sin(\omega n + \phi)$, where $\omega = 2\pi f$. $f(t)$ is called a continuous (or analog) time function, and $x(nT)$ and $x(n)$ are called discrete (or discontinuous) time functions.

28.4 ■ THE FREQUENCY DOMAIN

We have already seen many examples of spectra, which typically show either amplitude or phase values as a function of frequency. Just as time can be continuous or discontinuous, depending on whether we are speaking of an analog or digital signal, frequency can be either continuous or discontinuous as well.

We already know from the sampling theorem that the maximum frequency a digital signal can represent is $R/2$ hertz, where R is the sampling rate (any higher frequencies sampled at rate R will be aliased to some frequency less than $R/2$ according to Eq. 21.1). In other words, at least two samples are needed to digitally represent a frequency—the sampling theorem allows more than two samples to represent a given frequency but not less than two.* One way to understand discrete frequencies is then to consider periods of three samples, four samples, and so on, down to arbitrarily long periods (i.e., low frequencies). Another is to start with N samples of a digital signal and define the frequency corresponding to a period of N samples to be the fundamental frequency of a set of N harmonically related frequencies.

We can describe the digital signal's spectrum in terms of such harmonics, because Fourier's theorem assures us that we can represent any digital signal $x(n)$ as

$$x(n) = A_0$$

$$+ A_1 \cos\left(2\pi \frac{n}{N} + \phi_1\right)$$

$$+ A_2 \cos\left(2\pi \frac{2n}{N} + \phi_2\right)$$

$$+ \cdots \quad \text{(to } N \text{ terms)} \tag{28.1a}$$

$$= \sum_{k=0}^{k=N-1} A_k \cos\left(2\pi \frac{kn}{N} + \phi_k\right). \tag{28.1b}$$

Equations (28.1a) and (28.1b) are just two different ways of writing the same thing. Each term in (28.1a) is a sinusoid with amplitude A_k and phase offset ϕ_k. The fundamental

---

*Because a single frequency *must* have the shape of a sinusoid, the statement that only two samples per period can represent a frequency bothers some people. Two samples per period might seem to represent at best a square wave, or perhaps, up and down "impulses" of some sort. Recall, however, from Fig. 21.8 that the final step of digital-to-analog conversion involves *low-pass filtering at the Nyquist rate*. This smoothing filter (ideally) removes all energy above half the sampling rate, leaving only a pure sinusoid at the Nyquist rate. Similar reasoning applies to lower frequencies as well.

frequency is R/N, where N is the number of samples being considered. The second term has twice this frequency, the third term would have three times this frequency, and so on. Equation (28.1b) shows the sum notation commonly used in digital signal processing, where

$$\sum_{k=a}^{k=b} \xi_k$$

means that we write the expression ξ_k for every value of k starting with $k = a$, then $k = a + 1$, etc., until we reach $k = b$; then we add all the expressions. For example, we can express the sum of all the whole numbers from 50 to 100 by writing either

$$50 + 51 + \cdots + 100$$

or

$$\sum_{k=50}^{k=100} k$$

Equations (28.1a) and (28.1b) therefore show how N samples of a digital waveform can be represented as a sum of N harmonics with fundamental frequency of R/N hertz, where R

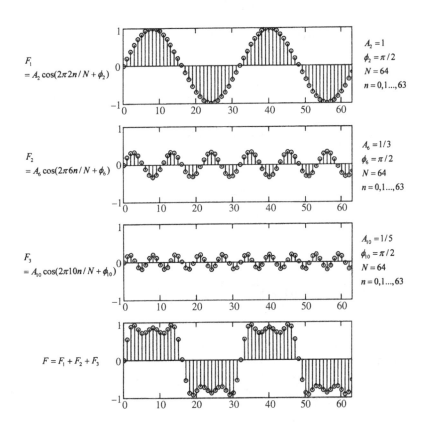

FIGURE 28.2
Building up a
digitized square
wave in terms of
Equations (28.1).
All amplitude and
phase terms not
mentioned are
equal to zero.

is the sampling rate. The amplitude and phase offset of the kth harmonic are A_k and ϕ_k, respectively (see Fig. 28.2).

The astute reader will have noticed that the frequencies start at zero and go up to a frequency of $(N - 1)R/N$ hertz (almost R), which exceeds $N/2$ hertz (the Nyquist rate) whenever N is greater than two. Zero hertz may be understood by reference to $A\cos(0)$, which is equal to A (a constant value) because $\cos(0) = 1$. (This is sometimes called dc, which stands for *direct current*. Frequencies different from zero are referred to collectively as ac, for alternating current.)

Frequencies between the Nyquist rate and the sampling rate can be understood (by reference to Equation 21.1) as negative frequencies (see Fig. 28.3). This is because the spectrum of a sampled signal is *periodic in frequency*, with an infinite number of periods centered about all multiples (positive and negative) of the sampling rate. Each "copy" of the spectrum ranges from $-R/2$ to $+R/2$ hertz. In most cases, the negative half of the frequency spectrum is just the left-right mirror image of the positive half of the frequency spectrum (possibly upside down as well). Because of this symmetry, it is common to show just the positive half of the frequency spectrum in a diagram in order to save space or to allow a more detailed view in a given space.

Although the sampling theorem forbids frequencies with magnitudes greater than the Nyquist rate, negative frequencies are perfectly permissible, so long as their magnitudes do not exceed the Nyquist rate. There is nothing more unusual about negative frequencies than a phase reversal in the case of so-called odd functions, such as $\sin(x)$ (because $\sin(-x) = -\sin(x)$); even functions such as $\cos(x)$ are not affected at all (because

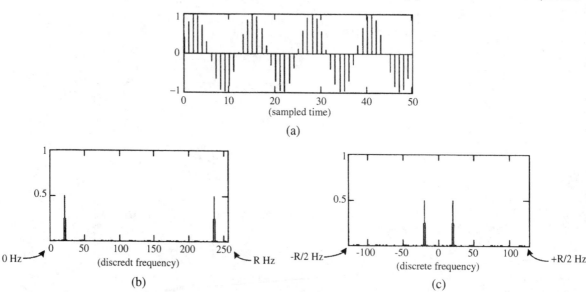

FIGURE 28.3   (a) shows the first 50 samples of $N = 256$ samples of a sampled waveform with a period of $N/20$ samples (its frequency is therefore 20 cycles in the time of N samples). (b) shows the magnitude of the (discrete) spectrum of this signal from 0 to R Hz. (c) shows the same spectrum, from $-R/2$ Hz to $+R/2$ Hz. Note that a sinusoidal signal in the time domain has symmetrical positive and negative components in the frequency domain, each with half the amplitude of the time domain signal.

$\cos(-x) = \cos(x)$). It is straightforward to show that *all* mathematical functions (and therefore all audio waveforms) are either purely odd, purely even, or the sum of odd and even parts.

Assuming N is an even number, then, Eqs. (28.1a) and (b) describe $x(n)$ in terms of N harmonics. The first half of these harmonics (of R/N hertz) range in frequency from 0 Hz to (almost) $+R/2$ hertz; the second half range from $-R/2$ hertz to (almost) 0 Hz, completing the range of possible frequencies in a digital signal.

When making plots of spectra, the frequency scale for digital signals can be given in hertz or in normalized frequency, which ranges from 0 to $\frac{1}{2}$ for positive frequencies, where $\frac{1}{2}$ refers to half of the sampling rate.

28.5 ■ THE DISCRETE FOURIER TRANSFORM

The time- and frequency-domain views of a signal are useful because we have a convenient way of transforming one into the other. This method is called the *Fourier transform* in the case of analog signals, and the *discrete Fourier transform* (DFT) in the case of digital signals. In order to do this transformation, we use the identity $A \sin(\omega n + \phi) = a \cos(\omega n) + b \sin(\omega n)$ to rewrite Eqs. (28.1a) and (b) as follows:

$$x(n) = a_0$$
$$+ a_1 \cos(\omega n) + b_1 \sin(\omega n)$$
$$+ a_2 \cos(2\omega n) + b_2 \sin(2\omega n) + \cdots \qquad (28.2a)$$
$$= \sum_{k=0}^{k=N-1} a_k \cos(k\omega n) + b_k \sin(k\omega n), \qquad (28.2b)$$

where a_k is equal to $A_k \sin(\phi_k)$, b_k is equal to $A_k \cos(\phi_k)$, and ω is equal to $2\pi/N$. We can also convert from Eq. (28.2) notation to Eq. (28.1) notation by noting that

$$A_k = \sqrt{a_k^2 + b_k^2} \qquad (28.3a)$$

and

$$\phi_k = \tan^{-1} \frac{a_k}{b_k} \qquad (28.3b)$$

Thus, although Eqs. (28.2a) and (b) do not describe the amplitudes and phases of the harmonics directly, they are entirely equivalent to Eqs. (28.1a) and (b).

Equations (28.2a) and (b) are useful because we have a very straightforward way to compute the values of the a_k and b_k coefficients. Once these coefficients are known, the values of the amplitudes and phases of the harmonics can also be readily computed (see Fig. 28.4).

The DFT of N samples of a digital signal $x(n)$ is calculated using the formulas

$$a_k = \frac{1}{N} \sum_{n=0}^{N-1} x(n) \cos(k\omega n) \qquad (28.4a)$$

and

$$b_k = \frac{1}{N} \sum_{n=0}^{N-1} x(n)\sin(k\omega n),$$ (28.4b)

where $k = 0, 1, \ldots, N-1$ and ω is equal to $2\pi/N$, as before. Once the a_k and b_k coefficients are known, we can use Eqs. (28.3a) and (b) to calculate the relevant amplitudes and phases offsets.

FIGURE 28.4
The Discrete Fourier Transform (DFT) of the noisy sinusoidal time domain signal (top) is given in two parts: the magnitude (amplitude) spectrum (middle) and the phase spectrum (bottom). Note how evident the frequency of the sinusoidal component is in the amplitude spectrum. Note also that the amplitude spectrum is symmetrical around 0 Hz, while the phase spectrum is roughly antisymmetrical (i.e., both backwards *and* upside down).

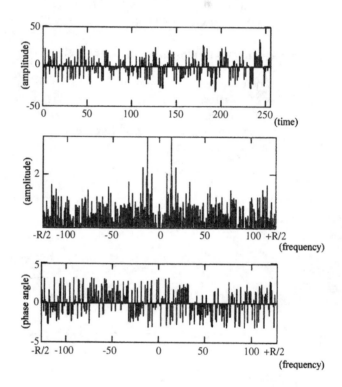

Another way to define the DFT is as follows.

$$\mathrm{DFT}[x(n)] = X(k)$$

$$= \frac{1}{N} \sum_{n=0}^{N-1} x(n)e^{-j\omega nk} \qquad 0 \le k \le N-1,$$ (28.5)

where j is the imaginary unit (i.e., the square root of -1), and $e^{j\theta} = \cos(\theta) + j\sin(\theta)$ (this is known as Euler's identity). Defined in this way, the DFT consists of N complex numbers, each of the form $a_k + jb_k$. The a_k and b_k values in this form are equal to the a_k and b_k values in Eqs. (28.2), (28.3), and (28.4).

Given this form of the DFT, the inverse DFT can be defined as follows:

$$\text{DFT}^{-1}[X(k)] = x(n)$$

$$= \sum_{k=0}^{N-1} X(k)e^{+j\omega nk} \qquad 0 \le n \le N - 1. \qquad (28.6)$$

Thus we can use Eq. (28.5) to compute the DFT of N samples of digital signal $x(n)$, yielding N samples of its spectrum, $X(k)$. Conversely, we can use Eq. (28.6) to compute N samples of the digital signal $x(n)$ from N samples of its spectrum, $X(k)$.

28.6 ■ THE FAST FOURIER TRANSFORM

Although it is a landmark in the field of computational algorithms, for our purposes the fast Fourier transform (FFT) is just a very efficient way to calculate the DFT. A discussion of the details of the FFT is beyond the scope of this book, but two simple observations can explain its usefulness.

First, the FFT yields exactly the same results as the DFT (Eqs. (28.5) and (28.6)). The reason it is almost universally preferred to the DFT is that the computation time for the FFT is proportional to $N \log N$, where N is the number of points being transformed, whereas the computation time for the DFT is proportional to N^2. For very small values of N, this difference is largely irrelevant. The larger N becomes, however, the greater the advantage of the FFT. When $N = 16$, for example, N^2 is about four times greater than $N \log N$. But when $N = 1024$, N^2 is more than 100 times greater than $N \log N$.

Second, the FFT usually requires that the value of N be restricted to "highly composite numbers," meaning that N has many prime factors (the least "composite" numbers are, therefore, prime). The *most* highly composite numbers are the powers of 2, so the FFT has its greatest speed advantage when N is restricted to powers of 2. Although it might seem that this imposes severe restrictions on the fundamental frequency used to compute the spectrum, these restrictions are relatively insignificant compared to the speed advantages of the FFT. It is, therefore, typical that most FFT-based systems—hardware or software— are restricted to use *only* powers of 2 (i.e., 2, 4, 8, 16, 32, 64, 128, 256, 1024, 2048, 4096, 8192, 16,384, etc.) for N.

28.7 ■ CONVOLUTION

Given the DFT—and its high-speed cousin, the FFT—for computing the spectrum of a waveform or (inversely) the waveform of a spectrum, we need to observe only two more mathematical facts before describing digital audio-signal processing in more detail. When waveforms are added (or mixed), the associated spectra are also added (or mixed). Saying the same thing mathematically, if $x_3(n) = x_1(n) + x_2(n)$ and $X_1(k)$ and $X_2(k)$ are the DFTs (or FFTs) of $x_1(n)$ and $x_2(n)$, respectively, then the DFT (or FFT) of $x_3(n)$ is equal to $X_3(k) = X_1(k) + X_2(k)$. Similarly, when two spectra are added, the waveform of the result is simply the sum of the waveforms associated with the two added spectra.

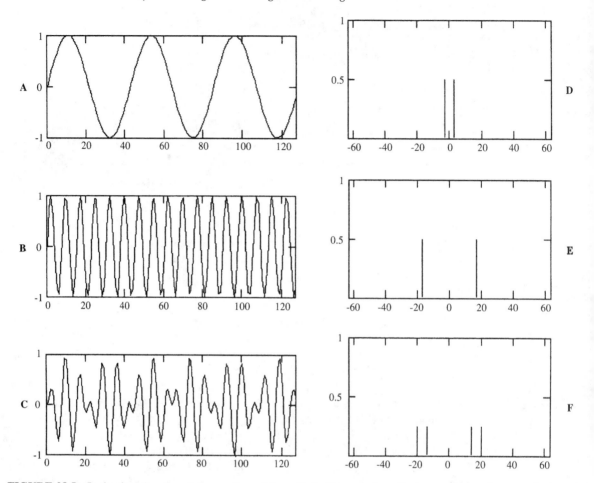

FIGURE 28.5 In the time domain, when waveform A is multiplied by waveform B, the result is product waveform C. In the frequency domain, we obtain the spectrum of waveform C by convolving the spectra of waveforms A and B. Note that the components of spectrum F are obtained by centering a scaled copy of spectrum D around each component of spectrum E. We see that, in this case, convolution illustrates the trigonometric identity: $\cos A \cos B = \frac{1}{2}[\sin(A - B) + \sin(A + B)]$.

When two waveforms are multiplied, however, the spectrum of the result is *not* the product of the two individual spectra. Rather, the spectrum of the product of two waveforms is found by *convolving the spectra* of the two individual waveforms (see Fig. 28.5). Similarly, when two spectra are multiplied, the waveform of the result is found by *convolving the waveforms* associated with the two individual spectra. In other words, *multiplication in the time or frequency domain is associated with convolution in the other domain.*

Just as point-by-point addition and multiplication are possible ways to combine two waveforms (or spectra), convolution is just another way to combine two functions, either analog or digital (we are concerned here only with the digital case). In order to convolve one digital function with another, we simply replace *each* sample of one function by *all* samples of the other function and add overlapping values, according to the formula

$$h(n) = f(n) \otimes g(n) = \sum_{m=0}^{n} f(m)g(n-m), \qquad (28.7)$$

where $\otimes$ denotes the convolution operation.

Convolution has a reputation for being hard to understand, but it is really quite straightforward. To convolve two digital signals, we simply substitute one signal for each sample of the other, multiplying—or scaling—by the substituted sample each time. Then we add all the (scaled) substitutions to get the final result. As a simple example, suppose one signal consists of the sample values $\{2, 2, 3, 3, 4\}$ and the other signal consists of the sample values $\{1, 1, 2\}$. Call the first signal $f(n)$ and the second $g(n)$. Equation (28.7) says that we convolve $f(n)$ with $g(n)$ by writing $g(n)$ starting at each position in $f(n)$, multiplying $g(n)$ by the value of $f(n)$ in that position each time. We then add all the overlapping sequences to get the result.

2	2	4					$g(n)$ times 2,
	2	2	4				$g(n)$ times 2,
		3	3	6			$g(n)$ times 3,
			3	3	6		$g(n)$ times 3,
				4	4	8	$g(n)$ times 4,
2	4	9	10	13	10	8	Result.

28.8 ■ DIGITAL FILTERING

One of the reasons that convolution is important in digital audio-signal processing is that filtering inherently involves multiplying the spectrum of a signal by another spectrum describing the operation of a filter. The frequency domain operation of a filter is called its *transfer function*, because it shows how an input signal is transferred to its output (see Fig. 28.6). The time domain version of the transfer function is therefore convolved with the input signal in order to compute the filter output. For reasons that we shall discuss, the time domain version of the digital filter's transfer function is called its *unit sample response*.*

If the unit sample response of a digital filter is $h(n)$, then the input signal $x(n)$ is converted by the filter to its output signal $y(n)$ according to the relation

$$y(n) = x(n) \otimes h(n), \qquad (28.8)$$

where $\otimes$ denotes the convolution operation (see Fig. 28.7).

---

*Because the unit sample function is the digital equivalent of the impulse function, the unit sample response is also sometimes informally called the digital filter's impulse response.

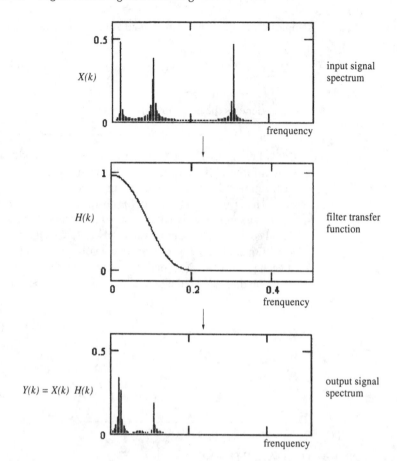

FIGURE 28.6
In the frequency
domain operation
of a digital filter,
the spectrum of the
input signal (top) is
multiplied by the
transfer function of
the filter (middle)
to produce the
spectrum of the
output signal
(bottom). The time
domain version of
the transfer
function is called
the "unit sample"
or "impulse"
response of the
filter.

FIGURE 28.7
Operation of a
digital filter in the
time domain: $h(n)$
is the "unit sample"
or "impulse"
response of the
filter. Note the
convolution
operation involved.

28.9 ■ THE z-TRANSFORM

Just as the DFT can be used to associate a spectrum in the frequency domain with a waveform in the time domain, the z-transform can be used to associate a "generalized" spectrum in the z domain with a waveform in the time domain.

The z-transform of a digital waveform $x(n)$ is defined as

$$X(z) = \sum_{n=-\infty}^{\infty} x(n)z^{-n}, \qquad (28.9)$$

FIGURE 28.8
Definition of a
complex value z_0.
When plotted on a
complex plane, it
has both real and
imaginary parts, or,
conversely, a
direction (angle)
and distance
(magnitude) from
the origin. The
relations between
these two forms of
representation are
essentially those of
Equations (28.3).

where z is an arbitrary complex variable (see Fig. 28.8). The quantity z^{-1} can be interpreted as the unit sample delay operator, because $x(n)z^{-1}$ is equal to $x(n-1)$ (see Fig. 28.9). The definition of the z-transform (Eq. (28.9)) also tells us that if $X(z)$ is the z-transform of $x(n)$, then the z-transform of $x(n-k)$ is $z^{-k}X(z)$.

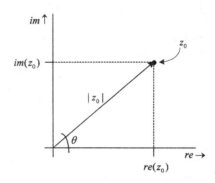

FIGURE 28.9
Definition of the
unit sample delay
operator. The
output is the same
as the input, but
delayed in time by
one sample period.

The z-transform actually includes the DFT as a special case, i.e., when we set $z = e^{j\omega}$, we have

$$X(e^{j\omega}) = \sum_{n=-\infty}^{\infty} x(n)e^{-j\omega n}, \qquad (28.10)$$

which is equivalent to Eq. (28.5) for values of n from 0 through $N - 1$.

Convolution is to the z-transform as it is to the DFT: multiplication in the time domain or z domain corresponds to convolution in the other domain. Because Eq. (28.8) shows that the output of a digital filter is the convolution of its input with its unit sample response, the z-transform of the output of a digital filter is the product of the z-transforms of its input and unit sample response, according to

$$Y(z) = X(z)H(z). \tag{28.11}$$

We see from Eq. (28.11) that $H(z)$ relates the input and output of a digital filter, because

$$H(z) = \frac{Y(z)}{X(z)}. \tag{28.12}$$

$H(z)$ is therefore called the *transfer function* of the digital filter.

28.10 ■ FILTER COEFFICIENTS

Many types of digital filters are possible, but by far the most common and useful are defined by the following equation:

$$y(n) = \sum_{i=0}^{M} a_i x(n - i) - \sum_{i=1}^{N} b_i y(n - i). \tag{28.13}$$

As before, $y(n)$ is the output signal and $x(n)$ is the input signal. The a_i values are constant coefficients that determine how much of the current and previous input signal samples contribute to the current sample of output (the current input, $x(n)$, is "weighted" by a_0, the previous input, $x(n - 1)$, by a_1, and so on, up to $x(n - M)$ and a_M). Similarly, the b_i values are constant coefficients that determine how much the previous output signal

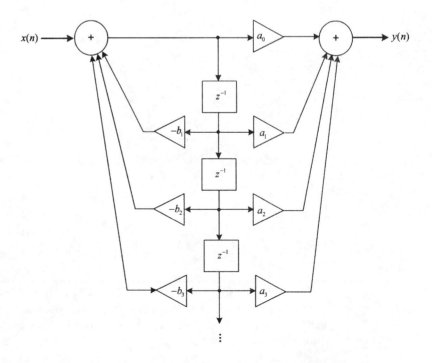

FIGURE 28.10
Direct implementation of Equation (28.13) in digital hardware requires only sample delays, multiply and accumulate (add) operations. This is the basis of almost all digital filtering operations.

samples contribute to the current sample of (the previous output, $y(n - 1)$, is weighted by b_1, and so on, up to $y(n - N)$ and a_N).

Equation (28.13) shows that the output of a digital filter at any sample index n depends on the current input and M previous inputs as well as N previous outputs (see Fig. 28.10). Note that the output can depend on the current input but not the current output, which is what the filter needs to compute. To do its job, a digital filter needs to keep track of M previous input values and N previous output values.

28.11 ◼ THE TRANSFER FUNCTION

The two halves of Equation (28.13) have far-reaching significance. To understand this, we can find the z-transform of Eq. (28.13) as follows.

$$Y(z) = \sum_{i=0}^{M} a_i z^{-i} X(z) - \sum_{i=1}^{N} b_i z^{-i} Y(z). \tag{28.14}$$

We can collect all the $Y(z)$ terms on the left and factor out common $Y(z)$ and $X(z)$ terms to see that

$$Y(z)\left(1 + b_1 z^{-1} + \cdots + b_N z^{-N}\right) = X(z)\left(a_0 + a_1 z^{-1} + \cdots + a_M z^{-M}\right) \tag{28.15}$$

Recalling Eq. (28.12), we now see that

$$H(z) = \frac{Y(z)}{X(z)} = \frac{a_0 + a_1 z^{-1} + \cdots + a_M z^{-M}}{1 + b_1 z^{-1} + \cdots + b_N z^{-N}} = \frac{N(z)}{D(z)} \tag{28.16}$$

Equation (28.16) is the basic definition of the transfer function of the basic digital filter defined by Eq. (28.13). The a_i and b_i values in Eq. (28.16) are the same as those in the basic definition.

28.12 ◼ POLES AND ZEROS

We see that the a_i values are the coefficients of a polynomial in the numerator, $N(z)$, of the transfer function, whereas the b_i values are coefficients of a polynomial in the denominator, $D(z)$. Recalling that the roots of a polynomial in z are the values of z that make the polynomial zero, we see that the roots of the numerator polynomial are the values of z that make transfer function $H(z)$ zero.

Similarly, the roots of the denominator polynomial are the values of z that make the transfer function arbitrarily large (i.e., infinite). The roots of $N(z)$ therefore define the *zeros* of the transfer function (the frequencies at which the filter's output is zero, regardless of its input). The roots of $D(z)$ similarly define the *poles* of the transfer function (the frequencies at which the filter can have an output whether the input is zero or nonzero). The poles of a filter are associated with *resonances*—frequencies at which the output of the filter is large. The zeros of a filer are associated with *antiresonances*—frequencies at which the output of the filter is nil (see Fig. 28.11).

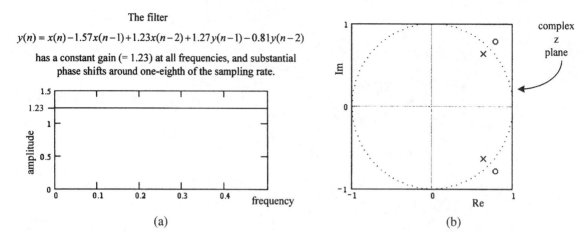

The filter

$$y(n) = x(n) - 1.57x(n-1) + 1.23x(n-2) + 1.27y(n-1) - 0.81y(n-2)$$

has a constant gain ($= 1.23$) at all frequencies, and substantial phase shifts around one-eighth of the sampling rate.

(a)   (b)

FIGURE 28.11   A particular digital allpass filter is achieved by placing poles and zeros at the same frequency, but symmetrically inside and outside the unit circle. The poles are marked with an **X**, the zeros are marked with an **O**. The indicate the roots of the denominator and numerator polynomials of the transfer function, respectively.

28.13 ■ FIR AND IIR FILTERS

According to Equation (28.13), a digital filter can have poles, zeros, or both. If a digital filter has only zeros, then $N = 0$, or, equivalently, all the b_i coefficients in Eq. (28.13) are equal to zero. This means that the filter output depends only on a weighted combination of present and past inputs (the a_i coefficients are the weights). Furthermore, the unit sample response of the filter will simply be the a_i coefficients themselves. Because there are just M such coefficients, it is clear that the unit sample response of the filter is of finite length. In other words, if a unit sample function (which is equal to $1, 0, 0, \ldots$) is input to the filter, the output will consist of the samples $a_0, a_1, a_2, \ldots, a_M$. Such a filter exhibits a finite impulse response (FIR) when excited by the digital form of an impulse signal.

FIR filters have many interesting properties, most especially that they are straightforward to implement directly in either digital hardware or software. Each output sample is the sum of the products of M coefficients with M input sample values. They also can be designed to have a *linear phase* characteristic, which guarantees that all frequencies are processed by the filter with an equal amount of delay (this prevents frequency-dependent phase misalignments). FIR filters often have no analog equivalent because they require pure delay (i.e., z^{-1}) operations for their implementation.

If any of the b_i coefficients in Eq. (28.13) are nonzero (i.e., $N \geq 1$), then each output sample can depend on previously computed output samples. This is the digital equivalent of feedback, because some of the output of the filter is being fed back into its subsequent output calculations. Even though N is a finite number, the response of the filter to a unit sample function (or equivalently, digital impulse function) can be in principle infinitely long. For example, suppose we have a very simple filter whose output is the sum of its input and nine-tenths of its previous output according to

$$y(n) = x(n) + 0.9y(n-1). \tag{28.17}$$

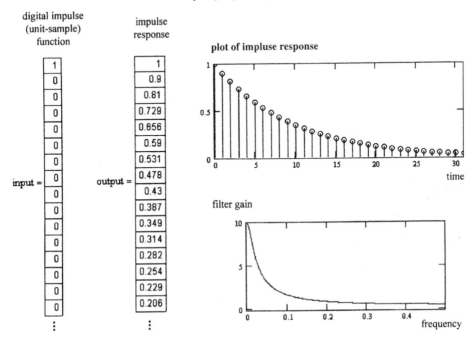

The Infinite Impulse Response (IIR) filter

$$y(n) = x(n) + 0.9y(n-1)$$

has the following impulse response and
frequency-dependent gain.

FIGURE 28.12

An IIR filter has an impulse response of theoretically infinite length. This is caused by inclusion of feedback terms in the filter equation.

If a signal consisting of 1 followed by all 0 values is input to this filter (the digital impulse, or unit sample, function), the response of this filter is a sequence of diminishing values: 1, 0.9, 0.81, ... Although this response eventually becomes very small, it theoretically just gets smaller and smaller, never really dying away completely. Such filters are therefore called infinite impulse response (IIR) filters (see Fig. 28.12).

IIR filters also have advantages, but they are complicated by the need to deal with calculated values that can become arbitrarily small (or large). Hardware and software implementations of IIR filters therefore need to take issues of numerical range and precision into account to prevent such filters from failing when the numbers get too small or large. IIR filters—which are digital equivalents to analog filters—also cannot have linear phase characteristics (though they may come close). That is, all IIR filters—like all analog filters—introduce some phase distortion into the signals they process, meaning that some frequencies pass through the filter in less time than others.

28.14 ■ PROCESSING AUDIO SIGNALS

Once signals can be converted to and from digital form and stored in long- and short-term memories, they can be operated upon by general-purpose computers and more specialized

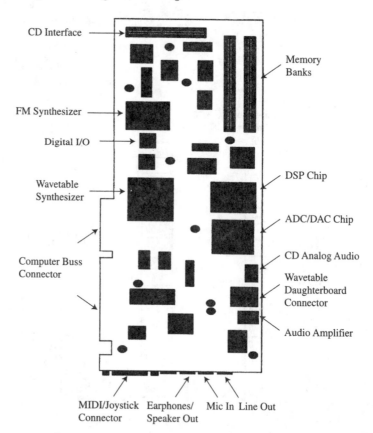

CD Interface

Memory Banks

FM Synthesizer

Digital I/O

DSP Chip

Wavetable Synthesizer

ADC/DAC Chip

CD Analog Audio

Computer Buss Connector

Wavetable Daughterboard Connector

Audio Amplifier

MIDI/Joystick Connector   Earphones/ Speaker Out   Mic In   Line Out

FIGURE 28.13
Anatomy of a sound card.

digital processors. To process audio signals with a general-purpose computer, we must first digitize an analog signal from either a microphone or a recording or other source.

Most desktop computers contain a "sound card" with a suitable analog-to-digital conversion system (see Fig. 28.13). Some sound cards also have provision for direct input of digital audio information from digital sources such as compact disc or digital audiotape players. Sound cards typically have playback as well as recording capability. In order to store and retrieve digital audio samples, they communicate with the central processing unit of the computer, of which they are a part under the supervision of an operating system. The operating system takes care of storing and retrieving the digital data in RAM or magnetic storage, and the sound card takes care of the more specialized task of converting the digital signals to and from analog signals that can be obtained from sources or played over loudspeakers.

In addition, a sound card will typically perform some digital processing on the signals, such as filtering (emphasizing and/or deemphasizing certain frequencies), adding echo, chorus, or reverberation, adjusting volume, and changing the sampling rate or encoding format. Finally, many sound cards also contain a digital sound-synthesis capability based on one or more of the computer music techniques described in the next section.

Typical sound cards accomplish these sound-conversion and processing tasks using a variety of specialized integrated circuits (ICs). There are standard ICs for performing commonly required functions, such as central processing units (CPU chips such as Pentium X, G_n, Athlon, etc.), memory (RAM, ROM, DRAM, EPROM, registers, etc.), and logic (AND, NOR, gate arrays, etc.). In addition there are special chips optimized around the requirements of digital signal processing (DSP). DSP chips typically cannot do anything that a CPU chip cannot do, but they use such techniques as parallel memory access to speed up DSP-specific operations such a forming a sum of products. Finally, many sound cards and similar specialized devices use custom-made application-specific integrated circuits (ASICs) designed around the specific requirements of particular computations that may be required in a particular manufacturer's patented algorithm for accomplishing especially advanced synthesis or spatialization algorithms for sound.

28.15 ■ SUMMARY

Digital audio signals are processed using mathematical techniques that correspond to the actions of inductors, capacitors, resistors and transistors on analog signals. Digital signals may be processed in either the time domain or the frequency domain. The Discrete Fourier Transform—or DFT—converts signals from the time to the frequency domain, while the Inverse DFT—or IDFT—converts signals from the frequency to the time domain. Both the DFT and IDFT are usually implemented using some variation of the Fast Fourier Transform (FFT) algorithm, which executes much more quickly than the simpler DFT and IDFT. Digital filters may be implemented in either the time domain or the frequency domain. In the time domain, the output of a digital filter is the convolution of its unit sample response with an input signal. In the frequency domain, the output of a digital filter is the product of its transfer function with the frequency domain version of the input signal. Digital filters may have either a Finite Impulse Response (FIR) or an Infinite Impulse Response (IIR), depending on whether they employ feedback. FIR filters can have a linear phase characteristic, meaning that all frequencies are processed in the same amount of time. IIR filters—which correspond to analog filters—may be quite efficient, but cannot have this linear phase characteristic. Digital audio signal processing software and hardware are typically associated with one or more sound cards found in a typical personal computer or workstation.

REFERENCES AND SUGGESTED READINGS

Moore, F. R. (1990). *Elements of Computer Music.* Englewood Cliffs, N.J.: Prentice Hall.

Steiglitz, K. (1996). *A Digital Signal Processing Primer with Applications to Digital Audio and Computer Music.* Reading, MA: Addison-Wesley.

Strawn, J., ed., with contrib. by J. W. Gordon, F. R. Moore, J. A. Moorer, T. L. Petersen, J. O. Smith, and J. Strawn.

(1993). *Digital Audio Signal Processing: An Anthology.* Madison, Wisc.: A-R Editions.

Strawn, J., ed. (1988). *Digital Audio Engineering.* Madison, Wisc.: A-R Editions.

Zölzer, U. (1997). *Digital Audio Signal Processing,* New York: John Wiley.

GLOSSARY

band limited signal An analog signal that has no frequencies higher than a known given frequency.

convolution A process for combining two functions, consisting of replacing each instant in one function with a scaled copy of the other function.

discrete Fourier transform (DFT) The discrete form of the Fourier transform, which associates with every time domain signal a frequency domain spectrum and vice versa.

fast Fourier transform (FFT) A particularly efficient way to compute the discrete Fourier transform (DFT).

frequency domain The representation of a signal as a function of frequency, such as frequency-dependent amplitude or phase.

poles and zeros Maximum and minimum values of a transfer function, associated with resonances and antiresonances, respectively.

sampling (Nyquist) theorem A theorem stating that a bandlimited signal containing frequencies up to F hertz must be sampled at least $2F$ times per second in order to avoid aliasing.

time domain The representation of a signal as a function of time, such as time-dependent amplitude of a sound waveform.

transfer function A function that represents how the input of a filter is transferred to its output, usually by multiplication in the frequency domain.

REVIEW QUESTIONS

1. Why is digital signal processing so mathematical?
2. What is meant by the terms *time domain* and *frequency domain*?
3. What is the spectrum of a digital signal?
4. Are negative frequencies like imaginary numbers?
5. Does the DFT compute the amplitude spectrum directly or indirectly?
6. What is the difference between an even function and an odd function? Are all functions even or odd?
7. What is the difference between a DFT and an FFT?
8. If you convolve two sound waveforms, what happens to their spectra?

9. What is the difference between an IIR and an FIR filter? Which can have linear phase?
10. If a particular filter has a strong resonance at 1000 Hz, would you expect to find a pole or a zero there?
11. Suppose the unit sample response of a particular digital filter is $\{0.1, -0.5, 0.1\}$. What is its transfer function?
12. What part of a typical home computer would likely perform digital filtering of audio signals?
13. What is an ASIC?

QUESTIONS FOR THOUGHT AND DISCUSSION

1. What would happen to the digital filter shown in Eq. (28.13) if the plus sign were changed to a minus sign?

EXERCISES

1. Setting z equal to $e^{j\omega}$ in a z-transform corresponds to evaluating the z-transform on the unit circle (a circle of radius 1). Can you explain how $e^{j\omega}$ describes a unit circle? (*Hint:* Use Euler's relation.)

2. Suppose a digital filter has the defining equation $y(n) = x(n) + x(n-1)$.
   (a) What are the filter coefficients in terms of Eq. (28.13)?
   (b) What is the gain of the filter at dc?
   (c) What is the gain of the filter at the highest possible frequency it can process?
   (d) What is the impulse response of the filter?
   (e) What is the transfer function of the filter?
   (f) Is the filter low-pass, high-pass, or something else?

3. Suppose one digital signal contains the frequencies 100 and 200 Hz. A second digital signal contains just a 10-Hz frequency. If the two signals are multiplied, what frequencies will be heard in the resulting signal? Explain your answer.

EXPERIMENTS FOR HOME, LABORATORY, AND CLASSROOM DEMONSTRATION

Home and Classroom Demonstrations

1. *Convolution and digital filtering* Write a computer program that demonstrates convolution by creating a digital filter with an impulse response consisting of the tune for *Frère Jacques*. Verify the impulse response by listening to it. Demonstrate convolution (and superposition) by exciting the filter with impulses spaced four beats apart, thus creating the round.

2. *Sound cards* Examine two or more sound cards, identifying as many components as you can.

Computer Music and
Virtual Acoustics

At first confined to research laboratories with large computing facilities, computer music and the associated digital audio technology have now become almost ubiquitous. Most of the music heard by people now comes from loudspeakers, and most of that music has been at least digitally processed and recorded, if not also digitally synthesized using techniques first explored in the field of computer music. One result of developments in the field of computer music has been an ever-expanding literature of some of the most imaginative works of music ever created. Another has been the development of digital synthesizers that are now possibly the most common type of musical instrument in the world. Linked with digital communications media such as the Internet, computer music techniques form the basis for the audible parts of virtual reality; therefore, it can only be called *virtual acoustics*.

In this chapter, you should learn:

- About the definition and origins of computer music;
- About the basic classifications for music synthesis;
- About the basic construction and operation of digital music synthesizers;
- About the basic concepts of the music instrument digital interface (MIDI);
- About the basic concepts of sound spatialization;
- About advances in digital techniques for encoding sound.

29.1 ■ COMPUTER MUSIC

Computer music is the sum total of ways in which computer technology can be applied to the creation of music (Moore 1990). More specifically, computer music is the computer-mediated analysis, processing and synthesis of musical scores and musical sounds. Any of these operations can occur in *real time*, i.e., in a manner that allows computer device(s) to be played like traditional musical instruments. Alternatively, some operations may occur either slower or faster than real time, according to the needs of the computer user and technological capabilities of the computing devices involved.

Computer Music Composition

Computer analysis of musical scores is widely used in the field of musicology, where several significant databases containing various encodings of musical scores have been developed. Statistical analysis of the works of Mozart, for example, has cast doubt on the authenticity of at least one of the wind serenades attributed to him.

Computer processing of musical scores is usually linked to score synthesis, which is just another name for musical composition. Computer composition of music is characteristically based on the work of human composers, although this is not necessarily the case.

The earliest significant application of computers to music was, in fact, the use of the computer to test stochastic models of musical compositions. Composer Lejaren A. Hiller (1924–1994), working in the early 1950s at the University of Illinois, used Markov chains and other methods borrowed from statistical theory and research chemistry (Hiller's other career) to construct statistical models of certain compositions by composers such as Palest-

		A	B	C#	D	E	F#	G	A'	total
	A'	2	0	0	0	0	0	2	0	4
	G	0	0	0	0	0	2	0	4	6
1ˢᵗ note	F#	0	0	0	0	4	6	4	0	14
	E	0	6	4	4	10	6	0	0	30
	D	6	2	12	6	2	0	0	0	28
	C#	2	16	2	12	10	0	0	0	42
	B	4	4	18	6	2	0	0	0	34
	A	2	6	5	0	3	0	0	0	16

FIGURE 29.1   Composing music by computer. Adjacent pairs of notes in this well-known melody by W. A. Mozart occur as many times as the table indicates (including the repeated sections, the note B is followed by C$^{\#}$ 18 times, for example). From this information it is straightforward to compute the corresponding note transition probabilities, such as those shown in the lower right diagram (for example, two-thirds of the time G$^{\#}$ is followed by high A, whereas one-third of the time G$^{\#}$ is followed by F$^{\#}$). Note that the sum of the probability arcs leaving any given note should add up to 1. By instructing a computer to choose notes based on these probabilities, a new melody with some of the same characteristics as Mozart's can be generated. Extensions of this technique were first investigated by L. A. Hiller in the early 1950s.

rina and Mozart (Hiller and Isaacson 1979). For example, a simple table can be constructed whose rows and columns represent all pitches in a given musical composition (see Fig. 29.1). Each entry in the table determines the probability that any note will follow any other note. The entries are made by counting all two-note sequences. A computer program can then generate a "new" composition based on the existing one in the sense that its note-to-note transitions obey the same statistics as the original (such statistical methods were originally used to study language in the work of the Russian linguist and mathematician Andrei Andreevich Markov (1856–1922)). Hiller's pioneering work resulted in several significant musical compositions, starting with the "Illiac Suite for String Quartet," created in 1956 in collaboration with Leonard Isaacson.

Work by UCSC music professor David Cope investigated the use of the computer to capture and extrapolate musical styles (Cope 1991, 1996). Some results of his work can be heard on a commercial recording entitled "Bach by Design." This recording includes several computer-stimulated piano works in the styles of classical composers Bach, Mozart, Chopin, Brahms, Joplin, Bartók, and Prokofiev, as well as a work in the style of Cope himself.

Computer Music Sound Synthesis

Digital sound analysis, processing, and recording have already been discussed in previous chapters. Max V. Mathews, a research engineer working in the Bell Telephone Laboratories, first accomplished synthesis of musical sound by digital computer in 1957. Mathews' original **MUSIC I** program ran on an IBM 704 computer, producing elementary sounds that had controllable pitch and little else. This modest beginning, however, has blossomed into a rich and unparalleled resource for musical expression. Mathews and his collaborators continually improved his music synthesis program, finally completing **MUSIC V** around 1968 (Mathews et al. 1969). The **MUSIC V** program has acted as a model for many other so-called acoustic compilers ever since, such as Barry Vercoe's **MUSIC360** and **Csound**, Structured Audio Orchestras Language (SAOL), Stanford University's **MUSIC 10**, and F. R. Moore's **cmusic** and **pcmusic**.

Using Mathews' basic design, all these programs translate a text description written in a special language of one or more sounds into one or more digital signals representing the described sound. The key concepts of acoustic compilers include *unit generators*, *stored functions*, and *input-output blocks*.

Unit generators are subprograms that simulate the operation of signal generators and processors such as those found in analog music synthesizers (see Chapter 27). One unit generator might produce an amplitude envelope signal with specifiable shape, duration, and attack and decay times. Other unit generators might combine signals in various ways, such as summation or multiplication. The acoustic compiler user makes selections from the library of available unit generators to build up virtual "instruments" by interconnecting unit generators using input-output blocks like the patch cords of analog synthesizers.

A particularly important unit generator is the signal source for most synthesis operations, called the *table lookup oscillator*, or *TLO* (see Fig. 29.2). This unit generator repeatedly scans a stored function that holds one cycle of a periodic waveform. In this context, the stored function is also called a *wavetable*. The content of the wavetable is completely arbitrary—it may result from measurement of a sound waveform coming from a traditional

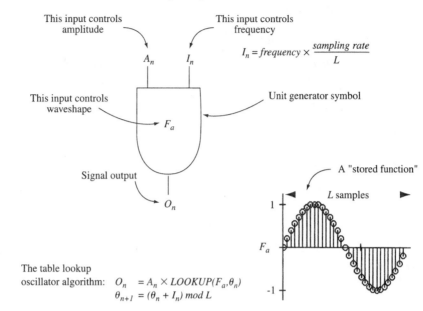

FIGURE 29.2
The *LOOKUP* function might use the integer part of the oscillator phase value to access the stored function, or it might use some form of interpolation on the waveform values stored in the table for greater accuracy. Output signal accuracy is also increased by increasing the table length, L.

The Table Lookup Unit generator

This input controls amplitude

This input controls frequency

$I_n = frequency \times \dfrac{sampling\ rate}{L}$

A_n I_n

This input controls waveshape

Unit generator symbol

F_a

Signal output

O_n

A "stored function"

L samples

F_a

1

-1

The table lookup oscillator algorithm:
$O_n = A_n \times LOOKUP(F_a, \theta_n)$
$\theta_{n+1} = (\theta_n + I_n)\ mod\ L$

musical instrument or it may be derived mathematically, such as by summing together one or more harmonically related sinusoids.

Acoustic Compilers

Acoustic compilers read *score files* (see Fig. 29.3) containing the definitions of one or more virtual instruments, one or more stored functions, and a time-ordered list of "notes" to be "played" on the defined instruments. Once a virtual instrument is defined, a typical acoustic compiler allows any number of overlapping notes to be played on it. The specification of how to play a given note depends on the definition of the instrument that plays the note. All notes must have a beginning time and duration, usually specified in seconds, although various provisions can be made to specify time in terms of beats or even samples, if desired. A simple instrument might further accept a simple pitch and volume specification. A more complex instrument might accept a pitch, volume, attack time, decay time, and brightness specification.

Assuming they all have the same duration, it generally takes an acoustic compiler ten times as long to compute the signal for ten notes on a given instrument as one note. This is true whether the notes are heard sequentially or simultaneously. An acoustic compiler can produce sounds of virtually any complexity, with the proviso that the more complex the specified sound, the longer it will take to compute. Thus, acoustic compilers typically do not operate in real time.

Notably, a specified sound is completely documented by a combination of its score file and the source code of the acoustic compiler program. Thus, a particular sound can be specified both precisely and repeatably. This makes acoustic compilers of great utility in studying both methods for synthesizing sounds of every description and the perception of those sounds.

Acoustic Illusions

Sound specifications for an acoustic compiler are quite arbitrary. Sounds may be specified that approximate natural sources or traditional musical instruments. Alternatively, the

```
{This score file might have a name like "mozart.sc"}
```
— (Score comments are enclosed in curly braces {}.)

```
#include <cmusic.h>
```
— (Reads in standard definitions.)

```
instrument 0 toot;
    seg     b2 p5 f2 d 0 ;
    osc     b1 b2 p6 f1 d ;
    out     b1 b1 ;
end;
```
(Defines an instrument named "toot" using three unit generators: "seg," "osc" and "out." bN is input-output block N, fN is stored function N, pN is note parameter N (see below))

```
SINE(f1) ;
```
— (Specifies that stored function f1 contains one cycle of a sine waveform.)

```
var 0 s3 "-c" ;
GEN3(f2) s3 0 1 1 1 1 0 ;
```
(Defines a simple trapezoidal amplitude envelope function.)

```
{
 In a note list, note parameter
 p1 specifies "note",
 p2 specifies the starting time of a note ("p2+p4"
 simply starts a note after the previous one),
 p3 specifies the instrument to be played, and
 p4 specifies the note duration (in seconds).
 The remaining note parameters depend on the
 instrument being played.  In this case,
 p5 specifies amplitude, and
 p6 specifies frequency.
}
```
(This long comment documents the following note list.)

{p1}	{p2}	{p3}	{p4}	{p5}	{p6}	
note	0	toot	2*3/16	-30dB	Cs(1)	;
note	p2+p4	toot	2*1/16	-30dB	D(1)	;
note	p2+p4	toot	2*1/8	-30dB	Cs(1)	;
note	p2+p4	toot	2*1/4	-20dB	E(1)	;
note	p2+p4	toot	2*1/8	-30dB	E(1)	;
note	p2+p4	toot	2*3/16	-30dB	B(0)	;
note	p2+p4	toot	2*1/16	-30dB	Cs(1)	;
note	p2+p4	toot	2*1/8	-30dB	B(0)	;
note	p2+p4	toot	2*1/4	-20dB	D(1)	;
note	p2+p4	toot	2*1/8	-30dB	D(1)	;

(This notelist specifies that the first measures of the Mozart melody from Figure 29.1 are to be played on instrument toot.)

```
    .
    .
    .
terminate ;
```

FIGURE 29.3   A sample score file for the pcmusic acoustic compiler.

sound specifications might not correspond to any naturally vibrating source. Studies in the latter category have led to the identification of a number of *acoustic illusions*. Psychologist Roger Shepard first synthesized tones that step up or down in pitch endlessly without ever getting anywhere. Using the **MUSIC V** program, composer/physicist Jean-Claude Risset demonstrated Shepard tones that glide up or down in pitch endlessly as well as tones that go up and down in pitch at the same time (Risset 1995). Using similar synthesis principles, Risset also created rhythms that constantly accelerate (or decelerate) without limit. Psychoacoustician Diana Deutsch has used **cmusic** to study such pitch paradoxes extensively, as well as to discover several others (Deutsch, Moore, and Dolson 1984, 1986).

29.2 ■ SOUND SYNTHESIS METHODS

Research using acoustic compilers and similar computer software has laid the foundations for a framework for sound synthesis methods. While they may be and often are combined in any given synthesis context, there are four basic methods for synthesizing sound: *additive synthesis*, *subtractive synthesis*, *nonlinear synthesis*, and *physical modeling*.

29.3 ■ ADDITIVE SYNTHESIS

Additive synthesis is based on the principle of adding together a number—possibly a *very large* number—of simple tone components such as sinusoids to produce an arbitrary complex tone. Additive synthesis is sometimes called *Fourier synthesis*, but it can be considerably more general than that. Fourier's famous theorem essentially states that any *periodic* waveform can be obtained by adding together a number—possibly an *infinite* number—of *harmonically related* sinusoids, each with a specific amplitude and phase. In additive synthesis, the components often are, but need not be, sinusoidal. Further, in additive synthesis the components may be, but need not be, harmonically related. Finally, the specification for the amplitude and phase of each component in additive synthesis might be, but typically is not, constant in time.

The synthesis structure for additive synthesis in an acoustic compiler is typically just a table lookup sinusoidal oscillator with time-varying amplitude and frequency specifications upon which many different notes can be played (see Fig. 29.2). To produce a Risset pitch glide in the upward direction, for example, the frequency of, say, four oscillators can be controlled so that they slowly sweep linearly upward on a log frequency scale five octaves in extent (this yields a constant number of semitones per second in upward pitch movement). As it sweeps from bottom to top of a five-octave range, the amplitude of each component increases from zero to a maximum value halfway through the sweep and then back to zero at the top of the sweep (see Fig. 29.4). The pitch of each of the four oscillators is initially one octave higher than the previous one, starting at the minimum frequency. After sweeping up one octave, the oscillator with the highest pitch is reset so that it enters at the lowest frequency, now an octave lower than the lowest of the remaining components. The frequency of the highest oscillator is repeatedly reset to the minimum frequency throughout the sound, and a precise octave separation among all components is maintained. Because the frequency resetting is done at zero amplitude, it is completely inaudible, and the aggregate sound seems to rise forever in pitch. This is the aural equivalent of the optical "barber-pole stripe" illusion, in which stripes appear to rise or fall forever.

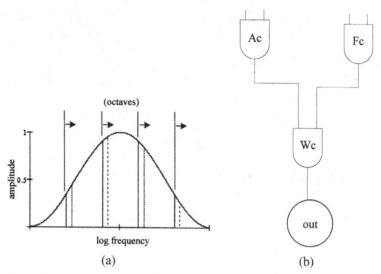

FIGURE 29.4 Rising-tone acoustical illusion due to Risset. To create this illusion, octave-separated sinusoidal components are made to sweep upward under control of an amplitude spectrum envelope that has a bell-like shape when plotted against log frequency. Components enter softly at a low frequency, become louder as they increase in frequency, and then die away as they reach higher frequencies. Just one acoustic compiler "instrument" playing overlapping notes is needed to synthesize all components, with stored function Ac providing amplitude control and stored function Fc providing frequency control (stored function Wc contains a sinusoidal waveform.

If one wishes to use additive synthesis to simulate the sound of a musical instrument, it is necessary to analyze the time-varying behavior of the sinusoidal components of that sound. Using time-varying spectrum measurements, psychologist John Grey concluded that as few as seven straight-line segments could be used to approximate the complicated amplitude behavior of each sinusoidal additive synthesis component for single notes played on several musical instruments (Grey 1975; Grey and Moorer 1977). Assuming that a straight-line segment can be specified in about two numbers (a slope and a duration, for example) implies that a given note on a trumpet, for example, might be encoded in as few as $14N$ numbers, where N is the number of additive synthesis components used in the note. A trumpet playing middle C produces at most around 76 audible harmonics (and probably far fewer). A trumpet playing an octave higher produces half this many audible harmonics. Multiplying these values by 14 suggests that a few hundred well-chosen numbers might fully encode single trumpet notes. This potentially represents between one and two orders of magnitude of data compression when compared to a CD-quality recording of the same notes.

The Phase Vocoder

In general, the problem of additive synthesis boils down to the problem of specifying proper time-varying amplitude and frequency values for the synthesis components. One technique for doing this involves an analysis/synthesis technique called the *phase vocoder* (Flanagan and Golden 1966). As its name implies, the phase vocoder was originally conceived as a form of *vocoder*, which is a contraction of *voice encoder*, named after Homer Dudley's pioneering *channel vocoder* (Dudley 1939).

Vocoders analyze their input signals in terms of multiple parallel channels, each of which describes activity in a narrow frequency band. Because the activity in each frequency band is usually much simpler than that of the input signal itself, each channel may be transmitted or encoded using much less information than is required for the original signal; in many cases, some channels may be omitted altogether.

A phase vocoder can therefore be described as a collection of filters that together cover all possible frequencies in the input signal. The output of each filter consists of two time-varying signals: one specifies amplitude behavior in a particular region of the spectrum, and the other specifies frequency behavior. If we connect the output of each analysis filter to a table lookup oscillator, the sum of all the resulting sinusoidal components can be identical to the original input (see Fig. 29.5).

By encoding the amplitude and frequency information for each channel and omitting channels according to certain criteria, we can use the phase vocoder to compress the input signal. Furthermore, we can manipulate the amplitude information in certain ways to effectively filter the input signal: certain frequency regions could be attenuated and other accentuated. Because the channel information also provides frequency information, we can alter the pitch of the input signal without affecting its duration. For example, if all frequencies were multiplied by $2^{1/12} \cong 1.059463$, the pitch of the resynthesized signal would be transposed upward by one semitone (we must be careful to avoid aliasing with upward pitch shifts).

If we can shift pitch without affecting duration, then by definition we can also alter duration without affecting pitch. Suppose, for example, we multiply all frequencies by a

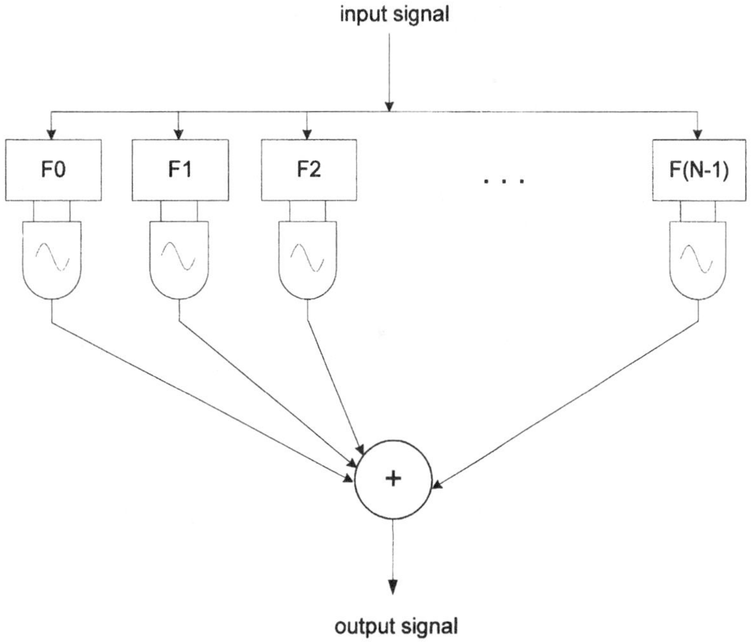

FIGURE 29.5

Basic diagram of a phase vocoder. Each of N parallel filters analyzes a small portion of the input signal spectrum, outputting time-varying amplitude and frequency information. These signals may be used to control a band of sinusoidal oscillators to resynthesize the signal. If the intermediate amplitude and frequency data are not modified, the output can theoretically be identical to the input.

factor of 2 (being careful to avoid aliasing), thus shifting the pitch of the output signal up by one octave. If we then play this signal at half speed (half the sampling rate for a digital signal), the original pitch would be restored, but the duration of the signal would double. We might also lose the top octave of frequencies in the original signal due to the necessity to avoid aliasing in the upward pitch shift. If, instead, we were to double the sampling rate of the original signal before processing it as before with the phase vocoder (we could design a phase vocoder to do this automatically), we could avoid losing any frequencies present in the original signal while doubling its duration.

Analysis-based additive synthesis gives the phase vocoder the ability to independently manipulate the spectrum, pitch, and duration of a digital audio signal. This makes it a powerful tool for both computer music and for data encoding and compression.

29.4 ■ SUBTRACTIVE SYNTHESIS

Although additive synthesis involves the summation of a number of relatively simple components to form a complex result, subtractive synthesis starts with a complex waveform from which certain frequencies are removed in order to produce a simpler—though still complex—result. The starting waveshape may be either periodic (harmonic), such as pulse, sawtooth, square, or triangular, or it may be random, such as white or pink noise. This waveform is then filtered in order to achieve the desired result (see Fig. 29.6). We often think of such a two-part system in terms of a source, or *excitation*, waveform followed by a modification, or *resonating*, system.

Many natural sound sources are well modeled, at least to a first approximation, by such an "excitation-resonator" system. For example, the bowed string on a violin provides an excitation waveform through the bridge to the body of the violin, which acts as a resonant system (filter) that modifies and radiates the resulting waveform. Similarly, a plucked string

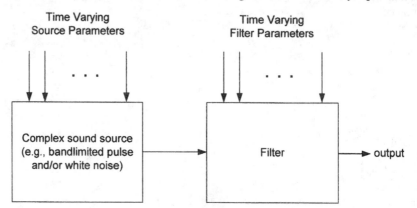

FIGURE 29.6   Processing model for subtractive synthesis. For general subtractive synthesis, the complex sources would include bandlimited pulse or sawtooth and noise sources as well as means for mixing these together in various proportions. If linear prediction is used to determine the time-varying filter parameters, the natural complex source is the error signal resulting from filtering a sound (typically speech) with the inverse of the time-varying filter used during synthesis (an *inverse filter* can be created simply by interchanging poles with zeros). The broadband error signal accounts for all parts of the original signal that are not accounted for by the filter.

on a string bass or guitar uses essentially the same process. The buzzing lips of a trumpet or horn player provide the excitation for a conical resonator of variable length (depending on the valves depressed by the player), whose radiation into the air is mediated by a bell. Turbulent airflow provides the source excitation for a flute (or a whistle or a bottle).

Significantly, human and animal vocalizations are also well modeled by an excitation-resonator system. In this case, the excitation is produced either via the pitched buzzing of the vocal folds or by turbulent airflow due to constrictions at various places in the throat or mouth. Resonances are created by the time-varying shape of the vocal tract, i.e., the positions of the throat, mouth, tongue, teeth, lips, and nasal passages (see Chapters 15–17).

Subtractive synthesis is thus applicable to the synthesis of a wide variety of musical sounds, including voices. One of the first significant results of computer music research was Jean-Claude Risset's finding that the brightness of a brass tone was proportional to its amplitude envelope (Risset and Mathews 1969). This finding gave theoretical support to a very practical method for synthesizing brass tones on early analog synthesizers: a sawtooth waveform feeding a voltage-controlled low-pass filter and amplifier whose cutoff frequency and gain followed the output of an ADSR envelope generator could produce convincing brasslike tones (see Fig. 27.13). Tone synthesized in this manner became so popular that they still persist on digital synthesizers under rubrics such as *analog brass* or *synth brass*.

Linear Prediction

Because the *source-excitation* model fits so many physical systems, the subtractive synthesis model can be interpreted as a first-order approximation to the synthesis technique known as *physical modeling*. Nevertheless, simple source-excitation modeling can be seen in terms of (1) the characteristics of the source (typically time-varying amplitude and possible pitch parameters), and (2) the filtering characteristics of the resonant system (typically time varying as well).

For speech, the linear prediction model (Makhoul 1975) boils down to the following basic idea: given N samples of a speech waveform, to what extent can we predict the next? It turns out that when the speech waveform is very regular (as during a vowel sound), it is quite possible to predict the next sample, given a fair number of previous values. When the speech waveform is very irregular (as with a turbulent airflow burst during a consonant), such prediction is nearly impossible because the waveform is nearly random. Fortunately, the hardest waveform to predict (the random one) is the easiest one to approximate, and vice versa.

Given N samples of a speech waveform, then, linear prediction attempts to find the best possible prediction of the next sample. It does this multiplying each of the N samples by a parameter value and adding together all the products to find the predicted value. Various parameter values are tried in order to find the best possible predictor set for the following sample.

The difference between the prediction and the actual next sample value is called the *prediction error*. If the error is small, the prediction is obviously pretty good. If the error is large, then the prediction is poor. The average size of the error over a few samples for a given set of prediction parameters is then a measure of the *predictability* of the signal.

When the predictability is large, the linear prediction process is working well, and vice versa.

More precisely, linear prediction involves modeling a signal, $y(n)$, in terms of some of its past values according to the equation

$$y(n) = e(n) + \sum_{i=1}^{M} b_i\, y(n - i). \tag{29.1}$$

$e(n)$ is called the error signal, because it represents the difference between $y(n)$ and what it was predicted to be by the sum. The problem of linear prediction is to find the b_i values (*prediction coefficients*) that makes the overall amplitude of the $e(n)$ signal as small as possible. By comparison with Eq. 28.15, we see that the prediction coefficients, b_i, define an Mth-order all-pole (IIR) filter that, when excited by input signal $e(n)$, results precisely in signal $y(n)$ (see Section 28.26).

The mathematical techniques for finding the best prediction coefficients are straightforward but involved, so we won't go into them here. Nevertheless we can see how the processing model shown in Fig. 29.6 is well matched by the linear prediction process described in Eq. (29.1).

In practice we find that speech signals are largely predictable. This is due to the fact that the shape and, therefore, the resonances of the vocal tract change relatively slowly, at least when compared with the rapid vibrations of the vocal folds. The prediction parameters specify an all-pole filter that represents the overall filtering function of the vocal tract, which is determined by the vocal tract shape as a function of time. When we say the word *bit*, the shape of the vocal tract changes rapidly during the initial *b* plosive, relatively slowly during the much longer *i* vowel, and quickly again during the final *t* plosive. During the vowel, the shape of the vocal tract is more or less constant, so the prediction parameters remain more or less constant as well (even though the pitch of the excitation function may change slightly during the vowel). The linear prediction parameters (filtering coefficients) can, therefore, remain nearly constant during the vowel, which is about 80% of the duration of the word. Coding of the excitation signal captures not only the pitch, but also the degree to which the speech sound is pitched at all, or noisy, or some combination of both.

Linear predictive coding (LPC) for speech includes coordinated information about the prediction parameters, which are essentially filter coefficients describing the position of the vocal tract and coding for the excitation function. Excitation coding typically includes information about the relative proportion and frequency of the "buzz" (vocal fold oscillations) versus the "hiss" (broadband excitations resulting from turbulent airflow).

29.5 ■ NONLINEAR SYNTHESIS

The signature of a linear time-invariant (LTI) system is that the frequencies coming out of it are the same as those going in, albeit with possibly different amplitudes and phases. A filter that changes the amplitude and phase of every frequency is therefore an LTI filter, because no frequency can come out of the filter that was not present in its input.

The signature of a nonlinear system is that frequencies can come out that did not go in. Thus, a device that outputs a 1000-Hz sinusoid whose amplitude matches that of its input regardless of the input signal frequency is inherently *nonlinear*. A system whose output amplitude and frequency are proportional to the input frequency and amplitude, respectively, is also nonlinear.

A simple example of a nonlinear system is an overdriven amplifier (see Fig. 29.7). Suppose an amplifier has a voltage gain of 100 and a maximum output voltage of ±10 V. As long as the input voltage remains below an absolute magnitude of $10/100 = 0.1$ V, the output of the amplifier is linear. But when the input voltage exceeds +0.1 V, the output "saturates" at its maximum of 10 V output. Similarly, an input of less than −0.1 V causes the output to remain constant at −10 V. Thus, any input with amplitudes exceeding ±0.1 V will cause the amplifier to "clip" its output to its maximum values. If the input is a sinusoid exceeding the linear input range, the output signal will be a sinusoid at the same frequency plus odd harmonics of the same frequency with amplitudes that depend on the extent of the clipping. Because frequencies come out of the amplifier that do not go in (the harmonics of the near-square waveform), the overdriven amplifier is operating in a nonlinear mode.

Frequency Modulation (FM) Synthesis

Amplitude modulation characteristically generates sum and difference frequencies, as we saw in Chapter 28. Frequency modulation (FM), where one signal drives the frequency variation of another, also generates new frequencies, although in this case the number of new frequencies generated is in principle infinite (see Fig. 29.8).

When a single sinusoidal oscillator produces a signal whose frequency is the sum of a constant and another sinusoidal oscillator, we see that the frequency of the modulated

FIGURE 29.7
Nonlinear gain in an overdriven amplifier. As long as the input signal has a magnitude of less than 0.2 V, the amplifier operates in its *linear* region. The input signal shown has a peak amplitude of ±0.2 V. Because the amplifier has a voltage gain of 100 but a maximum output amplitude of ±10 V, the output signal that is "clipped" (saturated) at the ±10-V level. Despite the sine-wave input, its nonlinear gain causes the amplifier output to resemble a square wave.

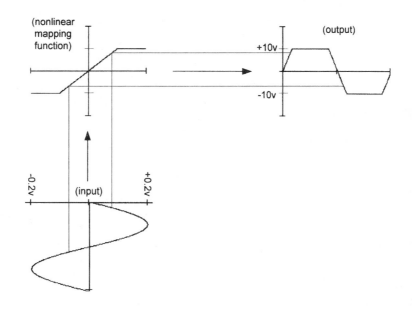

oscillator varies between $f_c - \Delta f$ and $f_c + \Delta f$, where f_c is the constant. Δf is the *frequency deviation*, which determines the extent to which the output signal varies above and below the carrier frequency. This variation occurs at a rate f_m Hz. f_m is called the *modulating frequency*, and f_c is called the *carrier frequency* by reference to FM radio.

The spectrum of a frequency-modulated signal contains an infinite number of *sidebands*, so named because they occur at intervals above and below the carrier frequency, separated by all possible integer multiples of the modulating frequency. That is, a frequency-

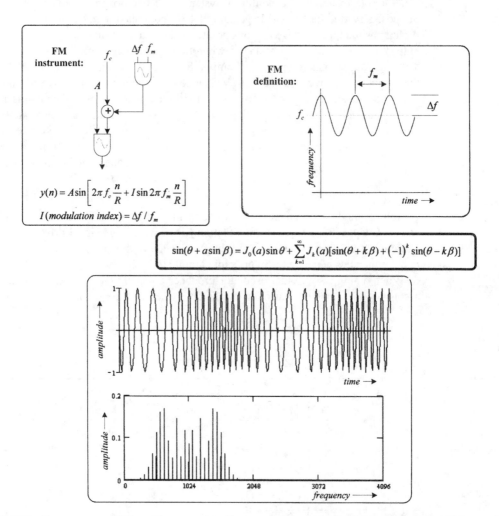

FIGURE 29.8 Frequency modulation. The upper two diagrams depict the basic FM instrument, in which one sinusoidal oscillator modulates the frequency of another and relations among carrier and modulating frequencies, frequency deviation, and modulation index I. The boxed formula shows how the FM process generates sidebands centered around the carrier frequency with amplitudes determined by Bessel functions. The bottom two diagrams plot the waveform and spectrum for an FM signal with $f_c = 1024$, $f_m = 64$, and $I = 8$ (note the symmetrical arrangement of sidebands around the carrier frequency).

modulated waveform contains an infinite number of frequencies f_c, $f_c \pm f_m$, $f_c \pm 2f_m$, $f_c \pm 3f_m$, ..., and so on. The amplitudes of the sidebands generally decrease as they move away from the carrier frequency (see Fig. 29.9). By "spreading" the spectrum of a modulated waveform to many different frequencies, an FM radio signal is far less susceptible to atmospheric interference than a similar AM signal. Due to its greater bandwidth, FM radio generally sounds better than AM radio, a fact first demonstrated in New York City by the inventor of FM radio, Edwin Howard Armstrong, on June 9, 1934.

When singers sing, various characteristics of their physiology interact to produce *vibrato*, which is partially a periodic variation of the pitch of their voices above and below the main pitch. In the late 1960s composer and computer music researcher John Chowning was experimenting at Stanford University with computer synthesis of the singing voice. An accidental specification of an extremely high vibrato rate and depth resulted in a sound that was entirely unexpected. Instead of producing the effect of a singing vibrato, Chowning had produced a frequency-modulated waveform in which the sidebands were distinctly audible, resulting in a complex timbre, even though the synthesis methods involved merely the modulation of the frequency of one sinusoid with another. After considerable research, Chowning discovered that by listening directly to the frequency-modulated waveform, one could produce in effect an infinite variety of time varying tone qualities merely by modulating the frequency of one sinusoid with another (Chowning 1973).

In its basic form, Chowning's FM synthesis method uses only two frequencies: the carrier frequency f_c and the modulating frequency f_m. Because an infinite number of sidebands normally result, FM is seen to be an inherently nonlinear form of waveform synthesis. If the modulation and carrier frequencies are the same, i.e., $f_c = f_m$, these sidebands will occur at f_c, $f_c \pm f_c$, $f_c \pm 2f_c$, $f_c \pm 3f_c$, ..., and so on. Because the spacing between all adjacent sidebands is equal to the carrier frequency, it can be seen that this is a harmonic spectrum with the same carrier frequency as fundamental frequency. Some of the frequencies are negative, in which case they either constructively or destructively interfere with their positive counterparts, depending on their phase. The component at 0 Hz may result in a constant bias for the entire waveform, depending on its amplitude and phase.

Many other relationships between f_c and f_m are possible. If $f_m = 2f_c$, the resulting spectrum corresponds to all the odd harmonics of f_c, whereas $f_c = 2f_m$ yields a spectrum consisting of all harmonics of f_m. Irrational relationships, such as $f_m = \sqrt{2}\, f_c$, can yield inharmonic spectra, which can be useful in the synthesis of bell-like and gong-like sounds.

The amplitudes of the sidebands depend on a quantity called the *modulation index*, I, which in the case of FM is equal to $\Delta f / f_m$. The amplitude of the nth sideband pair is given by $J_n(I)$, where $J_n(I)$ is the *Bessel function of the first kind of order n* evaluated at point I. Bessel functions of the first kind are similar in shape to damped sinusoids, are always less than or equal to 1.0, and are generally smaller as n increases. Thus, the sidebands of an FM waveform tend to get smaller in amplitude as they get further from the carrier frequency. Larger values of I (such as $I > 1$) tend to distribute energy more broadly to more sidebands, meaning components further from the carrier frequency have significant energy.

The relationship between the carrier and modulating frequency determine the *harmonicity* of the resulting waveform, whereas the relationship between the frequency deviation and modulating frequency control the *breadth of the spectrum*. These factors, together with independent control of the carrier frequency and amplitude of the FM waveform, provide an

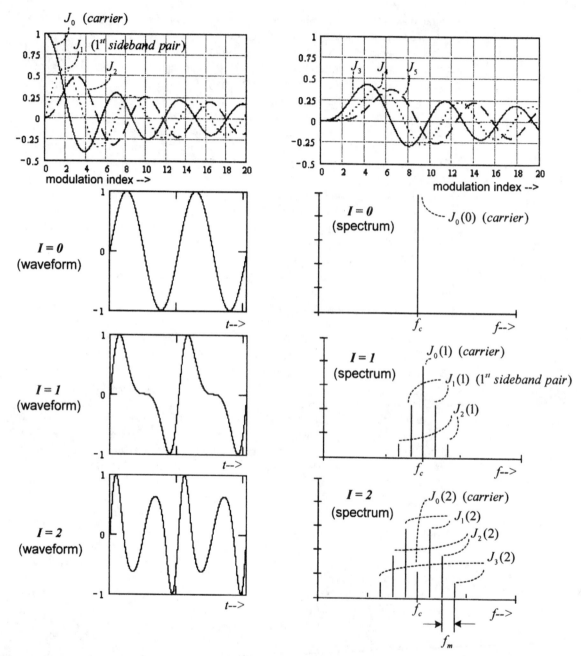

FIGURE 29.9 To find the amplitude of sideband pair n in FM synthesis, use the value $J_n(I)$, where J_n is the Bessel function of order n (the first six such function are shown at the top) and I is the modulation index ($\Delta f/f_m$, where Δf is the frequency deviation and f_m is the modulating frequency). Sidebands are spaced symmetrically around carrier frequency f_c at intervals of f_m Hz. The bottom three pairs of diagrams show representative waveforms and spectra for modulation indices of 0, 1, and 2.

extremely useful set of controls over the timbre of the resulting waveform. The conceptual simplicity of the FM synthesis technique allows it to be predicted and controlled in a time-varying manner much more easily than other methods. It is little wonder that FM became the first computer technique to be thoroughly exploited in a context of real-time digital music synthesis.

Waveshaping Synthesis

We can represent the nonlinear mapping function shown in Fig. 29.7 in the following way.

$$\text{clip}(x) = \begin{cases} +10, & x > 0.1, \\ 100x, & -0.1 \le x \le 0.1, \\ -10, & x < -0.1. \end{cases} \tag{29.2}$$

There are other interesting music synthesis–related possibilities for such nonlinear equations. One such technique involves using *Chebyshev polynomials* (LeBrun 1977). The first few Chebyshev polynomials are

$$T_0(x) = 1,$$
$$T_1(x) = x,$$
$$T_2(x) = 2x^2 - 1, \tag{29.3}$$
$$T_3(x) = 4x^3 - 3x,$$
$$T_4(x) = 8x^4 - 8x^2 + 1.$$

The recurrence rule here is $T_{n+1}(x) = 2x T_n(x) - T_{n-1}(x)$, so we see that, for example, $T_5(x) = 16x^5 - 20x^3 + 5x$. The reason Chebyshev polynomials are interesting from a

FIGURE 29.10
Nonlinear synthesis using Chebyshev polynomials. Here the nonlinear mapping function is a weighted sum of Chebyshev polynomials, each of which transforms a cosinusoidal input into a specified harmonic. This allows any specific periodic waveform to be synthesized exactly. Note that the amplitude of the input signal has a complicated effect on the output waveform.

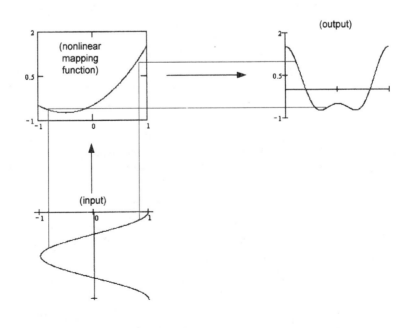

music synthesis standpoint is that, in general,

$$T_k[\cos(\theta)] = \cos(k\theta). \tag{29.4}$$

Therefore, if we feed a sinusoid into a nonlinear mapping function comprising a weighted sum of Chebyshev polynomials, a complex waveform will come out, where the amplitudes of the harmonics will equal the weights of the Chebyshev polynomials (see Fig 29.10). Unlike FM, this allows us to specify the exact composition of the complex waveform when the sinusoid has full amplitude.

For example, suppose we define $f(x)$ in the following way:

$$f(x) = T_1(x) + \frac{1}{2}T_2(x). \tag{29.5}$$

Using $f(x)$ as a nonlinear mapping function, we see that

$$f[\cos(\theta)] = \cos(\theta) + \frac{1}{2}\cos(2\theta). \tag{29.6}$$

Varying the amplitude of the input cosine wave has a complex effect on the spectrum of the output waveform, similar to varying the modulation index in the case of FM.

The possibilities for synthesizing complex waveforms by nonlinear means are limitless. Therefore, they represent a largely uncharted territory for future research.

29.6 ■ PHYSICAL MODELING SYNTHESIS

We have seen throughout this book how mathematics may be used to describe sound and, more specifically, the vibrating systems that give rise to sound. The basic notion of physical modeling is to write equations that describe how particular sets of physical objects vibrate and then to solve those equations in order to synthesize the resulting sound. We have examined equations that describe the vibrations of a mass suspended on a spring, a stretched string, open and closed columns of air, and so on. In order to understand such systems, we usually simplify the equations as much as possible so that we can study basic principles, such as how string tension or mass per unit length affect vibration frequency.

Much more precise equations can be written than the ones that have generally been discussed in this book. It is possible to describe a vibrating string in terms not only of its length, tension, and mass per unit length, but also of its frequency-dependent damping (which causes the string vibrations to die away) and its stiffness (which causes upper partials to have slightly higher frequencies than harmonic ones). The string might be bowed, plucked, or struck in any number of places and manners with soft or hard objects of various shapes. Multiple strings might be coupled together (as in the case of a piano), and they might be coupled to a resonator through a bridge. The bridge might have its own physical characteristics, such as stiffness, damping, mass, and so on. Resonators can have any number of shapes and sizes, as in the case of violins versus violas, viols, lutes, or guitars.

As one might imagine, the equations that describe the totality of such vibrating systems can be very complicated. Yet, if they can be written down and solved, then in principle the resulting waveform should sound just like the physically vibrating object. This is precisely the aim of physical modeling synthesis.

The Karplus-Strong Plucked-String Algorithm

Due to its mathematical complexity, an in-depth treatment of physical modeling synthesis is beyond the scope of this book. One method, however, is so delightfully simple that we can examine it in some detail. In 1983, the concept of a personal computer was very new, and the capabilities of such machines were extremely limited by modern standards. Random-access memory was measured in kilobytes, and computation speeds were limited to less than a million fixed-point operations per second. Some machines even lacked multipliers, simulating multiplication operations through repeated addition.

Remarkably, Kevin Karplus and Alex Strong devised a plucked-string synthesis technique in 1983 that not only sounded quite convincing but even ran in real time on such personal computers (Karplus and Strong 1983). The technique consists of filling a delay line with random values and then repeatedly feeding these values back through a simple filter (see Fig. 29.11).

Assuming the delay line contains p samples and ignoring the filter for a moment, it should be clear that the resulting sound would consist of p samples repeating over and over again. This periodic sample pattern should give rise to a pitch associated with a frequency of R/p hertz, where R is the sampling rate.

The filter in the feedback loop has two basic effects on the output signal. First, assuming the filter is a *weak* low-pass filter (one having a gain that rolls off gradually at high frequencies), it should be clear that the effect will be that the sound is robbed of some of its high-frequency content each time the samples circulate once around the loop. Because the strongest periodicity in the recirculating sound is the one associated with $R/(p+q)$ hertz (where q is the delay associated with the filter), the sound will eventually turn from a broadband noise into more or less a sinusoid at this frequency.

Karplus and Strong used the simplest possible low-pass filter in the feedback loop:

$$y(n) = \frac{x(n) + x(n-1)}{2}. \tag{29.7}$$

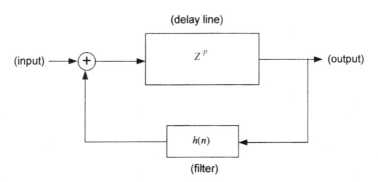

FIGURE 29.11   A simple physical model: the Karplus-Strong plucked string synthesis algorithm. The input fills the delay line with random values representing a broadband "pluck." The simple low-pass filter gradually removes high-frequency energy as the samples recirculate through the p-sample delay line, eventually leaving a sinusoid with a period of $p+q$ samples, where q is the delay through the filter. This simple structure emulates a plucked guitar or banjo string.

As we saw in Chapter 28, this is a weak low-pass filter with a frequency response defined by the first quadrant of a cosine curve. It also has a fixed time delay of one-half sample at all frequencies.

Despite its limitations, this synthesis technique illustrates several features of physical modeling. In a real plucked string, the string is initially displaced in a way that excites many frequencies in the string. The plucking place along the string will determine which frequencies are excited, as will the size and hardness of the object used to do the plucking (a broad, soft finger will elicit fewer high frequencies than a narrow, hard pick). The initial condition of the pluck is represented by the broadband noise used to fill the delay line. Once a real string is plucked, a traveling wave moves in both directions from the pluck point to the ends of the string (see Fig. 10.4), where it is partially absorbed and partially reflected in a frequency-dependent manner. In the Karplus-Strong algorithm, the wave moves only in one direction (around the feedback loop), with absorption represented by the action of the filter. In a real string, higher frequencies are damped more rapidly than lower ones, causing the string to "ring" for some time in a more or less sinusoidal manner once the initial "twang" of the pluck has died away. The low-pass filter simulates this effect.

The limitations of the Karplus-Strong algorithm are fairly severe. For example, the duration of the sound will be determined by how long it takes the recirculating samples to die away essentially to silence. Higher pitches die away more quickly than lower ones, just as in real plucked strings, but there is no way to adjust how long this takes. Even more serious is the fact that the available pitches become more coarsely defined with increasing frequency (done by decreasing p). This causes the algorithm to play more and more out of tune as higher pitches are played. David Jaffe and Julius Smith effectively addressed these and other limitations in the Karplus-Strong algorithm but at the cost of considerable increase in computational complexity (Jaffe and Smith 1983). Even the simple personal computers of 1983 could add two numbers and divide by two (using a right-shift operation), making the Karplus-Strong algorithm the first musically useful physical model (however simplified) to run in real time on personal computers.

Waveguide Synthesis

Modern computing power has greatly alleviated the need for oversimplification in physical models. Nevertheless, physical modeling is still in its infancy as a practical music synthesis technique. One of the most interesting approaches to physical modeling of musical instruments is based on digital waveguide models such as those developed in microwave engineering (Smith 1996, 2000). Instead of modeling a traveling wave with a single delay of p samples, as in the case of the Karplus-Strong algorithm, digital waveguide models take into account how the wave moves from point to point along the string. Thus, it is a collection of many individual delay units for each direction of wave propagation in the string, each of which can in principle be associated with frequency-dependent gain and other properties. In other words, a digital waveguide can model how energy is distributed along a string in a manner that is sampled in space as well as in time (see Fig. 29.12). This makes it possible to deal with such physical properties as time-varying nonuniformity in string tension, variable mass per unit length, frequency-dependent dispersion due to stiffness, nonrigid versus rigid terminations (such as bridges and nuts), different plucking, striking, or bowing positions and techniques, and so on.

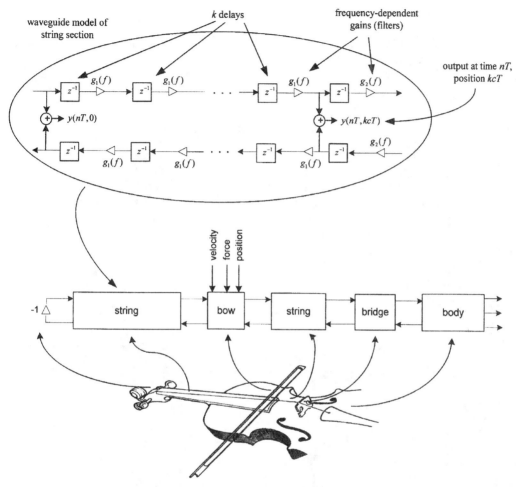

FIGURE 29.12   Physical modeling using waveguide techniques. Bidirectional waveguides model behavior of various parts of a violin in time-, position-, and frequency-dependent ways. Delay and filter properties are derived from equations describing the physical characteristics of instrumental subsections. Both right- and left-moving traveling waves in the string are modeled (the −1 multiplier at the nut implements a lossless, inverting reflection). Sometimes groups of similar parameters (such as the k delays shown with identical filtering) can be lumped together for computational efficiency. Output can be taken at any point by summing the two traveling waves. Ideally, changing a parameter such as bow position has the same effect on the synthesized sound as it would on a real instrument. (After Smith 2000.)

Unlike additive synthesis, which is based on the analysis of sound, physical modeling actually involves the mathematical and computational construction of a virtual musical instrument whose characteristics can match that of a real instrument as closely as desired or as computationally tractable. Physical models are not limited to reality, however, because improbable, impossible, and fanciful characteristics can be modeled just as readily as any other. It would be possible, for example, to listen to a violin whose wood is gradually turning into tin, in case one yearned to hear that.

29.7 ■ DIGITAL MUSIC SYNTHESIZERS

Computer music had its beginnings in the context of general-purpose computers programmed to compose music or synthesize digital waveforms representing musical sounds. In general, neither of these processes took place in real time, because the computation time necessary to compute most results exceeded the time it took to hear those results by a considerable margin.

If computer music is to be performed in real time, it is necessary to select a set of synthesis procedures, to optimize the manner in which these computations are carried out, and to place these computations in a context where interactive control is possible. Modern computer music workstations (see Fig. 29.13) mix real-time and non-real-time capabilities allowing the user to choose an operating mode. A typical computer music workstation includes one or more digital music synthesizers, which are, essentially, specialized computers connected to real-time controllers, such as keyboards, switches, knobs, and pedals. The first widely sold digital music synthesizers were introduced commercially in 1983 (see

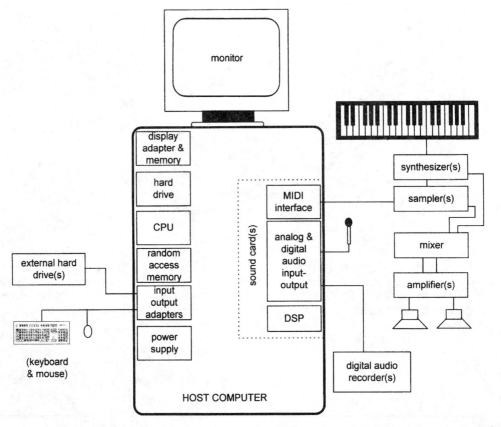

FIGURE 29.13 Typical computer music workstation components. A general-purpose computer is augmented with one or more sound cards providing for analog and digital I/O (including a MIDI interface) and digital signal processing (including digital music synthesis). External components include extra data and digital audio storage, as well as synthesizers, a sound system, and associated devices. Such a system provides highly flexible support for computer music applications.

The Yamaha DX-7

The Kurzweil 250

FIGURE 29.14 Two landmark digital music instruments. The Yamaha DX-7 digital music synthesizer introduced FM music synthesis. The Kurzweil 250 digital sampler introduced realistic ROM-based reproduction of piano, strings, choirs, drums, and other acoustic instruments. Both instruments also incorporated the MIDI interface and were first made commercially available in 1983.

Fig. 29.14). The extremely popular Yamaha DX-7 was based directly on the FM synthesis technique worked out by Chowning and extended by others.

Voices, Timbres, Channels, and Splits

Digital music synthesizers do a great many things besides sound synthesis. When a key goes down on a digital music synthesizer, a signal is sent to a control processor (essentially a microcomputer) indicating which key was depressed, as well as other information such as the key velocity, pressure, and so on (see Fig. 29.15). Early synthesizers were *monophonic*, which means that they can play only one note at a time. Other synthesizers are *polyphonic*, meaning that they can produce 8, 16, 32, 64, or more notes at once. Each depressed key must be associated with an available synthesis "voice," or "channel," if it is to be heard.

A synthesizer might be *polytimbral* as well, indicating that it can produce more than one tone quality at a time. This might be controlled using a keyboard split such that notes above, say, middle C would sound like flutes whereas notes below middle C would sound like string basses. Obviously it is possible to have more than one split in a keyboard, although the number of keys available for each sound quality is diminished by a large number of splits. A large number of splits can be useful, however, when synthesizing percussion instruments—each key of the entire keyboard might represent a sound of a different percussion instrument.

FIGURE 29.15
General architecture of a typical digital music synthesizer or sampler. The rightmost digital components are typically time multiplexed, performing their functions on a fixed schedule one or more times for each synthesis channel, or voice. The association of control inputs with particular voices is maintained by the CPU, which coordinates all processing in the instrument. In a typical synthesizer, the synthesis and processing engines might be combined; on a typical sampler they are separate.

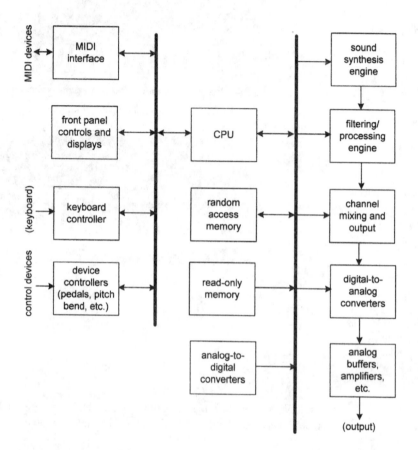

Synthetic sounds can also behave in a variety of ways. When a key is released, for example, the corresponding sound might stop more or less instantly, as in the case of a wind instrument or an organ. Some sounds may have extensive decays or releases, such as a piano (with the sustain pedal depressed) or a bell after it is struck and allowed to vibrate freely. The main control processor must keep track of which keys are depressed, which notes are sounding, and which synthesis channels are in use. When a synthesis channel has completed the release of a note, it can be reassigned to another task, such as playing a note with a different tone quality initiated by a key from a different portion of the keyboard.

Before the control processor can turn on a specific channel, all necessary information must be assembled, such as what key with what velocity initiated the note. To this must be added the type of sound to be synthesized. Additional information may include relevant transpositions (specified by the user or automatic), pitch bend information from a pitch wheel or similar device (a knob or slider that uses a spring to automatically "return to zero" after it is pushed one way or the other), and processing and routing information.

A synthesized sound might be further processed through a filtering system that brightens or smooths the overall sound. It might be "sweetened" through a multichannel reverberator according to various specifications. It might also be spatialized in various ways, such as playing it on the left or right stereo output, or panning it from left to right or along a more complex path.

The actual synthesis of the sound is typically carried out by a specialized processor, often called a *synthesis engine*. This unit typically feeds its output into a *processing engine*, which filters, reverberates, and mixes the sounds together according to the available output channels. Finally, the assembled output sounds are converted from digital to analog form, amplified, and sent to loudspeakers, which may or may not be part of the digital music synthesizer itself.

It is obviously necessary to keep the digital samples flowing at sufficient rates to avoid any break in the quality of the synthesized sound. What is perhaps less obvious is that it is also necessary to minimize and regularize the latency of responses to control inputs. A delay between the time a key is depressed and the time the associated sound begins is tolerable, provided it is small and relatively uniform; otherwise the instrument can become unplayable.

29.8 ■ MUSICAL INSTRUMENT DIGITAL INTERFACE (MIDI)

One of the features of the extremely popular DX-7 (introduced in 1983) was an interface that allowed the keyboard to control a different synthesizer, or the synthesizer to be controlled by a different keyboard, or even a computer. Called the musical instrument digital interface (MIDI), this communication standard allowed different synthesizers to be plugged together, augmenting greatly the capabilities that the player of a single keyboard could control.

The applications of basic MIDI interface (see Fig. 29.16) have since been expanded in a wide variety of ways. The MIDI interface eliminates the need for every digital synthesizer to have a keyboard, for example. It also eliminates the need for every MIDI keyboard to have a synthesizer. It provides a simple, standardized interface for a wide variety of control devices, such as breath controllers, drum pads, and microphones (via *pitch-to-MIDI* converters). It allows computer software (such as music notation software) to have a standard way of controlling a synthesizer or a standard way of being controlled by a synthesizer (e.g., automatic transcription of performed music). MIDI codes have been adapted to control a variety of nonmusical devices, such as audio mixing panels, theatrical lighting controllers, audio processing devices such as reverberators and "effects boxes," and, via SMPTE timecodes, to coordinate with video and graphics devices.

The MIDI protocol allows a device to send messages over 1 to 16 channels. Each channel corresponds to a different destination for the associated information; for instance, a multitimbral synthesizer may use different channels for notes of different timbres. Many MIDI systems allow more than one set of 16 channels to be used.

A MIDI message consists of one or more 10-bit words. Two of these bits are *start* and *stop* bits, leaving 8 bits for status and data. A status byte begins with a 1 and identifies a particular function, such as "note on on channel 3" or "note off on channel 6." A data byte begins with a 0 and provides a 7-bit value associated with a preceding status byte. For a typical synthesizer, when a key is played, three MIDI "words" are generated, specifying, for example, that a note turned on on channel 1, with a key number of 60 (middle C) and a velocity of 100 (loud). When the key is released, three more words are sent specifying

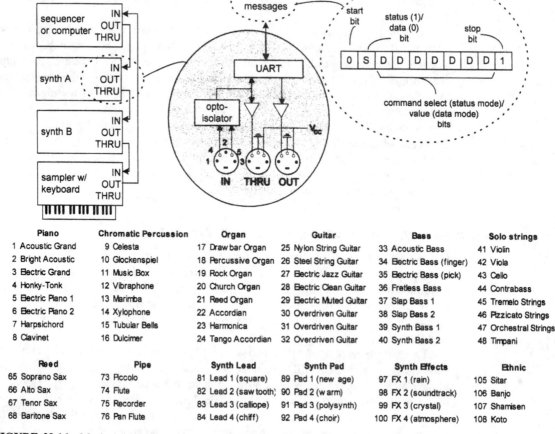

Piano	Chromatic Percussion	Organ	Guitar	Bass	Solo strings
1 Acoustic Grand	9 Celesta	17 Draw bar Organ	25 Nylon String Guitar	33 Acoustic Bass	41 Violin
2 Bright Acoustic	10 Glockenspiel	18 Percussive Organ	26 Steel String Guitar	34 Electric Bass (finger)	42 Viola
3 Electric Grand	11 Music Box	19 Rock Organ	27 Electric Jazz Guitar	35 Electric Bass (pick)	43 Cello
4 Honky-Tonk	12 Vibraphone	20 Church Organ	28 Electric Clean Guitar	36 Fretless Bass	44 Contrabass
5 Electric Piano 1	13 Marimba	21 Reed Organ	29 Electric Muted Guitar	37 Slap Bass 1	45 Tremelo Strings
6 Electric Piano 2	14 Xylophone	22 Accordian	30 Overdriven Guitar	38 Slap Bass 2	46 Pizzicato Strings
7 Harpsichord	15 Tubular Bells	23 Harmonica	31 Overdriven Guitar	39 Synth Bass 1	47 Orchestral Strings
8 Clavinet	16 Dulcimer	24 Tango Accordian	32 Overdriven Guitar	40 Synth Bass 2	48 Timpani

Reed	Pipe	Synth Lead	Synth Pad	Synth Effects	Ethnic
65 Soprano Sax	73 Piccolo	81 Lead 1 (square)	89 Pad 1 (new age)	97 FX 1 (rain)	105 Sitar
66 Alto Sax	74 Flute	82 Lead 2 (saw tooth)	90 Pad 2 (w arm)	98 FX 2 (soundtrack)	106 Banjo
67 Tenor Sax	75 Recorder	83 Lead 3 (calliope)	91 Pad 3 (polysynth)	99 FX 3 (crystal)	107 Shamisen
68 Baritone Sax	76 Pan Flute	84 Lead 4 (chiff)	92 Pad 4 (choir)	100 FX 4 (atmosphere)	108 Koto

FIGURE 29.16   Musical instrument digital interface (MIDI) characteristics. The MIDI connections shown in the upper left demonstrate how a hardware sequencer or computer can simultaneously control two synthesizers and a sample or the sampler's keyboard can control a sequencer/computer. The MIDI THRU signal is just a replica of MIDI IN. The MIDI universal asynchronous receiver/transmitter (UART) assembles MIDI messages into 10-bit packets, of which 8 are used: the first bit specifies whether the remaining 7 are status (command) or data bits. The table shows a portion of the general MIDI mode (GMM) standard widely used since about 1990. Most MIDI devices now implement some form of GMM, allowing a more or less standard association of MIDI program numbers with sounds while not restricting other MIDI possibilities.

that a key turned off on channel 1, with a key number and velocity. Other MIDI messages specify such things as key pressure (or aftertouch), pitch bend, position of a controller (such as a pitch wheel or lever), program change (e.g., select a new timbre for channel 14), or "all notes off."

It is important to understand that MIDI messages are sent over a single wire at the rate of 31,250 bits per second. Because each MIDI message takes 10 bits and typical MIDI commands take 1 to 3 words, we can easily calculate that somewhere between 50 to 150 commands per second can be sent to any one of 16 channels. MIDI is thus too slow to avoid many real-time performance artifacts. MIDI was designed to allow a keyboard to control a synthesizer and does a fair job of keeping up with what one keyboard player can do with 10 fingers. For most other purposes, however (such as controlling a synthesizer from a computer playing a moderately complex score written for, say, 10 instruments), MIDI can easily be overloaded. MIDI is particularly poor at handling continuously changing data, such as vibrato in a singing voice or violin (Moore 1988).

The slowness of MIDI can be ameliorated somewhat by the use of multiple MIDI ports. For example, sending each channel through a separate port can increase MIDI performance by about an order of magnitude. Because modern data connections run many hundreds or thousands of times faster than MIDI, it is also possible to multiplex many parallel MIDI channels over a single modern serial data connection. SMPTE timecodes can be useful for synchronizing several MIDI systems in such contexts.

29.9 ■ SAMPLERS

In addition to music synthesizers, digital audio technology has allowed the development of *samplers*. Although they may look like synthesizers, in a sampler the "synthesis engine" of the synthesizer is replaced by a digital library of recorded sounds in combination with a special processor for manipulating the duration and pitch of these recordings. There is potential for confusion here, because the recordings are called *samples*, just like the individual values in a digital signal.

Samplers are not really synthesizers because they start with sound recordings, not sound recipes. Many samplers allow new recordings to be placed in the waveform memory either from data sources such as memory modules or disk recordings or in real time from a microphone through an analog-to-digital conversion system. A recording may be played in its original form simply by pressing a key on the keyboard. The original pitch may be heard, for example, when the note middle C is depressed. When $C^\#$ is depressed, however, the pitch of the recording is transposed up one semitone (without affecting the duration of the recording). Arbitrary upward and downward transpositions are possible, sometimes within certain limits imposed by the pitch-shifting hardware.

In most cases, the more the pitch is shifted, the more the tone quality is also likely to shift in amusing or irritating ways. Upshifting the pitch of speech, for example, eventually creates the well known "chipmunk effect" associated with playing traditional recordings faster than normal. This is because *all* frequencies are shifted upward, including the frequencies of the speech formants. Shifting speech downward increases the apparent *size* of the speaker, even when the utterance does not slow down because of the downward pitch shift. Again this is due to the shift in spectrum envelope as well as pitch. To simply deepen the voice of a speaker, it would be necessary to somehow maintain the spectrum envelope while lowering the pitch of the excitation waveform. Such feats are possible, but not on typical samplers.

Because of undesirable timbre shifts associated with pitch shifts, it is usually necessary to sample (i.e., record) a musical instrument at many different pitches. Each of these samples becomes a separate recording in the waveform memory of a sampler. When a particular key is played, the recording (sample) with the closest pitch is selected for processing in order to avoid large pitch shifts. This is particularly important for certain instruments, such as the piano, that have both wide pitch and timbre ranges (timbre changes on a piano are so pronounced that some researchers have wondered how the ear accepts the piano as one instrument).

Pitch shifting allows a sampler to get by with many fewer recordings than possible notes. However, a good sampler may still contain many thousands of recordings. A piano, for example, needs to be recorded about every major third or so to avoid objectionable timbre shifts. Other instruments may be more forgiving, but for high-quality sound, a sampler needs to have high-speed access to a sizable waveform memory.

Looping

In addition to shifting the pitch, it is usually necessary to modify the duration of a recorded sample. For example, the recording of a piano tone in a sampler's waveform memory might have a duration of 0.25 s. If the recording matches the pitch of a played key, no pitch shifting is required. However, if the key is held down for more than 0.25 s, the duration of the note must somehow be extended.

The technique for extending a note is called *looping*. Looping consists of choosing two points during the recorded note as *loop points*. The first loop point normally occurs soon after the "attack" phase of the note has ended. The second loop point normally occurs some time later, but before any significant decay in the note's amplitude has begun. Repeating the section of the recorded note between the two loop points as often as needed can extend a note. As soon as the key is released, a transition is made immediately to the section of the note from the second loop point to the end of the recorded note.

It should be apparent that care must be taken in the transitions from end to beginning of the loop—and from the middle of the note to the release—to avoid discontinuities or other unevenness in the sound. Choosing good loop points is an art that often requires considerable experimentation.

Pitch Shifting

The real trick of making a sampler work is to achieve pitch shifting with high audio quality. Some of the original samplers did this by using multiple output channels, each with a separate digital-to-analog converter running at a different sampling rate. As the sampling rate of each channel was changed, so, too, was the cutoff frequency of the associated output analog low-pass filter. Making the cutoff frequencies of high-quality, high-performance, low-pass filters dynamically variable proved to be both difficult and expensive. Further, the pitch-shifted signals had to be mixed in the analog domain, and no further digital processing could be done.

More recent samplers achieve pitch shifting entirely in the digital (time) domain by high-resolution resampling of the recorded signal. Because all signals end up at the same sampling rate, further digital processing (including mixing) can readily be done. To shift the pitch of a recording down by a semitone, for example, each period of its waveform needs to occupy about 6% more time than the unshifted signal. The sampling rate of the signal therefore has to be *increased* by a factor of $2^{1/12}$ in order to shift its pitch *down* (by playing this greater number of samples at the original sampling rate).

Resampling a digital signal therefore involves *interpolation*. The waveform between the available (recorded) sample values must be estimated. Provided it is accurate enough, the estimated signal can then be resampled at any desired rate.

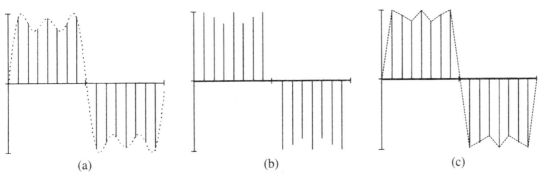

FIGURE 29.17 The limitations of linear interpolation. Plot (a) shows a waveform (dotted line) sampled at a rate of 16 times per period. Plot (b) shows the samples alone (this is how the original waveform is represented in digital form). Plot (c) shows an estimation of the original waveform(dotted line) based on linear interpolation. Note how poorly linear interpolation estimates the original waveform.

Simple *linear interpolation* is generally inadequate for high-quality results, although it is sometimes used because of its simplicity when high-fidelity sound is not required (see Fig. 29.17). *Ideal interpolation* would involve convolving the recording with a sinc function (see Eq. (29.8)), but this function is infinite in extent, making ideal interpolation infinitely expensive in terms of computation time and/or storage.

Convolving the signal with a special version of a sinc function can be used to achieve practical high-quality results (Smith and Gossett 1984). The sinc function, which normally extends forever (like the sine function, on which it is based), is typically tapered and then *truncated*, or cut off, so that it extends to only a finite extent in both directions. For example, four samples on either side of an interval between any two samples in the recording might be used to compute an interpolated value. This would mean that a total of $N = 8$ points of the original recording would participate in the calculation, which is far more accurate than using only $N = 2$ points (linear interpolation).

To shift the pitch of a recording down by one semitone, we need to resample the recording by a factor of $2^{1/12} \cong 1.059463$ times faster. In other words, if the original recording has a sampling rate of $R = 48,000$ Hz, we need to resample it at $R = 50,854.224$ Hz. When the resampled signal is played back at a rate of $R = 48,000$, all pitches will be lowered by a semitone.

We have the stored values of the recording, which we will call $x(n)$, where n is the usual integer-valued sample index. We need to estimate the values of this same recording at $x(0), x(1\Delta), x(2\Delta) \ldots$, where $\Delta = 2^{-1/12} \cong 0.94387$. That is, we need to estimate values of the waveform *in between* values of $x(n)$.

We can define the function sinc(p) as follows (see Fig. 29.18).

$$\text{sinc}(p) = \frac{\sin(\pi p)}{\pi p} \tag{29.8}$$

Note that sinc(p) is defined for all values of p, not just integer values. Defined in this way we see that sinc$(0) = 1$ and all zero crossings of sinc(p) occur at integer values of p, such as sinc$(-2) = $ sinc$(-1) = $ sinc$(1) = $ sinc$(2) = 0$. This definition makes it easy to relate sinc(p) to $x(n)$ when p happens to be an integer.

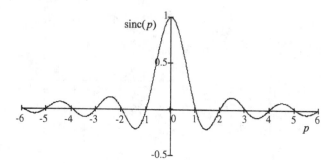

FIGURE 29.18
Plot of function
$\text{sinc}(p)$ defined in
Eq. (29.8) for
$-6 < p < +6$.

By definition, interpolation means estimating values of the form $x(n+\Delta)$, where n is an integer and Δ is a value between 0 and 1. We can estimate $x(n+\Delta)$ by directly convolving $x(n)$ with a shifted sinc function.

Why does this convolution "interpolate" the values of $x(n)$? Recall that $x(n)$ is a digital signal, which means that its spectrum contains infinitely many copies of its spectrum centered around all integer multiples of R hertz, the sampling rate. The underlying, smooth, "analog" form of the signal has no such copies. To access this underlying signal, we have

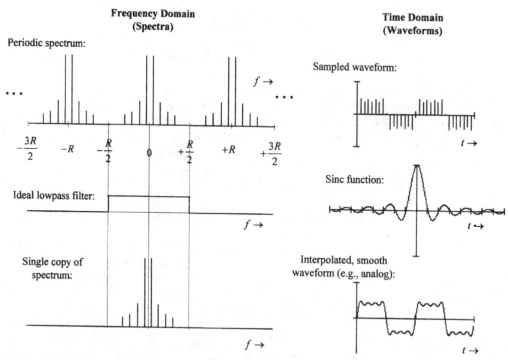

FIGURE 29.19   A sampled time function (upper right) theoretically contains infinitely many copies of its spectrum centered around all integer multiples of the sampling rate R (upper left). The "ideal" low-pass filter transfer function (middle left) rejects all frequencies greater than the Nyquist rate ($R/2$ hertz). The time-domain version of this rectangular boxcar function is the sinc function (middle right). The result of filtering the sampled function in this way is a smooth, interpolated waveform (lower right), which has only one copy of the associated spectrum (lower left).

only to remove all the copies of the spectrum with a low-pass filter—one that passes all frequencies up to half the sampling rate (recall that these frequencies can be either negative or positive). This ideal low-pass filter has a simple, rectangular "boxcar" transfer function in the frequency domain. If we simply multiply the boxcar by the spectrum of any digital signal, we remove all those periodic copies of the spectrum that make the difference between an analog and digital signal, leaving only the frequencies from $-R/2$ to $+R/2$ hertz (see Fig. 29.19).

But multiplication in the frequency domain, as we have learned, corresponds to convolution in the time domain. If we convert the ideal low-pass filter transfer function (the boxcar) from the frequency domain to the time domain using the Fourier transform, it turns out to be the sinc function. The reason the ideal low-pass filter cannot exist is seen to be that the sinc function—the impulse function of the ideal low-pass filter—is infinitely long. Further, it extends from minus to plus infinity, meaning that it begins to respond infinitely long *before* time $t = 0$, which is when the impulse occurs. It is impossible to build a real-time device that begins to respond to its input before it occurs, although such notions can exist mathematically. This also clarifies why we need to truncate the sinc function in order to use it.

To accomplish the interpolation, we first center the truncated copy of sinc(p) at position $n + \Delta$. $x(n + \Delta)$ is then just the sum of the nonzero products formed by $x(n)$ and sinc. Assuming that we use $N = 8$ points for an interpolation, the four participating values of $x(n)$ to the left of position $n + \Delta$ will be $x(n-3)$, $x(n-2)$, $x(n-1)$, and $x(n)$. The four participating values of $x(n)$ to the right will be $x(n+1)$, $x(n+2)$, $x(n+3)$, and $x(n+4)$. Because $x(n)$ lies Δ units to the left of position $n + \Delta$, it will be scaled by sinc$(-\Delta)$. Because $x(n-1)$ lies $\Delta + 1$ units to the left of position $n + \Delta$ it will be scaled by sinc$(-\Delta - 1)$. Extrapolating this procedure four points to the left and right allows us to write the following formula for the estimated value (see Fig. 29.20).

FIGURE 29.20
Interpolating with the truncated sinc function. Equation (29.9) says that to estimate the value of the waveform at point $n + \Delta$, multiply the heights of the white dots by the heights of the black dots and add the products together (several points are labeled in terms of the equation for reference).

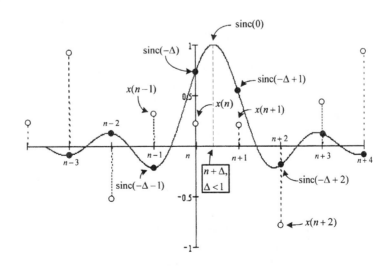

$$x(n + \Delta) = \text{sinc}(-\Delta - 3)x(n - 3) + \text{sinc}(-\Delta - 2)x(n - 2)$$
$$+ \text{sinc}(-\Delta - 1)x(n - 1) + \text{sinc}(-\Delta)x(n)$$
$$+ \text{sinc}(-\Delta + 1)x(n + 1) + \text{sinc}(-\Delta + 2)x(n + 2)$$
$$+ \text{sinc}(-\Delta + 3)x(n + 3) + \text{sinc}(-\Delta + 4)x(n + 4)$$
$$= \sum_{k=-3}^{4} \text{sinc}(k - \Delta)x(k). \tag{29.9}$$

We see from this formula that for $N = 8$, each interpolated value involves the sum of eight products involving eight adjacent samples of the recorded signal and eight values of $\text{sinc}(p)$.

Even though $x(n + \Delta)$ is just an estimate of the waveform value at position $n + \Delta$, it is a much better estimate for $N = 8$-point convolution-based interpolation than for linear interpolation. Whether 8 points are too few, too many, or just right depends on the sound quality needed and the speed and cost of the available hardware to do the necessary calculations (and memory accesses). If more accuracy is needed, N can be made larger. If lower cost is important, N can be made smaller.

29.10 ■ REAL-TIME SOFTWARE SYNTHESIS

Digital music synthesis can be implemented either through a computer program that runs on a general-purpose computer or by designing and building one or more specialized integrated circuits that perform the same algorithms. The great advantage of such hardware implementation is that hardware can be optimized around the requirements of a particular algorithm, such as additive or nonlinear synthesis, or looping and pitch shifting. Hardware designed for pitch shifting, however, is not likely to perform well (or even at all) if it is required to perform subtractive synthesis or any other function for which it was not designed. Hardware implementation greatly increases performance and typically comes at a great cost in flexibility.

The great advantage of software implementation is precisely that it is inherently much more flexible than hardware implementation (that is why standard computers are called general-purpose computers). To modify a software procedure or to add a new one, only new code needs to be written and linked into existing software. This is generally much easier than designing or modifying a piece of synthesis hardware (usually some form of VLSI circuit).

Software implementation can never match or even approach the performance level of optimized hardware. Nevertheless, digital circuits have become fast enough that even general-purpose computers are increasingly capable of real time performance. This allows the best of both the hardware and software worlds to overlap to a certain extent: we can simultaneously have real-time performance and relatively convenient software flexibility and extensibility.

Max and Pd

A current example of real-time software for computer music is the Pd program by UCSD music professor Miller Puckette. Based on his earlier program Max and similar to David Zicarelli's program MSP, Pd is described as a real-time graphical programming language for audio and graphics processing. The principal advantages of Pd (which means "pure data") over Max and MSP are that Pd is simpler and more portable (it runs on a variety of Linux, Windows, and other platforms). Pd can be combined with a program by Mark Danks called GEM to afford simultaneous computer animation (three-dimensional graphics and video) and computer audio.

Inputs and outputs to Pd can include real-time audio, MIDI, the computer's keyboard and display, and network messages. Because computers vary considerably in their processing capabilities, the performance of Pd is highly platform-dependent.

The Pd user controls real-time processing by designing a Pd *patch*, or *canvas*. The user graphically interconnects boxes with various functions such as reading a MIDI note message and converting it into suitable data to drive a software oscillator whose output signal is fed to the computer's sound output system. Such patches are similar to the instrument definitions used in acoustic compilers except that they read MIDI data instead of note statements and produce their outputs in real time. The user may also create new boxes from elements provided in Pd or add fundamentally new capabilities by adding new Pd "objects" written in C or other programming languages.*

29.11 ■ SPATIALIZATION

When we listen to sounds in their natural environment, we are aware of characteristics such as pitch, loudness, timbre, and temporal organization. Because sound waves exist in space as well as time, we are also aware of the spatial characteristics of many sounds that we hear. We can often decide to a fair degree of certainty from what direction a sound comes, how far away the source of the sound is, whether it is moving and how. We usually can tell something about the environment in which we are listening, such as whether it is an open or an enclosed space, and—if it is enclosed—something about the shape, size, and surface of the enclosure. Some enclosures help us hear sounds more clearly, whereas others tend to obliterate sound details. No room is optimum for listening to all types of sound. Some rooms are best suited to listening to speech, whereas others are better for chamber music; still others may be ideal for listening to opera, and still others for listening to large orchestras playing symphonies.

Many of these topics have been discussed in a general acoustical context elsewhere in this book. Computer music techniques provide many opportunities to study and exploit the spatial characteristics of sound for musical purposes. Computers allow us to move virtual sound sources through virtual sound spaces along virtual sound trajectories. Such capabilities provide new dimensions along which sound can be controlled and, therefore, composed and performed by musicians using computer technology.

---

*At the time of this writing, Pd and related software are available for free download from http://www.crca. ucsd.edu/~msp/software.html.

Virtual Acoustics

Sound spatialization is increasingly important in the ever-developing context of *virtual reality*. Because our experience of being in a virtual space can be affected as much by our ears as our eyes, for many virtual reality applications, synthesizing appropriate auditory cues can be as important as synthesizing appropriate visual cues (Begault 1994).

People listen to technologically mediated sound in several basic ways. The most common way is through one or two loudspeakers arranged more or less in front of the listener (*mono* or *stereo*). One may also listen to four speakers arranged in a rectangular or trapezoidal configuration, with the listener somewhere near the center (*quad* presentation). Another common approach is to use five loudspeakers: four in the corners of a quad box plus one centered in front of the listener. Such a system has its origins in movie theaters and is often called *home theater*, or 5.1 channel sound (if a subwoofer channel is used to provide the additional "0.1" part). Actual theater systems may include even more loudspeakers, arranged in virtually any geometry. One special auditory presentation space in San Francisco (the "Audium") arrayed dozens of loudspeakers on the inner surface of a large geodesic dome surrounding listeners who sat is circularly arranged chairs on a false floor, beneath which additional loudspeakers were present.

No matter how many loudspeakers are used to produce the sounds heard by a listener, we listen through only two ears. Albert Bregman (1990) brilliantly characterized the auditory space recognition problem in the following way.

> Imagine you are on the edge of a lake and a friend challenges you to play a game. The game is this: Your friend digs two narrow channels up from the side of a lake. Each is a few feet long and few inches wide and they are spaced a few feet apart. Halfway up each one, your friend stretches a handkerchief and fastens it to the sides of the channel. As waves reach the side of the lake they travel up the channels and cause the two handkerchiefs to go into motion. You are allowed to look only at the handkerchiefs and from their motions to answer a series of questions: How many boats are there on the lake and where are they? Which is the most powerful one? Which one is closer? Is the wind blowing? Has any large object been dropped suddenly into the lake?
>
>   Solving this problem seems impossible, but it is a strict analogy to the problem faced by our auditory systems. The lake represents the lake of air that surrounds us. The two channels are our ear canals, and the handkerchiefs are our ear drums. The only information that the auditory system has available to it, or ever will have, is the vibrations of these two ear drums. Yet it seems to be able to answer questions like the ones we were asked by the side of the lake: How many people are talking? Which one is louder, or closer? Is there a machine humming in the background? We are not surprised when our sense of hearing succeeds in answering these questions any more than we are when our eye, looking at the handkerchiefs, fails.

Thus we might spatialize sound either by placing a loudspeaker in every position from which we wish to hear a sound (effective but expensive) or, at the other extreme, by properly synthesizing the motions of Bregman's two handkerchiefs.

The best way to present information almost directly to the eardrums is with intraaural earphones (the kind that are placed in the ear canal). Even supraaural or circumaural earphones (which sit on or around the pinnae, respectively) can produce convincing spatial impressions if the appropriately processed binaural sound is fed into them (see Fig. 29.21). But earphones also have their limitations. Most people find earphones uncomfortable after

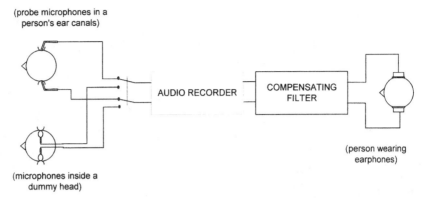

FIGURE 29.21 Binaural reproduction of sound. Microphones may be placed either in the ear canals of a human listener or inside a dummy head. The resulting signals may be recorded or listened to directly. In either case it is generally necessary to filter the signals to correct for amplitude and phase changes between the microphone position and the sound pressure at the outer ears, depending on the type of earphones used. Such a system is capable of reproducing all the spatial characteristics of sound.

a time, especially when they are worn inside the ear canal. And the listener's head movements usually cause the entire spatialized sound image to move.

Creating spatialized sound generally involves the creation of *spatial cues* that can be impressed on arbitrary sounds. Many of these cues have been discussed elsewhere in this book (see Section 5.5); for convenience a brief list is given here.

To spatialize a virtual sound source we must control its *direction* (made up of *azimuth* on the horizontal plane and *elevation* on the vertical plane) and *distance*, usually within a reverberant environment that portrays the spatial impression of a virtual listening space (see Fig. 29.22). Virtual sound sources may also exhibit *motion* of various kinds through virtual sound space.

Azimuth cues (horizontal direction) include *interaural time delay* (ITD) and *interaural intensity difference* (IID). The forward-facing *pinnae* (outer ears) also provide information for disambiguating whether a sound comes from in front of or behind a listener. At frequencies below about 1500 Hz, ITD provides the primary cue for azimuth because low frequencies can effectively diffuse around the head resulting in little, if any, IID. At frequencies above about 1500 Hz, ITD becomes ambiguous because the delay can represent many complete cycles of the waveform, but the "head-shadow" effect causes IID values up to 20 dB or so.

Other cues of importance for azimuth perception include the *precedence effect*, also known as the *Haas effect* or the law of the first wavefront (see Sections 5.5, 23.3, 25.3, and Blauert (1983)). This law states that if two more-or-less identical sounds arrive from different directions within about 35 ms of each other, they are heard as a single sound coming from the direction of the sound that arrives first. This can be true even when the second sound is slightly louder than the first.

Another important cue is the *Franssen effect*, or *interaural envelope difference* (IED), which states that the direction of a sound will be identified by high-frequency components associated with the onset of the sound, even if the preponderance of the energy is at low

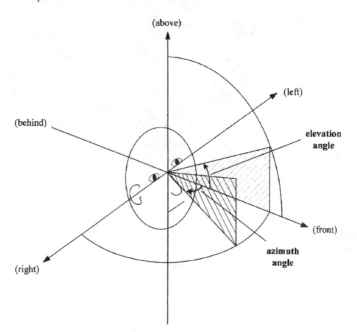

FIGURE 29.22
Head-related
coordinate system.
Azimuth is
normally measured
from a reference of
0° (front),
increasing to the
left. Elevation is
normally measured
from a reference of
0° (front),
increasing upward.

frequencies coming from a different direction (Hartmann and Rakerd 1989). This is why the location of a subwoofer is largely irrelevant in a multichannel loudspeaker system.

Elevation cues depend on subtle filtering effects of the pinnae. Although the origins of this filtering effect are not entirely understood, their effect can be measured. The pinnae introduce a significant "notch" in the frequency response of the ear. The center frequency of this notch goes from a bit less than 6000 Hz at elevations slightly below the horizontal plane to about 10,000 or so as the source moves directly overhead. The depth of the notch ranges from about 10 dB for sounds coming from the front to just a few decibels when the sound is directly overhead (it becomes deeper and lower in frequency again as the sound travels from overhead to behind (Blauert 1983)). All such observations are very sensitive to small changes in the size and shape of the pinnae and, therefore, vary significantly from person to person. Elevation effects thus tend to be much less reliable than azimuth effects.

Distance cues are different for familiar and unfamiliar sounds. For familiar sounds (such as speech), the direct sound basically drops off in accordance with the inverse square law, or about 6 dB per doubling of distance. When the sound source is of considerable extent (such as a freeway), intensity drops off less rapidly, often at about 3 dB per doubling of distance. When a sound source is unfamiliar (such as a novel synthesized tone), a doubling of distance is more likely to be associated with a halving of loudness, which is roughly 9 or 10 dB at middling loudness and frequencies.

For close sound sources (less than about 1 m from the listener) both azimuth parallax and IID will increase. Azimuth parallax occurs when the listener's head position moves slightly: for very close sound sources the change in azimuth will be greater than for distant sources. Similarly, the head-shadowing effect is greatest for very close sound sources, generally increasing IID. Sufficiently distant sound sources will also be affected by the

TABLE 29.1 Absorption due to air.

		Frequency in Hz				
		2000	4000	6000	8000	10000
Relative humidity	30%	14	49	102	168	241
	50%	10	30	62	105	159
	70%	9	23	46	78	118

Attenuation in dB/km
$T = 20°C$ (ISO 9613)

frequency-dependent attenuation through air that depends on temperature and humidity (see Table 29.1).

Reverberation cues include "early reflections" occurring from about 30 to 80 ms after the direct sound. Such reflections are typically due to the walls, floor, ceiling, and other prominent reflecting surfaces in a reverberant environment. After about 80 ms, second and higher-order reflections tend to accumulate so rapidly that *dense reverberation* results. We saw in Chapter 23 that the characteristics of such reverberation (intimacy, liveness, warmth, and so on) have a distinct effect on the listener's impression of the size, shape, and quality of the room in which sounds are heard.

Motion cues for sound sources include time-varying direction, distance, and *Doppler shift* (see Section 3.7). When apparent sound sources are made to move along virtual trajectories, Doppler shift may or may not be desirable. Doppler shifting may cause musical pitches to go out of tune, for example, but they may also add dramatically to the vividness of a motion effect. Doppler shift does not result from synthetic IID cues, but it does result from time-varying delays associated with either synthetic ITD cues or delays due to distance. When a delay becomes time varying, it is important to use interpolation in order to avoid discontinuities in the synthesized waveform (such discontinuities can and often do cause audible clicks and other distortions). As we have already seen in the case of pitch shifting, N-point interpolation based on convolution will always outperform simple linear interpolation, though the latter may suffice for slowly moving virtual sound sources.

There are several approaches to synthesizing virtual localization cues. One approach that works well for playback over loudspeakers for a sizable audience is to model the real listening space as a room embedded within a larger virtual listening space. This room-within-a-room approach is the basis for the space unit generator in the cmusic acoustic compiler (Moore 1990). See Fig. 29.23.

Head-Related Transfer Functions

Virtual reality applications typically concentrate on modeling an individual listener within a virtual acoustic space. In that case it is generally desirable to directly model the physical cues that the listener would hear if the sound source were actually located in a certain direction. This can be done by recording how a real source sounds when it is in a particular position. In order to capture pinna cues, the shadowing effect of the head, reflections from the shoulders, etc., it is necessary to make the recording inside the ear canal as close to the eardrum as possible.

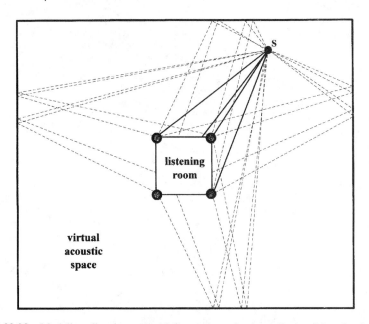

FIGURE 29.23   Modeling direct sound and first-order reflections. Each of four loudspeakers in a listening room of specified geometry receives direct sound (heavy lines) and a reflected sound (dashed lines). The size and geometry of the virtual space determine distances (and therefore time delays) for all the direct and reflected paths. In the cmusic space unit generator, the outside surfaces of the listening room are modeled as absorbing so that they "cut," or "shadow," certain reflections, which enhances the spatial effect.

Such recordings can then be used to derive a head-related transfer function (HRTF), which is essentially a filter through which any sound can be processed. HRTFs are usually made with the aid of a "dummy head," or mannequin, such as those manufactured for this purpose by such companies as Knowles Electronics, Brüel and Kjaer, and Neumann (see Fig. 29.24). Multiple HRTFs (for many individual positions) can be combined into a

FIGURE 29.24
Dummy head. This sketch of a dummy-head microphone might look like a real person, except for the places to remove the pinnae and the head. Such realism helps preserve spatial cues from acoustic shadowing effects of the body as well as pinna effects.

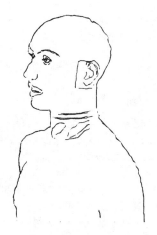

single, quite complex function of azimuth, elevation, and frequency that can be used for processing sounds in virtual-reality simulations.

29.12 ■ DIGITAL AUDIO COMPRESSION

A standard compact disk recording comprises two channels of digital audio, each containing 44,100 16-bit samples per second. The basic bit rate for CD-quality audio is therefore $2 \times 44,100 \times 16 = 1,411,200$ bits per second (bps). We can reduce this bit rate in one of three ways. First, we can lower the audio quality, for example, by reducing the sampling rate or number of bits per sample. Second, we can exploit the inherent predictability (i.e., redundancy) of audio waveforms. Finally, we can try to take advantage of certain limitations of human hearing, such as the tendency of one frequency to mask another. Reducing the bit rate means that we can store more audio information in the same amount of memory or transmit audio signals at reasonable and selectable audio quality using lower bandwidth channels.

Source Coding

Exploiting the redundant characteristics of the audio signal itself is called *source coding*. We have learned elsewhere, for example, that N bits can represent up to 2^N choices. The theory of information tells us that if all 2^N values are equally likely to occur, then we can do no better than to use equal-length codes. For example, to encode any four equally likely values, we could use the codes 00, 01, 10, and 11. If we transmitted these codes at the rate of 1000 values per second, we would then require a data-transfer rate (bandwidth) of 2000 bps.

Suppose, however, that one of the four values is much more likely than the others, occurring 91% of the time. We could lower the bandwidth by using a shorter code for the more likely value, even at the expense of using longer codes elsewhere. Assuming the remaining three values are still equally likely (each now occurring 3% of the time), we now have a situation in which it is possible to reduce the required data transfer rate to less than 2000 bps while still transmitting 1000 values per second. For example, we might use the four codes 0, 10, 110, and 111. A basic postulate of information theory states that we can now find the average length of a message by multiplying the lengths of the codes by the probability that they will occur and adding the results: $1 \times 0.91 + 2 \times 0.03 + 3 \times 0.03 + 3 \times 0.03 = 1.15$ bits per value (Abramson 1963). Note that even though the codes differ in length, any sequence of them is uniquely decodable (we would have trouble decoding a message comprising only the four code words 0, 00, 000, and 0000, for example). We could, therefore, use our new code to transmit 1000 values per second using an average bit rate of only 1150 bps.

Note that because the four values can be recovered from their encoded form perfectly, this encoding is *lossless*, meaning that no information is lost due to the encoding itself. Note also that although the *average* bit rate is assured, the *instantaneous* bit rate is not. If, for example, the first 9% of the message consisted only of the uncommon values, whereas the last 91% consisted of only the most common value, then during that first part of the message, the average code length would be $2/3 + 3/3 + 3/3 = 3.66\ldots$ bits per value. Even though the final 91% of the message would have utilized only 1 bit per value, if the

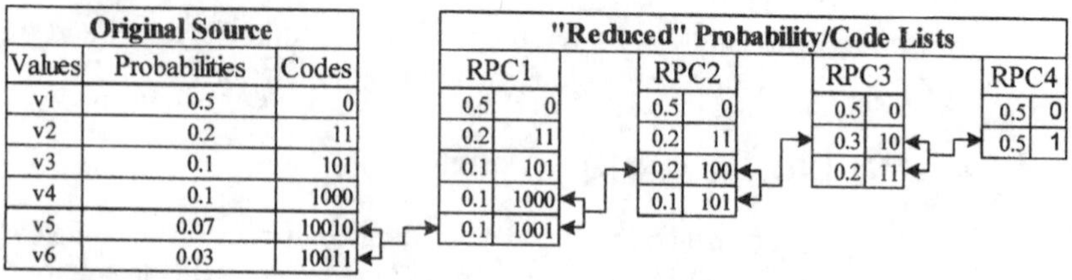

Original Source			"Reduced" Probability/Code Lists									
Values	Probabilities	Codes	RPC1		RPC2		RPC3		RPC4	
v1	0.5	0	0.5	0	0.5	0	0.5	0	0.5	0
v2	0.2	11	0.2	11	0.2	11	0.3	10	0.5	1
v3	0.1	101	0.1	101	0.2	100	0.2	11		
v4	0.1	1000	0.1	1000	0.1	101				
v5	0.07	10010	0.1	1001						
v6	0.03	10011								

$$Average\ Code\ Length = 0.5 \times 1 + 0.2 \times 2 + 0.1 \times 3 + 0.1 \times 4 + 0.07 \times 5 + 0.03 \times 5$$
$$= 2.1\ bits\ per\ value$$

FIGURE 29.25 Construction of a Huffman code. Values are listed in order of their probability. The smallest two probabilities on the list are repeatedly combined to form a shorter, sorted list. When only two items are left on the list, 0 and 1 are assigned as codes. The codes with combined probabilities are then expanded by appending 0 and 1 repeatedly, starting with the shortest list and working backward. Because 0 and 1 assignments may be interchanged at any point and because there are multiple ways to sort a list when two or more probabilities are the same, the code is not unique. However, all codes resulting from this procedure are uniquely decodable and have the shortest possible average length.

message were 100 min long, the first 9 min would have a significantly higher bit rate than the overall average.

In the field of information theory, lossless, optimum source codes based on the probabilities of encoded values are well understood. Assuming that we need to encode values *one at a time* and that the probability of each value is known, we can construct a *Huffman code* to achieve the smallest possible average code length in the following manner. Suppose we have n values to be encoded, $v_1, v_2, \ldots, v_n$ and that the probability of each value is given by $P_1, P_2, \ldots, P_n$. Assume also that the values are ordered so that $P_1 \geq P_2 \geq \cdots \geq P_n$. We can construct a Huffman code by combining the last two symbols on the list (and their probabilities) repeatedly until we end up with only two values (see Fig. 29.25). Clearly, the final two values can be encoded with only the binary codes 0 and 1. We can then work backward by placing a 0 and 1 after the code used for the combined probabilities. Although a Huffman code is not necessarily unique, it can be shown mathematically that no code has a shorter average length.

Many other source coding methods exist. In most cases, lossless coding must be used to encode a source when we don't know if any particular loss will be significant.

Receiver Coding

Lossy coding can be used when it is possible to determine the significance of any given loss. If an information loss is in some sense insignificant—and especially if it results in a significant data reduction—it can be accepted. For audio signals we can use psychoacoustic criteria to try to estimate the significance of certain types of losses. Because coding choices are based on characteristics of the receiver (the listener in this case), such techniques are called *receiver coding*. Receiver coding and source coding techniques can be, and often are, combined.

The main advantage of lossy coding is that it typically allows much better data compression than lossless coding. The main disadvantage is, of course, that some information being encoded is obliterated by the encoding method. In other words, it is impossible to obtain the exact original message from the encoded one.

MPEG/Audio Layer III (MP3)

Such data-compression schemes as MP3 and AC-3 use lossy receiver coding methods. MP3 stands for MPEG/Audio Layer III (MPEG stands for Motion Picture Experts' Group).* MP3 works by transforming a linear PCM signal into the frequency domain using a filter bank, much like the vocoders we discussed earlier (Pan 1995). Each of 32 equally spaced subband channels is analyzed according to a *psychoacoustic model* that tries to determine how best to apportion available bits to each channel. The 32 subband channels are then encoded and formatted into a compressed bitstream (the MP3 signal). The decoder reads this bitstream and transforms the signal back into the time domain at the original sampling rate. The decoded signal is represented in linear PCM format, but it is quite different from the input signal. If all goes well, however, the decoded signal should sound very much like the input signal because MP3 attempts to remove only "perceptually insignificant" information from the signal.

MP3 supports linear PCM audio sampling rates of 32, 44.1, and 48 kHz and four different modes:

1. A monophonic mode for one audio channel;
2. A dual-monophonic mode for two independent audio channels (functionally the same as stereo);
3. A stereo mode with bit-sharing between the channels (but no joint-stereo coding);
4. A joint-stereo mode that exploits either correlations between the channels, irrelevance of phase differences between the channels, or both.

The compressed (encoded) bitstream can have any of several fixed rates, ranging from 32 to 224 Kbps per channel, yielding compression factors between 2.7 and 24.

The MP3 encoder processes digital audio data in 1152-sample *frames*. The input signal is transformed into the frequency domain using a 32-channel *polyphase* filter bank, followed by a modified discrete cosine transform (MDCT), which is similar in its properties to the DFT discussed in Chapter 28 (the MDCT is typically implemented using FFTs). Each of the 32 channels has the same frequency bandwidth, equal (in hertz) to the Nyquist rate divided by the number of filters.

The input data are concurrently analyzed using a *psychoacoustic model* (see Fig. 29.26). The psychoacoustic model also transforms the audio data to the frequency domain using a fine-grained DFT (implemented as an FFT). For analysis purposes, the psychoacoustic model has considerably greater resolution in the frequency domain than the encoding process. The frequency domain representations are grouped into sections related to auditory critical bands (see Table 29.2). The psychoacoustic model then computes a *tonality index* for each band that characterizes whether it is relatively more tonelike or noiselike (tones and noise have different masking characteristics). The masking effect of a band is then estimated using a *spreading function*, which determines how energy in a given critical band affects its neighboring bands.

The sum total of all the masking effects on a given band, plus its signal content, determines how likely it is to be masked. This masking information is mapped onto the 32

---

*MPEG is standardized as ISO 11172-1, 11172-2, 11172-3, and 13818-3.

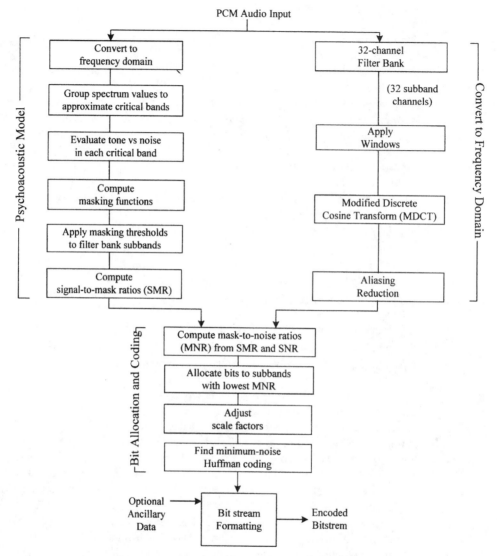

FIGURE 29.26   The MPEG/Audio Layer II (MP3) encoding process.

equal-width filter bank channels and used to apportion bits to each channel. The more likely a channel is to be masked, the fewer bits it receives. Channels with energy falling below the computed masking threshold may receive no bits at all.

Bits are shared among channels by using a *bit reservoir*. When a channel is found to need fewer bits than initially estimated, bits may be put into the reservoir; when a channel is found to contain highly significant data, it may take extra bits out of the reservoir. Bits are shared only within single frames, however; never among multiple frames.

MP3 has been found to perform quite well in a variety of situations, but it is important to understand some of its limitations. The MP3 standard specifies the bitstream format but

TABLE 29.2   Approximate critical-band boundary frequencies for MP3 (frequencies are at the top end of each band).

Band number	Frequency (Hz)	Band number	Frequency (Hz)	Band number	Frequency (Hz)
0	50	9	940	18	3,840
1	95	10	1,125	19	4,690
2	140	11	1,265	20	5,440
3	235	12	1,500	21	6,375
4	330	13	1,735	22	7,690
5	420	14	1,970	23	9,375
6	560	15	2,340	24	11,625
7	660	16	2,720	25	15,375

not the sound quality. The standard suggests various features that the psychoacoustic model might have but does not specify it precisely. MP3 psychoacoustic modeling is a topic of continuing research and is typically limited by such things as available real-time processing power. Thus, two different MP3 encoders might sound quite different when encoding the same input data. Also, certain types of audio signals might sound much worse than others. Sometimes increasing the bitstream rate will improve the MP3 sound quality dramatically. It is, therefore, important to monitor the sound quality carefully when using MP3 encoding.

29.13 ■ DIGITAL AUDIO AND THE INTERNET

MPEG-2/Audio handles multichannel audio (such as 5.1-channel home theater sound) and incorporates other enhancements such as multilingual audio support and lower sampling and bitstream rates. MPEG-2 is intended for use primarily with digital television systems and DVDs. AC-3 is a proprietary standard owned by Dolby Laboratories and widely used for encoding DVD audio, which is the first *de facto* standard for the distribution of audio with more than two channels. Other standards include MPEG-4, which is oriented toward multimedia for the Worldwide Web and mobile systems, and MPEG-7, a multimedia content description interface. MPEG-21 supports a "multimedia framework." Much of the motivation for such developments is the increasing need for high-quality audio in situations supported by limited bandwidth, such as real time streaming audio and video over the Internet, which is supported by vastly different bandwidths in different situations.

The Internet is continually evolving in many ways, with technological, social, and artistic consequences that are virtually impossible to predict. It seems clear that the Internet will augment the daily lives of most people—whether or not they are aware of it—on a more or less continuous basis. In addition to traditional computers, widespread Internet access will be multiplied by a variety of telecommunications devices, sensors, actuators, embedded processors, and "information appliances," many of them wireless.

Most Internet connections today involve connections with significant bandwidth restrictions. All technological indications are that maximum bandwidths will continue to increase, and all bandwidths will evolve in the direction of universal availability. Such developments support the implementation of distributed computing networks that could feasibly augment the "intelligence" of any device on an as-needed basis. Low-bandwidth connectivity, however, will continue to make data compression an important consideration for audio and video transmissions.

29.14 ■ SUMMARY

Perhaps the most important thing to understand about digital audio technology is that it is rapidly evolving. It is, therefore, important to understand the basic principles on which it is based, information that becomes obsolete much less rapidly than the technical details of a given implementation. Many current digital audio applications are based on notions first explored in the field of computer music, including artificial musical intelligence and the basic sound synthesis methods: additive, subtractive, nonlinear, and physical modeling. Digital music synthesis started in software and then moved to synthesizer and sampler hardware in order to achieve real-time performance, but it is gradually reverting to the much more flexible software context as processing speeds continue to improve. Advances in digital audio-signal processing currently support virtual-reality applications in a variety of ways, most notably in the areas of sound spatialization and data compression.

REFERENCES AND SUGGESTED READINGS

Abramson, N. (1963). *Information Theory and Coding*. New York: McGraw-Hill.

Begault, D. (1994). *3-D Sound: for Virtual Reality and Multimedia*. New York: Academic Press.

Blauert, J. (1997). *Spatial Hearing*. Cambridge, Mass.: MIT Press.

Bregman, A. (1990). *Auditory Scene Analysis*. Cambridge, Mass.: MIT Press.

Chowning, J. (1973). "The Synthesis of Complex Audio Spectra by Means of Frequency Modulation," *J. Audio Eng. Soc.*, **21**.

Cope, D. (1991). *Computers and Musical Style*. Madison, Wisc.: A-R Editions.

Cope, D. (1996). *Experiments in Musical Intelligence*. Madison, Wisc.: A-R Editions.

Danks, M. (1997). "Real-time Image and Video Processing in GEM," *Proc. International Computer Music Conf.* Thessaloniki, 220–223.

Deutsch, D., F. R. Moore, and M. Dolson (1984). "Pitch Classes Differ with Respect to Height," *Music Perception* **2**(2).

Deutsch, D., F. R. Moore, and M. Dolson (1986). "The Perceived Height of Octave-Related Complexes," *J. Acoust. Soc. Amer.* **68**.

Dudley, H. (1939). "The Vocoder," *Bell Labs Record* **17**.

Flanagan, J. J., and R. R. Golden (1966). "Phase Vocoder," *Bell Sys. Tech. J.* **45**.

Grey, J. M. (1975). *An Exploration of Musical Timbre*, Ph.D. dissertation, Department of Psychology, Stanford University.

Grey, J. M., and J. A. Moorer (1977). "Perceptual Evaluation of Synthesized Musical Instrument Tones," *J. Acoust. Soc. Amer.* **62**: 434–462.

Hartmann, W. M., and B. Rakerd (1989). "Localization of Sound in Rooms IV: The Franssen Effect," *J. Acoust. Soc. Amer.* **86**(4): 1366–1373.

Hiller, L. A., and L. M. Isaacson (1979). *Experimental Music: Composition with an Electronic Computer*. Westport, Conn.: Greenwood Press, c. 1959.

Jaffe, D., and J. Smith (1983). "Extensions of the Karplus-Strong Plucked String Algorithm," *Computer Music J.* **7**(2).

Karplus, K., and A. Strong (1983). "Digital Synthesis of Plucked String and Drum Timbres," *Computer Music J.* **7**(2).

LeBrun, M. (1977). "A Derivation of the Spectrum of FM with a Complex Modulating Wave," *Computer Music J.* **1**(4).

Makhoul, J. (1975). "Linear Prediction: A Tutorial Review," *IEEE Proc.* **63**: 561–580.

Mathews, M. V., F. R. Moore, J.-C. Risset, and J. R. Pierce (1969). *The Technology of Computer Music.* Cambridge, Mass.: MIT Press.

Moore, F. R. (1983). "A General Model for Spatial Processing of Sounds," *Computer Music J.* **7**(3).

Moore, F. R. (1988). "The Dysfunctions of MIDI," *Computer Music J.* **12**(1).

Moore, F. R. (1990). *Elements of Computer Music.* Upper Saddle River, N.J.: Prentice Hall.

Moorer, J. A. (1978). "The Use of the Phase Vocoder in Computer Music Applications," *J. Aud. Eng. Soc.* **24**.

Pan, D. (1995). "A Tutorial on MPEG/Audio Compression," *IEEE Multimedia J.* (Summer).

Puckette, M. (1997). "Pure Data," *Proc. International Computer Music Conf.* Thessaloniki, 224–227.

Risset, J.-C. (1995). *The Historical CD of Digital Sound Synthesis* [sound recording] Mainz, Germany: Wergo.

Risset, J.-C., and M. V. Mathews (1969). "Analysis of Musical Instrument Tones," *Phys. Today* **22**(2): 23.

Roads, C. (1996). *The Computer Music Tutorial.* Cambridge, Mass.: MIT Press.

Smith, J. O. (1996). "Physical Modeling Update," *Computer Music J.* **20**(2).

Smith, J. O. (2000). (Excerpts from) *Digital Waveguide Modeling of Musical Instruments*, draft version prepared for Acoustical Society of America Newport Beach meeting, 4 December. Also available at http://www.ccrma.stanford.edu/~jos/waveguide/.

Smith, J. O., and P. Gossett (1984). "A Flexible Sampling-rate Conversion Method," International Conference on Acoustics, Speech and Signal Processing (ICASSP) '84 *Proc.* **2** (March 19–21).

Strawn, J., ed., with contrib. By J. W. Gordon, F. R. Moore, J. A. Moorer, T. L. Petersen, J. O. Smith, and J. Strawn (1993). *Digital Audio Signal Processing: An Anthology.* Madison: Wisc.: A-R Editions.

Strawn, J., ed. (1988). *Digital Audio Engineering.* Madison, Wisc.: A-R Editions.

GLOSSARY

acoustic compiler A general-purpose computer program such as **MUSIC V** that translates a description of a sound written in a special language into the corresponding digital sound signal.

acoustic illusion By analogy to an optical illusion, a sound—typically synthetic—that "fools" the ear into believing it is hearing something impossible or contradictory (such as a pitch that rises and falls simultaneously).

additive synthesis The process of synthesizing sounds by adding together many simple components, such as sinusoids, each with typically time varying characteristics.

all-pole filter A digital filter that uses feedback to create resonances but does not use antiresonances (see Section 28.12).

Bessel functions One of a class of transcendental functions named for nineteenth century Prussian astronomer Friedrich W. Bessel that solve a particular differential equation. Bessel functions occur in many contexts, including in the theory of FM music synthesis.

convolution A way of combining two mathematical functions that substitutes the entirety of one function for each point of the other function. Convolution occurs in many contexts, including the theory of waveforms and spectra as well as digital filtering.

computer music The field of inquiry surrounding the application of computers and related technology to the synthesis, processing, and analysis of musical scores and sounds.

digital music synthesizer A digital device designed to synthetically create musical sounds.

head-related transfer function (HRTF) The name given to the filter transfer function that describes the manner in which sounds from particular positions in space are transmitted to the ear canals of a listener, taking into account the relative positions and frequency-dependent shadowing effects of the head, pinnae, nose, etc.

linear prediction A set of mathematical techniques designed to predict how a digital signal will behave based on a weighted sum of its most recent N values. The weights used in forming this sum are called *prediction coefficients*.

looping The process of extending the duration of a digitally recorded sound by repeatedly outputting a portion of it.

MIDI The musical instrument digital interface communications standard adopted widely in the music synthesizer industry in the early 1980s.

MP3 The common name given to digital audio that is encoded according to a particular standard known as the motion picture expert's group (MPEG)-1, Layer III. The perception-based encoding process compresses the digital representation greatly, allowing it to be readily transmitted or stored.

music synthesis Either computer-based composition of music or the synthesis of musical sounds based on computer or electronics technology.

nonlinear synthesis The process of synthesizing sounds through mathematically nonlinear processes that generate output frequencies different from those used as input, such as frequency modulation (FM).

phase vocoder A sound analysis-synthesis system based on measuring the time-varying amplitude and frequency characteristics of a complex sound. The intermediate data can be manipulated to independently shift the pitch and/or duration of the synthesized sound as compared to the original.

physical modeling synthesis The process of synthesizing sounds by deriving and operating mathematical models that describe the physical operation of real or imaginary sound sources.

pitch shifting The process of accurately *resampling* a digitally recorded sound at a new sampling rate, thus shifting the pitch when the resampled signal is played back at the same rate as the original recording.

real time The operation of a device or system in such a way that it responds to inputs without perceptually significant delay.

resampling The operation of changing the sampling rate of a digital signal by an arbitrary amount, which is necessary in *pitch-shifting*.

sampler A device designed to create musical sounds by digitally manipulating the pitches and durations of recorded sounds stored in a waveform memory.

score file The name given to the input file of an *acoustic compiler*; a score file typically consists of *instrument* and *stored function* definitions, followed by a list of notes to be played on the defined instruments.

sinc function The impulse response of an ideal low-pass filter.

sound encoding (lossless) The body of techniques typically used to reduce the number of bits needed to represent a sound signal in such a manner that the original signal can be recovered perfectly.

sound encoding (lossy) The body of techniques typically used to reduce the number of bits needed to represent a sound signal by discarding its perceptually irrelevant or unnecessary features. Because some information is discarded, the original signal cannot be perfectly reconstructed from its encoded form.

sound spatialization The use of technological means to create the impression that one or more sounds come from specified, possibly time-varying locations in a "virtual" acoustic space surrounding the listener.

stored functions The name given to data tables used to hold control and/or waveform data in an *acoustic compiler*.

subtractive synthesis The process of synthesizing sounds by passing time varying complex sounds through filters with time-varying characteristics.

table lookup oscillator The primary synthetic signal source in an *acoustic compiler*: it can typically output an arbitrary waveform (specified in a *stored function*) at an arbitrary amplitude, frequency, and phase.

unit generator The name given to subprograms that implement specific signal-processing functions (such as oscillating, filtering, or adding) in an *acoustic compiler*.

vocoder An abbreviation meaning *voice coder*, based on the analysis of speech sounds into many simultaneous frequency-spaced channels of information, with the information in each channel typically much simpler than the analyzed sound.

virtual acoustics Simulated acoustics that support the auditory aspect of virtual reality.

virtual reality Computer-based simulation of the real world by perceptual means intended to minimize the apparent artificial nature of those means.

REVIEW QUESTIONS

1. Define computer music.

2. What is real time?

3. How can a computer compose music?

4. What is an acoustic compiler? A unit generator?

5. What feature of an acoustic compiler corresponds to patch cords?

6. What increment value needs to be fed into a table lookup oscillator in order to obtain 440 Hz (assume a standard CD sampling rate and a wavetable length of 4096 samples)?

7. How many notes can an acoustic compiler produce at one time?

8. What computer music composer experimented with tones that rise and fall in pitch simultaneously?

9. Define additive synthesis.

10. How may the data for additive synthesis be reduced?

11. How can vocoders reduce data?

12. How does the phase vocoder allow the pitch to be changed without changing the duration of a sound?

13. What type of synthesis method uses an excitation-resonating model?

14. What part of a speech signal is most predictable? Least predictable?

15. How is linear prediction related to all-pole filtering?

16. If an amplifier clips, is it operating in its linear or non-linear mode? Why?

17. How are the frequency components in a frequency-modulated waveform arranged?

18. How can we determine the strength of the eight sideband pairs in an FM waveform?

19. Can FM produce inharmonic spectra? If so, how? If not, why not?

20. What Chebyshev polynomial is associated with the third harmonic?

21. For a delay length of 100 samples and a sampling rate of 10,000 Hz, what frequency would the Karplus-Strong plucked string algorithm produce?

22. Why do pitches produced by the Karplus-Strong plucked string algorithm become more out of tune as they get higher?

23. What is a keyboard split?

24. What is a polyphonic synthesizer?

25. How many channels can a basic MIDI interface control?

26. What is the difference between MIDI OUT and MIDI THRU?

27. What causes the chipmunk effect?

28. What complicates the looping process in a sampler?

29. What is a better alternative to linear interpolation for pitch shifting?

30. What are Max and Pd?

31. Name as many cues for sound localization as you can.

32. What is an HRTF?

33. Is MP3 a form of lossless or lossy encoding?

34. What is a Huffman code?

35. Assuming a 48,000-Hz sampling rate, what is the bandwidth of each filter bank channel for in MP3 encoding?

36. Do all MP3 encoders produce the same result?

37. How are wireless information appliances interconnected?

QUESTIONS FOR THOUGHT AND DISCUSSION

1. In what ways are current virtual-reality simulations limited, and how might those limitations be overcome?

2. Does the problem of programming a computer to compose music differ in any significant way from the general problem of programming computers to be artificially intelligent? If so, how?

3. What are the similarities and differences, if any, between subtractive synthesis and physical modeling synthesis?

4. What are the relative advantages and disadvantages of playing spatialized sounds over earphones versus loudspeakers?

EXERCISES

1. Using the information given in Fig. 29.1, devise a method for composing music with the same statistics as the Mozart melody shown. Explain how the resulting music is similar to Mozart's.

2. How many different ways can you think of to implement the *LOOKUP* function shown in Fig. 29.2? Describe their relative advantages and disadvantages.

3. At what musical tempo (dotted quarter notes per minutes) will the sound synthesized by the score in Fig. 29.3 be played?

4. How could the rising-tone acoustical illusion shown in Fig. 29.4 be modified to obtain a tone that goes up and down in pitch at the same time?

5. If the modulating and carrier frequencies of an FM wave are the same, at what frequencies will the sidebands be located? What if the modulating frequency is twice the carrier frequency?

6. In Fig. 29.17, assuming the signal comes from a CD player, what is the frequency of the waveform shown?

7. Estimate the value of $x(n + \Delta)$ as shown in Fig. 29.18 by performing the convolution as shown.

8. Assuming an ambient temperature of 20°C and 50% relative humidity, by how much would the sound of a 4000-Hz triangle be attenuated if it were 250 m away? If the original note had a loudness of 10 sones at 1 m, would it still be audible?

9. Devise a Huffman code for six equally likely values. What is its average code length?

10. Assuming that two sound events have to be separated by more than 30 ms for them to be perceived as occurring at different times, how many notes would you have to play at once on a MIDI synthesizer before they don't all seem simultaneous? Could the nonsimultaneous nature of sounds separated by less than 30 ms be perceived in another manner?

11. Suppose a given digital music sampler stores sounds at a rate of 48,000 samples per second and that a piano is represented by a $\frac{1}{2}$-s recording of every third note on its keyboard. About how many bytes of storage would such a sound library take? Could the same recordings be used to represent piano sound played with the *una corda* ("soft") pedal depressed? If not, why? If so, how?

12. Estimate how far away from the listener a musical instrument has to be before air absorption of sound plays a perceptually significant effect.

13. When a sound source is traveling toward you, its pitch is shifted upward due to Doppler shift. How fast does a sound source have to be moving before you would notice the pitch shift, assuming the pitches of a moving and stationary source are sounded (a) sequentially, and (b) simultaneously?

EXPERIMENTS FOR HOME, LABORATORY, AND CLASSROOM DEMONSTRATIONS

Home and Classroom Demonstrations

1. *Computer music*   Compare the performance of *Bicycle Built for Two* from the 1968 movie *2001: A Space Odyssey* and the 1963 Decca recording from Bell Telephone Laboratories entitled *Music from Mathematics*.

2. *Computer-composed music*   Compare the computer-composed music on the 1950s recording of *The Illiac Suite* by Lejaren Hiller and the 1990s recording *Bach by Design* by David Cope.

3. *Computer sound synthesis*   Using one or more personal computers and any convenient sequencer or notation software, compare the tone qualities of sounds produced by wavetable synthesis and FM synthesis. What are their relative similarities and differences?

4. *Physical modeling*   Write and execute a simple computer program implementing the Karplus-Strong plucked string algorithm. Verify its operation by listening to the result(s).

5. *MIDI*   Using one or more personal computers and any convenient sequencer or notation software, compare the tone qualities of sound produced by two or more different sound cards implementing General MIDI Mode (GMM). What are their relative similarities and differences?

6. *Looping*   Use any sampler with adjustable setting to demonstrate good and poor choices for looping points.

7. *Pitch shifting*   Write a simple computer program to demonstrate the difference between good and poor interpolation strategies for pitch shifting, including (a) no interpolation, (b) linear interpolation, and (c) convolution-based interpolation for 4, 6, 8, and 16 points.

8. *Spatialization*   Demonstrate and discuss the differences among different hall settings on a receiver that has sufficient documentation to allow the audible differences to be correlated with different reverberation algorithms.

9. *Binaural recording*   Obtain and compare various binaural recordings and listen to them over various types of earphones and loudspeakers.

10. *MP3 encoding*   Listen to and compare the same music encoded by as many different MP3 encoders as possible, discuss their differences, and try to determine the causes for the differing results.

Laboratory Experiment

A laboratory experiment can be made up using several of the above demonstration experiments, depending upon equipment available.

PART
VIII

Environmental Noise

Noise has been receiving increasing recognition as one of our critical environmental pollution problems. Like air and water pollution, noise pollution increases with population density; in our urban areas, it is a serious threat to our quality of life. Noise-induced hearing loss is a major health problem for millions of people employed in noisy environments. Beside actual hearing loss, however, human beings are affected in many other ways by high levels of noise. Interference with speech, interruption of sleep, and other physiological and psychological effects of noise have been the subject of considerable study recently.

Finding the technical solutions to many of our environmental noise problems requires the work of scientists and engineers with considerable knowledge of acoustics. On the other hand, many problems require social and political action rather than technical solutions. Thus it is important the the public be well informed about the basic principles of acoustics and noise control.

Chapters 30, 31, and 32 are intended as an introduction to environmental noise for the intelligent layperson, perhaps for future legislators or other governmental officials. Chapter 30 provides an introduction to the subject, Chapter 31 describes the way in which noise affects people, and Chapter 32 deals with the control of noise through both regulations and technical solutions.

30

Noise in the Environment

Music and speech are not the only contributions humans have made to the world of sound. With few exceptions, advances in technology, such as the development of labor-saving machines, have resulted in a steady increase in the amount of unwanted sound, which we call *noise*.

In this chapter, you should learn:

■ About various ways for rating noise levels;

■ How sound propagates outdoors;

■ About flow noise and machinery noise;

■ About motor vehicle noise and railroad noise.

If we try to think of all possible sound generators, we will probably find that they fall into one of the following categories:

1. Vibrating solid bodies, such as bars, membranes, plates, loud-speaker cones, etc.;

2. Vibrating air columns, such as those in musical instruments;

3. Flow noise in fluids due to turbulence, such as that which occurs in jet engines or from air leaking out of an air hose;

4. Interaction of a moving solid (such as a rotating propeller blade) with a fluid, or a moving fluid with a solid (such as air flowing in a duct or though a grill);

5. Rapid changes in temperature or pressure, such as the thunder caused by a lightning discharge or a chemical explosion;

6. Shock waves caused by motion or flow at supersonic speed.

In this chapter we will discuss some common noise sources that incorporate sound generators in one or more of these categories.

30.1 ■ SOUND POWER AND MECHANICAL POWER

Fortunately for the environment, even the noisiest machines convert only a small part of their total energy into sound. A modern jet aircraft, for example, may produce a kilowatt or more of acoustic power, but this is less than 0.02% of its mechanical output of 55,000 kW (Shaw 1975). Automobiles emit approximately 0.001% (one-thousandth) of their power as sound, and household appliances are comparable to this in "efficiency" as noise sources.

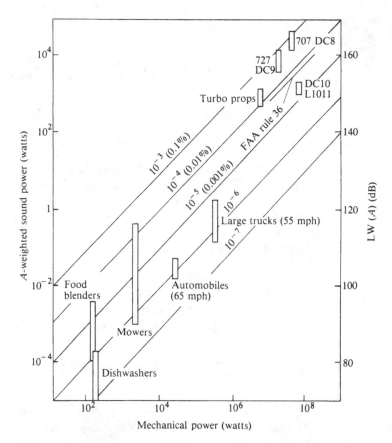

FIGURE 30.1
Sound power
compared to
mechanical power
for various
machines. Also
shown are the
corresponding
A-weighted sound
power levels
$L_w(A)$. (After
Shaw, 1975.)

Figure 30.1 compares the A-weighted sound power (see Section 6.2) of a number of large and small noise sources to their mechanical power. The appropriate values for the sound power level L_w are also given. The diagonal lines represent different percentages of mechanical power converted to sound power. The line labeled "FAA Rule 36" refers to the Federal Aviation Administration regulation on noise for new aircraft certified after 1969.

30.2 ■ NOISE LEVELS

The sound pressure level L_p decreases as we move away from the sound source. In a free field (which may exist outdoors or in an anechoic room), the sound pressure level decreases by 6 dB each time the distance from the source doubles (see Section 6.2). Thus, the sound pressure level one or two meters from a small appliance may exceed that due to jet aircraft at a distance of 10 km.

The most commonly used measure of noise is the A-weighted sound pressure level $L_p(A)$, which deemphasizes sounds of low frequency (see Fig. 6.5). Sound-level meters are nearly always equipped with a weighting network, so that the value of $L_p(A)$ can be read directly. A-weighted levels are frequently expressed as dB(A) or dBA.

Because the sound pressure level at a given location may fluctuate at various times, the use of some averaging procedure is desirable. One way to describe noise is to express

L_x, the level exceeded x percent of the time, for different values of x. Thus, L_{10} is the A-weighted sound level exceeded 10% of the time; L_{50} is the median level, since it is exceeded 50% of the time, and so on.

Another useful measure of noise is the *energy-equivalent level*, or *equivalent level L_{eq}*. It is defined as the decibel level of the steady noise that would give the same total energy over the same time period. For example, a single peak at 100 dB lasting 3.6 s is equivalent in energy to a whole hour of steady noise at 70 dB. The advantage of using L_{eq} rather than L_{50} or L_{90}, for example, in the description of noise is its sensitivity to short peaks of noise, which can be very annoying. Because noise that occurs at night is usually more annoying, a *day-night equivalent level L_{dn}* is frequently used to describe noise. In arriving at L_{dn}, one adds 10 dB to the sound level between 10:00 P.M. and 7:00 A.M. in the averaging process.

Still other noise levels may be found in the literature on noise control. Perceived noise level PNL (expressed in PNdB) and effective perceived noise level EPNL (expressed in EPNdB) are two criteria frequently used to describe aircraft noise. EPNL takes into account both the maximum loudness level and the duration of an event, such as an aircraft flyover. The noise-exposure forecast NEF, derived from EPNL, also takes into account the frequency of the events in a given neighborhood. Various noise levels are summarized in Table 30.1.

30.3 ■ SOUND PROPAGATION OUTDOORS

Although sound propagation outdoors may approach that of a free field when conditions are ideal, this rarely takes place in actual practice. Atmospheric turbulence, temperature and wind gradients, molecular absorption in the atmosphere, and reflection from the earth's surface all affect propagation and cause fluctuations in the sound intensity level at the receiver.

Attenuation is strongly influenced by the type of ground cover present. The attenuation of noise through a dense forest may be as great as 20 dB per 100 m. The attenuation through thick grass and shrubbery may be even greater. Mounds of earth placed beside highways are effective noise barriers. Depressing a highway by about 4 m typically reduces the sound level by 7 to 10 dBA.

TABLE 30.1   Noise levels used to rate noise.

$L_p(A)$	A-weighted sound pressure level
L_{10}	A-weighted sound level exceeded 10% of the time
L_{50}	A-weighted sound level exceeded 50% of the time
L_{90}	A-weighted sound level exceede 90% of the time
L_{eq}	Sound pressure level of steady noise that would give the same total energy as the noise being rated
L_{dn}	The same as L_{eq} with 10 dB added to noise measured between 10:00 P.M. and 7:00 A.M.
PNL	Perceived noise level (used to rate aircraft noise)
EPNL	Effective perceived noise level (used to rate aircraft noise) takes into account both maximum loudness and duration of noise events

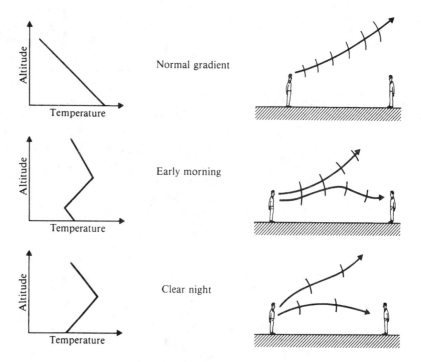

FIGURE 30.2
Refraction of sound
under different
conditions

Refraction, or the bending of sound by temperature and wind gradients, was described in Section 3.9. There it was shown that sound could be bent up away from the receiver (leading to attenuation in excess of the free field condition—a 6-dB loss when the distance from the source is doubled), or sound can be bent down toward the receiver, resulting in attenuation that is less than the free-field condition. Normally, the temperature decreases with altitude; thus there is an upward refraction, since sound travels faster in the warm air near the surface of the earth. Two examples of temperature inversion that will cause downward refraction are illustrated in Fig. 30.2. Mining companies, for example, carefully monitor the weather conditions to decide on the best time for blasting operations in order to minimize noise levels in surrounding communities.

Sound waves are weakly absorbed by the atmosphere itself. Atmospheric attenuation occurs mainly at high frequency, and depends rather strongly on humidity and temperature. Atmospheric attenuation in three different octave bands is shown in Fig. 30.3.

Effect of the Ground

When broadband sound propagates over the ground, certain frequencies are attenuated more than others. This gap in the spectrum can be heard, for example, when listening to the sound of a jet engine on the ground some distance away. Generally the gap occurs between 250 and 800 Hz, and the depth and width of the gap depends upon the nature of the ground cover.

Suppose the ground is hard, so that it is a good sound reflector. Interference occurs between the direct sound and the reflected sound, whose phase is reversed upon reflection

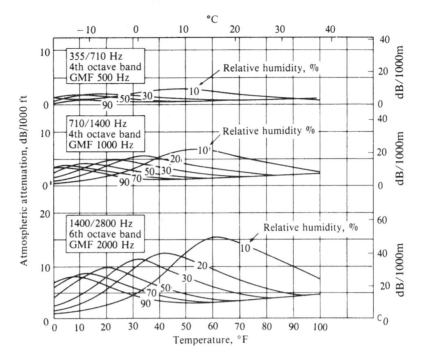

FIGURE 30.3
Atmospheric
attenuation of
sound at 40, 1000,
and 2000 Hz.
(From *Noise and
Vibration Control*,
revised. ©1988
Leo Beranek. Used
with author's
permission.)

(See Chapter 3). At certain frequencies this interference is destructive, and considerable
cancellation occurs. Typical paths are shown in Fig. 30.4, along with the excess attenuation
to be expected, in excess of the attenuation due to atmospheric absorption. The frequencies
at which this attenuation occurs depends upon the height of the source and the receiver
above the hard surface. This is not unlike the comb filtering experienced due to a hard
surface near a microphone.

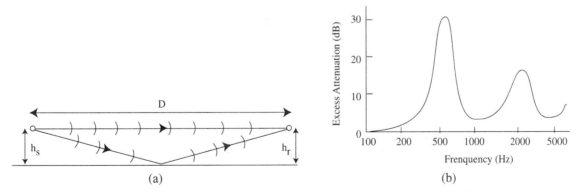

FIGURE 30.4   Excess attenuation due to interference between the direct sound and the sound reflected by a hard surface.
The frequency at which maximum attenuation occurs depends upon the height of the source h_s, the height of the receiver h_r,
and the distance D between them

If the surface is not highly reflecting, the situation is further complicated by the existence of a surface wave, a ground wave, or both. The excess attenuation due to the ground can reach 30 dB or more at certain frequencies (Piercy, Embleton, and Sutherland 1977). A porous ground surface can result in sound absorption and also in a change in the speed of sound due to sound penetration into pores. Predicting the access attenuation becomes quite difficult.

Outdoor Sound Propagation in the U.S. Civil War

An interesting example of the dramatic effects of outdoor sound propagation is related in a book *Trial by Fire: Science, Technology and the Civil War* (Ross 1999). Before electrical and wireless communications became available, the sound of battle was often the quickest and most efficient method by which a commander could judge the course of a battle. Troop dispositions were often made based on the relative intensity of the sounds from different locations on the battlefield. Acoustic shadows due to atmospheric absorption, wind shear, temperature inversions, ground effects, or all of these apparently influenced several key battles in the U.S. Civil War.

At the battle of Gettysburg, for example, Confederate General Ewell was apparently unable to hear the artillery of General Longstreet and hence did not move his troops. As a result, Union General Meade was able to shift his troops in the nick of time to defeat Longstreet's attack. On the previous day, Meade had been unable to hear the Gettysburg fighting from his position at Taneytown (12 mi away), yet the battle was clearly audible in Pittsburgh, 150 mi from Gettysburg (Ross 1999).

30.4 ■ FLOW NOISE

Noise associated with gas flow may be of three general types; an example of each type is shown in Fig. 30.5. The first type occurs when the flow of air is interrupted at a regular rate, as in a siren, giving rise to what is called a *monopole* source. The sound power radiated from such a source varies as the fourth power of the flow velocity. A second type of noise is generated when a moving stream of air strikes a solid object, forming a *dipole* source. The noise of air blowing through a grill, the sound generated by a moving fan blade, and the aerodynamic whistles described in Section 12.10 are examples of this type of noise. Dipole

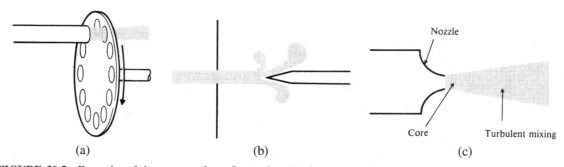

FIGURE 30.5 Examples of three types of gas flow noise; (a) siren (monopole source); (b) edgetone generator (dipole source); (c) gas jet (quadrupole source).

radiation is even more dependent on flow velocity, the power varying as the sixth power of velocity (v^6). Finally there is *quadrupole* radiation, which is weak at ordinary flow velocities but increases with v^8, so that at high flow velocities, it becomes appreciable. The noise from a gas jet (as in a jet airplane engine) is of this type.

From the strong dependence of flow noise on velocity, we can see that the best way to minimize flow noise is to utilize high-volume, low-velocity flow whenever possible. Doubling the flow velocity increases the strength of monopole sources 16 times, that of dipole sources 64 times, and that of quadrupole sources 256 times.

The flow of air is often maintained by fans, which are common sources of noise. Fan noise is of two types: rotation noise and aerodynamic noise. Rotation noise, common to all rotating machines, will be discussed in Section 30.5. Aerodynamic noise is generated by swirls of air, or *vortices*, caused by the blades moving through the air. Aerodynamic noise is amplified if there are stationary objects or guide vanes nearby with which the vortices interact.

Fans are of two types: axial and centrifugal (see Fig. 30.6). Axial fans have propeller blades that push air in the direction determined by their pitch. Centrifugal fans spin the air, and centrifugal force causes an outward flow of air that is funneled into the discharge duct. Centrifugal fans are noisiest at low frequency, which is less annoying. Axial fans, on the other hand, produce a more prominent noise, because the aerodynamic noise covers a wider range of frequency.

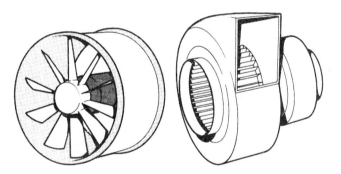

FIGURE 30.6
Axial fan and
centrifugal fan.

30.5 ■ MACHINERY NOISE

In Section 30.1, we pointed out that the sound power output of various machines is typically from 10^{-7} to 10^{-3} of the corresponding mechanical power output. Although these are small fractions the sound power from the large number of machines found in a factory (or even a home) can create a noisy environment. In order to control environmental noise, we must understand the main noise sources in machines.

Possible noise sources in machines include unbalance in rotating parts, friction, bearing noise, gear noise, hum induced by magnetic fields, fan noise, impact noise, and turbulence. The noise source itself may not emit much airborne sound, but it may act as an energy source for vibrations that are transmitted through the structure to be emitted as sound by some other part of the machine. This is especially noticeable if the driving source vibrates at the natural resonance frequency of a large radiating members, such as a metal cover panel. Airborne noise is usually caused by a vibrating surface or by turbulent air.

Noise reduction in machines may be reduced by one or more of the following courses of action: reducing the noise generated by the source; decoupling the source from its resonators; damping the resonators; reducing the noise radiation from a radiating source or resonator. The best course of action frequently cannot be determined until the main noise sources have been located through acoustical testing.

Often the frequency of a prominent noise gives away its origin. Magnetic hum from a motor or transformer will usually occur at the power-line frequency (60 Hz) or its harmonics. Fan noise will have a strong component at the blade-passage frequency. Vibrations generated by rotating parts will be related to the frequency of rotation.

Inspection often indicates the radiation members. Sometimes they can only be determined by attaching *accelerometers*, which generate electrical signals proportional to acceleration when they vibrate. Using an accelerometer enables one to determine what surfaces are vibrating strongly at the frequency of the most prominent components in the noise spectrum.

A rectangular panel or thin plate has a large number of resonance frequencies. If it is supported (but not rigidly clamped) at its boundaries, its resonance frequencies are given by

$$f_{mn} = 0.48 v_L h \left[\left(\frac{m}{l} \right)^2 + \left(\frac{n}{w} \right)^2 \right]$$

$v_L =$ speed of longitudinal (sound) waves in a bar (m/s) (see Table 13.1);
$h =$ thickness of plate (meters);
$l, w =$ length and width of plate (meters);
$m, n =$ integers (1, 2, 3...).

If the plate is clamped at its boundaries, the lowest resonance frequency is nearly doubled, but the others are raised by a smaller factor.

Noise reduction in machines is most effective when applied at the source. This may involve balancing rotating parts, repairing or redesigning bearings, replacing gears, etc. Isolation of the source from resonators may be accomplished by installing resilient mountings or even by enclosing a noisy part in a housing. Panels that radiate noise can be damped by covering them with special materials having a high internal friction.

30.6 ■ INDOOR NOISE

It is appropriate at this time to review Chapter 23 and 24, in which we discussed the acoustics of rooms. Let us for the moment look at Section 24.2, in which we focused on direct and reverberant sound fields, and in particular on Fig. 24.2, which shows the relationship between sound power level L_w and sound pressure level L_p in different types of rooms. This chart is useful in determining the amount of sound reinforcement needed in an auditorium, but it is also useful in determining the sound level to be expected from a noise source indoors.

The direct sound field for a nondirectional source ($Q = 1$) follows the $1/r$ relationship characteristic of a free field (see Section 6.2), and this is represented by the $R = \infty$ line in Fig. 24.2. The level of the direct sound decreases by 6 dB for each doubling of the distance, once we have moved out of the near field of the source. The sound level in a room is found to level off at a constant value as we move away from the source, and this is called the reverberant level. We have assumed sufficient reflecting surfaces in the room to distribute the reverberant sound eventually throughout the room; in many rooms, this will not be the case. In furnished residential rooms, for example, the rate of decrease appears to be 3 to 4 dB per doubling of distance throughout a large part of the room; in other words, there is a blending of the direct and reverberant fields. Similar behavior is noted in large, low rooms with distant sidewalls (Schultz 1980).

In order to describe completely the strength of a noise source, both the *sound power level* (or sound power) and the *directivity* must be known at all frequencies. In Eqs. 24.1 and 24.2, the directivity is expressed by a single value Q, which is the ratio of the sound intensity in front of a source to that averaged in all directions. In the case of a complex source such as a large machine, this is insufficient.

Ideally, the sound power level of a source should be measured in a reverberation room and the directivity in an anechoic room. Often this is not practical. Many sophisticated methods have been devised for determining both sound power level and directivity in real rooms. They usually involve considerable computation based on a number of samples of the sound field (Beranek 1971).

30.7 ■ MOTOR VEHICLES

With more than 100 million passenger cars in the United States traveling over 10^{12} miles annually, automobiles generate megawatts of acoustic power. Fortunately, much of this power is radiated to areas with low population density, but an appreciable portion is generated in urban areas. Figure 30.7 shows typical noise levels for automobiles, trucks and buses measured 50 feet from the highway as well as inside the vehicle itself. Sources of noise in an automobile or truck include the engine, the cooling fan, the drive train, the tires, aerodynamic turbulence, body vibrations, and the intake and exhaust systems.

The modern automobile, when kept in good repair, is a relatively quiet vehicle at low speeds. Engine noise and aerodynamic noise are low at these speeds, and most of the

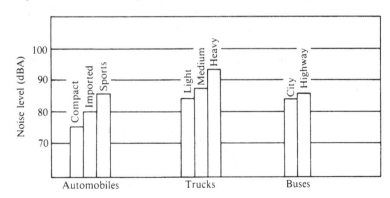

FIGURE 30.7
Typical noise level for motor vehicles measured at 50 feet from the highway.

noise is radiated from the exhaust system. Engine noise and aerodynamic noise increase with speed, however, and at high speeds tire noise also becomes an important factor. The average sound level, measured 50 ft from the center of the roadway, rises about 10 dB for each doubling of speed.

Although some work is being done on higher-performance mufflers, quieter tire treads, quieter fans, etc., for automobiles, it is unlikely that any great reduction in noise emission can be expected. The greatest problem with respect to automobile noise appears to be poor maintenance of equipment, especially muffler and exhaust systems.

Trucks generate about 10 times as much mechanical power as do automobiles, but they emit from 10 to 100 times as much acoustic power. Thus, they present a much more serious problem so far as road noise is concerned. Sources of truck noise are roughly the same as those of automobile noise, but the proportions assigned to each source are different. At low speeds, exhaust and fan noise are most important. At high speed, however, tire noise takes over completely. Tire noise at 55 mi/h for a single-chassis truck ranges from 75 to 95 dBA, depending on the design of the tread, as shown in Fig. 30.8.

The main noise source in motor vehicles is the combustion or explosion of the fuel-air mixture inside the cylinders. Fortunately, this very powerful noise source is buried deep inside the massive engine, and therefore is well attenuated. Some of the energy of combustion does appear as noise, however, due to vibration of the entire engine as well as individual parts. Furthermore, when the exhaust and intake valves open, loud sound of short duration are emitted, especially in the exhaust system, since the exhaust valve opens when the cylinder pressure is still quite high.

Diesel engines are noisier than gasoline engines, because the combustion is more sudden so that cylinder pressure rises more abruptly. Furthermore, a diesel engine produces nearly as much noise under no load as it does under full load.

Engine cooling fans produce a substantial amount of noise unnecessarily. Fans must be large enough to cool the engine in a stationary vehicle with the engine running at low speed. When the vehicle is moving on the highway, there is little need for forced air flow. Furthermore, fan noise increases rapidly with speed, and at high speeds it can be as noisy as the engine itself. Substitution of an electrically driven fan or use of a magnetic clutch

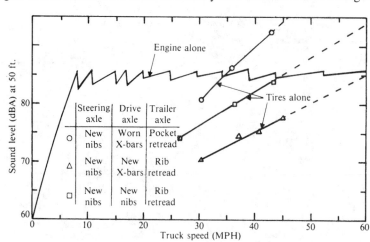

FIGURE 30.8
Engine and tire noise at various speeds for an 18-wheel tractor-trailer. Note that the truck has shifted gears 11 times. (From Close and Atkinson, 1973. Reprinted by permission of *Sound and Vibration*.)

so that the fan operates only when needed not only reduces noise but results in a saving of fuel.

Although regulation of motor vehicle noise is generally left to state and local governments, the Federal government does regulate the noise of new trucks, motor homes, motorcycles, and mopeds. Several states and cities set limits for passenger cars. Pass-by noise measurements generally follow procedures defined by the Society of Automotive Engineers in the appropriate standards (SAE J986). For example, noise from passenger cars is measured 15 m from the roadway center and 1.2 m above the ground while the vehicle is undergoing maximum acceleration, and the car is to attain maximum rated engine speed not more than 45 m past the microphone. Most states set the pass-by limit at 80 dBA. Noise limits in Europe are somewhat lower (Hickling 1997).

30.8 ■ RAILROADS AND AIRCRAFT

Noise from railroad operations is not as widespread as noise from highways. Nevertheless, for persons living near a railroad, it can be an annoyance. Major sources of noise are locomotives, rail-wheel interaction, whistles and horns, yard retarders, refrigerator cars, maintenance operations, and loading equipment. A recording of a freight train passby is shown in Fig. 30.9.

The Environmental Protection Agency has proposed the following limits on locomotive noise measured 100 ft from the track:

1. 96 dB(A) for locomotives manufactured before 1980;
2. 90 dB(A) for locomotives manufactured in 1980 or after;
3. 73 dB(A) when idling (70 dBA for manufacture after 1980).

Railcars (or combinations of them) may not emit more than:

1. 88 dB(A) at speeds of up to 45 mi/h;
2. 93 dB(A) at speeds above 45 mi/h.

Some of the possible ways in which railroad noise can be reduced include equipping diesel locomotives with mufflers (an approximate 6-dB reduction), use of welded rails

FIGURE 30.9
Sound level of a freight train passby. (From Close and Atkinson, 1973. Reprinted from permission of *Sound and Vibration*.)

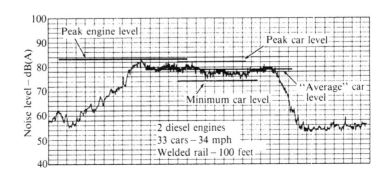

(an approximate 3-dB reduction), careful maintenance of rolling stock, and barrier walls around retarders in rail yards (up to a 20-dB reduction).

Aircraft noise is discussed in Sections 32.10 and 32.11.

30.9 ■ SUMMARY

Noise has been receiving increasing recognition as one of our critical environmental problems. Various machines have sound power outputs that are only 10^{-7} to 10^{-3} of their mechanical power output. Nevertheless, this results in a substantial noise level. Various noise levels take into account the frequency of occurrence of annoying sounds as well as their levels. Atmospheric turbulence, temperature and wind gradients, atmospheric absorption, and reflections all affect the propagation of sound outdoors.

Noise associated with gas flow may be described as originating from monopole, dipole, or quadrupole sources. Machinery noise can originate in a number of places, the airborne noise often being radiated by panels or other surfaces at some distance form the source. Sources of noise in motor vehicles include the engine, the cooling fan, the drive train, tires, aerodynamic turbulence, intake exhaust, and body vibrations. At high speeds, tire noises dominate, especially in the case of trucks. The EPA has written noise standards for both trucks and railroads.

REFERENCES AND SUGGESTED READINGS

Beranek, L. L., ed. (1971). *Noise and Vibration Control.* New York: McGraw-Hill.

Close, W. H., and T. Atkinson (1973). "Technical Basis for Motor Carrier and Railroad Noise Regulations," *Sound and Vibration* **7**(10): 28.

Hickling, R. (1997). "Surface Transportation Noise," in *Encyclopedia of Acoustics,* ed. M. J. Crocker, New York: John Wiley, Vol. 2, p. 1073.

Ingard, U. (1953). "The Physics of Outdoor Sound," *Proc. 4th National Noise Abatement Symposium.* (Reprinted in *Environmental Noise Control: Selected Reprints,* ed., T. D. Rossing, Am. Assn. Phys. Teachers, Stony Brook, N.Y., 1979).

Miller, T. D. (1976). "Machine Noise Analysis and Reduction," *Sound and Vibration* **1**(3): 8.

Piercy, J. E., T. F. W. Embleton, and L. C. Sutherland (1977). "Review of Noise Propagation in the Atmosphere," *J. Acoust. Soc. Am.* **61**: 1403.

Ross, C. D. (1999). *Trial by Fire: Science, Technology and the Civil War.* Shippensburg, Penn.: White Mane Publishing Co. See also "Outdoor Sound Propagation in the U.S. Civil War," *ECHOES* **9** (Winter).

Rossing, T. D. (1978). "Resource Letter ENC-1: Environmental Noise Control," *Am. J. Phys.* **46**: 444.

Schultz, T. J. (1980). "The Relation Between Sound Power Level and Sound Pressure Level Outside of Laboratories," *Noise Control Eng.* **14**: 24.

Shaw, E. A. G. (1975). "Noise Pollution—What Can Be Done?" *Physics Today* **28**(1): 46.

GLOSSARY

accelerometer A device used to measure vibration; its electrical output indicates its acceleration (see Section 1.6).

aerodynamic noise Noise generated by moving air or by the flow of air around a moving body (such as an automobile or truck).

day-night level L_{dn} Equivalent sound level that adds 10 dB to sounds that occur during the night-time hours of 10:00 P.M. to 7:00 A.M..

dipole source A noise source in which two halves are vibrating in opposite phase (e.g., two loudspeakers connected in opposite polarity, or the air at the leading and trailing edges of a fan blade).

equivalent level L_{eq} Sound pressure level that would give the same total energy as the noise being described.

monopole source A noise source in which the entire radiating surface vibrates in phase.

quadrupole source A noise source in which four parts vibrate alternately in phase.

refraction The bending of waves when the velocity changes (due to temperature and wind gradients, for example: see Section 3.9).

REVIEW QUESTIONS

1. Identify six categories of sound generators.

2. Which converts a higher percentage of mechanical power to sound power: automobiles, trucks, or airplanes?

3. What is meant by L_{10}? By L_{eq}?

4. What are several factors that influence the propagation of sound outdoors?

5. Explain how temperature inversion can increase the noise level at a point.

6. How did sound propagation affect the U.S. Civil War?

7. Give examples of monopole, dipole, and quadrupole sound sources.

8. What is an accelerometer?

9. What is the main source of truck noise at low speeds? At high speeds?

10. What is the maximum passby noise from passenger cars in most states?

11. What is the maximum noise level 100 ft from the tracks for locomotives manufactured after 1980?

12. What are some ways that railroad noise can be reduced?

QUESTIONS FOR THOUGHT AND DEMONSTRATION

1. Far more people are affected by truck noise than by airplane noise. Why are there fewer complaints, citizens' protest groups, etc., directed against highway noise?

2. Try to think of other examples of monopole, dipole, and quadrupole sound sources.

3. Describe the noise of a vacuum cleaner, and try to identify the two chief noise sources.

EXERCISES

1. How many watts of sound power and mechanical power are delivered by a typical automobile?

2. Compare the attenuation at 500 Hz and 2000 Hz for the following atmospheric conditions.
   (a) 70°F, 10% relative humidity
   (b) 0°C, 30% relative humidity.

3. Find the lowest resonance frequency of a steel panel ($v = 4905$ m/s) with dimensions $l = 1.5$ m, $w = 1.0$ m, and $h = 10^{-3}$ m.

4. Suppose that the velocity of air flow through a grill is cut in half but the grill area is doubled in order to maintain the same volume flow.
   (a) By what factor will the sound power be reduced?
   (b) By how many decibels will the sound power level of the source decrease?

(c) Will the sound level in the room change by the same amount?

5. (a) If a jet aircraft flying at an altitude of 10 km emits 1 kW of sound power, what will the sound level be on the ground? (Assume a free field; see Sections 6.2 and 24.2 and Eq. 24.1.)

   (b) If we add atmospheric attenuation of 30 dB/1000 m (see Fig. 30.3), what will the level be at the ground?

6. (a) If an automobile traveling at 60 mi/h emits 0.01 W of acoustical power (Fig. 30.1), estimate the average continuous power from 100 million automobiles, each of which travels 10,000 mi per year.

   (b) Make an estimate of the peak acoustical power. (How many automobiles might be traveling at a peak hour of the day?)

EXPERIMENTS FOR HOME, LABORATORY, AND CLASSROOM DISCUSSION

Home and Classroom Demonstrations

1. *Monopole, dipole, and quadrupole sources in a duct*   A wooden box is designed to fit snugly into a length of large ventilation duct, as shown. Four small loudspeakers are mounted in the box with four reversing switches so that they can form a monopole, a dipole, or a quadrupole source. Note the difference in the sound transmitted down the duct from the three different source types.

2. *Monopole, dipole, and quadrupole sources in a free field* Repeat Demonstration 1 without the duct. Note the decrease in low-frequency radiation in the dipole and especially the quadrupole configurations. (Radiated power from a monopole source varies as f^2, from a dipole source as f^4, and from a quadrupole source as f^6.)

3. *Feedback noise*   The Rijke tube illustrates how feedback from a resonator to a source results in tonal noise. The Rijke tube has one or more layers of wire gauze at about one-fourth of its length to produce turbulence. Heat gauze cherry red, and the tube sings when removed from the flame (as long as it is near vertical). To amuse the class, sound can be "poured out" into a bucket and then "poured back" into the tube.

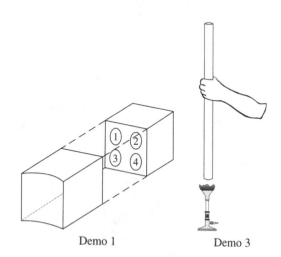

Demo 1   Demo 3

Laboratory Experiments

Noise ratings for small appliances (Experiment 40 in *Acoustics Laboratory Experiments*)

Monopole, dipole, and quadrupole sources (Experiment 41 in *Acoustics Laboratory Experiments*)

Sound power level of a stationary noise source (Experiment 43 in *Acoustics Laboratory Experiments*)

Measurement of vibration: the accelerometer (Experiment 45 in *Acoustics Laboratory Experiments*)

CHAPTER

31

The Effects of Noise on People

Noise affects people in many ways. In addition to causing temporary and permanent hearing loss, noise interferes with speech communication, interrupts sleep, reduces human efficiency, and is believed to produce other physiological and psychological effects.

It is interesting to note that many of these important effects on community mental health have only recently received careful study. Thus it is understandable that there are differences of opinion about the seriousness of some of the more subtle physiological and psychological effects. Differing opinions have also been expressed about the levels of noise that can be tolerated without adverse effect on community health and welfare.

In this chapter you should learn:

- About temporary and permanent hearing loss;
- About audiometers and audiograms;
- About ear damage due to noise;
- About hearing protectors;
- About psychological and physiological effects of noise.

31.1 ■ TEMPORARY HEARING LOSS

The ear has a remarkable ability to recover from conditions of overload if they are of short duration. Thus it is important to distinguish between noise-induced temporary threshold shift (NITTS or TTS) and noise-induced permanent threshold shift (NIPTS). Exposure to loud noise for a comparatively short time causes the ear to desensitize, especially at frequencies about 4000 Hz. However, if the overload is removed soon enough, the threshold will shift back almost to its normal level. If the stimulus continues or is repeated too frequently, however, a permanent shift can occur.

The primary measure of hearing loss is the hearing threshold level, the sound level of a tone that can just be detected. The threshold level at various test frequencies is usually measured with an *audiometer*. A pure-tone audiometer, shown in Fig. 31.1, allows tones of several frequencies to be applied at selected levels to one ear at a time. A plot of the threshold level at each frequency, compared to normal hearing, is called an *audiogram*. The reference threshold levels established for normal hearing with one type of earphone are also given in Fig. 31.1. Some audiometers automatically present the test tones and record the subject's responses on an audiogram.

Intense sounds, if sustained long enough, will produce *physiological fatigue*: a threshold shift that requires up to 16 h to disappear. The recovery is generally linear in the logarithm

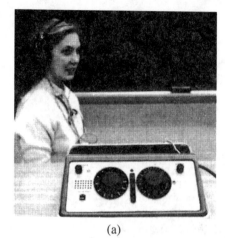

FIGURE 31.1
(a) A pure-tone audiometer.
(b) Pure-tone reference threshold levels for "normal" hearing.

Frequency (Hz)	dB (re 20 μN/m^2)
125	45
250	25.5
500	11.5
1000	7
1500	6.5
2000	9
3000	10
4000	9.5
6000	15.5
8000	13

(a)   (b)

of time. The first minute or so following intense sound generally involves a "rushing noise" tinnitus, which doesn't originate entirely in the cochlea but may involve higher centers as well (Ward 1977).

TTS that persists for more than 16 h is called *pathological fatigue* and is produced by more severe exposures. This TTS recovers linearly in time (i.e., a certain number of decibels per day) rather than in log time, suggesting a different correlate. As much as 3 weeks can be required for this delayed recovery to be complete (Ward 1977).

Temporary threshold shifts can vary in magnitude from a change in hearing sensitivity of a few decibels in a narrow band of frequencies to shifts so large that the ear is temporarily deaf. The time required for hearing sensitivity to return to near-normal levels can vary from a few hours to two or three weeks. In spite of considerable research on the subject, the laws describing temporary threshold shift have not been completely determined. Figure 31.2 shows the hypothetical growth of threshold shift after exposure to noise of varying levels and duration. These curves are based on data from several laboratories, and on extrapolations made on the basis of data from animals.

Although at this stage, curves of the type shown in Fig. 31.2 are hypothetical, they do indicate the way in which threshold shift depends on noise level and duration. For example, a shift of 34 dB, which is reached in one day at 80 dB, can be reached in about 10 min at 110 dB. These data represent a "worst case," because the noise, centered at 4000 Hz, is in the range to which the ear is most susceptible to damage; and the threshold is measured at 4000 Hz, where threshold shifts are often the greatest.

Hypothetical curves for recovery from noise-induced threshold shifts are shown in Fig. 31.3. These curves are for 7-day exposure to noise at varying levels, and hence they could be considered continuations of the curves shown in Fig. 31.2. Recovery from a given shift is more rapid if the exposure time is less than 8 h, as can be seen by the single dotted curve, which presents recovery from a 102-min exposure to noise at a level of 95 dB. For shifts above 40 to 50 dB, recovery is not always complete.

Although the physiological basis of temporary hearing loss is not yet understood completely, the following statements summarize what is known about the relationship between noise and TTS (see Ward 1969):

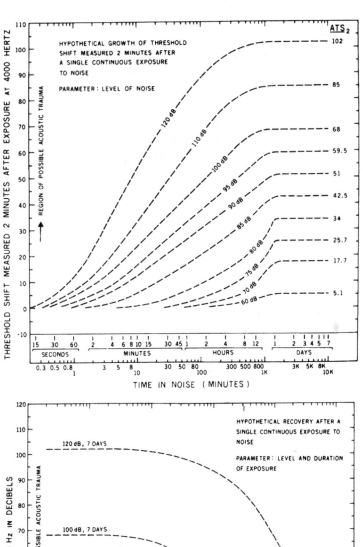

FIGURE 31.2
Hypothetical growth of temporary threshold shift due to exposure to noise of varying level and duration. (From Miller 1974.)

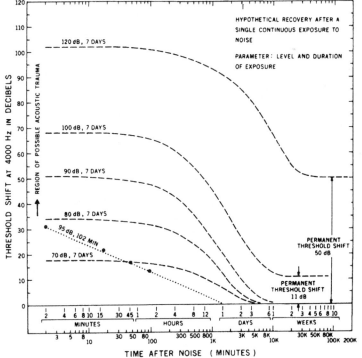

FIGURE 31.3
Hypothetical recovery from various threshold shifts. (From Miller 1974.)

719

1. The growth of TTS (in dB) is a nearly linear function of the logarithm of time. That is, doubling the exposure time tends to double the threshold shift, as can be seen in Fig. 31.2.

2. Moderate TTS recovers in time, recovery usually being complete in 16 hours. However, when TTS reaches 35 or 40 dB, recovery takes days or weeks, and above 50 dB recovery may never be complete.

3. Low-frequency noise produces less TTS than does high-frequency noise.

4. Narrowband noise produces a maximum TTS one-half to one octave above the noise band rather than in it.

5. An intermittent noise produces much less TTS than does a steady one. In fact, TTS is nearly proportional to the fraction of the time that the noise is present.

6. Neither growth nor recovery of TTS appears to be influenced by drugs, medications, time of day, hypnosis, or state of mind. Thus it appears to be entirely a physiological effect.

31.2 ■ PERMANENT HEARING LOSS

Noise-induced permanent threshold shift (NIPTS), like noise-induced temporary threshold shift (TTS), is usually greatest at frequencies around 4000 Hz. Figure 31.4(a) combined three audiograms to illustrate the progressive loss of hearing that might be found in a person who works in a noisy factory. In the early stages, hearing loss occurs mainly at frequencies between 2000 and 8000 Hz and does not interfere very much with ordinary speech (which is carried mainly by sounds in the "speech band," 300–3000 Hz). As the exposure to noise continues, loss of hearing of speech becomes more apparent. The fact that noise-induced hearing loss generally occurs first at frequencies above the speech band provides a means of detecting a dangerous condition in order to take preventive action before the ability to hear and understand speech is lost.

In analyzing noise-induced hearing loss, it is important to note that hearing becomes less sensitive with advancing age, even in the absence of damaging noise exposure. This effect, called *presbycusis*, is most prominent at the higher frequencies, as indicated in Fig. 31.4(b).

FIGURE 31.4
(a) Audiogram showing progressive loss of hearing for typical workers in a noisy factory. (b) Loss of hearing with age (presbycusis) based on the data of Schneider et al. (1970).

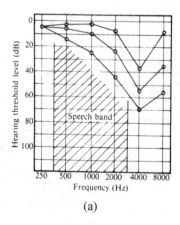

(a)

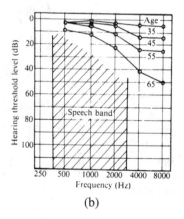

(b)

Actually, this process begins early in life; small children can usually hear up to 20,000 Hz, but adults can hear up to only 14,000 or 12,000 or 10,000 Hz (depending on age). By age 70, most people do not hear much above 8000 Hz, and the loss of hearing has a noticeable effect on frequencies within the speech band. Some television sets emit a fairly loud noise at 15,750 Hz, the frequency of the horizontal sweep oscillator. This noise is disturbing to young ears but goes completely unnoticed by people of middle age.

Two interesting hypotheses have been suggested as bases for predicting noise-induced hearing loss. The *equal-energy hypothesis* states that equal amounts of sound energy will cause equal amounts of NIPTS regardless of the distribution of the energy with time. The *equal temporary effect*, or TTS, *hypothesis* states that the temporary threshold shift measured two minutes after cessation of an 8-h exposure closely approximates the NIPTS that would result from a 10- to 20-year exposure to that same level. Neither of these hypotheses has been completely established, although there is a certain amount of evidence for each of them.

31.3 ■ EAR DAMAGE

Although the eardrum can be ruptured by extremely large sound pressures, such as those that occur during blasts, the outer ear, eardrum, and middle ear are not ordinarily damaged by exposure to noise. The primary site of ear damage due to noise is the receptor organ of the inner ear, the organ of Corti (see Section 5.2). From extensive examination of animal ears, and occasional post-mortem examinations of noise-damaged human ears, it has been fairly well established that excessive exposure to noise destroys the delicate hair cells in the organ of Corti and eventually the organ itself (Miller 1974). Figure 31.5 is a series of

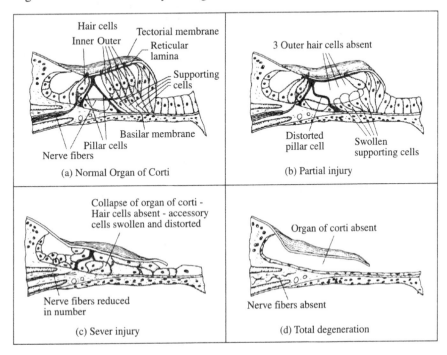

FIGURE 31.5
Drawings of the human organ of Corti that illustrate increasing degrees of noise-induced permanent damage. (From Miller 1974.)

drawings, based on photographs of damaged organs, that illustrates increasing degrees of injury.

The mechanism by which the hair cells are destroyed is not completely clear to us. During the reception of sound, movement of the basilar membrane bends the hairs at the top of the hair cells, and causes the cells to stimulate the auditory nerve fibers. One theory suggests that constant overexposure to sound forces the cells to work at too high a metabolic rate for too long a time, leading to the death of the cells. These delicate receptor cells do not regenerate, and if they die of overwork, they are lost for life.

The ear is an exceedingly well-designed organ, except that it lacks protective devices. The eye, which, like the ear, is sensitive to exceedingly small stimuli, has both an adjustable iris and an eyelid to protect it from damage due to overload. This may be due to the fact that through millions of years of evolution, humans have been exposed to the intense light of the sun. However, frequent exposure to loud sounds largely began with the industrial revolution and the invention of gunpowder. In another million years of so, human ears may very well be equipped with "earlids" or some other similar protective mechanism.

One protective mechanism the ear does have, however, is the *stapedius reflex* or acoustic reflex, which desensitizes the ear by as much as 20 dB. Loud noise triggers two sets of muscles that act in the middle ear. One set tightens the eardrum, and the other set draws the stirrup away from the oval window of the inner ear (see Section 5.2). The stapedius reflex may be triggered at sound levels between 80 and 95 dB in different individuals, depending on ear sensitivity. Unfortunately, it does not adequately protect the ear from impulsive or explosive sounds, because the reflex takes time. Although hearing sensitivity is measurably decreased in 30 or 40 ms, full protection may take as much as 200 ms. Thus in the case of a gunshot noise, the muscles "lock the barn door after the horse has already been stolen." It have been suggested that gun crews could be partly protected by stimulating their acoustic reflex with a loud tone a second or two before a gun is fired (Ward 1962).

An intense sound impulse, especially if it is too rapid for the stapedius reflex to act, can rupture the eardrum, causing a temporary loss of hearing and the hazard of middle-ear infection. The membrane will nearly always heal, but the heavier scar tissue may lower the sensitivity slightly for sounds of high frequency.

31.4 ■ HEARING PROTECTORS

Although it is much better to control noise at its source, people who are obliged to work in noisy environments can protect their hearing by the proper use of hearing protectors. Figure 31.6 shows the sound attenuation characteristics of various types of hearing protectors.

Cotton plugs are not very effective; cotton soaked in vaseline or soft wax is much better. Commercial ear plugs of the soft deformable plastic type, which make a good seal, can provide 20 to 30 dB of protection. Effective protection is offered by over-the-ear muffs, which are work by personnel working at airports. For maximum protection, both muffs and earplugs can be used (see Fig. 31.6).

For maximum protection, especially at low frequencies, hearing protectors must make an airtight seal with the walls of the ear canal or the cirumaural region around the pinna. Air-leakage paths can reduce attenuation by as much at 15 dB. Earmuffs are subject to vibration as a mass-spring system. Even if ear protectors were completely effective in

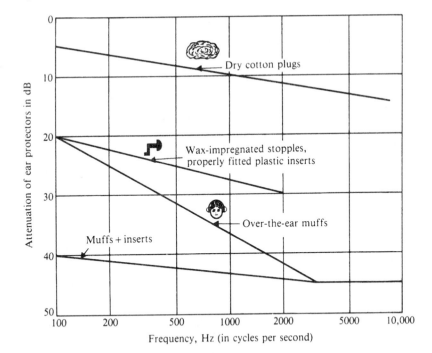

FIGURE 31.6
Sound attenuation
characteristics of
various types of ear
protectors. (From
Berendt, Corliss,
and Ojalvo 1976.)

blocking airborne sound, noise would still reach the inner ear by bone conduction. However, sound transmitted by bone conduction is generally 40 to 50 dB below that transmitted through an open ear canal, so it does not really compromise the effect of ear protectors.

Hearing protectors cannot really distinguish between noise and speech or music, so they do not really improve the speech-to-noise ratio. However, they may improve speech intelligibility slightly by reducing overload distortion in the cochlea. For normal hearers, hearing protectors have little degrading effect above 80 dbA but appear to cause considerable misunderstanding at lower levels, so they should not be worn at low levels where they are not needed (Berger and Casali 1997).

Efforts to improve communication under earplugs have involved the use of apertures or channels, sometimes opening into an air-filled cavity encapsulated by the earplug walls. For example, a small (0.5-mm) channel through an earplug creates a low-pass filter that passes speech frequencies up to 1000 Hz and proves up to 30 dB of attenuation at 8000 Hz. However much of the critical speech sounds in the 1000- to 4000-Hz band are missed. On the other hand, musicians want hearing protectors that have a uniform, or "flat," frequency response over the 100- to 8000-Hz range. Hearing protectors with a uniform 15-dB attenuation have been developed for musicians (Killian, DeVilbiss, and Stewart 1988).

31.5 ■ SPEECH INTERFERENCE

Speech communication is extremely important to human society. Therefore, interference with speech is a particularly disturbing effect of environmental noise pollution. Even though it is possible to communicate, in some noisy environments, by speaking louder and

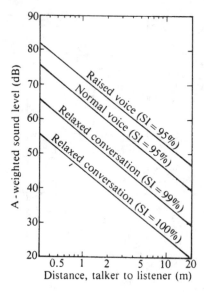

FIGURE 31.7
Maximum distances (outdoors) over which conversation is satisfactory. (From Environmental Protection Agency 1974.)

slower or by decreasing the distance from talker to listener, the need to do so can be quite annoying.

Accuracy of speech depends on many factors, such as noise level, vocal effort, distance between talker and listener, acoustical character of the room, speech clarity, message familiarity, and so on. Thus it is not easy to predict the dependence on any one of these factors, such as noise level. Nevertheless, considerable research on the problem has led to graphs of the type shown in Fig. 31.7, which illustrates the maximum distances over which conversation is considered satisfactory. Figure 31.8 shows the conditions under which speech

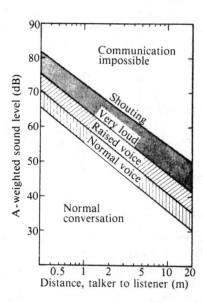

FIGURE 31.8
Maximum distances (outdoors) over which speech communication is possible. (After Environmental Protection Agency 1973.)

communication is possible at all. These graphs consider only direct sound, so they describe a free-field (outdoor) environment (see Section 6.2). In a room, reflected sound will reinforce the direct sound, and make communication possible over a greater distance.

A speaker generates a complicated series of sound waves that contain many acoustical cues to the sounds of spoken language, as we described in Chapter 16. The listener sorts out these cues, combines them with other available information (such as context, knowledge of the speaker, etc.), and translates them into an intelligible message, if possible. Most speech cues are carried by sounds with frequencies in the range of 300 to 3000 Hz, although a few cues occur as low as 100 Hz and as high as 8000 Hz. Speech ordinarily carries an abundance of extra information, so that it is still understandable even when some cues are lost. Lower noise levels are necessary if the talker has a poor articulation or speaks an unfamiliar dialect; young children, for example, have less precise speech than do adults. Also, the listener's ability to understand masked or distorted speech deteriorates with age (Palva and Jodinen 1970).

A number of ways to rate speech interference by noise have been suggested. Perhaps the most accurate is the *articulation index* (AI), which considers the average speech level and average noise level in each of 20 frequency bands over the range of 250 to 7000 Hz. Somewhat simpler to measure is the *speech interference level* (SIL), which is the average of the sound pressure levels in the octave bands centered at 500, 1000, 2000, and 4000 Hz. (Older versions used octave bands centered at 850, 1700, and 3400 Hz, or at 500, 1000, and 2000 Hz, and there is some sentiment voiced for averaging in a fourth octave band centered at 4000 Hz.) The simplest measure is the A-weighted sound level $L_p(A)$, and this is sufficiently precise for most purposes.

Thus far we have discussed only the effects of rather broad bands of steady noise. We will now briefly consider the speech interference due to narrow-band noise and fluctuating noise level. It is well known that tones of lower frequency can mask tones of higher frequency more effectively than the converse (see Section 6.10). Thus, narrow-band noise interferes with speech most effectively if its frequency is near the low end of the speech band. Certain bands of noise interfere with certain phonemes more than others.

The speech interference due to fluctuating noise levels depends on the nature of the fluctuations as well as the equivalent sound level. For a given equivalent level L_{eq}, intermittent noise will nearly always cause less interference with speech than will steady noise. In Section 6.10, we pointed out that loud intermittent noise can mask speech sounds that occur 20 ms or so after the noise ceases (forward masking) and can even mask speech sounds that occur 5 or 10 ms before the onset of the noise (backward masking).

31.6 ■ INTERFERENCE WITH SLEEP

From our everyday experience, we know that noise can interfere with sleep. Almost all of us have been awakened or kept from falling asleep by loud, strange, sudden sounds, and it is common to be awakened by an alarm clock. But it is also possible to become accustomed to sounds and sleep through them (even alarm clocks!). Apparently, unusual or unfamiliar sounds are most apt to disturb sleep.

The interference with sleep due to noise is not well understood at the present. Much of what is known is based on laboratory studies conducted on relatively few subjects. Field studies involving large numbers of persons are difficult to conduct. Among factors that appear to be important are the nature of the noise stimulus; the physical and emotional state of the subject; physiological differences such as sex and age; and the stage of sleep during which the noise occurs (see the box below).

There are several different ways of classifying the stages of sleep, but one way simply numbers them from 1 to 5. As one relaxes and enters a stage of drowsiness, the electroencephalogram (EEG) pattern changes from rapid, irregular waves to regular variations at 9 to 12 Hz, known as the *alpha rhythm*. The person is relaxed, but not yet asleep. In sleep stage 1, the alpha waves diminish in amplitude and frequency. In stage 2, the alpha rhythm gives way to bursts of waves (*spindle* waves) mixed with single slow waves. In stage 3, 30 to 45 minutes later, bursts of large, slow waves (*delta* waves) appear. In stage 4, the deepest sleep, delta waves occur over 50 percent of the time. An additional stage, called the *rapid eye movement* (REM) stage, exhibits many characteristics of stage 1. During a night's sleep, a person typically enters cycles from stage 4 to the REM-stage several times and may go through the entire cycle of stages several times. The time spent awake generally increases with age (EPA 1973).

A person's threshold for being awakened by noise appears to be lowest during the REM-stage and highest during stages 3 and 4. The amount of accumulated sleep time increases the probability of arousal, no matter what the stage of sleep. Stimuli of 50 dBA have been found to invoke some sort of response, either a change in sleep stage or an awakening, about half the time. When stimuli reach 70 dBA, awakening is the most likely response. In noise levels of 40 to 50 dBA, many subjects experience difficulty falling asleep and awaken rather easily.

Although further studies are needed to establish with any certainty the noise levels that interfere with sleep, it appears that indoor noise levels should not exceed 35 dBA in order to protect most of us from sleep interference.

31.7 ■ OTHER PSYCHOLOGICAL EFFECTS

The effect of noise on the performance of various tasks has been the subject of several investigations in the laboratory and in actual work situations. When mental or motor tasks do not involve auditory signals, the effects of noise on human performance have been difficult to assess.

Some general conclusions can be drawn about the effects of noise on performance. Steady noises below about 90 dBA do not seem to affect performance, but intermittent noise can be disruptive. Noise with appreciable strength around 1000 to 2000 Hz is more disruptive than is low-frequency noise. Noise is more likely to reduce the accuracy of work than to reduce the total quantity of work (Miller 1974).

Noise appears to interfere with the ability to judge the passage of time. Subjects in various experiments have judged time as passing too rapidly or too slowly, depending perhaps on the level of the noise they were hearing.

There is a general feeling that nervousness and anxiety are caused by exposure to noise or at least are intensified by it. Whether noise by itself causes a significant amount of stress is difficult to determine, because noise is often closely associated with events that involve fear and anxiety.

31.8 ■ THE PHYSIOLOGICAL EFFECTS OF NOISE

Sudden noises are startling. They trigger a muscular reflex that may include an eyeblink, a facial grimace, inward bending of arms and knees, etc. These reflexes prepare the body for defensive action against the source of the noise. Sometimes these reflexive actions interfere with other tasks; sometimes they even cause accidents.

Constriction of blood vessels, reduction of skin resistance, changes in heartbeat, changes in breathing rate, dilation of pupils, and secretion of saliva have been observed in human response to brief sounds (Davis, Buchwald, and Frankmann 1955). There is evidence that workers exposed to high levels of noise have a higher incidence of cardiovascular disorders; ear, nose, and throat problems; and equilibrium problems than do workers at lower levels of noise (Miller 1974).

Laboratory experiments with animals have shown that intense noise can cause pathological effects such as hypertrophy of the adrenal glands, developmental abnormalities of the fetus, brain damage, and sexual misfunction. These often occur at levels above those that humans normally encounter, however.

31.9 ■ SUMMARY

Noise affects people in many ways. One of the most serious effects is hearing loss, both temporary and permanent, due to noise exposure. Hearing threshold shifts usually occur first around 4000 Hz in amounts that depend both on noise level and duration. Ear damage centers in the organ of Corti, where prolonged exposure to noise results in the destruction of the hair cell sensors.

Because of the importance of speech communication, speech interference is a particularly annoying effect of noise. Most speech cues are carried by sounds in the range of 300 to 3000 Hz; thus, a speech interference level (SIL) is defined as the average of noise levels in the 500-, 1000-, and 2000-Hz octave bands. Noise can cause changes in a person's sleep stages or even awakening. It can also affect performance of mental and physical tasks, interfere with the judgment of time, intensify stress and anxiety, and trigger muscular reflexes.

REFERENCES AND SUGGESTED READINGS

American National Standards Institute (1969). "Specifications for Audiometers," ANSI S.3-1969. New York: American National Standards Institute (similar to International Standards Organization, ISO R839, 1964).

Berendt, R. D., E. L. R. Corliss, and M. S. Ojalvo (1976). *Quieting: A Practical Guide to Noise Control*. Washington, D.C.: National Bureau of Standards Handbook 119.

Berger, E. H., and J. G. Casali (1997). "Hearing Protection Devises," in *Encyclopedia of Acoustics*, vol. 2, ed. M. J. Crocker. New York: Wiley.

Davis, R. C., A. M. Buchwald, and R. W. Frankmann (1955). "Automatic and Muscular Responses and Their Relation to Simple Stimuli," *Psychol. Monogr.* **69**(20): 1.

Environmental Protection Agency (1973). *Public Health and Welfare Criteria for Noise*, Report 550/9-73-003, U.S. Environmental Protection Agency, Office of Noise Abatement and Control, Washington, D.C.

Environmental Protection Agency (1974). *Information on Levels of Environmental Noise Requisite to Protect Public Health and Welfare with an Adequate Margin of Safety*, Report 550/9-74-004, U.S. Environmental Protection Agency, Office of Noise Abatement and Control, Washington, D.C.

Killian, M., E. DeVilbiss, and J. Stewart (1988). "An Earplug with Uniform 15-dB Attenuation," *Hearing J.* **41:** 14.

Kryter, K. D. (1970). *Noise and Man.* New York: Academic Press.

Kryter, K. D. (1973). "Impairment to Hearing from Noise," *J. Acoust. Soc. Am.* **53:** 1211. (Several critiques of this paper also follow it.)

Lucas, J. S. (1975). "Noise and Sleep: A Literature Survey and a Proposed Criterion for Assessing Effect," *J. Acoust. Soc. Am.* **58:** 1232.

Miller, J. D. (1974). "Effects of Noise on People," *J. Acoust. Soc. Am.* **56:** 729.

Palva, A., and K. Jodinen (1970). "Presbycusis. V. Filtered Speech Tests," *Acta Oto-Laryngol.* **70:** 232.

Schneider, E. J., J. E. Mutchler, H. R. Hoyle, E. H. Ode, and B. B. Holder (1970). "The Progression of Hearing Loss from Industrial Noise Exposures," *Am. Ind. Hygiene Assoc. Jour.* **31:** 368.

Ward, W. D. (1962). "Studies on the Aural Reflex. III. Reflex Latency as Inferred from Reduction of Temporary Threshold Shift from Impulses," *J. Acoust. Soc. Am.* **34:** 1132.

Ward, W. D. (1969). "Effects of Noise on Hearing Thresholds," *Proc. of Conf. on Noise as a Public Health Hazard*, eds. W. D. Ward and J. E. Fricke. Washington: American Speech and Hearing Association.

Ward, W. D. (1977). "Effects of High-Intensity Sound," in *Encyclopedia of Acoustics*, vol. 3, ed. M. J. Crocker. New York: Wiley.

GLOSSARY

articulation index (AI) A means of rating speech interference by noise that considers the average speech level and average noise in each of 20 bands over the frequency range 250–7000 Hz.

audiogram A graph of a person's hearing threshold at several frequencies compared to thresholds for normal hearing.

audiometer An instrument that measures hearing thresholds of an individual at several audible frequencies.

electroencephalogram (EEG) A record of electrical potentials at several points in the brain (sometimes referred to as "brain waves").

equal energy hypothesis Postulates that risk of hearing loss is determined by the total amount of noise energy to which the ear is exposed each day, irrespective of its distribution in time.

equal temporary effect (TTS) hypothesis Postulates that risk of permanent hearing loss increases with average temporary loss (TTS).

hair cells Delicate sound-sensing cells in the organ of Corti that are destroyed by overexposure to noise.

permanent threshold shift (NIPTS) The amount that the threshold of hearing is raised irreversibly by exposure to noise.

presbycusis Gradual loss of hearing with age, especially at high frequency.

rapid eye movement (REM) state A stage of sleep during which a person can be awakened quite easily by noise (also the stage in which dreaming occurs).

speech cues Particular combinations of sounds or dynamic changes in sound by which a listener identifies phonemes (speech sounds).

speech interference level (SIL) The average of sound levels in the 500-, 1000-, and 2000-Hz octave bands.

temporary threshold shift (TTS) A reversible increase in the threshold of hearing that disappears in hours, days, or weeks depending on its severity (also called *auditory fatigue*).

REVIEW QUESTIONS

1. What is meant by NITTS or TTS? What is meant by NIPTS?
2. What is an audiogram?
3. How does TTS generally change with exposure time?
4. Which causes the most TTS: low-frequency noise or high-frequency noise?
5. How does the growth of TTS depends upon drugs?
6. What is the frequency of noise related to the horizontal sweep on a TV set in the United States?
7. What part of the organ of Corti is most likely to be destroyed by excess noise?
8. How does the stapedius reflex or acoustic reflex help to protect us from loud noises?
9. What is the maximum attenuation that can be normally obtained with ear plugs?
10. Does wearing earplugs in a noisy room make it easier to understand speech?
11. During what stage of sleep is a person most likely to be awakened by noise?
12. What are some physiological effects of noise on humans?

QUESTIONS FOR THOUGHT AND DISCUSSION

1. In what ways is the dependence of hearing damage to noise much like the effect of nuclear radiation on the body?
2. Do you expect that human beings will, in time, develop earlids or other noise protectors through evolution?
3. Have you ever had your sleep interrupted by a loud noise? What do you think the sound pressure level might have been?
4. What effects of noise in the environment are most costly?
5. Is the effect of noise on animals similar to the effect on people?

EXERCISES

1. Sound levels of 67 dB, 71 dB, and 68 dB are measured in the octave bands centered at 500, 1000, and 2000 Hz, respectively. What is the speech interference level?
2. The sound energy that the ear receives depends on the average intensity multiplied by the time.
   (a) Compare the intensities of sounds having levels of 85 dB and 110 dB.
   (b) Compare the energy received by the ear during exposures for 10 min at 110 dB and for 8 h at 85 dB (both of these exposures cause TTS of about 34 dB, according to Fig. 31.2).
3. If the A-weighted sound level outdoors is 50 dB, what is the maximum distance from talker to listener in order to carry on conversation with
   (a) normal voice;
   (b) very loud voice.
4. All the curves in Fig. 31.7 and 31.8 have essentially the same slope. How much does the allowed sound level decrease when the talker-to-listener distance doubles? Explain this result by referring to Section 6.2.

EXPERIMENTS FOR HOME, LABORATORY, AND CLASSROOM DEMONSTRATION

Home and Classroom Demonstration

1. Noise Levels Borrow or purchase an inexpensive sound-level meter and use it to record the ambient sound levels in different environments: restaurants, bars, concerts of different types, athletic events, etc. Report your findings to the class so that everyone gets a good feeling for different sound levels in our environment.

2. Effect of noise on performance Make your own test of the effect of noise on a mental task, such as the solution of a set of arithmetic problems of copying names and numbers from the telephone directory. Try it at medium to high noise levels (90 dBA and 100 dBA if you have access to a sound-level meter). A good source of broadband noise is an FM radio tuned between stations. Be careful that the sound does not become uncomfortably loud; do not exceed the level you would expect at a loud concert.

3. *Presbycusis*   Make a survey of acquaintances of various ages to determine the average age at which people no longer hear the 15,750-Hz sound emitted by many television sets in countries whose electrical power is transmitted at 60 Hz.

4. *Audiograms*   Many universities have free hearing clinics in which you can have your own audiogram recorded. Is there any evidence of noise-induced hearing loss?

Laboratory Experiments

Loudness level and audiometry (Experiment 13 in *Acoustics Laboratory Experiments*)

CHAPTER

32

The Control of Noise

The words *silencing* and *soundproof* are two very common, yet misleading, terms. Both suggest the elimination of noise, which generally is not possible (or even desirable). What is desirable is the reduction of noise to levels that are not injurious to the health and well-being of people who live and work in that particular environment.

In this chapter, you should learn:

- How to deal with noise-control problems;
- About environmental noise regulations;
- About noise attenuation by walls and barriers;
- About heating and air conditioning noise;
- About aircraft and airport noise;
- About active noise control.

32.1 ■ ANALYZING A NOISE PROBLEM: SOURCE-PATH-RECEIVER

A straightforward approach to solving a noise control problem is to divide it into its three basic elements: source, path, and receiver. The *source* may be a noisy machine or appliance, a jet aircraft, a noisy highway, a neighbor's loudspeaker or lawnmower, or any of a large number of mechanical noisemakers that are common in our society. The *path* may be a direct line-of-sight air path, a structural path through a building, or a complex path that includes propagation through ducts, windows, wall panels, etc. The *receiver* in which we are interested is a person or a group of persons receiving unwanted sound.

There is little doubt that the most effective place to control noise is at its source. The best policy is to select quiet machines or appliances initially. This would become easier if product noise labels were to become mandatory, as proposed by the Environmental Protection Agency at one time. Noisy machines can often be quieted by a few simple modifications such as the installation of resilient mounts, the damping of vibrating panels, the slowing down of air velocities, and so on. In other cases, major reengineering is required. Sound-attenuating housings around the machines are frequently helpful.

After careful attention has been paid to controlling noise at the source, the next line of defense is to cause the sound to be attenuated as much as possible along its path from source to receiver. This can be done by

1. Absorbing sound energy along the path;
2. Reflecting sound back toward the source by means of barriers;
3. Eliminating alternative paths for sound transmission.

It is usually more difficult to attenuate sound outdoors than indoors.

When efforts to reduce noise at its source or along the transmission path prove insufficient, one or more of the following measures may be taken to protect the receiver:

1. Using ear protectors (see Section 31.4);
2. Limiting the time of exposure to noise;
3. Operating noisy machines by remote control from a sound-attenuating enclosure; and
4. Making use of masking noise to reduce the annoyance due to pure-tone noise or nearby conversations.

32.2 ■ NOISE REGULATIONS

Like most aspects of environmental protection in the United States, noise regulation is shared by federal, state, and local governments. The federal government sets general noise standards, especially in areas where they will affect interstate commerce, including the health and hearing of workers employed by companies that engage in interstate commerce. State noise control measures have typically dealt with motor vehicle noise, airport noise, land use, and related aspects of noise control. Local governments regulate the use of noisy machines, the noise of vehicles (including recreational vehicles) that operate within their boundaries, nuisance noise, noise transmission within apartment buildings, and so forth.

The *Noise Control Act of 1972* was the first major piece of federal legislation in the field of noise control This act directed the United States Environmental Protection Agency (EPA) to develop and publish information on hazardous noise levels, to identify major noise sources, and to define permissible noise levels for them. The EPA was also directed to coordinate all federal noise research and control programs and to provide technical assistance to state and local governments.

Among the major sources of noise that were identified by the EPA are portable air compressors, medium and heavy trucks, motorcycles, buses, garbage trucks, jackhammers, railroad cars, snowmobiles, and lawnmowers. However, the plans that the EPA developed in the 1970s were effectively scrapped in 1980, and the EPA has had no national noise program since 1981. In 1981 Congress agreed to the Reagan Administration's proposal to cease funding for the EPA Office of Noise Abatement and Control (ONAC). However Congress did not repeal the Noise Control Act of 1972, and there have been several bills in the Congress to reestablish ONAC.

Other U.S. government agencies are also concerned with special areas of noise control. The *Federal Aviation Administration* (FAA) sets criteria and standards for aircraft noise.

The *Federal Highway Administration* has adopted noise control standards for motor vehicles, and the *Bureau of Motor Vehicle Safety* shares with state and local agencies the responsibility for enforcement. Both of these agencies, as well as the FAA, are in the Department of Transportation at this time.

TABLE 32.1 Permissible occupational noise exposure

Average sound level, dB (A-weighted, slow response)	Hours per day
90	8
92	6
95	4
97	3
100	2
102	$1\frac{1}{2}$
105	1
110	$\frac{1}{2}$
115	$\frac{1}{4}$

The *Occupational Safety and Health Administration* (OSHA) sets regulations designed to protect the hearing of workers in companies engaged in interstate commerce. Popularly known as *OSHA standards*, the current limits on noise exposure are given in Table 32.1.

The *Department of Housing and Urban Development* (HUD) has developed standards for sound-insulation characteristics of walls and floors in multifamily residences and other buildings that qualify for HUD mortgage insurance. It also sets guidelines for permissible noise levels at sites of housing developments.

State and Local Noise Regulations

Unlike the Federal government, which has been relatively inactive in regulating the noise environment, many states and cities have enacted noise-control ordinances. These generally address maximum levels for motor vehicles and maximum levels at property boundaries. In Minneapolis, for example, the maximum noise level for passenger cars measured 50 ft from a street ranges from 69 dBA in areas where the speed limit is 30 mi/h to 74 dBA where the speed limit if 45 mi/h. Heavy trucks in a 35 mi/h speed zone are limited to 75 dBA during daytime, but this reduces to 65 dBA at night.

In Madison, Wisconsin, the maximum sound level 50 ft from a residential property line is set at 75 dBA during the day and 70 dBA at night. In San Diego the 1-h average sound level at property boundaries is set at 50 dBA during the day and drops to 40 dBA at night. Minneapolis specifies the maximum sound level in each of nine different octave bands.

32.3 ■ EXPOSURE TO OCCUPATIONAL NOISE

Some of the noisiest environments in the world are found within factories in which thousands of workers earn their living. Although hearing loss among factory workers has been common, only recently has occupational noise been regulated.

The Occupational Safety and Health Act of 1970 set limits on permissible noise exposure for workers based on the amount of time they are exposed to these levels. These standards have precipitated some rather heated discussion, and there is considerable feeling that the limits should be lowered.

The combinations of exposure times and sound levels given in Table 32.1 are considered the limit of daily dose that will *not* produce disabling loss of hearing in more than 20% of a population exposed throughout a working lifetime of 35 years. For exposure at two or more different levels, effects are combined according to the relationship

$$\text{Fraction of allowed dose} = C_1/T_1 + C_2/T_2 + \cdots + C_n/T_n,$$

where C_n represents the actual exposure time at a given sound level and T_n is the time permitted if exposure were all at that level.

When the sound level exceeds the maximum permissible combination of dose and time given in Table 32.1, the company should

1. Take steps to reduce the noise at its source;
2. Provide hearing protection; and
3. Carry out a program to test and conserve hearing.

32.4 ■ PRODUCT LABELING

There are indications that some people prefer to purchase a quiet product even if it costs slightly more than a noisier one. Phrases such as "whisper quiet" often appear in product advertising. Nevertheless, it is difficult for the prospective buyer to compare the noise output from different brands.

The American National Standards Institute published a method for rating noise emission by small stationary sources to form the basis for product noise labeling, but the proposed method was never adopted and has been replaced by a method based on A-weighted sound power level (see Section 6.5).

32.5 ■ WALLS AND FLOORS

When an airborne sound wave strikes a solid wall, the largest part is reflected, whereas smaller portions are absorbed and transmitted through the wall. The coefficients of *reflection*, *absorption*, and *transmission* are determined by the physical properties of the wall and by the frequency of the sound and its angle of incidence to the wall.

The *transmission coefficient* τ is defined as the fraction of the acoustic power incident on the wall that is transmitted through the wall to the other side. The principle of conservation of energy tells us that the transmitted power is that part of the incident power that is neither reflected nor absorbed. This fraction is expressed as *transmission loss* (TL) in decibels:

$$\tau = \frac{W_{\text{transmitted}}}{W_{\text{incident}}};$$

$$\text{TL} = 10 \log \frac{1}{\tau}.$$

Sound waves striking a wall can bend it, shake it, or both. (These motions may be described as flexural and compressional waves, respectively, in the wall.) At low frequency, the sound transmission loss in a solid wall follows a *mass law*; it increases with increasing frequency f and mass M of the wall. Formulas for calculating the transmission loss at low frequency are given in the following box.

For waves that approach a wall of large dimensions with normal (perpendicular) incidence, the transmission loss is

$$(\text{TL})_0 = 10 \log \left(1 + \frac{\pi M f}{400} \right),$$

where M is wall area mass (in kg/m^2) and f is frequency (in Hz). In a room, it is a good approximation to assume the sound waves of low frequency to be randomly distributed over all angles from 0 to 80°. This decreases the transmission loss by about 5 dB:

$$\text{TL} = 10 \log \left(1 + \frac{\pi M f}{400} \right) - 5.$$

From these formulas, two facts are clear:

1. To reduce transmission of sound between adjoining rooms, the common wall should be as heavy as possible;
2. Low-frequency sounds are the most difficult to block (this should be clear if you recall that sound will "leak through" from a hi-fi system playing in the next room).

Transmission loss for a wall may fall considerably below that predicted by the mass law. This may be due to any of the following effects:

1. Wall resonances that occur at certain frequencies;
2. Excitation of bending waves at the *critical frequency*, where they travel at the same speed as certain sound waves in air;
3. Leakage of sound through holes and cracks.

The transmission losses for walls of several materials are shown in Fig. 32.1. Note the dip in TL at the critical frequency, which is different for each material.

Leakage of sound through small holes or cracks in walls tends to be underestimated all too often in building construction. Openings around pipes and ducts and cracks at the ceiling and floor edges of walls allow the leakage of airborne sound. Common causes of leakage in party walls separating apartments may include back-to-back electrical outlets or medicine cabinets. Cracks under doors are especially bad. Figure 32.2 illustrates the effect of holes of various sizes on the transmission loss of walls.

It is often desirable to provide a single-number rating of a wall for purposes of comparison. This is done by measuring the transmission loss of a wall sample at 16 different frequencies, and comparing these values to standard curves. The resulting single number is called the *sound transmission class* (STC) of the wall. Table 32.2 gives the sound transmission class for various wall structures, and also indicates the degree of privacy that they

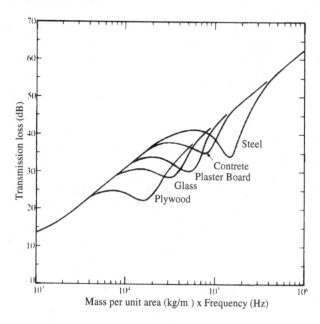

FIGURE 32.1

Transmission loss (TL) of a wall as a function of mass and frequency. Note the drop in TL near the critical frequencies for exciting bending waves

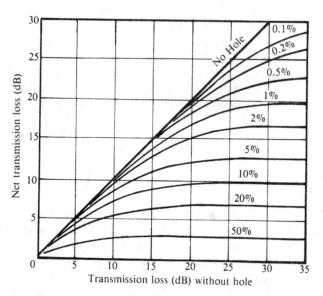

FIGURE 32.2

The effect of a hole on transmission loss TL. The horizontal axis is the transmission loss without the hole; the vertical axis is the transmission loss with the hole. The numbers that label the individual curves give the hole area as a percentage of total wall area.

offer. The Department of Housing and Urban Development recommends at least STC 55 for walls between two apartments (party walls) in nonurban areas.

In the case of floor-ceiling combinations, the transmission of impact noise is generally more important than the transmission of airborne sound. The impact-sound transmission level (ISTL) is measured with the help of a *standard tapping machine*, and the resulting data are used to determine the *impact isolation class* (IIC) of a floor-ceiling combination.

TABLE 32.2 Sound transmission class (STC) for various wall structures

STC Rating	Privacy Afforded	Wall Structure
25	Normal speech easily understood	$\frac{1}{4}$-in. wood panels nailed on each side of 2×4 studs
30	Normal speech audible but not intelligible	$\frac{3}{8}$-in. gypsum wallboard nailed to one side of 2×4 studs
35	Loud speech audible and fairly understandable	$\frac{3}{8}$-in. gypsum wallboard nailed to both sides of 2×4 studs
40	Loud speech audible but not intelligible	Two layers of $\frac{5}{8}$-in. gypsum wallboard nailed to both sides of 2×4 studs
45	Loud speech barely audible	Two sets of 2×3 studs staggered 8 in. on centers fastened to 2×4 base and head plates with two layers of $\frac{5}{8}$-in. gypsum wallboard nailed on the outer edge of each set of studs
50	Shouting barely audible	2×4 wood studs with resilient channels nailed horizontally to both sides with $\frac{5}{8}$-in. gypsum wallboard screwed to channels on each side
55	Shouting not audible	$3\frac{5}{8}$-in. metal studs with 3-in. layer of glass fiber blanket stapled between studs. Two layers of $\frac{5}{8}$-in. gypsum wallboard attached to each side of studs.

Source: Berendt, Corliss, and Oljavo (1976).

One of the easiest ways to increase the impact-isolation of a floor-ceiling structure is to cover the upper surface with a thick carpet over a resilient pad.

32.6 ■ BARRIERS—INDOORS AND OUTDOORS

Sound barriers, which block the direct sound path from source to receiver, can result in appreciable noise reduction, both indoors and outdoors. Unfortunately, their potential is often overestimated, and the actual noise reduction achieved from barriers is disappointing.

Sound transmission *through* a barrier is generally less important than sound transmission *around* a barrier. In Fig. 32.2, for example, we can see that as the open area increases in percentage of the total area, the transmission loss becomes more or less independent of the TL of the wall itself. Thus, the primary attention should be directed at the "flanking" paths that allow sound transmission over or around a barrier.

A typical situation in an indoor office is shown in Fig. 32.3. There are three types of transmission paths to be considered: transmission through the barrier (path SCR), diffraction around the barrier (path SBR), and reflection from the ceiling (paths SAR, SDER, etc.).

Transmission through the barrier is similar to that through a full wall of the same construction; unless the barrier is flimsy, direct transmission will be much less than will transmission by diffraction and reflection.

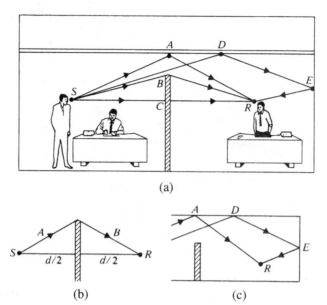

(a)

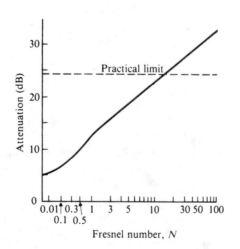

FIGURE 32.3
(a) Transmission
paths through and
around a barrier.
(b) Diffraction
around the barrier.
(c) Reflection paths
around a barrier.

(b)   (c)

The diffraction of sound of a given frequency around a barrier depends on the Fresnel number N, which is expressed as

$$N = \frac{2}{\lambda}(A + B - d),$$

where λ is the wavelength of sound and lengths A, B, and d are as shown in Fig. 32.3(b). Figure 32.4 illustrates the attenuation that can be achieved with a barrier when diffraction is the only type of transmission. This curve assumes a fairly large open area above the barrier. If the barrier extends nearly to the ceiling, the attenuation will be increased.

FIGURE 32.4
Attenuation of
sound by a barrier
if diffraction were
the only mode of
transmission.

Transmission by reflection depends on the acoustic properties of the ceiling, the size of the opening above the barrier, and the nature of the walls on the source and receiver sides of the barrier. A highly absorbent ceiling is essential in an open-plan office or school.

Even when carefully designed, barriers in open-plan offices and schoolrooms cannot provide anywhere near the privacy or noise reduction of full-size walls. It has frequently been found that to obtain any degree of privacy in an open-plan environment, work areas must be spread out more than is necessary for enclosed work spaces, thus offsetting the economy anticipated from open planning. Even so, the nonacoustical advantage of open planning may outweigh the acoustical disadvantages in some situations.

Barriers are frequently used outdoors to attenuate sound from a noisy highway or railroad, for example. The same physical laws that govern the performance of indoor barriers serve for outdoor barriers as well but the design considerations are different. Most of the sound transmission outdoors is due to diffraction over the barrier (see Fig. 32.4). Because source-receiver distances are greater outdoors, achieving a large Fresnel number $(A + B - d)$ requires a very high barrier. Also, refraction of sound caused by wind or temperature gradients may lead to additional transmission of sound over the barrier (see Fig. 3.18(b)). Sound barriers are much more effective in attenuating high-frequency noise than low-frequency noise.

32.7 ■ ENCLOSURES

Consider the demonstration experiment shown in Fig. 32.5 (better yet, perform the experiment yourself). First, a small noise source, such as a battery-operated doorbell or buzzer, is allowed to operate on a wooden table. The sound level is measured at some nearby location. The source is now covered or enclosed with acoustic tile or other absorbing material (Fig. 32.5(b)); the reduction in sound level is minimal. Next, a small piece of the same material is placed under the source to decouple it from the table (c); a larger reduction in sound level results. Now a lightweight solid enclosure (e.g., a small wastebasket or cookie can) is placed over the source (d); the sound level is reduced by 6 dB (half the sound pressure); the can and pad together (3) give 11 dB. Finally, the absorbing material is placed *inside* the enclosure (f), and the sound level is further reduced by about 3 dB.

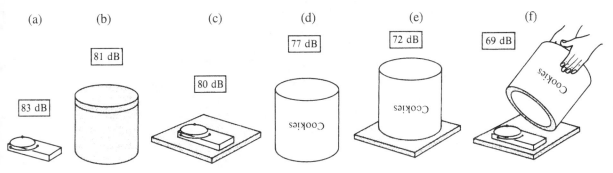

FIGURE 32.5 demonstration of noise reduction with an enclosure; (a) noise source; (b) covered with absorbing material; (c) placed on a resilient pad; (d) covered with a metal can; (e) can and pad; (f) can lined with absorbing material.

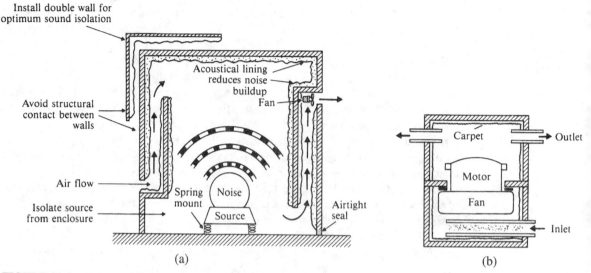

FIGURE 32.6   (a) A noise-reducing enclosure requiring air circulation. (Berendt, Corliss and Olijavo, 1976). (b) A home-made air blower that is 10 to 15 dB quieter than vacuum cleaner blowers.

If the metal can in Fig. 32.5 makes a reasonably good seal, it reduces the sound to some 10 or 11 dB below its level inside the can. If there is insufficient absorption inside the can, however, the reverberant level builds up inside the can, so the net reduction is only 6 dB (Fig. 32.5(d)). Placing absorbing material on the inside will absorb the reverberant sound so that the sound attenuation by the can results in a greater reduction in the noise level outside. Absorbing material on the outside of the can would be of little use.*

If the noise source needs ventilation, so that the enclosure cannot be made airtight, less noise reduction is possible. A muffler can be included in the air inlet and outlet ducts to reduce noise transmission, especially if the noise source emits sound at certain frequencies more than others. If mufflers are not practical, the air inlet and outlet ducts should be lined with absorbing material, and they should make as many bends as possible. Vented enclosures are shown in Fig. 32.6.

32.8 ■ SOUND ABSORBERS

Absorption of sound requires the conversion of sound energy to heat. One of the most effective ways to absorb airborne sound is to use a porous material made of many small fibers or cells. The rush of air particles back and forth among the fibers, due to passage of a sound wave, generates heat and absorbs the sound wave. Because the energy of a sound wave is small, the temperature rise in the material is not normally measurable.

Porous absorbers are more effective at high frequencies than at low frequencies (see Table 23.1). Also, porous absorbers are more effective if mounted away from a reflecting

---

*To constantly remind me of this fact, I have a noisy duct from an exhaust fan passing through my office. Some well-meaning person carefully installed absorbing material on the outside rather than lining the inside of the duct.

surface where the air-flow velocity is greater. (Recall from Chapter 3 that at the hard surface of a reflector, the sound pressure is a maximum but the particle velocity is zero.)

A second type of absorber is a panel that is set into vibration by an airborne sound wave. Sound energy is converted into heat by internal friction in the panel. A panel absorber is effective mainly at low frequencies (see Table 20.1). The frequency of maximum absorption can be controlled by choice of panel weight and by the depth of air space behind the panel, and the absorption can be increased by filling the air space behind the panel with a porous absorbing material.

An absorbing surface that combines the principles of both porous and panel absorbers is a perforated panel backed by porous material. The holes act as resonators, very similar to Helmholtz resonators, without cavity walls (see Section 2.3). The frequency of maximum absorption can be controlled by choice of panel thickness, the size and spacing of the holes, and the depth of the space behind the panels. Panel absorbers with and without holes are shown in Fig. 32.7.

FIGURE 32.7
(a) Panel absorber spaced away from a reflecting wall (the air space may contain a porous absorber).
(b) Perforated panel backed by porous material.

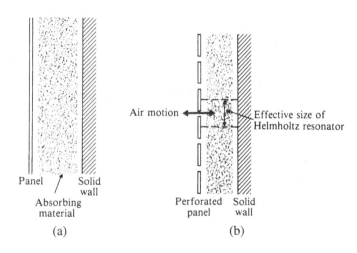

(a)   (b)

Perforated tiles, or acoustical tiles, are commonly used as absorbers. The front surface is porous with openings that penetrate into the interior. If the openings are regular in size and spacing, the material will show an absorption maximum, typically near a frequency of 1000 Hz. If the holes are random, the absorption maximum will be broader. Acoustical tiles of several designs are shown in Fig. 32.8.

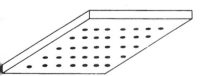
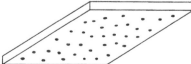
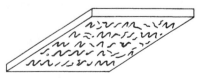

FIGURE 32.8 "Acoustical" tiles with porous surfaces.

32.9 ■ HEATING AND AIR CONDITIONING NOISE

Heating, ventilating, and air-conditioning systems are a most annoying source of noise in many homes, offices, classrooms, and even concert halls. In nearly all cases, the annoyance is completely unnecessary. In some cases, the level of the noise indicates a malfunction that may also reduce efficiency.

Several different types of noise may be encountered in air-handling systems. These include the following:

1. Noise from motors, blowers, and compressors that is transmitted through the air ducts. Unlined ducts act as "waveguides," delivering sound as well as fresh air efficiently throughout the building.
2. Noise from air flow and turbulence within the duct, especially at inlet and outlet grills.
3. Mechanical noise, as in (1), transmitted through the building structure, which results in a low-pitched rumble.
4. Noise from fresh air intake and exhaust ducts and compressors outside the building that radiate noise throughout the neighborhood. Window-mounted air conditioners may be particularly annoying to neighbors who wish to sleep with their windows open (or even closed).

In planning new construction, residential or otherwise, it is most important to select quiet heating and ventilating equipment, and to isolate it as much as possible from the quietest parts of the building. Centrifugal or squirrel-cage fans are usually less noisy than are axial or propeller fans. Motors, blowers, and compressors should be mounted on resilient pads to isolate vibrations. Heavy equipment in large buildings should be bolted to a concrete slab that is isolated from the rest of the building by resting either on vibration isolators or on its own foundation. The blower should be mechanically isolated from supply and return ducts by a flexible "boot."

Noise transmitted through ducts can be reduced by installing duct liners. A 1-in.-thick lining will reduce high-frequency noise by as much as 10 dB per meter, although low-frequency noise may require a thicker lining for effective absorption. Both supply and return ducts should be lined; lining is most effective when installed near the open or grill end of the duct. In difficult cases, expansion or plenum chambers may be necessary.

Aerodynamic noise generated in ducts and grills can be minimized by keeping air velocities low (see Section 30.4). Sharp corners, ragged joints, and dampers can cause noise-generating turbulence in air ducts, as shown in Fig. 32.9. If removing a grill results in a substantial reduction in noise, replacement with a quieter grill should be considered. Quiet grills or diffusers are made of heavy-gauge metal with widely spaced streamlined deflectors devoid of sharp corners and edges.

Many practical suggestions for quieting home heating, air-conditioning, and plumbing systems are given in an inexpensive Bureau of Standards (now NIST) handbook (Berendt, Corliss, and Oljavo 1976).

32.10 ■ AIRCRAFT NOISE

Control of aircraft noise is one of the most challenging of urban environmental issues. Airlines transport approximately 80% of all intercity passenger traffic traveling by common

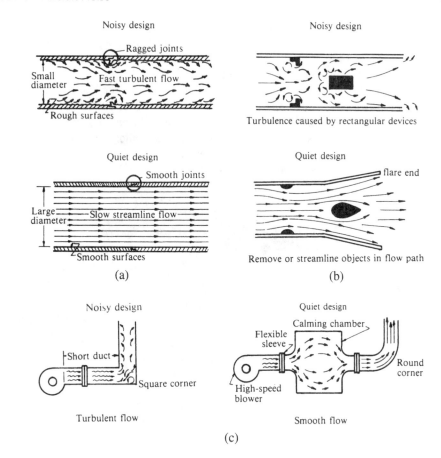

FIGURE 32.9
Examples of duct designs that reduce noise from air turbulence. (From Berendt, Corliss, and Ojalvo, 1976.

carrier in the United States. In spite of efforts to develop high-speed ground transport, the number of aircraft takeoffs and landings near major cities continues to grow. Furthermore, as land prices rise, residential dwellings encroach on noise buffer zones near airports in increasing numbers.

The modern jet aircraft is actually a rather inefficient noise source, radiating less than 0.02% of its total power as sound. Nevertheless, this may exceed 1 kW of sound power because of the prodigious amount of mechanical power generated by the engine. The development of the turbofan engine in 1960 led to greater efficiency and somewhat less total sound power, but it added a new source of annoyance: a sirenlike whine from the fan. Sound power radiation from early long-range jets, such as the Boeing 707 and the Douglas DC-8, exceeded 10 kW, which meant that A-weighted sound levels at 1000 ft often exceeded 100 dB. In the mid-1950s, considerable work was begun on the quieting of large jet engines.

A turbofan jet engine produces two main types of noise. The first is due to the turbulence created when the high-velocity jet of gas reacts with the quiescent atmosphere. This noise, which has a considerable low-frequency component, dominates during takeoff and climb. The second type of noise is the high-pitched whine of the fan, which becomes dominant during a landing approach with reduced power. Two major engineering developments have spearheaded the attack on jet engine noise: the development of acoustic linings for engine nacelles and the *high-bypass-ratio* engines, now used in the widebody 747, DC-10, L-1011, 767, and A-300 aircraft.

A cutaway view of the high-bypass-ratio engine used in the DC-10 is shown in Fig. 32.10. In this type of engine, a fan supplies intake air to the engine but also blows a stream of bypass air, which surrounds the primary jet and mixes with it. The new acoustical linings, some of them capable of functioning at temperatures of 700°F and sound levels up to 170 dB, are especially designed to attenuate fan whine and noise in the frequency range between 1000 and 5000 Hz.

FIGURE 32.10
Cutaway view of high bypass engine used in the DC-10 (From "Air Transport Noise Reduction" by Robert J. Koenig from *Noise Control Engineering Journal* (May/June 1977). Reprinted by permission.)

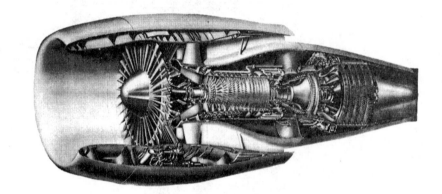

A quantity called the *effective perceived noise level* (EPNL), expressed in decibels, is used in Federal Aviation Regulation (FAR) Part 36 for certification of aircraft noise. (The EPNL combines the sound pressure levels in 24 one-third-octave bands and integrates over the duration of an aircraft flyover). FAR Part 36 imposes noise-level limits on aircraft certified after 1969. Stage 1 aircraft, certified before 1969, are being phased out. Figure 32.11 shows the maximum EPNL limits during approach and takeoff for stage 2 aircraft, certified after 1969, and for stage 3 aircraft, certified after 1975. Stage 1 aircraft with a takeoff weight of more than 75,000 lb have not been allowed to operate at U.S. airports since 1985.

FIGURE 32.11
EPNL limits during approach and takeoff for Stage 2 and Stage 3 aircraft.

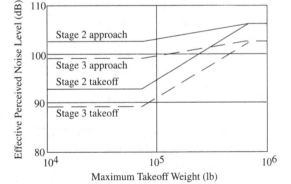

Many modern aircraft operate well below the stage 3 limits in Fig. 32.11. Older aircraft have been retrofitted with new quiet engines or with "hushkits" for engine-noise reduction. Many airports restrict operations by stage 2 aircraft during certain hours, and stage 2 aircraft weighing more than 75,000 lb are being phased out altogether. The FAA estimates that this phaseout of stage 2 aircraft will reduce the number of people living in neighborhoods where the L_{dn} (day-night sound level, see Table 30.1) exceeds 65 dB by 85% from earlier estimates (Eldred 1997).

32.11 ■ SUPERSONIC AIRCRAFT

Although the United States decided to terminate its program to develop a civilian SST in the mid-1960s, the Russian TU-144 and the Anglo-French Concorde were put into service. Supersonic transports present unique noise problems around those airports that they serve.

Both the TU-144 and the Concorde use afterburners for increased thrust during takeoff. Afterburners increase the noise emission, especially at low frequency where atmospheric absorption is very low. These low-frequency sounds are also more apt to excite building vibration and rattle. On the other hand, the afterburners can be cut out shortly after takeoff to reduce noise by sacrificing rate of climb. Effective perceived noise levels on takeoff of the Concorde, the 707-302B (not retrofitted), and the 747 are shown in Fig. 32.12.

The Concorde is only slightly noisier than a 707. However, if we compare it to a DC-10 with high bypass-ratio engines, we find that it emits substantially higher levels.

A *sonic boom* is a pressure transient of short duration that occurs during the flyover of an aircraft at a speed that exceeds the speed of sound (approximately 770 mi/h or 343 m/s at low altitude). A sonic shock wave is analogous to the bow wave produced by a boat moving through water. At ground level, a momentary overpressure of 10 to 100 N/m^2 occurs, followed a moment later by a similar underpressure as the pressure fronts (Mach cones) pass by (see Fig. 32.13). For the Concorde, flying at an altitude of 40,000 ft, the overpressure on the ground is about 20 N/m^2 (about 10^{-3} atm), and the time between overpressure and underpressure is approximately 0.2 to 0.3 s.

The *boom carpet*, a term applied to the zone affected by the sonic boom, is roughly 1 mi wide for each 1000 ft of altitude of the aircraft. Thus, a flyover at very high altitude produces a less intense sonic boom but a wider boom carpet. Regulations of the FAA forbid sonic booms over land areas in United States territory by civilian aircraft, although military craft continue a limited number of supersonic operations over land.

FIGURE 32.12
Effective perceived
noise levels of
several aircraft.
(From Wesler,
1975).

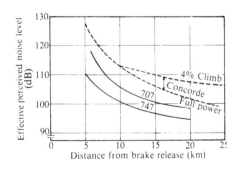

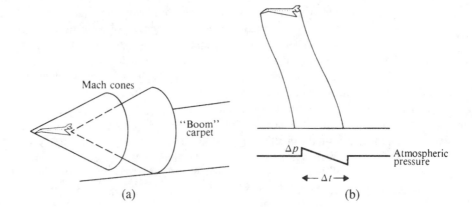

FIGURE 32.13
Structure of the sonic boom generated by aircraft flying at supersonic speed.

The increasing use of helicopters, short-takeoff and landing (STOL) aircraft, and even vertical-takeoff and landing (VTOL) aircraft will present new noise problems, because these craft will operate closer to areas of high population density. Minimizing their noise emission will require careful engineering.

32.12 ■ ULTRASOUND AND INFRASOUND

Ultrasound is sound with frequencies higher than the audible range (greater than 20,000 Hz). *Infrasound* is sound with frequencies lower than the audible range (less than 16 Hz). There are indications that at high enough sound pressure levels, both types of inaudible sound can affect people adversely.

Ultrasonic waves are emitted by jet engines, high-speed drills, cleaning devices, etc. There have been reports of people experiencing nausea, headache, and changes in blood sugar, due to exposures to ultrasound of high intensity. Fortunately, ultrasonic waves are absorbed very strongly by air, so that ultrasound does not travel very far from the source. At sound pressure levels below 105 dB, no adverse effects have been reported (Environmental Protection Agency 1973).

Infrasound of moderate intensity is radiated by a number of natural sources, including waterfalls, volcanoes, ocean waves, earthquakes, wind, and thunder. There are also a number of sources, which radiate higher intensities, such as jet aircraft, air-conditioning systems, and other machinery. Headaches, coughing, blurred vision, and nausea are among symptoms reported from excess exposure to high levels of infrasound (Environmental Protection Agency 1973).

32.13 ■ ACTIVE NOISE CONTROL

Active noise control is sound-field cancellation by electroacoustical means. In its simplest form, an electronic controller drives a loudspeaker or other electromechanical actuator to produce a sound field that is a mirror image of the offending sound in order to reduce the sound. Active noise control complements *passive* noise-control methods such as barriers, enclosures, absorptive treatments, or vibration mounts. Active noise control works well at low frequencies where passive techniques are less effective.

Active noise control works best when the sound field is of relatively low frequency and has a simple geometry. The classical active control problem is a low-frequency tone traveling through a duct, as shown in Fig. 32.14. A microphone samples the sound so that a controller can drive a loudspeaker and create *antisound*. An error microphone samples the resulting signal and adds a correction that can be used to minimize the sound.

Headphones with active noise control, which combine a noise-canceling signal with the desired signal, are used to listen to music or speech in noisy airplanes. They are especially effective in helicopters or propeller aircraft, because blade-passage noise is mainly of low frequency. Speakers in aircraft wall panels can also reduce noise generated as the propeller tips rotate past the aircraft fuselage. As costs are reduced, such systems will probably appear in automobiles as well.

FIGURE 32.14
Reduction of noise in a duct by means of active noise cancellation. A microphone samples the sound so that a controller can drive a loudspeaker and create "anti-sound."

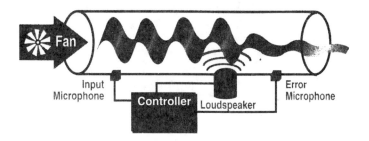

32.14 ■ SUMMARY

Noise-control problems can be dealt with by considering the *source*, the *path*, and the *receiver*. The most effective approach is to quiet the source by damping of vibrating panels, reducing air velocity, etc. Regulation of noise is shared by federal, state, and local governments. Occupational noise should be limited under the provisions of the Occupational Safety and Health Act. Labeling of certain noisy appliances and hearing protectors has been proposed by the EPA.

Noise transmission through walls decreases as the frequency of the noise and the wall mass increase; attenuation of low-frequency noise required heavy walls. Wall resonances, excitation of bending waves (at the critical frequency), and leakage through holes can cause excessive sound transmission through the walls. Sound barriers are used as substitutes for walls in open-plan rooms and as aids in the reduction of outdoor noise. Sound reflected and diffracted around a barrier reduces its attenuation substantially below that of a comparable airtight wall, however. Enclosures are frequently used to reduce the noise from machines. Porous surfaces absorb well at high frequency, whereas free panels absorb well at low frequency. Perforated panels and similar materials combine both principles of absorption.

Heating, ventilating, and air-conditioning systems distribute noise of aerodynamic and mechanical origin. Aircraft noise is being reduced by technological advances and by government regulation, but this noise reduction is offset to some extent by the increase in the number of flight operations per day at major airports. Thus, aircraft noise continues to be one of our major environmental noise problems. Sonic boom is a pressure transient that occurs during the flyover of an aircraft at supersonic speed. The term ultrasonic and infrasonic refer to sound above and below the audible range of frequencies, respectively.

REFERENCES AND SUGGESTED READINGS

American National Standards Institute (1975). "Method for Rating the Sound Power Spectra of Small Stationary Noise Sources," ASA Std. 4-1975 (ANSI 3.17-1975) New York: Acoustical Society of America.

Bazley, E. N. (1966). *The Airborne Sound Insulation of Partitions*. London: Her Majesty's Stationary Office.

Berendt, R. D., E. L. R. Corliss, and M. S. Oljavo (1976). *Quieting: A Practical Guide to Noise Control*, Handbook 119. Washington, D.C.: National Bureau of Standards.

Beranek, L. L. (1971). *Noise and Vibration Control*. New York: McGraw-Hill.

Eldred, K. M. (1997). "Airport Noise," in *Encyclopedia of Acoustics*, Vol. 2, p. 1059, ed. M. J. Crocker. New York: Wiley.

Environmental Protection Agency (1973). *Public Health and Welfare Criterion for Noise*, Report 550/9-73-003. U.S.

Environmental Protection Agency, Office of Noise Abatement and Control, Washington, D.C.

Magrab, E. G. (1975) *Environmental Noise Control*. New York: Wiley.

Rossing, T. D. (1979). *Environmental Noise Control, Selected Reprints*, Stony Brook, N.Y.: American Association of Physics Teachers.

Ruckman, C. E. (1996). "Active Noise Control," *ECHOES* **6** (Winter).

U.S. Gypsum (1972). *Sound Control Insulation*. Chicago: U.S. Gypsum Company.

Wesler, J. E. (1975). "Noise and Induced Vibration Levels from Concorde and Subsonic Aircraft," *Sound and Vibration* **9**: 10, 18.

GLOSSARY

aerodynamic Having to do with the flow of air and its interaction with objects in its path.

critical frequency The frequency of bending (flexural) waves in a panel that can be excited by sound waves traveling at the same speed.

FAR Part 36 A regulation by the Federal Aviation Administration that sets standards for noise from new aircraft.

Fresnel number The parameter that determines the sound diffracted around a barrier.

IIC (impact isolation class) A number that describes the effectiveness of a ceiling-floor structure in attenuating impact sound.

infrasonic Having a frequency below the audible range.

OSHA (Occupational Safety and Health Agency) The agency that publishes industrial safety standards set by the Occupational Safety and Health Act.

sonic boom Pressure transient that occurs during the flyover of an aircraft faster than the speed of sound.

STC (sound transmission class) A number that describes the effectiveness of a wall structure in attenuating airborne noise.

supersonic Having a speed greater than that of sound (approximately 340 m/s or 770 mi/h).

TL (transmission loss) A number that describes the reduction in the sound transmitted through a wall relative to the incident sound.

turbofan engine A type of jet aircraft engine that uses a large fan to drive air into or through the engine.

ultrasonic Having a frequency above the audible range.

waveguide A device that transmits waves (e.g., sound, light, or radio waves) over a particular path minimizing their tendency to propagate in all directions.

REVIEW QUESTIONS

1. What are three basic elements of a noise-control problem?

2. Where is the most effective place to control noise?

3. What is the maximum allowable noise exposure for a worker during an 8-h day?

4. Why is there a dip in the transmission loss through a wall at the *critical frequency*?

5. What is the transmission loss for an enclosure that has a hole with an area equal to 2% of the total wall area?

6. In what two ways can sound travel around a barrier?

7. In the experiment in Fig. 32.5, why did enclosing the noise source with a sound-absorbing material reduce the sound level by only 2 dB?

8. What are two sources of noise in a turbofan jet engine?

9. What is meant by a high-bypass-ratio engine?

10. What is meant by a stage 3 aircraft?

11. What is infrasound?

12. What is ultrasound?

13. Explain how active noise control works.

14. Is it possible to soundproof a room? How?

QUESTIONS FOR THOUGHT AND DISCUSSION

1. Why is an enclosure made of a thick blanket of an absorbing material ineffective in attenuating noise?

2. Room A in the following figure contains a noisy machine and Room B is an office. You have insufficient absorbing material to cover the walls and ceiling of both rooms. Would you use the available material in Room A, in Room B, or part of it in each?

3. In response to your complaints about the noise from the apartment above you, the landlord offers to install acoustical tile on your ceiling. Would you expect this to be effective in reducing the noise? Why? What measures for reducing the noise would you suggest?

EXERCISES

1. Sound pressure from a dipole source, such as a fan, is proportional to the sixth power of the air velocity (Section 30.4). If the velocity is doubled, how much will the sound pressure increase? How much will the sound pressure level (in dB) increase?

2. A common wall between two rooms has a transmission loss of 30 dB. If the sound level on the source side is 60 dB, how much soundpower is actually transmitted through the wall, if it has an area of 20 m^2?

3. Suppose the wall described in Problem 1 has an opening with an area of 0.2 m^2. What is the transmission loss of the wall with the hole? Is more sound power transmitted through the wall or through the hole?

4. Calculate the total force on one wall of a typical house due to an overpressure of 20 N/m^2 in a sonic boom.

5. If workers are exposed to an A-weighted sound level of 97 dB for 1 h, how long can they then be exposed to 92 dB before they reach the maximum exposure allowed?

6. Using Fig. 32.1, estimate the critical frequency for

    (a) A plywood wall with an area mass density of 10 kg/m^2;

    (b) A glass window with an area mass density of 5 kg/m^2;

    (c) A concrete wall with an area mass density of 500 kg/m^2.

7. Using Fig. 32.1, estimate the transmission loss (TL) at 100 Hz and at 1000 Hz for each of the three walls described in Exercise 6.

EXPERIMENTS FOR HOME, LABORATORY, AND CLASSROOM DEMONSTRATION

Home and Classroom Demonstrations

1. *Noise attenuation with an enclosure*   A foam cylinder is placed over a noise source (such as a doorbell) (Fig. 32.5(b)). Then the source is placed on a foam slab (c). Placing a metal cookie can over the source (d) is slightly better than the foam cylinder. Placing the can over the insulating foam further reduces the noise level (e), whereas placing an absorbing lining in the can (f) gives the lowest level of all. The foam does little to attenuate the sound (b), but it effectively reduces the reverberant field in the enclosure (f).

2. *Leakage in an enclosure*   The effect of a small hole in an enclosure can be demonstrated by enclosing a noise source in a box with holes that can be opened or closed.

Laboratory Experiments

Reduction of noise with an enclosure (Experiment 42 in *Acoustics Laboratory Experiments*)

Mufflers (Experiment 44 in *Acoustics Laboratory Experiments*)

APPENDIX

A

A.1 ■ UNITS: SI AND OTHER SYSTEMS

The preferred system of units, based on the meter, kilogram, and second, is called the *Système International*, or *SI*. It is also referred to as the *MKS system*. It is recommended for all scientific publications and will become more familiar as the United States converts to metric measure. The SI system is used throughout this book, although occasionally the English system is used as well.

The *English system*, in use for many years in the United States, uses the foot as a unit of length, the pound as a unit of force, and the second as a unit of time. It has a number of disadvantages, the greatest of which is that it is not used in most countries of the world. It is a clumsy system and is generally avoided in scientific work.

Another metric system that is in common use is the *cgs system*, based on the centimeter, the gram, and the second. Units in this system are considerably smaller than the corresponding SI units; conversions between the two are relatively simple.

Let us compare these three systems by referring to Newton's law: force = mass × acceleration.

System	Force unit	Mass unit	Acceleration unit
SI (MKS)	newton (N)	kilogram (kg)	meters/second^2 (m/s^2)
cgs	dyne (dyn)	gram (g)	centimeters/second^2 (cm/s^2)
English	pound (lb)	slug (slug)	feet/second^2 (ft/s^2)

Other physical quantities and units are given in Table A.1.

One of the features of the metric system is that units are related by powers of ten. New units can be derived by using standard prefixes. For example, a kilogram is 1000 grams; a centimeter is $\frac{1}{100}$ meter, etc. Table A.2 gives the standard prefixes; Table A.3 gives factors for converting from one system of units to another.

A.2 ■ POWERS OF TEN AND LOGARITHMS

It is convenient to use *scientific notation* to express large and small numbers. The number 30,000 can be obtained by multiplying $30 \times 10,000$, or $3 \times 10 \times 10 \times 10 \times 10$. This is written as 3×10^4 or 3.0×10^4; on some electronic calculators, this appears as 3.0 E4. Table A.4 further illustrates scientific notation.

TABLE A.1 Physical quantities and units

Quantity	Symbol	SI unit	Abbrev.	cgs Unit	Abbrev.	English Unit	Abbrev.	Formula
Distance, length	d, L	meter	m	centimeter	cm	foot	ft	
Mass	m	kilogram	kg	gram	g	slug	—	
Time	t	second	s	second	s	second	s	
Speed	v	(meters/s)	m/s	(centimeters/ second)	cm/s	(feet/second)	ft/s	$v = d/t$
Force	F	newton	N	dyne	dyn	pound	lb	$F = ma$
Energy,	E	joule	J	erg	erg	(foot-pound)	ft-lb	
work	$\mathcal{W}$	joule	J	erg	erg	(foot-pound)	ft-lb	$W = FY$
Pressure	p	pascal (N/m^2)	Pa	(dyne/cm^2)	dyn/cm^2	(pounds/inch^2)	lb/in.^{2}	$p = F/A$
Power								
mechanical	$\mathcal{P}$	watt	W	(ergs/second)	erg/s	horsepower	hp	$\mathcal{P} = \mathcal{W}/t$
sound	W	watt	W					$W = pU$
electrical	P	watt	W					$P = I^2R = V^2/R$
Potential	V	volt	V					
Current	I	ampere	A					$I = V/R$
Resistance, impedance	R, Z	ohm	Ω					
Frequency	f	hertz	Hz					$f = 1/T$

TABLE A.2 Standard prefixes

Prefix	Symbol	Factor
terra	T	10^{12}
giga	G	10^9
mega	M	10^6
kilo	k	10^3
milli	m	10^{-3}
micro	μ	10^{-6}
nano	n	10^{-9}
pico	p	10^{-12}

Other prefixes are peta (10^{15}), exa (10^{18}), deci (10^{-1}), centi (10^{-2}), femto (10^{-15}), and atto (10^{-18}).

TABLE A.3 Conversion of units

1 meter = 39.37 in. = 3.281 ft
1 kilogram = 0.0685 slug
1 newton = 10^5 dyne = 0.225 lb
1 joule = 10^7 ergs = 0.738 ft-lb
1 pascal (N/m^2) = 10 dyn/cm^2 = 1.45×10^{-4} lb/in^2 = 9.872×10^{-6} atm
1 foot = 0.3048 m
1 pound = 4.448 N
1 horsepower = 746 W

EXAMPLE A.1

$$23{,}500 = 2.35 \times 10^4 \text{ or } 2.35 \text{ E4};$$

$$0.084 = 8.4 \times 10^{-2} \text{ or } 8.4 \text{ E} - 2;$$

TABLE A.4   Scientific notation

$1,000,000 = 10^6$ or E6
$100,000 = 10^5$ or E5
$10,000 = 10^4$ or E4
$1,000 = 10^3$ or E3
$100 = 10^2$ or E2
$10 = 10^1$ or E1
$1 = 10^0$ or E0
$.1 = 10^{-1}$ or E − 1
$.01 = 10^{-2}$ or E − 2
$.001 = 10^{-3}$ or E − 3
$.0001 = 10^{-4}$ or E − 4
$.00001 = 10^{-5}$ or E − 5
$.000001 = 10^{-6}$ or E − 6

$$0.005307 = 5.307 \times 10^{-3} \text{ or } 5.307 \text{ E} - 3;$$

$$186,000 = 1.86 \times 10^5 \text{ or } 1.86 \text{ E5}.$$

Arithmetic in Scientific Notation

To *multiply* in scientific notation, we add the exponents; to *divide*, we subtract the exponents. For example:

$$(3.1 \times 10^4) \times (4.2 \times 10^5) = 13.02 \times 10^9 = 1.302 \times 10^{10};$$

$$(3.1 \times 10^4) \times (2.3 \times 10^{-2}) = 7.13 \times 10^2;$$

TABLE A.5   Logarithms of some numbers

x	$\log x$	x	$\log x$	x	$\log x$
1	0	10	1.000	$\frac{1}{2}$	−0.301
2	0.301	20	1.301	$\frac{1}{3}$	−0.477
3	0.477	30	1.477	$\frac{1}{4}$	−0.602
4	0.602	40	1.602	$\frac{1}{5}$	−0.699
5	0.699	50	1.699	0.1	−1.000
6	0.778	60	1.778	0.2	−0.699
7	0.845	70	1.845	0.3	−0.523
8	0.903	80	1.903	0.4	−0.398
9	0.954	90	1.954	0.5	−0.301
10	1.000	100	2.000		

TABLE A.6   Identities

$\log AB = \log A + \log B$
$\log A/B = \log A - \log B$
$\log A^n = n \log A$
$\log 1/A = -\log A$

$$(6.4 \times 10^5)/(4.1 \times 10^3) = 1.56 \times 10^2;$$
$$(7.35 \times 10^3)/(3.2 \times 10^{-2}) = 2.30 \times 10^5.$$

Logarithms

The logarithm to the base 10 of a number x is the power to which 10 must be raised to equal x. That is, if $x = 10^y$, then $y = \log x$. Logarithms of some numbers are given in Table A.5, and several useful identities involving logarithms are listed in Table A.6.

A.3 ■ SOLVING EQUATIONS

To solve an equation for an unknown quantity, perform the same arithmetic operations to both sides of the equation, step by step, until the unknown quantity stands alone.

$6t = 30$ Divide by 6 to get $t = 5$.

$5x + 7 = 32$ Subtract 7: $5x = 25$;
Divide by 5: $x = 5$.

$4p^2 + 8 = 44$ Subtract 8: $4p^2 = 36$;
Divide by 4: $p^2 = 9$.
Take the square root: $p = 3$ (strictly speaking,
$p = \pm 3$, because $(-3)(-3) = 9$ also).

A.4 ■ GRAPHS

Graphs are very useful in showing how one quantity relates to another. Often one of the quantities is time. A graph of pressure versus time, for example, which represents the waveform of a sound wave, conveys a great deal of information about the sound.

A Linear Graph

If two quantities are related by the equation $y = 2x + 3$, the graph of y versus x will be a straight line having a slope of 2, as shown in Fig. A.1.

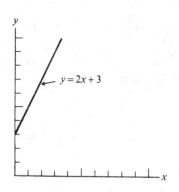

FIGURE A.1

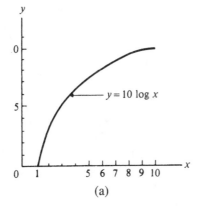

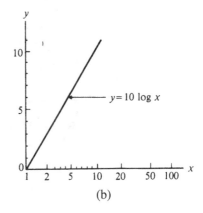

FIGURE A.2   (a)   (b)

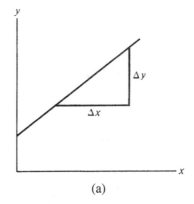

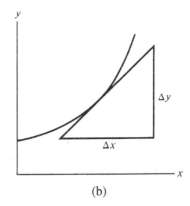

FIGURE A.3   (a)   (b)

A Logarithmic Graph

If two quantities are related by the equation $y = 10 \log x$, the graph of y versus x will not be a straight line, unless the scale for x is logarithmic, as shown in Fig. A.2.

Determining the Slope of a Curve

If a curve is a straight line, its slope is the number of units of *rise* for each unit of *run*. The slope can be calculated by using any two points on the curve and determining Δy and Δx between those two points. If a curve is not a straight line, a tangent should be drawn to the curve at a particular point and Δy and Δx determined for two points on the tangent line, as shown in Fig. A.3(b).

A.5 ■ DECIBELS

Sound pressure level L_p, sound intensity level L_I, and sound power level L_W are all measured in decibels. The formulas are as follows:

$$L_p = 20 \log p/p_0, \qquad p_0 = 2 \times 10^{-5} \text{ N/m}^2 = 20 \ \mu\text{Pa} \qquad \text{(micropascals)};$$

$$L_I = 10 \log I/I_0, \qquad I_0 = 10^{-12} \text{ W/m}^2 \qquad \text{(watts/meter}^2\text{)};$$

$$L_W = 10 \log W/W_0, \quad W_0 = 10^{-12} \text{ W} \qquad \text{(watts)}.$$

A.6 ■ A PROPOSAL FOR THREE NEW CLEFS

Music is most often written on two staffs, designated by treble and bass clefs. The staff with the bass clef extends from G_2 to A_3 (98 to 220 Hz), and the staff with the treble clef extends from E_4 to F_5 (330 to 698 Hz). However, the piano keyboard extends from A_0 to C_8 (27.5 to 4186 Hz), and the range of human hearing may extends to C_{10} (16,744 Hz) and beyond. Notes above and below the staff are written by using either ledger lines (extra lines added above or below the staff) or notations such as "8va," "15va," etc. Figure A.4 shows notations commonly used for the notes C_6 (1047 Hz) and C_7 (2093 Hz).

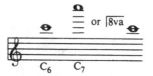

FIGURE A.4

Adding two new staffs above the treble clef and one new staff below the bass clef avoids the use of ledger lines and the various "-va" notations. The five staffs together allow notation of over nine octaves (F_0 to G_9) without ledger lines. The audible overtones of all musical instruments, as well as the fundamental notes, can be written on musical staffs. The entire piano range (A_0 to C_8) requires only the lower four staffs, the upper one being for overtones. The new clefs are shown in Fig. A.5.

I have named these clefs *sub-bass* (SB), *super-treble* (ST), and *supra-super-treble*, (SST). The SB clef is two octaves below the bass clef, so its staff has the same note names;

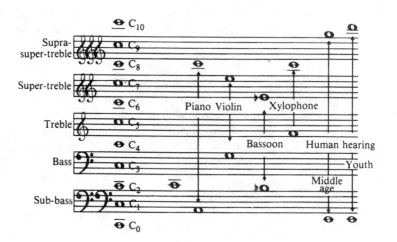

FIGURE A.5

the ST and SST clefs are two and four octaves above the treble clef, respectively, so their staffs have the same note names as the treble. Note that whereas there is one line (middle C) between the treble and bass staffs, there are two lines between each of the others.

A.7 ■ NOTATION USED FOR NOTES

Throughout this book we have used the U.S. standard note designation, which is widely used in world literature as well. However, many European musicians use a different notation for octaves and designate sharps and flats by "is" and "es" respectively, following the note name. B_3 is denoted by "h" and $B_3^\flat$, by "b."

FIGURE A.6

U.S.A. standard	C_0	C_1	C_2	C_3	C_4	$F_4^\sharp$	$A_4^\flat$	$B_4^\flat$	B_4	C_5	C_6	C_7	C_8
European	C_2	C_1	C_0	c	c^1	fis^1	aes^1	b^1	h^1	c^2	c^3	c^4	c^5
Some organ music	CCCC	CCC	CC	C	c^1					c^2	c^3	c^4	c^5

A.8 ■ AUDIO COMPRESSION

Digital audio signals typically consist of 16-bit samples recorded at a sampling rate more than twice the actual audio bandwidth (44.1 kHz for CDs, 48 kHz for DATs). Thus, representing 1 s of stereo music takes more than 1.4 megabits (0.175 Mbytes). Digital storage media have huge capacities (one CD stores 782 Mbytes, for example). However, transmitting digital audio can be a problem. A 28.8 kbps modem, for example, would take around 49 min to transmit each minute of music. Thus, the great interest in digital audio compression is not at all surprising.

As pointed out in Section 29.4, there are three ways to reduce the bit rate: by reducing the sampling rate or number of bits per sample; by mathematically reducing redundancy; or by using psychoacoustic principles to assign to each sample only the number of bits that can be heard. Most often we do not wish to compromise the sound quality by reducing the sampling rate or number of bits per sample, so we examine the other two possibilities.

Reducing the redundancy is done by using a mathematical algorithm such as a Huffman code, as explained in Section 29.4. This is termed a *lossless* compression, because no data are actually lost. Unfortunately the reduction in bit rate for representing music is generally less than a factor of 2. Adding preprocessing using linear prediction (such as LPAC, linear predictive audio compression) to eliminate statistical dependence within the signal may increase the bit rate reduction to between 2 or 3. Nevertheless, lossless compression is generally combined with some type of *lossy* compression.

Most lossy compression systems make use of psychoacoustic principles to decide which features of a sound are audible and, therefore, need to be coded. Quantization noise, which

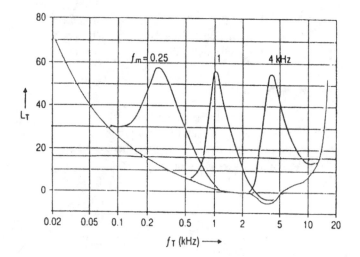

is a necessary by-product of digital sound and is inversely related to the number of bits by which samples are represented (see Section 21.5), is allowed up to a level that just fails to be heard.

Compression Using Psychoacoustics

An early compression system that used psychoacoustic principles was MUSICAM (masking pattern for adapted universal coding and multiplexing), which allowed the reduction of CD data by a factor of about 4 without noticeable loss of sound quality. Sony adopted a similar technique (ATRAC) for its MiniDisc technology.

Figure A.7 shows how the hearing threshold is raised in the presence of a masking sound, in this case three narrow bands of noise centered around 250, 1000, and 4000 Hz, but it could just as easily represent a musical sound with these three prominent frequencies. Note the evidence of asymmetrical masking; the raised threshold falls off more slowly at frequencies above the masker (see Section 6.10). Broadband quantization noise is represented by a horizontal line in Fig. A.7.

It is easy to see that if too few bits are used to encode the digital sound, quantization noise may become audible in the deep valleys between the tone frequencies. This can easily occur if 8-bit or 12-bit codes are used (in which cases the quantization noise will be 48 or 72 dB below the maximum sound level, respectively), but with the 16-bit code used in CD and DAT (which reduces quantization noise to 90 dB below the maximum level), this will hardly ever be audible.

If we could shape the spectrum of quantization noise according to the spectrum of the signal, we could allow much greater amounts of noise without it actually being heard. MUSICAM and its descendants achieve this by dividing the signal up into a series of subbands by means of band-pass filters. The optimal way to choose these filters would be to emulate the critical bands of hearing (see Section 6.7). The outputs from each of the filters would then be coded to a number of bits such that the quantization noise stays below the masking threshold, as shown in Fig. A.8.

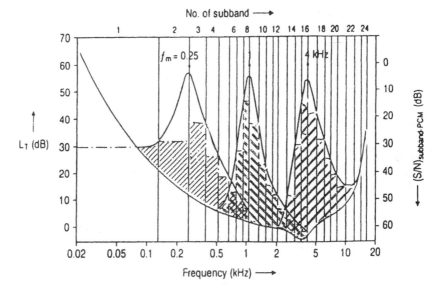

FIGURE A.8
This is the same as in Fig. A.7, but quantization noise is allowed in 24 subbands (Houtsma 1992).

In practice, however, it is much easier to use digital filters having a constant bandwidth, even though more filters are required. MUSICAM uses 32 filters of equal bandwidth, which comes out to be about 700 Hz (half the sampling rate divided by 32).

Dynamic Bit Allocation

The spectrum of the actual signal to be recorded, of course, will not be static, as shown in Figs. A.7 and A.8, but rather will be a rapidly changing one. In our ears, spectral analysis is performed for a sample that extends over about the past 5 to 15 ms. MUSICAM, on the other hand, computes the spectrum every 24 ms and allocates to each subband the necessary number of bits to make the quantization noise inaudible. Because the spectral analysis, threshold computation, and bit allocation are done for very short signal segments, the coding system is dynamic and can keep up with temporal and spectral details (including transients) almost as well as our auditory system can.

Comparison of Compressed and Uncompressed Sound

Scientists at the Institute for Perception Research in the Netherlands performed some psychoacoustical experiments to find out what effect a reduced bit rate would have on sound quality (Houtsma 1992). Listeners heard two sequential music fragments, one taken directly from a CD and the other with a reduced bit rate; they were asked to respond whether the CD version came first or second. The resulting *psychometric functions* for one listener, showing the percentage of correct responses as a function of bit rate, is shown in Fig. A.9. The bit rate for a 75% score, generally taken as the discrimination threshold, is a little below 3 bits/sample. The required bit rate was found to depend on the character of the music; a tenor and orchestra required 2.48 bits/sample, on the average, whereas a viola sound requires 3.16 bits/sample. An average bit rate of 4 bits/sample appeared to be adequate to make the compressed signal virtually indistinguishable from the original one.

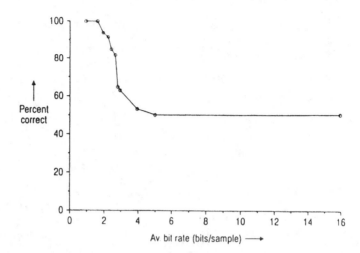

FIGURE A.9
Psychometric function for one listener to a fragment of Mozart's *Requiem*, showing the percentage of correct responses as a function of bit rate (Houtsma 1992).

The MPEG Family

MPEG (The Moving Picture Experts Group) was established in 1988 as a working group of the International Standards Organization (ISO) to develop standards for coded representation of digital audio and video. MUSICAM became the basis for MPEG Audio Layers I and II, but with the finalization of MPEG-1 in 1992, the original MUSICAM algorithm was modified. Later standards include MPEG-2, which is aimed at digital television; however, the audio part addresses 5.1 channel home theater sound (see Section 20.10), MPEG-4, and MPEG-7 (see Section 29.4).

In both MPEG-1 and MPEG-2, three layers are defined. These layers represent a family of coding algorithms whose complexity and efficiency increase from layer I to layer III. Layer I results in a data reduction of about 4; layer II results in a reduction of 6 to 8; layer III results in a reduction of 10 to 12. Layer III requires the lowest bit rate, or for a given bit rate it achieves the highest sound quality. MPEG-1 layer III, which has come to be known as MP3, has become very popular for coding music to be transmitted on the Internet.

An improvement in efficiency compared to MP3 is MPEG-2 AAC (advanced audio coding). It offers a slightly better compression ratio than MP3, and it can include up to 48 audio channels, so it is suitable for surround-sound coding.

REFERENCES AND SUGGESTED READINGS

Houtsma, A. J. M. (1992). "Psychophysics and Modern Digital Audio Technology," *Philips J. Res* **47**: 3.

Rossing, T. D. (1994). "Digital Techniques for Sound Recording," *Phys. Teach.* **32**: 458.

A.9 ■ HOW TRANSISTORS WORK

The nucleus of a hydrogen atom, the simplest atom known, has a single proton, which carries an electric charge of $+e$. A single electron, which has a charge of $-e$, can be thought of as orbiting around the nucleus. According to the quantum theory, the atom can have only discrete amounts of energy, given by $E_n = -13.6/n^2$ eV, where $n = 1, 2, 3 \ldots, \infty$. One electron volt (eV) equals 1.6×10^{-16} J, the amount of energy an electron would acquire in moving through a potential difference (voltage) of 1 V, and it is a convenient way for expressing energy on the atomic scale. When $n = 1$, $E_n = -13.6$ eV, and the atom is in its *ground state*. As $n \to \infty$, $E_n \to 0$, and the electron is said to be a *free electron*, i.e., it has enough energy to escape from the attractive force of the nucleus. The energy levels of a hydrogen atom are shown in Fig. A.10.

Crystalline solids are made up of many nuclei arranged in a regular array. Electrons experience electrical forces from not only their own nuclei but also neighboring nuclei as well. Thus, the energy levels of a single isolated atom give way to energy *bands*. Each energy band is made up of many, many energy levels so close together that they are hardly distinguishable. Between the bands there may be gaps, which represent electron energies not permitted by quantum mechanics. The highest band is the *conduction band*, and electrons in this band are free to move around the crystal when an electric field is applied. The lowest band is the *valence band*, and electrons in this band are tightly bound to their nucleus.

The electrical conductivity of a solid is largely determined by its energy bands. In an insulator there is a large gap between the valence band and the conduction band, and the conduction band is nearly empty. If the valence and conduction bands are close together (or even overlap), as in metals, the solid will be a good conductor. If the band gap is of the order of 1 eV or so, the solid will be a *semiconductor*. In a pure semiconductor, the number of electrons in the conduction band is small, and it depends strongly upon temperature: the greater the temperature, the more electrons that are able to break loose from their parent nucleus and join the conduction band. The bandgap in silicon, for example, is 1.1 eV, whereas the average energy of an electron at room temperature is about 0.03 eV, so only a few unusually energetic electrons have enough energy to break loose.

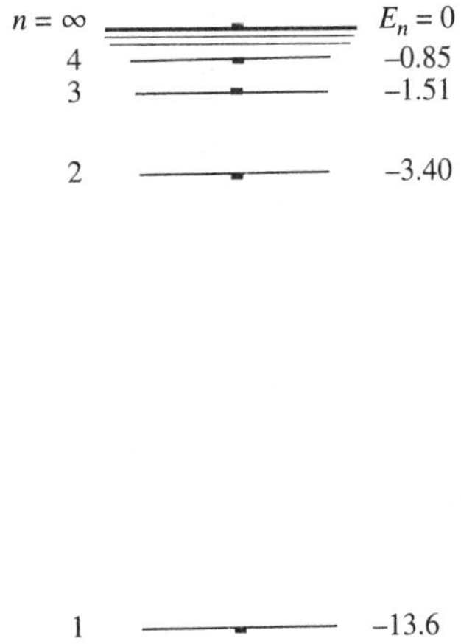

$$
\begin{array}{lll}
n = \infty & \text{———————} & E_n = 0 \\
4 & \text{———————} & -0.85 \\
3 & \text{———————} & -1.51 \\
2 & \text{———————} & -3.40 \\
1 & \text{———————} & -13.6
\end{array}
$$

FIGURE A.10

Energy levels in a hydrogen atom; n is the principle quantum number and E_n is the energy level relative to that of a free electron ($E = 0$).

n- and p-type Semiconductors

When impurities are added to pure silicon (or other semiconductors), their behavior changes rather dramatically. For example, replacing a silicon atom with one of phosphorus, which has five electrons in its outer shell compared to the four in silicon, results in a rather loosely held electron in the crystal lattice. This electron is in an energy level only a fraction of an electron volt below the conduction band, so it is easily freed from its parent nucleus by applying an electrical voltage or by heating the solid. Thus it can move around the lattice and transport electrical charge, creating an electrical current. The number of free electrons (in the conduction band) depends upon the concentration of impurity atoms as well as the temperature. Silicon that has been *doped* with phosphorus, or other pentavalent impurity is called an *n-type* semiconductor because it conducts by the movement of negative charge carriers (electrons).

If the silicon crystal is doped with boron (or similar material with only three electrons in its outer shell), it becomes a *p-type* semiconductor. The vacancy in the lattice would result in a new energy level above the valence band, which electrons from other atoms could easily occupy. Thus, the vacancy, or hole, could move around the lattice, transporting a net positive charge from place to place; hence the name *p*-type. The hole behaves rather much as an electronlike "particle" carrying a positive charge. The electrical conductivity of *n*-type and *p*-type silicon is much greater than that of pure silicon.

pn Junction

When *p*-type and *n*-type semiconductors are placed in intimate contact, electrons diffuse across the junction, creating a positive charge on the *n*-side and a negative charge on the *p*-side, as shown in Fig. A.11. This results in a potential "hill," which governs the number of electrons and holes that can diffuse across it. Forward-biasing the junction (e.g., connecting the positive terminal of a battery to the *p*-side of the junction) lowers the hill and allows more carriers to cross, thus creating a forward current. Reverse-biasing the junction (by connecting the negative terminal of a battery to the *p*-side) raises the hill and results in the depletion of charge carriers near the junction, as shown in Fig. A.12. A *pn* junction diode thus allows current to flow easily in the forward direction but not in the reverse direction. If the reverse voltage becomes too large, it is possible for breakdown to occur, and the diode may be destroyed by passing too much current and heating up.

Although charge carriers (electrons and holes) are generally created by applying a voltage to a *pn* junction, they can also be created by shining light on the junction. A photodi-

FIGURE A.11
(a) Energy bands in *p*-type and *n*-type semiconductors. Acceptor and donor levels appear in the bandgap.
(b) Diffusion of electrons across a *pn* junction results in a potential hill.

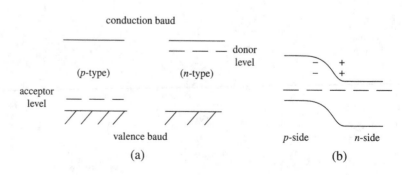

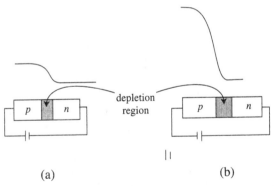

FIGURE A.12   (a) Forward-biasing a junction lowers the potential hill and narrows the charge depletion region, so more carriers cross; backward-biasing raises the hill and widens the charge depletion region, so few charge carriers cross. (b) Current increases rapidly with increasing forward voltage, but the reverse current remains small until the breakdown voltage is reached.

ode, or photo cell, converts light energy to electrical energy in this way. A photon (quantum, or *particle*, or light) "knocks" an electron out of an atom, thus creating an electron-hole pair. This creates a small potential difference across the *pn* junction, which can be used to power a calculator, for example. On a larger scale, photodiodes are the basis of solar energy panels that create electrical power. A light-emitting diode (LED) uses the same principle in reverse. Forward current in a diode creates a free electron on the *p*-side of a junction and a hole on the *n*-side. When these carriers recombine, a photon is emitted.

Junction Transistor

A bipolar junction transistor consists of two *pn* junctions back to back. One of the junctions (the emitter-base junction in Fig. A.13) is forward-biased, whereas the other (the collector-base junction) is reverse-biased. The base region is made very thin, so that carriers introduced into the base region by the emitter-base current will quickly diffuse across the base region, and most of them will flow to the collector. In Fig. A.13(a), the majority carriers are holes, and the holes travel from emitter to collector, both of which are of *p*-type

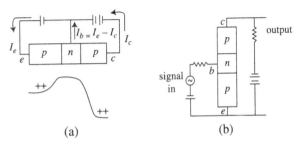

FIGURE A.13   (a) In a bipolar junction transistor, the emitter-base junction is forward-biased, whereas the collector-base junction is reversed-biased. (b) In a common-base amplifier, the input signal is applied to the base, whereas the output is from the oscillator.

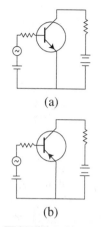

(a)

(b)

FIGURE A.14
(a) A common-emitter
circuit with an *npn*
transistor; (b) a
common-emitter
circuit with a *pnp*
transistor.

material. The small base current I_b is the difference between the emitter current I_e and the collector current I_c.

The most common arrangement of a transistor in a circuit is the common-emitter arrangement shown in Fig. A.13(b). The emitter junction is forward-biased and the collector junction is reverse-biased, as before, but in addition to the bias voltage, a signal voltage is applied to the base. This signal voltage modulates the base current, which, in turn, controls the collector current. Because the collector current is so much larger than the base current, however, a small change in the base current produces a much larger change in the collector current. Thus current amplification results. By making the load resistor in the collector circuit large, a large voltage gain can result as well.

Figure A.14(b) shows a simple common-emitter amplifier circuit using a *pnp* transistor, and Fig. A.14(a) shows a common-emitter amplifier using an *npn* transistor. Note the opposite polarities of the batteries in the base and collector circuits used with the two types of transistors. Having two types of junction transistors gives the circuit designer flexibility in designing circuits. The output circuits in audio amplifiers, for example, often use *complementary pairs* of *npn* and *pnp* transistors.

Field-Effect Transistor

The field-effect transistor (FET) is a device for modifying the flow of current between two terminals (the *source* and the *drain*) in a slab of semiconductor material under the control of an electric field induced by a third terminal (the *gate*). Whereas a junction transistor is basically a current amplifier with a rather low input impedance, a field-effect transistor is a voltage amplifier with a high input impedance. The control mechanism is the creation and control of a depletion layer, as in all reverse-biased junctions. As shown in the *n*-channel FET in Fig. A.15, a current flows in the channel of *n*-type material, whereas the gate is of *p*-type material. Symbols for *n*-channel and *p*-channel FETs are shown in Fig. A.16 (compare with the symbols for the junction transistors in Fig. 18.21).

Most field-effect transistors in common use are either metal-semiconductor field-effect transistor (MESFET) type or the metal-oxide-semiconductor (MOSFET) type rather than the junction (JFET) type shown in Fig. A.15, but the basic principle remains the same.

Although there are several other types of transistors, bipolar junction transistors and field-effect transistors are the principal types of transistors used in analog and digital audio circuits.

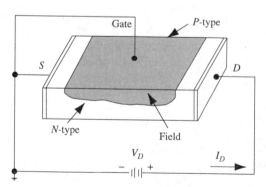

FIGURE A.15
An *n*-channel FET
with a gate of
p-type material.

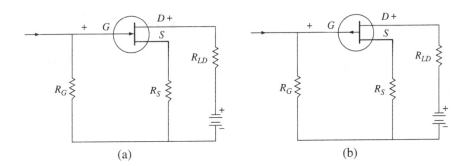

FIGURE A.16
(a) *n*-channel
field-effect
transistor circuit;
(b) *p*-channel
field-effect
transistor circuit.

A.10 ■ A LITTLE BIT OF TRIGONOMETRY

In Section 8.2 we pointed out that simple harmonic motion is often referred to as sinusoidal motion, because the projection of circular motion on the *y*-axis is given analytically by a trigonometric function called the sine (sin); i.e.,

$$y = A \sin \phi = A \sin 360 \frac{t}{T} \qquad (A.1)$$

where 360 is the number of degrees in a complete circle, t/T is the fraction of a complete circle subtended by the angle ϕ in time t, T is a complete period, and sin is the abbreviation for sine. It was pointed out that the language of trigonometry would rarely be used in this book. However, on occasion it is advantageous to use it, and so we will explain some basic principles of this language.

The trigonometric functions of an angle are defined as ratios to the lengths of the sides of a right angle erected on this angle (see Fig. A.17). The *sine*, *cosine*, and *tangent* of angle ϕ are defined as

$$\sin \phi = y/r; \qquad (A.2)$$

$$\cos \phi = x/r; \qquad (A.3)$$

$$\tan \phi = y/x. \qquad (A.4)$$

■ ─────────────────────

EXAMPLE A.2   Find the sine, cosine, and tangent for angles of 0, 90°, and 45°.

FIGURE A.17
The trigonometric
functions of an
angle are defined as
ratios of the lengths
of the sides of a
right triangle with
this angle.

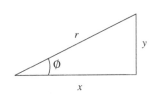

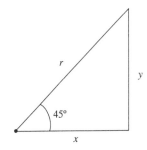

FIGURE A.18
Right angle with
two 45° angles and
two equal sides.

Solution For an angle of 0, the opposite side is zero ($y = 0$), and the adjacent side coincides with the hypotenuse ($x = r$). Hence, $\sin 0 = 0$; $\cos 0 = 1$; and $\tan 0 = 0$.

For an angle of 90°, the adjacent side is zero ($x = 0$), and the opposite side coincides with the hypotenuse ($y = r$). Hence $\sin 90° = 1$; $\cos 90° = 0$; and $\tan 90° = 1/0 = \infty$.

For an angle of 45° (see Fig. A.18), the adjacent and the opposite sides have the same length ($x = y$) and the hypotenuse has a length $\sqrt{2}$ times the length of either side ($r = \sqrt{2}x = \sqrt{2}y$). Hence $\sin 45° = 1/\sqrt{2}$; $\cos 45° = 1/\sqrt{2}$; and $\tan 45° = 1$.

The definitions of sin, cos, and tan are also valid for angles greater than 90°, such as the angle in Fig. A.19, where the quantities x and y are interpreted as the rectangular coordinates of the point P. For angles larger than 90°, however, one or both of the coordinates x and y are negative, so some of the trigonometric functions will also be negative. For example:

$$\sin 135° = \sin 45° = 1/\sqrt{2}; \qquad \cos 135° = -\cos 45° = -1/\sqrt{2};$$

$$\tan 135° = -1.$$

From definitions A.2–A.4 it is easy to see that $\tan\phi = \sin\phi/\cos\phi$. Also, because $x^2 + y^2 = r^2$ (according to the Pythagorean theorem), we can substitute $x = r\cos\phi$ and $y = \cos\phi$ to obtain $r^2\cos^2\phi + r^2\sin^2\phi = r^2$. Dividing each term by r^2, we obtain

$$\cos^2\phi + \sin^2\phi = 1 \tag{A.5}$$

Inverse Trigonometric Functions

Given a trigonometric function, what is the associated angle? We frequently wish to answer this question, so most scientific calculators have inverse trigonometric function keys labeled either *arcsin* or *sin*$^{-1}$ (similarly for *cos* and *tan*) to perform the calculation for us. On some calculators one presses the INV key followed by the trigonometric function key to obtain the inverse function.

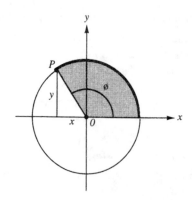

FIGURE A.19
Trigonometric functions of an angle greater than 90°.

A.11 Designing Loudspeaker Enclosures **767**

EXAMPLE A.3

(a) Find $\sin^{-1} 0.5$.
(b) Find $\cos^{-1} 0.4$.

Solution

(a) Key in 0.5; then press INV followed by sin or whatever keys give $\sin^{-1}$. The answer is
    $30°$.
(b) Key in 0.4, followed by $+/-$ to make it negative, followed by INV and cos. The answer
    is $114°$. (The cosine is negative between $90°$ and $180°$).

Angles in Radians

The angle in Fig. A.17 is expressed in *degrees*. However, angles are sometimes expressed in
radians (rad). A radian is an angle that subtends a portion of the circumference equal to the
radius. Just as there are $360°$ in a circle, there are 2π radians in a circle. Any angle given in
radians is an appropriate fraction of 2π. For example, $90°$ (one-fourth of a complete circle)
is $\pi/2$ rad and $45°$ (one-eighth of a complete circle) is $\pi/4$ rad. You can generally tell from
the context whether angles are being expressed in degrees or radians; you should learn to
be comfortable with either one. (When you turn on a scientific calculator, it will generally
find trigonometric functions of numbers keyed in degrees. You can generally switch it to
read radians by pressing an appropriate key.)

Graphs of Trigonometric Expressions

Figure 8.2 shows how simple harmonic motion can be represented as the projection of a
vertical axis of a point P moving around a circle at a uniform rate. The resulting curve is
really a graph of $\sin 360t/T$ (angle in degrees) or $\sin 2\pi t/T$ (angle in radians) from $t = 0$
to T. If the graph began one-fourth of a cycle (period) later, it would be a graph of $\cos 2\pi t$.
The only difference between graphs of the sine and cosine of a quantity is the initial phase.
Graphs of $\sin\phi$, $\cos\phi$, and $\tan\phi$ are shown in Fig. A.20. Often, the term *sine wave* is used
to describe a graph of $\sin 2\pi f t$ or, in other words, a pure tone of single frequency f.

Is it necessary to understand trigonometry in order to understand the use of sine and
cosine functions in acoustics or music? Probably not, but it certainly make the functions
much more meaningful!

A.11 ■ DESIGNING LOUDSPEAKER ENCLOSURES

Designing an Air-Suspension Enclosure

The design of loudspeaker enclosures is aided by using Thiele-Small parameters developed in the early 1960s by two Australian engineers. The simplest enclosure is a closed
box. Loudspeaker manufacturers commonly include the following Thiele-Small parameters with loudspeaker drivers:

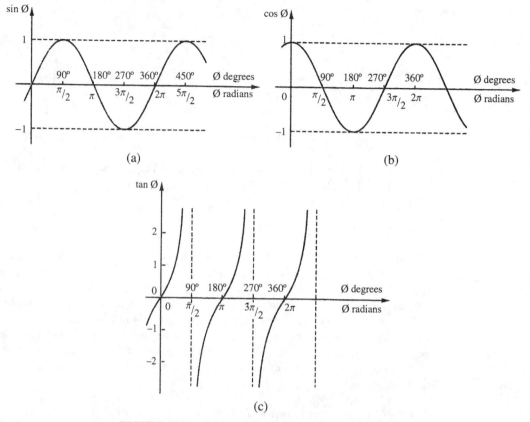

FIGURE A.20   Graphs of (a) $\sin\phi$, (b) $\cos\phi$, and (c) $\tan\phi$.

f_d the driver's free-air resonance frequency

Q_d the driver's Q factor

V_d the driver's compliance (stated as equivalent air volume as discussed above)

The design process involves finding three other Thiele-Small parameters that characterize the loudspeaker system, then building a closed-box enclosure to meet these parameters:

f_s the system resonance

Q_s the system Q factor

V_B the volume of the enclosure

Because placing a loudspeaker driver in a closed box increases the resonance frequency and Q, it is important to have a driver with low enough resonance frequency for its Q to adapt well to a closed-box enclosure. This is determined by calculating an F factor (the resonance frequency of a driver in a closed box with Q_d of 1), where

$$F = \frac{f_d}{Q_d}. \tag{A.6}$$

For best performance, the F factor for a loudspeaker in a closed box should be around 50. If the F factor is too high, the driver should be used with a reflex enclosure.

Once an appropriate loudspeaker driver has been selected, the volume of the box and system resonance frequency are calculated. Start by selecting a system Q factor (Q_s) based on the desired bass response of your system. Next calculate the alpha of the system where

$$\alpha = \left[\frac{Q_s}{Q_d}\right]^2 - 1. \qquad (A.7)$$

This alpha is the ratio of the driver compliance to the box volume so

$$V_B = \frac{V_d}{\alpha}. \qquad (A.8)$$

This is the volume of the enclosure. Because the resonance frequency is proportional to the driver's Q factor, the frequency of the system can be determined from

$$f_s = f_d\left(\frac{Q_s}{Q_d}\right). \qquad (A.9)$$

EXAMPLE A.4 Designing an Air-Suspension Enclosure.

A 0.30 m (12-in.) poly-paper cone woofer is selected with the following Thiele-Small parameters: $f_d = 30$ Hz; $Q_d = 0.6$; and $V_d = 0.127$ m^3 (4.5 ft^3). Let's first check the F factor:

$$F = \frac{f_d}{Q_d} = \frac{30}{0.6} = 50 \text{ Hz}.$$

This driver should work well for a closed-box enclosure. Next we desire a smooth roll off in the bass so we pick a $Q_s = 0.7$. The alpha is

$$\alpha = \left[\frac{Q_s}{Q_d}\right]^2 - 1 = \left[\frac{0.7}{0.6}\right]^2 - 1 = 0.36,$$

giving a box volume of

$$V_B = \frac{V_d}{\alpha} = \frac{0.127}{0.36} = 0.354 \text{ m}^3 (12.5 \text{ ft}^3).$$

The resulting resonance frequency is

$$f_s = f_d\frac{Q_s}{Q_d} = 30\left(\frac{0.7}{0.6}\right) = 35 \text{ Hz}.$$

Designing a Bass-Reflex Enclosure

Designing bass-reflex enclosures is more difficult than designing air-suspension enclosures. In addition to finding the box volume V_B, you must find the box resonance frequency f_B, the system cutoff frequency (where the response is down 3 dB) f_3, and the level of the "hump" (increase in response at the resonance frequency for different values of Q) in dB (see Fig. 19.15). Using Thiele's papers, D. B. Keele, Jr. developed a pocket-calculator method for designing bass-reflex enclosures (details are given in Weems, 1990). There are two sets of equations used by Keele for calculating the necessary parameters for a bass-reflex enclosure: The first set is used when the compliance and Q of the loudspeaker driver result in an enclosure volume of a reasonable size; the second set is used when you need to adjust the size of the enclosure. The loudspeaker parameters f_d, Q_d, and V_d are the same as those used in the air-suspension design.

The first step is to determine the volume of the box

$$V_B = 15Q_d^{2.87} V_d. \tag{A.10}$$

If the volume calculated is a reasonable size then the following set of equations are used to calculate the system cutoff frequency (f_3) and the box resonance frequency (f_B).

$$f_3 = 0.26Q_d^{-1.4} f_d \tag{A.11}$$

$$f_B = 0.42Q_d^{-0.9} f_d. \tag{A.12}$$

Because this is an optimal design there is no need to calculate the hump in the frequency response. If the volume calculated for V_B is not reasonable (too large or too small) you would select a new volume and use the following set of equations to calculate the parameters:

$$f_3 = f_d \sqrt{\frac{V_d}{V_B}} \tag{A.13}$$

$$f_B = f_d \left(\frac{V_d}{V_B}\right)^{0.32} \tag{A.14}$$

$$\text{hump(dB)} = 20 \log \left[2.6Q_d \left(\frac{V_d}{V_B}\right)^{0.35} \right]. \tag{A.15}$$

The bass-reflex enclosure acts as a Helmholtz resonator so the area and length of the port can be determined from the enclosure volume and resonance frequency (see Section 2.3):

$$f_B = \frac{v}{2\pi} \sqrt{\frac{A}{V_B L}}, \tag{A.16}$$

where the box frequency f_B and volume V_B are calculated above and the duct area A and length L are to be determined. A good rule of thumb is to divide the piston area by seven to determine the port area. The diameter of the port is then the piston area divided by $\sqrt{7}$. The length of the port can then be calculated from the Helmholtz resonator equation,

$$L = \frac{A}{V_B} \left(\frac{v}{2\pi f_B}\right)^2. \tag{A.17}$$

EXAMPLE A.5 Designing a Bass-Reflex Enclosure.   The 0.30 m (12-in.) woofer from
Example A.4 is to be used in a bass-reflex enclosure. The Thiele-Small parameters are:

$$f_d = 30 \text{ Hz}; \quad Q_d = 0.6; \quad \text{and} \quad V_d = 0.127 \text{ m}^3 (4.5 \text{ ft}^3);$$

$$V_B = 15(0.6)^{2.87}(0.127 \text{ m}^3) = 0.440 \text{ m}^3 (15.5 \text{ ft}^3).$$

This is a box about 0.61 m (2 ft) on each side. If this is a reasonable sized enclosure we
use the first set of equations

$$f_3 = 0.26(0.6)^{-1.4}(30) = 16 \text{ Hz},$$

$$f_B = 0.42(0.6)^{-0.9}(30) = 20 \text{ Hz}.$$

The overall system cutoff frequency will be reduced from 30 Hz of the driver to 16 Hz. The
box frequency will be 20 Hz with a volume of 0.440 m³ (15.5 ft³). For a 0.30 m (12-in.)
diameter woofer the port diameter would be $0.30/\sqrt{7} = 0.11$ m (4.5 in.) that has an area
of 0.01 m² (0.11 ft²). The length of the port is

$$L = \left(\frac{0.01 \text{ m}^3}{0.44 \text{ m}^3} \right) \left(\frac{343 \text{ m/s}}{2\pi 20} \right)^2 = 0.17 \text{ m (about 7 in.)}.$$

If the same loudspeaker driver were used, but the enclosure volume were reduced from
0.44 m³ (15.6 ft³) to 0.28 m³ (10.0 ft³) the design would be adjusted as follows:

$$f_3 = 30 \sqrt{\frac{0.44}{0.28}} = 38 \text{ Hz}$$

$$f_B = 30 \left(\frac{0.44}{0.28} \right)^{0.32} = 35 \text{ HZ}$$

$$\text{hump} = 20 \log \left[(2.6)(0.6) \left(\frac{0.44}{0.28} \right)^{0.35} \right] = 5.2 \text{ dB}.$$

The overall system frequency will be increased from 30 Hz of the driver to 38 Hz. The box
frequency will be 32 Hz. The area of the port would still be 0.01 m² (0.11 ft²) with a length
of

$$L = \left(\frac{0.01 \text{ m}^3}{0.44 \text{ m}^3} \right) \left(\frac{343 \text{ m/s}}{2\pi 35} \right)^2 = 0.06 \text{ m (about 2.2 in.)}.$$

The consequence of making a smaller enclosure is less bass response and a 5.2 dB hump
in the response curve.

Answers to Odd-Numbered Exercises

Chapter 1

3. 1000 kg/m^3, 920 kg/m^3

5. 1.86 m/s^2 = 0.19 g

9. For cylinder 2 m long, 1 m in circumference,
$F = 2 \times 10^5$ N

11. 83%, dissipated as heat (core loss, friction, etc.)

Chapter 2

1. (a) 49 N/m; (b) 1.11 Hz

3. 0.99 m

5. 0.061 J

7. 3.05 Hz and 1.56 Hz, compared to 3.15 Hz and 1.58 Hz

Chapter 3

1. (a) 3×10^6 Hz; (b) 4×10^{14} Hz; (c) 6×10^{14} Hz; (d) 10^{10} Hz

3. From $t = 23°$ to $37°$, $\Delta v = 8.4$ m/s, $\Delta f/f = 2.4\%$

5. 260 m/s

7. $t = 2.42 \times 10^{-3}$ s

9. 1029 m

11. 349.9 m/s and 349.3 m/s

Chapter 4

1. 15 Hz

3. (a) 35 Hz, 70 Hz; (b) 18 Hz, 53 Hz

5. 2.5 Hz

7. 2550 Hz, 5100 Hz, 7650 Hz

Chapter 5

1. 2858 Hz

3. 5.5×10^{-7} N

5. 8.7×10^{-4} s for $L_2 - L_1 = 30$ cm

7. (a) 8.0×10^{-3}; (b) 30; (c) 1.73×10^3; (d) $4.19946 \times 10^2 \simeq 420$

9. (a) $x = 2$; (b) $x = 1000$; (c) $x = 20$; (d) $x = \frac{1}{2}$

Chapter 6

1. 43 dB, 25 dB, 6 dB, 4 dB, -2 dB, 18 dB

3. 56 dB

5. $L_W = 117$ dB; $L_p = 106$ dB, 94 dB

7. 6.3×10^{-3} N/m^2 (or Pa); 10^{-7} W/m^2

9. 74 phons; 110 phons

Chapter 7

1. At 21.7 cm from either end

3. c.b./jnd = 30, 31, 50, 38

5. 100 s^{-1}, 200 s^{-1}, 300 s^{-1}; 100 Hz

7. 130; 270; 380

9. 572 Hz

11. 300, 900, 1500, 2100 Hz

Chapter 8

1.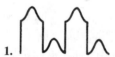

5. 60*, 120, 180*, 480, 600, 720, 900

7. America

9. 799 kHz, 801 kHz

Chapter 9

3. B_0, B_1, B_2, B_3, B_4, B_5, B_6, B_7, B_8, B_9

5. ET $= 2^{7/12} \cdot 2^{5/12} = 2$; $2^{9/12} \cdot 2^{3/12} = 2$
Just: $\frac{3}{2} \cdot \frac{4}{3} = 2$; $\frac{5}{3} \cdot \frac{6}{5} = 2$
Pyth: $\frac{3}{2} \cdot \frac{4}{3} = 2$; $(1.687)(1.184) = 2$

7. $\frac{32}{27}$

9. 37.6¢ flat

Chapter 10

1. Turn the figure upside down and read from bottom to top.

3. 0.21 mm; 2.3 m/s

5. G$_3$ and D$_4$; viola resonances are somewhat higher (about A$_3^\sharp$ and F$_4$)

7. 373, 746 and 520, 1039 compared to 384, 795 and 584, 980
Air in the guitar cavity has resonances like two closed pipes equal to length and width (i.e., nearly plane waves propagate back and forth between ribs).

Chapter 11

1. 8.1 cm

3. 100, 255, 380, 495, 610, 730, 840, 960 Hz
Bugle notes: 233, 349, 466, 587, 698, 831, 932
(Lower resonances are sharp by nearly a semitone)

5. 1.07, 1.18, 1.20, 1.38, 1.41

Chapter 12

1. (a) 128 Hz; (b) Frequency of D$_3$ is 17% lower; (c) Flare of bell shortens effective pipe length and reed appears capacitive (played below its resonance frequency)

3. 73%

5. (a) 408 Hz; (b) 14% lower; (c) end correction (very large at mouth end; see Sect. 14.9)

Chapter 13

1. 1 : 2.71 : 5.15 : 8.43 : 12.21 compared to 1 : 2.76 : 5.40 : 8.93 (thin bar)
Glockenspiel bar shows thick bar behavior.

3. $\dfrac{f_t}{f_L} = \dfrac{9\pi t}{4L\sqrt{12}}$; 0.0858 compared to 0.0888

5. 1.77

7. 2440, 3180, 3980, 4180 Hz; 802 Hz, G$_5$ + 39¢

Chapter 14

1. 0.0156 m or 16 mm; not very practical

3. Piano is longer; 1.054

5. 1.0038; 6.3¢

7. 34.5 ft, 17.2 ft, 8.6 ft; compared to 32, 16, 8 ft.

Chapter 15

1. 780 Hz, 2239 Hz, 3898 Hz; compare 690, 2610, 3570 Hz

5. 536 Hz, 1608 Hz, 2680 Hz, 3752 Hz

7. 1426, 4279, 7132 Hz

9. 546 Hz

Chapter 16

1. 81 Hz male

3. di: 300 Hz in 0.05 s; da: −300 Hz in 0.05 s; du: −700 Hz in 0.1 s

Chapter 17

1. Male: 32–311 Hz, 330–440 Hz, 494 Hz up; Female: 175–370 Hz, 392–659 Hz, 698–1046 Hz

3. /u/ 300, 750, 2500 vs 350, 640, 2550
/ɑ/ 600, 1000, 2500 vs 700, 1200, 2600
/i/ 300, 1800, 2500 vs 300, 1950, 2750

5. (a) 0.01 W; (b) 0.1 W; (c) 1.6 W

Chapter 18

1. (a) 2.5 A; (b) 1 A; (c) 4.5 A; (d) 1 A

3. (a) 2906 Hz; (b) 3.2 Hz

5. (a) 2.31 Ω; (b) 4 Ω

7. 0.56 W; 0.63 W

9. 7958 Hz (low pass); 318 Hz (low pass)

11. 5000, 37 dB

13. (a) Positive feedback at f_0; (b) 712 Hz

Chapter 19

1. 10 dB greater

3. (a) 1 m, 20 cm; (b) 172 Hz, 860 Hz

5. (a) 0.86 m; (b) 1.53 m

7. 1.43 m

9. 8.7×10^{-7} cm/dyne; 0.85 mm; yes

Chapter 20

1. A gives twice the output of B
3. 4 Ω
5. 1.94 A; 2 A

Chapter 21

1. (a) 1100; (b) 10110; (c) 100100
5. 0 1111110 00000000000000000000000
7. 8192; $20 \log 8192 = 78.3$
9. 144 dB

Chapter 22

1. 6 μm
3. 0.63; 0.14; 2/53

Chapter 23

1. (a) 14 m^2 vs 8 m^2; (b) 4 m^2 vs 71 m^2
3. $t_1 = 23$ ms; side
5. 2.0 s, 1.7 s, 1.5 s, 1.5 s

Chapter 24

1. 900 Hz, 6482 Hz
3. 707 Hz, 354 Hz
5. 69 Hz; yes
7. 62 dB; 59 dB

Chapter 25

1. 17.2 Hz, 34.4 Hz, 34.4 Hz

Chapter 26

3. 6,351,155

5. (a) 0.35 m/s; (b) 0.1%; (c) 1/60 of a semitone; (d) It would probably not be noticeable

Chapter 27

1. (a) exponential; (b) 3.58 V

Chapter 29

3. 80 beats/min (80 M.M.)
5. f_c and all its harmonics
7. 0.24
9. 2.667 bits per value
11. 288,000 bytes
13. (a) 1.8 m/s; (b) 172/f_s (0.39 m/s for A440)

Chapter 30

1. 15–30 mW of sound power and 25 kW of mechanical power
3. 1.1 Hz
5. (a) 59 dB; (b) −241 dB

Chapter 31

1. 68.7 dB
3. (a) 4 m; (b) 20 m

Chapter 32

1. 64 times, 18 dB
3. 19 dB; through the hole
5. 4 hr
7. (a) 14 dB, 24 dB; (b) 8 dB, 25 dB; (c) 37 dB, 56 dB

Index